T0133900

Trends and Perspectives in Modern Computational Science

Lecture Series on Computer and Computational Sciences
Editor-in-Chief and Founder: Theodore E. Simos

Volume 6

Trends and Perspectives in Modern Computational Science

Lectures presented at the International Conference of Computational Methods in Sciences and Engineering 2006 (ICCMSE 2006). Chania, Crete. Greece

Recognised Conference by the European Society of Computational Methods in Sciences and Engineering (ESCMSE)

Editors:

George Maroulis and Theodore Simos

CRC Press is an imprint of the
Taylor & Francis Group, an **informa** business

First published 2006 by Koninklijke Brill NV

Published 2021 by CRC Press
Taylor & Francis Group
6000 Broken Sound Parkway NW, Suite 300
Boca Raton, FL 33487-2742

First issued in hardback 2021

ISBN 13: 978-1-138-41297-2 (hbk)
ISBN 13: 978-90-04-15541-1 (pbk)

Publisher's Note
The publisher has gone to great lengths to ensure the quality of this reprint but points out that some imperfections in the original copies may be apparent.

Visit the Taylor & Francis Web site at
http://www.taylorandfrancis.com

and the CRC Press Web site at
http://www.crcpress.com

A C.I.P. record for this book is available from the Library of Congress

COVER DESIGN: ALEXANDER SILBERSTEIN

Brill Academic Publishers
P.O. Box 9000, 2300 PA Leiden
The Netherlands

Lecture Series on Computer and Computational Sciences
Volume 6, 2006, pp. I-II

Trends and Perspectives in Modern Computational Science

Lectures presented in the
International Conference of Computational Methods in Science and Engineering
(ICCMSE 2006)
Chania, Crete
Greece
27 October – 1 November 2006

Preface

We have collected in this volume the scientific papers presented in the International Conference of Computational methods in Science and Engineering (ICCMSE 2006), held in Chania (Crete), Greece. Included herein are the

A. The Highlighted Lectures
B. Invited Papers
C. Talks of the Symposium Organizers
D. Special talks by Keynote Symposium Speakers

Three eminent scientists were invited to give the Highlighted Lectures this year: Professor **A.D.Buckingham** (Cambridge, United Kingdom), Professor **B.Roos** (Lund, Sweden) and Professor **W.Kutzelnigg** (Bochum, Germany).

The Invited Speakers are **T.Bancewicz, S.Canuto, M.Cho, C.Cramer, M.Heaven, H.Herrmann, K.Hirao, J.Jellinek, P. Jørgensen, I.Kaplan, B.Ladanyi, J. Leszczynski, P.Mezey, M.Nakano, P. Pyykkø, J.Sauer, H.F.Schaefer III, N.S.Scott, M.Urban** and **K.Yamaguchi**.

The following Symposium Organizers have contributed to this volume: **A.Avramopoulos, M.G.Papadopoulos, H.Reis, F.Batzias, D.Bonchev, B.Champagne, L.Gagliardi, U.Hohm, Jin Yong Lee, M.Sugimoto, G.Maroulis, S.Nikolic, C.Pouchan, P.Schwerdtfeger, A.J.Thakkar, G.Verros** and **M.J.Wójcik**.

The Keynote Speakers are **I.Grant, M.Chardonnet, C.Puzzarini, S.Farantos, B.M.Rode, K.Szalewicz, C.Ciobanu, P.Calaminici** and **Y.R.Shen**.

This collection continues the tradition of excellence established by the ICCMSE. Invited Speakers, Symposium Organizers and Keynote Symposium Speakers are expected to present important work and the text of their talk is included in this volume for the benefit of a wider audience of specialists. We have especially in mind the educational character of these papers for young scientists, so our speakers have to present their results with the widest possible purview in mind and to link their work with the fronts of modern Computational Science.

Professor **George Maroulis**
Department of Chemistry
Faculty of Science
University of Patras
GR-26500 Patras
GREECE

Professor **Theodore Simos**
Department of Computer Science and Technology
Faculty of Sciences and Technology,
University of the Peloponnese,
GR-221 00 Tripolis
GREECE

Archytas of Taras

On harmonics

Mathematicians seem to me to have excellent discernment, and it is not at all strange that they should think correctly about the particulars that are; for inasmuch as they can discern excellently about the physics of the universe, they are also likely to have excellent perspective on the particulars that are. Indeed, they have transmitted to us a keen discernment about the velocities of the stars and their risings and settings, and about geometry, arithmetic, astronomy, and, not least of all, music. These seem to be sister sciences, for they concern themselves with the first two related forms of being [number and magnitude].

Brill Academic Publishers
P.O. Box 9000, 2300 PA Leiden
The Netherlands

Lecture Series on Computer and Computational Sciences
Volume 6, 2006, pp. III-IV

European Society of Computational Methods in Sciences and Engineering (ESCMSE)

Aims and Scope

The ***European Society of Computational Methods in Sciences and Engineering (ESCMSE)*** is a non-profit organization. The URL address is: **http://www.uop.gr/escmse/**. Soon we will have a new URL address: **http://www.escmse.org/**

The aims and scopes of ***ESCMSE*** is the construction, development and analysis of computational, numerical and mathematical methods and their application in the sciences and engineering.

In order to achieve this, the ***ESCMSE*** pursues the following activities:

• Research cooperation between scientists in the above subject.
• Foundation, development and organization of national and international conferences, workshops, seminars, schools, symposiums.
• Special issues of scientific journals.
• Dissemination of the research results.
• Participation and possible representation of Greece and the European Union at the events and activities of international scientific organizations on the same or similar subject.
• Collection of reference material relative to the aims and scope of ***ESCMSE.***

Based on the above activities, ***ESCMSE*** has already developed an international scientific journal called **Journal of Numerical Analysis, Industrial and Applied Mathematics (JNAIAM)**. The copyright of this journal belongs to the ESCMSE.

JNAIAM is the official journal of ***ESCMSE.***

Categories of Membership

European Society of Computational Methods in Sciences and Engineering (ESCMSE)

Initially the categories of membership will be:

• **Full Member (MESCMSE):** PhD graduates (or equivalent) in computational or numerical or mathematical methods with applications in sciences and engineering, or others who have contributed to the advancement of computational or numerical or mathematical methods with applications in sciences and engineering through research or education. Full Members may use the title MESCMSE.

• **Associate Member (AMESCMSE):** Educators, or others, such as distinguished amateur scientists, who have demonstrated dedication to the advancement of computational or numerical or mathematical methods with applications in sciences and engineering may be elected as Associate Members. Associate Members may use the title AMESCMSE.

• **Student Member (SMESCMSE):** Undergraduate or graduate students working towards a degree in computational or numerical or mathematical methods with applications in sciences and engineering or a related subject may be elected as Student Members as long as they remain students. The Student Members may use the title SMESCMSE

• **Corporate Member:** Any registered company, institution, association or other organization may apply to become a Corporate Member of the Society.

Remarks:

1. After three years of full membership of the European Society of Computational Methods in Sciences and Engineering, members can request promotion to Fellow of the European Society of Computational Methods in Sciences and Engineering. The election is based on international peer-review. After the election of the initial Fellows of the European Society of Computational Methods in Sciences and Engineering, another requirement for the election to the Category of Fellow will be the nomination of the applicant by at least two (2) Fellows of the European Society of Computational Methods in Sciences and Engineering.

2. All grades of members other than Students are entitled to vote in Society ballots.

3. All grades of membership other than Student Members receive the official journal of the ESCMSE Applied Numerical Analysis and Computational Mathematics (ANACM) as part of their membership. Student Members may purchase a subscription to ANACM at a reduced rate.

We invite you to become part of this exciting new international project and participate in the promotion and exchange of ideas in your field.

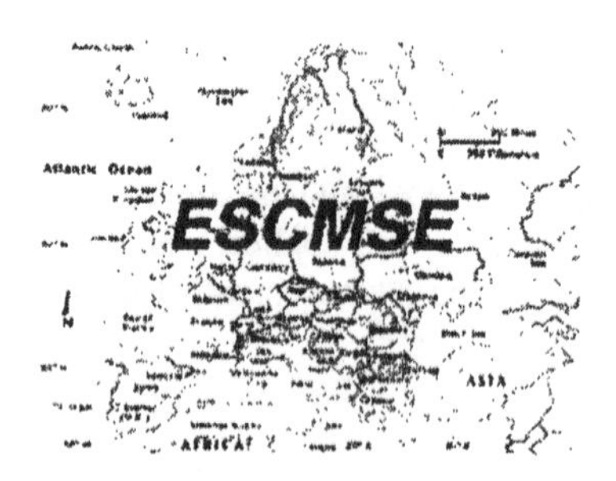

Brill Academic Publishers
P.O. Box 9000, 2300 PA Leiden
The Netherlands

Lecture Series on Computer and Computational Sciences
Volume 6, 2006, pp. V-VII

Table of Contents

Brill Academic Publishers
P.O. Box 9000, 2300 PA Leiden
The Netherlands

Lecture Series on Computer and Computational Sciences
Volume 6, 2006, pp. 1-5

Electric Polarization Induced by Nuclear Spins

A D. Buckingham

Department of Chemistry, Cambridge University, Cambridge CB2 1EW, U.K

Received 3 July, 2006; accepted 10 July, 2006

Abstract: It is well known that a nuclear magnetic dipole induces a magnetic moment in the surrounding electronic cloud and that a nuclear quadrupole moment can induce an electric dipole moment. We investigate whether a nuclear magnetic dipole can induce an electric dipole in its surroundings and note that it would breach both time-reversal and parity symmetries. However, the strong magnetic field of an NMR spectrometer does break time-reversal symmetry, so we expect electric polarization to be induced by coherently precessing nuclear spins located in a non-centrosymmetric position, as in many crystals. Also, in a chiral liquid, there will be a weak rotating electric polarization induced by the rotating nuclear magnetization following a $\pi/2$ pulse. The largest electric polarization induced by nuclear spins is likely to arise in the case of a ferromagnetic or ferrimagnetic sample.

Keywords: Nuclear magnetic resonance, nuclear quadrupole resonance, chirality, electric polarization.

1. Introduction

In addition to its mass and charge, a nucleus can have a magnetic dipole moment (for $I \geq 1/2$), an electric quadrupole moment (for $I \geq 1$), a magnetic octopole moment (for $I \geq 3/2$), and an electric hexadecapole moment (for $I \geq 2$). The parity symmetry operator $\hat{P}$ rules out the existence of odd-parity properties such as an electric dipole moment or magnetic quadrupole moment [1]. The most important of these spin-dependent properties is the nuclear magnetic dipole moment which is responsible for NMR spectroscopy. The nuclear quadrupole moment interacts with the electric field gradient at the nucleus and contributes to the hyperfine structure of atomic and molecular spectral transitions.

2. Nuclear Magnetic moments

In NMR spectroscopy the nuclear magnetization $\langle M_z \rangle$ is rotated into the xy-plane by a resonant radiofrequency $\pi/2$ pulse and the coherent precession of the nuclei about the magnetic field B_z is recorded by an inductor coil on the x-axis. The rotating magnetization induces a voltage in the coil that is recorded over time. The spectrum is the Fourier transform of the voltage. The magnetic field at the nucleus N is affected by the current induced in the electron cloud by the applied field $\boldsymbol{B}$:

$$B_\alpha^{(\mathrm{N})} = B_\beta(\delta_{\alpha\beta} - \sigma_{\alpha\beta}^{(\mathrm{N})}) \tag{1}$$

where $\sigma_{\alpha\beta}^{(\mathrm{N})}$ is the nuclear magnetic shielding tensor. The spin Hamiltonian is

$$H = -\sum_{\mathrm{N}} m_\alpha^{(\mathrm{N})}(\delta_{\alpha\beta} - \sigma_{\alpha\beta}^{(\mathrm{N})})B_\beta + h\sum_{\mathrm{N<N'}} J_{\alpha\beta}^{(\mathrm{NN'})} I_\alpha^{(\mathrm{N})} I_\beta^{(\mathrm{N'})} \tag{2}$$

where $\boldsymbol{m}^{(\mathrm{N})} = \hbar\gamma^{(\mathrm{N})}\boldsymbol{I}^{(\mathrm{N})}$ is the magnetic dipole moment of nucleus N, $\boldsymbol{J}^{(\mathrm{NN'})}$ is the nuclear spin-spin coupling tensor (in hertz) of the pair of nuclei N and N′, $\hbar\gamma^{(\mathrm{N})}\boldsymbol{I}^{(\mathrm{N})}$ is the spin angular momentum of N and $\gamma^{(\mathrm{N})}$ its magnetogyric ratio.

For isotropic media, equation (2) becomes

$$H = -\sum_{\mathrm{N}} m_z^{(\mathrm{N})}(1-\sigma^{(\mathrm{N})})B_z + h \sum_{\mathrm{N<N'}} J^{(\mathrm{NN'})} \boldsymbol{I}^{(\mathrm{N})} \bullet \boldsymbol{I}^{(\mathrm{N'})} \tag{3}$$

where $\sigma^{(\mathrm{N})} = \frac{1}{3}\sigma_{\alpha\alpha}^{(\mathrm{N})} = \frac{1}{3}(\sigma_{xx}^{(\mathrm{N})} + \sigma_{yy}^{(\mathrm{N})} + \sigma_{zz}^{(\mathrm{N})})$ is the nuclear shielding constant and $J^{(\mathrm{NN'})} = \frac{1}{3} J_{\alpha\alpha}^{(\mathrm{NN'})}$ the spin-spin coupling constant.

From equation (2), we see that the dimensionless nuclear magnetic shielding tensor $\sigma_{\alpha\beta}^{(\mathrm{N})}$ has two meanings: it can give the extra magnetic field $\Delta \boldsymbol{B}^{(\mathrm{N})}$ at N due to the current induced in the electron cloud surrounding N by the applied field $\boldsymbol{B}$:

$$\Delta B_\alpha^{(\mathrm{N})} = -\sigma_{\alpha\beta}^{(\mathrm{N})} B_\beta \tag{4}$$

or it gives the magnetic dipole moment $\Delta \boldsymbol{m}^{(\mathrm{N})}$ induced in these surroundings by the nuclear magnetic moment $\boldsymbol{m}^{(\mathrm{N})}$:

$$\Delta m_\alpha^{(\mathrm{N})} = -\sigma_{\beta\alpha}^{(\mathrm{N})} m_\beta^{(\mathrm{N})} \tag{5}$$

3. Nuclear Quadrupole Interactions

A nuclear electric quadrupole moment $Q_{\alpha\beta}^{(\mathrm{N})}$ interacts with the electric field gradient $-q_{\alpha\beta}^{(\mathrm{N})}$ at N produced by the surrounding electrons and nuclei to give hyperfine structure. In pure quadrupole resonance in solids application of a resonant pulse causes the nuclear spins to oscillate around the principal axis q_{zz} of the field gradient and the coherent oscillation of the resonant spins about the z-axis is detected as an oscillation of the nuclear magnetization as in NMR [2].

An electric quadrupole produces an electric field in its vicinity that decreases with distance r as r^{-4} and this field can induce an electric dipole moment $\Delta \boldsymbol{\mu}^{(\mathrm{N})}$ in the atom or molecule:

$$\Delta \mu_\alpha^{(\mathrm{N})} = C_{\alpha,\beta\gamma}^{(\mathrm{N})} Q_{\beta\gamma}^{(\mathrm{N})} \tag{6}$$

The electric polarization induced by ^{35}Cl nuclei in a $NaClO_3$ single crystal was observed by Sleator et al. [3]. The magnitude of the dipole induced in a ClO_3^- anion was estimated as 10^{-4} D and the rotating electric polarization $|P_y|$ as $\sim 10^{-13}$ Cm^{-2} at 4 K.

Ab initio computations of the electric multipole moments induced by a nuclear quadrupole have been carried out [4].

4. Electric Dipoles Induced by Nuclear Magnetic Moments

Can a rotating nucleus N of spin 1/2, and therefore having zero electric quadrupole moment, induce an electric dipole in the molecule? If the nuclear magnetic moment $\boldsymbol{m}^{(\mathrm{N})}$ is precessing at the angular frequency ω we can write [5-7]

$$\Delta \mu_\alpha^{(\mathrm{N})} = \xi_{\alpha\beta}^{(\mathrm{N})} m_\beta^{(\mathrm{N})} + \xi_{\alpha\beta}^{\prime(\mathrm{N})} \dot{m}_\beta^{(\mathrm{N})} / \omega \tag{7}$$

Time-dependent perturbation theory gives the following formulas for $\xi^{(\mathrm{N})}$ and $\xi^{\prime(\mathrm{N})}$ [5,6]:

$$\xi_{\alpha\beta}^{(\mathrm{N})} = -\frac{\mu_0}{4\pi}\frac{2e}{m_e}\sum_{p\neq 0}\frac{\omega_p \mathrm{Re}\{\langle 0|\mu_\alpha|p\rangle\langle p|\Sigma_j l_{j\beta}/r_j^3|0\rangle\}}{\omega_p^2-\omega^2} \tag{8}$$

$$\xi_{\alpha\beta}^{\prime(\mathrm{N})} = \frac{\mu_0}{4\pi}\frac{2e\omega}{m_e}\sum_{p\neq 0}\frac{\mathrm{Im}\{\langle 0|\mu_\alpha|p\rangle\langle p|\Sigma_j l_{j\beta}/r_j^3|0\rangle\}}{\omega_p^2-\omega^2} \tag{9}$$

where $\hbar \boldsymbol{l}_j = \boldsymbol{r}_j \times \boldsymbol{p}_j$ is the orbital angular momentum operator of electron j about the origin of coordinates at the nucleus N, $\hbar\omega_p$ is the excitation energy from the ground state $\langle 0|$ to the excited state $\langle p|$. Consideration of time-reversal symmetry $\hat{T}$ shows that $\xi^{(\mathrm{N})}$ is odd under $\hat{T}$ while $\xi'^{(\mathrm{N})}$ is even. Thus $\xi^{(\mathrm{N})}$ can exist only (i) for molecules in a magnetic field $\boldsymbol{B}$ or (ii) for magnetic samples.

For a closed-shell molecule in the presence of a magnetic field $\boldsymbol{B}$

$$\xi_{\alpha\beta}^{(\mathrm{N})}(\boldsymbol{B}) = \xi_{\alpha\beta\gamma}^{(\mathrm{N})}B_\gamma + \mathrm{O}(B^3) \tag{10}$$

Properties of the third-rank nuclear magnetic polarizability $\xi_{\alpha\beta\gamma}^{(\mathrm{N})}$ have been explored [5,6]. Its isotropic part $\bar{\xi}^{(\mathrm{N})}\varepsilon_{\alpha\beta\gamma}$, where $\varepsilon_{\alpha\beta\gamma}$ is the unit skew-symmetric tensor and the pseudoscalar $\bar{\xi}^{(\mathrm{N})} = \frac{1}{6}\varepsilon_{\alpha\beta\gamma}\xi_{\alpha\beta\gamma}^{(\mathrm{N})} = \frac{1}{6}(\xi_{xyz}^{(\mathrm{N})} - \xi_{xzy}^{(\mathrm{N})} + \xi_{yzx}^{(\mathrm{N})} - \xi_{yxz}^{(\mathrm{N})} + \xi_{zxy}^{(\mathrm{N})} - \xi_{zyx}^{(\mathrm{N})})$ vanishes for achiral molecules and is equal and opposite for enantiomers [5].

A remarkable identity (for $\omega = 0$) linking $\xi_{\alpha\beta\gamma}^{(\mathrm{N})}$ to the electric field dependence of the nuclear magnetic shielding polarizability $\sigma_{\alpha\beta}^{(\mathrm{N})}(\boldsymbol{E})$ in the presence of an electric field $\boldsymbol{E}$ [8] has been established [6]:

$$\sigma_{\alpha\beta}^{(\mathrm{N})}(\boldsymbol{E}) = \sigma_{\alpha\beta}^{(\mathrm{N})} + \sigma_{\alpha\beta\gamma}^{(1)(\mathrm{N})}E_\gamma + \frac{1}{2}\sigma_{\alpha\beta\gamma\delta}^{(2)(\mathrm{N})}E_\gamma E_\delta + \ldots \tag{11}$$

$$\xi_{\alpha\beta\gamma}^{(\mathrm{N})} = -\sigma_{\beta\gamma\alpha}^{(1)(\mathrm{N})} \tag{12}$$

Equation (12) provides a convenient route to the computation of $\xi_{\alpha\beta\gamma}^{(\mathrm{N})}$ since programs are available for determining the nuclear magnetic shielding tensor in the presence of a finite electric field (E_x, E_y, E_z) [9,10].

Calculations of the pseudoscalar $\bar{\xi}^{(\mathrm{N})}$ for the protons in the chiral species hydrogen peroxide HOOH have been reported as a function of the dihedral angle ϕ and are of the order of 10^{-17} $\mathrm{V^{-1}}$ m (it is zero for $\phi = 0$ or 180°) [6]. Rapid inversion motion of the protons prevents the enantiomers from being isolated in HOOH. The heavier nuclei in CHFCl(COOH) have $\bar{\xi}^{(\mathrm{N})}$ values one or two powers of ten larger than that for protons [6]. Thus the e.m.f. induced in a capacitor on the y-axis following application of a $\pi/2$ pulse to a chiral liquid would be of the order of 1 nV for protons [6].

Results for the full nuclear magnetic shielding tensor $\sigma_{\alpha\beta\gamma}^{(\mathrm{F})} = -\xi_{\gamma\alpha\beta}^{(\mathrm{F})}$ in FOOH are shown in Table 1. It is clear that there is much cancellation in the resulting pseudoscalar $\bar{\xi}^{(1)(\mathrm{N})}$ and that is to be expected since the antisymmetric part $\sigma_{\alpha\beta}^{(\mathrm{N})} - \sigma_{\beta\alpha}^{(\mathrm{N})}$ is generally small compared with the symmetric part. There are therefore advantages to be gained by looking at crystalline samples, although in that case electric polarization induced by precessing nuclear spins can be expected from achiral, as well as chiral, samples.

Table 1. Hartree-Fock calculations, using DALTON [9], of the geometry and $\sigma^{(1)(F)}_{\alpha\beta\gamma}$ tensor of the chiral species *S*-FOOH. The geometry was optimised using a cc-pVDZ basis set to give a dihedral angle of 82.8°. The shielding tensor was evaluated using the Ahlrichs VTZ basis. $\bar{\sigma}^{(1)(F)} = \frac{1}{6}(\sigma^{(1)(F)}_{xyz} - \sigma^{(1)(F)}_{xzy} + \sigma^{(1)(F)}_{yzx} - \sigma^{(1)(F)}_{yxz} + \sigma^{(1)(F)}_{zxy} - \sigma^{(1)(F)}_{zyx}) = 6.99 \times 10^{-6}$ a.u. $= -\bar{\xi}^{(F)}$

Molecular geometry (a.u.):

	x	y	z
F	0	0	0
O	0	0	2.5975
O	2.4457	0	3.2765
H	2.9240	-1.7331	3.1834

The 27 elements of the $\sigma^{(1)(F)}_{\alpha\beta\gamma}$ tensor (10^{-6} a.u.):

$\gamma = x$	x	1928.22	-26.98	-748.44
	y	-77.79	3567.98	25.18
	z	109.35	-17.01	363.84
$\gamma = y$		-475.00	-120.55	42.44
		-21.03	-413.18	278.28
		-37.46	42.80	-106.96
$\gamma = z$		8359.19	-166.26	116.77
		-162.03	8405.92	-29.98
		-1.63	11.68	189.27

5. Electric Polarization Induced by Nuclei in Crystalline Samples

The tensors $\xi^{(N)}_{\alpha\beta}$ and $\xi^{(N)}_{\alpha\beta\gamma}$ are odd under the parity operator $\hat{P}$ and therefore can only exist for nuclei in non-centrosymmetric positions in molecules or crystals. The smallness of $\bar{\xi}^{(N)}$ provides a serious experimental challenge to the direct detection of chiral effects in NMR. Much larger signals are to be expected from anisotropic crystals, but the effect would not be restricted to chiral samples.

Consider the component $\xi^{(N)}_{yyz}$: it exists for the non-centrosymmetric unit cells of all the seven crystal systems except the cubic – see Birss [11]. For example, if the trigonal axis of a crystal of symmetry class 3 is aligned in the direction of the magnetic field B_z, and a $\pi/2$ pulse applied in the x-direction the resonant nuclei will precess around the z-axis and the y-component of the rotating nuclear magnetization will induce an electric polarization in the y-direction that may be ~ 10^2 times larger than that from a pure chiral liquid.

Still larger electric polarization should result from rotating nuclear magnetization in a ferromagnetic or ferrimagnetic crystal where an effect is induced by the axial second-rank tensor element $\xi^{(N)}_{yy}$ rather than the third-rank polar tensor element $\xi^{(N)}_{yyz}B_z$. The challenge is to find a magnetic crystal containing an abundant nucleus of spin 1/2 in an appropriate symmetry class (it must be non-centrosymmetric since $\xi^{(N)}_{\alpha\beta}$ (and $\xi^{(N)}_{\alpha\beta\gamma}$) is odd under $\hat{P}$). Such a magnetic crystal is likely to give an oscillating electric polarization 10^4-10^5 times that of a pure chiral liquid. Because electronic spin-orbit coupling increases rapidly with nuclear charge, the effect is more likely to be found using heavier nuclei having a spin of 1/2 in good abundance, e.g. Si, P, Y, Rh, Ag, Cd, Sn, Te, Yb, W, Pt, Hg, Tl and Pb.

Acknowledgement

The author is grateful to Dr Peer Fischer for helpful comments and for computing the shielding tensor of FOOH.

References

[1] C. H. Townes and A. L. Schawlow, *Microwave Spectroscopy.* McGraw-Hill, London 1955, p. 142.

[2] T. P. Das and E. L. Hahn, *Nuclear Quadrupole Resonance Spectroscopy.* Academic Press, New York, 1958.

[3] T. Sleator, E. L. Hahn, M. B. Heaney, C. Hilbert and J. Clarke, Nuclear-quadrupole induction of atomic polarization, *Phys. Rev. B* **38** 8609-8624 (1988).

[4] A. D. Buckingham, P. W. Fowler, A. C. Legon, S.A. Peebles and E. Steiner, A distributed electrostatic model for field gradients at nuclei in Van der Waals molecules. Application to complexes of HCl, *Chem. Phys. Letters* **232** 437-444 (1995).

[5] A. D. Buckingham, Chirality in NMR spectroscopy, *Chem. Phys. Letters* **398** 1-5 (2004).

[6] A. D. Buckingham and P. Fischer, Direct chiral discrimination in NMR spectroscopy, *Chem. Phys.* **324** 111-116 (2006).

[7] P. Lazzeretti and R. Zanasi, Electric and magnetic nuclear shielding tensors. A study of the water molecule, *Phys. Rev. A* **33** 3727-3741 (1986).

[8] A. D. Buckingham, Chemical shifts in the nuclear magnetic resonance spectra of molecules containing polar groups, *Canadian J. Chem.* **38** 300-307 (1960).

[9] T. Helgaker et al. DALTON. A molecular electronic structure program. Release 1.2.1 (2005).

[10] A. Rizzo, T. Helgaker, K. Ruud, A. Barszczewicz, M. Jaszunski and P. Jørgensen, Electric field dependence of magnetic properties. Multiconfigurational self-consistent field calculations of hypermagnetizabilities and nuclear shielding polarizabilities of N_2, C_2H_2, HCN and H_2O, *J. Chem. Phys.* **102** 8953-8966 (1995).

[11] R. R. Birss, *Symmetry and Magnetism.* North-Holland, Amsterdam, 1966, Tables 4a and 4e.

Brill Academic Publishers
P.O. Box 9000, 2300 PA Leiden,
The Netherlands

Lecture Series on Computer and Computational Sciences
Volume 6, 2006, pp. 6-22

On the nature of the metal-metal multiple bond

Laura Gagliardi[1] and Björn O. Roos

Department of Physical Chemistry
Sciences II University of Geneva
30, Quai Ernest Ansermet
CH-1211 Geneva 4 Switzerland
and
Department of Theoretical Chemistry
Chemical Center, P.O.B. 124
S-221 00 Lund, Sweden

Received 11 July, 2006; accepted 17 July, 2006

Abstract: The quantum chemical description of covalent chemical bonds with bond orders larger than three has been a challenge for quantum chemistry since the such bonds were first detected in 1964. We illustrate in this contribution how modern multiconfigurational quantum chemistry can be applied to describe such bond and to define an effective bond order for them. Examples are given from metal-metal bonds through the entire periodic table, starting with the chromium dimer and ending with the multiple bond between two uranium atoms. A number of complexes containing a multiply bonded metal dimer as the central unit have also been studied and illustrate how this building block might be used to form complex ions and molecules also for actinide dimers.

Keywords: metal-metal multiple bonds, CASSCF/CASPT2, Cr_2, U_2, PhCrCrPh

Mathematics Subject Classification: Here must be added the AMS-MOS or PACS Numbers

PACS: Here must be added the AMS-MOS or PACS Numbers

1 Introduction

A covalent chemical bond between two atoms is in molecular orbital (MO) theory described by a bonding orbital occupied by two electrons. The total wave function for a number of such bonds is the closed shell Hartree-Fock determinant. The situation becomes, however, more complicated when we want to describe the bonding along the potential curve where the bond distance increases towards dissociation. Each molecular fragment in a dissociated system will contain one electron in a localized orbital. The wave function for such a system contains two determinants (the Heitler-London wave function). Alternatively, it can be described using two delocalized MOs, the bonding and the antibonding orbital. The wave function is then a linear combination of two configurations, one with the bonding orbital and one with the antibonding orbital doubly occupied. The occupation number of the two orbitals are 2-x and x, where x is one at the dissociation limit. Such a wave function will actually give a qualitatively correct representation of the electronic structure for all distances. In the bonding region, x will be small (typically less than 0.05) and the wave function

[1]Corresponding author E-mail: Laura.Gagliardi@chiphy.unige.ch

will be close to the Hartree-Fock determinant. We can use x to define an effective bond order that will be valid along the whole potential curve. If we define the effective bond order to be 1-x, we will get a value close to one in the bonding region and zero at dissociation.

Why is this interesting? Well, assume that the the bond formed between the two fragments is weak for some reason, for example, due to steric hindrance. Then, the value of x may be quite different from zero. Say, that it is 0.5. We can then say that the bond is only half way formed and the effective bond order is 1-x=0.5. In multiply bonded systems, the different orbitals forming the bonds may have different overlaps and x may vary considerably from bond to bond.

How can we extend this definition of the effective bond order to many electron systems and multiple bonds. The wave function will in such cases be more complicated and contain many electronic configurations, if we wish to be able to describe the full potential curve. We can then use the natural orbitals (NO) and their occupation numbers to define the bonding. These are quite stable properties of a wave function, which do not change much when we increase the accuracy, once we have defined a multiconfigurational wave function that adequately describes the bonding pattern in terms of the bonding and antibonding orbitals. One condition is of course that the NOs are localized to the bonding region, which normally is no problem in the case of metal-metal bonds. Several examples will be given below to illustrate the point. For multiple bonds we can then define the effective bond order as the sum for each individual bond, that is:

$$BO = \sum_i (\eta_i - \eta^*)/2$$

where η_i and η_i^* are the occupation numbers for the bonding and antibonding NOs i.

It was for a long time assumed that the highest bond order that could exist between two atoms was three, as illustrated for example by the nitrogen molecule. It was not until 1964, when Cotton *et al.* reported on the crystal structure of $K_2[Re_2Cl_8]{\cdot}2H_2O$, that the idea of a quadruple bond between two transition-metal atoms was introduced. The $Re_2Cl_8^{2-}$ ion has since then become the prototype for this type of complexes. A new era of inorganic chemistry was born. Over the years we have studied several metal-metal multiple bond compounds and recently we have extended this concept to the whole periodic table, including actinides. In this article we shall give a number of illustrations of such studies, which are all based on multiconfigurational wave functions where the effective bond order can easily be computed. This gives us the possibility to answer the following question: if three is not the highest bond order of the periodic table, what is it? In the following we shall describe the methods that we use and then we will report some of the systems containing a metal-metal multiple bonding that we have recently studied. We shall describe the $Re_2Cl_8^{2-}$ ion, the Cr_2 dimer, the U_2 dimer and some inorganic systems containing a U_2 central unit.

2 Computational methods

The electronic structure of the compounds studied have been obtained using multiconfigurational wave functions. We have used the Complete Active Space (CAS) SCF method [1] with dynamic correlation effects included using multiconfigurational second order perturbation theory, CASPT2 [2, 3], This approach allows us to obtain a qualitatively correct representation of the electronic structure at the CASSCF level of theory, which can be used to analyze the multiplicity of the bond, while the CASPT2 method allows a determination of the energetics.

A crucial feature of the approach is the choice of the active space, which must include the metal-metal bonding and antibonding orbitals and in appropriate cases also ligand orbitals that interact with the metal d- or f-type orbitals. The choice of the active space is particularly challenging for metal-metal bonds. One reason is the large number of atomic orbitals that are involved in the bond formation. The uranium atom, for example uses the 7s, 6d, and 5f orbitals to form the bond to a second U atom. This results in an active space of 26 orbitals with 12 active electrons, a clearly impossible case for the CASSCF method. Here the problme was solved by making three of the bonds inactive because the occupation numbers for the bonding orbitals turned out to be close to two at equlibrium geometry. The price to pay was that only potential curves close to the equilibirum geometry could be studied and not the full dossociation path.

Another example is the Cr dimer, where the 3d and 4s orbitals are used to form the multiple bond. Thus, the active space is 12 electrons in 12 orbitals, which is quite feasible. It tourns out, however, that such an active space leads to severe intruder state problems in the CASPT2 treatment. An extension to 16 orbitals (including 4pπ) was necessary to cope with this problem. The choice of the active space is usually simpler for ionic systems because fewer atomic orbitals are involved in the bonding. More details about the choice of active orbitals and electrons will be given below in connection with the different examples presented.

Atomic natural orbital basis sets (ANO-RCC) have been used, which includes scalar relativistic effects using the Douglas-Kroll-Hess Hamiltonian [4, 5]. Scalar relativistic effect are thus included at all levels of theory. For the heavy atom systems we have also included spin-orbit coupling as a last step in the calculation, using a recently developed approach where the CASSCF wave functions comprise the basis for a spin-orbit CI calculation with energies shifted to include dynamic correlation [6] The calculations have been performed using the MOLCAS suite of programs [7].

3 The mutiple metal-metal bond in $Re_2Cl_8^{2-}$ and related systems

In 1965 F. A. Cotton and C. B. Harris reported the crystal structure of $K_2[Re_2Cl_8]\cdot 2H_2O$ [8]. A surprisingly short Re-Re distance of 2.24 Å was found. This was the first example of a multiple bond between two metal atoms and the $Re_2Cl_8^{2-}$ ion (Figure 1) has since then become the prototype for this type of complexes. Cotton analyzed the bonding using simple Molecular Orbital (MO) theory and concluded that a quadruple Re-Re bond was formed [8, 9]. Two parallel $ReCl_4$ units are connected by the Re-Re bond. The $d_{x^2-y^2}$, p_x, p_y, and s orbitals of the valence shell of each Re atom form the σ bonding to each Cl atom. The remaining d_{z^2} and p_z orbitals with σ symmetry relative to the Re-Re line, the d_{xz} and d_{yz} with π symmetry, and the d_{xy} with δ symmetry form the quadruple Re-Re bond: one σ bond, two π bonds and one δ bond. Because there are eight electrons to occupy these MOs, the ground state configuration will be $\sigma^2\pi^4\delta^2$. The presence of the δ bond explains the eclipsed conformation of the ion. In a staggered conformation, the overlap of the δ atomic orbitals are zero and the δ bond disappears. The visible spectrum was also reported in these early studies. The notion of a quadruple bond is based on the inherent assumption that four bonding orbitals are doubly occupied. Today we know that this is not the case for weak inter-metallic bonds. The true bond order depends on the relation between the occupation of the bonding and antibonding orbitals, respectively. Such a description is, however, only possible if a quantum chemical model is used that goes beyond the Hartree-Fock model. The CASSCF model will be used here and we shall demonstrate that the true bond order between the two Re atoms is closer to three than to four.

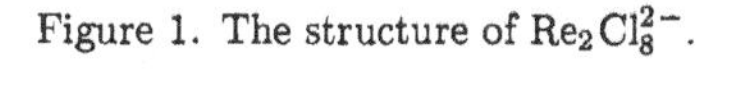

Figure 1. The structure of $Re_2Cl_8^{2-}$.

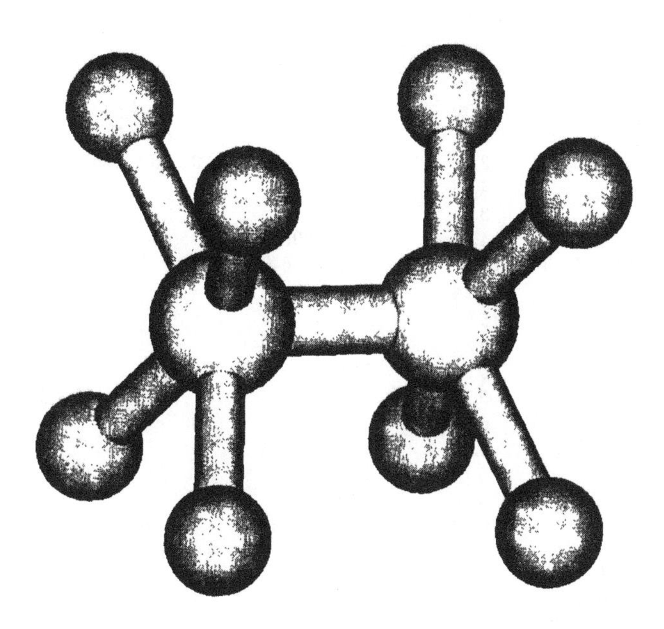

Because the $Re_2Cl_8^{-2}$ ion is such an important entity in inorganic chemistry, we decided to study its structure and electronic spectrum using multiconfigurational quantum chemistry [10] . Relativistic effects and also spin-orbit coupling were included in this study. The geometry of $Re_2Cl_8^{-2}$ was obtained at the CASPT2 level of theory and several excited states were calculated at this geometry. The calculations were performed using the active space formed by 12 active electrons in 12 active orbitals. They are nicely paired such that the sum of the occupation numbers for the bonding and antibonding orbitals of a given type is almost exactly two. The two bonding Re-Cl orbitals are mainly located on Cl as expected, while the antibonding orbitals have large contributions from $3d_{x^2-y^2}$. The occupation is low and these orbitals are thus almost empty and may be used as acceptor orbitals for electronic transitions.

The strongest bond between the two Re atoms is the $\sigma-$bond, with an occupation number of $\eta_b = 1.92$ of the bonding and $\eta_a = 0.08$ of the antibonding natural orbital. We could estimate the effective bond order as $(\eta_b - \eta_a)/(\eta_b + \eta_a)$. For the $\sigma-$bond we obtain the value 0.92. The corresponding value for the $\pi-$bond is 1.74. The δ pair gives an effective bond order of only 0.54. Adding up these numbers results in a total effective bond order of 3.20 for $Re_2Cl_8^{-2}$. The main reduction of the bond order from 4.0 to 3.2 is due to the weakness of the $\delta-$bond.

Vertical excitation energies and oscillator strengths, calculated at the CASPT2 level with and without the inclusion of spin orbit coupling have been calculated. While we refer to the original manuscript for the details of the calculations, we here report only some information about the most significant features of the spectrum. The most relevant transitions are reported in Table 1.

Table 1: Spin-free excitation Energies in Re_2Cl_8 (in eV) calculated at the CASSCF (CAS) and CASPT2 (PT2) level. Oscillator strengths are given within parentheses. Q(Re) gives the Mulliken charge on one Re atom.

State	ΔE(CAS)	ΔE(PT2)	Expt[a].	Q(Re)
$\delta \to \delta^*$, $^1A_{2u}$	3.08	2.03(0.0037)	1.82(0.023)	1.03
$\delta \to \pi^*$, $^1A_{1g}$	2.90	2.29(f)	2.19(weak)	1.03
$\pi \to \delta^*$, ^1E_g	3.41	2.70(f)	2.60($\epsilon = 65$)	1.04
$\delta \to \pi^*, \pi \to \delta^*$, ^1E_g	3.87	3.10(f)	2.93(very weak)	1.04
$\delta \to \sigma^*$, $^1B_{1u}$	4.47	3.10(f)		1.00
$\delta \to \delta_{x^2-y^2}$, $^1A_{2g}$	3.96	3.37(f)		1.11
$(\delta,\pi) \to (\delta^*)^2$, ^1E_u	4.20	3.38(0.29E-03)	3.35($\epsilon = \epsilon_1$)[b]	1.04
$Cl(3p) \to \delta^* LMCT$, ^1E_u	6.37	3.56(0.60E-04)	3.48($\epsilon = \epsilon_2$)[b]	0.84
$\delta \to \delta^*_{x^2-y^2}$, $^1A_{1u}$	4.24	3.59(f)		1.13
$\pi \to \pi^*$, $^1A_{1u}$	5.02	3.76(f)		1.04
$(\delta,\pi) \to (\delta^*)^2$, ^1E_u	4.81	3.80(0.92E-04)		1.04
$(\delta,\pi) \to (\delta^*\pi^*)$, $^1A_{1g}$	5.01	3.91(f)		1.05
$\pi \to \pi^*$, $^1B_{1u}$	5.17	4.00(f)		1.05
$Cl(3p) \to \delta^*, LMCT$, ^1E_u	6.54	4.08(0.08)	3.83(intense)	0.88
$\sigma \to \delta^*, \pi \to \pi^*$, $^1B_{1u}$	6.01	4.13(f)		1.05
$\pi \to \delta_{x^2-y^2}$, ^1E_u	6.15	4.17(0.009)		1.08
$(\delta,\pi) \to (\delta^*\pi^*)$, $^1B_{2g}$	5.66	4.30(f)		1.04
$\delta\pi \to \delta^*\sigma^*$, ^1E_u	6.79	4.40(1.0E-04)	4.42(complex)	1.03
$\sigma \to \sigma^*, \pi \to \pi^*$, $^1A_{2u}$	6.66	4.56(0.015)	4.86(intense)	1.04

[a] From Ref. [11].
[b] $\epsilon_1 + \epsilon_2 = 400$ [11].

The lowest band detected experimentally peaks at 1.82 eV (14 700 cm^{-1}) with a oscillator strength, of 0.023. It has been assigned to the $\delta \to \delta^*$ ($^1A_{1g} \to {}^1A_{2u}$) transition. Our 12/12 calculation predicts an excitation energy of 2.03 eV at the CASPT2 level with an oscillator strength equal to 0.004. We performed calculations also with enlarged active spaces, like, for example, with 16 electrons in 14 orbitals (16/14). These various calculations predict a CASPT2 excitation energy that varies between 1.68 and 1.74 eV. The oscillator strength varies between 0.007 and 0.092, which shows that this quantity is sensitive to the active space. The low energy of this transition is of course a result of the weak $\delta-$bond, which places the δ^*-orbital at low energy.

In the region of weak absorption between 1.98–3.10 eV, (16000–25000 cm^{-1}), the first weak peak occurs at 2.19 eV (17675 cm^{-1}) and has been assigned to a $\delta \to \pi^*$ ($1\,^1A_{1g} \to 2\,^1A_{1g}$) transition mostly located on Re with little ligand character. We predicted it at 2.29 eV and it is a forbidden transition.

Two bands with a total ϵ of 400 have been assigned to charge transfer, CT, states. They occur at 3.35 and 3.48 eV respectively. It was suggested that they correspond to two A_{2u} spin-orbit components of two close-lying ^3E_u states [12]. We have not studied the triplet ligand to metal charge transfer (LMCT) states, but our first singlet CT state was predicted at 3.56 eV, corresponding to a $Cl(3p) \to \delta^*$ ($^1A_{1g} \to {}^1E_u$) LMCT transition. Thus, it seems natural to assign the upper of the two bands to this transition. The peak at 3.35 eV has been assigned to a metal localized transition above.

A $(\delta,\pi) \to (\delta^*,\pi^*)$ ($^1A_{1g} \to {}^1A_{1g}$) transition is predicted at 3.91 eV, and a $\pi \to \pi^*$ ($^1A_{1g} \to {}^1B_{1u}$) transition at 4.00 eV. No corresponding experimental bands could be found. An intense CT state is found in the experimental spectrum at 3.83 eV and it is assigned to the $Cl(3p) \rightarrow \delta^*$ ($^1A_{1g} \rightarrow {}^1E_u$) transition that we predict at 4.08 eV with an oscillator strength of 0.08.

Togler *et al.* have suggested that the complex band found at 4.42 eV should be a mixture of two LMCT transitions. We find no evidence of this in the calculations but a weak ^1E_u state is found at 4.40 eV and there are other forbidden transitions nearby.

A rather intense band is found at 4.86 eV with a tentative assignment $\pi \rightarrow \pi^*(^1A_{1g} \rightarrow {}^1A_{2u})$. We agree with this assignment and compute the state to occur at 4.56 eV with an oscillator strength of 0.015.

The spectrum was recomputed with the inclusion of spin-orbit coupling. This effect does not in general change any of the qualitative features of the spectrum. There is a small shift in the energies and intensities, but we do not see any new states with intensities appreciably different from zero. We may, however, have lost some information because we have not studied the LMCT triplet states and the corresponding effects of spin-orbit splitting.

More recently, four compounds containing a metal-metal quadruple bonds, the $[M_2(CH_3)_8]^{2n-}$ ions, where M = Cr, Mo, W, Re and n = 4, 4, 4, 2, respectively, have been studied theoretically [13] using the same methods as in the $Re_2Cl_8^{2-}$ case. The molecular structure of the ground state of these compounds has been determined and the energy of the $\delta \rightarrow \delta^*$ transition has been calculated and compared with previous experimental measurements. The high negative charges on the Cr, Mo and W complexes lead to difficulties in the successful modeling of the ground-state structures, a problem that has been addressed by the explicit inclusion of four Li^+ ions in these calculations. The ground-state geometries of the complexes and $\delta \rightarrow \delta^*$ transition have been modeled with either excellent agreement with experiment (Re) or satisfactory agreement (Mo, Cr and W).

Our primary goal when we started this study was to provide a theoretical understanding of the apparently linear relationship between metal-metal bond length and $\delta \rightarrow \delta^*$excitation energy for the octamethyldimetallates of Re, Cr, Mo, and W. As our results demonstrated, these seemingly simple anionic systems represent a surprising challenge to modern electronic structure methods, largely because of the difficulty in modeling systems that have large negative charges without electronegative ligands. Nevertheless, by using the CASPT2 method with Li^+ counterions for the Mo, Cr and W cases, we have been able to model the ground-state geometries of the complexes with either excellent agreement with experiment (Re) or satisfactory agreement (Mo, Cr and W). This multiconfigurational approach, which is critical for the calculation of excited-state energies of the complexes, does a fairly good job of modeling the trends in the $\delta \rightarrow \delta^*$ excitation energy with the metal-metal bond length, although the accuracy is such that we are not yet able to explain fully the linear relationship discovered by Sattelberger and Fackler [14]. Future progress on these systems will require better ways to accommodate the highly negative charges. These efforts are ongoing.

4 The Cr-Cr multiple bond

The chromium atoms has a ground state with six unpaired electrons ($3d^54s$, ^7S). Forming a bond between two Cr atoms could therefore in principle result in an hextuple bond. It is not surprising that the chromium dimer has become a challenging test for various theoretical approaches to chemical bonding. Almost all existing quantum chemical methods have been used. The results are widely varying in quality (see Ref. [15] for references to some of these studies). It was not until the CASSCF/CASPT2 approach was applied to Cr_2 that a consistent picture of the bonding was achieved. The most recent study resulted in a bond energy (D_0) of 1.65 eV, a bond distance of 1.66 Å, and an ω_e value of 413 cm^{-1} (experimental values are 1.53+-0.06 eV [16], 1.68 [17], and 452 cm-1 [18], respectively.

Does the two chromium atoms form an hextuple bond? The calculations quoted above give the following occupations of the bonding and antibonding orbitals: $4s\sigma_g$ 1.90, $4s\sigma_u$ 0.10, $3d\sigma_g$ 1.77, $3d\sigma_u$ 0.23, $3d\pi_u$ 3.62, $3d\pi_g$ 0.38, $3d\delta_g$ 3.16 eV, $3d\delta_u$ 0.84, yielding a total effective bond order of 4.46. We notice that the δ bond is weak and could be considered as an intemediate between a chemical bond and four antiferromagnetically coupled electrons. The chromium dimer could thus also be described as a quadruply bonded system with the δ electrons localized on the separate atoms and coupled to a total spin of zero. The molecular orbitals are shown in Fig. 2 together with their occupation numbers.

Figure 2. The active molecular orbitals for the Cr dimer at equilibirum geometry (contour line 0.04). The Natural Orbital, NO, occupation numbers are given below the orbitals. Only one of the two components of the π and δ orbitals are shown.

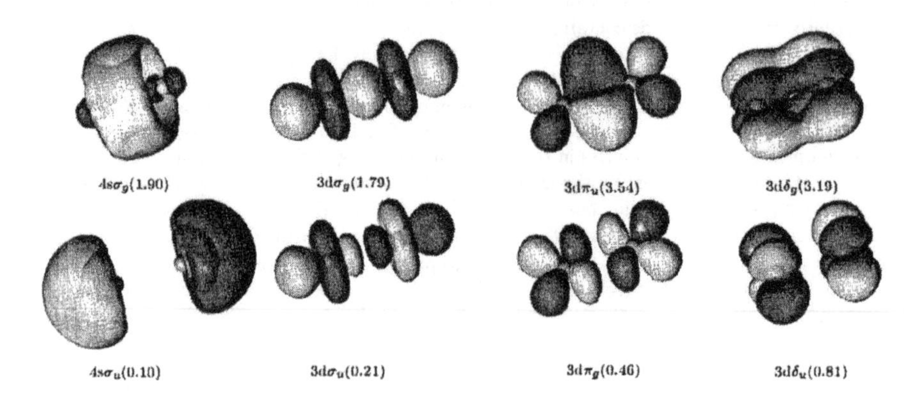

The difficulty to form all six bonds is mainly due too the large difference in size between the 3d and 4s orbitals. When the 3d orbitals reach an effective bonding distance is the 4s already far up on the repulsive part of the potential. This explains why the bond energy is so small in spite of the high bond order. This difference in size decreases for heavier atoms. Thus, is the 5s orbital of Mo more similar in size to the 4d orbital. Even more pronounced is the effect for W where the relativistic contraction of the 6s orbital and the corresponding expansion of the 5d orbital makes the two orbitals very similar in size. The result is a much stronger bond with a bond energy above 5 eV and an effective bond order of 5.19 [19]. The tungsten dimer can thus be described as a truly hextuply bonded system. The occupation numbers of the bonding orbitals are never smaller than 1.8. This is the highest bond order found among any dimer of the periodic table.

Nguyen *et al.* synthesized in 2005 a dichromium compound with the general structure ArCr-CrAr where Cr is in oxidation state 1+ [20]. This s the first example of such a compound. A bond distance of 1.83 Å was determined for the Cr-Cr bond and it was concluded that a quintuple bond was formed. Subsequent CASSCF/CASPT2 calculations on the model compound PhCrCrPh (Ph=Phenyl) essentially confirmed this picture [21]. The NO occupation numbers were found to be: $3d\sigma_g$ 1.79, $3d\sigma_u$ 0.21, $3d\pi_u$ 3.54, $3d\pi_g$ 0.46, $3d\delta_g$ 3.19 eV, $3d\delta_u$ 0.81, very similar to the chromium dimer with again a weak δ bond. The total effective bond order is 3.52, so the bond is intermediate between a triple bond with four antiferromagnetically coupled δ electrons and a true quintuple bond. The bond energy was estimated to be about 3.3 eV, which is twice as much as for the chromium dimer. The reason is the absence of the 4s electron in the Cr(I) ion.

Dichromium(II) compounds have been known for a long time. In particular, the tetracarboxylates have been extensively studied since the first synthetic work of Cottone *et al.* [22]. The Cr-Cr bond length varies extensively depending on the donating power of the bridging ligands and the existence of additional axial ligands. The shortest bond length, 1.966 Å was found for $Cr_2(O_2CCH_3)_4$ in a gas phase measurment [23]. A CASSCF calculation at this bond distance yields the following natural orbital occupation numbers: 3dσ 1.68, 3dσ^* 0.32, 3dπ 3.10, 3dπ^* 0.90, 3dδ 1.21, 3dδ^* 0.79, which gives an effective bond order of only 1.99. This is far from a quadruple bond, which explains the great flexibility in the bond length dependng on the nature of the ligands. Another festure of these compounds is temperature dependent paramegnetism, which is explained by the existence of low lying triplet excited states, arising from a shift of the spin of the weakly coupled δ electrons [24].

A general picture of the Cr-Cr multiple bond emerges from these studies. Not unexpectedly, fully developed bonds are formed by the 3dσ and 3dπ orbitals, while the 3dδ orbitals are only weakly coupled. The notion of an hextuple bond in the Cr_2 system, a quintuple bond in ArCrCrAr, and a quadruple bond in the Cr(II)-Cr(II) complexes is therefore an exaggeration. The situation will be different for the corresponding compounds with the heavier atoms Mo and W where more fully developed multiple bonds can be expected in all three cases.

5 The uranium-uranium multiple bond in the U_2 molecule

The uranium atom (atomic number 92) has 6 valence electrons and the ground-state electronic configuration $(5f)^3(6d)^1(7s)^2$. The 5f and 6d electrons are unpaired, giving rise to a quintet ground state. However, it is not energetically costly to unpair the 7s electrons by forming hybrids between that orbital and, for example, the 6d or 7p orbitals. Uranium thus has in principle 6 electrons available with which to form chemical bonds. How then are these electrons used to form a bond to another uranium atom? In a Lewis-like formalism, one would expect all of the electrons to combine in electron-pair bonds that would give rise to a hextuple bond between the two atoms and a singlet ground state. By analogy, this kind of bonding occurs in the chromium dimer where the six valence electrons reside in the 3d and 4s orbitals of the Cr atom, as explained above. However, the situation in the uranium molecule is considerably more complex. The U atom has seven 5f, five 6d, one 7s, and three 7p orbitals, 16 in total all of which are quite close to one another in energy. As a result, all 16 orbitals may be considered to be valence orbitals and thus to be equally available for use in forming the chemical bond. There is no simple rule that can guide us in predicting the nature of the bond between two uranium atoms. The strength of a covalent bond depends on two factors: the energy of the atomic orbitals on the two different centres involved and the overlap between them. We can predict that the overlap between two 7s orbitals will be large, as is also the case for three out of the five 6d orbitals (forming one s- and two p-type orbitals, respectively). The 5f orbitals, on the other hand, will all have smaller overlap. However, the 5f atomic orbital energy level is lower than those of the 6d and 7s levels, and on this basis we may presume that the system would prefer not to deviate too much from the electronic configuration of the free atom in forming the dimer. The uranium molecule, U_2, has been studied theoretically in the past [25], but the technology available in those days did not allow for a conclusive treatment of such a complex chemical bond. It has also been detected in the gas phase [26, 27], but it has not yet been isolated. The only way to address this challenge from the computational point of view is to perform as accurate a quantum chemical calculation as possible, using a method that allows all of the valence orbitals to combine freely in order to form the most stable chemical bond. We have recently performed such a calculation. The approach has been used in a number of earlier studies of actinide compounds with results that are in agreement with experiment both for structural properties and relative energies (see for example the recent study of the electronic spectrum of the

UO_2 molecule [28]). Based on this experience we estimate that bond distances are computed with an accuracy of better than 0.05 Å for U_2. It is more difficult to give an estimate of the uncertainty in the binding energy because a so complicated chemical bond has never been treated before and it is not easy to obtain a balanced treatment of the molecule and the free atoms. It is without doubt that the molecule is bound with an appreciable binding energy. All low lying energy levels correspond to the same general electronic structure so it is certain that our description of the binding mechanism is correct.

The molecular orbitals involved in the bonding can be described as follows. The lowest energy doubly occupied orbital is a $7s\sigma_g$ type, and corresponds to a typical single σ bond. The next two degenerate doubly occupied orbitals are of covalent π-type and are formed from combinations of 6d and 5f atomic orbitals. Two singly occupied orbitals, of σ ($6d\sigma_g$) and δ ($6d\delta_g$) type, respectively, give rise to one-electron bonds between the two atoms, and two other singly occupied orbitals, of δ and π type, respectively, ($5f\delta_g$ and $6d\pi_u$) give rise to two additional weak one-electron bonds. Finally, there are two electrons in what may be considered to be fully localised $5f\phi$ orbitals. In sum, U_2 is predicted to have three strong normal electron-pair bonds, two fully developed one electron-bonds, two weak one-electron bonds, and two localised electrons. The singly occupied orbitals add up to a 5f shell on each atom with an occupation number of two and the electron spins parallel. In such a situation one normally would expect the two pairs of electrons to couple in such a way that the total spin becomes zero (spin up on one atom and spin down on the other) because this antiferromagnetic coupling provide some additional bonding, even if small. Here, however, all spins are predicted to be parallel (ferromagnetic coupling). This counterintuitive behaviour derives from the interaction between the non-bonding 5f electrons and the two one-electron bonds. It is then energetically more favourable for all open-shell electrons to have the same spin. This "exchange stabilisation" is larger than the small gain in bond energy that would otherwise be obtained by an antiferromagnetic coupling of the 5f electrons.

The calculations also show that all open shell electrons have parallel spin (Hund's rule 1), resulting in a total spin angular momentum of 3, a septet state. The U_2 chemical bond is thus unique and more complex than any other diatomic bond in the periodic system. Summing up the bonding electrons, the molecule turns out to have a quintuple bond. Of particular novelty, strong bonding and strong ferromagnetism coexist in the U_2 molecule. We are now exploring the possibility that this new type of bond may exist in other diactinide compounds as well. One may also speculate about new chemistries based on a central U_2 unit, with the multi-radical nature of the diatomic permitting it to form chemical bonds with a variety of ligands. Having six electrons available for binding, could U_2 form species like NU_2N or $Cl_3U_2Cl_3$? The molecules OU_2O [26, 27] and $H_2U_2H_2$ [29] have already been detected. In addition, the orbital angular momentum about the molecular axis is unique. The large number of orbitals available to the unpaired electrons permits this molecule to have a very large total orbital angular momentum, adding up to 11 as shown above. However, the spin and orbital angular momenta cannot be approximated as independent quantities in a heavy-atom system. Instead, the combination of the two must be considered and the calculations show that the total angular momentum around the molecular axis is 14 (11 from the orbitals plus 3 from the total spin). This value is unique for a chemical bond. High values for the angular momentum have been found in lanthanide compounds, but in those cases the 4f electrons that give rise to such values are localised and they do not participate in chemical bonding. Here, all valence electrons do.

Quantum chemists have been exploring the nature of the chemical bond for almost 80 year, beginning with Heitler and London's valence bond description of the hydrogen molecule in 1927 [30]. Quantum chemistry has been used to explore the bonding between pairs of atoms throughout the periodic table. Most molecules have bonds made up of electron pairs, one for each bond. Some molecules also share single electrons; typical examples are B_2 and O_2. Such one-electron bonds become more abundant in transition metals. High spin states have been found in the lanthanides because of the non-bonded 4f electrons. In this work we have described the electronic structure among the heaviest atoms in the periodic table, the actinides. Unless the chemistry of the as yet undiscovered super-heavy elements proves to be accessible, we appear to have reached the end of the road.

If two of the twelve valence electrons of U_2 are removed, some simplification of the electronic structure could be expected. Indeed, preliminary DFT or CAS calculations[31] suggested for U_2^{2+} a singlet, predominantly $(\sigma_g)^2(\pi_u)^4(\delta_g)^4$ state with bond lengths, as short as 2.15 Å.

On the experimental side it is interesting to note that a mass-spectroscopic study using an Au-U liquid-alloy ion source for uranium with 20% ^{235}U and 80% ^{238}U showed no fingerprint for the mixed diatomic dication, while the mixed-isotope U_2^+ was seen [32]. The authors point out, however, that the strong electric field in the experiment might destabilize the heteronuclear species.

We have studied the U_2^{2+} species, at a comparable level as that used for U_2 [33]. The most stable spin-free electronic state was found to be a singlet state and to have a total orbital angular momentum, Λ, equal to 10 atomic units, (a ^{1}'N' state). The equilibrium bond distance is 2.30 Å. without the inclusion of spin-orbit coupling and the harmonic vibrational frequency is 300 cm^{-1}. A triple-bond covalent radius of 1.18 Å has recently been proposed for uranium [34]. For a triple U-U bond this would give 2.36 Å, close enough to the calculated 2.30 Å. The U_2^{2+} system is metastable and lies 1.59 eV higher in energy than two U^+ ions.

6 Diuranium inorganic chemistry

Actinide chemistry poses a formidable challenge for chemists, both from an experimental and a theoretical perspective. Actinide compounds are not easy to handle in a laboratory. Still, a good understanding of their chemistry is important in a number of areas. Our study on the U_2 molecule, suggested that the U_2 unit could form the framework for a diuranium chemistry.

As already discussed in the previous section, the uranium atom has six valence electrons and the U-U bond in U_2 is composed of three normal two-electron bonds, four electrons in different bonding orbitals and two non-bonded electrons leading to a quintuple bond between the two uranium atoms. Multiple bonding is also found between transition metal atoms. The Cr, Mo, and W atoms have six valence electrons and a hextuple bond is formed in the corresponding dimers, even if the sixth bond is weak. The similarity between these dimers and the uranium dimer suggests the possibility of an inorganic chemistry based on the latter. A number of compounds with the M_2 (M=Cr, Mo, W, Re, etc.) unit are known. Among them are the chlorides, for example, Re_2Cl_6, $Re_2Cl_8{}^{2-}$ [8] and the carboxylates, for example $Mo_2(O_2CCH_3)_4$ [35, 36]. The simplest of them are the tetraformates, which in the absence of axial ligands have a very short metal-metal bond length [37].

Figure 3. The structure of U_2Cl_6.

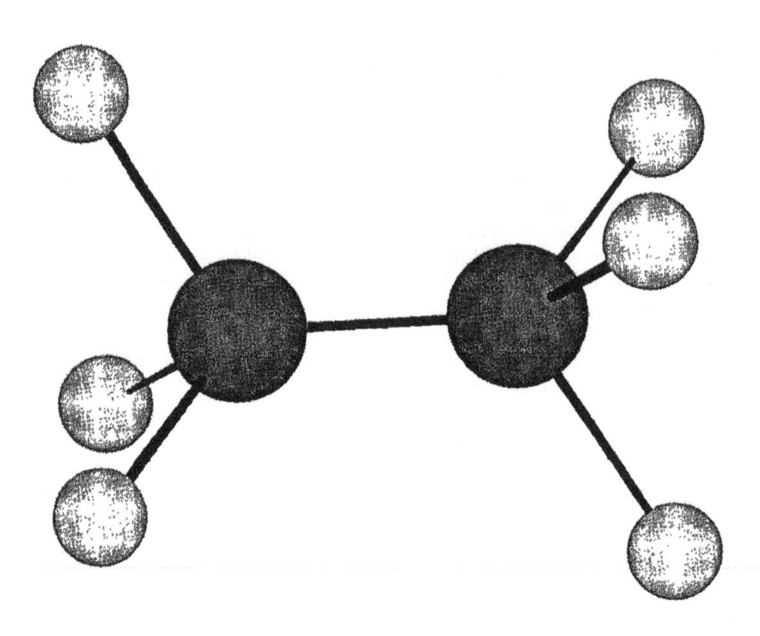

We have recently shown that corresponding diuranium compounds are also stable with a multiple U-U bond and short bond distances [38]. We have studied two chlorides, U_2Cl_6 and $U_2Cl_8{}^{2-}$, both with U(III) as the oxidation state of uranium (See Figure 3), and three different carboxylates (See Figure 4), $U_2(OCHO)_4$, $U_2(OCHO)_6$, and $U_2(OCHO)_4Cl_2$. All species have been found to be stable with a multiply bonded U_2 unit.

In the diuranium chlorides, the formal charge of the uranium ion is +3. Thus six of the twelve valence electrons are available and a triple bond can in principle be formed. U_2Cl_6 can have either an eclipsed or a staggered conformation. Preliminary calculations have indicated that the staggered conformation is about 12 kcal/mol lower in energy than the eclipsed form. We have thus focus our analysis on the staggered structure.

The U-U and U-Cl bond distances and the U-U-Cl angle have been optimized at the CASPT2 level of theory. The ground state of U_2Cl_6 is a singlet state with the electronic configuration $(\sigma_g)^2(\pi_u)^4$. The U-U bond distance is 2.43 Å, the U-Cl distance 2.46 Å and the U-U-Cl angle 120.0 degrees. At the equilibrium bond distance, the lowest triplet lies within 2 kcal/mol from the singlet ground state. The two states are expected to interact via the spin-orbit coupling Hamiltonian. This will further lower the energy, but will have a negligible effect on the geometry. The dissociation of U_2Cl_6 to 2 UCl_3 has been studied. UCl_3, unlike U_2Cl_6, is known experimentally [39] and has been the object of previous computational studies [40]. Single point CASPT2 energy calculations have been performed at the experimental geometry, as reported in Ref. [40], namely a pyramidal structure with a U-Cl bond distance of 2.55 Å and a Cl-U-Cl angle of 95 degrees.

Figure 4. The structure of $U_2(OCHO)_4$.

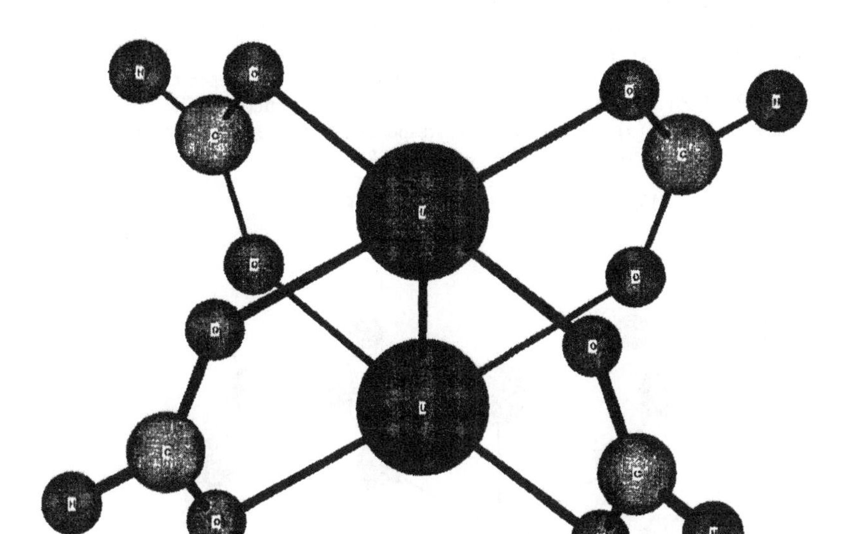

$U_2Cl_8^{2-}$ is the analogue of $Re_2Cl_8^{2-}$. We have optimized the structure for $U_2Cl_8^{2-}$ using an active space formed by six active electrons in thirteen active orbitals, assuming D_{4h} symmetry. As in the U_2Cl_6 case, the molecular orbitals are linear combinations of U 7s, 7p, 6d, and 5f orbitals with Cl 3p orbitals. The ground state of $U_2Cl_8^{2-}$ is a singlet state with an electronic configuration of $(5f\sigma_g)^2(5f\pi_u)^4$. The molecule presents a U-U triple bond. The U-U bond distance is 2.44 Å, the U-Cl bond distance is 2.59 Å and the U-U-Cl angle is 111.2 degrees. $U_2Cl_8^{2-}$ is different compared to $Re_2Cl_8^{2-}$ in terms of molecular bonding, in the sense that the bond in $Re_2Cl_8^{2-}$ is formally a quadruple bond, even though the δ_g bond is weak, because Re^{3+} has four electrons available to form the metal-metal bond. Only a triple bond can form in $U_2Cl_8^{2-}$ because only three electrons are available on each U^{3+} unit.

Based on several experimental reports of compounds in which the uranium is bound to a carbon atom, we have considered the possibility that a CUUC core containing two U^{1+} ions could be incorporated between two sterically hindered ligands [41] We have performed a theoretical study of a hypothetical molecule, namely PhUUPh (Ph = phenyl), the uranium analogue of the PhCrCrPh compound. We have chosen to mimic the bulky terphenyl ligands, which could be potentially promising candidates for the stabilization of multiply bonded uranium compounds, using the simplest phenyl moiety. We demonstrate that PhUUPh could be a stable chemical entity with a singlet ground state. The CASSCF method was used to generate molecular orbitals and reference functions for subsequent CASPT2 calculations. The structures of two isomers were initially optimized using DFT, namely the bent planar PhUUPh isomer (Isomer A Figure 5) and the linear isomer (Isomer B Figure 6). Starting from a trans bent planar structure, the geometry optimization for isomer A predicted a rhombic structure (a bis(μ-phenyl) structure), belonging to the D_{2h} point group and analogous to the structure of the experimentally known species U_2H_2 [29]. Linear structure B also belongs to the D_{2h} point group. CASPT2 geometry optimizations for several electronic states of various spin multiplicities were then performed on selected structural parameters, namely the U-U and U-Ph bond distances, while the geometry of the phenyl fragment was kept fixed. The most relevant CASPT2 structural parameters for the lowest electronic states of the isomers A and B, together with the relative CASPT2 energies, are reported in Table 1. The ground state of PhUUPh is a ^1A_g singlet, with a bis(μ-phenyl) structure (Figure 1-A), and an electronic configuration $(\sigma)^2(\sigma)^2(\pi)^4(\delta)^2$, which corresponds to a formal U-U quintuple bond. The effective bond order between the two uranium atoms is 3.7.

Figure 5. The bent planar PhUUPh isomer.

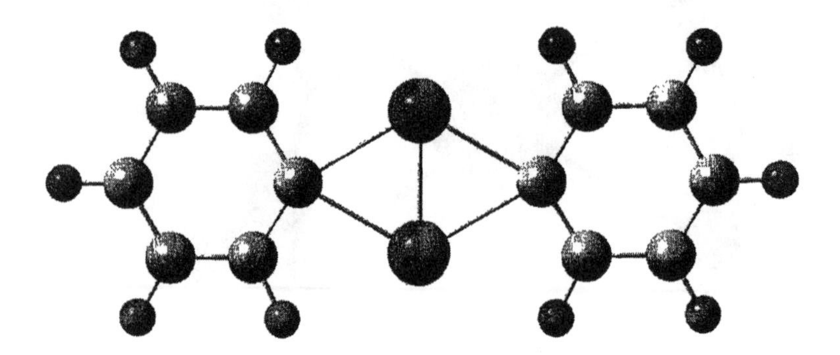

Figure 6. The linear PhUUPh isomer.

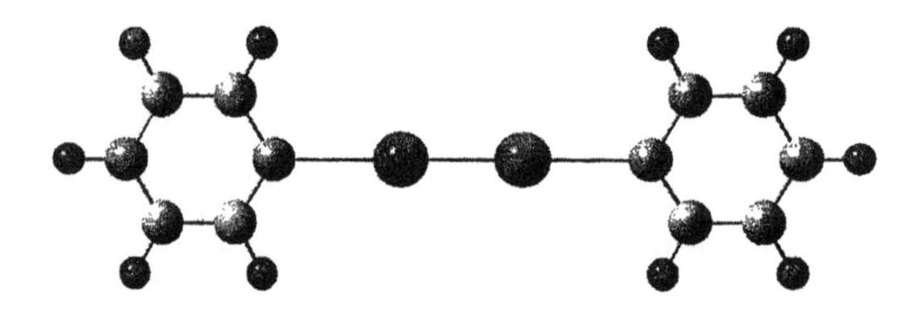

It is interesting to investigate briefly the difference in the electronic configurations of the formal U_2^{2+} moiety in PhUUPh and that of the bare metastable U_2^{2+} cation [33]. The ground state of U_2^{2+} has an electronic configuration $(\sigma)^2(\pi)^4(\delta_g)^1(\delta_u)^1(\phi_g)^1(\phi_u)^1$, corresponding to a triple bond between the two U atoms and four fully localized electrons. In PhUUPh, the electronic configuration is different, due to the fact that the molecular environment decreases the coulomb repulsion between the two U^{1+} centers thus making the U-U bond stronger than in U_2^{2+}. The corresponding U-U bond distance, 2.29 Å, is also slightly shorter than in U_2^{2+} (2.30 Å). A single bond is present between the U and C atoms.

Table 2: CASPT2 most significant structural parameters (distances in Å, angles in degrees) and relative energies (kcal/mol) for the lowest electronic states of isomer A and B of PhUUPh.

Isomer	El. State	R(U-U)	R(U-Ph)	UPhU	PhUPh	ΔE
A	^1A_g	2.286	2.315	59.2	120.8	0
A	^3A_g	2.263	2.325	58.3	121.8	+0.76
A	$^5B_{3g}$	2.537	2.371	64.7	115.3	+4.97
A	$^5B_{3u}$	2.390	2.341	61.4	118.6	+7.00
A	$^3B_{3g}$	2.324	2.368	58.8	121.2	+7.00
A	$^1B_{3g}$	2.349	2.373	59.3	120.7	+7.14
B	$^3B_{3g}$	2.304	2.395		180	+19.67
B	^3A_g	2.223	2.430		180	+22.16
B	$^1B_{3g}$	2.255	2.416		180	+27.62

Inspection of Table 2 shows that the lowest triplet state, ^3A_g, is almost degenerate with the ground state, lying only 0.76 kcal/mol above. Several triplet and quintet states of various symmetries lie 5-7 kcal/mol above the ground state. The lowest electronic states of the linear structure (Figure B) lie about 20 kcal/mol above the ground state of the bis(μ-phenyl) structure. Since the ^1A_g ground state and the ^3A_g triplet state are very close in energy, they may be expected to interact via the spin-orbit coupling operator. To evaluate the impact of such interaction on the electronic states of the PhUUPh, the spin-orbit coupling between several singlet and triplet states was thus computed at the ground state (^1A_g) geometry. The ordering of the electronic states is not affected by the inclusion of spin-orbit coupling. To assess the strength of the U-U bond in PhUUPh, its bonding energy was computed as the difference between the energy of the latter and those of the two unbound PhU fragments. PhUUPh is lower in energy than two PhU fragments by about 60 kcal/mol, with the inclusion of basis set superposition error correction. The question that one would like to answer is how to make PhUUPh and analogous species. PhUUPh could in principle be formed in a matrix - analogous to the already detected diuranium polyhydride species[9,10] - by laser ablation of uranium and co-deposition with biphenyl in an inert matrix. The phenyl ligand might however be too large to be made, and its reactions controlled, in a matrix and species like for example CH_3UUCH_3 may be more feasible.

7 Conclusions

To explore the nature of the chemical bond has been a central issue for both theoretical and experimental chemistry since the dawn of quantum chemistry eighty years ago. Today, we have a detailed understanding of what the electrons are doing in the molecules both on a qualitative and quantitative basis. We have quantum chemical methods that allow us to compute with high accuracy the properties of chemical bonds, such as bond energies, polarities, charge transfer, back bonding, etc. In later years it has been possible to extend thses methods to cover also bonding involving atoms from the lower part of the periodic table.

Until 1964 the general consensus was that the largest multiplicity that could be found in a covalent bond was three, illustrated by the strong chemical bond of the N_2 and CO molecules with bond energies close to 230 kcal/mol. The synthesis in 1964 of the formally quadruply bonded ion $(Re_2Cl_8)^{2-}$ therefore came as a surprise to the inorganic chemistry community. Since then a large number of similar compounds involving several first, second and third row transition metals have been synthetized.

In this overview we have illustrated how modern quantum chemistry can be used to explore the nature of such chemical bonds. A classical case is the Cr-Cr bond ranging from the quadruply bonded Cr(II)-Cr(II) moiety to the formal hextuple bond between to neutral chromium atoms. It is clear from this analysis that the bonding between the $3d\delta$ electrons is weak and should be considered as an intermediate between two pairs of antiferromagnetically coupled localized 3d electrons and a true chemical bond. The Cr-Cr case also illustrates that there is no simple relation between bond order and bond energy. The energy of the bond in the Cr(I) compound PhCrCrPh is twice as large as in the Cr(0) in spite of the decreased bond order. On the other hand, would the Cr(II)-Cr(II) moiety hardly be bound at all without the help of bridging ligands such as the carboxylate ions.

We have extended the analysis of the metal-metal multiple bond to cover also actinide atoms, in particular uranium. A fromal quintuple bond was found for the uranium dimer with a unique electronic structure involving six one-electron bonds with all electrons ferromagnetically coupled resultin in a high spin ground state. We have also explored if the U_2 unit could possibly be used as a central moiety in inorganic complexes similar to those explored by Cotton *et al.* for transition metal dimers. Corresponding chlorides and carboxylates were found to be stable units with multiply bonded U(III) ions. It might even be possible to use the elusive U(I) ion in metal-metal bonding involving protective organic aryl ligands in parallel to the recently synthetized ArCrCrAr compound.

The work reported here gives an almost complete picture of the metal-metal multiple bond ranging from first row transition metals to actinides. We have reported that the highest covalent bond order found in the periodic table is that between two tungsten atoms, which is a true hextuple bond with an effective bond order of 5.18 [19]. We have also shown that multiple bonding exists between actinide atoms. Recent work has shown that this is not only true for uranium but also for the other early actinide where in particular the thorium dimer exhibits a strong quadruple bond [42].

Acknowledgment

The authors thank all the developers of MOLCAS, whose effort has been essential in order to be able to study such an exciting chemistry! In particular we would like to thank Roland Lindh, Per-Åke Malmqvist, Valera Veryazov and Per-Olof Widmark. LG thanks the Swiss National Science Foundation and BOR the Swedish Foundation for Strategic Research for financial support.

References

[1] Roos, B. O. In Lawley, K. P., ed., *Advances in Chemical Physics; Ab Initio Methods in Quantum Chemistry - II.* John Wiley & Sons Ltd., Chichester, England, 1987 399.

[2] Andersson, K.; Malmqvist, P.-Å.; Roos, B. O.; Sadlej, A. J.; Wolinski, K. *J. Phys. Chem.* **1990**, *94*, 5483.

[3] Andersson, K.; Malmqvist, P.-Å.; Roos, B. O. *J. Chem. Phys.* **1992**, *96*, 1218.

[4] Roos, B. O.; Lindh, R.; Malmqvist, P.-Å.; Veryazov, V.; P.-O-Widmark. *J. Phys. Chem. A* **2005**, *109*, 6575.

[5] Roos, B. O.; Lindh, R.; Malmqvist, P.-Å.; Veryazov, V.; P.-O-Widmark. *Chem. Phys. Letters* **2005**, *295-299*, 409.

[6] Roos, B. O.; Malmqvist, P.-Å. *Chem. Phys. Phys. Chem* **2004**, *6*, 2919.

[7] Karlström, G.; Lindh, R.; Malmqvist, P.-Å.; Roos, B. O.; Ryde, U.; Veryazov, V.; Widmark, P.-O.; Cossi, M.; Schimmelpfennig, B.; Neogrady, P.; Seijo, L. *Computational Material Science* **2003**, *28*, 222.

[8] Cotton, F. A.; Harris, C. B. *Inorg. Chem.* **1965**, *4*, 330.

[9] Cotton, F. A. *Inorg. Chem.* **1965**, *4*, 334.

[10] Gagliardi, L.; Roos, B. O. *Inorg. Chem.* **2003**, *42*, 1599.

[11] Trogler, W. C.; Gray, H. B. *Acc. Chem. Res.* **1978**, *11*, 232.

[12] Trogler, W. C.; Cowman, C. D.; Gray, H. B.; Cotton, F. A. *J. Am. Chem. Soc.* **1977**, *99*, 2993.

[13] Ferrante, F.; Gagliardi, L.; Bursten, B. E.; Sattelberger, A. P. *Inorg. Chem.* **2005**, *44*, 8476.

[14] Sattelberger, A. P.; Fackler, J. P. *J. Am. Chem. Soc.* **1977**, *99*, 1258.

[15] Roos, B. O. *Collect. Czech. Chem. Commun.* **2003**, *68*, 265.

[16] Simard, B.; Lebeault-Dorget, M.-A.; Marijnissen, A.; ter Meulen, J. J. *J. Chem. Phys.* **1998**, *108*, 9668.

[17] Casey, S. M.; Leopold, D. G. *J. Phys. Chem.* **1993**, *97*, 816.

[18] Hilpert, K.; Ruthardt, K. *Ber. Bunsenges. Physik. Chem.* **1987**, *91*, 724.

[19] Borin, A. C.Private Communication.

[20] Nguyen, T.; Sutton, A. D.; Brynda, M.; Fettinger, J. C.; Long, G. J.; Power, P. P. *Science* **2005**, *310*, 844.

[21] Brynda, M.; Gagliardi, L.; Widmark, P.-O.; Power, P. P.; Roos, B. O. *Angew. Chem. Int. Ed.*. **2006**, *45*, 3804.

[22] Cotton, F. A. *Chem. Soc. Rev.* **1975**, *4*, 27.

[23] Ketkar, S. N.; Fink, M. *J. Am. Chem. Soc.* **1985**, *107*, 338.

[24] Andersson, K.; Bauschlicher, Jr., C. W.; Persson, B. J.; Roos, B. O. *Chem. Phys. Letters* **1996**, *257*, 238.

[25] Pepper, M.; Bursten, B. E. *J. Am. Chem. Soc.* **1990**, *112*, 7803.

[26] Gorokhov, L. N.; Emel'yanov, A. M.; Khodeev, Yu. S. *Teplofiz. Vys. Temp.* **1974**, *12*, 1307.

[27] Gorokhov, L. N.; Emel'yanov, A. M.; Khodeev, Yu. S. *High Temp.* **1974**, *12*, 1156.

[28] Gagliardi, L.; Heaven, M. C.; Krogh, J. W.; Roos, B. O. *J. Am. Chem. Soc.* **2005**, *127*, 86.

[29] Souter, P. F.; Kushto, G. P.; Andrews, L.; Neurock, M. *J. Am. Chem. Soc.* **1997**, *119*, 1682.

[30] Heitler, W.; London, F. *Z. Physik* **1927**, *44*, 455.

[31] Pyykkö, P.; Runeberg, N.; Hess, B. A. **2004**. as quoted by L. Gagliardi, P. Pyykkö, **2004** *Angew. Chem. Int. Ed.* *43*, 1573-1576; *Angew. Chem.* **2004**, *116*, 1599-1602.

[32] van de Walle, J.; Tarento, R. J.; Joyes, P. *Surf. Rev. Lett.* **1999**, *6*, 307. and private communication to P. Pyykkö.

[33] Gagliardi, L.; Roos, B. O. *Nature* **2005**, *433*, 848.

[34] Pyykkö, P.; Riedel, S.; Patzschke, M. *Chem. Eur. J.* **2005**. electronically published 14 April.

[35] Lawton, D.; Mason, R. *J. Am. Chem. Soc.* **1965**, *87*, 921.

[36] Stephenson, T. A.; Bannister, E.; Wilkinson, G. *J. Chem. Soc.* **1964**, 2538.

[37] Cotton, F. A.; Hillard, E. A.; Murillo, C. A.; Zhou, H.-C. *J. Am. Chem. Soc.* **2000**, *122*, 416.

[38] Roos, B. O.; Gagliardi, L. *Inorg. Chem.* **2006**, *45*, 803.

[39] Bazhanov, V. I.; Komarov, S. A.; Sevast'yanov, V. G.; Popik, M. V.; Kutnetsov, N. T.; Ezhov, Y. S. *Vysokochist. Veshchestva* **1990**, *1*, 109.

[40] Joubert, L.; Maldivi, P. *J. Phys Chem. A* **2001**, *105*, 9068.

[41] Macchia, G. L.; Brynda, M.; Gagliardi, L. *Angew. Chem. Int. Ed.* **2006**, *xx*, xx.

[42] Roos, B. O.; Malmqvist, P.-Å.; Gagliardi, L.In preparation.

Brill Academic Publishers
P.O. Box 9000, 2300 PA Leiden,
The Netherlands

Lecture Series on Computer and Computational Sciences
Volume 6, 2006, pp. 23-62

Density Functional Theory (DFT) and ab-initio Quantum Chemistry (AIQC). Story of a difficult partnership.

Werner Kutzelnigg[1]

Lehrstuhl für Theoretische Chemie,
Ruhr-Universität Bochum, D-44780 Bochum, Germany

Received 29 July, 2006; accepted in revised form 2 August, 2006

Abstract: The history of the relations, the competition, the cooperation and the irritations between Density Functional Theory (DFT) and ab-initio Quantum Chemistry (AIQC) is critically reviewed. We start with a discussion of the the main source of density functional theory (DFT), namely the Thomas-Fermi-model. An analysis of the Hartree method and the Hartree Fock method follows, with an emphasis on the role of the self-interaction of the electrons. The two methods are compared and the advantages and drawbacks of the Hartree-Fock scheme are discussed. In connection with Slater's $X\alpha$-method it is explained why with an approximate local exchange the bond dissociation is better described than with the exact delocalized Hartree-Fock exchange. The Hohenberg-Kohn (HK) theorem is presented in terms of a sequence of two Legendre transformations. The essential message of the HK theorem is the existence of a variation principle in terms of a trial density. Problematic aspects of the HK paper are not ignored, especially claims without any physical meaning such as that a state is entirely characterized by its density. After an analysis of a system of non-interacting particles the Kohn-Sham recipe is presented and analyzed. The closeness of the Kohn-Sham approach to the much older Slater exchange is stressed. The final part of this paper deals with new trends both in DFT and AIQC. This includes orbital functional theory, exchange-only DFT, and density matrix functional theory (DMFT). We review paradigms of ab-initio theory, that are partially a reaction to the challenge of DFT methods, especially the minimal parametrization of a state. We stress the importance of the concept of separability and the need to formulate the theory in terms of additively separable quantities. In this context the cumulants of reduced density matrices deserve special interest. Methods for the direct calculation of the two-particle density matrix and of density cumulants are discussed. Speculations about a possible convergence of DFT and AIQC towards each other close this paper.

Keywords: DFT, Thomas-Fermi model, Hartree, Hartree-Fock, $X\alpha$, Hohenberg-Kohn theorem, Kohn-Sham, Legendre transformation, DMFT, density cumulants, Nakatsuji theorem, Nooijen conjecture, generalized normal ordering.

Mathematics Subject Classification: PACS: 31.10.+z, 31.15.Ar, 31.15.Ew, 31.25.-v

PACS: 31.10.+z, 31.15.Ar, 31.15.Ew, 31.25.-v

1 Introduction

There is no doubt that the communities of ab-initio Quantum Chemistry (AIQC) and Density Functional Theory (DFT) are *partners*. They cooperatate in the design of program packages for

[1]Corresponding author. E-mail: werner.kutzelnigg@rub.de

theoretical calculations on atoms, molecules, clusters or solids. They have even shared the Nobel price 1998, given to J.A. Pople as a representative of AIQC and W. Kohn as his counterpart in DFT.

It would, nevertheless, be exaggerated, or at least premature, to classify the relation between AIQC and DFT as a *friendship*. It is rather dominated by competition and mutual prejudices, even hostility. Many ab-initio quantum chemists accept DFT only because it often leads to good results at low costs, but have, at the same time, the feeling that the good cost-performance ratio can only be due to dubious tricks or to a cancellation of errors. DFT people, on the other hand, often claim that DFT has made wave function theory obsolete, and regard AIQC as a waste of time, and ab-initio quantum chemists as dinosaurs. There is a current discussion whether DFT methods belong to the general class of *ab-initio theory*, or whether they should rather be classified as *semi-empirical*, since they require a calibration with empirical (or accurate ab-initio) data. A standard characterization of the difference between AIQC and DFT has been that in AIQC there is a *hierarchy of approximations*, such that it is, at least in principle, always possible to achieve any desired accuracy, while there is usually no prescription how one should proceed if a certain DFT method fails.

In recent years the communication between AIQC and DFT has much improved and one can even observe a mutual convergence. It may be that the way from a partnership to a friendship is not too far. To open this way it is mandatory to analyze the *origins of the very reluctant attitude of the two communities towards each other*, and this requires to have a look at the history, as we shall do in this review. One of the reasons for the poor communication between the two communities is certainly that DFT has its origin in *solid state physics*, and uses in part a language that differs from that current in chemistry. A large role has been the importance given in DFT to the Hohenberg-Kohn theorem, the real meaning of which is *not easily understood*, and which has, more or less willingly, often been mystified. This is manifest in the fact that the papers by Hohenberg and Kohn [1] as well as by Kohn and Sham [2] are still frequently quoted, which would not have been happened if these fundamental papers had, in due time after their publication, been *canonized* and become textbook knowledge, for which the original papers are usually not quoted.

Some confusion has been due to the fact that the vast majority of calculations with the label DFT were based on the Kohn-Sham recipe, and that the names 'DFT' and 'Kohn-Sham' were used almost synonymously. If one reserves the name 'DFT' to methods in which the *entire internal energy* E_{int} is formulated as a functional of the density, Kohn-Sham methods do not belong to density functional theory in a strict sense, because the kinetic energy is evaluated in term of a *density matrix* (with a correction, see sec. 5.7), rather than a density.

It has been clear for quite a while that Kohn-Sham type DFT methods often perform rather well, *not because of the HK theorem*, but because they are sufficiently similar to Hartree-Fock like mean field theories, though with some flexibility to avoid some drawbacks of the Hartree-Fock method. Unfortunately a systematic analysis on these lines has never been done, although this would have been in order some 30 years ago already. AIQC should have learned from the success of DFT, that it is possible to get acceptable results by simple-minded approaches and should have tried to derive theories with a similar structure in a strict ab-initio framework. It ought to have reacted to the challenge of DFT, rather than become simply overwhealmed by it.

We try here such an analysis, and start by a historical review about DFT and AIQC and their encounters.

2 From the 'free' electron gas to Thomas-Fermi theory

2.1 The 'free' electron gas.

In a way DFT is older than AIQC. It goes back to the earliest quantum mechanical study of an n-fermion system, namely the so-called 'free' electron gas.

Let us consider n non-interacting electrons in a square box of length A, ignoring the problem, how one could confine non-interacting particles to a finite volume. One could take infinitely repulsive boundaries, but it is easier and leads to the same result, to use periodic boundary conditions, i.e. to require that the one-electron eigenfunctions φ satisfy

$$\varphi(\vec{r}) = \varphi(\vec{r} + \{A,0,0\}) = \varphi(\vec{r} + \{0,A,0\}) = \varphi(\vec{r} + \{0,0,A\}) \quad (1)$$

The (non-relativistic) Hamiltonian

$$H = -\frac{\hbar^2}{2m}\nabla^2 \quad (2)$$

only consists of the kinetic energy and its normalized one-electron eigenfunctions, which satisfy the boundary condition (1) are plane-wave states:

$$\begin{aligned} \varphi_{\vec{k}} &= A^{-\frac{3}{2}} e^{i\vec{k}\vec{r}}; \int |\varphi_{\vec{k}}| d^3r = 1 & (3) \\ \vec{k} &= \frac{2\pi}{A}\vec{m};\ \{m_x, m_y, m_z\} = 0,1,2... & (4) \\ H\varphi_{\vec{k}} &= \varepsilon_k \varphi_{\vec{k}} & (5) \end{aligned}$$

Any possible eigenfunction $\varphi_{\vec{k}}$ is characterized by a *wave vector* $\vec{k}$. The corresponding eigenvalue, representing the kinetic energy is

$$\varepsilon_k = \frac{\hbar^2 k^2}{2m};\ k = |\vec{k}| \quad (6)$$

It depends only on the absolute value k of $\vec{k}$. We take care of the Pauli principle in filling the $\varphi_{\vec{k}}$ in order of increasing energy with two electrons each (one for α and one for β spin) up to the highest occupied energy level ε_{k_F}. More precisely we construct the n-electron wavefunction Ψ as a Hartree product or a Slater determinant built up from the energetically lowest one-electron functions. Since the $\varphi_{\vec{k}}$ are orthonormal, a Hartree product and a Slater determinant yield the same one-particle energy. The k-value k_F corresponding to the highest occupied energy ε_{k_F} is called the *Fermi momentum* (in units of $\hbar$) and ε_{k_F} the *Fermi energy*. The number of electrons corresponding to some k_F is equal to twice the number of $\vec{k}$ points in a sphere of radius k_F, the Fermi sphere. For a sufficiently large number of electrons in the box, we can replace the summation over $\vec{k}$-values by an integral and get so for the number of particles n, and the density ϱ, corresponding to the maximum occupied energy ε_{k_F}:

$$n = 2\left(\frac{A}{2\pi}\right)^3 4\pi \int_0^{k_F} k^2 dk = \frac{A^3 k_F^3}{3\pi^2};\ \rho = \frac{n}{A^3} = \frac{k_F^3}{3\pi^2} \quad (7)$$

We can invert this relation to get the Fermi momentum and the Fermi energy as a function of the density ϱ, i. e. of the number of electrons per Volume.

$$k_F = (3\pi^2\rho)^{\frac{1}{3}};\ \varepsilon_{k_F} = \frac{\hbar^2 k_F^2}{2m} \quad (8)$$

The total energy E is then

$$E = 2\left(\frac{A}{2\pi}\right)^3 4\pi \int_0^{k_F} \frac{\hbar^2 k^2}{2m} k^2 dk = \frac{k_F^5 A^3 \hbar^2}{10 m\pi^2} \quad (9)$$

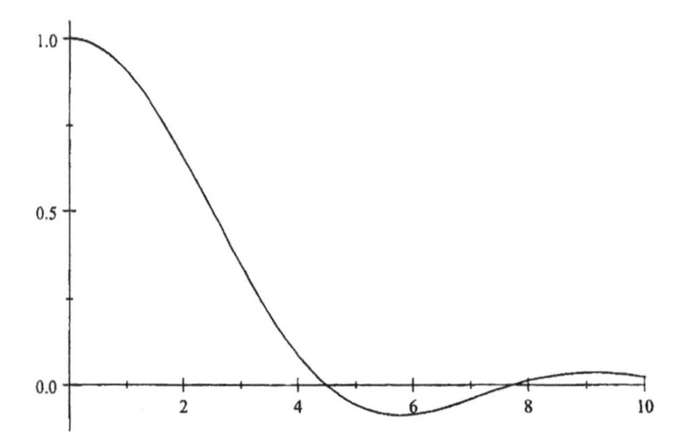

Figure 1: The density matrix $\Gamma(\vec{r};\vec{r}')$ of a free-electron gas. Γ/ρ is plotted as a function of $k_F|\vec{r}-\vec{r}'|$.

and the *energy density* ε, i.e. the energy per volume is

$$\varepsilon = \frac{E}{A^3} = \frac{k_F^5\hbar^2}{10m\pi^2} = \frac{(3\rho)^{\frac{5}{3}}\pi^{\frac{4}{3}}\hbar^2}{10m} \tag{10}$$

We note that for constant ρ, E is proportional to A^3 (or to n), i. e. the energy is *extensive*. We further see that the *energy density* is not proportional to the density, but goes as $\rho^{\frac{5}{3}}$, i.e. for twice the density the energy increases by a factor $2^{\frac{5}{3}} \approx 3$.

Both results come from the fact that the electrons are *fermions*, i.e. that one one-electron state can at most be occupied by two electrons. There is only kinetic energy.

The volume in phase space occupied by our n-electron system is equal to the product of the volume A^3 in position space and the volume of the Fermi sphere in momentum space,

$$A^3\frac{4\pi}{3}(\hbar k_F)^3 = \frac{4\pi}{3}(A\hbar k_F)^3 = \frac{1}{2}h^3 n, \tag{11}$$

i.e. an electrons requires the volume element $\frac{1}{2}h^3$ in phase space. This is a fundamental relation of Fermi statistics. There is no such limitation for bosons, which can all go into the same element of phase space.

The spinfree one particle density matrix for the free electron gas

$$\Gamma(\vec{r},\vec{r}\,') = \sum_{occ} \varphi(\vec{r})\varphi^*(\vec{r}\,') \tag{12}$$

is evaluated by replacing the summation over the occupied states by an integral over the Fermi sphere:

$$\begin{aligned}
\Gamma(\vec{r},\vec{r}\,') &= 2\frac{A^3}{(2\pi)^3}2\pi\int_0^{k_F}dk\int_0^{\pi}\sin\theta d\theta\frac{e^{-ik|\vec{r}-\vec{r}\,'|\cos\theta}}{A^3}k^2 \\
&= \frac{1}{\pi^2}\frac{\int_0^{k_F}k\sin(k|\vec{r}-\vec{r}\,'|)dk}{|\vec{r}-\vec{r}\,'|} \\
&= \frac{\sin(k_F|\vec{r}-\vec{r}\,'|)-k_F|\vec{r}-\vec{r}\,'|\cos(k_F|\vec{r}-\vec{r}\,'|)}{\pi^2|\vec{r}-\vec{r}\,'|^3} \qquad (13)
\end{aligned}$$

$$\Gamma(\vec{r},\vec{r}) = \varrho(\vec{r}) = \varrho = \frac{k_F^3}{3\pi^2} \qquad (14)$$

$\Gamma(\vec{r},\vec{r}\,')$ is actually a function of $|\vec{r}-\vec{r}\,'|$ only. The 'diagonal' element $\Gamma(\vec{r},\vec{r})$ is equal to the density ϱ, that we have derived earlier.

A plot of $\Gamma(|\vec{r}-\vec{r}\,'|)/\varrho$ is seen on fig. 1. Apparently Γ decays rather fast with $k_F|\vec{r}-\vec{r}\,'|$. This is a result of the integration over $k \le k_F$. The individual terms in the sum (12) have much longer range.

2.2 The jellium model and the exchange density

The assumption that electrons are non-interacting, is rather unrealistic. It is even more unrealistic to consider a system of electrically charged particles with the same sign of the charge. In reality electrons move in the field of positively charged nuclei. If one introduces the potential created by nuclear charges into the Hamiltonian, one destroys the homogeneity of the electron distribution. This can be preserved, if one introduces instead a uniform background of positive charge, such that the net electric charge vanishes. This defines the jellium model. In this model the repulsions between the electrons and of the positive background with itself cancels the Coulomb attraction between the electrons and the uniformly positive background. Neither of these interactions would be extensive individually, but rather scale with the square of the size of the system, so it is essential that these terms cancel.

However, the cancellation is not complete. Since there is no self-interaction between the electrons, each electron interacts with $n-1$ rather than with n electrons, such that the energy of interaction between the electrons has (unlike the two other interaction terms, which involve a genuine positive continuum charge) to be multiplied by a factor $(n-1)/n$. This factor, which has nothing to do with the fermionic character of the electrons [3, 4], and which is unavoidable if one uses a Hartree product wave function, goes to 1 for large n, and is usually ignored. We come back to this in subsection 3.5.

We now use explicitly a single-Slater-determinant wave function. This is not exact, but usually a good approximation. What one neglects in this approximation, is called the correlation energy.

The probability density $\varrho^{(2)}(\vec{r}_1,\vec{r}_2)$ for two electrons to be found at the positions $\vec{r}_1$ and $\vec{r}_2$ respectively, based on a single Slater determinant wave function, is expressible via the density ϱ and the (spinfree) one-particle density matrix $\Gamma(\vec{r}_1,\vec{r}_2)$ as

$$\begin{aligned}
\varrho^{(2)}(\vec{r}_1,\vec{r}_2) &= \varrho(\vec{r}_1)\varrho(\vec{r}_2)-\frac{1}{2}|\Gamma(\vec{r}_1,\vec{r}_2)|^2 \\
&= \varrho^2\left\{1-\frac{9}{2}\left[\frac{\sin(k_Fr_{12})-k_Fr_{12}\cos(k_Fr_{12})}{k_F^3r_{12}^3}\right]^2\right\} \qquad (15)
\end{aligned}$$

There is an *exchange contribution* [5] plotted on fig. 2

$$-\frac{1}{2}|\Gamma(\vec{r}_1,\vec{r}_2)|^2 \qquad (16)$$

with the properties

$$-\frac{1}{2}\int|\Gamma(\vec{r}_1,\vec{r}_2)|^2 d^3|\vec{r}_1-\vec{r}_2| = -\varrho \tag{17}$$

$$-\frac{1}{2}\int|\Gamma(\vec{r}_1,\vec{r}_2)|^2 d^3r_1d^3r_2 = -n \tag{18}$$

$$-\frac{1}{4}\int\frac{|\Gamma(\vec{r}_1,\vec{r}_2)|^2}{|\vec{r}_1-\vec{r}_2|}d^3|\vec{r}_1-\vec{r}_2| = -\frac{3}{4}(\frac{3}{\pi})^{\frac{1}{3}}\varrho^{\frac{4}{3}} \tag{19}$$

The last expression is the *exchange energy density*. The *exchange energy* turns out to be extensive (like the kinetic energy). It increases, in absolute value, more strongly than linearly with the density, but not as fast as the kinetic energy. Unlike the latter it is negative (attractive).

The *pair density* $\varrho^{(2)}$ integrates to $n(n-1)$, indicating that the number of pairs of electrons is $n(n-1)$ rather than n^2. By taking care of exchange, one mainly removes the *unphysical self-pairing* of the electrons. However, this looks here different than in the case of atoms or molecules. For these the Fermi hole mainly reflected the shell structure, provided that one used a localized representation, in which the shell structure is manifest. For an electron gas there is no obvious procedure to construct localized orbitals, this system is inevitably *delocalized*. If we describe this by means of the Hartree approximation (in terms of delocalized orbitals, although the Hartree approximation is – strictly speaking – only acceptable in a localized representation), we find

$$\varrho^{(2)}(\vec{r}_1,\vec{r}_2) = \frac{n-1}{n}\varrho(\vec{r}_1)\varrho(\vec{r}_2) \tag{20}$$

i.e. the electrons appear statistically independent. The 'hole' corresponding to removal of the unphysical self-pairing is completely delocalized. Conversely the Fermi-hole, which manifests here genuine exchange, (and not a shell structure) is of rather short-range type. This is an indication that the exchange hole of the free electron gas is *by no means a representative model* for atomic or molecular systems. This did not prevent its use as a model of Fermi correlation in atoms or molecules, first in Slater's *X-α method* [6] and later in Kohn-Sham approaches of LDA-type [2].

2.3 The Thomas-Fermi atom.

We consider an atom with charge Z and $n = \int \varrho(\vec{r})d^3r$ electrons. We assume that the density of the kinetic energy is the same functional of the density as for a free electron gas. Then we start with the approximate expression for the energy expectation value.

$$\begin{aligned} E &= \int T(\varrho)d^3r - Z\int\frac{\varrho}{r}d^3r + \frac{1}{2}\int\frac{\varrho(\vec{r}_1)\varrho(\vec{r}_2)}{|\vec{r}_1-\vec{r}_2|}d^3r_1d^3r_2 \\ T(\varrho) &= \frac{(3\varrho)^{\frac{5}{3}}\pi^{\frac{4}{3}}\hbar^2}{10m} \end{aligned} \tag{21}$$

This is obviously a functional of the density ϱ. If we minimize the energy with respect to variation of ϱ, subject to its normalization, we obtain the Thomas-Fermi equation [7, 8], as the simplest possible approach to the ground state of all neutral atoms, as well as for all atomic ions.

The Thomas-Fermi model describes all neutral atoms by a *universal function*. It predicts that the energy of any atomic ground states is a negative constant times $Z^{7/3}$. This is in error by more than 50% for Z=1, but only by 15% for Z=100. It describes, in a way, the non-relativistic limit for large Z. In this model the electron density decays too slowly for $r \to \infty$, namely as $\sim r^{-6}$ rather than $\sim \exp(-ar)$. A *serious drawback* is that the Thomas-Fermi model does not account for the *shell structure of atoms*. One can object to this model that only the Coulomb interaction

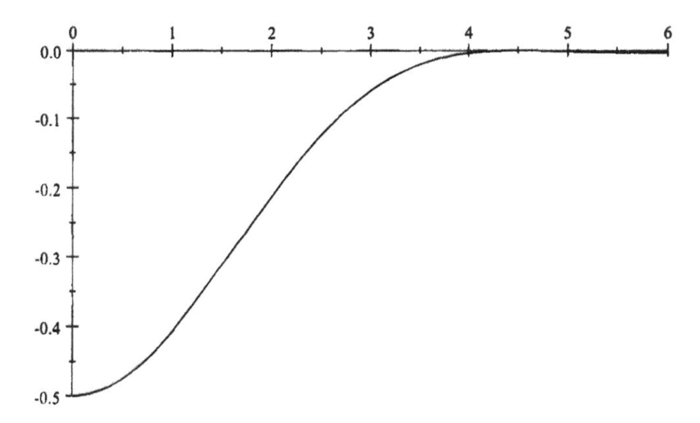

Figure 2: The Fermi hole of a free-electron gas as function of $k_F|\vec{r} - \vec{r}\,'|$.

between the electrons is considered, but the exchange is ignored. It is relatively easy to include the exchange energy as a functional of the density, taking again that derived for the jellium model. This leads us to the Thomas-Fermi-Dirac model, in which there is an additional contribution to the energy.

$$E_x = \int X(\varrho)d^3r; \; X(\varrho) = -\frac{3}{4}\left(\frac{3}{\pi}\right)^{\frac{1}{3}} \varrho^{\frac{4}{3}} \tag{22}$$

Unfortunately, the corrections introduced by the inclusion of exchange go into the wrong direction, they make the agreement between theory and experiment worse, rather than better. This is why the Dirac exchange [5] hardly became popular in Thomas-Fermi theory. A substantial improvement of the Thomas-Fermi model came with the Weizsäcker correction [9], in which the dependence of the kinetic energy density on the *gradient* of the density is taken care of to the leading order. Actually the density in an atom is very far from homogeneous, such that the use of an expression that is valid for the homogeneous electron gas, is certainly far from realistic.

However, even with the Weizsäcker correction *no shell structure* of the atoms was found.

Moreover Teller 1962 [10] has shown that there is *no chemical bond* within the Thomas-Fermi model. The failure of this model lies obviously in the difficulty to express the kinetic energy (which is mainly determined by the Fermi statistics, i.e. the antisymmetry of the wave function) as a functional of the density. This works satisfactorily only for the trivial case that the density is a constant as in the jellium model. As we shall see later, it is always possible to express the kinetic energy as a functional of the *one-particle density matrix*.

3 Hartree-Fock and the like

3.1 Hartree's theory of atomic structure

What could not be achieved by means of the Thomas-Fermi model, namely a quantitative theory of atoms, became possible with Hartree's studies on atomic structure.

Hartree [11] regarded the electrons in a atom as *distinguishable* and argued that any electron (describable by a one-electron function) moves in the effective Coulomb field of the *other* electrons. So he had to solve a one-electron Schrödinger equation for the kth electron with the Hamiltonian (in Hartree units)

$$h_k = -\frac{1}{2}\nabla_k^2 - \frac{Z}{r_k} + \sum_{l \neq k} \int |\varphi_l(\vec{r}_l)|^2 \frac{1}{|\vec{r}_l - \vec{r}_k|} d^3 r_l \tag{23}$$

Since h_k depends on the unknown φ_l of the other electrons, the various eigenvalue equations had to be solved iteratively, until *self-consistency*. Hartree took care of the Pauli principle by occupying each one-electron state by at most two electrons. In the sum in (23) up to two of the φ_k may be the same.

It is essential for the Hartree model that there is *no unphysical interaction of an electron with itself*. This implies that different one-electron functions φ_k (now called *orbitals*) are eigenfunctions of different Hamiltonians. The electron interaction energy of the Hartree method is:

$$E_{ee} = \sum_{k<l} \frac{|\varphi_k(1)|^2 |\varphi_l(2)|^2}{r_{12}} d^3 r_1 d^3 r_2 \tag{24}$$

In the language which is now current, one would call the Hartree scheme an *orbital functional theory*. The energy expectation value is a functional of the occupied orbitals φ_k.

3.2 Hartree-Fock theory

A formulation of Hartree's concept in the language of many-electron quantum mechanics was given by Fock [12]. He started from the expectation value of the n-electron Hamiltonian in terms of what is now called a *Slater determinant*, built up from the spin-orbitals ψ_k, which are of the form either $\varphi_k \alpha$ or $\varphi_k \beta$, with α and β the two spin functions.

$$\begin{aligned} E &= \sum_k \langle \psi_k | -\frac{1}{2}\nabla_k^2 - \frac{Z}{r_k} | \psi_k \rangle + \sum_{k<l} \int \frac{|\psi_k(1)|^2 |\psi_l(2)|^2 - \psi_k^\dagger(1)\psi_l(1)\psi_l^\dagger(2)\psi_k(2)}{r_{12}} d\tau_1 d\tau_2 \\ &= \sum_k \langle \psi_k | -\frac{1}{2}\nabla_k^2 - \frac{Z}{r_k} | \psi_k \rangle + \frac{1}{2} \sum_{k,l} \int \frac{|\psi_k(1)|^2 |\psi_l(2)|^2 - \psi_k^\dagger(1)\psi_l(1)\psi_l^\dagger(2)\psi_k(2)}{r_{12}} d\tau_1 d\tau_2 \\ &= \sum_k \langle \psi_k | -\frac{1}{2}\nabla_k^2 - \frac{Z}{r_k} | \psi_k \rangle + \frac{1}{2} \int \frac{|\varrho(1)|^2 |\varrho(2)|^2 - |\gamma(1,2)|^2}{r_{12}} d\tau_1 d\tau_2 \end{aligned} \tag{25}$$

$$\gamma(1,2) = \sum_k \psi_k(1)\psi_k^\dagger(2) \tag{26}$$

$$\varrho(1) = \gamma(1,1) = \sum_k \psi_k(1)\psi_k^\dagger(1) = \sum_k |\psi_k(1)|^2 \tag{27}$$

with $\gamma(1,2)$ the one particle density matrix, and $\varrho(1)$ the one-particle density (both depending on spin). The volume element $d\tau = d^3 r ds$ symbolizes integration in ordinary space and summation over spin.

Variation of the spin orbitals ψ_k occupied in the Slater determinant Φ, subject to the orthonormality of the ψ_k (which is no loss of generality) then lead to the Hartree-Fock equations, i.e.

one-electron pseudo-eigenvalue equations with the Fock operator

$$\begin{aligned} f_k &= -\frac{1}{2}\nabla_k^2 - \frac{Z}{r_k} + \sum_{l\neq k} \int |\varphi_l(\vec{r}_l)|^2 \frac{1}{|\vec{r}_l - \vec{r}_k|} d\tau_l - \sum_{l\neq k} K_{kl} \\ &= -\frac{1}{2}\nabla_k^2 - \frac{Z}{r_k} + \int \varrho(\vec{r}_l) \frac{1}{|\vec{r}_l - \vec{r}_k|} d\tau_l - K_k \end{aligned} \tag{28}$$

with K_k the *non-local*, so-called exchange operator, which is complicated in configuration space, but easy in Fock space (see section 7.6).

The Hartree-Fock energy differs from the Hartree energy in the presence of an *exchange interaction*, and the Fock operator from the Hartree operator in the presence of the *exchange operator*. From a quantitative point of view the *exchange terms are not very important*. However the very presence of exchange makes an important simplification possible. Unlike in the original Hartree equations, where the restriction $k \neq l$ is essential, it does in the Hartree-Fock equations *not make a difference* whether or not one imposes this restriction. The *unphysical Coulomb self-interaction* cancels exactly with the equally *unphysical self-exchange*. Taking advantage of this cancellation and omitting the summation restriction, one achieves that all ψ_k become eigenfunctions of the same *Fock operator*. This automatically guarantees that all ψ_k are automatically orthogonal to each other, which considerably simplifies the formalism. It also follows that the individual ψ_k have no direct physical meaning, since the theory is invariant under a unitary transformation among the occupied ψ_k. A Hartree-Fock state is fully characterized by its one-particle density matrix γ_1 (in terms of spin-orbitals). So it can be classified as a *density matrix functional theory*.

3.3 Merits of Hartree-Fock theory

Hartree-Fock theory certainly has a few merits:

(a) It is both conceptually and operationally rather simple

(b) It is variational, i. e. provides an upper bound for the ground state energy

(c) The Hartree-Fock wave function is the 'best' approximation, in the sense of the *variation principle*, in the form of a single Slater-determinant.

(d) It is a good approximation, if the exact γ_1 is close to idempotent, i.e. $\gamma_1 \approx \gamma_1^2$ or equivalently if the exact two-particle cumulant λ_2 is close to 0 (see section 7.5).

(e) It cannot be improved by inclusion of 'single substitutions' (Brillouin theorem)

(f) It is a convenient starting point for more refined calculations, such as Møller-Plesset perturbation theory or coupled-cluster theory.

(g) It can easily be generalized to be applicable to molecules and clusters (keywords MO- and MO-LCAO theory), and even to solids, at least to insulators or semiconductors.

(h) According to Koopmans' theorem the canonical orbital energies are acceptable approximations to ionization potentials.

(i) Hartree-Fock theory is essentially a theory in terms of the one-particle density matrix.

3.4 Drawbacks of Hartree-Fock theory

One cannot deny, however, that Hartree-Fock has also a few drawbacks, namely:

(a) It is a theory for a single state, and hardly gives informations on other states, except, in view of Koopmans' theorem for singly ionized states.

(b) In particular the virtual (unoccupied) orbitals have no physical meaning, and are not related to excited states.

(c) The energy gap between the highest occupied molecular orbital (HOMO) and the lowest unoccupied MO (LUMO) has no relation to an observable band gap.

(d) The exchange potential is *non-local.* This is serious only for a theory in real space, not in a basis representation.

(e) Hartree-Fock theory does not describe bond dissociation.

(f) It fails in situations of near-degeneracy.

(g) It gives the wrong density of states at the Fermi level for metals.

(h) So-called Hartree-Fock instabilities may occur.

(i) To cover various cases, variants of HF theory have to be defined, such as RHF (restricted Hartree-Fock), UHF (unrestricted Hartree-Fock) etc. Ambiguities are possible.

(j) Hartree-Fock is a problematic starting point for a qualitatively correct derivation of a one-electron theory [13]. This is mainly related to the inability of HF to describe bond dissociation.

(k) Hartree-Fock leads to wrong electron distribution in transition metal complexes like MnO_4^- [14].

(l) The Hartree-Fock exchange appears too delocalized. There is a too high probability for two electrons to come close to each other.

To be fair, one should not blame HF theory for not accounting for phenomena, for which is was not designed. It was certainly designed as a theory for *one state,* and not as a theory to describe spectral excitations.

Of course, most defects of HF theory disappear if one climbs up the hierarchy of approximations, that can be built upon HF. However, one looses then the simplicity of a mean-field theory. One should like to overcome these defects *without loosing the simplicity* of the HF scheme. DFT has achieved this goal to a certain extent. Most DFT approaches that work, especially those of Kohn-Sham type, are basically mean-field theories (although this is often not admitted). Within the DFT frame corrections to Hartree-Fock theory were possible, that were *prohibited in the strict ab-initio framework* (e.g. by the insistence on the variation principle or the strict satisfaction of the Pauli principle). One of these corrections is the replacement of delocalized by *local exchange,* which attenuates some of the drawbacks listed above. It is on these lines where one ought to have searched in order to understand the sometimes unexpected success of DFT methods.

3.5 Hartree vs. Hartree-Fock

Let us now reconsider the relation between Hartree and Hartree-Fock and ask the somewhat provocative question: *Can Hartree be better than Hartree-Fock?*

There are two equivalent expressions for the electron interaction in Hartree-Fock theory,

(a) excluding self-interaction and self-exchange

$$E_{ee} = \sum_{k<l} \int \frac{|\psi_k(1)|^2|\psi_l(2)|^2 - \psi_k^\dagger(1)\psi_l(1)\psi_l^\dagger(2)\psi_k(2)}{r_{12}} d\tau_1 d\tau_2 \tag{29}$$

(b) including self-interaction and self-exchange

$$\begin{aligned} E_{ee} &= \frac{1}{2}\sum_{k,l} \int \frac{|\psi_k(1)|^2|\psi_l(2)|^2 - \psi_k^\dagger(1)\psi_l(1)\psi_l^\dagger(2)\psi_k(2)}{r_{12}} d\tau_1 d\tau_2 \\ &= \frac{1}{2}\int \frac{\varrho(1)\varrho(2) - \gamma(1,2)\gamma(2,1)}{r_{12}} d\tau_1 d\tau_2 \end{aligned} \tag{30}$$

It is not a very crude simplification if one ignores the exchange terms in the restricted sum (29). This namely is just the Hartree approximation with

$$E_{ee} = \sum_{k<l} \int \frac{|\psi_k(1)|^2 |\psi_l(2)|^2}{r_{12}} d\tau_1 d\tau_2 \tag{31}$$

To ignore the exchange terms in the unrestricted sum (29) is, by no means, justified, because it amounts to adding an *unphysical self interaction* to the Hartree approximation.

In the Hartree approximation different electrons move in different effective fields. This has the disadvantage that the orbitals are not automatically orthogonal, and also that one cannot eliminate the orbitals in favor of the density or the density matrix. However, one only deals with local operators.

Hartree-Fock theory is invariant with respect to *unitary transformations* between the occupied spin orbitals. One can, however, try to find among all possible equivalent choices of the occupied orbitals, those between which the *genuine exchange interaction* is absolutely a minimum (or the self-exchange a maximum). These orbitals turn out to be the best localized orbitals according to the criterion of Edmiston and Ruedenberg [15]. For these orbitals the best agreement between Hartree and Hartree-Fock is achieved. These orbitals should also come out if one attempts to minimize the Hartree energy.

The best Hartree orbitals should not only minimize the total energy but also reduce the orbital repulsion. There is more flexibility in the Hartree than the Hartree-Fock method In a Hartree theory the virtual orbitals are more meaningful than in Hartree-Fock.

Studies in the frame of the Hartree method in a post-Hartree-Fock context have hardly been performed, probably because one does not respect the Pauli principle in a strict sense, but it might have been worthwhile. In the spirit of what has been said at the end of the last subsection one can argue that Hartree can be regarded as a correction to Hartree-Fock, that is physically meaningful but prohibited in an ab-initio context.

4 From Slater exchange to Kohn-Sham

4.1 Slater exchange ($X\alpha$)

Slater [6] was concerned with one drawback of Hartree-Fock theory, namely the *non-locality of the exchange potential.* He proposed a local approximation to the latter, in the same spirit as Dirac's exchange potential [5] in Thomas-Fermi-Dirac theory, with the difference that unlike in Thomas-Fermi-Dirac theory the *kinetic energy* is evaluated exactly, and not approximated by a functional of the density.

The name $X\alpha$ contains two *factors*, the X stands for (local) exchange, the α for an adjustable linear parameter.

Slater's motivation was a simplification of the Hartree-Fock method, such that it could be applied to solid states, for which straight Hartree-Fock is too complicated. The use of the Dirac type local exchange worked rather nicely for solids, but less so for atoms or molecules. This is probably related to the fact that the Dirac exchange is appropriate for systems, which are highly delocalized and for which the density does not not vary too much. For systems with strongly localized (or rather localizable) orbitals, neglect of exchange as in the Hartree method might have been preferable, but was not tried. The nonlocality of the exchange operator in Hartree Fock theory hardly causes problems if one introduces a basis and a *matrix representation* of the terms in the Hamiltonian. An unconventional access to molecules, based on the *muffin tin* potential, a concept that had played a big role in solid state theory, and which is partially a theory in real space, was greatly simplified if one used a local exchange potential. So the appearance of Slater exchange in molecular theory was mainly accompanied by the muffin tin approximation [16]. In spite of a great effort of the people involved in this combination to *conquer* molecular theory, it did not turn out to be very convincing, and was almost entirely abandoned in molecular theory around 1975. Both the muffin tin potential and Slater exchange survived in solid state theory.

Possibly the overselling and the final disillusion about the combination of muffin tin with Slater exchange, did some harm to the latter, which it not really deserved. It had hardly been noticed that a combination of Slater exchange with standard techniques of AIQC is possible, and led to acceptable results [17].

Anyway the general feeling in the AIQC community around 1975 was not only that it had been cured from density functional theory (after all, Slater exchange is, in some sense, a density functional theory), but that it had become immunized for ever.

Actually it took roughly 15 years, before the next *attack* of DFT on atomic and molecular theory started, and this time it was much more successful.

In the meantime some interesting observations were made with respect to Slater exchange calculations. Although Slater exchange was introduced as a simplification of Hartree-Fock, the results were often *better than those of genuine Hartree-Fock theory.* The results of $X\alpha$ calculations were therefore often compared directly with experiment rather than with those theoretical results that they were meant to approximate.

It had, so to say, been forgotten that Slater exchange was introduced as a simplification of Hartree-Fock, and it was interpreted as an approximation to post-Hartree-Fock n-electron theory. That an approximation could be better than the theory that it was designed to approximate, was at best regarded as a curiosity. One hardly cared for an explanation. Some rationalization of the unexpected good performance of Slater exchange came much later with the work of Kohn and Sham [2], in which a similar ansatz was proposed, but this time not as an approximation to Hartree-Fock, but rather as a simplification of density functional theory (see subsection 5.7). The point is that as a first approximation the same functional form can be taken for the exchange and the correlation functional, such that the exchange functional multiplied by an adjustable factor can also be used to describe *exchange and correlation at the same time.*

4.2 Bond dissociation in MO theory

The following considerations are not directly related to Slater exchange or DFT, but they illustrate that a brute-force replacement of delocalized by localized exchange may lead to a substantial improvement of MO theory, that is especially useful if one wants to build a qualitative theory of the chemical bond on the concept of MO theory. In doing so, one is faced with the problem that MO theory does not correctly dissociate chemical bonds. The full theory has been outlined elsewhere [13]. We shall here only consider the simple case of the H_2 molecule.

Take the example of H_2 and choose the simple MO-LCAO description in a minimal basis consisting of the AOs a and b at the respective centers

$$\Psi(1,2) = \varphi(\vec{r}_1)\varphi(\vec{r}_2) \tag{32}$$

$$\varphi(\vec{r}) = \frac{1}{\sqrt{2(1+S)}}[a(\vec{r}) + b(\vec{r})]. \tag{33}$$

We use the standard notation for the relevant one- and two-electron matrix elements

$$H_{aa} = \langle a|h|a\rangle;\ H_{ab} = \langle a|h|b\rangle;\ S = \langle a|b\rangle \tag{34}$$

$$h = -\frac{1}{2}\Delta - \frac{1}{r_a} - \frac{1}{r_b};\ (a:bb) = \int \frac{1}{r_a} b(\vec{r}_1)^2 d^3r \tag{35}$$

$$(pq|rs) = \int \frac{p(1)q(1)r(2)s(2)}{r_{12}} d^3r_1 d^3r_2. \tag{36}$$

Let us make the *Mulliken approximation* for two-electron integrals.

$$(ab|cd) \approx \frac{1}{4} S_{ab} S_{cd} [(aa|cc) + (aa|dd) + (bb|cc) + (bb|dd)] \tag{37}$$

This is a very good approximation and leads to significant simplifications. We actually get the following expression for the MO-LCAO energy:

$$E = 2\frac{H_{aa} + H_{ab}}{1+S} - \frac{1}{R} + \frac{1}{2}[(aa|aa) + (aa|bb)] \tag{38}$$

The incorrect bond dissociation is related to the fact that the electron interaction energy is obtained as $\frac{1}{2}[(aa|aa) + (aa|bb)]$, i.e. one half of the one-center and one half of the two-center repulsion, consistent with the picture that in MO theory there is equal probability for two electrons to be near the the same or near different nuclei.

Actually the electron repulsion energy of the MO approximation consists of a Coulomb term

$$E_{coul} = (aa|aa) + (aa|bb) \tag{39}$$

and an exchange energy term

$$E_{ex} = -\frac{1}{2}(aa|aa) + (aa|bb) \tag{40}$$

The *exchange hole* is *delocalized* over the *entire molecule*.

Let us now, in a brute-force way, replace the delocalized exchange energy by a localized one!

$$E_{lex} = -(aa|aa) = -(bb|bb) \tag{41}$$

Then we get for the total electron interaction energy simply $(aa|bb)$ i.e. the correct electron repulsion energy, and the molecule dissociates correctly.

The total energy of H_2 becomes:

$$E = 2\frac{H_{aa} + H_{bb}}{1+S} + (aa|bb) - \frac{1}{R} \tag{42}$$

This is *not a variational energy* and does not correspond to a wave function. It is an energy expression composed of a one-electron part $2\frac{H_{aa}+H_{bb}}{1+S}$ like in MO-theory – as it is valid for H_2^+ – and a two-electron part like in VB theory (neglecting terms of higher order in the overlap).

If one wants to improve the MO ansatz in a rigorous way to account for *left-right correlation* one can either use the *Heitler London* ansatz

$$\Psi = \frac{1}{\sqrt{2(1+S^2)}}[a(1)b(2) + b(1)a(2)] \tag{43}$$

or use a CI mixing of the MO configurations σ_g^2 and σ_u^2

$$\Psi = c_1\sigma_g(1)\sigma_g(2) - c_2\sigma_u(1)\sigma_u(2) \tag{44}$$

The *one-particle contribution* to the chemical bond is changed from

$$2\frac{H_{aa} + H_{ab}}{1+S} = 2\alpha + 2\frac{\beta}{1+S} - 2(b:aa);\ \beta = H_{ab} - SH_{aa} \tag{45}$$

to

$$2\alpha + 2\frac{\beta S}{1+S^2} - 2(b:aa) \tag{46}$$

This means that if one wants to establish the correct dissociation in a rigorous way, one must not only modify the electron interaction energy, but also the one-electron terms. One then reduces the *interference*, which is responsable for the chemical bond [18]. If one ignores this *penalty*, and corrects only the electron interaction, one overestimates the binding energy, as is characteristic for methods with local exchange like the $X\alpha$ theory.

The MO theory fails for dissociation because on one hand the exchange is too delocalized, on the other hand because the one-particle *density matrix* has the wrong dissociation limit. The electron density ϱ in the dissociation limit is at least qualitatively correct.

For H_2 the MO *density* $\varrho_{MO} = \frac{1}{2(1+S)}[a^2(1) + b^2(2) + 2a(1)b(1)]$ has e. g. the correct limit for $R \to \infty$.

This makes plausible why density functional methods have less problems with bond dissociation than Hartree-Fock type theories.

4.3 The various generations of DFT

In a paper with the title *'Inhomogeneous electron gas'* Hohenberg and Kohn (HK) [1] laid the ground for what is now called density functional theory (DFT). We shall discuss in a later section (5.3) what Hohenberg and Kohn have actually shown. This is somewhat controversial. It suffices here to mention that Hohenberg and Kohn were mainly involved in *existence theorems*. They did not give a prescription for how to construct a density functional. This was done roughly a year later by Kohn and Sham [2], who proposed a Hartree-Fock like mean field theory with a local exchange correlation potential, that in its simplest version, the local density approximation (LDA) was hardly distinguishable from Slater's $X\alpha$ theory. The point was that to the leading order correlation had the same dependence on ϱ as exchange, such that one could simply switch from an exchange functional to an exchange-correlation functional, simply by varying a numerical factor.

The two papers [1, 2] had certainly a great immediate impact on solid-state theory, but hardly received much attention in the community of atomic or molecular theorists. This community had just been cured from the 'attack' by $X\alpha$ and remained immune for a while to similar approaches. It was noticed that the DFT calculations current at that time were slightly superior to Hartree-Fock, but not competitive w ith post-Hartree-Fock methods such as Møller-Plesset (MP) or coupled-cluster (CC). While Hartree-Fock underestimated binding energies, LDA calculations overestimated binding energies considerably, and were far from what had been defined as *chemical accuracy*.

The situation changed drastically, when DFT methods of the *second generation* of the type of the generalized gradient approximation (GGA) entered the market [19, 20], that no longer suffered from the overbinding of LDA, and that were often close to chemical accuracy, at much lower cost than ab-initio post-Hartree-Fock methods.

One usually refers to the first option within the Kohn-Sham ansatz, the so-called *local density approximation* (LDA) as the *first generation of DFT*. It makes then sense to call the very similar $X\alpha$ theory as the *zeroth generation of DFT*, and its precursor Thomas-Fermi as $(-1)^{st}$ *generation* of DFT. A breakthrough came with the *2nd generation* of the type of *generalized gradient approximation* (GGA). More recently DFT methods of the *3rd generation* of the type of the *optimized potential methods* (OPM) or *orbital functional theory* entered the scene [21]. These are, in some sense, closer to AIQC than to original Kohn-Sham DFT (see section 6).

DFT methods of the 0th and $(-1)^{st}$ generation gave slightly better results than Hartree-Fock, but were not competitive with post-Hartree-Fock methods. They mainly suffered from *overbinding*.

In those of the 2nd generation the overbinding was corrected, and they became competitive with post-Hartree-Fock methods and hence attractive for chemistry.

The attitude of quantum chemists in this matter was very pragmatic. DFT methods of the 2nd generation were implemented into quantum chemical program packages. This new implementation, taking advantage of the knowhow of AIQC lead to better numerical stability and better computational performance than earlier implementations in the DFT community. The success of quantum chemical calculations with DFT methods of GGA type was so spectacular, that in the Nobel price 1998 given to numerical quantum chemistry, the share of DFT was especially honored by choosing W. Kohn at the side of J. A. Pople as a Nobel laureate.

5 The Hohenberg-Kohn theorem and the Kohn-Sham recipe

5.1 The exact energy expression

Let us consider a Hamiltonian for clamped nuclei (i.e. making implicitly the Born-Oppenheimer approximation).

$$H = \hat{T} + V_{en} + V_{ee} + V_{nn}; \; V_{en} = \sum_{k=1}^{n} v(\vec{r}_k) \tag{47}$$

with $\hat{T}$ the operator of the kinetic energy, and V_{en}, V_{ee}, V_{nn} the various potential energy terms. The expectation value for an arbitrary eigenstate Ψ is

$$E = E_{ext} + E_{int} \tag{48}$$

$$E_{ext} = \langle\Psi|V_{en}|\Psi\rangle = \int v(\vec{r})\varrho(\vec{r})d^3r \tag{49}$$

$$E_{int} = \langle\Psi|T + V_{ee} + V_{nn}|\Psi\rangle \tag{50}$$

for Ψ normalized as $||\Psi|| = 1$, where ϱ is the *electron density*, defined as

$$\varrho(\vec{r}) = n \int \Psi^*(\vec{r}, \vec{r}_2, ...\vec{r}_n)\Psi(\vec{r}, \vec{r}_2, ...\vec{r}_n)d^3r_2.....d^3r_n ds \tag{51}$$

and where the summation over all spin variables is formally written as an integral over the variable s. One part of the energy expectation value, the *external energy* E_{ext}, is expressible through the density, but the other part, the internal energy E_{int} is not. It requires the one-particle density matrix γ_1 (with matrix elements γ_p^q) for the kinetic energy T, and the pair density (or the two-particle density matrix γ_2) for the electron interaction V_{ee}. The kinetic energy is most conveniently evaluated in a matrix representation in an orthonormal basis $\{\chi\}$.

$$\langle\Psi|T|\Psi\rangle = T_q^p\gamma_p^q; \; T_q^p = \langle\chi_q|T|\chi_p\rangle \tag{52}$$

$$\gamma_p^q = \langle\Psi|a_p^q|\Psi\rangle; \; a_p^q = a_q^\dagger a_p \tag{53}$$

For more details on this notation see subsect. 7.1. A formulation in terms of the density ϱ like (49) is only possible for local operators such as the external potential.

So to make use of the variational principle, we must make E stationary with respect to variations of ϱ, γ_1 and γ_2 in such a way, that these variations *correspond to variations of the normalized n-electron wave function* Ψ. This requirement is usually referred to as *(fermion) n-representability.*

It would be nice if it were possible to express the entire energy as a functional of the density ϱ only. This is *not possible* because γ_1 and γ_2 contain more information than ϱ. It is always possible to obtain ϱ from γ_1 or γ_2, but *not* vice versa – unless one gets the *missing information from another source.*

This is the problem that HK wanted to solve.

Let us temporarily consider that we have minimized the ground state energy expectation value E_0 for a certain Hamiltonian H_0 characterized by an external potential v_0. Let us then try to get the ground state energy E for a Hamiltonian H with the external potential v, that differs *infinitesimally* from v_0. Perturbation theory then tells us that

$$E - E_0 = \int (v - v_0)\varrho_0 d^3r \tag{54}$$

In the limit of an infinitesimal change of the external potential the change of the energy is entirely determined by the change of v and is expressible through the density alone. The evaluation of the ground state energy of all possible Hamiltonians H that differ infinitesimally from H_0, requires γ_1 or γ_2 a single time, and only ϱ for each particular H. This is so, because we have assumed that the external potential is *local*. This observation is a vague hint that, although the construction of E requires γ_1 and γ_2, the *dependence* of E on v is monitored by ϱ alone.

5.2 The basis of the Hohenberg-Kohn theorem

The Hohenberg-Kohn theorem (with all what it implies) and even the entire density functional theory (DFT) is based on the following simple observation:

If the ground state densities ϱ_1 and ϱ_2 for two Hamiltonians $H(v_1)$ and $H(v_2)$ are the same, the potentials v_1 and v_2 can only differ by a constant. So, specifying potentials only up to an arbitrary constant, there is a *unique mapping* from ϱ to v, and v can be regarded as a functional of ϱ, symbolized $v(\varrho)$. This mapping is *anticausal*, because v is the cause, and ϱ is the effect.

This observation, which hardly deserves the name of a theorem, let us call it the *HK lemma*, is an immediate consequence of the variation principle and the choice of local potentials. We start from

$$E(v_1) - E(v_2) \leq \int (v_1 - v_2)\varrho(v_1) d^3r \tag{55}$$

which becomes a strict inequality if $\Psi(v_1) \neq \Psi(v_2)$. The assumption that $v_1 \neq v_2$, and $\Psi(v_1) \neq \Psi(v_2)$, and at the same time $\varrho(v_1) = \varrho(v_2)$, leads to a contradiction. Hence $\varrho(v_1) = \varrho(v_2)$ for $v_1 \neq v_2$ is only possible, if $\Psi(v_1) = \Psi(v_2)$, which is the case if v_1 and v_2 differ simply in a constant, which specifies the gauge. We can here not comment on a recent criticism of the *reductio ad absurdum* contained in this proof, in terms of Kato's cusp conditions, and a new proof based on these very conditions by Kryachko [22].

Is this simple lemma strong enough to carry all the burden of the HK theorem and of DFT? There is an apparent problem with the concept of a functional. It has been used in a very generous way in the DFT context. Hohenberg and Kohn (HK) run into a dilemma, when they state that the ground state energy E and wave function Ψ are functionals of v. They state this, because v determines H (different H differ only in v), and if one knows H, one has *just to solve the n-electron Schrödinger equation*, in order to get E and Ψ. Combination of the two mappings then almost trivially lead to the claim that E and Ψ are functionals of ϱ.

The dilemma is, that the purpose of DFT is to *avoid the solution of the n-electron Schrödinger equation*. So any prescription: 'solve the n-electron Schrödinger equation' should be absolutely tabu in the DFT context. More serious is that the information contained in v or ϱ is not sufficient for the construction of Ψ from ϱ, one needs information about the entire Hamiltonian.

To formulate a DFT one should rather try to find a counterpart of the Schrödinger equation or the associated variation principle by something, in which the density ϱ replaces the wave function, such that one can construct the ground state energy E and the corresponding density ϱ *without worrying about the wave function*. HK were actually able to formulate such a variation principle in terms of the density. It involves a functional $F(\varrho)$, which represents the *internal energy* as a functional of ϱ. Unfortunately, on the way to show that $F(\varrho)$ exists, HK argue that Ψ is a functional of ϱ. This is problematic and ought to be avoided. The question arises, whether the existence of $F(\varrho)$ can be proved without any explicit reference to the wave function, making only use of quantities that are meaningful in a DFT context. This is, in fact, the case, if one uses the theory of Legendre transformations, as introduced by E. Lieb [23]. This is one of the reasons why the derivation of the basic theorems of DFT is preferable in this framework. Only well-characterized functionals enter the theory. Another reason for preferring the Legendre transformation is that in this way a rigorous discussion of the domains for v and ϱ is straightforward.

In the context of the Legendre transformation, wave functions never appear. However, we have to start from the energy $E(v)$ as a functional of the potential v. This knowledge is not available, if one has not solved a set of Schrödinger equations beforehand. So there is nothing free, and one can only show that $F(\varrho)$ exists and one cannot indicate a simple way how to construct it. As far as the wave function and thus the complete information about a state is concerned, it remains that it is only obtained by solution of the Schrödinger equation, or something equivalent.

5.3 The Hohenberg-Kohn by means of a Legendre transformation

Let us now assume that we know $E(v)$ for a family of Hamiltonians $H(v)$, characterized by the potential v. This is not a very realistic assumption. Why should we assume that the more difficult problem to know $E(v)$ for a family of v is solved, if we are interested in the much simpler problem to obtain $E(v)$ for a single v? Moreover, why should we not just use the information contained in $E(v)$ trivially by specifying v? Let us assume, nevertheless, that we know $E(v)$ and let us use this to construct the Legendre transform $-F(\varrho)$ of $E(v)$ [23]

$$-F(\varrho) = \inf_v \{ \int \varrho(\vec{r}) v(\vec{r}) d^3r - E(v) \} \tag{56}$$

To make this precise one must specify a *domain* of functions $\varrho(\vec{r})$, for which one wants to consider the above expression, and a *domain* of $v(\vec{r})$ over which the infimum has to be taken. Conditions for the existence of $F(\varrho)$ are that the domains for v and ϱ are *dual spaces*, which implies that

$$\int \varrho(\vec{r}) v(\vec{r}) d^3r < \infty \tag{57}$$

and that $E(v)$ is a concave functional, i.e. that

$$E(\lambda v_1 + [1-\lambda] v_2) \geq \lambda E(v_1) + [1-\lambda] E(v_2); \ 0 \leq \lambda \leq 1 \tag{58}$$

This is easily shown as a consequence of the variation principle, very much like the HK lemma (55).

In the special case that the infimum is a stationary minimum, we conclude from (56) that

$$\varrho = \frac{\partial E(v)}{\partial v} \tag{59}$$

i.e that the admitted ϱ are functional derivatives of the energy with respect to the potential. This is valid for a non-degenerate state. One can then write the Legendre transform $-F(\varrho)$ in the more traditional form

$$-F(\varrho) = \int \frac{\partial E(v)}{\partial v} v(\vec{r}) d^3r - E(v); \ v = v(\varrho) \tag{60}$$

which requires to *invert the mapping* from v to ϱ. The formulation (56) is preferable, not only because the inversion of the mapping (59) is not always possible, but also since the functional derivative is only unique for non-degenerate states.

Anyway (60) illustrates that $F(\varrho)$ is just the internal energy (50), written as a functional of (ϱ).

A nice feature of the Legendre transformation is, that it is, in some sense, self-inverse. If we iterate the Legendre transformation, we get back the original function – with some reservations, related to the fact that the *domains* do not necessarily transform in a self-reflective way.

Let us now suppose that we know the functional $F(\varrho)$ and let us consider the (second) Legendre transformation

$$\tilde{E}(v) = \inf_{\tilde{\varrho}} \mathcal{E}(v, \tilde{\varrho}) = \inf_{\varrho} \{ \int \tilde{\varrho}(\vec{r}) v(\vec{r}) d^3r + F(\tilde{\varrho}) \} \tag{61}$$

We now fix the potential v and we search for the infimum of $\mathcal{E}(v, \tilde{\varrho})$ over a trial set of densities $\tilde{\varrho}$. We find that the $\tilde{E}(v)$ so constructed is, in fact the ground state energy for the chosen v, and it is achieved as the infimum of a *functional of the trial density*.

So, if this second Legendre transformation exists, it appears in fact possible to formulate a variation principle for the energy (for fixed v) as a functional of the trial density $\tilde{\varrho}$. Condition for the existence of $\tilde{E}(v)$ is that $F(\varrho)$ is a convex functional. This property of $F(\varrho)$ is less easily shown than the concavity of $\tilde{E}(v)$, and requires a careful study of the domains of the functionals considered [23, 24].

The conclusion is that a variation principle for the energy as a functional of the trial density exists. If one knew $F(\varrho)$, the minimum of $\mathcal{E}(v, \tilde{\varrho})$ would – for a non-degenerate state – be achieved for the particular $\tilde{\varrho}$ that is related to v as (59) and be the exact ϱ of the ground state.

Unfortunately this *Hohenberg-Kohn variation principle* only holds for the exact $F(\varrho)$, that one will usually not know, because $F(\varrho)$ contains roughly as much information as $E(v)$ and one cannot expect that this is a simple expression.

The main message of this section is, that there is some justification for replacing a *known* functional of γ_1 and γ_2 by an *unknown* functional of ϱ. This is the basis of density functional theory (DFT). The theory does not give any hint as to a possible construction of $F(\varrho)$, but establishes at least its existence.

We have here not discussed the domains which have to be chosen for v and for ϱ. This involves a few non-trivial aspects [23, 24, 25], but does not affect the general conclusions. If the domains are correctly chosen, there is neither a v-representability problem, nor an n-representability problem for ϱ. However, the n-representability is shifted to the functional $F(\varrho)$. It is not easily checked whether the chosen $F(\varrho)$ is compatible with the antisymmetry of the n-electron wave function.

5.4 Is an n-electron state fully characterized by the electron density?

The derivation of the HK theorem that we have sketched in the last section, which goes back to E. Lieb [23] is very compact, and leads in two steps, without any detour, to the essential message, namely the *existence of a variational principle in terms of a trial density.*

Unfortunately this very elegant derivation is much less popular than the original one by Hohenberg and Kohn [1], which is more pedestrian and in which the importance of the domains for v and ϱ is ignored. The original derivation has the disadvantage that claims are made on the way, that are *not necessary for the proof of the final result*, namely the existence of a HK variation principle, and which are in part responsible for some current mystification of the HK theorem.

The most popular of these claims is that the *electron density determines an n-electron state entirely.* It is based on the correct observation that the mapping between the potential v and the density ϱ is unique in the direction $\varrho \to v$. Hohenberg and Kohn [1] then argue that if one knows the density, one also knows the potential, and hence the Hamiltonian, and that one has then nothing to do but *solve the Schrödinger equation* in order to get the wave function Ψ for the ground state (and all other states as well) and from the wave function all properties. So the wave function and any property is a *functional of the density* ϱ.

This claim is not even genuinely wrong. One can define the term *functional* so generously that the existence of a mapping from A to B allows one to call B a functional of A. This remains meaningless unless one knows at least something about the mathematical properties of the mapping $A \rightarrow B$. The claim that Ψ is a *functional of the density*, has (unlike that of the existence of $F(\varrho)$) no consequences and therefore no physical meaning. It is simply an *empty statement.*

Although the potential can be regarded as a functional of the density, the problem that ϱ is given, but v unknown, *never arises.* One will always *start from a given* v and the corresponding Hamiltonian and will *never want* to construct H from ϱ. The Hamiltonian is always given at the outset.

It is hard to understand why the just outlined, at best curious argument has so often been repeated and can be found even in modern textbooks [26, 27].

A problem with the persistence of the empty statement about $\Psi(\varrho)$ is probably that Hohenberg and Kohn first argued that Ψ is a functional of ϱ, and used afterwards this claim to show that $F(\varrho)$ is a functional of ϱ. It is an advantage of the derivation by means of the Legendre transformation, that the proof of the existence of $F(\varrho)$ is not based on the existence of a functional $\Psi(\varrho)$.

The way from H to Ψ is highly non-trivial, and the possibility that there may be a sufficiently simple shortcut directly from ϱ to Ψ is not more than a speculation.

In their reply to a polemic by Kryachko [28], who questioned that the electron density characterizes a state entirely, Levy and Perdew [29] discuss the possibility of such a shortcut, based on Levy's *constrained variation* [30]. They suggest, so to say, to split the variational solution of the Schrödinger equation into two parts. One first determines ϱ, say from the HK variation principle (i.e. assuming that one knows $F(\varrho)$). One then restricts the set of variational wave functions Ψ to those, which in the sense of (51) integrate to the given ϱ. All these Ψ have the same E_{ext}, such that we need only find the minimum for

$$\langle\Psi|T + V_{ee}|\Psi\rangle \tag{62}$$

over all Ψ, which have the same ϱ. (This minimum is, with some precautions, equal to $F(\varrho)$). It is not obvious whether this *constrained variation* is simpler than a straight variational calculation without *prescreening*. At least it represents a *Gedanken-Experiment* that associates a Ψ with every (exact) ϱ. The existence of a mapping from ϱ to Ψ in a purely formal sense is, however, uncontested anyway (see subsection 5.2). The essential point is, that this mapping is not possible from the *information contained in* ϱ, (the loss of information on the way (51) cannot be gained back), but it requires knowledge of the Hamiltonian, and a step, that is essentially equivalent to the solution of the n-electron Schrödinger equation.

In using the constrained variation one need not assume that one knows $F(\varrho)$ to start with. One can first consider an arbitrary set of trial densities $\tilde{\varrho}$ and construct for each of these the corresponding trial wave functions and trial internal energies $F(\tilde{\varrho})$. So one gets automatically a functional $F(\tilde{\varrho})$, in a kind of alternative to the approach via the Legendre transformation, which is of as little use as the latter for a *practical construction.*

One often hears that the expression for the *energy* in DFT is a functional of the density alone, because the potential and hence the Hamiltonian is a functional of the density. In claiming this, one overlooks that the energy expression contains a term which depends *both* on v and on ϱ, namely the external energy. One would never want to eliminate v in favor of ϱ, especially not since in the expression of the HK variation principle we have the exact potential combined with a trial density $\tilde{\varrho}$. The trial density $\tilde{\varrho}$ does certainly *not* determine the exact potential v, but rather a physically meaningless trial potential $\tilde{v}$.

Hohenberg and Kohn call $F(\varrho)$ a *universal functional* and even claim that it is *independent of*

v. This is problematic. In view of the mapping between v and ϱ any functional of ϱ is necessarily a functional of v and vice versa. Behind this misleading claim is, of course, the fact that $E(v)$ and $-F(\varrho)$ are Legendre transforms of each other, which implies that v is the *natural* variable for E, and ϱ that for $-F$. So $E(v)$ and $-F(\varrho)$ are related to each other somewhat like $U(V)$ and $H(p)$ in thermodynamics or $L(x, \dot{x})$ and $H(x, p_x)$ in classical mechanics.

In the derivation of the HK theorem by means of the Legendre transformation all quantities depend parametrically on the particle number n, while HK [1] consider functionals without a parametric particle-number dependence. Eschrig [24] has shown that a functional, in which the particle number only enters via the normalization of ϱ can be derived by means of a switch to a grand canonical ensemble, that involves an additional Legendre transformation, in which the particle number is replaced by the chemical potential.

5.5 Take-home lessons in connection with the Hohenberg-Kohn theorem

The content of the HK theorem is, that in the exact variational expression for the ground state energy one can replace the *known* expression for the *internal energy* as a functional of γ_1 and γ_2 by an *unknown* functional of ϱ. The construction of this functional is, in principle - but not in practice, possible by means of a sequence of two Legendre transformations, which requires the knowledge of the energy $E(v)$ as a functional of all v in the considered domain.

There is a variation principle for the energy in terms of the given potential v, and a trial density $\tilde{\varrho}$. Unfortunately the formulation of this variation principle requires the knowledge of an unknown functional $F(\varrho)$, that is defined as the Legendre transform of $E(v)$, and does therefore exist (under some assumptions about the domains of v and ϱ), but about the construction of which the HK theorem gives no information. While for the construction of E in terms of ϱ at least the existence of an algorithm can be proven, there is no way to arrive from the density at the wave function (and at properties, that are not immediately derivable from the density), except via the solution of the n-electron Schrödinger equation, or a procedure essentially equivalent to this solution.

The fact that DFT does not determine an n-electron state fully, does not appear to have caused serious problems. Most people who used DFT codes were apparently only interested in the energy or in properties directly derivable from the energy or the density. In the Kohn-Sham variant of DFT (see subsect. 5.7) the output contains a wave function, that can be used as an approximation to the exact Ψ, if one is interested in this.

At this point one should be aware that a DFT strictly in the spirit of the HK theorem would have little chance to work. The classical DFT is Thomas-Fermi theory, and any genuine DFT has to be some generalization of the latter. The main problem with such a theory is that it is very hard to express the *kinetic energy as a functional of the density*. This is why Kohn and Sham [2], shortly after Hohenberg and Kohn [1] have proposed a method in which the formulation of the kinetic energy as a functional of the density is completely avoided. This led to the paradoxical situation that the entire field of what is now called DFT is not a DFT in the *genuine* sense.

To prepare the recipe of Kohn and Sham, we try to formulate a system of non-interacting electrons in the language of DFT.

5.6 A system of non-interacting electrons

Let us now have a look at the hypothetical n-electron system *without electron interaction*. In this situation the functional $F(\varrho)$ becomes the functional $T_0(\varrho)$ of the *kinetic energy*. It is obvious that this functional $T_0(\varrho)$ must fully *take care of the Pauli principle*. The potential term in the energy expression (61) does not discriminate between fermions and bosons.

To formulate *the kinetic energy* and the Pauli principle approximatively via a functional of the density is (except for the homogeneous electron gas [7, 8]) a rather hopeless task, and is certainly much more difficult than to represent the electron interaction in terms of the density.

Let us now make the simplifying assumption (which is not necessary) that we deal with a closed shell state, that is correctly described by a single Slater determinant, with the orbitals φ_k occupied with α and β spin each. In this case we can solve the Schrödinger equation exactly. We attempt, however, to formulate this exact solution in the *language of DFT*.

Even in this simple case the functional $T_0(\varrho)$ is *unknown*, and so is the functional derivative $\frac{\delta T_0(\varrho)}{\delta \varrho}$ that we need in order to formulate the Euler-Lagrange equations of the HK variation principle (61)

$$\{v + \frac{\delta T_0(\tilde{\varrho})}{\delta \tilde{\varrho}} - \lambda\}\delta\tilde{\varrho} = 0 \tag{63}$$

We are here in the fortunate situation that we can express ϱ through the occupied orbitals φ_k

$$\varrho = 2\sum_{k=1}^{\frac{n}{2}} |\varphi_k|^2 \tag{64}$$

and that we also know the expectation value $\langle \hat{T}_0 \rangle$ of the kinetic energy as a functional of the (doubly) occupied φ_k

$$\langle \hat{T}_0 \rangle = -\sum_{k=1}^{\frac{n}{2}} \langle \varphi_k^* | \nabla^2 | \varphi_k \rangle \tag{65}$$

One could phrase this also by saying that we know T_0 as a functional of the one-particle *density matrix* γ_1, which, in this case, happens to be idempotent (in terms of spin orbitals).

There are two ways of expressing the variation of the kinetic energy with respect to variation of the φ_k, either via the unknown $\frac{\delta T_0(\varrho)}{\delta \varrho}$ and the dependence of ϱ on the φ_k, or directly from the known expression (65). The two results must be identical. Furthermore, one sees that stationarity of the variational expression with respect to variation of the φ_k is sufficient for stationarity with respect to variation of ϱ. One could also say that stationarity with respect to variation of γ_1 implies that with respect to variation of ϱ. This means, that if we satisfy

$$\{v - \frac{1}{2}\nabla^2 - \varepsilon_k\}\varphi_k = 0;\ E = \sum_k \varepsilon_k \tag{66}$$

i.e. if we solve the Schrödinger equation exactly, we also satisfy the equations (63), i.e. we get a solution of the HK variation principle.

We are so in the curious situation, that we are able to solve the HK variational equations, *without even knowing* anything about the functional $F(\varrho)$, nor its derivative (except that they exist). This is, of course, only possible since we are able to solve the problem exactly, without any need to refer to the HK theorem.

We can, of course, phrase the content of this subsection in a very simple way. We do not know the kinetic energy as a functional of the density. However, if it were known, we would from the HK variation principle get the exact density and energy. Fortunately, we are, for a system of non-interacting electrons, able to solve the Schrödinger equation exactly, so we simply perform the exact calculation and call this DFT. This is a very generous interpretation of the concept of DFT.

We have said earlier (sec. 5.2) that it should be tabu in the DFT context to solve an n-electron Schrödinger equation. Apparently one can break this tabu, if this solution is as easy as in the present case.

Note that in this section we have only assumed that the density is a functional of the occupied orbitals, which is certainly correct, but *not* that the orbitals are functionals of the density.

5.7 The Kohn-Sham recipe

The previous subsection has been the preparation for the discussion of a recipe proposed by Kohn and Sham [2], in a paper shortly following that of Hohenberg and Kohn [1], on a reformulation of the HK variation principle, in which the need to consider the functional $T_0(\varrho)$ is completely circumvented. This approach followed operationally, though not conceptually earlier work of Slater [6]).

One arrives at the Kohn-Sham recipe in the following way. The unknown functional $F(\tilde{\varrho})$ may be decomposed into two parts

$$F(\tilde{\varrho}) = T_0(\tilde{\varrho}) + I(\tilde{\varrho}) \tag{67}$$

where T_0 is the kinetic energy of a system of n non-interacting electrons, and I all the rest, i.e. the electron interaction energy, plus the difference between the actual kinetic energy T and T_0. Like in the example of the preceding subsection we reformulate the HK variation principle, such that T_0 does not enter. We start from

$$0 = \{v + \frac{\delta T_0(\tilde{\varrho})}{\delta \tilde{\varrho}} + \frac{\delta I(\tilde{\varrho})}{\delta \tilde{\varrho}} - \lambda\}\delta\tilde{\varrho} \tag{68}$$

We can define the *effective one-electron potential*

$$v_{eff} = v + \frac{\delta I(\tilde{\varrho})}{\delta \tilde{\varrho}} \tag{69}$$

and rewrite (68) as

$$0 = \{v_{eff} + \frac{\delta T_0(\tilde{\varrho})}{\delta \tilde{\varrho}} - \lambda\}\delta\tilde{\varrho} \tag{70}$$

This *can be interpreted* as the variation of the energy of an n-electron system formally *without electron interaction*, but in the effective (ϱ-dependent) external potential V_{eff}. Since this is a system formally without electron interaction, we can treat it is as in the preceding subsection. We assume again that it is of closed shell type and describable by a single Slater determinant. The generalization to the case of degeneracy is nontrivial, but possible, because any n-electron density can be formulated as that of a single Slater determinant [23]. By the same argument as in the previous subsection we conclude that stationarity with respect to variation of the occupied orbitals φ_k is sufficient for stationarity of the variation with respect to ϱ, and we get so the Kohn-Sham equations

$$\{v_{eff} - \frac{1}{2}\nabla^2 - \varepsilon_k\}\varphi_k = 0 \tag{71}$$

for the construction of the KS orbitals φ_k, that are used for the construction of the density according to (64), which is then used to update v_{eff}, and so forth. Again we have completely circumvented to worry about the dominant part T_0 of the functional F, and achieved that only $I(\varrho)$ enters. This can further be decomposed as

$$I(\varrho) = \frac{1}{2}\int \varrho(\vec{r}_1)\frac{1}{|\vec{r}_1 - \vec{r}_2|}\varrho(\vec{r}_2)d^3r_1 d^3r_2 + E_{xc} \tag{72}$$

where the first term is the Coulomb interaction, including the self-interaction, and where E_{xc}, called *exchange-correlation functional*, also contains $T - T_0$, but is dominated by self-exchange.

Although, in the spirit of DFT, we are only interested in E and ϱ, we also get, so to say for free, a wave function, the *Kohn-Sham determinant*, which is, however, not the exact n-electron wave function Ψ (it only has the same density as Ψ), or equivalently an idempotent approximation $\gamma^{(0)}$ to the one-particle density matrix γ, and a set of orbital energies. The question arises whether these auxiliary quantities have any physical meaning.

It is a curious aspect of the history of what is now called DFT, that the search for approximate density functionals $E_{xc}(\varrho)$ (including a correction to the kinetic energy) became only worthwhile after the search for the complementary functional $T_0(\varrho)$ for the kinetic energy had been completely abandoned.

5.8 Does the hypothetical Kohn-Sham system have a physical reality?

There has been some controversy whether the Kohn-Sham orbitals and the Kohn-Sham determinant are just intermediates on the way towards a construction of the density (and the energy), or whether they have an immediate physical meaning. The matter appears to be settled in the sense of the second interpretation [31, 32]. If one does not want to cheat oneself, one must admit that methods based on the Kohn-Sham recipe are *mean-field methods* (belonging to the same family as Hartree-Fock) and that the central quantity is not the density ϱ, but an idempotent approximation $\gamma^{(0)}$ to the exact density matrix γ with the special requirement that the density $\varrho^{(0)}(\vec{r}) = \gamma^{(0)}(\vec{r}, \vec{r})$ corresponding to $\gamma^{(0)}$ agrees with the exact $\varrho(\vec{r}) = \gamma(\vec{r}, \vec{r})$ corresponding to γ.

The only non-trivial aspect is that it is possible [23] to fulfill this requirement to select one particular among the many possible idempotent approximations to γ.

It results from the derivation of the Kohn-Sham recipe that the *exchange correlation potential* does not purely describe exchange and correlation, but contains also the difference between the (unknown) density functionals of the exact kinetic energy and that for non-interacting electrons.

There has been a discussion whether the so-called *Kohn-Sham system* has any physical reality. This hypothetical system is defined as a system of non-interacting electrons moving in an *external field* (69). The *electron density* of this *Kohn-Sham system* should be the same as that of the *exact system.*

Actually the *external* potential is typically that of a mean-field theory, containing an effective electron interaction. It is somewhat an abuse of the concept of a non-interacting system to call the system described by the Kohn-Sham equations *non-interacting.* This system is describable by a single Slater-determinant wave function, because the energy expression has been so decomposed that one part of it is the kinetic energy of non-interacting particles, while the remainder of the kinetic energy is shifted into the electron interaction part. So it is much more straightforward to interpret the Kohn-Sham recipe as an especially defined mean field theory. The Kohn-Sham system interacts in a mean-field way, rather than consisting of non-interacting electrons.

If one accepts this interpretation of the Kohn-Sham equations, it becomes hard to make a difference between the real system and an artificial Kohn-Sham system. Then also the formal introduction of a *coupling parameter*, which interpolates between the Kohn-Sham system and the real system, in the sense of an *adiabatic connection*, no0 longer makes sense.

5.9 Take-home lessons in connection with the Kohn-Sham recipe

The Kohn-Sham recipe consists in *avoiding the most difficult part* of a density functional theory, namely the expression of the kinetic energy as a functional of the density. It evaluates the kinetic energy in terms of a *one-particle density matrix* like in AIQC. So it is by far superior to any *genuine* density functional theory like that of Thomas-Fermi.

It results from the derivation of the Kohn-Sham recipe that the *exchange correlation potential* does not purely describe exchange and correlation, but contains also the difference between the (unknown) density functionals of the exact kinetic energy and that for non-interacting electrons.

Although the Kohn-Sham recipe can be derived from the HK variation principle, the *exchange correlation functional* cannot be formally derived by means of a Legendre transformation. One must use physical plausibility arguments to derive approximations to it, like e.g. the popular B3LYP [33].

As already mentioned earlier, the working equations of Kohn-Sham theories are very similar to those of Slater's $X\alpha$ theory. Differences are only in the physical interpretation.

6 Recent trends in density functional theory and density matrix functional theory.

There has, obviously, been a lot of progress in Kohn-Sham DFT, since the original work [2]. Only a limited amount of this progress is relevant in the present context. In methods of the *first generation* so-called *local* exchange correlation functionals dominated. They had the essential drawback of *overbinding*, i.e. of overestimating binding energies.

In DFT methods of the 2nd generation, density functionals that also depend on the gradient of the density (keyword GGA for *generalized gradient approximation*) entered the scene. They are mainly associated with the names of Becke [19] and Perdew [20]. These functionals no longer suffered from overbinding, and became therefore attractive for chemical applications. In methods of this generation the introduction of adjustable parameters into the functionals became quite common. This often left the impression that DFT was basically a *semi-empirical* theory.

So-called *hybrid functionals* were introduced, which are simply linear combinations of DFT and ab-initio functionals, mainly of Hartree-Fock type, in which both local potentials, typical for DFT, and non-local potentials (typical for Hartree-Fock) appeared. A particularly popular choice has e.g. been the use of a one-to-one mixture of LDA and Hartree-Fock. Since LDA is overbinding and HF underbinding, there appears to be a good chance for a cancellation of errors.

Now there are so many density functionals on the market, often especially designed for special purposes, that it has become hard to make statements about DFT in general. There are as many theories as there are functionals and each of it has merits and drawbacks.

On the methodologic side most current work in DFT is in the spirit of DFT of the *third generation*. While methods of the first and second generation still cared for functionals of the density, those of the third generation mainly deal with *orbital functionals* [21]. It is often not admitted, that this is a real *change of paradigm*, because it has been a dogma of DFT that all information should be in the density. Strictly speaking this dogma has already been abandoned in the Kohn-Sham recipe, because there the central quantity is an *approximate one-particle density matrix*. The DFT past is often reminiscent in orbital functional theories, when orbitals are regarded as functionals of ϱ, and the chain rule is formally used for differentiation.

The motivation for an *orbital functional theory* has been the observation that the main defect of traditional Kohn-Sham type DFT lies in the presence of the unphysical Coulomb self-repulsion of the electrons, that is – in theories with local exchange – only incompletely cancelled by the (also unphysical) self-exchange of the electrons, while this cancellation is exact in Hartree-Fock and other ab-initio theories.

There is the problem that self-interaction and self-exchange can be defined in a clean way only in an orbital picture, in which each electron is associated with an electron (see our earlier discussion of Hartree-Fock theory in subsec. 3.2, but also the rigorous ab-initio definition in subsec. 7.6). These concepts cannot be defined in terms of the density. Fortunately the Kohn-Sham recipe provides orbitals, and it is, in terms of these orbitals, like in Hartree theory, possible to eliminate the self-interaction of the electrons. This requires, of course, to eliminate the self-exchange also from the exchange-correlation potential, which has formally been the bulk contribution to the latter. One obviously has to redefine the remaining exchange correlation functional, which becomes 'smaller', such that the final result is less sensitive to this residual functional.

It has sometimes been claimed that an orbital functional theory can still be called a density functional theory, because *the density determines everything*, and hence the Kohn-Sham orbitals are also functionals of the density. If one realizes that *the density does not determine everything* (at least not without solving the n-electron Schrödinger equation), one must admit that one has left the ground of DFT, but the question *whether or not an orbital functional theory is DFT* has hardly any relevance. Note that there is no objection to the complementary claim that the orbitals determine the density (see eqn. (64)). One can, of course, argue, that if one knows an approximate one-particle density matrix anyway, namely that of Kohn-Sham, and use this for the evaluation of the kinetic energy, why not use it for the bulk of the exchange energy as well. This leads us to a DFT with *exact exchange.* The residual functional of the density then consists of the *full correlation* and of *corrections to both the kinetic energy and the exchange.* It then becomes essentially a *correlation functional* rather than an exchange-correlation functional.

A first step on these lines is the *exchange-only DFT* [32]. This sounds, at first glance, a little curious, since an exact exchange-only theory is available in the Hartree-Fock method. Nevertheless the exchange-only DFT has also some ingredients of DFT, in particular the mean field is represented as a *local* operator. Exchange-only DFT is, so to say, the best local approximation to the Hartree-Fock scheme.

If one combines exchange-only DFT with a correlation functional, taken from traditional DFT, the results are rather disappointing. A completely new approach to the correlation functional appears to be necessary.

An interesting rather new branch of research is *ab-initio DFT* [31, 34]. It is based on the observation that the (virtually) *exact* exchange correlation potential in the sense of the Kohn-Sham recipe can, for any particular system, be derived from an accurate ab-initio calculation. One can construct an exchange-correlation potential, such that the solution of the corresponding Kohn-Sham equations duplicates the ab-initio result. Such a study does not give any physical information, that is not already available from the ab-initio calculation on which it is based. It is further not competitive with original Kohn-Sham DFT, as far the computational requirements are concerned. However, it is, on this way, possible to compare the virtually exact exchange correlation *potentials* with those that are derived from exchange correlation *functionals*, that are available on the market. The comparisons made so far show very little similarities between the exchange correlation potentials derived from the two sources and rather support the view that in order to get acceptable Kohn-Sham results, it is *not necessary to use the correct exchange correlation potentials.* This is further evidence for the importance of error cancellation in DFT.

Many recent applications, especially on properties, are based on *time-dependent DFT* [35]. The mathematical basis of this theory is actually being disputed [36].

There has also been increased interest in density matrix functional theory (DMFT), more precisely in a *one-particle-density-matrix* functional theory [37]. In this theory one starts from the same energy expression (48) as before, but one only tries to eliminate the dependence on the two-particle density matrix γ_2. By means of a sequence of two Legendre transformations, similar to (56) and (61) one can show that it is possible to replace the *electron interaction energy as a known functional* of γ_2 by an *unknown* one of γ_1. A large part of this unknown functional is actually known, namely the Coulomb and exchange energy, such that only the *correlation energy* is unknown.

DMFT has the advantage over DFT, that the kinetic energy and a large part of the electron interaction energy is a known functional of γ_1, such that the residual unknown functional of γ_1 only plays a minor role, but the disadvantage, that the variation of the energy must be made stationary subject to the conditions for n-representability of γ_1. These conditions are relatively harmless and one must only require that the eigenvalues of γ_1 are between 0 and 1 (and sum up to n). In DFT no n-representability problem arises, because all correctly normalized densities are n-representable. The essential difference between DMFT and Kohn-Sham DFT is that in DMFT one cares for the exact one-particle density matrix γ_1, in Kohn-Sham DFT for the best idempotent approximation to γ_1.

To rationalize DMFT we start from the following energy expression (for details of the notation see subsect. 7.1.

$$E = h_q^p \gamma_p^q + \frac{1}{2} g_{rs}^{pq} \gamma_{pq}^{rs}; \; h_q^p = T_q^p + v_q^p \tag{73}$$

We consider the Legendre transform

$$-F(\gamma_1) = \inf_{\mathbf{v}} \{E(\mathbf{v}) - v_q^p \gamma_p^q\} = \inf_{\mathbf{v}} \{T_q^p \gamma_p^q + \frac{1}{2} g_{rs}^{pq} \gamma_{pq}^{rs}\} \tag{74}$$

$$= \inf_{\mathbf{v}} \{T_q^p \gamma_p^q + \frac{1}{2} g_{rs}^{pq} (\gamma_p^r \gamma_q^s - \gamma_q^r \gamma_s^p) + \frac{1}{2} g_{rs}^{pq} \lambda_{pq}^{rs}\} \tag{75}$$

If the infimum is a stationary minimum, we get

$$\gamma_q^p = \frac{\partial E}{\partial v_p^q} \tag{76}$$

i.e. γ is the functional derivative of E with respect to $\mathbf{v}$. If we assume that the potential, represented by the matrix $\mathbf{v}$ with elements v_q^p, is *nonlocal*, the Legendre transform is automatically a functional of the nonlocal one-particle density matrix γ_1 with elements γ_p^q.

In the functional F only $g_{rs}^{pq} \lambda_{pq}^{rs}$, i. e. the *correlation energy* is unknown.

The *kinetic energy is exact.* Ignoring λ_2 we are led to Hartree-Fock.

There has also been some recent work in the spirit of the Legendre transformation [38] and on the borders between DFT and DMFT [39].

7 Recent trends in ab-initio theory

7.1 The paradigm of minimal parametrization

Let us now mention some recent work in AIQC that is in part a reaction to the challenge by DFT. A paradigm of *modern n-electron theory* is that one does not care about the wave function, because it contains too much irrelevant information, but rather for *reduced density matrices*, or even the *cumulants* of the reduced density matrices, in which the information is reduced to its relevant part. This leads us to search for a parametrization in terms of a minimal number of parameters [40]. A further step would lead us to *subminimal parametrization* such as terms of the electron density only. For the latter unknown functionals must be guessed, while for minimal parametrization there are no unknown functionals, but instead unknown n-representability conditions.

We start with the formulation of the *Fock space Hamiltonian* in a finite basis [41].

Let us choose an orthonormal one-electron spin-orbital basis $\{\psi_p\}$. We define annihilation and creation operators a_p and $a^p = a_p^\dagger$ respectively for the spin orbital ψ_p, and in terms of these the one-and two particle excitation operators a_q^p and a_{rs}^{pq}.

$$a_q^p = a^p a_q = a_p^\dagger a_q \tag{77}$$

$$a_{rs}^{pq} = a^q a^p a_r a_s = a_q^\dagger a_p^\dagger a_r a_s \tag{78}$$

Excitation operators are *particle-number conserving.*

Let the Hamiltonian in configuration space

$$H(1,2,3\ldots.n) = \sum_{k=1}^{n} h(k) + \sum_{k<l=1}^{n} \frac{1}{r_{kl}} \tag{79}$$

be given, and let us define the matrix elements

$$h^p_q = \langle \psi_q | h | \psi_p \rangle; \qquad O(m^2) \tag{80}$$

$$g^{pq}_{rs} = \langle \psi_r(1)\psi_r(2) | \frac{1}{r_{12}} | \psi_p(1)\psi_q(2) \rangle; \quad O(m^4) \tag{81}$$

Their number is of $O(m^2)$ and $O(m^4)$ respectively. Then the *Fock space Hamiltonian* in a finite basis

$$H = h^p_q a^q_p + \frac{1}{2} g^{pq}_{rs} a^{rs}_{pq} \tag{82}$$

with the *Einstein summation convention* implied, is equivalent to (78) in the sense that the two Hamiltonians have the same matrix elements with respect to n-electron Slater determinants. The eigenstates of this Hamiltonian are the *full-CI states*.

It is reasonable to require that n-electron states should not be specified by more parameters than those which characterize the Fock space Hamiltonian. Their number is of $O(m^4)$, where m is the dimension of the basis. We refer to this as *minimal parametrization* [40].

That this is possible has been shown in *many-body perturbation theory (MBPT)*, which can be formulated entirely in terms of the matrix elements in the Hamiltonian (82) [42]. Full CI contains too many parameters, of $O(m^n)$.

7.2 Reduced density matrices

Consider now a state described by the n-electron wave function Ψ. We define the k-particle density matrices in the following way [4]:

$$\gamma = \gamma_1 : \gamma^p_q = \langle \Psi | a^p_q | \Psi \rangle \tag{83}$$

$$\gamma_2 : \gamma^{pq}_{rs} = \langle \Psi | a^{pq}_{rs} | \Psi \rangle \tag{84}$$

$$\gamma_3 : \gamma^{pqr}_{stu} = \langle \Psi | a^{pqr}_{stu} | \Psi \rangle \tag{85}$$

The *spinfree* k-particle density matrices Γ_k are obtained by summation over spin.

$$\Gamma_1 : \quad \Gamma^P_Q = \gamma^{p\alpha}_{Q\alpha} + \gamma^{p\beta}_{Q\beta} \tag{86}$$

$$\Gamma_2 : \quad \Gamma^{PQ}_{RS} = \gamma^{P\alpha Q\alpha}_{R\alpha S\alpha} + \gamma^{P\alpha Q\beta}_{R\alpha S\beta} + \gamma^{P\beta Q\alpha}_{R\beta S\alpha} + \gamma^{P\beta Q\beta}_{R\beta S\beta} \tag{87}$$

The counterparts of these matrices in *in configuration space* are the integral kernels

$$\gamma(\vec{r};\vec{r}') = \sum_{p,q} \psi_p(\vec{r})\gamma^p_q\psi^*_q(\vec{r}') = \sum_\nu n_\nu \chi_\nu(\vec{r})\chi^*_\nu(\vec{r}')$$

$$\Gamma_1(\vec{r};\vec{r}') = \sum_{P,Q} \varphi_P(\vec{r})\gamma^P_Q\varphi^*_Q(\vec{r}')$$

with χ_ν the natural spin orbitals (NSOs) with occupation numbers n_ν. The *electron density* ϱ is

$$\varrho(\vec{r}) = \Gamma(\vec{r};\vec{r})$$

We then get the following energy expectation value

$$E = \langle \Psi | H | \Psi \rangle = h^p_q \gamma^q_p + \frac{1}{2} g^{pq}_{rs} \gamma^{rs}_{pq} \tag{88}$$

The one-particle density matrix γ_1 is diagonal in a natural spin orbital (NSO) basis:

$$\gamma^p_q = n_p \delta^p_q; \ \sum_p n_p = n \tag{89}$$

The definitions of the γ_k is easily generalized to *ensemble states*. The γ_p are normalized such that

$$\begin{aligned} \mathrm{Tr}\gamma_k &= n!/(n-k)! = O(n^k) && (90)\\ \mathrm{Tr}\gamma_1 &= n && (91)\\ \mathrm{Tr}\gamma_2 &= n(n-1) && (92) \end{aligned}$$

There are two necessary n-representability conditions for $\gamma = \gamma_1$. The following matrices must be non-negative [43]

$$\gamma \geq 0;\ \eta \geq 0 \qquad (93)$$

$$\begin{aligned} \gamma_p^p &= \langle\Psi|a^p a_p|\Psi\rangle = \langle a_p\Psi|a_p\Psi\rangle \geq 0 && (94)\\ \eta_p^p = 1-\gamma_p^p &= \langle\Psi|a_p a^p|\Psi\rangle = \langle a^p\Psi|a^p\Psi\rangle \geq 0 && (95) \end{aligned}$$

This implies

$$0 \leq n_p \leq 1 \qquad (96)$$

This condition is also sufficient for ensemble n-representability.

There are analogous necessary n-representability conditions for γ_2. The following matrices must be non-negative:

$$\gamma_2 \geq 0; \eta_2 \geq 0; \beta_2 \geq 0 \qquad (97)$$

For historical reasons these non-negativity requirements are referred to as P (or D), Q, and D conditions respectively [43, 44]. This implies for the diagonal elements:

$$\begin{aligned} \gamma_{pq}^{pq} &= \langle a_p a_q\Psi|a_p a_q\Psi\rangle \geq 0;\ (P,D) && (98)\\ \eta_{pq}^{pq} &= \langle a^p a^q\Psi|a^p a^q\Psi\rangle \geq 0;\ (Q) && (99)\\ \beta_{p,q}^{q,p} &= \langle \tilde{a}_q^p\Psi|\tilde{a}_q^p\Psi\rangle \geq 0;\qquad (G) && (100) \end{aligned}$$

These inequalities can be formulated as inequalities for the elements of γ_2. There are further Cauchy-Schwarz-type inequalities like

$$\gamma_{rs}^{pq}\gamma_{pq}^{rs} \leq \gamma_{pq}^{pq}\gamma_{rs}^{rs} \qquad (101)$$

which imply relations between γ_{rs}^{pq} and γ_{pq}^{pq}.

Recently two more conditions (T_1 and T_2) going back to Erdahl [45] and Weinhold-Wilson [46] have received interest,

The following matrices must be non-negative:

$$\gamma_3 + \eta_3 \geq 0\ (T_1);\ \beta_3 + \beta_3' \geq 0\ (T_2) \qquad (102)$$

$$\begin{aligned} \gamma_{stu}^{pqr} &= \langle a_p a_q a_r\Psi|a_s a_t a_u\Psi\rangle && (103)\\ \eta_{stu}^{pqr} &= \langle a^s a^t a^u\Psi|a^p a^q a^r\Psi\rangle && (104)\\ \beta_{s,tu}^{pq,r} &= \langle a^s a_p a_q\Psi|a^r a_t a_u\Psi\rangle && (105)\\ \beta_{st,u}^{p,qr} &= \langle a^s a^t a_p\Psi|a^q a^r a_u\Psi\rangle && (106) \end{aligned}$$

There are again diagonal and Cauchy-Schwarz type conditions.

7.3 Direct calculation of the two-particle density matrix

In view of the partial trace relation

$$\sum_p \gamma_{pq}^{ps} = (n-1)\gamma_q^s \tag{107}$$

γ_1 is expressible through γ_2 and the energy expectation value (88) is entirely expressible in terms of γ_2 and a *reduced (2-electron) Hamiltonian* $\bar{H}$ [47].

$$\bar{H} = \frac{1}{2}\bar{H}_{rs}^{pq} a_{pq}^{rs} \tag{108}$$

$$\bar{H}_{rs}^{pq} = \frac{1}{[n-1]}(h_r^p \delta_s^q + \delta_r^p h_s^q) + g_{rs}^{pq} \tag{109}$$

$$E = \bar{H}_{rs}^{pq} \gamma_{pq}^{rs} \tag{110}$$

E is stationary with respect to variations of γ_2, if $\bar{\mathbf{H}}$ and γ_2 commute, i.e. have common eigenfunctions.

Let us first consider the simpler case of a *Hamiltonian without electron interaction*

$$H_0 = h_q^p a_p^q; \; E_0 = h_q^p \gamma_p^q \tag{111}$$

E_0 is stationary with respect to variation of γ if $[\mathbf{h}, \gamma] = 0$. In a basis of the eigenstates of $\mathbf{h}$, the *off-diagonal elements* of γ must vanish.

$$H_0 = \varepsilon_p a_p^p; \; \gamma_p^q = 0 \; for \; \varepsilon_p \neq \varepsilon_q; \; E_0 = \varepsilon_p n_p \tag{112}$$

There is *no information* from the stationarity condition on the n_p. We use the n-representability to get these. We know that

$$0 \leq n_p \leq 1; \; \sum_p n_p = n \tag{113}$$

We get the state of lowest energy if we fill the lowest one-electron states ε_p with electrons in the spirit of the *aufbau principle.*

Bopp [47] was probably the first who tried to minimize the energy with respect to a variation of γ_2. He assumed that the eigenvalues of γ_2 lie between 0 and 2, (which is plausible but incorrect) and filled the $n(n+1)/2$ lowest eigenstates of $\bar{H}$ with one electron pair each. The result was a *complete failure.*

The point is that one is only allowed to vary γ_2 under the subsidiary condition that it is *(fermion) n-representable*, i.e. that γ_2 is *derivable from an n-fermion wave function.* We have earlier mentioned a few necessary conditions for n-representability. There are some known sufficient conditions, which are too restrictive. Unfortunately *no simple complete set of necessary and sufficient conditions* has been available [43, 44, 48].

A recent paper by Coleman [49] on the solution of the n-representability problem, based on the concept of the *Kummer variety* [50], is both impressive and disappointing. The message turns out to be that, when formulated in a finite basis γ_2 is n-representable if it derived from a full-CI wave function [51]. This is by no means surprising, but highly frustrating. If one cares for exact n-representability, one obviously does not gain with respect to traditional quantum chemistry. One only has a chance if one is satisfied with approximate n-representability conditions.

At this point one can recall an old paper by Garrod et al. [52], in which only the three classical non-negativity conditions (97) were imposed. The results were encouraging, but had, for decades, no real consequences. Only rather recently this concept was taken up again, and was implemented together with the numerical technique of *semi-definite programming*. It turned out that if one took care of the T_1 and T_2 conditions (102) in addition to the P, Q and G conditions, results of a quality comparable to that of CCSD(T), which is a kind of standard of higher level coupled-cluster theory, could be achieved, though at significantly higher computational cost [53].

So the direct calculation of two-particle density matrices is still from being competitive with accurate methods of traditional AIQC, but the n-representability problem is no longer an unsurmountable barrier.

7.4 Separablity

Suppose that the space of one-electron basis functions can be divided into two subspaces A and B, such that all matrix elements h^p_q and g^{pq}_{rs} vanish, unless p and q, or p, q, r, s, belong to the same subspace A or B.

Then the Fock space Hamiltonian is additively separable [4]:

$$H = H_A + H_B \tag{114}$$

The solution of the Schrödinger equation is, in some sense, equivalent to the *block diagonalization* of H by means of a *similarity transformation* with the wave operator W.

$$L = W^{-1}HW \tag{115}$$

$$L_A = W_A^{-1}H_AW_A;\; L_B = W_B^{-1}H_BW_B \tag{116}$$

$$L = L_A + L_B;\; W = W_AW_B \tag{117}$$

The transformed operator L is additively separable, while the wave operator W is multiplicatively separable. We prefer to write (in the spirit of coupled-cluster theory)

$$W = \exp(S) \tag{118}$$

$$W_A = \exp(S_A);\; W_B = \exp(S_B);\; S = S_A + S_B \tag{119}$$

such that the *cluster amplitude* S is *additively separable*.

Separability implies extensivity (size-consistency) and connected-diagram theorem.

The combination of separability and minimal parametrization implies *linear scaling* with the number of atoms in a molecule.

7.5 Density cumulants

One can define the *cumulants of the k-particle density matrices*, for short *density cumulants*, which have a few advantages with respect to the k-particle density matrices [54, 55].

The *cumulant* $\boldsymbol{\lambda}_2 = \{\lambda^{pq}_{rs}\}$ of the two-particle density matrix γ^{pq}_{rs} is the *difference* between γ^{pq}_{rs} and *what one expects for independent particles* - that obey Fermi statistics; $\boldsymbol{\lambda}_3 = \{\lambda^{pqr}_{stu}\}$ is defined in an analogous way.

$$\lambda^{pq}_{rs} = \gamma^{pq}_{rs} - \gamma^p_r\gamma^q_s + \gamma^p_s\gamma^q_r \tag{120}$$

$$\begin{aligned}
\lambda^{pqr}_{stu} &= \gamma^{pqr}_{stu} - \gamma^p_s\lambda^{qr}_{tu} - \gamma^q_t\lambda^{pr}_{su} - \gamma^r_u\lambda^{pq}_{st} + \gamma^p_t\lambda^{qr}_{su} \\
&+ \gamma^p_u\lambda^{qr}_{ts} + \gamma^q_s\lambda^{pr}_{tu} + \gamma^q_u\lambda^{pr}_{st} + \gamma^r_s\lambda^{pq}_{ut} + \gamma^r_t\lambda^{pq}_{su} \\
&- \gamma^p_s\gamma^q_t\gamma^r_u - \gamma^p_t\gamma^q_u\gamma^r_s - \gamma^p_u\gamma^q_s\gamma^r_t \\
&+ \gamma^p_s\gamma^q_u\gamma^r_t + \gamma^q_t\gamma^p_u\gamma^r_s + \gamma^p_t\gamma^q_s\gamma^r_u
\end{aligned} \tag{121}$$

$\boldsymbol{\lambda}_2$ is a measure of pair correlation, in the form a *correlation increment*, preferable to the more common *correlation factor*. Cumulants of any *particle rank* be defined via a *generating function*[54]. A more compact formulation is in terms of the Grassmann (or wedge) products [55]

$$\boldsymbol{\lambda}_2 = \gamma_2 - \gamma_1 \wedge \gamma_1 \tag{122}$$

$$\boldsymbol{\lambda}_3 = \gamma_3 - \gamma_1 \wedge \boldsymbol{\lambda}_2 - \gamma_1 \wedge \gamma_1 \wedge \gamma_1 \tag{123}$$

$$\boldsymbol{\lambda}_4 = \gamma_4 - \gamma_1 \wedge \boldsymbol{\lambda}_3 - \boldsymbol{\lambda}_2 \wedge \boldsymbol{\lambda}_2 - \gamma_1 \wedge \gamma_1 \wedge \boldsymbol{\lambda}_2 - \gamma_1 \wedge \gamma_1 \wedge \gamma_1 \wedge \gamma_1 \tag{124}$$

The density cumulants have a few interesting properties.
Hermiticity: $\lambda^{pq}_{rs} = (\lambda^{rs}_{pq})^*$
Antisymmetry: $\lambda^{pq}_{rs} = -\lambda^{qp}_{rs} = -\lambda^{pq}_{sr} = \lambda^{qp}_{sr}$
Separability: For $\Psi = \mathcal{A}\{\Psi_A(1,2,\ldots n_A)\Psi_B(n_A+1,\ldots n_A+n_B)\}$ with Ψ_A and Ψ_B *strongly orthogonal*, $\lambda^{pq}_{rs} = 0$, unless *all labels* refer either to subsystem A or B.
The density cumulants are (unlike the reduced density matrices for $k > 1$) *additively separable.*

$$\lambda^{pq}_{rs} = (\lambda_A)^{pq}_{rs} + (\lambda_B)^{pq}_{rs} \tag{125}$$

$$\lambda^{pq}_{rs} = 0, \text{ in NSO} - \text{basis if any } n_p = 0 \; or = 1 \tag{126}$$

Trace relations:

$$\mathrm{Tr}(\gamma_1) = \Sigma_k n_k = n \tag{127}$$

$$\mathrm{Tr}(\boldsymbol{\lambda}_2) = \mathrm{Tr}(\gamma_1^2 - \gamma_1) = \Sigma_k(n_k^2 - n_k) = O(n);\; \mathrm{Tr}(\gamma_2) = O(n^2) \tag{128}$$

$$\mathrm{Tr}(\boldsymbol{\lambda}_3) = \mathrm{Tr}(-4\gamma_1^3 + \gamma_1^2 - 2\gamma_1) = 6\Sigma_k n_k(n_k - \frac{1}{2})(n_k - 1) = O(n) \tag{129}$$

Partial trace relations

$$\lambda^{pr}_{qr} = -\gamma^p_q + \gamma^p_r\gamma^r_q = (\gamma^2 - \gamma)^p_q \tag{130}$$

$$\lambda^{prt}_{qst} = 2\lambda^{pr}_{qs} - \gamma^p_t\lambda^{rt}_{sq} - \gamma^r_t\lambda^{pt}_{qs} - \gamma^t_q\lambda^{pr}_{ts} - \gamma^t_s\lambda^{pr}_{qt} \tag{131}$$

$$\lambda^{prt}_{qrt} = (-2\gamma^3 + 4\gamma^2 - 2\gamma)^p_q - \gamma^r_t\lambda^{tp}_{rq} - \gamma^t_r\lambda^{pr}_{qt} \tag{132}$$

Particle-hole symmetry/ Hole density matrices

$$\eta^p_q = <\Phi|a_q a^p|\Phi> = \delta^p_q - \gamma^p_q \tag{133}$$

$$\begin{aligned}\eta^{pq}_{rs} &= <\Phi|a_s a_r a^p a^q|\Phi> \\ &= \delta^p_r\delta^q_s - \delta^p_s\delta^q_r - \delta^p_r\gamma^q_s - \gamma^p_r\delta^q_s + \delta^p_s\gamma^q_r + \gamma^p_s\delta^q_r \\ &+ \gamma^{pq}_{rs}; etc.\end{aligned} \tag{134}$$

$$\eta^p_q = \delta^p_q(1 - n_p) \; in \; NSO - basis \tag{135}$$

These $\boldsymbol{\eta}_m$ matrices have the *same cumulants* as the corresponding $\boldsymbol{\gamma}_m$ matrices, just with γ^p_q replaced by η^p_q and with some sign changes, e. g.

$$\eta^{pq}_{rs} = \lambda^{pq}_{rs} + \eta^p_r\eta^q_s - \eta^p_s\eta^q_r \tag{136}$$

From the non-negativity conditions

$$\gamma_2 \geq 0;\; \boldsymbol{\eta}_2 \geq 0;\; \boldsymbol{\beta}_2 \geq 0;\; \gamma_3 + \boldsymbol{\eta}_3 \geq 0;\; \boldsymbol{\beta}_3 + \boldsymbol{\beta}'_3 \geq 0 \tag{137}$$

some important inequalities for the diagonal matrix elements of of $\boldsymbol{\lambda}_2$ can be derived

$$0 \leq \gamma^{pq}_{pq} = \lambda^{pq}_{pq} + \gamma^p_p\gamma^q_q - \gamma^p_q\gamma^q_p \tag{138}$$
$$0 \leq \eta^{pq}_{pq} = \lambda^{pq}_{pq} + \eta^p_p\eta^q_q - \eta^p_q\eta^q_p \tag{139}$$
$$0 \leq \beta^{p,q}_{q,p} = \lambda^{qp}_{pq} + \eta^q_q\gamma^p_p = -\lambda^{pq}_{pq} + \eta^q_q\gamma^p_p \tag{140}$$
$$0 \leq \beta^{q,p}_{q,p} = \lambda^{qp}_{pq} + \eta^p_p\gamma^q_q = -\lambda^{pq}_{pq} + \eta^p_p\gamma^q_q \tag{141}$$

Especially in an NSO basis we obtain

$$\lambda^{pq}_{pq} \geq \max\{-n_p n_q, -(1-n_p)(1-n_q)\} \geq -\frac{1}{4} \tag{142}$$

$$\lambda^{pq}_{pq} \leq \min\{n_p(1-n_q), n_q(1-n_p)\} \leq +\frac{1}{4} \tag{143}$$

In terms of Møller-Plesset perturbation theory one gets [56]:

$$\lambda^{ij}_{ij} = \frac{1}{2}\lambda^{ab}_{ij}\lambda^{ij}_{ab} + O(\mu^3) \tag{144}$$
$$\lambda^{ab}_{ab} = \frac{1}{2}\lambda^{ij}_{ab}\lambda^{ab}_{ij} + O(\mu^3) \tag{145}$$

where μ is the perturbation parameter.

That the T_1 condition is a condition for $\boldsymbol{\lambda}_2$ rather than $\boldsymbol{\lambda}_3$ can be seen in the following way.

$$\gamma_3 = \boldsymbol{\lambda}_3 + \boldsymbol{\gamma}\wedge\boldsymbol{\lambda}_2 + \boldsymbol{\gamma}\wedge\boldsymbol{\gamma}\wedge\boldsymbol{\gamma} \tag{146}$$
$$\eta_3 = -\boldsymbol{\lambda}_3 + \boldsymbol{\eta}\wedge\boldsymbol{\lambda}_2 + \boldsymbol{\eta}\wedge\boldsymbol{\eta}\wedge\boldsymbol{\eta} \tag{147}$$
$$\gamma_3 + \eta_3 = \boldsymbol{\delta}\wedge\boldsymbol{\lambda}_2 + \boldsymbol{\gamma}\wedge\boldsymbol{\gamma}\wedge\boldsymbol{\gamma} + \boldsymbol{\eta}\wedge\boldsymbol{\eta}\wedge\boldsymbol{\eta} \tag{148}$$

7.6 The energy in terms of the density cumulants

In terms of the density cumulants we get the following expression for the energy expectation value [54]:

$$E = h^p_q\gamma^p_q + \frac{1}{2}g^{rs}_{pq}(\gamma^p_r\gamma^q_s - \gamma^p_s\gamma^q_r) + \frac{1}{2}g^{pq}_{rs}\lambda^{rs}_{pq}$$
$$= \frac{1}{2}(h^p_q + f^p_q)\gamma^p_q + \frac{1}{2}g^{pq}_{rs}\lambda^{rs}_{pq} \tag{149}$$
$$f^p_q = h^p_q + \bar{g}^{pr}_{qs}\gamma^s_r; \; \bar{g}^{pr}_{qs} = g^{pr}_{qs} - g^{pr}_{sq} \tag{150}$$

One can try to make this stationary with respect to variations of $\boldsymbol{\gamma}$ and $\boldsymbol{\lambda}_2$, subject to the respective n-representability conditions. This looks easier than the energy expression in terms of $\tilde{H}$ and γ_2, and is *separable*. It is the basis of *density cumulant functional theory*.

One sees that the electron interaction energy consists of three parts, the Coulomb energy (including self interaction) [4]

$$E_{coul} = \frac{1}{2}g^{rs}_{pq}\gamma^p_r\gamma^q_s = \frac{1}{2}\int\frac{\varrho(1)\varrho(2)}{r_{12}}d^3r_1d^3r_2 \tag{151}$$

the exchange energy (including self-exchange)

$$E_{ex} = \frac{1}{2}g^{rs}_{pq}\gamma^p_s\gamma^q_r = -\frac{1}{4}\int\frac{\gamma_1(1,2)\gamma_1(2,1)}{r_{12}}d\tau_1 d\tau_2 \tag{152}$$

and the correlation energy

$$E_{corr} = \frac{1}{2} g_{pq}^{rs} \lambda_{pq}^{rs} = -\frac{1}{2} \int \frac{\lambda_2(1,2;1,2)}{r_{12}} d\tau_1 d\tau_2 \tag{153}$$

$$\lambda_2(1,2;1',2') = \gamma_2(1,2;1',2') - \gamma_1(1;1')\gamma_1(2;2') + \gamma_1(1;2')\gamma_1(2;1')$$

The explicit elimination of the *self interaction* is only possible for a single Slater determinant.

The definition of the Coulomb energy agrees with that used in DFT. In DFT a different definition of the *exact* exchange energy is used, namely with the exact γ_1 replaced by its idempotent approximation in terms of the Kohn-Sham orbitals. Consequently the correlation energy of DFT differs as well and contains contributions from exchange and the kinetic energy. The definition of the correlation energy current in AIQC (going back to Löwdin) [57] is different again. It is based on two separate calculations, one with and the other without correlation, and does hence not correspond to any expectation value.

Interesting are the following trace relations (implying summation over spin)

$$\mathrm{Tr}\,\varrho = \mathrm{Tr}\,\gamma_1 = n = \sum_\nu n_\nu \tag{154}$$

$$\mathrm{Tr}\,\gamma_2 = n(n-1) \tag{155}$$

$$\mathrm{Tr}\,\{\varrho(1)\varrho(2)\} = n^2 \tag{156}$$

$$\mathrm{Tr}\,\{\gamma_1(1,2)\gamma_1(2,1)\} = -\frac{1}{2}\mathrm{Tr}\,\{(\gamma_1)^2\} = -\sum_\nu n_\nu^2 \tag{157}$$

$$\mathrm{Tr}\,\lambda_2 = \frac{1}{2}\mathrm{Tr}\,\{\gamma_1(\gamma_1-2)\} = \sum_\nu n_\nu(n_\nu-1) \tag{158}$$

$$\mathrm{Tr}\{\lambda_2 - \gamma_1(1,2)\gamma_1(2,1)\} = -n \tag{159}$$

Here n_ν is the occupation number of the NSO χ_ν. We can distinguish two extreme situations: (a) γ is close to idempotent, i.e. the n_ν are either close to 1 or to 0. Then the exchange density integrates to something close to n and the correlation density to something close to 0. (b) There are many n_ν far from both 0 and 1. Then the traces of the exchange and the correlation hole can have the same order of magnitude, and the trace of the correlation hole can even dominate.

7.7 Contracted Schrödinger equations CSE_k and k-particle Brillouin conditions BC_k

One can try to *avoid the n-representability problem* by starting from the energy expectation value

$$\langle\Psi|H|\Psi\rangle = 0 \tag{160}$$

and making it stationary with respect to *arbitrary* (not norm-conserving) *variations* of Ψ, with X an excitation operator and E a Lagrange multiplier [58]

$$\Psi \to (1+Z)\Psi \tag{161}$$

$$\langle\Psi|Z^\dagger(H-E)|\Psi\rangle = 0 \tag{162}$$

$$\langle\Psi|(H-E)Z|\Psi\rangle = 0 \tag{163}$$

According to the particle rank k of Z in (161) we get the k-particle *contracted Schrödinger equations* CSE_k [60, 59, 58]

$$\langle\Psi|a^p_q(H-E)|\Psi\rangle = 0 \tag{164}$$

$$\langle\Psi|a^{pq}_{rs}(H-E)|\Psi\rangle = 0 \text{ etc.} \tag{165}$$

The explicit form of (164) is e.g.

$$h^r_s\gamma^{qs}_{pr} + h^r_p\gamma^q_r + \frac{1}{2}\bar{g}^{rs}_{pt}\gamma^{qt}_{rs} + \frac{1}{2}\bar{g}^{rs}_{tu}\gamma^{qtu}_{prs} = 0 \tag{166}$$

From these equations one can express γ_1 through γ_2 and γ_3, γ_2 through γ_3 and γ_4 etc. There is no obvious truncation of the k-particle hierarchy.

Valdemoro [61] has proposed a *'reconstruction'*, in approximating the higher-order γ_k in terms of those of lower orders and so derived a hierarchy of approximations characterized by the particle rank at which the system of equations is truncated. This was refined by Nakatsuji and Yasuda [62] and Mazziotti [55].

An alternative formulation, in which no reconstruction is necessary, was given by Kutzelnigg and Mukherjee [58] in terms of stationarity conditions in generalized normal ordering as a hierarchy of equations in which the density cumulants appear directly instead of the reduced density matrices (see subsect 7.9).

If one considers stationarity of the energy (160) with respect to norm-conserving variations,

$$\Psi \to \exp(Z)\Psi : \; Z^\dagger = -Z \tag{167}$$

one is led to the k-particle Brillouin conditions [63]

$$\langle\Psi|[H,Z]|\Psi\rangle = 0 \tag{168}$$

For a comparison of the k-particle Brillouin conditions with the k-particle contracted Schrödinger equations see the original literature [58].

7.8 Nakatsuji theorem and Nooijen conjecture

Nakatsuji [59] has made an interesting observation that is now known as Nakatsuji theorem.

It is *sufficient* for Ψ to be an exact full-CI wave function that the CSE_2 (which imply the CSE_1) are satisfied. We prove this for a somewhat more general theorem. Let H be expanded in an operator basis U_p, and let the CSE be satisfied for all U_p in this basis

$$H = \sum_p H_p U_p; \; \langle\Psi|(H-E)U_p|\Psi\rangle = 0 \tag{169}$$

Multiply by H_p and sum over p

$$\langle\Psi|(H-E)H|\Psi\rangle = 0 \tag{170}$$

subtract

$$E\langle\Psi|(H-E)|\Psi\rangle = 0 \tag{171}$$

The result is that the variance of the energy expectation value vanishes.

$$\langle\Psi|(H-E)^2|\Psi\rangle = \langle(H-E)\Psi|(H-E)\Psi\rangle = 0 \tag{172}$$

This is only possible if $(H-E)\Psi = 0$ (in the full-CI basis). The proof uses explicitly that H is a two-particle operator.

Nakatsuji [64] and Nooijen [65] have independently considered the CCGSD (*Generalized coupled cluster with single and double excitations*) ansatz for the wave function.

$$\Psi = e^T\Phi;\; T = \sum_{\mu=1}^{d} T_\mu U_\mu \;=\; T^p_q a^q_p + \frac{1}{2} T^{pq}_{rs} a^{rs}_{pq} \tag{173}$$

$$\langle\Phi|e^{T^\dagger}(H-E)U_\mu e^T|\Phi\rangle \;=\; 0;\; \mu = 1,2,...,d \tag{174}$$

This is a nonlinear system of as many (d) equations as there are unknowns T_μ - or more specifically T^p_q and T^{pq}_{rs}:

The number of equations to be satisfied corresponds to minimal parametrization.

If this set has a solution, then via the Nakatsuji theorem the wave function $e^T\Phi$ is equivalent to a full-CI function.

Nooijen has conjectured that this set of equations has a solution. There is strong evidence that this conjecture is not rigorously true, however it appears to hold to high degree of approximation [40].

7.9 Generalized normal ordering. Irreducible contracted Schrödinger equations $ICSE_k$ and k-particle Brillouin conditions IBC_k

For a refined formulation of many-electron quantum mechanics it is recommendend to introduce the concept of *generalized normal ordering* [66]. We can here not go into details. The so-called *particle-hole picture* can be regarded as a special case of generalized normal ordering, limited to a *single Slater determinant reference state*. The basic excitation operators in normal order with respect to the arbitrary reference function Ψ are symbolized by a tilde, like $\tilde{a}^{pq}_{rs}$. They are expressible in terms of the traditional excitation operators like a^{pq}_{rs} (78), which are in normal order with respect to the *genuine vacuum* e.g.

$$\tilde{a}^{pq}_{rs} \;=\; a^{pq}_{rs} - \gamma^p_r \tilde{a}^q_s - \gamma^q_s \tilde{a}^p_r + \gamma^p_s \tilde{a}^q_r + \gamma^q_r \tilde{a}^p_s - \gamma^{pq}_{rs} \tag{175}$$

These operators have two important properties.

1. For an operator in normal order with respect to Ψ, the expectation value in terms of Ψ vanishes, e.g.

$$\langle\Psi|\tilde{a}^{pq}_{rs}|\Psi\rangle = 0 \tag{176}$$

2. A product of two operators in generalized normal ordering can – in the sense of a generalized Wick theorem – always be written as a sum of operators in normal order, and possibly a constant, where the terms in this sum are the *normal product* and all possible *contractions*. There are particle contractions, hole contractions and all combinations of them, including *full contractions* (which are single numbers), but also contractions that involve the density cumulants of Ψ. For a single Slater determinant reference state all density cumulants vanish, and there are only particle and hole contractions and their combinations, as in the traditional particle-hole picture. An example is

$$\tilde{a}^p_q \tilde{a}^r_s = \tilde{a}^{pr}_{qs} + \eta^r_q \tilde{a}^p_s - \gamma^p_s \tilde{a}^r_q + \eta^r_q \gamma^p_s + \lambda^{pr}_{qs} \tag{177}$$

A consequence of these properties is that to the expectation value of a product of operators in generalized normal order only all full contractions (including those involving cumulants) contribute, e.g.

$$\langle\Psi|\tilde{a}^p_q \tilde{a}^r_s|\Psi\rangle = \eta^r_q \gamma^p_s + \lambda^{pr}_{qs} \tag{178}$$

In normal order with respect to the eigenfunction Ψ with eigenvalue E of the Hamiltonian H, the latter can be written as

$$H = E + f_q^p \tilde{a}_p^q + \frac{1}{2} g_{rs}^{pq} \tilde{a}_{pq}^{rs}; \; f_q^p = h_q^p + \bar{g}_{qs}^{pr} \gamma_r^s \tag{179}$$

If, in the contracted Schrödinger equations (164, 165) or the k-particle Brillouin conditions one replaces the operators without a tilde by those with a tilde (in normal order with respect to Ψ) one gets the irreducible contracted Schrödinger equations (ICSE$_k$) [58]

$$\langle \Psi | \tilde{a}_q^p (H - E) | \Psi \rangle = 0; \; CSE_1 \tag{180}$$

$$\langle \Psi | \tilde{a}_{rs}^{pq} (H - E) | \Psi \rangle = 0; CSE_2 \tag{181}$$

and the irreducible k-particle Brillouin conditions respectively

$$\langle \Psi | [H, \tilde{a}_q^p] | \Psi \rangle = 0 : \; IBC_1 \tag{182}$$

$$\langle \Psi | [H, \tilde{a}_{rs}^{pq}] | \Psi \rangle = 0 : \; IBC_2 \tag{183}$$

These lead directly to expressions in terms of the density cumulants. The IBC$_1$ expresses γ_1 through γ_1 and λ_2. All these conditions are additively *separable* and expressible through connected diagrams only. Unlike for the BC$_k$ and the CSE$_k$, a truncation at any particle rank is possible for the IBC$_k$ and the ICSE$_k$. So one automatically arrives at a hierarchy of equations for the successive determinantion of the $\boldsymbol{\lambda}_k$ without any need of a reconstruction.

IBC_1: $\gamma_1 = F(\gamma_1, \lambda_2)$

IBC_2: $\lambda_2 = F(\gamma_1, \lambda_2, \lambda_3)$

$ICSE_1$: $\gamma_1 = F(\gamma_1, \lambda_2, \lambda_3)$

$ICSE_2$: $\lambda_2 = F(\gamma_1, \lambda_2, \lambda_3, \lambda_4)$

The explicit form of the IBC$_1$ is

$$f_q^p \gamma_q^s - \gamma_q^p f_q^s + \frac{1}{2} \bar{g}_{tu}^{pr} \lambda_{qv}^{tu} - \frac{1}{2} \lambda_{rs}^{pu} \bar{g}_{qu}^{rs} = 0. \; f_q^p = h_q^p + \bar{g}_{qs}^{pr} \gamma_r^s \tag{184}$$

If we ignore $\boldsymbol{\lambda}_2$ we are led to Hartree-Fock.

There are still a few unsolved problems [56], but in principle an iterative construction of the λ_k with increasing particle rank is possible on these lines.

8 Conclusions

DFT based methods will continue to play a role in quantum chemical program packages. Whether their share is going to increase or decrease is hard to predict. It is likely that DFT will be used for routine calculations, whereas AIQC will dominate the realm of highly accurate computations. The potential for an improvement of the cost performance ratio is much larger in AIQC. DFT methods will develop further in the direction of AIQC and the border between DFT and AIQC may even disappear.

An evolution of AIQC in the direction of DFT is less likely. The main paradigm of AIQC has been that there are certain rules that must be strictly observed, e.g. that the wave function must be strictly antisymmetric, or that uncontrolled error cancellations should be avoided. DFT was much less subject to such rules and used this flexibility to 'improve' the performance. The still unsatisfactory cost-performance ratio of AIQC is in part due a overparametrization, i.e. to a description of a state in terms of a redundant set of parameters, where this redundancy can be genuine or numerical. AIQC can learn from DFT how one can be successful even with a subminimal parametrization.

As long as the HK theorem is not canonized and textbook stuff, people will continue to cite the original work, but it may happen that a pragmatic attitude takes over, and people will realize that there has been little correlation between attempts of a theoretical justification and the practical performance of DFT. The traditional conjuration of the HK theorem at the beginning of the presentation of DFT calculations is likely to disappear, together with the unsustainable claim that there are two equivalent alternative approaches to the n-electron problem, namely wave function theory and density functional theory. It is interesting that in the early days of the partnership between DFT and AIQC the main issue has been to understand why DFT often performs surprisingly well, while at present the shortcomings of DFT become more and more obvious such that there is more concern with how to improve DFT rather than to explain its success. DFT has still the advantage to be based on a subminimal parametrization of a state, but as soon as AIQC has achieved at least minimal parametrization, a new round of the competition may start. The partnership between DFT and AIQC is already more relaxed than it used to be, further atmospheric improvement can be expected.

Acknowledgment

This work has been supported by Deutsche Forschungsgemeinschaft (DFG) and Fonds der Chemie (FCh). The author thanks R. Jaquet and W. H.E. Schwarz fr helpul comments.

References

[1] P. Hohenberg, W. Kohn, *Phys. Rev. B* **136**, 864 (1964)

[2] W. Kohn, L. S. Sham, *Phys. Rev. A* **140**, 1133 (1965)

[3] W. Kutzelnigg, *Top. Curr. Chem.* **41**, 31 (1973)

[4] W. Kutzelnigg, *Theory of electron correlation*, in J. Rychlewski ed. *Explicitly correlated wave functions in chemistry and physics*, Kluver, Dordrecht 2003.

[5] P. A. M. Dirac, *Proc. Cambr. Phil. Soc.* **26**, 376 (1930); **27**, 240 (1931)

[6] J. C. Slater, *Phys. Rev.* **81**, 385 (1951); *J. Chem. Phys.* **435**, 228 (1965)

[7] L. H. Thomas, *Proc. Cambr. Phil. Soc.* **23**, 542 (1927)

[8] E. Fermi, *Rend. Accad. Linc.* **6**, 602 (1927)

[9] C. F. von Weizsäcker, *Z. Phys.* **96**, 431 (1935)

[10] E. Teller, *Rev. Mod. Phys.* **34**, 627 (1962)

[11] D. R. Hartree, *Proc. Cambr. Phil. Soc.* **24**, 111, 426 (1928)

[12] V. Fock. *Z. Phys.* **61**, 126 (1930)

[13] W. Kutzelnigg, Einführung in die Theoretische Chemie, Vol. 2. Die Chemische Bindung, Wiley-VCh, 1978, 1994, Last edition 2003 in one volume.

[14] M. A. Buijse and E. J. Baerends, *Mol. Phys.* **93**, 4129 (1990)

[15] C.Edmiston and K. Ruedenberg *Rev. Mod. Phys.* **35**, 457 (1963); *J. Chem. Phys.* **43**, 597 (1965)

[16] K. H. Johnson,*Adv. Quant. Chem.* **7**, 143 (1973)

[17] E. J. Baerends, P. Ros *Int. J. Quantum Chem.* **S12**, 169 (1978)

[18] K. Ruedenberg, *Rev. Mod. Phys.* **39**, 326 (1962);

[19] A. D. Becke, *Phys. Rev. A* **38**, 3098 (1988)

[20] J. Perdew and W. Yue, *Phys. Rev. B* **33**, 8000 (1986)

[21] J. B. Krieger, Y. Lie, G. J. Jafrate, *Phys. Rev. A* **46**, 5453 (1992); **47**, 165 (1993)

[22] E. S. Kryachko, *Int. J. Quantum Chem.* **103**, 118 (2005), **106**, 1795 (2006)

[23] E. H. Lieb, *Int. J. Quantum Chem.* **24**, 243 (1983)

[24] H. Eschrig, *The Fundamentals of Density Functional Theory* (Teubner, Stuttgart, 1996); A second edition (revised and extended) has been published by EAG.LE, Leipzig 2003, ISBN 3-937219-04-8

[25] W. Kutzelnigg, *J. Mol. Struct. THEOCHEM*

[26] R. M. Dreizler and E. K. U. Gross, *Density Functional Theory*, Springer, Berlin, 1990

[27] W. Koch and M. C. Holthausen, *A chemist's guide to density functional theory*, Wiley-VCh, Weinheim 2000

[28] E. S. Kryachko, *Int. J. Quantum Chem.* **18**, 1029 (1980)

[29] M.Levy and J. P. Perdew, *Int. J. Quantum Chem.* **21**, 511 (1982)

[30] M. Levy,*Proc. Natl. Acad. Sci. USA* **76**, 6062 (1979): Phys. Rev. A **26**, 1200 (1982)

[31] E. J. Baerends and O. V. Grisenko, *J. Phys. Chem. A* **101**, 5383 (1997); E. J. Baerends, *Phys. Rev. Lett.* **87**, 133004 (2001)

[32] A. Görling, M. Levy, *Phys. Rev. A* **50**, 196 (1994); **53**, 3140 (1996); **50**, 196 (1994) A. Goerling,*Phys. Rev. Lett.* **83**, 5459 (1999); **85**, 4229 (2000)

[33] C. Lee, W. Yang, and R. G. Parr *Phys. Rev. B* **37**, 785 (1988)

[34] I. Grabovski, S. Hirata, S. Ivanov, and R. J. Bartlett,*J. Chem. Phys.* **116**, 4415 (2002); I. Grabovski, S. Hirata, S. Ivanov, and R. J. Bartlett, *J. Chem. Phys.* **118**, 461 (2003); A. Beste and R. J. Bartlett, *J. Chem. Phys.* **120**, 8395 (2004); R. J.Bartlett, I. Grabovski, S. Hirata, and S. Ivanov, *J. Chem. Phys.* **122**, 034104 (2005); R. J. Bartlett, V. F. Lotrich, and I. V. Schweigert, *J. Chem. Phys.* **116**, 4415 (2002)

[35] E. Runge, E. K. U. Gross, *Phys. Rev. Lett.* **52**, 997 (1984)

[36] J. Schirmer, A. Dreuw, *Phys. Rev. A* submitted

[37] G. Zumbach and H.Maschke, *J. Chem. Phys.* **82**, 5604 (1975); T. L. Gilbert, Phys. Rev. B **12**, 2111 (1975); A. Mïler, *Phys. Lett. A* **105**, 866 (1984); M.A. Buijse and E. J. Baerends, *Mol. Phys.* **100**, 401 (2002); S. Goedecker and C. Umrigar, *Phys. Rev. Lett.* **81**, 866 (1998); J. Cioslowski and K.Pernal, *J. Chem. Phys.* **111**, 3396 (1999); M. Piris, *Int. J. Quantum Chem.* **106**, 1093 (2006)

[38] F. Colonna and A. Savin, J. Chem. Phys. **110**, 2828 (1999); W. T. Yang, P.W. Ayers, and Q. Wu, Phys. Rev. Lett. **92**, 146404 (2004); P. W. Ayers and M. Levy, J. Chem. Sci. **117**, 507 (2005); P. W. Ayers, Phys. Rev. A **73**, 012513 (2006); P. W. Ayers, S. Golden, J. Chem. Phys. **124**, 054101 (2006)

[39] J. K. Percus, *J. Chem. Phys.* **122**, 234103 (2005)

[40] W. Kutzelnigg and D. Mukherjee, *Phys. Rev. A* **71**, 022502 (2005)

[41] W. Kutzelnigg, *J. Chem. Phys.* **77**, 3081 (1982);**82**, 4166 (1984)

[42] W. Kutzelnigg, *The many-body perturbation theory of Brueckner and Goldstone*, in Applied many-body methods in spectroscopy and electronic structure, D. Mukherjee ed. Plenum, New York, 1992

[43] A. J. Coleman, *Rev. Mod. Phys.* **35**, 668 (1963)

[44] C. Garrod and J. K. Percus, *J. Math. Phys.* **5**, 1756 (1964)

[45] R. M. Erdahl, *Int. J. Quantum Chem.* **13**, 697 (1978)

[46] F. Weinhold, E. B. Wilson, *J. Chem. Phys.* **47**, 2298 (1967)

[47] F. Bopp, Z. Phys. **156**, 348 (1959)

[48] A. J. Coleman and V. I. Yukalov, *Reduced Density Matrices, Lecture Notes in Chemistry* (Springer, Berlin, 2000)

[49] A. J. Coleman, Phys. Rev. A **66**, 022503 (2002)

[50] H. Kummer, *Int. J. Quantum Chem.* **12**, 1033 (1977)

[51] A. Beste, K. Runge, and R. J. Bartlett, *Chem. Phys. Lett.* **335**, 263 (2002)

[52] C. Garrod, M. V. Mihalovic and M. Rosina, *J. Math. Phys.* **16**, 868 (1975)

[53] M. Nakata, H. Nakatsuji, M. Ehara, M. Fukuda, K. Nakata, K. Fujisawa, *J. Chem. Phys.* **114**, 8282 (2001); D. A. Mazziotti, *Phys. Rev. Lett.* **93**, 213001 (2004); *J. Chem. Phys.* **121**, 10957 (2004); Z. J. Zhao, B. J. Braams, M. Fukuda, M. L. Overton, and J. K. Percus, *J. Chem. Phys.* **120**, 2095 (2004)

[54] W. Kutzelnigg and D. Mukherjee, *J. Chem. Phys.* **110**, 2800 (1999)

[55] D. Mazziotti, *Chem. Phys. Lett.* **289**, 419 (1998); *Int. J. Quantum Chem.***70**, 557 (1998); *Phys. Rev. A* **57**, 4219 (1998); *Phys. Rev. A* **60**, 4396 (1999)

[56] W. Kutzelnigg and D. Mukherjee, *J. Chem. Phys.* **120**, 7350 (2004)

[57] P. O. Löwdin, *Adv. Chem. Phys.* **22**, 207 (1959)

[58] D. Mukherjee and W. Kutzelnigg, *J. Chem. Phys.* **114**, 2047 (2001); Erratum **114**, 8226 (2001)

[59] H. Nakatsuji, *Phys. Rev.* **14**, 41 (1976)

[60] L. Cohen and C. Freshberg, *Phys. Rev.* **13**, 927 (1976)

[61] C. Valdemoro, *Phys. Rev.* **31**, 2114 (1985)

[62] H. Nakatsuji and K. Yasuda, *Phys. Rev. Lett* **76**, 1039 (1996)

[63] W. Kutzelnigg, *Chem. Phys. Lett.* **64**, 383 (1979); *Int. J. Quantum Chem.* **18**, 3 (1980)

[64] H. Nakatsuji, *J. Chem. Phys.* **113**, 2949 (2000)

[65] M. Nooijen, *Phys. Rev. Lett.* **84**, 2108 (2000)

[66] W. Kutzelnigg and D. Mukherjee, *J. Chem. Phys.* **107**, 432 (1997)

Brill Academic Publishers
P.O. Box 9000, 2300 PA Leiden,
The Netherlands

Lecture Series on Computer and Computational Sciences
Volume 6, 2006, pp. 63-79

Collisional *ab initio* hyperpolarizabilities in computing hyper-Rayleigh spectra of noble gas heterodiatomics

W. Głaz°, T. Bancewicz°[1], J.-L. Godet◇, G. Maroulis•[2] and A. Haskopoulos•

°Nonlinear Optics Division, Faculty of Physics, Adam Mickiewicz University, 61-614 Poznań, Poland

◇Laboratoire des Propriétés Optiques des Matériaux et Applications, Université d'Angers, 2 boulevard Lavoisier, 49045 Angers, France

and

•Department of Chemistry, University of Patras, GR-26500 Patras, Greece

Received 27 July, 2006; accepted 2 August, 2006

Abstract: We review some recent studies on the theoretical and computational aspects of collision-induced nonlinear properties and phenomena, such as: the first hyperpolarizability of inert gas heteroatomic pairs (He-Ne, He-Ar, N-Ar and Kr-Xe) and spectral properties of hyper-Rayleigh scattered light in such systems. The hyperpolarizability data obtained by means of a variety of methods of quantum chemistry were applied to calculate spectral HR profiles. Computations are performed both quantum-mechanically and classically for two temperature values—T=95 K and T=295 K. Resulting tensorial properties and spectra are presented and discussed.

Keywords: Collisional ab initio hyperpolarizabilities, hyper-Rayleigh scattering, collisional spectra, MP2, SCF

PACS: 33.80.Gjxa, 34.30.th

1 Introduction

When a molecular microsystem is interacting with the electric field of a light beam of a high photon density, a nonlinear three-photon scattering effect may occur: the molecule 'absorbs' simultaneously two photons (of circular laser frequency ω_L) and 'emits' a photon of the doubled frequency $2\omega_L$ in a process of spontaneous three-photon scattering referred to as hyper-Rayleigh (HR) scattering. For many years the HR scattered intensities have been attributed to the first hyperpolarizability tensor **b** of individual molecules. However, interactions between monomers of a system may result in the so-called collisional hyperpolarizability yielding its contribution to the HR scattered light. centrosymmetric scattering microsystems cannot possess the permanent first hyperpolarizability tensor b_{ijk} due to the symmetry requirements, so they do not participate in nonlinear processes such as HR scattering. Though, it is possible that intermolecular interactions can suppress the symmetry limitations and the process, forbidden in the case of a single atom or a pair of identical atoms, will emerge at the level of the binary regime (i.e. for dissimilar atomic pairs).

Inevitably the theoretical analyses of such processes have to refer to the tensorial quantities responsible for inducing the nonlinear effects. In this review we are going to present results of our

[1]Corresponding author for the theoretical spectroscopy part. E-mail tbancewi@ zon12.physd.amu.edu.pl
[2]Corresponding author for the *ab initio* computational part. E-mail maroulis@upatras.gr

project in which the collision-induced hyper-Rayleigh light scattering processes in supermolecular systems of noble gas heterodiatoms were analyzed both theoretically and numerically [1, 2, 4, 5].

The basic quantity to be considered in this field of interest is the 3rd rank first hyperpolarizability tensor **b** . Accurate values of of its components or, more precisely, their dependence on the interatomic distance, are needed in order to obtain reliable theoretical predictions.

Fortunately in recent years, a number of theoretical and numerical concepts belonging to the field of quantum chemistry have emerged and been refined. This high-level computational chemistry has made a substantial contribution to the efforts by predicting accurate properties of the interacting monomers. The detailed philosophy on which these methods are based is presented in detail in previous work [6]; the more specific methodology and results related to the diatomic systems into which the HR research was carried out within the framework of the reviewed project (He-Ne, He-Ar, Ne-Ar nd Kr-Xe) were developed and reported in [7, 8, 9, 10]. In this study we shall present both the quantum chemistry aspects of calculations performed as well as the theoretical and computational background of HR analyses. Moreover, the results obtained on the grounds of the methods will be shown and discussed.

The paper is organized as follows. The basic concepts and details of the theory, both quantum mechanical and classical, have been given in Section 2 and 3. Sections 3.1 and 3.2 contain theoretical and numerical considerations on the quantum and classical methods of computing of the HR spectra. In 3.3 the *ab inito* quantum chemistry theories and procedures of evaluating collisional-induced molecular properties are briefly described, including the details of the chosen basis sets and methodological principles of the different tools used (SCF, MP2, CCSD and B3LYP). Section 4 is devoted to the discussion of the results obtained for all the noble gas atomic pairs studied. The dependence of the first hyperpolarizability components (b_1 and b_3) on the interatomic distance is illustrated and analyzed with regard to its influence on HR spectral properties. The spectral profiles obtained with these **b** tensor values are shown and compared with the linear scattering effect.

2 Basic definitions and concepts

Most of the quantities characterizing interactions of electromagnetic radiation with matter (e.g. light scattering or absorption), namely: intensities, differential crossections and absorption coefficients are in quite obvious way related to the electrical properties of molecular systems. From theoretical point of view the latter are expressed in terms of tensorial entities, such as for instance: dipole moments, polarizabilities and hyperpolarizabilities of different orders. In our earlier work [1, 2, 4, 5] we decided that spectral features of light nonlinearly scattered in molecular media can be in the most convenient manner (both theoretically and experimentally) expressed via the so-called double differential intensity, defined as the intensity (per supermolecule; diatom, in other words) of light irradiated within a frequency range $d\omega$ into a solid angle $d\Omega$ and normalized to the square of the incident intensity. Denoted hereafter as *normalized double differential intensity* (NDDI) it assumes the form of:

$$\mathcal{J}_{AZ}^{2\omega_L}(\omega) \equiv \left(\frac{\partial^2 I_{AZ}^{2\omega_L}}{\partial\Omega\partial\omega}\right)_{H-R} / I_0^2 = \frac{\pi}{2c} k_s^4 \sum_{i,i'} \rho_i \left|\left\langle i' | b_{AZZ} | i \right\rangle\right|^2 \delta(\omega - \omega_{i'i}), \tag{1}$$

where $\hbar\omega_{ii'} = E_{i'} - E_i$, ρ_i is the density matrix element of the initial state i, k_s stands for the wave-vector of the H-R scattered light, whereas b_{AZZ} is a Cartesian component of the first hyperpolarizability tensor—the quantity responsible for nonlinear radiative transitions between states i and i'.

The choice of the Cartesian indices in Eq. 1 is basically dependent on the geometry of an experimental setup used. Assuming that the incident light propagates along the Y axis, while the scattered beam is observed in the direction of X, two polarizability states of the latter can be considered: vertical (polarized component) with $A \equiv Z$ and horizontal (depolarized) one, when $A \equiv Y$.

The first hyperpolarizability tensor **b** is the one that in phenomena considered ties the square of an electric field applied with the induced dipole moment of a molecular system from which the nonlinear scattering effects actually emanate:

$$\mu_i^{2\omega_L} = b_{ijk} \, E_i \, E_k. \tag{2}$$

From mathematical point of view this is exactly the some tensor that is involved in the expressions determining the energy of interaction of an uncharged molecule with a homogeneous weak electric field:

$$E^p = E^0 - \mu_\alpha F_\alpha - \frac{1}{2} a_{\alpha\beta} F_\alpha F_\beta - \frac{1}{6} b_{\alpha\beta\gamma} F_\alpha F_\beta F_\gamma - \frac{1}{24} c_{\alpha\beta\gamma\delta} F_\alpha F_\beta F_\gamma F_\delta + \dots \tag{3}$$

where F_α , ... is the field, E^0 the energy of the free molecule, μ_α the dipole moment, $a_{\alpha\beta}$ the dipole polarizability, $b_{\alpha\beta\gamma}$ the first dipole hyperpolarizability, and $c_{\alpha\beta\gamma\delta}$ the second dipole hyperpolarizability. It must be stressed, however, that the physical nature of b_{ijk} and $b_{\alpha\beta\gamma}$ should be considered as different in this respect that in general b_{ijk} is a dynamic, frequency dependent quantity, while $b_{\alpha\beta\gamma}$ used in the *ab initio calculations* is a static entity which does not depend on ω. Nonetheless, it can be shown, that in the conditions assumed in the reported studies the dynamical hyperpolarizability may be replaced with its static approximation with relatively meaningless inaccuracy of the results obtained.

Having recourse to the Cartesian notations is seemingly natural at the first glance; however, when dealing for example with the angle dependent quantities, applying a reduced number of the so-called spherical invariants can lead to more treatable formulae and procedures. The number of such components vastly depend on the symmetry involved in the situation studied. For example, for the systems such as these discussed in this review (linear heterodiatoms), the ordinary polarizability can be expressed in terms of two components:

$$\begin{aligned} \bar{a} &= \frac{1}{3}\left(a_{zz} + 2\,a_{xx}\right), \\ \Delta a &= a_{zz} - a_{xx}, \end{aligned} \tag{4}$$

and quite similarly the first hyperpolarizability converts into terms defined as

$$\begin{aligned} b_1 &= \frac{3}{5}\left(b_{zzz} + 2b_{zxx}\right), \\ b_3 &= b_{zzz} - 3b_{zxx}, \end{aligned} \tag{5}$$

which, moreover, can be directly related to spherical tensor components [11, 12] $b_{10} = -\sqrt{5/3}\, b_1$ and $b_{30} = \sqrt{2/5}\, b_3$. It should be noted here that b_1 and b_3, according to their mathematical and physical nature, are often referred to as the *vector* and the *septor* part of the hyperpolarizability. In what follows we will be using both the spherical invariants as well as the spherical tensor approach and relevant terminology according to which is more convenient and appropriate in a particular context.

3 Methodology of computing

The general strategy which is to bring forth numerically computed nonlinear spectra consists of a number of combined specific methods. In the next sections we shall present an account of the procedures, both quantum as well as classical, and several examples of the results obtained with the tools they provide. It is not our intention, however, to get to far into the tedium of an elaborate outline here—a more detailed description, if needed, can be found in our previous works and the references therein [2, 4, 1].

3.1 Spectral calculations; quantum approach

The double differential intensity defined in Eq. 1 may be evaluated to a more numerically doable shape by means of the spherical tensor algebra. Indeed, the wave functions of the relative motion of two atoms involved in the formula can be factorized into the form of

$$|i\rangle = |n\,l\,m\rangle = Y_{lm}(\hat{\mathbf{R}})\,\frac{R_i(R)}{R}, \tag{6}$$

where the angular part of the product is given by the spherical harmonic $Y_{lm}(\hat{\mathbf{R}})$ and is fully detached from $R_i(R)$ which is the radial wave-function for state i; additionally, R denotes the interatomic separation [13, 14].

On applying the transformation and orthogonality principles typical of the spherical tensors we arrive at the following form of NDDI expression for the polarized component [2]:

$$\begin{aligned}\left(\frac{\partial^2 I^{2\omega_L}_{ZZ}}{\partial\Omega\,\partial\omega}\right)/I_0^2 &= \frac{\pi}{2c}k_s^4\sum_{i,i'}\rho_i\left\{(2l+1)\left[\frac{1}{5}H(1)_l^{l'}\left|(b_{10})_i^{i'}(E,\omega)\right|^2\right.\right.\\ &\left.\left.+\frac{2}{35}H(3)_l^{l'}\left|(b_{30})_i^{i'}(E,\omega)\right|^2\right]\right\}\delta(\omega-\omega_{ii'}),\end{aligned} \tag{7}$$

whereas for the depolarized component we obtain

$$\begin{aligned}\left(\frac{\partial^2 I^{2\omega_L}_{YZ}}{\partial\Omega\,\partial\omega}\right)/I_0^2 &= \frac{\pi}{2c}k_s^4\sum_{i,i'}\rho_i\left\{(2l+1)\left[\frac{1}{45}H(1)_l^{l'}\left|(b_{10})_i^{i'}(E,\omega)\right|^2\right.\right.\\ &\left.\left.+\frac{4}{105}H(3)_l^{l'}\left|(b_{30})_i^{i'}(E,\omega)\right|^2\right]\right\}\delta(\omega-\omega_{ii'}.)\end{aligned} \tag{8}$$

The formulae look slightly convoluted at first sight. Noticeably though, quite neatly, there are two distinguished basic terms included that can be evaluated separately. As it can be easily found, the intensity factors $H(k)_l^{l'}$ may be expressed in terms of the 3-j Wigner symbols:

$$H(k)_l^{l'} = (2l'+1)\begin{pmatrix} l' & k & l \\ 0 & 0 & 0 \end{pmatrix}^2, \tag{9}$$

and the task of calculating their numerical values does not pose any difficulty whatsoever [11, 12]. Note that k is in the above expressions is directly linked to the rank of the tensorial property involved; in the case of the H-R effect discussed in this review k=1 for the vector and k=3 for the septor components of the spectrum. On the other hand, evaluating numerical values of the other crucial constituent element of the NDDI formula, $b_{k\,0}(R)$, is by no means a trivial endeavor. By its definition,

$$(b_{k\,0})_i^{i'}(E,\omega) = \int_0^\infty \Psi_{i'}^*(R)\,b_{k\,0}(R)\,\Psi_i(R)\,dR, \tag{10}$$

it is simply a radial matrix element of the hyperpolarizability tensor component b_{k0}, with $\Psi_i(R)$ being the radial wave function of energy E and the angular momentum quantum number l and $\Psi_{i'}(R)$ is the wave function corresponding to E' and l'. Naturally, these wave functions are solutions of the Schrödinger equation in the form relevant to the system pondered. To be more precise, the matrix element is eventually involved in a more complex expression that appears during the final stage of evaluating the NDDI formula. Following to some extent the treatment developed in [13, 14] and performing the averaging introduced into the expressions in Eqs. 7 and 8 via the density-matrix ρ_i one can eventually obtain

$$\begin{aligned} \mathcal{J}_{YZ}^{2\nu_L}(\nu) = V\left(\frac{\partial^2 I_{YZ}^{2\nu_L}}{\partial\Omega\,\partial\nu}\right)/I_0^2 &= \frac{\pi}{2}k_s^4\,h\,L_0^3\sum_l(2l+1)\left\{\frac{1}{45}\left[H(1)_l^{l+1}\,(\mathcal{B}_{10})_l^{l+1}(\nu)\right.\right. \\ &+ \left. H(1)_l^{l-1}\,(\mathcal{B}_{10})_l^{l-1}(\nu)\right] + \frac{4}{105}\left[H(3)_l^{l+1}\,(\mathcal{B}_{30})_l^{l+1}(\nu) + H(3)_l^{l-1}\,(\mathcal{B}_{30})_l^{l-1}(\nu)\right. \\ &+ \left.\left. H(3)_l^{l+3}\,(\mathcal{B}_{30})_l^{l+3}(\nu) + H(3)_l^{l-3}\,(\mathcal{B}_{30})_l^{l-3}(\nu)\right]\right\}, \end{aligned} \tag{11}$$

where L_0 is the thermal de Broglie wavelength of the relative motion of two atoms, $\nu = \omega/(2\pi)$ is the frequency shift, and V is the active scattering volume. The b_{k0} related terms, $(\mathcal{B}_{k0})_l^{l'}$,

$$(\mathcal{B}_{k0})_l^{l'}(\nu) = \int_0^\infty dE\,e^{-\beta E}\left|(b_{k0})_l^{l'}(E,\nu)\right|^2 \tag{12}$$

are the ones to be computed so that we could obtain the desired H-R spectral shapes. Unfortunately, the available analytical methods of solving Schrödinger equations for relatively realistic physical situations are not in abundance, and very often we can cope with the problem by having recourse to methods of different 'genre'. In the series of works reported here the numerical approach was chosen in order to create final spectral profiles both quantum and classical. The quantum computing procedure was based on an algorithm that one was proven to be exceptionally effective in light absorption calculations [15] and within the framework of the H-R project was upgraded so that it could perform also the nonlinear scattering calculations. Applying the Numerov method of solving differential equations [16] it additionally moves all the computing into the complex plane in order to bypass the harmful influence of the cancellation error, especially severe for high frequency region of spectral lines. A vast number of the details concerning the technicalities of the procedure can be found in previous works [2, 4, 5].

Additionally, our studies on the hyper-Rayleigh scattering reviewed here were supplemented with an analyze of the spectral moments of the computed spectra, mainly treated as a benchmarking tool for previous calculations. These moments, associated with the particular components of the hyperpolarizability tensor can be obtained from a spectral profile, according to their definition [14]:

$$\mathcal{M}_{2n}^{b_{k0}} = \frac{2c}{\pi}\left(\frac{1}{2\pi}\frac{\lambda_L}{2}\right)^4\int_{-\infty}^{+\infty}\mathcal{J}_{k0}^{2\nu_L}(\nu)\,(2\pi c\nu)^{2n}d\nu. \tag{13}$$

where λ_L denotes the wave length of the laser light (in our computations we assumed $\lambda_L = 514.5$ nm), and the intensities $\mathcal{J}_{k0}^{2\nu_L}$ are functions of the polarized $\mathcal{J}_{ZZ}^{2\nu_L}(\nu)$ and the depolarized $\mathcal{J}_{ZY}^{2\nu_L}(\nu)$ components of the experimental spectra:

$$\mathcal{J}_{10}^{2\nu_L}(\nu) = 6\mathcal{J}_{ZZ}^{2\nu_L}(\nu) - 9\mathcal{J}_{ZY}^{2\nu_L}(\nu), \tag{14}$$

$$\mathcal{J}_{30}^{2\nu_L}(\nu) = -\frac{7}{2}\mathcal{J}_{ZZ}^{2\nu_L}(\nu) + \frac{63}{2}\mathcal{J}_{ZY}^{2\nu_L}(\nu), \tag{15}$$

Alternatively the moments, expressed in $\mathrm{cm}^{12}\mathrm{s}^{-2n}\mathrm{erg}^{-1}$, may be calculated by means of the formulas applying the pair correlation function $g(R)$. According to Refs. [2] and [14], the appropriate expression for the zeroth moment (sum rule) is of the shape:

$$M_0^{b_{k0}} = 4\pi \int_0^\infty b_{k0}^2(R)\, g(R)\, R^2\, dR. \tag{16}$$

The expressions defining the moments of higher order are rather tangled and we have found it not so very much useful to present them here; moreover, if necessary, they can be found published elsewhere [17]. The equilibrium pair correlation function can be derived within the quantum approach by expression:

$$g(R) = \frac{V}{\mathcal{Z}} \langle \boldsymbol{R} | \exp(-\beta \mathcal{H}) | \boldsymbol{R} \rangle, \tag{17}$$

where $\mathcal{Z}$ is the partition function for the relative motion of the atoms of the supermolecule in a box of volume V and $\mathcal{H}$ is the Hamiltonian of the system:

$$\mathcal{H} = \frac{p^2}{2\mu} + \mathcal{V}(R), \tag{18}$$

with μ denoting the reduced mass of two colliding particles. On applying appropriate procedures of the quantum mechanics, we can derive from these equations an expression determining the pair correlation function in a form suitable for numerical computations:

$$g(R) = \frac{L_0^3\, \mu^{3/2}}{2^{1/2}\, \pi^2\, \hbar^3} \sum_{l=0}^{\infty} \int_0^\infty dE\, E^{1/2} \exp(-\beta E) \mathcal{R}_{El}^*(R) \mathcal{R}_{El}(R), \tag{19}$$

where $\mathcal{R}_{El} = \Psi_{El}(R)/R$, and $\Psi_{El}(R)$ is a solution of the radial Schrödinger equation of the system considered. Fortunately, a precise code needed in order to obtain the wave functions $\mathcal{R}_{El}$, can be derived with far less difficulty than the one which is used for calculating the spectral profiles [16] since now we deal with merely the $\mathcal{R}_{El}$ squared and, as a result, the problem of the error due to the cancellation-oscillation effect becomes meaningless.

3.2 Spectral calculations; classical approach

The classical calculation of the CIH-R spectra due to the free dimers X-Y (colliding atoms coming from infinite) can been carried out by using $\mathcal{V}$, an appropriate interatomic potential, according to a procedure commonly used for Rayleigh spectra [13, 18]. Considering an immobile target and a projectile of reduced mass μ, three parameters must be taken into account: s, the velocity of the projectile when it is located at the infinity; q, the impact parameter of the projectile relatively to the target; t, the algebraic time relatively to the moment of the closest approach. In our calculations, we select a tight grid of points (q, s) and determine the polar coordinates $(R(t), \theta(t))$ for each point (q, s). Making a transposition of a formula proposed by Posch for Rayleigh spectra [19], we are then able to compute for any value of ν, the frequency shift, the Fourier transforms of the hyperpolarizability tensor components ($B_{10}(\nu, q, s)$ for the vector $b_{10}(R)$ and $B_{30}(\nu, q, s)$ for the septor $b_{30}(R)$):

$$B_{10}(\nu, q, s) = 4 \left(C_{1,1}(\nu, q, s) + S_{1,1}(\nu, q, s) \right), \tag{20}$$

$$B_{30}(\nu, q, s) = \frac{5}{2} \left(C_{3,3}(\nu, q, s) + S_{3,3}(\nu, q, s) \right) + \frac{3}{2} \left(C_{3,1}(\nu, q, s) + S_{3,1}(\nu, q, s) \right), \tag{21}$$

where (for $j = 1$ or 3):

$$C_{k,j} = \left(\int_0^\infty b_{k0}\,[R(t,q,s)]\,\cos(2\pi\nu t)\,\cos[j\theta(t,q,s)]\,dt \right)^2, \tag{22}$$

$$S_{k,j} = \left(\int_0^\infty b_{k0}\,[R(t,q,s)]\,\sin(2\pi\nu t)\,\sin[j\theta(t,q,s)]\,dt \right)^2. \tag{23}$$

In absolute units ($cm^8\,s\,erg^{-1}$), the contributions of these Fourier transforms to the hyper-Rayleigh spectra scattered by a scattering volume V are of the form (for the vector, $k = 1$, and the septor, $k = 3$, parts):

$$\mathcal{J}_{k0}^{2\nu_L}(\nu) = \frac{\pi}{2}\,k_s^4\,G(\nu) \int_0^\infty s\,f_{MB}(s,T)\,4\pi s^2 ds \int_0^\infty B_{k0}(\nu,q,s)\,2\pi q\,dq, \tag{24}$$

where $f_{MB}(s,T)\,4\pi\,s^2 ds$ stands for the Maxwell-Boltzmann factor whereas $G(\nu) = \exp\left(h\nu/(2k_BT)\right)$ is the detailed balance correction.

The reliability of these classical results have to be considered from two points of view. First, it is useful to verify that the obtained absolute intensities and spectral line shapes are compatible with the forecasts of the method of the spectral moments. Second, attention must be paid to the relative importance of purely quantal contributions.

The calculation of the classical spectral moments at the zeroth-order allows to check that the integrated intensities of the CIH-R spectra are correct and, therefore, that the intensities are provided in good units. Nevertheless, these zeroth-order moments give only information on low frequencies because the contribution of the high-frequency intensities to the zeroth-order moments is very small. Information on spectral line shapes is rather provided by higher-order moments. The $2n^{th}$-order moments corresponding to the vector ($k = 1$) and the septor ($k = 3$) components calculated from the spectral profiles are this time defined by:

$$\mathcal{M}_{2n}^{b_{k0}} = \frac{4c}{\pi}\left(\frac{1}{2\pi}\frac{\lambda_L}{2}\right)^4 \int_0^{+\infty} \mathcal{J}_{k0}^{2\nu_L}(\nu)G(\nu)^{-1}\,(2\pi c\nu)^{2n} d\nu. \tag{25}$$

They must be equal to the corresponding classical canonical averages [20]:

$$M_0^{b_{k0}} = V\,x_0^k\,\langle b_{k0}^2(R)\rangle, \tag{26}$$

$$M_2^{b_{k0}} = V\,x_2^k\left(\frac{k_BT}{\mu}\right)\left\langle \frac{k(k+1)\,b_{k0}^2}{R^2} + {b'_{k0}}^2 \right\rangle, \text{ etc.} \tag{27}$$

Here, $\langle \dots \rangle$ denotes the low density mean value:

$$\langle f(R)\rangle = \frac{1}{V}\int_0^\infty f(R)\,g_0(R)\,4\pi R^2\,dR, \tag{28}$$

where $g_0(R) = \exp(-V/(k_BT))$ is the classical pair correlation function. Besides, the dimensionless portions x_{2n}^k stand for the parts of the averages due to the free dimers. They can be calculated by using the classical method of Levine [21]. At high temperatures and/or if the well of the interatomic potential is shallow, the intensity contribution due to bound and metastable dimers (atoms trapped in the well of the effective potential) is negligible and the x_{2n}^k are roughly equal to one. Conversely, at low temperatures and/or if the well of the potential is deep, the x_{2n}^k can be significantly lower than one and a contribution of the bound and metastable dimers is expected at low frequencies (generally below 20 cm^{-1}). In Table 1, the examples of the He-Ne dimer moments at $T = 295$ K and $T = 95$ K are considered by using computed CIH-R intensities published in Ref. [4]. A good agreement between spectral moments and corresponding canonical averages is obtained, particularly at room temperature, showing the internal coherence of the classical calculations.

Table 1: The spectral 0th and 2nd moments of the b_1 contribution to the H-R spectra of He-Ne; classical (class) and quantum (quant) values calculated both from the spectrum profiles ($\mathcal{M}_{2n}$) as well as from the pair-correlation function (M_{2n}). $\mathcal{M}_{2n}^{sym}(quant)$ denotes the quantum moments obtained from symmetrized spectral lines.

T	$2n$	$M_{2n}(class)$	$\mathcal{M}_{2n}(class)$	$M_{2n}(quant)$	$\mathcal{M}_{2n}(quant)$	$\mathcal{M}_{2n}^{sym}(quant)$
95 K	0	$2.645 \cdot 10^{-88}$	$2.634 \cdot 10^{-88}$	$2.604 \cdot 10^{-88}$	$2.565 \cdot 10^{-88}$	$2.367 \cdot 10^{-88}$
	2	$2.582 \cdot 10^{-62}$	$2.669 \cdot 10^{-62}$	$3.192 \cdot 10^{-62}$	$3.133 \cdot 10^{-62}$	$2.223 \cdot 10^{-62}$
295 K	0	$3.986 \cdot 10^{-88}$	$3.960 \cdot 10^{-88}$	$3.909 \cdot 10^{-88}$	$3.923 \cdot 10^{-88}$	$3.845 \cdot 10^{-88}$
	2	$9.730 \cdot 10^{-62}$	$9.644 \cdot 10^{-62}$	$9.870 \cdot 10^{-62}$	$1.001 \cdot 10^{-62}$	$9.258 \cdot 10^{-62}$

For light dimers and/or at low temperatures, purely quantal contributions can have a substantial influence on the absolute intensities as well as on the spectral profiles of the corresponding H-R spectra. A way to foresee the relative importance of these quantal effects is to consider the Wigner expansion in powers of the Planck constant of the pair correlation function [22]:

$$g_2(R) = g_0(R)\left\{1 + \frac{\hbar^2}{12\mu\,(k_BT)^2}\left(\frac{\mathcal{V}'^2}{2k_BT} - \frac{2\mathcal{V}'}{R} - \mathcal{V}''\right)\right\}, \tag{29}$$

where the symbols $'$ and $''$ stand for the first and the second R-derivatives, respectively. This Wigner expansion up to $\hbar^2$ remains valid for corrections of a few per cents only. However, the comparison between $g_0(R)$ and $g_2(R)$ allows to estimate quickly in which measure the calculation can be considered as classical or not. For example, the ratio of the difference $|g_2(R) - g_0(R)|$ at its maximum, at the distance R_M, to $g_0(R_M)$ provides an order of magnitude of the quantum mechanical corrections introduced by Eq. (29). For the heavy dimer Kr-Xe, this ratio is lower than 0.3% at room temperature: classical and quantum mechanical calculations of CIH-R spectra give similar results [2]. Conversely for the light dimer He-Ne, this ratio is ten times bigger at room temperature and greater than 10% below 120 K: classical and quantum mechanical calculations of CIH-R spectra differ significantly, especially at low temperature [4].

3.3 *Ab initio* computed collisional hyperpolarizabilities

Collisional properties of molecular systems are expressed in terms of quantities that can be evaluated by means of a number of analytical and numerical methods. Interaction related tensorial invariants may be given in a convenient analytical form, for instance on the grounds of the well known multipolar approximation treated as a *first approach* model [23, 24], within which the tensorial components of interest to us are expressed as:

$$
\begin{aligned}
{}^{(\alpha B)}b_{zzz} &= 9\,(B^A\,\alpha^B\;-\;B^B\,\alpha^A)\,R^{-4},\\
{}^{(\alpha B)}b_{xxz} &= -\,\frac{9}{2}\,(B^A\,\alpha^B\;-\;B^B\,\alpha^A)\,R^{-4},
\end{aligned}
\tag{30}
$$

where $\alpha^{(i)}$ is the ordinary dipole^2 polarizability of the unperturbed species i whereas $B^{(i)}$ denotes its dipole^2-quadrapole hyperpolarizability. It is noteworthy that within the multiple model ${}^{(\alpha B)}b_1$ vanishes, whereas for ${}^{(\alpha B)}b_3$ we obtain:

$$
{}^{(\alpha B)}b_3 = \frac{45}{2}\,(B^A\,\alpha^B\;-\;B^B\,\alpha^A)\,R^{-4}.
\tag{31}
$$

The multipolar approximation, however, is of limited value for the inert gas diatomic systems considered here as, by its nature, it is generally valid for rather complex molecules interacting at relatively long intermolecular separations. As a consequence, in the spectral analyzes we present in this review, we have found it inevitable to apply the pair hyperpolarizability values obtained by having recourse to the so-called *ab initio* numerical procedures of quantum chemistry.

For decades a vast number of such methods was developed and refined, although they were mainly related to lower rank tensors describing radiative processes, such as the light absorption or the linear light scattering, i.e. collisional dipole moments and polarizabilities. Only relatively recently have a number of works emerged reporting values of higher order molecular properties, among them —the first order collisional hyperpolarizabilities for supermolecular noble gas hetero-diatom entities [25, 26] .

Quantum chemistry calculations are based on several standard theoretical approaches and algorithms derived from them. In their work Maroulis, Haskopoulos and Xenides utilized a number of such procedures to calculate linear and nonlinear properties of colliding atomic pairs [6, 7, 8, 9]; worth mentioning is also López-Cacheiro's et al contribution to this field [10]. Of all possible models of different merits and accuracy the tools chosen and applied in these studies belong to a 'short' list which comprises those denoted as: SCF, MP2, CCSD and B3LYP. Thus, the variety ranges from relatively simple self-consistent field (SCF) approach through a group of post-Hartree-Fock procedures [27], such as: second-order Møller-Plesset Perturbation Theory (MP2), coupled-cluster theory with singles and double (CCSD), coupled-cluster theory with single, doubles and perturbatively linked triple excitations (CCSD)(T), as well as density functional theory (B3LYP) [28]. Anticipating results presented in the sections to follow it might be useful here to remark that the $b_k(R)$ dependences stemming from calculations based on the above-mentioned procedures are not very much different with regard to the shape of the curves, with merely few local discrepancies observed; the only exception being the B3LYP case, which values are systematically lower than those of the other profiles for $R > 6a_0$. Naturally, we may expect that the discrepancies would have their influence on the shape of the spectral H-R lines, which therefore could possibly serve as a benchmarking tool to infer on distinguishing features of the particular $b_k(R)$ models. In the next chapters we shall refer back to some of the results reported in [1] and modified in this work with a more realistic potential and present H-R spectra computed by means of all the discussed methods for the Ne-Ar heterodiatom. As far as the remaining supermolecular pairs are concerned, we shall restrain ourselves to two specific schemes, namely—MP2 and SCF [29] intuitively found as the most trustworthy of all. Therefore, afterwards, a more detailed outline of the main principles of these particular approaches is to be elaborated on.

In our quantum chemical computations the electrical properties of interacting atoms A and B are based on Boys-Bernardi counterpoise-correction (CP) method [30]. The tensorial invariants considered are defined in Eqs. 5. In general the form of such interaction quantities are expressed as

$$P_{\text{int}}(A\cdots B)(R) = P(A\cdots B)(R) - P(A\cdots X)(R) - P(X\cdots B)(R), \tag{32}$$

where the symbol P(A···B) denotes the property P for the A-B, P(A···X) is the value of P for the subsystem A in the presence of the ghost orbitals of subsystem B, and R is the internuclear separation. Meticulously treated large and flexible Gaussian-type basis functions (GTF) were of near Hartree-Fock quality; and [6s4p3d1f] was the basis set for He, [9s6p5d1f] for Ne (40 and 59 contracted GTFs) [6], [8s6p5d3f] for Ar [8], and [8s7p6d5f/9s8p7d5f] for Kr-Xe. Details of their construction were given in a previous paper [6]. Their quality is evidenced by the dipole polarizabilities calculated at the SCF level for the two atoms. $\alpha_{\text{He}} = 1.322$ and $\alpha_{\text{Ne}} = 2.368\ e^2a_0^2E_h^{-1}$. to be compared with the numerical Hartree-Fock [31] values of 1.32223 and 2.37674, respectively. The calculations were performed entirely with *Gaussian 98* [32]. Within the framework of the computing procedure all electrons were correlated at the MP2 level, although it seems that, apart from this short notice, elaborating more on the size of correlation is beyond the scope of this work. Besides, relevant discussion on this question can be found elsewhere (see e. g. [8]). It should be added, however, that one of the advantages of the performed calculations is the range of the interatomic distances considered, especially the low level limit, which reaches exceptionally small values. Within this region, the correlation effects together with the triples to the CCSD treatment play an essential role [33]. The question is of some importance then since, as we shall show in the sections further on, that close encounters of the atoms influence significantly the shape of the spectral profiles.

4 Results and discussion

The methodology presented in the previous sections brings forth a tool for calculating electrical properties of colliding inert gas atoms and related spectral phenomena, in our case—hyper-Rayleigh light scattering. The key quantity here is the collision-induced first hyperpolarizability tensor, more specifically—its functional dependence on the interatomic separation. The quantum chemistry theoretical and computational techniques used are an efficient means of calculating these tensors for a variety of molecular systems. As a consequence, we are provided with data for relatively complete set of diatomic noble gas atoms ranging from the lightest He-Ne with the reduced mass $\mu = 3.34$ a. u. and slightly heavier He-Ar ($\mu = 3.64$) through the far end of the mass scale—Kr-Xe ($\mu = 51.15$ a. u.), with Ne-Ar falling in between ($\mu = 13.33$ a. u.). Spectral properties, quite obviously, coincide with the mass values, mainly via the dynamics of the colliding systems or, for instance, deciding to what extent the effects studied reveal their quantum nature. Indeed, some regular evolution of the properties, such as line half-widths, slopes and intensities throughout the miscellany of supermolecules can be observed, although certain irregularities in purely monotonic trends are also noticed. The latter comes as no surprise, since no doubt other mechanisms—electrical properties, potentials—must also have their share in shaping the profiles. In what follows we shall discuss the most characteristic spectral features of the inert gas diatomic pairs and how they are determined by the above-mentioned concurrently acting factors.

4.1 He-Ne

This is the lightest of all considered diatomic supermolecules and this fact may be reflected in properties of its H-R spectra, the other factor of most importance being the collisional hyperpolarizability. In Figs. 1, the **b** tensor dependences on the interatomic distance are shown.

In our calculations these functions were studied within a relatively large range of separations including those extremely short. As a consequence some characteristic behaviour of $b(R)$ can be

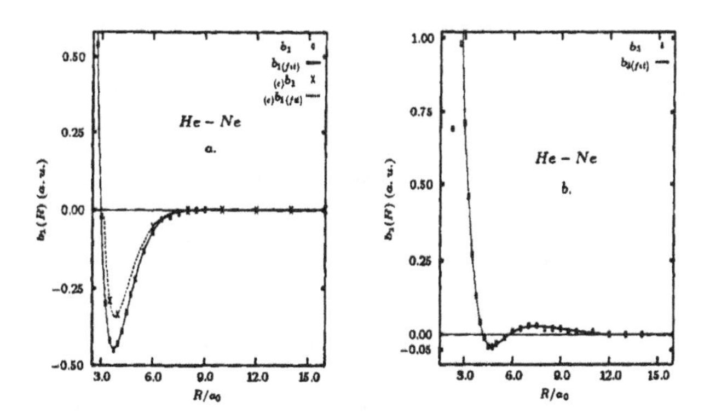

Figure 1: The interatomic distance dependence of b_1 and b_3 components of the hyperpolarizability tensor of He-Ne.

observed. Let us notice that the $b_1(R)$ profile shows in this region a profound negative regular well, whereas the $b_3(R)$ component goes through a characteristic pattern of minimum/maximum shape. To some extent this kind of $b(R)$ functions can be also met in the case of other atomic supermolecules (e.g. in Kr-Xe). Not surprisingly, these features find their reflection in the shapes of related spectral lines. Figure 2 illustrates the polarized component of the hyper-Rayleigh spectrum of He-Ne supermolecules.

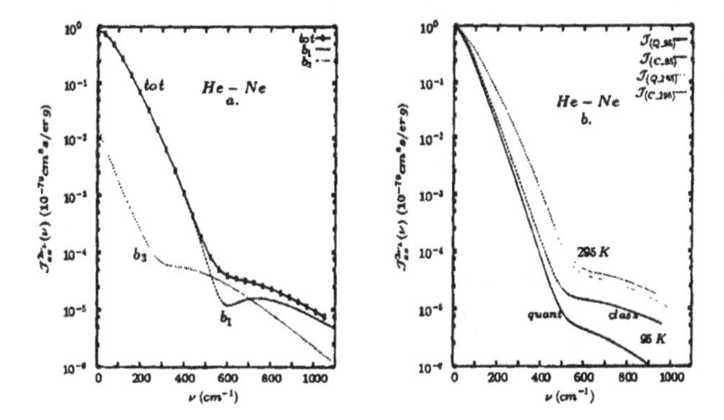

Figure 2: The polarized H-R spectra of He-Ne. The b_1 and b_3 contribution to the total profile are shown (a.), and the quantum and classical lines are compared for T=295 K and T=95 K (b.).

One of most striking features visible here are the bents 'spoiling' the monotonic decay of the intensities. According to our analysis, the bents apparently originate from the short and mid-range properties of $b_k(R)$, mainly from the minima and maxima which are there within the small distances available for the colliding atoms to penetrate. Furthermore, there are some other aspect of the spectral lines worth mentioning. Several parameters characterizing the lines are of significant importance, when one would like to find out the differences between spectral shapes. For example, the so-called slope of the line within the frequency region where it may be regarded as roughly exponential is equal to 48 cm^{-1}. It is the parameter which value will be compared with its counterparts calculated for the other noble gas diatomics. The He-Ne system is the one that, due to the smallest masses of its constituent atoms, is the most convenient object to judge on the influence of the quantum effect of the phenomena occurring therein. Having compared the He-Ne spectra in Fig. 2b., we find that the discrepancy between their properties determined within the quantum and the classical approach cannot be treated as meaningless. Although at the room temperature the profiles of classical and quantum origin basically seem almost the same—especially when plotted in the semi-logarithmic scale—more thorough numerical analysis reveals that the difference between them is in fact systematic and growing from the center of the lines, where the deviation is of the order of approximately 2.5% (both for the b_1 and the b_3 components), then grows to 6.0% and 4.5%, for b_1 and b_3, respectively, at $\nu \approx 250$ cm^{-1}, and eventually reaches $\approx 20\%$ for $\nu \approx 900$ cm^{-1}. All the more so is the tendency conspicuous at lower temperature (95 K): the difference ranges approximately from ≈10% (for the b_1 component) and 5% (b_3) at the line center to 33 and 25%, respectively at 200 cm^{-1} towards 400 and 500% at above 900 cm^{-1} (see Fig. 2b.)! This aspect of spectral distributions can be also considered on the grounds of the spectral moments analyze. In order to cut this review short we present in Tab. 1 only a limited example of the spectral moment values—the zeroth and the second rank quantities for the most domineering contribution to the total intensities linked to the $b_{10}(R)$ component of the hyperpolarizability (Eq. ??). Nevertheless, it does not violate the generality of our conclusions, especially that other order values, if needed, can be found in literature [4, 34]. In Table 1 we have both the moments received by integrating the obtained spectral profiles (Eq. 25) as well those evaluated from the sum rules formulae (Eqs. 16 and 27). It is easy to notice that the values are in remarkably good agreement. On the other hand, when the classical and quantum results are compared they show significant discrepancies ranging from ≈ 3% and 4% cent for the zeroth and the second moments, respectively at T=295 K towards ≈ 10% and ≈ 17% at 95 K. Thus, although in the higher temperatures region the classical treatment of the spectra of He-Ne can be accepted in many applications, for lower temperatures it seems to be inevitable to employ the fully quantum theory such as the one mentioned earlier in this article.

4.2 He-Ar

As not being significantly more massive than the previously discussed pair, this diatomic entity could be expected to show a great deal of similarity to He-Ne. Surprisingly though, there are a few crucial differences between the two cases. In Figs. 3 the hyperpolarizability components do not resemble those of He-Ne: $b_1(R)$ decays asymptotically to zero with no extremum present, and $b_3(R)$ has only one such feature. This has its immediate impact on the spectral properties visualized in Fig. 4a., where the b_1 line has no bent feature on it.

On this occasion it might be also worth noticing that, like in all other H-R spectra of the noble gas diatoms, the b_1 related part of them plays definitely the most important role in shaping the total spectrum in the low and mid-frequency region, whereas the b_3 contribution is visible only in certain sections of the line, mainly at wings (see also Fig. 2). Not only in this respect is the behaviour of the He-Ar surprising. The calculated slop of its H-R spectrum is greater than for the He-Ne case, $\Delta \approx 58$ cm^{-1}, while intuitively one might have expected lower values for the more massive supermolecules (indeed, for Ne-Ar $\Delta \approx 40$ cm^{-1} and for Kr-Xe $\Delta \approx 16$ cm^{-1}). This seemingly contradictory result can be attributed to the rather obvious fact that not only is the dynamics of the molecular system responsible for the spectral line shape, but also the electrical properties must have their share in it.

In Fig. 4b. the classical and quantum profiles are compared, analogously to the situation illustrated in Fig. 2b. Now, there is no surprise: the discrepancy between the to types of lines is

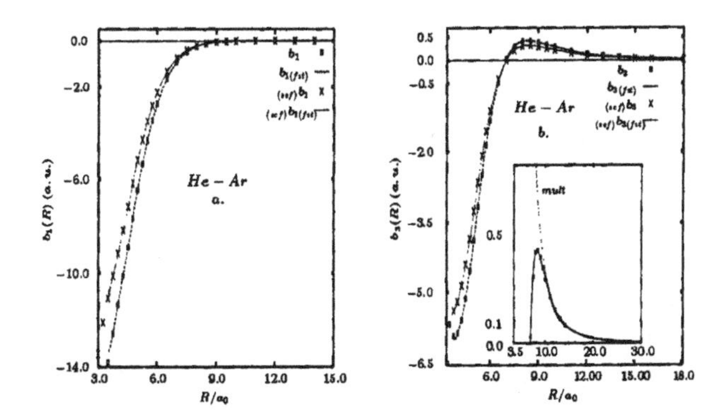

Figure 3: The same as in Fig. 1 for He-Ar diatoms. In the inset $b_3(R)$ calculated on the grounds of the multipolar approach is visualized (b.).

visible, yet it can be shown that it is lower than for the lighter He-Ar pair; the more pronounced is this tendency in heavier Ne-Ar and Kr-Xe systems ,where the difference for higher temperatures is barely noticeable, if at all.

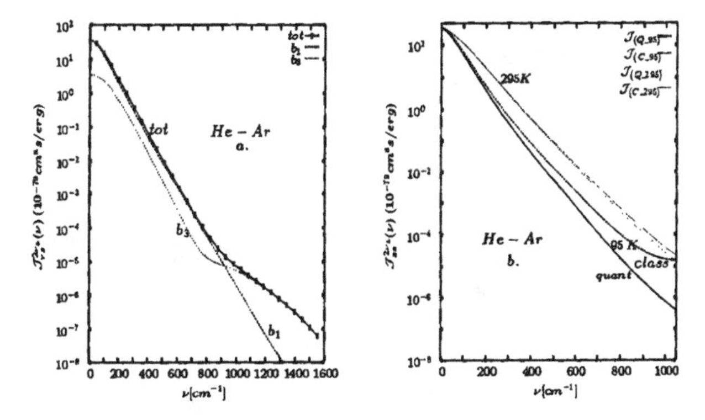

Figure 4: The same as in Fig. 2 but for the depolarized intensity of the He-Ar spectra(a.). The polarized quantum and classical lines are also compared (b.)

4.3 Ne-Ar

This supermolecule will serve us to present how the different models and theories used in the *ab initio* calculations of the molecular properties can affect the final form of the calculated H-R spectra.

In the previous sections the lines obtained with the MP2 methodology and occasionally with SCF were discussed; here we supplement the considerations with two other methods: CCSD and B3LYP. As we stated it previously the functional dependence $b(R)$ of all the methods, except

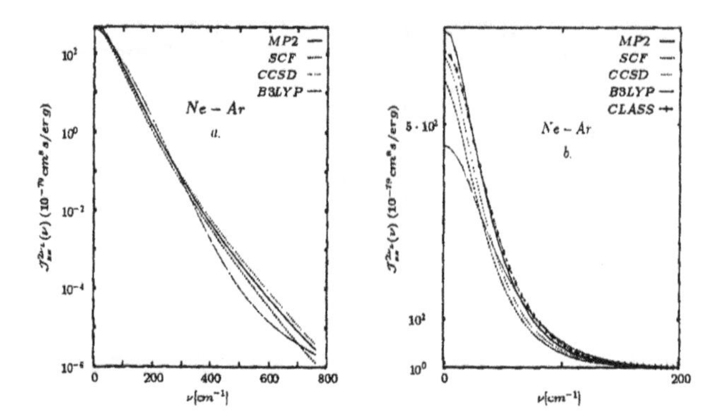

Figure 5: The depolarized spectra of Ne-Ar calculated by means of different methods plotted in semi-log (a.) and linear (b.) scale. Additionally the classical MP2 based profile is shown (b.) for comparison.

for B3LYP, follow more or less the same pattern akin to that revealed by He-Ar quantities [5]. Therefore when analyzing the spectral profiles plotted in Fig. 5 in semi-log scale we find that qualitatively their characteristics are close to each other (though, apart from the 'odd man out', B3LYP).

We must be aware, however, that a closer look at the lines based on more quantitative assessments can show some more pronounced differences, as it can be seen in Fig. 5b., where the linear scale is applied.

4.4 Kr-Xe

This is the heaviest noble gas diatom of all considered with all the consequences of this fact. Although the shapes of the $b(R)$ lines resembles those typical of the He-Ne system, both the parameters characterizing the spectra (slope, half-width, peak-to-wing ratio) as well as the computing problems it brings about are extreme. We discussed that question partially in the sections above, so herein we would like to consider briefly another interesting aspect concerning the H-R spectra: the differences between them and the collision-induced Rayleigh profiles. The *ab initio* methods presented before in this review can also provide values of a variety of tensorial quantities other than the first hyperpolarizability, among them polarizabilities or the multipole moments. The polarizability anisotropy dependence on R for Kr-Xe supermolecules is depicted in Fig. 6b. and the Rayleigh spectra in Fig. 6a., compared to their H-R counterparts (normalized to the same peak value). Quite similarly to what we found in our previous study [1, 5] concerning the Ne-Ar and He-Ar systems, the linear scattering feature is less diffuse with the peak-to-wing ratio of orders of magnitude higher than the nonlinear effects. As we discussed it before this kind of behaviour can be attributed to the electric characteristics of the diatoms, more precisely—to the range of the effective influence of the tensorial properties involved.

5 Conclusions

We have presented a review on the theoretical and numerical research into the subject of the nonlinear optical properties of diatomic noble gas heterodiatomic systems. The main attention was

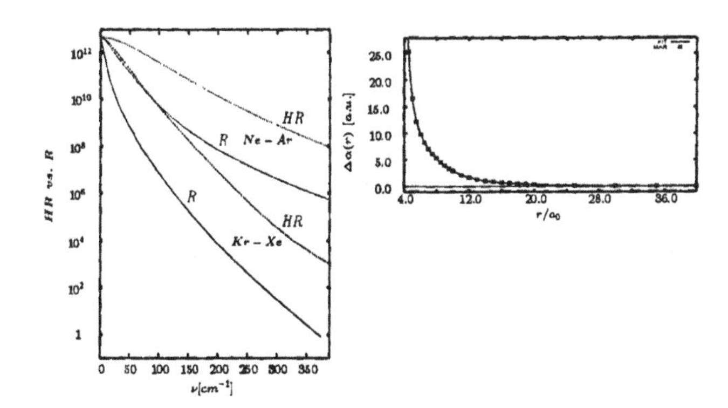

Figure 6: The depolarized Rayleigh (R) collision-induced light scattering spectrum for the Kr-Xe pair (a.) computed with the polarizability anisotropy presented in (b.). The hyper-Rayleigh (H-R) spectrum computed for the same temperature and for the same model is also presented. The R spectrum has been adjusted to fit the H-R one at the initial intensity. For comparison, the Ne-Ar spectra are added.

concentrated on the hyper-Rayleigh light scattering and the molecular property responsible for the occurrence of this effect, namely—the first hyperpolarizability. A number of quantum chemistry methods of evaluating the values of the components of the hyperpolarizability tensor were discussed. The spectral collision-induced HR spectra of several inert gas pairs have been obtained on the grounds of both classical and quantum-mechanical approach. We expect the combined quantum-chemical and theoretical approach to pave the way to more systematic experimental investigations. Admittedly though, to the best of our knowledge, to date only the allowed monomer hyper-Rayleigh (MHR) spectra have been measured. According to our estimations the effects discussed are substantially more subtle and thus require adequately more powerful methods of detection than those applied in MHR experiments. Nevertheless, in view of the incredible development in the domain of experimental molecular spectroscopy, allowing measurements of extremely small intensities [35], we may expect in foreseeable future to have at hand a set of data to which the methods presented here might be successfully applied.

Acknowledgment

This work has been supported in part by grant No. 1 PO3B 08 30 of the Polish Ministry of Sciences.

References

[1] W. Głaz and T. Bancewicz, J. Chem. Phys **118**, 6264 (2003).

[2] W. Głaz and T. Bancewicz, J. Chem. Phys **118**, 6264 (2003).

[3] W. Głaz, T. Bancewicz, and J.-L. Godet, J. Chem. Phys **122**, 224323 (2005).

[4] W. Glaz, T. Bancewicz, J.-L. Godet, G. Maroulis and A. Haskopoulos, Phys. Rev. A **73**, 042708 (2006).

[5] G. Maroulis, A. Haskopoulos, W. Glaz, T. Bancewicz and J.-L. Godet, Chem. Phys. Lett. (2006), to be published.

[6] G. Maroulis, J. Phys. Chem. A **104**, 4772 (2000).

[7] G. Maroulis and A. Haskopoulos, Chem. Phys. Lett. **349**, 335 (2001).

[8] G. Maroulis and A. Haskopoulos, Chem. Phys. Lett. **358**, 64 (2002).

[9] A. Haskopoulos, D. Xenides, and G. Maroulis, Chem. Phys. **309**, 271 (2005).

[10] J. López-Cacheiro, B.Fernandez, D.Marchesan, S.Coriani, C.Haettig and A.Rizzo, Mol. Phys. **102**, 101 (2004).

[11] A. R. Edmonds, *Angular Momentum in Quantum Mechanics* (Princeton University Press, Princeton, 1957).

[12] D. A. Varshalovich, A. N. Moskalev, and V. K. Khersonskii, *Quantum theory of angular momentum* (World Scientific, Singapore, 1988).

[13] L. Frommhold, Adv. Chem. Phys. **46**, 1 (1981).

[14] L. Frommhold, *Collision-induced Absorption in Gases* (Cambridge University Press, Cambridge, 1993).

[15] W. Głaz and G. C. Tabisz, Can. J. Phys. **79**, 801 (2001).

[16] W. Głaz, J. Yang, J. D. Poll, and C. G. Gray, Chem. Phys. Lett. **218**, 183 (1994).

[17] M. Moralid, A. Borysow, and L. Frommhold, Chem. Phys. Lett. **86**, 339 (1984).

[18] N. Meinander, J. Chem. Phys. **99**, 8654 (1993).

[19] H. Posch, Mol. Phys. **46**, 1213 (1982).

[20] T. Bancewicz, Chem.Phys.Lett. **213**, 363 (1993).

[21] H. B. Levine, J. Chem. Phys. **56**, 2455 (1972).

[22] E. Wigner, Phys. Rev. **40**, 749 (1932).

[23] A. D. Buckingham, E. P. Concannon, and I. D. Hands, J. Phys. Chem. **98**, 10455 (1994).

[24] X. Li, K. L. C. Hunt, J. Pipin, and D. M. Bishop, J. Chem. Phys. **105**, 10954 (1996).

[25] *Phenomena Induced by Intermolecular Interactions, NATO ASI Series* (Plenum, New York, 1985), edited by G Birnbaum.

[26] *Collision-and Interaction-Induced Spectroscopy*, Vol. 452 of *NATO ASI Series C: Mathematical and Physical Sciences* (Kluwer Academic Publishers, Dordrecht, 1995), edited by G.C. Tabisz and M.N. Neuman.

[27] A. Szabo and N. S. Ostlund, *Modern Quantum Chemistry* (MacMillan, New York, 1982).

[28] A. D. Becke, J. Chem. Phys. **98**, 5648 (1993).

[29] T.Helgaker, P. Jørgensen, and J. Olsen, *Molecular Electronic-Structure Theory* (Wiley, Chichester, 2000).

[30] S. Boys and F. Bernardi, Mol. Phys. **19**, 55 (1970).

[31] J. Stiehler and J. Hinze, J. Phys B **28**, 4055 (1995).

[32] M. J. Frisch, G. W. Trucks, and H. B. Schlegel at al, GAUSSIAN 98, Revision A7 , Gaussian, Inc., Pittsburgh PA (1988).

[33] P. Karamanis and G. Maroulis, Comp. Lett. **1**, 117 (2005).

[34] W. Głaz, T. Bancewicz, and J.-L. Godet, in preparation.

[35] F. Rachet, M. Chrysos, Ch. Guillot-Noel, and Y. Le Duff, Phys. Rev. Lett., **84**, 2120 (2000).

Brill Academic Publishers
P.O. Box 9000, 2300 PA Leiden
The Netherlands

Lecture Series on Computer and Computational Sciences
Volume 6, 2006, pp. 80-90

Molecular Polarization in Liquid Environment

R. C. Barreto, S. Canuto[1], K. Coutinho and H. C. Georg

Instituto de Fvsica,
Universidade de Syo Paulo,
CP 66318, 05315-970, Syo Paulo, SP, Brazil

Received 25 July, 2006; accepted 1 August, 2006

Abstract: The polarization of organic molecules in different liquid environments is considered. Two important aspects are discussed. First, the electronic polarization of the solute by the solvent and the corresponding changes in spectroscopic properties. Second, the structural changes induced by the solvent into the solute. This includes both geometrical aspects as well as more subtle chemical changes. For the electronic polarization we use the sequential Monte Carlo/quantum mechanics methodology. The specific case of acetone in water and homogeneous liquid acetone at normal conditions are considered. MP2/aug-cc-pVDZ calculations of the dipole moment of the reference solute molecule in the solvent environment are performed on statistically uncorrelated structures extracted from isothermic-isobaric MC simulations. To probe the reliability of the results this solute polarization is used to obtain the solvatochromic shift in UV-Vis absorption spectra. For the structural changes, two interesting cases are presented. First we show that geometry changes of *ortho*-betaine in water, is responsible for the major part of the very large solvatochromic shift. Second we show that the phenol blue possibly changes the chemical structure from a normal to a zwitterion state and this change could explain the solvent dependence of the hyperpolarizability observed experimentally.

Keywords: QM/MM, liquid environment, electronic polarization, molecular spectra, solvent effects

Mathematics SubjectClassification: PACS: 31.70.Dk, 31.10.+z, 31.15.Bs, 33.20.Kf

1. Introduction

The last decade has seen an enormous interest in the study of molecular systems in a liquid environment [1]. There are several reasons for this. First, understanding solvent effects is crucial for rationalizing several phenomena in physics, chemistry and biochemistry. Most experiments in organic chemistry laboratories are made in solution. This includes ultra-violet and visible (UV-Vis) spectra that give information on the electronic structure. Second, the water environment is essential for all processes related to life and regulates the biological activity. The ongoing computer revolution, both in software and hardware, is allowing the study of increasingly complex molecular systems. In this direction, the computer simulation of liquids is now not only possible but also desirable to deal with the natural statistical aspect of liquid systems. In fact a liquid is characterized by a large number of possible configurations at a certain temperature. Hence both molecular dynamics and Monte Carlo methods [2] can be used to generate liquid structures. As a consequence, in addition to the traditional self-consistent reaction-field, the use of computer simulation is gaining rapid use for the study of solvent effects. In special, the use of a hybrid methodology using both quantum mechanics and molecular mechanics (QM/MM) [3] is advancing with success in both spectroscopy and solvation. One

[1] Corresponding author. E-mail: canuto@if.usp.br, Fax: +55.11.3091-6831

variant of interest is the sequential use of MM and QM, where the classical simulation generates structures for subsequent quantum mechanical calculations [4]. This two-step procedure has the great advantage that after the simulation all statistical information is available permitting an efficient protocol for the more expensive QM calculations [5]. In particular the information obtained from the statistical correlation allows that the average results obtained be statistically convergent [6]. In fact, only statistically converged results can be of interest. In this work we use the sequential procedure with Monte Carlo simulations to study the effect of the polarization of the solvent environment in a reference solute molecule. Understanding the polarization effect in liquid systems is important in connection with several problems in solution chemistry [7]. In fact, we pay attention to the dipole moment, as the leading electrostatic term, of molecular systems in a liquid solvent environment, as compared to the dipole moment of the isolated molecule. The dipole moment of liquid systems is not a quantity that is easily amenable to experimental evaluation, although indirect results can be inferred, for instance, using integrated infrared intensities [8]. In this work we will consider the in-solution dipole moment of some representative organic molecules, like acetone, acetonitrile and water polarized by different and similar solvents. To probe the accuracy of these results the calculated electrostatic moments are used to evaluate solvatochromic shifts in UV-Vis absorption spectra. An additional and equally important consequence of the solvent polarization is the possible structural changes induced into the solute. This may be a simple geometrical change or a chemical deformation changing a solute to a *zwitterion* structure. Or even an *eno* → *keto* change of chemical structure. Zwitterion structures have increased dipole moment because of the charge separation and hence a possible increase in the stability of the molecule in polar solvents. Examples discussed below consider these aspects. Hence we first analyze electronic polarization and consequences in spectroscopic properties. Next we analyze structural changes. For the structural changes, we first consider betaine molecules where geometrical changes imply large spectroscopic modifications. Second, we consider phenol blue where the solvent may induce chemical changes with important consequences in the response properties such as the dipole hyperpolarizabilty.

2. Method of calculation and preliminary consideration

In this contribution results from several previous studies will be discussed. So we give a general description of typical simulations. Standard Monte Carlo (MC) Metropolis [2] simulations are carried out to generate the structures of liquid. The canonical (*NVT*) and the isobaric-isothermal ensemble (*NpT*) have been used and, as usual, periodic boundary conditions with the minimum image method in a cubic box are used. The simulations are performed with the DICE program [9] for a system consisting of the reference solute plus typically 500-1000 solvent molecules at T = 298 K and pressure of 1 atm. The interactions are described by the 6-12-1 potential composed of the usual Lennard-Jones 6-12 function plus the Coulomb interaction.

The simulations consist of a thermalization stage followed by an averaging stage typically of 5.0 X 10^7 MC steps. After the simulation the auto-correlation function of the energy is calculated to obtain information about the statistical correlation [4-6]. Next, configurations are sampled with less than 15% of statistical correlation. The quantum mechanical calculations are typically performed with augmented correlation-consistent basis set in the second-order Møller-Plesset perturbation (MP2/aug-cc-pVXZ, X=2 or 3) theory, as implemented in the Gaussian 03 program [10]. The QM calculations are performed on the structures composed of the central reference molecule and all others within a given solvation shell, these represented by simple point charges. Typically, to get converged results, calculations are made with around 100 configurations, implying 100 MP2/aug-cc-pVDZ calculations. The table 1 below illustrates the calculated in-solution dipole moment of liquid acetonitrile [11]. We can note convergence with respect to the number of solvent molecules used. Using 253 solvent molecules the dipole moment of acetonitrile changes by 0.71± 0.19 D, with respect to the gas phase value (3.93 D using MP2/aug-cc-pVDZ and 3.94 using MP2/aug-cc-pVTZ). Using the largest structure composed of the central acetonitrile surrounded by 253 acetonitrile molecules (represented by point charges) we obtain a dipole moment of 4.65 ± 0.19 D at the MP2/aug-cc-pVTZ level, in agreement with the experimental estimate of 4.5 ± 0.1 D [11]. Figure 1 illustrates the statistical distribution of calculated values. We note that the MP2/aug-cc-pVDZ method also gives good results. In general, the results have shown a good convergence with the correlation-consistent basis set. These are published results and a more thorough discussion can be found in the ref. [11]. A similar study for the dipole moment of liquid water is also published [12]. Our converged value gives an induced dipole moment of 0.74 ± 0.14 D, leading to a dipole moment of liquid water corresponding to 2.60 ± 0.14 D.

Table 1: Calculated changes (⟨Δμ⟩ ± σ) in the dipole moment (Debye) using MP2 methods for different solvation shells (radii in E). See ref. [11].

Radius of solvation	System	MP2	
		aug-cc-pVDZ	aug-cc-pVTZ
4.35	1 + 12	0.647 ± 0.259	0.648 ± 0.259
6.3	1 + 30	0.633 ± 0.216	0.634 ± 0.219
9.0	1 + 70	0.677 ± 0.195	0.677 ± 0.196
13.0	1 + 175	0.699 ± 0.193	0.700 ± 0.193
15.0	1 + 253	0.708 ± 0.189	0.709 ± 0.189

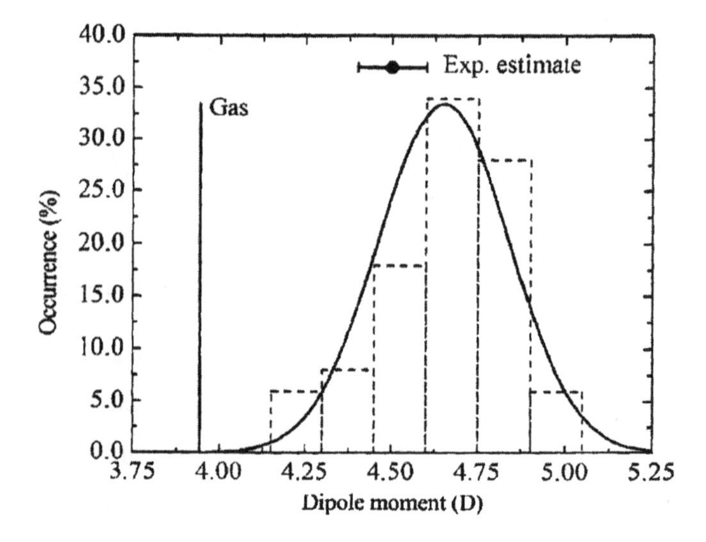

Figure 1: Statistical distribution of calculated dipole moments of liquid acetonitrile. Also shown are the gas phase calculated value and the experimentally inferred value.

A very important aspect of the simulation is the selection of the classical force field. Several force field parameters are available. As the sequential QM/MM adopted here has the classical part uncoupled of the QM part the pre-determined electrostatic parameters are not necessarily adapted to the specific solute-solvent situation. Normally, the atomic charges are selected such as to have an implicit polarization and a typical value is 20%. That is, the dipole moment of the solute is increased by 20% with respect to the gas phase dipole moment. When calculating the heterogeneous situation of a solute in the solvent field it is to be expected that the electrostatic field responds to different solute-solvent situation. The solute polarizes differently for different solvents and different physical-chemical circumstances (temperature, density, etc.). Hence it seems important to adjust the parameters to the specific situation in case. To obtain the electrostatic equilibrium between the solute and the solvent an iterative procedure is used in our group [13]. This is made according to the chart below. This strategy is similar to that adopted in the RISM [14] and ASEP/MD [15] methods. The basic difference is that our iterative procedure seeks converged value for the electrostatic field of the solute and this is used as the empirical classical potential for performing the MC simulations to be used in the subsequent QM calculations of the average property of interest. At this stage we have a three-step situation. First, the determination of the electrostatic in-solution parameters; second, using these in a MC simulation to generate liquid structures; and third, using the statistically uncorrelated structures sampled from the MC simulation to perform QM calculations and obtain the average property of interest.

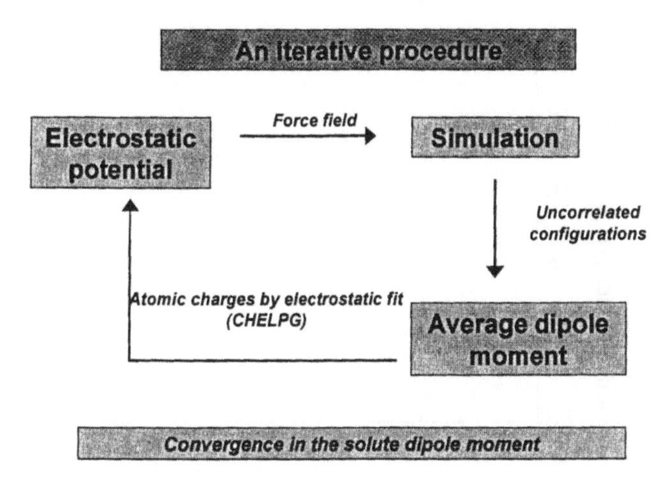

Chart 1: Iterative procedure to obtain in-solution electrostatic potential. Convergence is verified in the dipole moment.

3. Further results and discussion: molecular polarization in solution

3.A. Electronic polarization

Several systems have been considered but because of space limitation we discuss the case of acetone in the environments of water and acetone molecules. In the latter case, we are considering the case of homogeneous liquid acetone. For acetone we use the same potential as in ref. [16] and the potential for water is the SPC [17]. We now discuss three different procedures for obtaining the in-solution dipole moment of acetone. In the first case we calculate the average dipole moment of one acetone molecule in the electrostatic field of 200 solvent water molecules represented by simple point charges. This leads to a natural increase in the dipole moment corresponding to an estimate of the in-solution value. Second, we update this first estimate of the in-solution dipole moment in the electrostatic part of the classical potential to perform another MC simulation. The procedure is iterated till the convergence of the QM dipole moment with the classical potential (see Figure 2). At this point the in-solution dipole moment is in electrostatic equilibrium with the solvent [13].

The third case considered is the case of homogeneous liquid acetone, where the in-solution value is obtained also by iteration, but in this case we update both the solute and the solvent molecules, as they are in fact the same molecule (acetone). Every result reported is the average of 100 MP2/aug-cc-pVDZ calculations. Table 2 shows the results. The dipole moment of isolated acetone is obtained as 2.98 D, in good agreement with the experimental value [18] of 2.93 D. For the in-water case the calculations made without updating the potential (iteration 1) gives the value of 4.05 ± 0.03 D. Finally, iterating till electrostatic equilibrium we obtain the converged in-solution dipole moment of acetone in water as 4.80 ± 0.04 D, corresponding to a large polarization (increase of 60%) compared to the gas phase value.

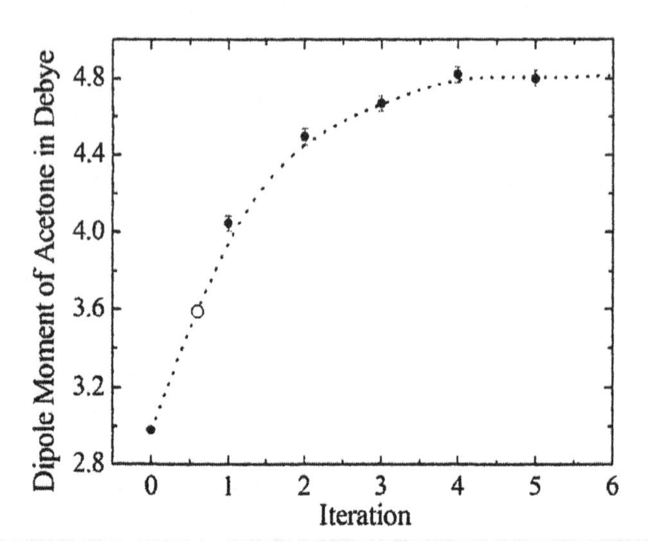

Figure 2: Convergence of the calculated in-solution dipole moment of acetone in water.

This value is in very good agreement with previous QM/MM methods, in special the value of 4.90 D obtained from Car-Parrinello molecular dynamics [19], corresponding also to a 60% increase compared to the gas-phase value.

Table 2: Calculated gas phase and in-solution dipole moment of acetone. Calculations are made at the MP2/aug-cc-pVDZ level using 100 statistically uncorrelated configurations generated by Monte Carlo simulations.

	Present	Ref [19]	Experiment.
μ_{Gas}	2.98	3.08	2.93 [18]
$\mu_{In\text{-}Water}$	4.80	4.90	-
$\mu_{Liquid\ acetone}$	3.56	-	-

Table 2 also shows the calculated dipole moment of liquid acetone. The converged value of 3.56 D represents a polarization of 20% with respect to the gas phase value. This shows the considerable effect of water in polarizing solute molecules. Using the in-water calculated electrostatic moment in the MC classical potential, configurations were generated, and sampled, for calculating the solvatochromic shift of the n - π^* transition. The corresponding gas to in-water shift is calculated with INDO/CIS method implemented in the ZINDO program with the spectroscopic parametrization [20]. The converged average result of 1650 ± 42 cm^{-1} is in excellent agreement with the experimental result of 1500-1700 cm^{-1} [21]. Using TD-DFT (B3LYP/6-311++G(d,p)) with the solvent molecules represented by simple point charges we obtain a larger value of 2200 cm^{-1} for the solvatochromic shift. Figure 3 shows the statistical convergence of the solvatochromic shift.

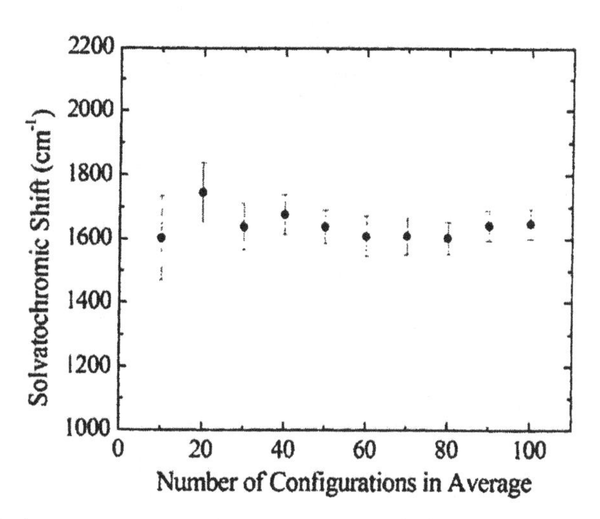

Figure 3: Statistical convergence of the calculated solvatochromic shift of the $n \rightarrow \pi^*$ transition of acetone in water. Converged value is 1650 ± 42 cm^{-1}.

These results suggest the strong polarization of the solute molecule in the presence of liquid solvents. Several examples are available for liquid systems [22] including homogeneous acetonitrile, water and even large solute molecules such as benzophenone in water.

3.B. Structural and chemical changes induced by the solvent

Structural changes of molecules in liquid solvent environment are a common effect in biomolecules interacting with the aqueous environment. On the spectroscopic side there is an enormous interest in betaine systems. These dyes are of great importance in probing solvent polarity from solvatochromic shifts [23]. Of special interest in this context is the so-called $E_T(30)$ Reichardt betaine, which has one of the largest measured solvatochromic shifts, ranging from a maximum at 453 nm in water to 810 nm in diphenyl ether, that is, a solvatochromic shift of 357 nm (a huge 9730 cm^{-1} of shift). It is for this high sensitivity that $E_T(30)$ is used a spectroscopic indicator of a normalized solvent polarity [23]. We consider next another betaine. The so-called *ortho*-betaine (Chart 2). It is simpler but it still shows all the essential character related to the dependence of the solvatochromic shift with the solvent. The shift of *ortho*-betaine in changing from toluene to water is ca. 7500 cm^{-1} [24]. Using the sequential MC/QM procedure we sampled configurations for QM calculations of the $n \rightarrow \pi^*$ transition. The calculated shift was computed as a disappointingly low value of ca. 1600 cm^{-1} [25] using in water the same structure of *ortho*-betaine as in gas phase. However, as Figure 4 shows, whereas in gas phase the torsion angle is 30°, in water the free energy calculation indicate that this angle changes considerably to a value of 60°, but the potential is floppy between 60-90[7]. Figure 4 shows the free energy potential curve, calculated using thermodynamic perturbation theory [26] implemented in the DICE program [9].

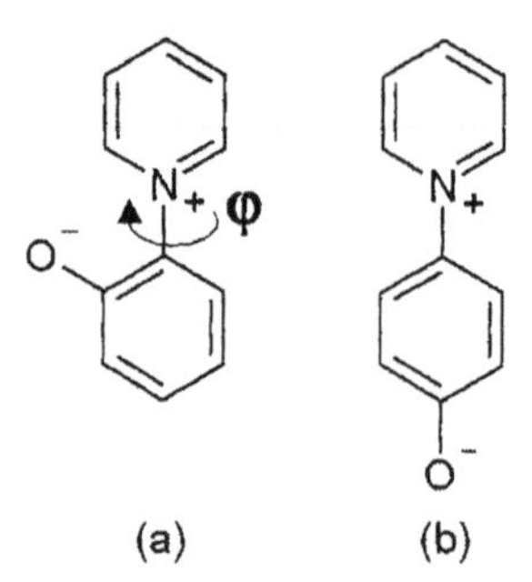

Chart 2: *Ortho-* and *para-*betaine.

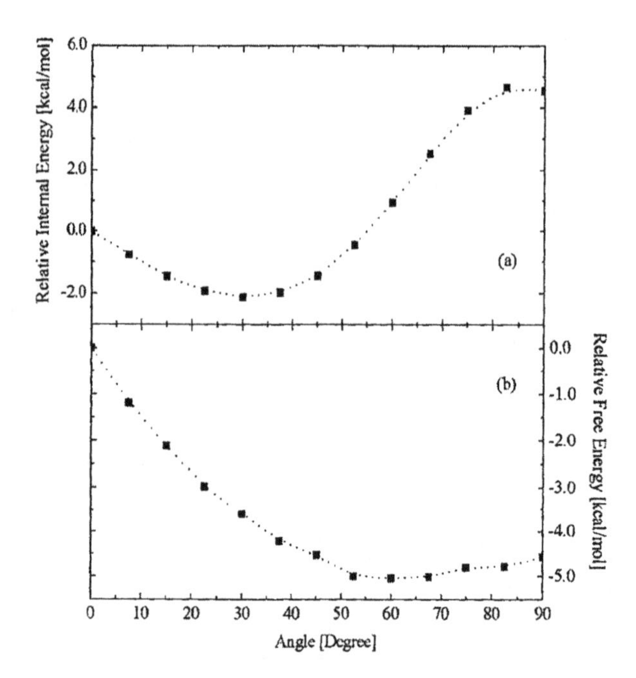

Figure 4: Helmholtz free energy as a function of the torsion angle using ab initio calculation with MP2/6-31G and Thermodynamic Perturbation Theory in the MC simulation, respectively.

Figure 4 shows that there are significant qualitative as well as quantitative differences between the internal rotation in gas phase and in water. The most stable conformer is shifted from approximately 30° in the gas phase to near 60° in water. But also, the maximum in the potential energy corresponds to 90° in gas phase whereas in water it is shifted to 0°. The minimum of the Helmholtz energy profile in water is very flat compared to the potential energy minimum in gas phase. The barrier to internal rotation changes from approximately 29 kJ/mol in gas phase to around 21 kJ/mol in water. These results are significant since it is common practice to just include the solvent effects into the gas phase potential energy surface or even use only the most stable conformer in gas phase for simulations in solutions. This twist angle has been noted in the $E_T(30)$ Reichardt betaine [27]. In the present case it is clear that there is a soft inter-ring rotation mode associated to the flat potential of *ortho*-betaine in water. Figure 5 shows the implication of this in the calculated solvatochromic shift. We can see that the conformational change responds to ca. 3000 cm^{-1} of the total shift. Incorporating the influence of the outer solvation shells and the conformation change a total solvatochromic shift of ca. 6100 cm^{-1} is obtained [25].

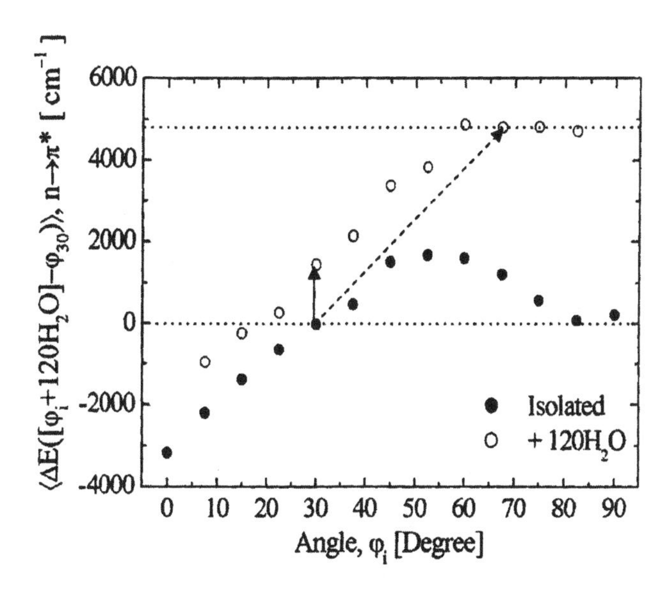

Figure 5: The calculated $n \rightarrow \pi^*$ transition (in cm^{-1}) with respect to the conformation with $\varphi = 30°$ as a function of the torsion angle. The filled circles are the results obtained for the isolated *ortho*-betaine and the open circles are the averaged results obtained to 61 configurations of one *ortho*-betaine surrounded by 120 water molecules.

This final calculated solvatochromic value of 6100 cm^{-1} for the $n \rightarrow \pi^*$ transition is in reasonable agreement with the observed value of ca. 7500 cm^{-1}, obtained as the shift of the maximum in the absorption spectra of the *ortho*-betaine in water and toluene [24]. A detailed discussion of the solvatochromic shift of *ortho*-betaine can be found in ref. [25].

A more subtle change due to the solvent is the chemical change in *keto*-type structure to the *eno*-type structure. As the charge-separated species has a larger dipole moment it is expected to be more stable in polar solvents. One example is the phenol blue molecule (Figure 6) where an interesting study [28] has experimentally investigated the dipole hyperpolarizability as a function of the solvent, showing a maximum. This is rationalized on the basis of two possible configurations, the *eno* and *keto* forms. The continuous variation between the two forms can be studied theoretically [29] by defining an order parameter for the transition, that in this case is the bond length alternation χ, defined on the caption of Figure 6. The variation of the calculated hyperpolarizability [29] exhibits the same aspect as the experimental [28] result suggesting that solvent effects in phenol blue may induce structural modifications leading to strong changes in molecular properties. One of these is the dipole hyperpolarizability that changes with the solvent polarity. Whether this is corroborated by additional, even if indirect, observations is not clear. NMR chemical shifts of indo-anilines in different solvents conclude that only the *keto*-like structure exists [30]. Phenol blue was not considered, though. However, resonance Raman spectra [31] seem to suggest a possible structure change. Our study [29] considered also the first electronic absorption transition that showed marked changes with the solvent polarity and therefore also in the hyperpolarizability.

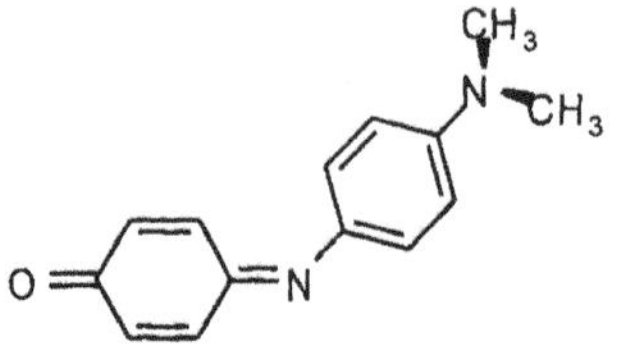

Phenol blue in the *keto*-like form

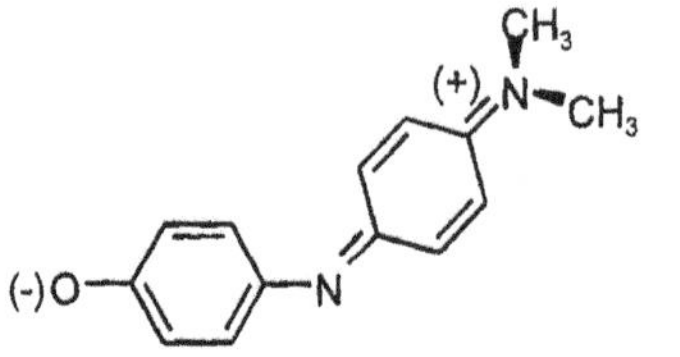

Phenol blue in the charge-separated *eno* form.

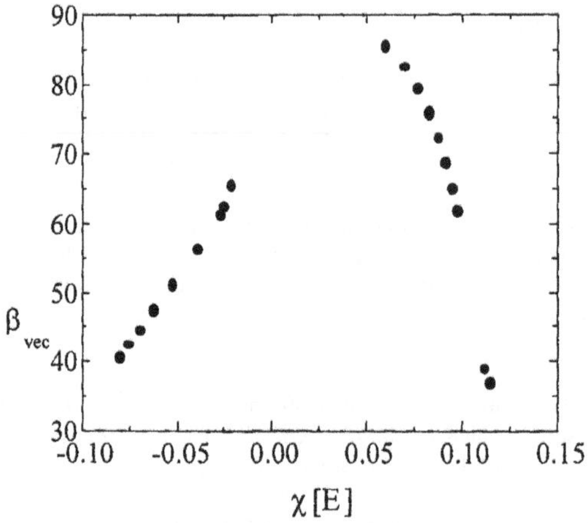

Figure 6: Dipole hyperpolarizability as a function of bond-length parameter χ = R(N-C1) – R(N-C2), where C1 e C2 are the C atoms closest to N. If χ is negative the top structure prevails. Otherwise, it prevails the charge-separated structure.

4. Summary and conclusions

The combined use of molecular mechanics and quantum mechanics in an important alternative to study molecular systems interacting with a liquid environment. Computer simulation of liquids is used to generate structures for subsequent quantum mechanics calculations. This sequential procedure has many advantages and flexibility over conventional QM/MM methodologies. Most of all, it allows an efficient protocol ensuring statistical convergence of the quantum mechanical calculated property [4-6]. This is a consequence of using statistically uncorrelated structures. Several applications have indeed shown the efficacy of the procedure. However, this sequential procedure has one disadvantage over conventional on-the-fly QM/MM. This is that the classical simulation is uncoupled from the quantum mechanics. Hence the solute polarization is treated in some approximate way. As an alternative we devised an iterative procedure where a prior step is made with the determination of the solute polarization in-solution. Examples include acetone in water where the acetone polarization in-water has improved the result for the solvatochromic shift of the $n \rightarrow \pi^*$ transition.

Some additional considerations were made regarding structural changes induced by the solvent. One class is the change in geometry that may be responsible for marked changes in spectroscopic and other molecular properties. The example considered belong to the cathegory of betaine systems. We show that the change in the torsion angle between the two molecular rings is responsible for a shift of ca. 3000 cm^{-1} to the total solvatochromic shift of the $n \rightarrow \pi^*$ transition of *ortho*-betaine in water. Another important structural change has been discussed for phenol blue. In this case the solvent is more important in inducing chemical changes rather than the direct solvent field effect. The change from a *keto-* to *eno*-like form changes the dipole hyperpolarizabilty by a factor of two and may be held responsible for the qualitative behavior in solvents of different polarity.

Acknowledgments

This work has been partially supported by CNPq and FAPESP (Brazil).

References

[1] J. Tomasi, Theor. Chem. Accounts, 112 (2004) 184.

[2] M. P. Allen, D. J. Tildesley, *Computer Simulation of Liquids*, Clarendon Press, Oxford, 1987.

[3] A. Warshel and M. Levitt, J. Mol. Biol. 103 (1976) 227; J. Gao, J. Am. Chem. Soc. 116, 9324 (1994); M. J. Field, P. A. Bash and M. Karplus, J. Comp. Chem. 11 (1990) 700.

[4] K. Coutinho and S. Canuto, Adv. Quantum Chem. 28 (1997) 89.

[5] K. Coutinho and S. Canuto, J. Chem. Phys. 113 (2000) 9132.

[6] T. Malaspina, K. Coutinho and S. Canuto, J. Chem. Phys. 117 (2002) 1692.

[7] J. Gao and X. Xia, Science 258 (1992) 631.

[8] T. Ohba, S. Ikawa, Mol. Phys. 73 (1991) 985.

[9] K. Coutinho and S. Canuto, *DICE: A Monte Carlo program for molecular liquid simulation* (University of Syo Paulo, Syo Paulo, 2003).

[10] M. J. Frisch, et al. Gaussian 03 (Revision B.04); Gaussian, Inc.: Pittsburgh, PA, 2003.

[11] R. Rivelino, B. J. Costa Cabral, K. Coutinho and S. Canuto, Chem. Phys. Lett, 407 (2005) 13.

[12] K. Coutinho, R. C. Guedes, B. J. Costa Cabral and S. Canuto, Chem. Phys. Lett, 369 (2003) 345.

[13]] H. C. Georg, K. Coutinho and S. Canuto, Chem. Phys. Lett, submitted.

[14] K. Naka, A. Morita and S. Kato, J. Chem. Phys. 111 (1999) 481.

[15] M. E. Martin, M. L. Sσnchez, F. J. Oliveira del Valle and M. A. Aguilar, J. Chem. Phys. 113 (2000) 308.

[16] K. Coutinho and S. Canuto, J. Mol. Struct. (Theochem) 632 (2003), 235.

[17] H. J. C. Berendsen, J. P. M. Postma, W. F. van Gunsteren and J. Hermans, in *Intermolecular Forces*, edited by B. Pullman (Reidel, Dordrecht, 1981), p. 331.

[18] R. Peter and H. Dreizler, Z. Naturforsch. A 20 (1965) 301.

[19] U. F. Rφhrig, I. Frank, J. Hutter, A. Laio, J. Vandevondele and U. Rothlisberger, Chem. Phys. Chem. 4 (2003) 1177.

[20] M. C. Zerner; *ZINDO, A semi-empirical program package*, University of Florida, Gainesville, FL 32611 (2000).

[21] W. P. Hayes and C. J. Timmons, Spectrochim. Acta 21 (1965) 529; N. S. Bayliss and E. G. McRae, J. Phys. Chem. 58 (1954) 1006; N. S. Bayliss and G. Wills-Johnson, Spectrochim. Acta, Part A 24 (1968) 551.

[22] R. C. Barreto, H. C. Georg, K. Coutinho and S. Canuto, in preparation.

[23] C. Reichardt, *Solvent Effects in Organic Chemistry*, Verlag Chemie, Weinheim, New York, 1979.

[24] D. Gonzalez, O. Neilands and M. C. Rezende, J. Chem. Soc. Perkin Trans., 2 (1999) 713.

[25] M. Z. Hernandez, R. Longo, K. Coutinho and S. Canuto, Phys. Chem. Chem. Phys., 6 (2004) 2088.

[26] R. W. Zwanzig, J. Chem. Phys., 22 (1954) 1420.

[27] S. R. Mente and M. Maroncelli, J. Phys. Chem. B, 103 (1999) 7704; J. Lobaugh and P. Rossky, J. Phys. Chem. A 103 (2000) 899; W. Bartkowiak and J. Lipinski, J. Phys. Chem. A, 102 (1998) 5236.

[28] S. R. Marder, D. N. Beratan and L. T. Cheng, Science 252 (1991) 103.

[29] A. Serrano and S. Canuto, Int. J. Quantum Chem., 87 (2002) 275.

[30] J. O. Morley and A. L. Fitton, J. Phys. Chem. A 103 (1999) 11442.

[31] T. Yamaguchi *et al*, J. Chem. Phys. 109 (1998) 9084; F. Terenziani, A. Painelli and D. Comoretto, J. Phys. Chem. 104 (2000) 11049.

Brill Academic Publishers
P.O. Box 9000, 2300 PA Leiden
The Netherlands

Lecture Series on Computer and Computational Sciences
Volume 6, 2006, pp. 91-111

Coherent Two-Dimensional Optical Spectroscopy

Minhaeng Cho[1]

Department of Chemistry and Center for Multidimensional Spectroscopy
Korea University, Seoul 136-701, Korea

Received 5 July, 2006; accepted 10 July, 2006

Abstract: Two-dimensional (2D) vibrational and electronic spectroscopy has been paid a great deal of attention recently because it provides spatio-temporal information on complicated molecular systems such as proteins and coupled multi-chromophore systems in subpicosecond time scale. In this review, a theoretical description of coherent 2D optical spectroscopy is presented. Developing a coupled multi-chromophore model, we discuss several examples of 2D spectroscopic studies of peptide solution structure determination and excitation transfer process in electronically coupled multi-chromophore system. A few concluding remarks and perspectives of this research area are given.

Keywords: Nonlinear optical spectroscopy, Two-dimensional spectroscopy, Molecular dynamics simulation, Protein structure and dynamics

PACS: 42.65.Ky, 42.65.An, 33.15.Mt

1. Introduction

Most of the conventional linear spectroscopies, though they are still extremely useful for studying structural and dynamic properties of complex molecules, can only provide highly averaged information. Therefore, novel spectroscopic techniques with much higher information content has been sought and tested continuously. In NMR spectroscopy research, such effort led to developing a variety of 2D NMR techniques such as NOESY (Nuclear Overhauser Enhancement Spectroscopy) and COSY (correlation spectroscopy) methods among many others, and they have revolutionized the application of NMR to both structural and dynamical problems of proteins in solution.[1-3]

As an optical analog of 2D NMR, two-dimensional (2D) optical spectroscopy utilizing multiple ultrafast coherent laser pulses have been used to study protein structure and dynamics,[4-10] femtosecond solvation dynamics,[11] hydrogen-bonding dynamics,[12-15] solute-solvent complexation,[16] and excitation migration process in a photosynthetic light harvesting complex.[17] Due to dramatic advent of recent laser technology, femtosecond laser systems operating in infrared and visible frequency range have been commercially available so that we have seen a wide range of applications utilizing such ultrafast nonlinear spectroscopic techniques.

There exist some analogies between 2D NMR and 2D optical spectroscopies. In the case of the 2D NMR, nuclear spin-spin coupling constant that carries information on the 3D molecular structure can be measured. Similarly, the 2D vibrational spectroscopy is of use to extract information on the vibrational coupling constant between two spatially separated vibrational modes and the 2D electronic spectroscopy can be of use to determine the magnitudes of electronic coupling constant between coupled chromophores. Despite that the coupling mechanism between two different vibrational (electronic) degrees of freedom differs from that between two nuclear spins, the coupling strengths in both cases are sensitively dependent on the 3D molecular structure. However, due to the differences in the time scale of the system-bath interaction-induced transition frequency fluctuation and coupling mechanisms, one cannot directly use the same theoretical framework developed for 2D NMR to describe 2D optical spectroscopic measurement.

[1] Corresponding author. E-mail: mcho@korea.ac.kr.

Furthermore, advantages of optical domain multi-dimensional spectroscopy are quite clear because of the dramatic gain in time resolution (~ subpicosecond scale) possible and because of the ability to directly observe and quantify the couplings between quantum states involved in molecular dynamical processes. For the 2D optical spectroscopy, the existence of cross peaks can be direct evidence on the vibrational or electronic coupling.[18,19] In addition, the time-dependent changes of te diagonal and off-diagonal cross peak amplitudes can provide information on the kinetic network among different quantum states in real time.

In this review, brief accounts of theoretical works reported by the author's group recently will be presented and discussed mostly. We will first discuss a general prescription of ultrafast vibrational or electronic excitation and probing methods utilizing femtosecond laser pulses and the linear and nonlinear response function formalism of optical spectroscopy. Secondly, a coupled multi-chromophore model for polypeptides and electronically coupled complexes, within the Frenkel exciton theory, will be briefly described. We then present several examples of 2D vibrational and electronic spectroscopy of peptides and light-harvesting complex. A few concluding remarks and perspectives will be given in the last section.

2. Ultrafast excitation and probing methods

Numerous time-domain excitation and probing methods have been used to directly probe time-evolution of quantum population and coherence states of molecules in condensed phases. Depending on the molecular nonlinear response of interest, one can probe the relaxation of the diagonal (population state) or off-diagonal (coherent state) density matrix elements selectively.[20] There are in principle quite a number of different excitation methods, but we specifically consider five different *vibrational* excitation methods: (E1) Absorption of a resonant IR photon, (E2) Stimulated Raman excitation by electronically non-resonant optical fields, (E3) Stimulated hyper-Raman excitation by electronically non-resonant optical fields, (E4) Absorption of a resonant circularly polarized IR photon, (E5) Stimulated Raman excitation by electronically non-resonant (circularly polarized) optical fields.

One of the most straightforward methods to create a vibrational coherence state is to use a resonant infrared pulse. The IR-field-matter interaction is, within the electric dipole approximation, $-\mu E(t)$ so that the force exerted on the *j*th vibrational degree of freedom is $-(\partial\mu/\partial Q_j)E(t)$, where μ is the electric dipole moment of a molecule in its electronically ground state. Due to the finite spectral bandwidth of femtosecond IR pulse, $\tilde{E}(\omega)$, which is the Fourier-transform of the temporal envelope of the IR pulse, the excitation process involves a range of vibrational excitations of modes of which frequencies are within the envelope of the pulse spectrum, $\tilde{E}(\omega)$. The second method (E2) is to use a stimulated Raman process to vibrationally excite the electronically ground state molecule. By using an ultrafast optical pulse whose frequency is electronically non-resonant with respect, Raman-active vibrational modes can be excited when the corresponding vibrational frequencies are within the spectral bandwidth of the short pulse. In this case, the effective field matter interaction is approximately given as $-\alpha\,|\,E(t)\,|^2$, where α is the molecular polarizability, so that the force exerted on the *j*th vibrational degree of freedom is $(\partial\alpha/\partial Q_j)\,|\,E(t)\,|^2$. Various stimulated Raman scattering experiments involve such kind of vibrational excitation processes. The third method (E3) involves higher-order field-matter interaction, hyper-Raman process, so that the effective field-matter interaction in this case is $-\beta E_1(t)E_2(t)E_3^*(t)$, where β is the first molecular hyperpolarizability and three injected fields are denoted as $E_j(t)$. Then, the *j*th mode experiences the force, $(\partial\beta/\partial Q_j)E_1(t)E_2(t)E_3^*(t)$, when the molecule is exposed to these electronically off-resonant fields. Often, the vibrational selection rule of the hyper-Raman process differs from that of the Raman process so that the stimulated hyper-Raman excitation method can be of use to excite different vibrational degrees of freedom in comparison to the conventional impulsive stimulated Raman scattering method. The first three methods utilize linearly polarized beams, but one can employ circularly polarized (CP) beams and measure the difference between the left-CP and right-CP signals to extract information on the molecular optical activity. As theoretically proposed recently, time-resolved optical activity measurement can be achieved by using CP light beams in the sum-frequency-generation or four-wave-mixing spectroscopic schemes.[21-23] These novel spectroscopic methods can be of effective use in studying ultrafast dynamics of chiral molecules in condensed phases. The fourth and fifth methods, E4 and E5 in Fig.3, utilize CP IR and CP optical pulsed fields to create a vibrational coherence state.

Time-dependent evolution of created vibrational coherence states can be probed by employing suitable ultrafast probing methods. Much like the ultrafast excitation (vibrational coherence state generation) process, one can use the same types of coherent transition processes and detect the spontaneous or stimulated emission field amplitude or intensity.

Designing a coherent multidimensional spectroscopy. Choosing a proper set of ultrafast excitation methods in combination with a chosen excitation method, one can measure a variety of different molecular properties such as linear or nonlinear dipole, polarizability, or hyperpolarizability response functions. As an example, the coherent Raman scattering spectroscopy utilizing femtosecond laser pulses is to measure Raman active mode relaxation, $\sim e^{-\gamma_j t}\sin\omega_j t$, where γ_j and ω_j are the vibrational dephasing constant and frequency of the *j*th mode. Then, by combining two or more excitation pulses that are separated in time, it will be possible to create temporally (and/or spatially) multi-dimensional vibrational (or electronic) transient gratings. Thus, two- and three-dimensional spectroscopies are essentially to measure the temporal profiles of thus created two- and three-dimensional transient gratings. In practice, one measures the two- or three-dimensional signal, $S(t_1, t_2)$ or $S(t_1, t_2, t_3)$, in time domain, where t_j are the experimentally controlled delay times, and its multi-dimensional Fourier transform gives the corresponding 2D or 3D spectrum.

As outlined above, there are different ways to create coherent states. Consequently, there could be a number of different ways to create a 2D TG. However, only a few of them have been experimentally explored so far. In Scheme 1, we show some combinations of exciton methods. If the first three methods in Fig.4 are only considered, there are 9 distinctly different ways to create a 2D TG, i.e.,

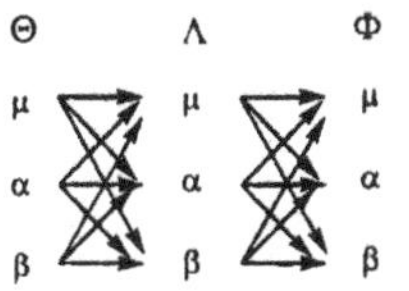

Θ Λ Φ

μ μ μ

α α α

β β β

Scheme 1

Then, the corresponding nonlinear response function is defined as

$$R^{(2)}(t_2,t_1,t_0)=\left(\frac{i}{\hbar}\right)^2\langle[[\Phi(t_2),\Lambda(t_1)],\Theta(t_0)]\rho(t_0)\rangle, \tag{1}$$

where the thermal equilibrium density operator was denoted as $\rho(t_0)$. In the case of three-dimensional spectroscopy, there are 27 different ways to create 3D transient grating (see Scheme 2).

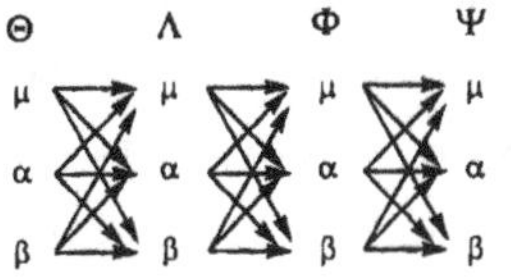

Θ Λ Φ Ψ

μ μ μ μ

α α α α

β β β β

Scheme 2

The third-order nonlinear response function describing molecular response to the series of field-matter interaction in the above figure is defined as

$$R^{(3)}(t_3,t_2,t_1,t_0)=\left(\frac{i}{\hbar}\right)^3\langle[[[\Psi(t_3),\Phi(t_2)],\Lambda(t_1)],\Theta(t_0)]\rho(t_0)\rangle, \tag{2}$$

where Ψ , Φ, Λ, and Θ are molecular operators, such as dipole (μ), polarizability (α), first hyperpolarizability (β) etc. The four-wave-mixing photon echo spectroscopy involves three or four laser pulses that are separated in time and its field-matter interaction sequence is given as μ→μ→μ→μ.[24,25] The final pulse in the case of the heterodyne-detected photon echo experiment acts like a time-gating pulse. Thus, the photon echo technique belongs to the case of 3D TG experiment in general. However, if, during the second delay time t_2, the system is on the population state (diagonal density matrix), the photon echo can be considered to be one of the coherent two-dimensional spectroscopy, i.e., a reduced 2D vibrational spectroscopy.[9]

3. Linear and Nonlinear Response Functions

Note that the *n*th-order polarization, which is linearly proportional to the signal electric field and the measured quantity, can be written as [26]

$$\mathbf{P}^{(n)}(\mathbf{r},t)\sim N\int_0^\infty dt_n\cdots\int_0^\infty dt_2\int_0^\infty dt_1\; R^{(n)}(t_n,\cdots,t_2,t_1)\mathbf{E}(\mathbf{r},t-t_n)\mathbf{E}(\mathbf{r},t-t_n-t_{n-1})$$

$$\times \cdots \times \mathbf{E}(\mathbf{r}, t - t_n - \cdots - t_2 - t_1) \tag{3}$$

where N is the number of chromophores. The measured nth-order signal field amplitude is related to the nth-order as $\mathbf{E}^{(n)} \sim i\omega \mathbf{P}^{(n)}$. Now, by using the eigenstate representation and taking into account the system-bath interaction properly, the linear and nonlinear response functions can be obtained and written as [27-29]

$$R^{(1)}(t_1) = \left(\frac{i}{\hbar}\right) \theta(t_1) \sum_{ab} P_a(t_0) \left\{ \Lambda_{ab} \Theta_{ba} \exp(-i\omega_{ba} t_1) F_{ba}(t_1) - c.c \right\} \tag{4}$$

$$R^{(2)}(t_2, t_1) = \left(\frac{i}{\hbar}\right)^2 \theta(t_2)\theta(t_1) \sum_{abc} P_a(t_0) \left[\Phi_{ab} \Lambda_{bc} \Theta_{ca} \exp(-i\omega_{ba} t_2 - i\omega_{ca} t_1) G_{bc}^{(1)}(t_2, t_1) \right.$$
$$\left. - \Lambda_{ab} \Phi_{bc} \Theta_{ca} \exp(-i\omega_{cb} t_2 - i\omega_{ca} t_1) G_{bc}^{(2)}(t_2, t_1) \right] + c.c. \tag{5}$$

$$R^{(3)}(t_3, t_2, t_1) = \left(\frac{i}{\hbar}\right)^3 \theta(t_1)\theta(t_2)\theta(t_3) \sum_{\alpha=1}^{4} \left[H_\alpha(t_3, t_2, t_1) - c.c. \right], \tag{6}$$

where the auxiliary functions are

$$F_{ba}(t_1) = \exp\left\{ -\int_0^{t_1} d\tau_2 \int_0^{\tau_2} d\tau_1 \xi_{bb}(\tau_2 - \tau_1) \right\} \tag{7}$$

$$G_{bc}^{(1)}(t_1, t_2) = \exp\left\{ -\left[\int_0^{t_2} d\tau_1 \int_0^{\tau_1} d\tau_2 \xi_{bb}(\tau_1 - \tau_2) + \int_0^{t_1} d\tau_1 \int_0^{\tau_1} d\tau_2 \xi_{cc}(\tau_1 - \tau_2) + \int_0^{t_2} d\tau_1 \int_0^{t_1} d\tau_2 \xi_{bc}(\tau_1 + \tau_2) \right] \right\} \tag{8}$$

$$G_{bc}^{(2)}(t_1, t_2) = \exp\left\{ -\left[\int_0^{t_2} d\tau_1 \int_0^{\tau_1} d\tau_2 \xi_{bb}^*(\tau_1 - \tau_2) + \int_0^{t_1+t_2} d\tau_1 \int_0^{\tau_1} d\tau_2 \xi_{cc}(\tau_1 - \tau_2) - \int_0^{t_2} d\tau_1 \int_0^{t_1+t_2} d\tau_2 \xi_{bc}(\tau_2 - \tau_1) \right] \right\} \tag{9}$$

$$H_1(t_3, t_2, t_1) = \sum_{abcd} P_a(t_0) \Lambda_{ad} \Phi_{dc} \Psi_{cb} \Theta_{ba} \exp\{ i\omega_{cb} t_3 + i\omega_{db} t_2 - i\omega_{ba} t_1 \}$$
$$\times \exp\{ -g_{dd}^*(t_2) - g_{cc}^*(t_3) - g_{bb}(t_1 + t_2 + t_3) - g_{dc}^*(t_2 + t_3) + g_{dc}^*(t_2) + g_{dc}^*(t_3)$$
$$+ g_{db}(t_1 + t_2) - g_{db}(t_1) + g_{db}^*(t_2 + t_3) - g_{db}^*(t_3) + g_{cb}(t_1 + t_2 + t_3) - g_{cb}(t_1 + t_2) + g_{cb}^*(t_3) \} \tag{10}$$

$$H_2(t_3, t_2, t_1) = \sum_{abcd} P_a(t_0) \Theta_{ad} \Phi_{dc} \Psi_{cb} \Lambda_{ba} \exp\{ i\omega_{cb} t_3 + i\omega_{db} t_2 + i\omega_{da} t_1 \}$$
$$\times \exp\{ -g_{dd}^*(t_1 + t_2) - g_{cc}^*(t_3) - g_{bb}(t_2 + t_3) - g_{dc}^*(t_1 + t_2 + t_3) + g_{dc}^*(t_1 + t_2) + g_{dc}^*(t_3)$$
$$+ g_{db}(t_2) + g_{db}^*(t_1 + t_2 + t_3) - g_{db}^*(t_1) - g_{db}^*(t_3) + g_{cb}(t_2 + t_3) - g_{cb}(t_2) + g_{cb}^*(t_3) \} \tag{11}$$

$$H_3(t_3, t_2, t_1) = \sum_{abcd} P_a(t_0) \Theta_{ad} \Lambda_{dc} \Psi_{cb} \Phi_{ba} \exp\{ i\omega_{cb} t_3 + i\omega_{ca} t_2 + i\omega_{da} t_1 \}$$
$$\times \exp\{ -g_{dd}^*(t_1) - g_{cc}^*(t_2 + t_3) - g_{bb}(t_3) - g_{dc}^*(t_1 + t_2 + t_3) + g_{dc}^*(t_1) + g_{dc}^*(t_2 + t_3)$$
$$+ g_{db}^*(t_1 + t_2 + t_3) - g_{db}^*(t_1 + t_2) - g_{db}^*(t_2 + t_3) + g_{db}^*(t_2) + g_{cb}(t_3) + g_{cb}^*(t_2 + t_3) - g_{cb}^*(t_2) \} \tag{12}$$

$$H_4(t_3, t_2, t_1) = \sum_{abcd} P_a(t_0) \Psi_{ad} \Phi_{dc} \Lambda_{cb} \Theta_{ba} \exp\{ -i\omega_{da} t_3 - i\omega_{ca} t_2 - i\omega_{ba} t_1 \}$$
$$\times \exp\{ -g_{dd}(t_3) - g_{cc}(t_2) - g_{bb}(t_1) - g_{dc}(t_2 + t_3) + g_{dc}(t_2) + g_{dc}(t_3)$$
$$- g_{db}(t_1 + t_2 + t_3) + g_{db}(t_1 + t_2) + g_{db}(t_2 + t_3) - g_{db}(t_2) - g_{cb}(t_1 + t_2) + g_{cb}(t_1) + g_{cb}(t_2) \} \tag{13}$$

Here, $P_a(t_0)$ is the Bolzmann probability of finding the quantum state $|a\rangle$ at time t_0 at temperature T. The heavy-side step function is denoted as $\theta(t)$. The complex conjugate is denoted as $c.c.$. The transition matrix element, for example Θ_{ba}, is defined as $\Theta_{ba} = \langle b | \Theta | a \rangle$. The entire response functions are determined by the *complex* auto- and cross-correlation functions of the fluctuating transition frequencies defined as

$$\xi_{xy}(\tau_1, \tau_2) \equiv \langle \delta\omega_{xa}(\tau_1) \delta\omega_{ya}(\tau_2) \rangle. \tag{14}$$

In summary, the linear and nonlinear response functions can be expressed in terms of the correlation functions of the bath degrees freedom that are coupled to the molecular vibrational or electronic transition.

4. Coupled multi-chromophore model

Both polypeptides and molecular complexes with a number of vibrational or electronic chromophores that are coupled to each other electronically can be successfully modeled by using the

Frenkel exciton theory, where each monomeric chromophore is either a two-level system or an anharmonic oscillator system. In the case of the amide I vibrational modes, each individual amide I local oscillator can be approximated as an anharmonic oscillator with just three low-lying vibrational states. Note that most of the 2D vibrational spectroscopic techniques based on the four-wave-mixing method such as 2D IR pump-probe and photon echo involve vibrational transitions up to the second excited states that are either overtone or combination bands. Thus, one can assume that a vibrational chromophore in this case is a three-level system with the overtone anharmonicity correctly taken into consideration. On the other hand, for coupled multi-chromophore system such as photosynthetic light-harvesting complex or *J*-aggregates, where the associated optical transition is electronic in nature, one can assume that the monomer is nothing but a two-level system. However, due to the electronic couplings among the two-level monomeric chromophores, the entire multi-chromophore system consists of not only the one-exciton states but also the two-exciton states. Nevertheless, the model Hamiltonians for these two different classes of systems are little different from each other.

A. Model Hamiltonian for coupled multi-chromophore systems

In order to describe nonlinear optical properties of the coupled two-level chromophore systems, the Frenkel exciton Hamiltonian has been extensively used.[30] Denoting a_m^+ and a_m to be the creation and annihilation operators of an electronic excitation at the mth chromophore, the zero-order Hamiltonian can be written as

$$H_0 = \sum_{m=1}^{N} \hbar\omega_m a_m^+ a_m + \sum_{m}^{N} \sum_{n \neq m}^{N} \hbar J_{mn} a_m^+ a_n + H_{ph} \quad (15)$$

where the excited state energy of the mth chromophore energy, electronic coupling constant between the mth and n^{th} chromophores, and the phonon bath Hamiltonian were denoted as $\hbar\omega_m$, J_{mn}, and H_{ph}, respectively. If each monomeric chromophore is an anharmonic oscillator instead of a two-level system, the zero-order Hamiltonian should be written as

$$H_0 = \sum_{m=1}^{N} \hbar\omega_m a_m^+ a_m + \sum_{m}^{N} \sum_{n \neq m}^{N} \hbar J_{mn} a_m^+ a_n + \sum_{m}^{N} \sum_{n}^{N} \hbar\Delta_{mn} a_m^+ a_n^+ a_m a_n + H_{ph} \quad (16)$$

Note that the third term on the right-hand side of Eq.(16) describes the quartic anharmonicities.

The chromophore-bath interaction and the changes of inter-chromophore distance and orientation induce fluctuations of site energies and coupling constants. Thus, the general system-bath interaction Hamiltonian is written as

$$H_{SB} = \sum_{m} \sum_{n} \hbar q_{mn}(\mathbf{Q}) a_m^+ a_n \quad (17)$$

where $q_{mn}(\mathbf{Q})$ is an operator of bath coordinates, $\mathbf{Q}$, and it is assumed that its expectation value calculated over the bath eigenstates, $\langle q_{mn}(\mathbf{Q})\rangle_0$, is zero. Then, the total Hamiltonian is written as

$$H = H_0 + H_{SB} + H_B \,. \quad (18)$$

B. Transformation of the Hamiltonian in the delocalized exciton representation

For any general four-wave-mixing spectroscopy, we need to consider three well-separated quantum state manifolds: the ground state, N one-exciton states, and $\sim N^2$ two-exciton states. The one- and two-exciton eigenvalues and eigenvectors can be obtained by diagonalizing the one- and two-exciton Hamiltonian matrices obtained in the site representation, and these two matrices are denoted as $\tilde{H}_1$ and $\tilde{H}_2$. Then we have

$$U^{-1}\tilde{H}_1 U = \hbar\tilde{\Omega}$$
$$V^{-1}\tilde{H}_2 V = \hbar\tilde{W} \,, \quad (19)$$

where the one- and two-exciton eigenvalues are the diagonal matrix elements of $\hbar\tilde{\Omega}$ and $\hbar\tilde{W}$, respectively. The one- and two-exciton states can be expanded as $|e_j\rangle = \sum_m U_{jm}^{-1} |m\rangle$ and $|f_k\rangle = \sum_{m=1}^{N-1} \sum_{n=m+1}^{N} v_{mn}^{(k)} |m,n\rangle$, where $|m\rangle = a_m^+|0\rangle$ and $|m,n\rangle = a_m^+ a_n^+ |0\rangle$. The eigenvector elements of the jth one-exciton and the kth two-exciton states were denoted as U_{jm}^{-1} and $v_{mn}^{(k)}$, respectively. The matrix elements of $v^{(k)}$ correspond to the elements of the kth row of the matrix V^{-1}.

Due to the chromophore-bath interaction, the corresponding matrices $H_{SB}^{(1)}$ and $H_{SB}^{(2)}$ in the site representation can be transformed into the delocalized exciton state representation, i.e.,

$\hbar\tilde{\Xi}_{SB}^{(1)}(\mathbf{Q}) = U^{-1}\tilde{H}_{SB}^{(1)}(\mathbf{Q})U$ and $\hbar\tilde{\Xi}_{SB}^{(2)}(\mathbf{Q}) = V^{-1}\tilde{H}_{SB}^{(2)}(\mathbf{Q})V$. The diagonal matrix elements, $\left[\tilde{\Xi}_{SB}^{(1)}(\mathbf{Q})\right]_{jj}$ and $\left[\tilde{\Xi}_{SB}^{(2)}(\mathbf{Q})\right]_{kk}$, describe the energy fluctuations of the jth one-exciton and the kth two-exciton states, respectively, which are induced by chromophore-bath interaction. The off-diagonal matrix elements of $\tilde{\Xi}_{SB}^{(1)}(\mathbf{Q})$ and $\tilde{\Xi}_{SB}^{(2)}(\mathbf{Q})$ will induce exciton relaxations within the one- and two-exciton state manifolds, respectively.

C. Transition dipoles

Once the eigenvectors of the one- and two-exciton states are determined, the exciton transition dipole matrix elements can be expressed as linear combinations of transition dipoles of each chromophore, i.e., $\mu_{e_j} = \sum_m U_{jm}^{-1} d_m$ and $\mu_{e_j f_k} = \sum_{m=1}^{N-1}\sum_{n=m+1}^{N} v_{mn}^{(k)}(U_{jn}^{-1}d_m + U_{jm}^{-1}d_n)$, where d_m is the transition dipole vector of the mth chromophore, i.e., $d_m \equiv <0|\hat{\mu}|m>$, where $\hat{\mu}$ is the electric dipole operator. These transition dipole matrix elements are used to calculate various 2D spectroscopic response functions of coupled multi-chromophore systems, such as polypeptides and light-harvesting complexes.

D. Frequency-frequency correlation functions

For a multi-level system such as an N coupled chromophore system, we need both the auto and cross-correlation functions of the one- and two-exciton transition frequencies to eventually calculate the 2D spectroscopic nonlinear response functions. We have found that the time-dependent correlation between any given two one-exciton transition frequencies is given as [18]

$$\left\langle \delta\Omega_j(t)\delta\Omega_k(0)\right\rangle = \left(\sum_m U_{mj}^2 U_{mk}^2\right)C(t) \tag{20}$$

where $\delta\Omega_j(\mathbf{Q}) = \Omega_j(\mathbf{Q}) - \tilde{\Omega}_{jj}$ and $\delta\Omega_j(t) = \exp(iH_{ph}t/\hbar)\delta\Omega_j(\mathbf{Q})\exp(-iH_{ph}t/\hbar)$. $C(t)$ is the site energy fluctuation correction function, i.e., $C(t) = \left\langle q_{mm}(t)q_{mm}(0)\right\rangle$ for all m. Using spectral density representing the spectral distribution of the chromophore-bath coupling constants, one can rewrite the real and imaginary parts of $C(t)$, denoted as $a(t)$ and $b(t)$, as

$$a(t) = \int_0^\infty d\omega\ \rho(\omega)\coth\left[\frac{\hbar\omega}{2k_BT}\right]\omega^2\cos\omega t, \tag{21}$$

$$b(t) = \int_0^\infty d\omega\ \rho(\omega)\omega^2\sin\omega t. \tag{22}$$

Note that the absolute magnitude of the spectral density is determined by the solvent reorganization energy as $\lambda = \hbar\int_0^\infty d\omega\ \omega\rho(\omega)$,[24] and λ is one-quantum energy averaged over the spectral density.

In order to experimentally measure the auto-correlation function, $<\delta\Omega_j(T)\delta\Omega_j(0)>$, one can use the photon echo peak shift measurement method. It was shown that the photon echo peak shift changes with respect to the delay (population evolution) time T, i.e.,[24,31]

$$\tau^*(T;\omega_1 = \omega_3 = \bar{\omega}_{e_j g}) = \frac{\mathrm{Re}[<\delta\Omega_j(T)\delta\Omega_j(0)>]}{\sqrt{\pi} <\delta\Omega_j^2>^{3/2}}, \tag{23}$$

where the center frequencies of the femtosecond laser beams, ω_1 and ω_3, are tuned to be identical to the average transition frequency of the one-exciton $|e_j>$ state. Also, as shown recently,[32] the two-color photon echo peak shift measurement method, where the two different frequencies of the incident laser beams are simultaneously resonant with the jth and k^{th} one-exciton states, can be used to study the cross-correlation function, $<\delta\Omega_j(T)\delta\Omega_k(0)>$, i.e.,

$$\tau_{two}^*(T;\omega_1 = \bar{\omega}_{e_j g}, \omega_3 = \bar{\omega}_{e_k g}) = \frac{\mathrm{Re}[<\delta\Omega_j(T)\delta\Omega_k(0)>]}{\sqrt{\pi} <\delta\Omega_j^2><\delta\Omega_k^2>^{1/2}}. \tag{24}$$

Now, the correlation functions between any given two two-exciton transition frequencies are found to be

$$\left\langle \delta W_j(t)\delta W_k(0)\right\rangle \cong \left(\sum_{m=1}^{N-1}\sum_{n=m+1}^{N}\left(v_{mn}^{(J)}\right)^2\{P_m^{(k)} + P_n^{(k)}\}\right)C(t) \tag{25}$$

where $P_m^{(k)} \equiv \sum_{j=1}^{m-1}\left(v_{jm}^{(k)}\right)^2 + \sum_{j=m+1}^{N}\left(v_{mj}^{(k)}\right)^2$. In addition to the correlation functions between two one-exciton transition frequencies and between two two-exciton transition frequencies, the cross-correlation functions between $\delta\Omega_j(t)$ and $\delta W_k(0)$ are required in calculating the nonlinear response functions. They are

$$\langle \delta\Omega_j(t)\delta W_k(0)\rangle \cong \left(\sum_{m=1}^{N} U_{mj}^2 P_m^{(k)}\right) C(t) \tag{26}$$

In the present section, we have described the nature of one- and two-exciton states, $|e_j\rangle$ and $|f_k\rangle$, and also presented the expressions for transition dipoles and various frequency-frequency correlation functions. Then, using the theoretical formula for 2D spectroscopic nonlinear response functions given in section 3 and carrying out 2D Fourier transformation of the resultant nonlinear response function, one can calculate the 2D vibrational or electronic spectrum of any coupled multi-chromophore system.

5. Two-dimensional vibrational spectroscopy of polypeptides

In the case of 2D vibrational spectroscopy, due to the doubly-vibrationally-resonant condition, two different IR fields, for example, can be simultaneously in resonance with two different vibrational degrees of freedom. Furthermore, in order for the signal not to be zero, the vibrational coupling between two modes via mechanical or electric anharmonicities should not vanish. This is only possible when the two vibrational chromophores interact with each other via through-bond or through-space interaction. This shows why the 2D vibrational spectroscopy can be a useful method to determine whether a given two vibrational chromophores are close to each other or not, i.e., the *two-body interaction.*[20] This is precisely why 2D vibrational spectroscopy can provide much more detailed information on three-dimensional molecular structure.

Much like the 2D vibrational spectroscopy, the 3D vibrational spectroscopy when the three different vibrational degrees of freedom are excited by the three beams with different frequencies can provide through-bond and/or through-space connections among the three modes. This method therefore is a useful tool for measuring *three-body interaction*[20] and analogous to multi-dimensional hetero-nuclear NMR spectroscopy.

In general, in order for 2D vibrational spectroscopic techniques to be of use in probing structures and dynamics of proteins in solution, there are a number of issues to be addressed as summarized in Fig.1. For a given target vibrational degree of freedom, solvation can induce both solvatochromic frequency shift and time-dependent frequency fluctuation. For example, the hydrogen-bonding effects on vibrational properties are to be understood to quantitatively establish protein structure-spectra relationships. In addition to the solute-solvent interactions, the intramolecular vibrational interactions between different local vibrational degrees of freedom should also be understood. For instance, the amide I mode localized at a given peptide bond in a protein backbone can interact with a neighboring peptide vibration, which also induces frequency shift and fluctuation as the protein structure changes in time.

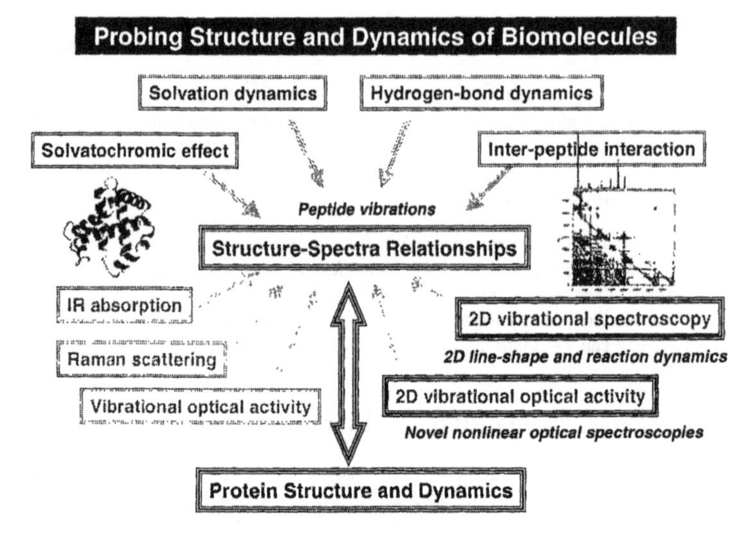

Figure 1. A schematic diagram showing issues to be studied and methods to be used to establish the structure-spectrum relationships.

Among a number of spectroscopic methods that can be used to probe structure and dynamics of biomolecules (Fig.1), the conventional 1D IR absorption and Raman scattering methods have been extensively used. Also, a few vibrational optical activity measurement methods such as vibrational circular dichroism and Raman optical activity spectroscopies were found to be quite useful and informative. Only recently, the 2D vibrational (IR) spectroscopy utilizing either IR photon echo or IR pump-probe scheme has been paid much attention. Furthermore, it was theoretically proposed that 2D vibrational optical activity spectroscopy would provide much detailed information on structure of chiral molecules such as proteins and other biomolecules. In this chapter, we will present brief accounts on the theoretical studies of protein vibrational dynamics.

A. *N*-methylacetamide: A prototype peptide model system

Hydration effects on the molecular structure as well as the amide I mode frequency of the prototype peptide molecule, *N*-methylacetamide (NMA), when it is solvated by a few water molecules, were investigated by carrying out *ab initio* calculations for a number of randomly chosen NMA-water clusters.[33] The linear relationship between the structural distortion of peptide bond, i.e., C=O bond length, and the amide I mode frequency was described by using the theory presented in Ref.[34]. The harmonic frequency shift of the amide I mode in NMA-$n$$D_2O$ (n=1~5) complex was found to originate from the combination of the molecular cubic anharmonicity and displacement of the amide I coordinate when the NMA is hydrated.[34] Using a multivariate least square fitting method, the effective transition charges of six NMA sites were determined. Then, this empirical model was successfully used to quantitatively describe solvatochromic frequency shift of the NMA amide I mode in solution.

Once the amide I mode frequency shift of NMA in NMA-water clusters is quantitatively determined, one can use the theoretical model to describe amide I mode frequency shift and fluctuation of NMA in *liquid water* in combination with molecular dynamics simulations of an NMA in H_2O and D_2O solutions. The ensemble averaged amide I mode frequency shift was found to be -78 cm^{-1} in comparison to that of the gas-phase NMA molecule, which was found to be in excellent agreement with the experimental value of -81 cm^{-1}.[35] Similar to the solvation correlation function of a polar solute in liquid water, the amide I mode frequency-frequency correlation function exhibits a bimodal decaying pattern and both the hindered translational and the librational motions of the water molecules directly hydrogen-bonded to the NMA were found to play critical roles in the pure dephasing of the amide I mode.[36] The pure dephasing constant was estimated to be 11 cm^{-1}. It was shown that the vibrational broadening mechanism is mainly determined by the motional narrowing process. The vibrational Stokes shift of the amide I mode was estimated to be as small as 1.2 cm^{-1}. The amide I IR absorption spectrum thus calculated without any adjustable parameters except for the lifetime of the first excited state has a full width at half maximum of 26.9 cm^{-1} and is found to be in good agreement with the experiment.[37-39]

We then used this method to numerically simulate 2D IR spectrum of NMA in liquid water, which was experimentally studied by Woutersen et al.[37] and Zanni et al..[38] We presented a theoretical description of two-dimensional (2D) IR pump-probe spectroscopy of a three-level system by taking into account the system-bath interaction properly.[40] There are six different nonlinear optical transition pathways. The transient (excited state) absorption contribution involves a transition from the first excited state to the second excited state so that its contribution to the pump-probe signal is negative. On the other hand, the ground-state bleaching and stimulated emission contributions to the signal are positive. Thus, the quantum interference at the amplitude level is critical in this nonlinear vibrational spectroscopy and the lack of perfect destructive interference due to finite vibrational anharmonicities is the key for interpreting the non-zero 2D IR pump-probe signals from an anharmonic oscillator system such as amide I vibration of NMA molecule.

By using the correlation function of the fluctuating amide I mode frequency of NMA in D_2O, which was obtained by combining both *ab initio* calculations with MD simulation results, the time-resolved 2D pump-probe spectra as a function of pump-probe pulse delay time (*T*) were calculated and compared with experimental results.[40] It was found that the vibrational dephasing becomes homogeneous in 2 ps time scale, which is a bit faster than the experimental result. Also, it was shown that the degree of slant of 2D contours, $\sigma_{PP}(T)$, is inversely proportional to the correlation function of the fluctuating amide I mode frequency, i.e., $1/\sigma_{PP}(T) = \langle \delta\omega(T)\delta\omega(0)\rangle / \langle \delta\omega^2 \rangle$, where $\sigma_{PP}(T)$ is the slope of the tangential line in the node of 2D spectrum and $\delta\omega(T)$ is the fluctuating amide I mode frequency.[11] This suggests that the 2D IR spectroscopy can provide direct information on the vibrational frequency fluctuation dynamics and on the magnitude of static inhomogeneity.

Although the carbonyl oxygen atom of the NMA in liquid water was found to maximally form H-bonds with surrounding water molecules, that in liquid methanol can have two different solvation structures differing from each other by the number of H-bonded methanol molecules (see Fig.2). Due to the presence of these two differently solvated NMA molecules in MeOD, the amide I IR band exhibits a doublet feature.[12,13] From the molecular dynamics simulations, it was found that the lower frequency amide I peak is associated with the amide I mode of NMA with two H-bonded methanol molecules at the carbonyl oxygen site, whereas the high-frequency amide I peak is with that having one H-bonded methanol molecule. Carrying out *ab initio* calculation studies of NMA-$(CH_3OD)_n$ clusters and MD simulations,[13] we were able to quantitatively describe the solvatochromic vibrational frequency shift induced by the hydrogen-bonding interaction between the NMA and the solvent methanol molecules. The amide I mode frequency distribution was found to be notably non-Gaussian and it can be decomposed into two Gaussian bands that are associated with two distinctively different solvation structures (Fig.2). The ensemble-average-calculated linear response function associated with the IR absorption is found to be oscillating, which is in turn related to the doublet amide I band shape. Numerically calculated infrared absorption spectrum was directly compared with experiment and the agreement was found to be excellent. By using the Onsager's regression hypothesis, the rate constants of inter-conversion process between the two solvation structures in Fig.2 were obtained. Then, the nonlinear response functions associated with 2D IR pump-probe spectroscopy were simulated and successfully compared with experimental result by Woutersen et al..[12]

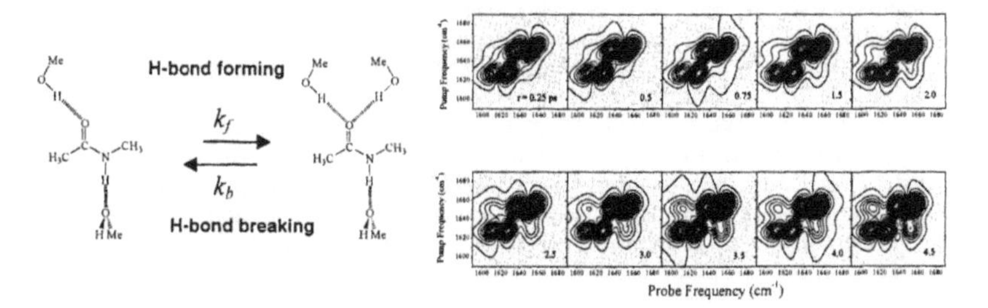

Figure 2. NMA molecule in methanol can have two different solvation structures. The two differ from each other by the number of hydrogen-bonded methanol molecules to the NMA carbonyl group, and they are surrounded by the remaining solvent methanol molecules. These two species are in thermal equilibrium and they can be separately identified in the IR absorption spectrum. The rate constants of hydrogen-bond association and dissociation processes can be estimated by measuring the time-

dependent cross peak amplitudes in the 2D IR spectrum. Time-resolved 2D IR pump-probe spectra of NMA in methanol. The waiting time varies from 0.25 to 4.5 ps. Adapted from [13].

As can be seen in Fig.2, as the waiting (population evolution) time increases from 0.25 to 4.5 ps, the cross peak amplitudes increase. As theoretically described in detail in Refs.[13,14], this time-dependent cross peak amplitude change is directly related to the H-bond formation and dissociation dynamics and they can be quantitatively described by the conditional probability functions obtained by solving the relevant kinetic equations.

B. Dipeptide: *N*-acetylproline amide

Structure-spectrum relationship of a dipeptide that is the simplest model for polypeptides has been extensively studied by using a variety of spectroscopic methods such as linear absorption,[41,42] vibrational circular dichroism,[43] non-resonant and resonant Raman scattering,[44] and 2D IR spectroscopies. Among various polypeptide vibrations, the amide I band has been known to be highly sensitive to the secondary structure of polypeptide. There are a number of literatures and works that aim at determining solution structure of small peptides such as alanine dipeptide, acetylproline amide (AP), etc.. Over the years, these experimental studies utilizing IR, Raman, VCD, CD, and NMR methods show that the solution structure of these dipeptides are not random coil but close to the left-handed threefold helical polyproline II (P_{II}) structure. Han et al.[45] carried out quantum chemistry calculation studies of alanine dipeptide with a few solvated water molecules and showed that the water bridge connecting the two carbonyl group in the dipeptide plays a critical role in stabilizing the dipeptide in its P_{II} structure. However, a few recent experimental studies suggest that such solvent-mediated H-bonding interaction may not be the determining factor for the stabilization of dipeptide in the P_{II} conformation, but rather the intramolecular steric repulsive interaction plays the role.[46]

To elucidate the essential role of H-bond network forming solvents, we studied the N-acetylproline amide by using a few different spectroscopic methods. Before we discuss these works, it should be mentioned that Hochstrasser and coworkers have extensively studied the same dipeptide (AP) dissolved in either water or chloroform, employing the time-resolved 2D IR photon echo spectroscopic method.[47-49] They showed that the vibrational coupling constant measured with such 2D IR method can be a good constraint that is of use to determine the solution structure of AP.

The AP contains two peptide bonds and their amide I local mode frequencies are separated from each other by about 30 cm^{-1} so that the amide I modes are relatively localized on one of the two peptide groups.[50] Nevertheless, the two anharmonic amide I local modes can couple to each other to form two slightly delocalized amide I normal modes and therefore cross peaks revealing the existence of coupling were observed. In the case of the aqueous AP solution, the two peptide groups form hydrogen bonds with water molecules in the first solvation shell. On the other hand, the chloroform solvent molecule cannot make a direct hydrogen bond with the AP peptide groups and, as suggested before, the two peptide groups form a direct intramolecular hydrogen bond to make its C_{7eq} structure to be the global energy minimum conformation in non-polar aprotic solvent. Therefore, the three-dimensional conformation of AP strongly depends on solvent, particularly on the hydrogen-bonding ability of the solvent molecules. Therefore, the AP molecule has been considered to be an ideal dipeptide system for detailed investigations. Zanni et al.[47] were able to determine the absolute 3D conformations of AP in liquid water as well as chloroform by separately measuring the cross peak intensities of both parallel- and perpendicular-polarization photon echo signals.

In order to study the ensemble averaged structures of AP in solutions independently from the above mentioned 2D IR spectroscopic studies, we have carried out both experimental and theoretical studies on the molecular structure of the AP in D_2O solution. In order to determine its aqueous solution structure, IR and vibrational circular dichroism spectra of both L- and D-form AP solutions were measured. Molecular dynamics simulations with two different force fields (*ff99* and *ff03* parameters in AMBER7 and AMBER8 programs, respectively) and density functional theory calculations for *trans*- and *cis*-rotamers of AP were performed to numerically simulate those spectra.[51]

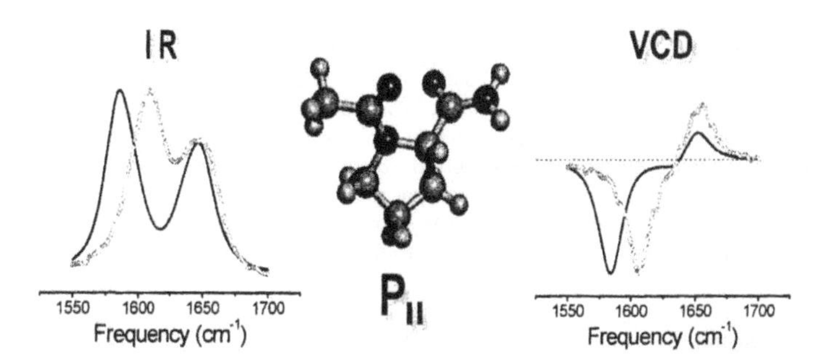

Figure 3. The N-acetylproline amide adopts polyproline II structure in water. Experimentally measured IR and VCD spectra (open circles) are directly compared with simulated ones (solid curves).

Comparisons between experimentally measured and computationally simulated spectra directly suggest that the AP in water adopts a polyproline II-like conformation (see Fig.3) and that the force field parameter *ff03* in AMBER8 suite of program is more realistic and reliable in predicting molecular structure of AP in water than the *ff99* in AMBER7.

C. α-helix

Chain length- and site-dependencies of amide I local mode frequencies of α-helical polyalanines were theoretically studied by carrying out semiempirical quantum chemistry calculations recently. A theoretical model that can be used to quantitatively predict both the local amide I mode frequencies and coupling constants between two different local amide I modes was developed.[52] By using this theoretical model and performing molecular dynamics simulation of an α-helical polyalanine in water, the conformational fluctuation and hydrogen-bonding dynamics of an α-helix in water were investigated by monitoring the amide I frequency fluctuations and by analyzing the 2D spectral line shape change.

The instantaneous normal mode analysis method was used to obtain densities of states of the amide I one- and two-exciton bands and to examine the extent of delocalization of the instantaneous amide I normal modes. Also, by introducing a novel concept of the so-called weighted phase-correlation factor, the symmetric natures of the delocalized amide I normal modes were elucidated and also it was shown that there is no unique way to classify any given amide I normal mode of the α-helical polyalanine in water to be either A-mode-like or E_1-mode-like. From the ensemble-averaged dipole strength spectrum and density of one-exciton states, the amide I IR absorption spectrum was numerically calculated and its asymmetric line shape was theoretically described. Considering both transitions from the ground state to one-exciton states and those from one-exciton states to two-exciton states, we calculated the 2D IR pump-probe spectra and directly compared them with experimental results presented by Hamm and coworkers.[53]

Amide I IR, VCD, and 2D IR spectra of various *isotope-labeled* α-helical polyalanines in water were calculated by combining semiempirical quantum chemistry calculation, Hessian matrix reconstruction, fragmentation approximation, and molecular dynamics simulation methods.[54] The solvation-induced amide I frequency shift was found to be about -20 cm^{-1}. Properly taking into account the motional narrowing effect on the vibrational dephasing and line broadening process, we showed that the simulated IR, VCD, and 2D IR spectra can be quantitatively predicted. Depending on the relative positions of the ^{13}C and/or ^{13}C=^{18}O labeled peptides in a given α-helix, the IR absorption line shape, IR intensity distribution, positive-negative VCD pattern, and diagonal/off-diagonal 2D IR spectral features were found to change dramatically. It was shown that these different spectroscopic observations can be described in a consistent manner by using the developed simulation method and coupled exciton model outlined in section 4. Therefore, properly designed isotopomers and their IR, VCD, and 2D IR spectrum analyses have been shown to be of use for extracting incisive information on the (local) three-dimensional polypeptide structure and dynamics.

D. β-hairpin

Amide I IR and 2D IR photon echo spectra of a model β-hairpin in aqueous solution were also simulated.[55] In order to fully take into account the motional and exchange narrowing processes and cross correlations of fluctuating exciton state energies in the calculations of various vibrational spectra, we used time-correlation function formalism for the linear and nonlinear response functions associated with IR absorption and 2D IR photon echo spectroscopies. Numerically simulated IR absorption and 2D IR spectra were found to be largely determined by the amide I normal modes delocalized on peptides in the two anti-parallel β-sheet strands. When the five peptides in the turn region are $^{13}C{=}^{16}O$ labeled, the isotope peaks appear to be broad and featureless, whereas the β-strand region with six isotope-labeled peptides produces a well-resolved isotope peak in the IR absorption and both diagonal and cross peaks in the 2D IR spectra. This site-specific isotope-labeling method in combination with the 2D vibrational spectroscopic technique was shown to be quite useful in studying structures and (folding or unfolding) dynamics of the β-hairpin in solution.

E. Anti-parallel and parallel β-sheet polypeptides

Amide I local mode frequencies and vibrational coupling constants in various multiple-stranded anti-parallel β-sheet polyalanines were calculated by using the semiempirical calculation and Hessian matrix reconstruction methods. The amide I local mode frequency is strongly dependent on the number of hydrogen bonds to the peptides. Vibrational couplings among amide I local modes in the multiple-stranded β-sheets were shown to be fully characterized by eight different coupling constants.[56] The intra-strand coupling constants were found to be much smaller than the inter-strand ones. Introducing newly defined inverse participation ratios and phase-correlation factors, the extent of 2D delocalization and vibrational phase relationship of amide I normal modes were elucidated. The A-E_1 frequency splitting magnitude was found to be strongly dependent on the number of strands but not on the length of each strand. A reduced one-dimensional Frenkel exciton model was therefore used to describe the observed A-E_1 frequency splitting phenomenon.

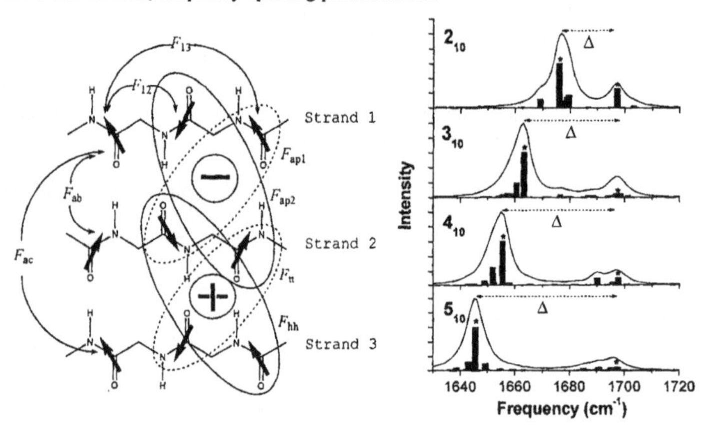

Figure 4. Different vibrational coupling constants required in the numerical simulation of vibrational spectra of anti-parallel β-sheet polypeptides. The simulated IR absorption spectra of M_{10} anti-parallel β-sheet polypeptides, where M is the number of strands involved in the anti-parallel β-sheet and each strand contains 10 peptide bonds, are plotted in this figure. The frequency splitting amplitude is denoted as Δ. Note that Δ increases as the number of strands, M, increases. Adapted from [56].

On the right-hand side of Fig.4, the amide I IR bands of M_{10} anti-parallel β-sheet polyalanines, where M is the number of involved strands, are plotted and the frequency splitting magnitude Δ is shown to increase as M increases. It was found that $\Delta(M)$ can be successfully fitted by using the following stretched-exponentially rising function,[56]

$$\Delta(M) = \Delta(\infty) - (\Delta(\infty) - \Delta(2))\exp\{-k(M-2)^{\alpha}\}. \tag{27}$$

In the case of this model system, the corresponding values in Eq.(27) were estimated to be $\Delta(\infty) = 57$ cm^{-1}, $\Delta(2) = 20$ cm^{-1}, k=0.47, and α=1.17. Although the model anti-parallel β-sheet polyalanines considered were ideal systems, there is likely a similar relationship between the frequency splitting magnitude and the number of strands in real anti-parallel β-sheet polypeptides.

The anti-parallel and parallel β-sheets are two of the most abundant secondary structures found in proteins. Despite that various spectroscopic methods have been used to distinguish these two different structures, the linear spectroscopic measurements couldn't provide incisive information for distinguishing an anti-parallel β-sheet from a parallel β-sheet. After carrying out quantum chemistry calculations and model simulations, we showed that the polarization-controlled 2D IR photon echo spectroscopy can be useful for distinguishing these two different β-sheets.[57] Particularly, the ratio between the diagonal peak and the cross peak was found to be strongly dependent on the quasi-2D array of the amide I local mode transition dipole vectors that are determined by the relative alignment of the participated strands. Although the 2D IR photon echo spectrum of the model anti-parallel β-sheet peptide shows some distinctive differences from that of the parallel β-sheet peptide, we showed that the spectral difference is fairly small so that it is not possible to determine the absolute configuration of β-sheet peptide. However, it turns out that the relative amplitudes of the cross peaks in the 2D difference spectrum, which is defined as $S^{diff}(T)=S^{ZZZZ}(T)-3S^{ZXXZ}(T)$, of an *anti-parallel* β-sheet are significantly larger than those of the diagonal peaks, whereas the cross peak amplitudes in the 2D difference spectrum of a *parallel* β-sheet are much weaker than the main diagonal peak amplitudes.

We found that, assuming the excited state absorption contributions to the 2D difference spectrum are not strongly overlap with the ground state bleaching and stimulated emission contributions in the 2D difference spectrum, the cross peak amplitude is approximately determined by the product of the two corresponding dipole strengths and angle factor, i.e.,[57]

$$S_{kj}^{diff}(T)=S_{kj}^{ZZZZ}(T)-3S_{kj}^{ZXXZ}(T)\propto \mu_k^2\mu_j^2\sin^2\theta_{kj}. \quad (28)$$

Here, the transition dipole matrix elements associated with the vibrational transitions of the *j*th and *k*th modes are denoted as μ_j and μ_k. The angle between the two transition dipole vectors is denoted as θ_{kj} $(=\vec{\mu}_k\cdot\vec{\mu}_j/|\vec{\mu}_k||\vec{\mu}_j|)$. A detailed examination of the delocalized amide I modes showed that the low frequency A-mode has a transition dipole vector that is perpendicular to that of the high-frequency E_1-mode, when the β-sheet forms an anti-parallel structure. Thus, the cross peak amplitude in the 2D difference spectrum can be very large. In contrast, there are no such high-frequency modes of which transition dipole vectors are large in magnitude and perpendicular to that of the low-frequency A-mode, for parallel β-sheet. This clearly shows why the 2D spectroscopic method is a better tool to determine a global structure of polypeptide that cannot be easily determined by other linear vibrational spectroscopic methods.

6. Two-dimensional electronic spectroscopy of coupled multi-chromophore systems

As a molecular system becomes increasingly complex, such as photosynthetic complexes, molecular aggregates of quantum dots or nanoparticles, etc., conventional optical spectroscopic methods such as time- or frequency-resolved absorption spectroscopy or spontaneous emission spectroscopy are of limited use to extract direct information neither on the molecular properties such as electronic couplings among chromophores nor on structures. In this regard, the 2D electronic spectroscopy based on heterodyne-detected photon echo spectroscopic technique can provide far more detailed information and be a useful method for efficient data collection. Despite that there exist quite a number of 2D IR spectroscopic investigations of peptides and proteins over the last decade, only a few experimental and theoretical studies with 2D electronic spectroscopy were reported just recently.[10,17-19] Interestingly, except for the fact that the electronic chromophore can be approximately modeled as a two-level system, one can use the same theoretical method used for polypeptides to numerically calculate the 2D electronic spectra of coupled multi-chromophore systems.

For an effective two-level system with considerable static inhomogeneous broadening, the transition frequency is widely distributed, though each individual two-level system has the same frequency in both time periods. Thus, this inhomogeneity can lead to elongation of the 2D signal along the diagonal $(\omega_\tau=\omega_t)$. However, if there are dynamical processes, such as solvent relaxation, conformational fluctuation, excitation or electron transfer, etc., that they can scramble the optical frequencies of the individual chromophores, the correlation magnitude between the frequencies in the first and third time periods decreases in time T and the 2D spectrum at sufficiently large T becomes symmetric. The transient behavior of 2D line shape can therefore provide information on the time scale of the system-bath interaction-induced decoherence process, and the slope of the diagonally elongated peak decreases from 1 at T=0 to 0 at T=∞ if there is no truly static inhomogeneity. Using the theory presented in Ref.[11], one can show that this slope is linearly proportional to the transition frequency-transition frequency correlation function, i.e.,[11] $\sigma_{2D-PE}(T) = \langle \delta\omega(T)\delta\omega(0)\rangle / \langle \delta\omega^2 \rangle$, where $\sigma_{2D-PE}(T)$ is the slope of the diagonally elongated 2D photon echo contour.

In Ref.[18], when there are excitation transfers among the excited states we showed that the 2D photon echo spectrum is given by a sum of five distinctively different contributions as

$$S^{(3)}(\Omega_1,T,\Omega_3) = GB + SE(j\to j) + SE(j \to k) - EA(j\to j) - EA(j\to k) \quad (29)$$

where *GB*, *SE*, and *EA* represent the ground-state bleaching, stimulated emission, and excited state absorption terms, respectively (see Eq.(33) in Ref.[18]). In the above equation, "$(j\to k)$" means that the initially created population state on the *j*th one-exciton state undergoes a transition to the *k*th state during *T*, and summations over all *j* and *k* should be performed. Therefore, the last four terms in Eq.(29) should properly include such population transfer processes. In the case when the excitation transfer processes are incoherent, one can solve the Master equation for conditional probability functions, i.e.,

$$\frac{d}{dt}G_{kj}(t) = \sum_{\ell\neq k} K_{k\ell}\, G_{\ell j}(t) - G_{kj}(t)\sum_{\ell\neq k} K_{\ell k}\,. \quad (30)$$

Here, $G_{kj}(t)$ is the conditional probability of finding the population on the *k*th exciton state at time *t*, when it was on the *j*th state at time zero. The $(j\to k)$ transition rate constant was denoted as K_{kj}. As shown earlier, the transition rate constant between two one-exciton states is linearly proportional to the spatial overlap between $|\psi_j|^2$ and $|\psi_k|^2$, i.e., $K_{kj} \propto \langle \delta\Omega_j \delta\Omega_k \rangle$. It should be noted that $\langle \delta\Omega_j \delta\Omega_k \rangle$ is the correlation magnitude between the two fluctuating transition frequencies and in principle is a measurable quantity by using the two-color photon echo peak shift measurement method (see Eq.(24)).

Now let us consider off diagonal (cross) peaks. These peaks arise only when the electronic states of the individual molecules comprising the complex interact (mix) and their amplitudes and locations provide extra information on the molecular structure and coupling strength of the coupled multi-chromophore system. Although the detailed theoretical expression for the cross peak in a given 2D spectrum is highly complicated due to the multi-dimensional line-broadening factor, the essential physics behind the time-dependent changes of cross peak amplitudes is rather simple. It was shown that the amplitude of the cross peak $S_{kj}(t)$ at $(\omega_\tau = \omega_{jg}, \omega_t = \omega_{kg})$ is approximately proportional to the product of the associated conditional probability function and transition dipole factors, i.e.,

$$S_{kj}(T) \propto G_{kj}(t)\left[\langle \mu_{gk}^2 \mu_{gj}^2 \rangle - \langle \mu_{gj} \mu_{kf}^2 \mu_{gj} \rangle\right] \quad (30)$$

where μ_{gk} is the transition dipole from the ground state to the kth one-exciton state and μ_{kf} from the kth one-exciton state to the fth two exciton state and $\omega_{kf} \cong \omega_{gk}$. Eq.(31) shows that the appearance and disappearance of cross peaks are directly related to the excitation transfer within the one-exciton state manifold. This demonstrates that the 2D electronic spectroscopy is a powerful method to directly obtain a detailed picture on the kinetic network among coupled chromophores.

FMO complex. Recently, Tobias et al. performed 2D photon echo experiments on the Fenna-Mathews-Olson (FMO) complex consisting of seven bacteriochlorophyll molecules. In Ref.[18], a detailed theoretical description of the time-evolved 2D photon echo spectra and deduced excitation transfer mechanism were presented. Using the theoretical model developed for this FMO complex, we numerically simulated the 2D spectra and plot the 1 ps snapshot spectrum in Fig.5(f). In addition to the main diagonal peaks that are directly associated with the peaks in the absorption spectrum, several cross peaks are clearly observed. The total 2D spectrum is essentially a sum of five different sets of nonlinear optical transition pathways so that the quantum interference at the amplitude level is critical in determining the overall 2D line shape in general. We calculated these contributions separately and plotted them in Fig.5(a)~(e). Adding these five spectra to get the total spectrum in Fig.5(f) means that these different quantum pathways constructively or destructively interfere to produce the total 2D spectrum. The cross peak amplitudes in the bottom half region of Fig.5(f) were observed to increase in time and their time-dependent changes were shown to directly provide state-to-state population transfer pathways and rates.

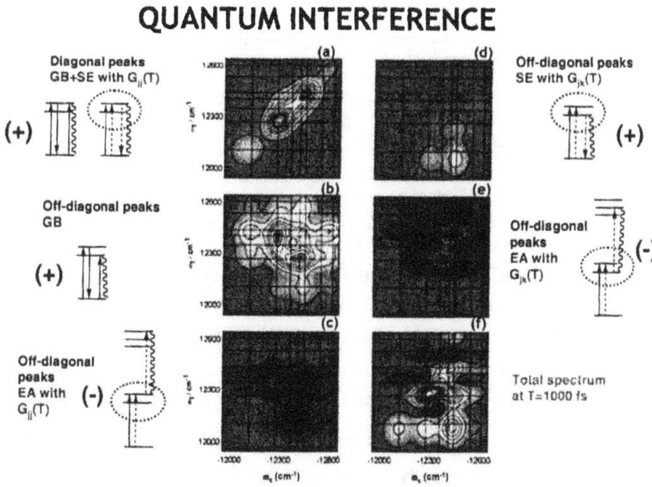

Figure 5. Numerically simulated 2D electronic spectrum of FMO complex is plotted in Figure (f), where the waiting time T is 1 ps. Five different sets of nonlinear optical transition pathways produce distinctively different 2D spectra (figures (a)~(e)). Adding these five spectra gives the total 2D spectrum in figure (f). Adapted from [18].

In Fig.5, we also plot the representative energy level diagrams (nonlinear optical transition pathways). The figures 5(a), (b), and (d) are associated with the GB and SE contributions and they all are positive. If one ignores the population transfer processes completely and considers the GB and SE terms, the 2D spectrum would appear approximately diagonally symmetric in shape. However, due to the existence of population transfer processes in the one-exciton state manifold and of the EA contribution (see Figs.5(c) and 5(e)), the total 2D electronic spectrum in Fig.5(f) becomes highly asymmetric. Nevertheless, the lower part of the spectrum in Fig.5(f) is largely dictated by the SE involving population transfer from upper-lying one-exciton states to lower-lying states. As T increases, not only the cross peak amplitudes but also the amplitude and detailed contour line shapes of diagonal peaks change in a very complicated way, but they can be quantitatively described once the electronic couplings and site energies are accurately determined.

Dimer. Although we have discussed a rather complex molecular system, e.g., FMO light-harvesting protein, one can learn a great deal of the underlying physics of 2D spectroscopy by considering a simple excitonically coupled dimer system. Here, again each monomer is modeled as a two-level system so that for this dimer system there are two one-exciton states $|e_1>$ and $|e_2>$, and a single two-exciton state $|f>$. When the population transfer between the two one-exciton states is relatively slow and when the two local transition dipoles d_1 and d_2 have the same magnitudes, i.e., $d=d_1=d_2$, one can estimate the maximum (or minimum) amplitudes of the diagonal and cross peaks, denoted by S^0_{jk}, as [32]

(1) Diagonal peak at ($\omega_1 = \bar{\omega}_{e_1 g}$, $\omega_3 = \bar{\omega}_{e_1 g}$)

$$S^0_{11} = \frac{2\pi}{<\delta\Omega_1^2>}\left\{2\left\langle \mu_{e_1}^2 \mu_{e_1}^2 \right\rangle\right\} = \frac{2\pi|d|^4}{<\delta\Omega_1^2>}\left\{12\left(1+2\kappa\cos\phi\right)^2\right\} \tag{31}$$

(2) Diagonal peak at ($\omega_1 = \bar{\omega}_{e_2 g}$, $\omega_3 = \bar{\omega}_{e_2 g}$)

$$S^0_{22} = \frac{2\pi}{<\delta\Omega_2^2>}\left\{2\left\langle \mu_{e_2}^2 \mu_{e_2}^2 \right\rangle\right\} = \frac{2\pi|d|^4}{<\delta\Omega_2^2>}\left\{12\left(1-2\kappa\cos\phi\right)^2\right\} \tag{32}$$

(3) Off-diagonal peak at ($\omega_1 = \bar{\omega}_{e_1 g}$, $\omega_3 = \bar{\omega}_{e_2 g}$)

$$S^0_{12} = \frac{2\pi}{\sqrt{<\delta\Omega_1^2><\delta\Omega_2^2>}}\left\{\left\langle \mu_{ge_1}^2 \mu_{ge_2}^2 \right\rangle - \left\langle \mu_{ge_1}\mu_{e_1 f}\mu_{fe_1}\mu_{e_1 g}\right\rangle\right\}$$

$$= -\frac{2\pi|d|^4}{\sqrt{<\delta\Omega_1^2><\delta\Omega_2^2>}}\left\{8\left(2\kappa^2+3\kappa\cos\phi+4\kappa^2\cos^2\phi\right)\right\} \tag{33}$$

(4) Off-diagonal peak at ($\omega_1 = \bar{\omega}_{e_2g}$, $\omega_3 = \bar{\omega}_{e_1g}$)

$$S_{21}^0 = \frac{2\pi}{\sqrt{<\delta\Omega_1^2><\delta\Omega_2^2>}}\left\{\left\langle\mu_{ge_2}^2\mu_{ge_1}^2\right\rangle - \left\langle\mu_{ge_2}\mu_{e_1f}\mu_{fe_2}\mu_{e_2g}\right\rangle\right\}$$

$$= -\frac{2\pi|d|^4}{\sqrt{<\delta\Omega_1^2><\delta\Omega_2^2>}}\left\{8\left(2\kappa^2-3\kappa\cos\phi+4\kappa^2\cos^2\phi\right)\right\}. \tag{34}$$

Here, κ is a measure of the delocalization of the excited states and is defined as $\kappa \equiv \cos\theta\sin\theta$, where the mixing angle θ is determined as $\tan 2\theta = 2J_{12}/(\omega_1-\omega_2)$. The electronic coupling constant between the two monomers is J_{12} and the electronic transition frequencies of the two monomers are ω_1 and ω_2. The angle between the two monomer transition dipole vectors is ϕ.

Note that the cross peak amplitude S_{12}^0 is given by the difference between $\left\langle\mu_{ge_1}^2\mu_{ge_2}^2\right\rangle$ and $\left\langle\mu_{ge_1}\mu_{e_1f}\mu_{fe_1}\mu_{e_1g}\right\rangle$. That is to say, the two different transition pathways interfere to produce the cross peak. However, these two terms exactly cancel out when the two monomers do not electronically interact. Therefore, the existence of cross peaks is definite evidence of non-zero electronic coupling. It is also interesting to note that the cross peak amplitude can provide critical information on the relative orientation of the two monomers.

Overall, it was demonstrated that 2D optical heterodyne-detected photon echo spectroscopy enables couplings, relaxation pathways and rates, and spatial relationships between exciton states to be measured. Also, by tracking the cross peak amplitudes in time, one can directly follow the energy flow on a molecular length scale with femtosecond time resolution. Thus, 2D electronic spectroscopy should provide insights into all systems with electronic band structures.

7. A few concluding remarks and perspectives

Ultrafast 2D vibrational and electronic spectroscopies have been paid a lot of attention because they can provide detailed and highly dense information on molecular structure of peptides and proteins, molecular interactions and dynamics, nucleic acid structures, and semiconductor dynamics. Perhaps, the most important advantage of this spectroscopic technique is its unprecedented ultrafast time-resolution so that it will trigger a number of new researches and investigations of chemical reaction dynamics and biochemical processes involving transient species. Conventional applications of time-resolved spectroscopy such as pump-probe method mainly focus on measurements of life-times of an excited state and of radiationless transition rates among different quantum states. In order to elucidate the entire kinetic network and rates, one should perform quite a number of two-color pump-probe measurements by varying pump and probe field frequencies separately. This experiment can be quite tedious and time-consuming. Furthermore, due to time-energy uncertainty, it is not possible to achieve both ultrafast time-resolution and high frequency-resolution simultaneously. On the other hand, since the 2D spectroscopy utilizes femtosecond laser pulses with broad spectral bandwidths, coherent quantum states of which energies are within the pulse spectral bandwidth can be created simultaneously and probed in time by using yet another femtosecond laser pulses.

A. Chemical reaction dynamics in condensed phases

One of the most important applications of ultrafast 2D spectroscopy would be to study chemical reaction dynamics in condensed phases including solutions, surfaces, and interfaces. The spatial connectivity between any two different vibrational (or electronic) degrees of freedom (chromophores) via through-bond and through-space interactions is the key information that can be extracted from the time-resolved multidimensional spectra. Thus, in general, by properly selecting two vibrational (electronic) degrees of freedom directly associated with reactive species, one can in principle follow the chemical reaction dynamics by monitoring the cross peak amplitude changes in time.

The role of the solvent molecules in chemical reactions has been the central research theme of physical chemistry. In addition, the solvent plays a critical role in stabilizing protein structure in its native conformation. However, there is still a lack of experimental method to obtain direct information on how each individual solvent molecule participates in a given chemical reaction. Most of previous spectroscopic investigations were to follow the dynamics and spectral changes of probing solute mode to indirectly infer an unknown time-scale and dynamics of interacting solvent molecules. Now, the ultrafast 2D spectroscopy might be of use to simultaneously follow the dynamics of both solute and solvent modes. Suppose that there is a characteristic vibrational chromophore in reactive species and that its vibrational properties such as frequency and dipole strength are strongly coupled to a particular solvent vibrational motion. These two mode frequencies are now denoted as ω_{solute} and $\omega_{solvent}$. If the laser pulse frequencies, ω_1 and ω_2, are tuned to be resonant with ω_{solute} and $\omega_{solvent}$, i.e., $\omega_1 \approx \omega_{solute}$ and $\omega_2 \approx \omega_{solvent}$, only when the two characteristic modes are vibrationally coupled to each other does the 2D vibrational spectroscopic signal, i.e., cross peaks, not vanish. The cross peak at ($\omega_1 \approx \omega_{solute}$, $\omega_2 \approx \omega_{solvent}$) is therefore a direct indicator revealing how the solvent mode is participated in the course of this chemical reaction. Two-dimensionally displayed spectra in time will thus give a detailed picture on the solvent dynamics during the chemical reaction. A few potentially interesting chemical reactions are photo-dissociation, photo-induced electron transfer, excited state isomerization, proton transfer induced by a photo-dissociation or photo-excitation, etc.

The 2D spectroscopy has been shown to be exceptionally useful in studying hydrogen-bond formation and dissociation processes and van der Waals complexation dynamics. An advantage of this application is that this technique does not need any external ultrafast perturbation such as T-jump, pressure-jump, concentration change, etc. Note that the cross peak amplitude changes in time are directly related to population changes of chemical species (reactants) initially pumped at time zero to product species of which characteristic mode frequencies are in different spectral (probing) window. However, thus far the 2D spectroscopic technique has been used to study relatively weak intermolecular interaction such as H-bonding and dispersion interactions. One can however use this method to directly follow the chemical reaction dynamics by examining the cross peak, where the two chromophores are specifically belonging to reactants and products exclusively. Then, maybe it will be possible to follow the entire kinetic network even for chemical reactions with multiple intermediates in femtosecond time scale.

B. Solvation structure and dynamics

Solvation dynamics has been extensively studied over the last two decades. Among different experimental methods, the fluorescence Stokes shift measurement has been one of the most effective methods. The other widely used method is the photon echo peak shift (PEPS) measurement.[24] PEPS was shown to be useful not only to study ultrafast solvation dynamics but also to quantitatively measure the inhomogeneous width. Nevertheless, it is still desired to have an experimental method providing direct information on how the surrounding solvent molecules participate in the solvation dynamics and in the formation of the local structure around it. The IR pump-probe and photon echo methods have shown to be useful for such a purpose. However, these works have focused on the vibrational dynamics of the solute only, that is to say, the solute-solvent interaction-induced changes of the vibrational properties of the *solute* (not both solute and solvent) were only observed. Then, are there any experimental methods that can be used to study solvation dynamics by watching the vibrational dynamics of both solute and solvent? As emphasized in this review, 2D vibrational spectroscopy can provide direct information on the coupling between two *spatially separated* but coupled vibrational modes. Let's assume that each of the two external field frequencies is adjusted to be resonant with one of the two vibrational chromophores, when one of the two belongs to the solute and the other to the solvent. Then, the time-dependent change of the cross peak amplitude will tell us about the dynamics of the solute-solvent interaction, i.e., microscopic aspect of the solvation dynamics.

C. Biological applications

The 2D IR spectroscopy has already been proven to be a useful method for determining local structure of peptides. Other types of 2D vibrational spectroscopies and dual frequency IR photon echo spectroscopies will serve as critical tools for this purpose. In addition, one can combine vibrational and electronic spectroscopy, e.g., triply resonant 2D vibrational spectroscopy or vibrational/electronic four-wave-mixing spectroscopy. The correlation between two modes that are coupled to both vibrational and electronic transitions of the chromophore, such as a peptide bond, is likely to be sensitive to the 3D structure of the polypeptide backbone. Alternatively, the electronically resonant fifth-order Raman spectroscopy and fifth-order three pulse scattering spectroscopy might be of use to get information about the 3D structure of proteins.

As emphasized in this review, the key advantage of the ultrafast multi-dimensional spectroscopy over the 2D NMR is its experimentally accessible time-scale. Although protein folding process occurs in a wide range of time scales, from picoseconds to seconds, solution NMR cannot be used to study the early part of the protein folding process due to its limited time resolution. In this respect, the ultrafast multi-dimensional spectroscopy utilizing IR or visible pulses has a clear advantage over the other techniques. Recently, the nonlinear spectroscopy utilizing circularly polarized beams was theoretically proposed and is expected to be useful to study protein folding dynamics since it is an ultrafast optical activity measurement technique.

Another interesting application of the ultrafast multi-dimensional spectroscopy is to investigate the substrate-enzyme interaction. In order for a given enzyme to catalyze a biochemical reaction, the substrate (or ligand) should form a complex with the catalytic site of the enzyme. Tuning the two external field frequencies to be in resonance with the characteristic vibrational modes of the substrate and enzyme, one can directly measure the formation and dissociation processes of the ES complex. Similarly, protein-DNA (or protein-RNA) complexes, antibody-antigen complexes etc can be other interesting targets to be studied with the ultrafast multi-dimensional spectroscopy.

In this review, we have summarized our recent theoretical investigation results demonstrating a wide range of applications of coherent 2D optical spectroscopy. It is believed that this relatively young spectroscopic method has a great potential and is highly useful in studying chemical and biological processes in real time.

Acknowledgments

This work was supported by the Creative Research Initiatives Program of KOSEF (MOST, Korea).

References

[1] K. Wuthrich, *NMR of Proteins and Nucleic Acids*; Wiley-Interscience: New York, 1986.

[2] R. Ernst, G. Bodenhausen, and A. Wokaun, *Principles of Nuclear Magnetic Resonance in One and Two Dimensions*; Clarendon: Oxford, 1987.

[3] J. K. M. Sanders and B. K. Hunter, *Modern NMR Spectroscopy*; Oxford University Press: New York, 1994.

[4] P. Hamm, M. Lim, W. F. DeGrado, and R. M. Hochstrasser, *Proc. Natl. Acad. Sci. USA* **1999**, *96*, 2036.

[5] M. Khalil, N. Demirdöven, and A. Tokmakoff, *J. Phys. Chem. A* **2003**, *107*, 5258.

[6] S. Woutersen and P. Hamm, *J. Phys. Cond. Matt.* **2002**, *14*, R1035.

[7] J. C. Wright, *Int. Rev. Phys. Chem.* **2002**, *21*, 185.

[8] M. Cho, In *Advances in Multi-photon processes and spectroscopy*; Lin, S. H.; Villaeys, A. A.; Fujimura, Y., Eds.; World Scientific: Singapore., 1999, Vol. 12, p 229.

[9] S. Mukamel, *Annu. Rev. Phys. Chem.* **2000**, *51*, 691.

[10] D. M. Jonas, *Annu. Rev. Phys. Chem.* **2003**, *54*, 425.

[11] K. Kwac and M. Cho, *J. Phys. Chem. A* **2003**, *107*, 5903.

[12] S. Woutersen, Y. Mu, G. Stock, and P. Hamm, *Chem. Phys.* **2001**, *266*, 137.

[13] K. Kwac, H. Lee, and M. Cho, *J. Chem. Phys.* **2004**, *120*, 1477.

[14] K. Kwac and M. Cho, *J. Raman Spectrosc.* **2005**, *36*, 326.

[15] Y. S. Kim and R. M. Hochstrasser, *Proc. Natl. Acad. Sci. U.S.A.* **2005**, *102*, 11185.

[16] J. Zheng, K. Kwak, J. Asbury, X. Chen, I. R. Piletic, and M. D. Fayer, *Science.* **2005**, *309*, 1338.

[17] T. Brixner, J. Stenger, H. Vaswami, M. Cho, R. E. Blankenship, and G. R. Fleming, *Nature.* **2005**, *434*, 625.

[18] M. Cho, H. Vaswani, T. Brixner, J. Stenger, and G. R. Fleming, *J. Phys. Chem. B.* **2005**, *109*, 10542.

[19] M. Cho, T. Brixner, I. Stiopkin, H. Vaswani, and G. R. Fleming, *J. Chinese Chem. Soc.* **2006**, *53*. 15.

[20] M. Cho, *PhysChemComm.* **2002**, *5*, 40.

[21] M. Cho, *J. Chem. Phys.* **2002**, *116*, 1562.

[22] M. Cho, *J. Chem. Phys.* **2003**, *119*, 7003.

[23] S, Cheon and M. Cho, *Phys. Rev. A* **2005**, *71*, 013808.

[24] G. R. Fleming and M. Cho, *Annu. Rev. Phys. Chem.* **1996**, *47*, 103.

[25] D. A. Wiersma, W. P. de Boeij, and M. S. Pshenichnikov, *Annu. Rev. Phys. Chem.* **1998**, *49*, 99.

[26] S. Mukamel, *Principles of Nonlinear Optical Spectroscopy*; Oxford University Press: New York, 1995.

[27] J. Sung and M. Cho, *J. Chem. Phys.* **2000**, 113, 7072.

[28] J. Sung, R. J. Silbey, and M. Cho, *J. Chem. Phys.* **2001**, *115*, 1422.

[29] M. Cho, *J. Chem. Phys.* **2001**, *115*, 4424.

[30] W. M. Zhang, T. Meier, V. Chernyak, and S. Mukamel, *J. Chem. Phys.* **1998**, *108*, 7763.

[31] M. Cho, J.-Y. Yu, T. Joo, Y. Nagasawa, S. A. Passino, and G. R. Fleming, *J. Phys. Chem.* **1996**, *100*, 11944.

[32] M. Cho and G. R. Fleming, *J. Chem. Phys.* **2005**, *123*, 114506.

[33] S. Ham, J.-H. Kim, H. Lee, and M. Cho, *J. Chem. Phys.* **2003**, *118*, 3491.

[34] M. Cho, *J. Chem. Phys.* **2003**, *118*, 3480.

[35] G. Eaton and M. C. R. Symons, *J. Chem. Soc. Faraday Trans.* **1989**, *85*, 3257.

[36] K. Kwac and M. Cho, *J. Chem. Phys.* **2003**, *119*, 2247.

[37] S. Woutersen, R. Pfister, P. Hamm, Y. G. Mu, D. S. Kosov, and G. Stock, *J. Chem. Phys.* **2002**, *117*, 6833.

[38] M. T. Zanni, M. C. Asplund, and R. M. Hochstrasser, *J. Chem. phys.* **2001**, *114*, 4579.

[39] M. F. DeCamp, L. DeFlores, J. M. McCracken, A. Tokmakoff, K. Kwac, and M. Cho, *J. Phys. Chem. B* **2005**, *109*, 11016.

[40] K. Kwac and M Cho, *J. Chem. Phys.* **2003**, *119*, 2256.

[41] *Infrared Spectroscopy of Biomolecules*; H. H. Mantsch and D. Chapman, Eds.; Wiley-Liss: New York., 1996.

[42] *Infrared and Raman Spectroscopy of Biological Materials*; H.-U. Gremlich and B. Yan, Eds.; Marcel Dekker: New York., 2000.

[43] *Circular Dichroism: Principles and Applications*; N. Berova, K. Nakanishi, and R. W. Woody, Eds.; Wiley-Liss: New York, 2000.

[44] F. Eker, K. Griebenow, and R. Schweitzer-Stenner, *J. Am. Chem. Soc.* **2003**, *125*, 8178.

[45] W.-G. Han, K. J. Jalkanen, M. Elstner, and S. Suhai, *J. Phys. Chem. B* **1998**, *102*, 2587.

[46] A. N. Drozdov, A. Grossfield, and R. V. Pappu, *J. Am. Chem. Soc.* **2004**, *126*, 2574.

[47] M. T. Zanni, S. Gnanakaran, J. Stenger, and R. M. Hochstrasser, *J. Phys. Chem. B* **2001**, *105*, 6520.

[48] N.-H. Ge, M. T. Zanni, and R. M. Hochstrasser, *J. Phys. Chem. A* **2002**, *106*, 962.

[49] I. V. Rubtsov and R. M. Hochstrasser, *J. Phys. Chem. B* **2002**, *106*, 9165.

[50] S. Hahn, H. Lee, and M. Cho,*J. Chem. Phys.* **2004**, *121*, 1849.

[51] K.-K. Lee, S. Hahn, K.-I. Oh, J. S. Choi, C. Joo, H. Lee, H. Han, and M. Cho, *J. Phys. Chem. B* **2006**. in press.

[52] S. Ham, S. Hahn, C. Lee, T.-K. Kim, K. Kwak, and M. Cho, *J. Phys. Chem. B* **2004**, *108*, 9333.

[53] S. Woutersen and P. Hamm, *J. Chem. Phys.* **2001**, *115*, 7737.

[54] J.-H. Choi, S. Hahn, and M. Cho, *Int. J. Quantum Chem.* **2005** , *104*, 616.

[55] S. Hahn, S. Ham, and M. Cho, *J. Phys. Chem. B* **2005**, *109*, 11789.

[56] C. Lee and M. Cho, *J. Phys. Chem. B* **2004**, *108*, 20397.

[57] S. Hahn, S.-S. Kim, C. Lee, and M. Cho, M. *J. Chem. Phys*. **2005**, *123*, 084905.

Brill Academic Publishers
P.O. Box 9000, 2300 PA Leiden
The Netherlands

Lecture Series on Computer and Computational Sciences
Volume 6, 2006, pp. 112-139

SM*x* Continuum Models for Condensed Phases

Christopher J. Cramer[1] and Donald G. Truhlar[1]

Department of Chemistry and Supercomputing Institute,
University of Minnesota,
207 Pleasant St. SE,
Minneapolis, MN 55455-0431, USA

Received 21 June, 2006; accepted 25 June, 2006

Abstract: The SM*x* continuum models are designed to include condensed-phase effects in classical and quantum mechanical electronic structure calculations and can also be used for calculating geometries and vibrational frequencies in condensed phases. Originally developed for homogeneous liquid solutions, the SM*x* models have seen substantial application to more complicated condensed phases as well, e.g., the air-water interface, soil, phospholipid membranes, and vapor pressures of crystals as well as liquids. Bulk electrostatics are accounted for via a generalized Born formalism, and other physical contributions to free energies of interaction between a solute and the surrounding condensed phase are modeled by environmentally sensitive atomic surface tensions associated with solute atoms having surface area exposed to the surrounding medium. The underlying framework of the models, including the charge models used for the electrostatics, and some of the models' most recent extensions are summarized in this report. In addition, selected applications to environmental chemistry problems are presented.
Keywords: Solvation; Polarization; Partitioning; Solubility; Thermodynamics; Vapor pressure; Electrochemistry.

1. Introduction and Underlying Physics

Many excellent reviews of the general theory and development of continuum solvation models are available [1-13], so this contribution will not attempt to provide yet another comprehensive overview of these powerful techniques. Instead, we focus specifically on the history and present status of the SM*x* models, which have also been reviewed [14-18], but not recently enough to include the latest developments included here. The "*x*" in SM*x* contains information about the model. Any number standing alone (e.g., 1, 2, 3, or 4) or preceding a decimal point (e.g., the "5" in SM5.42) indicates the generation of the model, and generations have typically been defined by a substantial change in the algorithmic approach undertaken for electrostatics, surface tensions, parameterization strategies, or some combination thereof as outlined in more detail below. Any number or numbers following a decimal point (e.g., the "42" in SM5.42) generally provide information about the charge models used to represent the solute charge distribution (this is also discussed in more detail below).

At the foundation of the SM*x* models is a partitioning of the free energy of transfer $\Delta G^{\circ}_{\text{tr}}$ from the gas phase to the condensed phase into two components [19]

$$\Delta G^{\circ}_{\text{tr}} = \Delta G^{\circ}_{\text{ENP}} + G^{\circ}_{\text{CDS}} \tag{1}$$

where the first term on the right-hand-side is, at the quantum mechanical level, computed as

$$\Delta G^{\circ}_{\text{ENP}} = \left\langle \Psi^{(1)} \middle| H + \frac{1}{2}V \middle| \Psi^{(1)} \right\rangle - \left\langle \Psi^{(0)} \middle| H \middle| \Psi^{(0)} \right\rangle \tag{2}$$

where $\Psi^{(0)}$ and $\Psi^{(1)}$ are, respectively, the wave functions that minimize the expectation values of the gas-phase Hamiltonian H and the Hamiltonian in solution; the latter adds to H one half the reaction field operator V, which represents the field acting on the solute due to the polarization of the

[1] Corresponding authors. E-mail: cramer@umn.edu; truhlar@umn.edu.

surrounding dielectric medium by the charge distribution of the solute. The factor of ½ comes from linear response theory [5] and accounts for the cost of polarizing the medium. Thus, $\Delta G^{\circ}_{\text{ENP}}$ accounts for the polarization (P) of the medium, for changes in the solute electronic (E) structure, and, if geometry reoptimization is undertaken, for changes in the nuclear (N) coordinates.

If the concentration of the solute is different in the liquid and vapor phases, eq. (1) needs another term to account for the associated entropy change. Our standard procedure is to use 1 M as the concentration in both phases, use eq. (1) as written, and then change to any other standard state that may be desired, e.g., to 1 atm standard pressure for the vapor.

In the SM*x* models, the reaction field operator is represented using the generalized Born (GB) equation [6,8,19-28]. In the GB approach, the charge distribution of the solute is represented by an atom-centered distribution of monopoles (i.e., partial atomic charges) and at each atom *k* the reaction field is defined as

$$V_k = \left(1 - \frac{1}{\varepsilon}\right) \sum_{k'}^{\text{atoms}} q_{k'} \gamma_{kk'} \tag{3}$$

where ε is the dielectric constant of the medium, q_k is a partial atomic charge, and $\gamma_{kk'}$ is an effective Coulomb integral first suggested by Still et al. [25]

$$\gamma_{kk'} = \left(r_{kk'}^2 + \alpha_k \alpha_{k'} e^{-r_{kk'}^2 / d_{kk'} \alpha_k \alpha_{k'}} \right)^{-1/2} \tag{4}$$

where $r_{kk'}$ is the interatomic distance, α is an effective atomic Born radius, $d_{kk'}$ is a parameter chosen by Still et al. [25] to be 4 and by us to be either 4 or some value rather close to 4 and α is an effective Born radius.

The effective Born radius may be calculated in a number of different ways [19,25,28-36] which will not be reviewed in detail here. Conceptually, it is the radius that a spherical monatomic ion—having a charge equal to the partial charge of the atom in the solute—would need to have in order to have the same solvation free energy in the medium as does the solute atom embedded in the full, otherwise uncharged solute (the volume of which, by displacing the dielectric medium, descreens the particular atom in question). The free energy of solvation of the atom embedded in the solute is computed by solving the Poisson equation for the charged species in the dielectric medium. The necessary integration for this solution is accomplished either numerically or by using an analytical quadrature approach, and one aspect particularly worthy of attention here is that some lower limit for the integration must be chosen; in SM*x* models this limit for a monatomic ion is referred to as the intrinsic or atomic Coulomb radius ρ_k and the determination of robust atomic ρ values is a key aspect in parameterization.

The second term on the r.h.s. of eq. (1) refers to the free energy of cavitation (C), dispersion (D), and changes in the otherwise homogeneous solvent structure (S) induced by the solute. The SM*x* models assume that these various free energies may be partitioned into atomic contributions, and that each atom's contribution will depend on its atomic number, its intramolecular environment, and the extent to which it is exposed to the surrounding medium. Thus, we compute

$$G^{\circ}_{\text{CDS}} = \sum_{k}^{\text{atoms}} \sigma_k(\mathbf{Q}) A_k \tag{5}$$

where A_k is the solvent accessible surface area [28,37,38] of atom *k* and σ_k has units of surface tension and is not necessarily a simple constant for all atoms of a given atomic number but instead a function of the nuclear coordinates **Q** and, in SM2 and SM3, on functions of the electronic density matrix as well.

We next proceed to detail how the SM*x* models have evolved within the general framework described by eqs. (1)–(5). We pay special attention to pointing out the details that distinguish one SM*x* model from another.

2. A Brief History of the SM*x* Models for Homogeneous Liquid Solutions

2.1. SMx models for semiempirical Hamiltonians. The first SM*x* models to be reported were SM1 and SM1a [19]. These two models were the first quantum mechanical continuum solvation models to be parameterized against an extensive set of experimental aqueous free energies of solvation (141 neutral compounds and 27 ions; most prior models had tended to focus on specific components of the solvation free energy that are not physical observables, or had focused on only one or a very small number of free energies of solvation for related molecules). Parameterization against increasingly large experimental data sets has been a hallmark of all SM*x* model development.

Both SM1 and SM1a were based on the semiempirical Austin Model 1 (AM1) Hamiltonian [39]. In these first generation models, the partial atomic charges used to define the reaction field were taken calculated from the electronic density matrix by the population analysis of Mulliken (which reduces to the description of Coulson within the context of a zero diatomic overlap model like AM1) [40]. Thus, Fock matrix elements in the self-consistent reaction field (SCRF) process were defined as

$$F_{\mu\nu}^{(1)} = F_{\mu\nu}^{(0)} + \delta_{\mu\nu}\left(1-\frac{1}{\varepsilon}\right)\sum_{k'}\sum_{\nu\in k'}(Z_{k'} - P_{\nu\nu})\gamma_{kk'} \tag{6}$$

where **F** is the Fock matrix indexed over basis functions μ and ν, δ is the Kronecker delta, Z is the AM1 valence nuclear charge, and **P** is the density matrix. The Coulomb integral $\gamma_{kk'}$ was defined as in eq. (4) except for OO and NH atom pairs, for which an additional term was added to effectively increase the screening between these two charges.

Another feature of SM1 and SM1a was that the atomic Coulomb radii were made to vary as a function of partial atomic charge. In particular, we used

$$\rho_k = \rho_k^{(0)} - \rho_k^{(1)}\left[\frac{1}{\pi}\tan^{-1}\left(\frac{q_k + q_k^{(0)}}{q^{(1)}}\right) + \frac{1}{2}\right] \tag{7}$$

where $\rho_k^{(0)}$, $\rho_k^{(1)}$, and $q_k^{(0)}$ are atom-specific parameters, and $q^{(1)}$ was a universal parameter set to 0.1. The idea of charge-dependent radii has some intuitive conceptual appeal, and such a treatment persisted through SM4. However, the approach was abandoned in favor of constant Coulomb radii beginning with SM5-type models. There were several reasons for this choice. First, atomic partial charges tend to cluster in discrete, narrow regions, making it difficult to develop switching functions between those regions that are not fairly arbitrary. Unfortunately, while such arbitrariness may not affect typical stable molecules, there may be much more significant effects on transition-state structures, where atoms may be passing between standard hybridizations and charges, so the consequences for reaction dynamics may be large. In addition, because SM1 and SM1a were designed for use at the semiempirical level, it was not terribly expensive to carry out numerical geometry optimizations. However, in order to compute analytic derivatives it is quite inconvenient to have the Coulomb radii, and thus the effective Born radii, depend on density matrix elements in a particularly complicated way. In unpublished work preceding the development of the SM6 solvation model, we (C. Kelly, C. J. Cramer, and D. G. Truhlar) revisited the issue of whether more accurate models could be obtained by letting the intrinsic Coulomb radii depend linearly or quadratically on particular atomic charges, but the gain in accuracy was insignificant.

With respect to $G_{\text{CDS}}^{\text{o}}$, SM1 used the simplest possible approach for defining σ_k in eq. (6): the surface tension depended only on the atomic number of atom k. SM1a, on the other hand, parameterized surface tensions based on atom "types", much as in molecular mechanics, and was designed to be used only for neutral molecules. Thus, the user would assign a type for each atom (e.g., ether oxygen or nitrile nitrogen) and the surface tension would be chosen accordingly. Over the parameterization training set of neutrals, SM1 and SM1a exhibited root-mean-square (rms) errors of 1.52 and 0.78 kcal/mol, respectively. Over the ionic training set, the rms error for SM1 was 4.4 kcal/mol. The models were defined for molecules containing H, C, N, O, F, S, Cl, Br, and I.

The next two SM*x* models to be reported, SM2 [26,41] and SM3 [26,42], were also aqueous solvation models based on semiempirical Hamiltonians. SM2 was designed for use with AM1 and SM3 for use with the Parameterized Model 3 (PM3) of Stewart [43]. The key difference between SM2 and SM3 and their precursor SM1 is in the determination of G°_{CDS}. In essence, recognizing the significant improvement SM1a offered over SM1, some of the heuristic principles by which a chemist assigns atom types were encoded into algorithms that assigned surface tension not only on the basis of atomic number, but also based on the number of attached hydrogen atoms, which were taken to have zero atomic radius in the computation of G°_{CDS}. In essence, then, SM2 and SM3 are united-atom models where heavy-atom properties are modified based on the hydrogen atoms attached to them.

Heavy atom surface tensions were defined as

$$\sigma_k = \sigma_k^{(0)} + \sigma_k^{(1)}\left[f\left(B_{k\text{H}}\right) + g\left(B_{k\text{H}}\right)\right] \tag{8}$$

where $\sigma_k^{(0)}$ and $\sigma_k^{(1)}$ are atom-specific parameters and f and g are functions of the bond order matrix **B** [44,45], the first being defined as

$$f\left(B_{k\text{H}}\right) = \tan^{-1}\left(\sqrt{3}B_{k\text{H}}\right) \tag{9}$$

and the second being a function targeted only to oxygen and nitrogen and designed to correct for otherwise systematic errors in H_2O, NH_3, and primary oxonium and ammonium ions. Note that the bond order matrix is a function of the electronic density matrix [45].

Eq. (8) permits substantial distinction between different functional groups, and its adoption caused SM2 and SM3 to be much more accurate for the prediction of the free energies of solvation for hydrophobic species. The parameterization set for SM2 and SM3 was also extended by 9 molecules compared to SM1, which permitted P to be added to the list of allowed atoms in solute molecules, although parameterizations for P were not entirely reliable until SM5.42. Over the neutral set, SM2 and SM3 had rms errors of 0.9 and 1.3 kcal/mol, respectively, and over the ionic set these errors were 3.9 and 5.6 kcal/mol, respectively. The larger errors exhibited by SM3 are associated in part with the now well known poor quality of PM3 partial charges on N atoms [26].

Because of their ready availability in the free code AMSOL [46] and the commercial code SPARTAN [47], the SM2 and SM3 models saw substantial use after their introduction [48-92]. Subsequent to their original development, improvements were made in the numerical stability of certain aspects of the calculation (e.g., the quadrature scheme used for integration in determining effective Born radii, the analytical approach used for the computation of molecular surface area, and the updating scheme used in construction of the Fock matrix including the reaction field operator), and more robust versions of SM2 and SM3 were produced and called SM2.1 and SM3.1, respectively [28]; results from these models were practically unchanged compared to their precursors. The numerical improvements introduced into SM2.1 and SM3.1 have been incorporated into all subsequent SM*x* models as well. SM2, SM3, SM2.1, and SM3.1 are no longer recommended since several models in the SM5 series discussed below are parameterized more broadly and more accurately for AM1 and PM3 as well as the equally inexpensive Modified Neglect of Diatomic differential Overlap (MNDO) model.

The SM4 models were the first to introduce a significant change in the calculation of $\Delta G^{\circ}_{\text{ENP}}$. These models continued to be founded on the semiempirical Hamiltonians AM1 and PM3, but instead of using Mulliken partial atomic charges in the reaction field operator, they made use of Charge Model 1 (CM1) to compute Class IV partial atomic charges [93]. The CM1 partial atomic charge is defined as

$$q_k^{\mathrm{CM1}} = q_k^{(0)} + B_k \Delta q_k - \sum_{k' \neq k} B_{kk'} \Delta q_{k'} \tag{10}$$

where $q_k^{(0)}$ is a reference charge (e.g., a semiempirical Mulliken charge), $B_{kk'}$ is the covalent bond order (also called bond index) [45] between atoms k and k', B_k is the sum of all bond orders from atom k to all other atoms, and Δq_k is defined as

$$\Delta q_k = c_k q_k^{(0)} + d_k \tag{11}$$

where c and d are an empirically optimized scaling parameter and offset, respectively, for atom k. The form of eq. 10 ensures that total molecular charge is conserved in the CM1 mapping procedure. The parameter vectors **c** and **d** for all atoms are optimized so that molecular dipole moments computed from the distributed partial atomic charges best fit an experimental data set for neutral molecules, and partial atomic charges computed from electrostatic potentials for select charged molecules [93]. CM1 charges are much more accurate then semiempirical Mulliken charges for the computation of molecular electrical moments, leading to a more physical partition of the total solvation free energy into ENP and CDS components when they are used in the GB procedure (note that by the nature of the parameterization of the surface tension terms, errors in the ENP component are perforce absorbed—as well as possible—into the CDS component).

The first SM4 model to be reported was not for water as solvent, but instead for n-hexadecane [94] (a solvent for which many data are available because it can be used as a stationary phase in capillary gas chromatography). For the hexadecane model, another new feature was also introduced into the SM4 model, namely, the use of different *solvent* radii to compute solvent accessible surface areas associated with different physical components of the CDS term. In particular, a CD area and a CS area were identified separately and the full CDS term was computed as

$$G_{\mathrm{CDS}}^{\mathrm{o}} = \sigma^{\mathrm{CS}} \sum_k A_k^{\mathrm{CS}} + \sum_k \sigma_k^{\mathrm{CD}}(\mathbf{Q}) A_k^{\mathrm{CD}} \tag{12}$$

where σ^{CS} is an empirically optimized molecular surface tension associated with a large solvent probe radius to generate the molecular surface area A^{CS} and σ^{CD} is an atom-specific surface tension that may either be an empirically optimized constant, or a function of empirical constants and bond orders between atom k and other atoms. In the latter case, the spirit of the approach taken in SM2 and SM3 was preserved, but the united atom model was abandoned in favor of a situation in which each H atom had a surface tension influenced by the atom attached to it, rather than vice versa. Atomic CD surface areas were computed with smaller probe radii, under the assumption that dispersion interactions occur over a much shorter range than do changes in the solvent structure when a solute is introduced.

Preliminary AM1-SM4 and PM3-SM4 models for n-hexadecane were parameterized against data for 153 neutral molecules. However, their parameters and performances were so similar to one another (because the CM1 charge mapping results in charge distributions that are effectively independent of the underlying Hamiltonian) that a single set of parameters was optimized for use with either Hamiltonian and taken to define SM4-hexadecane. The rms error of this model over the 306 data derived from combining the AM1 and PM3 calculations was 0.41 kcal/mol (for comparison, the dispersion in the reference data was 2.05 kcal/mol).

Subsequent work extended the SM4 model to other alkanes [95]. A data set of 350 solute/solvent combinations, where the solvents spanned 16 different alkanes, was constructed and added to the n-hexadecane data set. Since the alkanes would be expected to behave very similarly with one another when it comes to dispersion interactions with a solute, and since the dielectric constants are known, it was assumed that the only model parameter that one might assume to vary as a function of solvent would be σ^{CS} in eq. (12). The free energy of cavitation/structural rearrangement was assumed to be proportional to the *macroscopic* surface tension γ of the *solvent* and empirical optimization established that for all alkanes

$$\sigma^{CS} = 0.03332\gamma + 15.95 \text{ cal/mol} \bullet \text{Å} \tag{13}$$

provided an excellent fit to all of the experimental data (eq. (13) was constrained to reproduce SM4-hexadecane exactly). Values of σ^{CS} computed from eq. (13) differed from values optimized on a solvent-by-solvent basis by no more than 0.02 cal/mol•E. Again, a single set of parameters was found to be equally applicable to both the AM1 and PM3 Hamiltonians, and the rms error over all data was 0.45 kcal/mol.

An aqueous SM4 model for water was never developed. In two cases, preliminary work was reported for solutes containing C, H, and O atoms. However, those cases were not general parameterizations, but were instead restricted to certain classes of compounds for use in modeling aqueous solvation effects on the Claisen rearrangement [14] and sugar conformational analysis [96]. Careful analysis of work accomplished up to that point indicated that the dependence of surface tensions and Coulomb radii on bond orders and partial atomic charges led to instabilities associated with the difficulty of properly including these terms in Fock matrix updates as part of the SCRF equations. A decision was taken to alleviate this problem by eliminating such dependencies, and the resulting new protocol was referred to as SM5.

In practice, the term "SM5" does not refer to any specific model, but instead represents a general functional form wherein (i) all Coulomb radii are taken to be independent of partial atomic charge, (ii) Eq. (4) holds for all pairwise atomic combinations without exception, (iii) the van der Waals radii of Bondi [97] together with a solvent probe radius are used for the calculation of solvent-accessible surface area for use in computing G^{o}_{CDS}, and (iv) all atomic surface tensions are taken to depend only on atomic number and local solute geometry, *not* on bond orders. Except for H, all Coulomb radii are taken to be constant. For H atoms bonded to N and O, a somewhat smaller radius is assigned based on a geometric function that identifies such bonding. The dependence of atomic surface tensions on molecular geometry is, ultimately, simply a more computationally convenient and unambiguous way of encoding the heuristic rules so successfully used for assigning atomic types in the SM1a model, but with the advantage that the rules are smooth and differentiable functions of atomic coordinates, making quantum calculations on unusual species and reaction coordinates entirely feasible.

With respect to specific models, SM5 models are always identified by additional characters in their names. When CM1 Class IV charges are used in the computation of ΔG^{o}_{ENP}, and surface tensions are optimized for the corresponding G^{o}_{CDS}, the model is referred to as SM5.4. When Class II semiempirical Mulliken charges and the resulting associated surface tensions are used, the model is designated SM5.2. When ΔG^{o}_{ENP} is ignored completely, and the total solvation free energy is assumed to be computable entirely from eq. (5), the model is named SM5.0. In addition, during the course of the development of *all* of the SM*x* models, it was observed that reoptimization of geometries typically contributes at most 5% to a total solvation free energy. As a result, computation of SCRF solvation free energies at gas-phase geometries is an efficient alternative, and when geometries are restricted to the gas phase, the model may be identified by appending an "R" to the model name, e.g., SM5.0R. Later work at Hartree-Fock (HF) and density functional theory (DFT) levels, however, has tended to adopt instead the usual "double-slash" notation (where levels for energies are specified prior to the double slash and geometries are specified afterwards, e.g., SM*x*/HF/6-31G//HF/6-31G).

Another suffix that may be added to the SM5 notation is PD, for "pairwise descreening" [30,98]. At semiempirical levels, solution of the SCF equations is sufficiently fast that the computation of effective Born radii can become the slowest step in the SCRF calculation. Pairwise descreening approximates the difficult integral involved in computing the atomic solvation free energy with a parametrically scaled combination of pairwise, analytic integrals between the atom in question and all other atoms. This scheme speeds up the calculation considerably, and indeed the PD approach[2] has been widely employed in *classical* GB schemes (where atomic partial charges are force field parameters) in order to take advantage of its greater speed, especially for calculations on biopolymers [99,100]. At the HF and DFT levels the computation of effective Born radii no longer comprises a significant fraction of the

[2] This is sometimes referred to in the literature as the HCT approximation, for Hawkins, Cramer, and Truhlar.

effort, so most modern SM*x* calculations do not adopt the PD approach, although it remains an efficient scheme for large- or multi-scale molecular dynamics calculations. The original report [30] of the PD approach reoptimized surface tensions for SM2.1 within the context of PD descreening and named a preliminary version of this method SM2.2; subsequent work completed this parameterization and adopted the PD suffix [98].

Given that overall framework, the first SM5 model to be published [101] described the SM5.4A and SM5.4P aqueous solvation models (designed for use with AM1 and PM3, respectively); their rms errors over an expanded test set of 215 aqueous solvation free energies of neutral molecules were 0.72 and 0.62 kcal/mol, respectively. Over 34 ions their rms errors were 5.7 and 5.4 kcal/mol, respectively. A key advance accomplished at the SM5 level was the extension of the model to organic solvents (briefly called OSM5.4 to emphasize its "organic" nature, although this nomenclature is no longer used) [102,103]. This extension was done in a general way such that the model was (and still is) referred to as being "universal". As in all SM5 models, the ENP term is computed using the normal GB formalism and the appropriate dielectric constant for the solvent. The CDS term, on the other hand, is computed from eq. (12) using solvent probes of 1.7 and 3.4 E for the CD and CS surface areas, respectively, but the CD and CS surface tensions themselves are made to be linear functions of macroscopic *solvent* descriptors. In particular

$$\sigma_k^{\mathrm{CD}} = \sum_{\lambda=n,\alpha,\beta} \sigma_k^{\mathrm{CD},\lambda}\lambda \tag{14}$$

where all parameters implicit in each $\sigma_k^{\mathrm{CD},\lambda}$ are empirically optimized over a data set comprising free energies of solvation into multiple organic solvents, n is the solvent index of refraction, α is Abraham's hydrogen bonding acidity $\sum\alpha_2^{\mathrm{H}}$ [104], and β is Abraham's hydrogen bonding basicity $\sum\beta_2^{\mathrm{H}}$ [104]. For the molecular surface tension, we took

$$\sigma^{\mathrm{CS}} = \sum_{\lambda=n,\gamma} \sigma^{\mathrm{CS},\lambda}\lambda \tag{15}$$

where all symbols have been described above.

Over a data set of 1786 free energies of solvation for 206 different solutes in 90 different organic solvents, the SM5.4/AM1 and SM5.4/PM3 models achieved mean unsigned errors of under 0.5 kcal/mol after parameterization. For chloroform [105] and benzene and toluene [103], refined models were described where solvation free energies for solutes into these solvents were heavily overweighted in the optimization of the parameters implicit in equations (14) and (15). Systematic errors in these solvents were eliminated in this fashion, although the magnitude of these errors was less than 1 kcal/mol in terms of mean unsigned error.

Aqueous solvation models using the SM5 functional forms for surface tensions but combining these with $\Delta G_{\mathrm{ENP}}^{\mathrm{o}}$ computed from Class II Mulliken charges in combination with pairwise descreening were reported for the AM1 and PM3 Hamiltonians and referred to as SM5.2PD models [98]. Aqueous pairwise descreening models SM5.4PD were also developed [98]. All of the models achieved similar overall accuracies, but the partitioning between ENP and CDS components is presumably more realistic with Class IV charges and without adopting the PD approximation. Subsequently, SM5.2 organic models were developed for AM1, PM3, MNDO, and MNDO/d [106-109], again achieving accuracy similar to SM5.4 analogs at reduced cost (primarily because of simpler coding) but with some loss of physicality.

At the extreme end of sacrificing physicality for speed, it was discovered that eliminating the computation of $\Delta G_{\mathrm{ENP}}^{\mathrm{o}}$ altogether and computing the full aqueous free energy of solvation using only SM5 surface tension functionals led to a model, SM5.0, that, was essentially as accurate as any of the other SM5 aqueous models for neutral solutes [110]. The SM5.0 model does not require any electronic structure calculations, so it is extraordinarily fast, although molecular geometries must be chosen in some fashion, of course. The SM5.0 model was subsequently extended in a universal way to the organic parameterization set [111] and a freely distributed code, OMNISOL [112], was created incorporating these models. Another model, SM5.05, was also described [110] that was designed to permit an extension of aqueous SM5.0 to include the charged groups found in protein side chains at neutral pH, e.g., imidazolium ions, carboxylate anions, primary ammonium cations, and guanidinium ions.

Finally, SM5 surface tension functional forms and parameters have been optimized for use with $\Delta G^{\text{o}}_{\text{ENP}}$ values computed by means other than the GB formalism. In particular, SM5C [113] uses the SM5 surface tension functional form but computes $\Delta G^{\text{o}}_{\text{ENP}}$ at the AM1, PM3, MNDO, and MNDO/d level from the conductor-like screening model (COSMO) of Klamt and co-workers [114]. A key difference between the GB and COSMO approaches is that the latter uses the full electron density to represent the solute charge distribution instead of atom-centered point charges, and there may be instances where this more complete representation yields a more physical partitioning between electrostatic and non-electrostatic components of the full free energy of solvation. In practice, however, the accuracy of the SM5C model over the full SM5 training set is about the same as that observed for any of the other SM5-type models.

2.2. SMx models for ab initio Hartree-Fock theory and density functional theory. All of the quantum mechanical model development described for the SM*x* models thus far relied on underlying semiempirical QM levels of theory. Early efforts to extend the models to the HF and DFT levels made it clear, however, that the CM1 mapping scheme was not particularly well suited to these more complete levels of theory: Mulliken charges and bond orders show strong basis-set dependence and take on very unphysical values as basis sets become more saturated. In order to create a better charge mapping scheme for use at the HF and DFT levels, Charge Model 2 (CM2) was developed [115,116]. The CM2 mapping may be regarded as a more flexible and simpler extension of the CM1 model with the simultaneous adoption of pairwise specific parameters describing charge transfer between bonded atoms. The CM2 partial atomic charge is defined as

$$q_k^{\text{CM2}} = q_k^{(0)} + \sum_{k' \neq k} B_{kk'}\left(D_{kk'} + C_{kk'}B_{kk'}\right) \tag{16}$$

where the reference charge $q_k^{(0)}$ is now taken to be a Løwdin charge [117] (to reduce basis-set dependence), $B_{kk'}$ is the Mayer bond order between atoms k and k' [118,119], and $C_{kk'}$ and $D_{kk'}$ are empirically optimized parameters that change sign upon reversal of k and k'. Over a database of 211 polar molecules, CM2 models specific to level of theory/basis set combinations predict dipole moments with rms errors on the order of 0.2 D.

Subsequent SM*x* model development using CM2 charges led to the creation of SM5.42 models, where the "4" again denotes Class IV charges and the final "2" indicates that the Class IV charges come from CM2. At the HF level, the first SM5.42 model [120] was an aqueous model parameterized for the MIDI! basis set [121,122] based on gas-phase HF/MIDI! geometries. In short order, aqueous *and* organic SM5.42 models were described for a large number of additional HF and DFT levels /basis set combinations [123,124]. For organic solvents, eq. (15) was modified so that the summation runs over 4 solvent descriptors: the macroscopic surface tension (γ) the square of Abraham's hydrogen bonding basicity (β^2) the square of the fraction of solvent heavy atoms that are either F, Cl, or Br (ϕ^2), and the square of the fraction of solvent heavy atoms that are aromatic carbon atoms (ψ^2). Over all solvents and solutes, mean unsigned errors in neutral solvation free energies were found typically to be on the order of 0.5 kcal/mol and the errors for ions in aqueous solution were about an order of magnitude larger. Solutes containing Si were added to the training set so that parameters for this atom could be determined [125]; predictive accuracies for the training set molecules containing Si are similar to all others.

In addition to *ab initio* levels of theory, a CM2 model was developed [126] for the INDO/S semiempirical Hamiltonian [127,128] based on a training set containing both ground- and excited-state charge distributions. An SM5.42/INDO model for ground-state molecules [129] was built upon this CM2 model, as was a two-time-scale electrostatics-only vertical excitation model (VEM4.2) designed to predict solvatochromism in UV absorption spectra [130]. This model, augmented by dispersion and hydrogen bonding terms, was used very successfully to predict the solvatochromic shifts of acetone in nine solvents.

During the course of this development effort, analytic derivatives were derived for SM5.42 solvation free energies [131], permitting efficient optimization of molecular geometries in solution. As of this date, analytic second derivatives have not yet been derived, although vibrational frequencies may be determined from numerical differentiation of the analytic first derivatives.

The SM5.42 models may be regarded as being reasonably mature, and they continue to be useful for modern research. Nevertheless, they are not the most modern members of the SM*x* family. In order to improve further upon the electrostatics in the GB treatment, a new Charge Model 3 [132-135] was developed that did not differ from CM2 in functional form, but was parameterized over a much more diverse test set roughly double the size of that used for CM2. In addition, a charge renormalization scheme was adopted for use with basis sets containing diffuse functions since with such basis sets Løwdin charges and Mayer bond orders were otherwise found to be less reliable [136].

SM5-type models making use of CM3 charges are referred to as SM5.43 models [137,138]. Aqueous and organic models with analytic gradients are available for the HF and B3LYP [139-142] levels with the 6-31G(d) basis set [143]. They are also available for the MPW*X* density functional, which corresponds to mixing *X*% of Hartree-Fock nonlocal exchange with (100–*X*)% of local *m*PW exchange and 100% of PW91 density functional correlation, with *X* taking on any value from 0 to 100% [144-146], with any of the MIDI!, 6-31G(d), 6-31+G(d), or 6-31+G(d,p) basis sets. The models are coded in the freely distributed software packages GAMESSPLUS [147], HONDOPLUS [148], and SM*x*GAUSS [149], and their inclusion in other codes is ongoing (updates may be found on the comp.chem.umn.edu website). Comparisons of select SM5.43 models to some of the most recent versions of the polarized continuum model (PCM; [150-155]), which is also widely used for continuum solvation calculations, are presented in Tables 1 and 2 for aqueous and organic solvents, respectively. The superior performance of the SM5.43 models is noteworthy, particularly for organic solvents.

The most recently developed SM*x* model is SM6, which is presently available only for water [156]. Charge model 4 (CM4) [156] was developed as part of the SM6 parameterization. The primary difference between CM3 and CM4 is that the former was judged, after detailed analysis, to sometimes predict C–H bonds to be too polar. The CM4 parameterization therefore began by optimizing the C_{CH} and D_{CH} parameters so as best to reproduce the Optimized Potentials for Liquid Simulations (OPLS; [157]) partial atomic charges for 19 hydrocarbons.[3] Subsequent to that, all other parameters were optimized in the same fashion as for CM3. SM6 also has a neutral training set with some additional organic functionality not considered for SM5.43 and previous models. More importantly, however,

Table 1: Mean unsigned error of the aqueous free energy of solvation (kcal/mol) of various solute classes calculated by various continuum solvation models using HF/6-31G(d).

solute class	no. data	SM5.42	C-PCM[a]	D-PCM[a]	IEF-PCM[a]	SM5.43
neutral H, C, N, O, F[b]	171	0.57	0.79	1.21	0.76	0.51
Cl, Br, S, and P neutrals[c]	86	0.49	1.64	1.69	1.64	0.48
all neutrals	257	0.54	1.07	1.37	1.06	0.50
H, C, N, O, F ions[b]	32	5.20	7.66	9.18	7.63	4.91
Cl, Br, P, S ions[c]	15	4.03	2.16	13.12	2.18	4.10
all ions	47	4.83	5.90	10.44	5.89	4.65

[a]As coded in *Gaussian 03*

[b]Solutes containing at most the five listed elements

[c]Solutes containing at least one of these elements plus, in most cases, elements from the previous row

[3] Although the OPLS partial charges are optimized for liquid phases, where dielectric screening makes bonds more polar than in the gas phase, the difference is expected to be small for C–H bonds, and we elected to use these charges as target values for gas-phase calculations.

Table 2: Mean unsigned error of the free energy of solvation (kcal/mol) of solutes in the indicated solvent calculated by various continuum solvation models with the 6-31G(d) basis set.

		B3LYP				*m*PW1PW91
solute class	no. data[a]	C-PCM[b]	D-PCM[c]	IEF-PCM[c]	SM5.43	SM5.43
acetonitrile	7	5.22	5.13	5.22	0.37	0.41
aniline	9	7.55	7.54	7.73	0.50	0.50
benzene	68	4.81	5.09	5.15	0.62	0.63
carbon tetrachloride	72	4.53	4.76	4.80	0.48	0.49
chlorobenzene	37	3.90	4.01	4.12	0.51	0.55
chloroform	96	4.56	4.75	4.82	0.53	0.55
cyclohexane	83	2.30	2.52	2.53	0.40	0.41
dichloroethane	38	3.85	3.94	4.05	0.43	0.44
diethyl ether	62	3.26	3.49	3.68	0.62	0.63
dimethyl sulfoxide	7	3.42	3.31	3.43	0.57	0.61
ethanol	8	2.17	2.45	1.87	1.30	1.35
heptane	60	2.21	2.46	2.49	0.33	0.34
methylene chloride	11	2.01	3.16	3.26	0.45	0.47
nitromethane	7	2.83	2.64	2.84	0.67	0.71
tetrahydrofuran	7	3.12	3.10	3.20	0.32	0.33
toluene	49	3.43	3.67	3.75	0.40	0.41
16 above solvents	621	3.68	3.89	3.96	0.49	0.51
all other 74 solvents	1359	n.a.[d]	n.a.[d]	n.a.[d]	0.50	0.52
self-solvation energies[e]	76 (16)[f]	(3.94)	(4.14)	(4.20)	0.50 (0.49)	0.53 (0.51)

[a]Number of experimental data in this solvent
[b]As coded in *Gaussian 98*
[c]As coded in *Gaussian 03*
[d]not available
[e]Standard-state free energy of solvation of a solute in a pure liquid of the solute (i.e., equivalent to a vapor pressure).
[f]The PCM-type continuum solvation models are explicitly defined for 16 of the 76 solvents used to compute the self-solvation energies and the MUEs for these 16 solvents are given in parentheses.

SM6 employs an ionic training set that is more than double the size of that used for prior models. In addition, ions clustered with one water molecule are included in the training set. Because ionic solvation free energies are so sensitive to atomic Coulomb radii, a good ionic training set is very important for setting this critical parameter. By using ionic cluster data, it proved possible to do a much more thorough job of determining physically realistic Coulomb radii. The performance of the SM6 model is marginally better than SM5.43 for neutral molecules, but significantly better for ions, which makes the model more accurate for properties such as those described in more detail in Section 3 of this article, e.g., pK_a values and oxidation and reduction potentials. Development of organic solvation models within the SM6 framework is presently ongoing.

Another new frontier that is the subject of exploration is extension of the SM6 model to temperatures other than 298 K. A preliminary temperature-dependent water model that is applicable to compounds containing C, H, and O has been described [158]; the results are very encouraging.

2.3. Vapor pressures and solubilities. Another partitioning that is of special interest is that of a solute between a solvent and a pure phase of the solute itself, i.e., pure liquid or solid solute. This is of particular interest because one may relate this free energy of partitioning of a liquid or solid solute to that solute's solubility in a given phase. If we first consider a liquid species $A_{(l)}$ in equilibrium with its own vapor $A_{(g)}$

$$A_{(g)} \Leftrightarrow A_{(l)} \tag{17}$$

With a 1 molar standard-state and ideal behavior in both phases, the standard-state free energy of this process is

$$\Delta G_1^\circ = RT \ln \frac{P_A^\bullet}{P^\circ M_A^l} \tag{18}$$

where R is the universal gas constant, T is the temperature, $P_A^\bullet$ is the equilibrium vapor pressure of A over pure A, P° is the pressure (24.45 atm) of an ideal gas at 1 molar concentration and 298 K, and M_A^l is the equilibrium molarity of the pure liquid solution of A, which is obtained from the liquid density of A. By rearrangement of eq. (18) one may compute the vapor pressure using models that predict ΔG_1°. The universal SM5-type models have been used for this purpose for a large number of cases where the necessary solvent descriptors are already available [137,159,160]; accuracies within a log unit are routine (cf. last line of Table 2).

Let us now consider the equilibrium between pure liquid A and an aqueous solution of A (we specify aqueous because of the importance of water, but the below analysis may be generalized to any solvent)

$$A_{(l)} \Leftrightarrow A_{(aq)} \tag{19}$$

assuming that all activity coefficients are unity, the standard-state free energy for the phase transfer is

$$\Delta G_2^\circ = -RT \ln \frac{M_A^{aq}}{M_A^l} \tag{20}$$

where M_A^{aq} is the equilibrium aqueous molarity of solute A, i.e., its solubility in molarity units, also denoted as S. Combining eqs. (17) and (19) gives

$$A_{(g)} \Leftrightarrow A_{(aq)} \tag{21}$$

The standard-state free energy change in this process is the standard-state aqueous free energy of solvation of solute A, $\Delta G_{S(aq)}^\circ$; and we can calculate it by adding eqs. (18) and (20), which yields

$$\Delta G_{S(aq)}^\circ = RT \ln \frac{P_A^\bullet}{P^\circ} - RT \ln M_A^{aq} \tag{22}$$

Equation (22) is based on the assumption that the aqueous solution of A obeys Henry's law, i.e., that the saturated solution behaves as though infinitely dilute. In general, then, we may predict solubility from

$$S \equiv M_A^{aq} = \left(P_A^\bullet / P^\circ \right) \exp\left[-\Delta G_{S(aq)}^\circ / RT \right] \tag{23}$$

where the SMx models are used to compute $P_A^\bullet$ and $\Delta G_{S(aq)}^\circ$. In the case of 70 organic liquids and 13 organic solids, we obtained MUEs (log units) of less than 0.4 when using the SM5.42 model for all calculations with either AM1, HF/MIDI!, or B3LYP/MIDI! as the underlying Hamiltonian [160]. The successful prediction of the solid solubilities is particularly noteworthy as we treated these compounds essentially as supercooled liquids. That is, we estimated the necessary "solvent" descriptors for use with the universal SM5.42 model and then computed vapor pressures for the solutes in equilibrium with their solid form. The solids were for the most part aromatic hydrocarbons, and this approximation is likely to be less useful for solutes that have more specific intermolecular interactions in their crystalline solid form. Nevertheless, it suggests that the computation of solid/liquid partitioning and solubility may be reasonably started from an SMx foundation.

3. SMx Models for Other Condensed Phases

Homogeneous liquid solutions are arguably the simplest condensed phases to represent with a dielectric continuum model. We may, however, expect that other condensed phases having liquid-like properties, e.g., a fluid-phase lipid membrane, might be addressable with the SM*x* modeling approach. Inasmuch as the SM5 organic parameterizations are designed to be universal, the data required to model *any* phase are simply the most appropriate values of ε, n, γ, α, and β (and φ and ψ if an SM5.42 or SM5.43 model is employed). Typically, of course, values for these quantities are not available. However, they may be estimated by any number of approaches (e.g., on the basis of molecular functionality also found in other molecules for which solvent parameters are available). A still more pragmatic approach for dealing with such poorly characterized systems is to treat the unknown "solvent" descriptors as being fitting parameters themselves. For all but the dielectric constant, this is particularly simple insofar as the solvation free energy depends upon them linearly. Thus, a typical approach is to make an educated guess at the dielectric constant, compute $\Delta G^{\text{o}}_{\text{ENP}}$ for solutes for which solvation free energies or partition coefficients (which are differences of solvation free energies between two phases, as described further below) are known, and determine the optimal values of the remaining descriptors that minimize errors in the remaining term $G^{\text{o}}_{\text{CDS}}$ through multilinear regression. A final model may be determined by trial-and-error optimization of ε.

The first example of this approach was described for a phosphatidyl choline bilayer [17]. Available experimental data were partition coefficients P between the bilayer and surrounding aqueous solvent

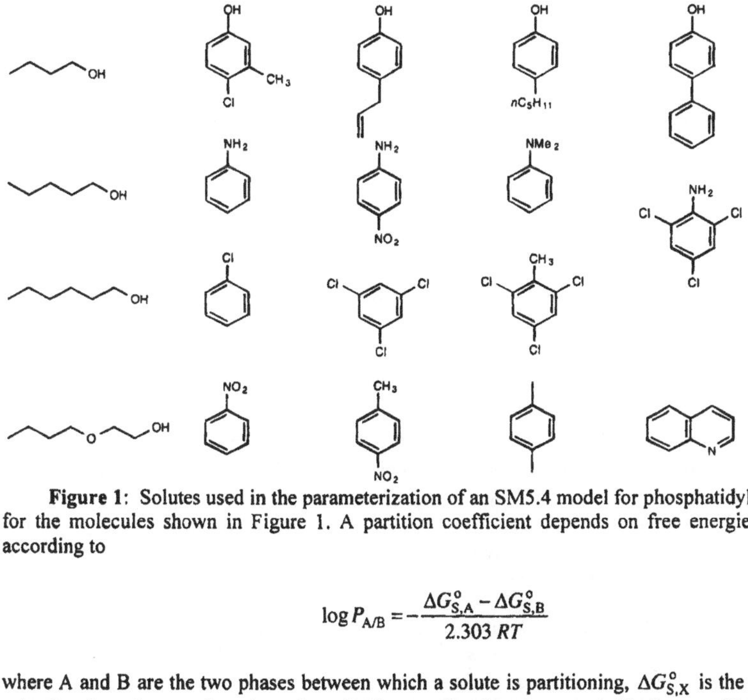

Figure 1: Solutes used in the parameterization of an SM5.4 model for phosphatidyl choline.

for the molecules shown in Figure 1. A partition coefficient depends on free energies of solvation according to

$$\log P_{\text{A/B}} = -\frac{\Delta G^{\text{o}}_{\text{S,A}} - \Delta G^{\text{o}}_{\text{S,B}}}{2.303\,RT} \tag{24}$$

where A and B are the two phases between which a solute is partitioning, $\Delta G^{\text{o}}_{\text{S,X}}$ is the free energy of solvation from the gas phase into phase X (the quantity directly computable from an SM*x* model), R is the universal gas constant, and T is temperature. Using either experimental or computed aqueous solvation free energies provides target values for the solvation free energy into the phosphatidyl choline bilayer by rearrangement of eq. (17). Assuming the dielectric constant of the bilayer to be 5.0 (an estimate based on the dielectric constant of octanol), and assuming the α value to be 0.0 (as there is no significant hydrogen bond donating functionality in phosphatidyl choline), an SM5.4 model was developed by regressing the target free energies of solvation on the n, γ, and β solvent parameters. The regression provided values of 1.40, 27, and 1.15 for these parameters, which quantities are very much in the range of physically realistic values that one might have expected based on analogous molecules as solvents. In this case, the regression was also allowed to have a constant term, which was calculated to be 0.59 log units. The quality of the regression, with an R^2 value of 0.80 (Figure 2), is in the range of those typically employed for predictions of bioavailability of druglike molecules, and the specific-range-parameter (SRP) SM5.4 model may be expected to be useful when used to make predictions about molecules having functionalities not too different from those included in the training set. Of course, the introduction of significantly different functionality would likely require retraining to ensure maximum predictive accuracy.

A similar phase parameterization has been accomplished for soil (i.e., dirt) [161], using a training set much larger than the phosphatidyl choline case. From an environmental perspective, an important physical parameter affecting the fate of ecosystem contaminants is the soil-water partition coefficient. Because this often depends primarily on the soil's organic carbon content, measured values are usually normalized for the organic carbon (OC) content of soil, in which case the soil sorption equilibrium constant is expressed as [162]

$$K_{OC} = \frac{C_{soil} / C_{soil}^{o}}{C_{w} / C_{w}^{o}} \tag{25}$$

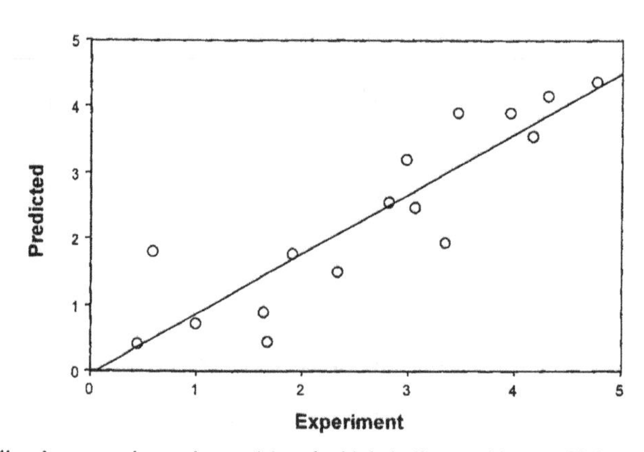

Figure 2: Predicted vs. experimental water/phosphatidyl choline partition coefficients ($\log_{10}$ units).

where C_{soil} is the concentration of solute per gram of carbon in a standard soil, and C_w is the concentration of solute per volume of aqueous solution. The standard state concentrations C_{soil}^{o} and C_{w}^{o} are typically chosen as 1 μg solute per g organic carbon for soil and 1 μg solute per mL for aqueous solution. Although this quantity is often called the soil/water equilibrium constant for simplicity, one should note that it is actually a measure of partitioning between soil *organic matter* and water, normalized to organic carbon content.

In this instance, a database of 387 molecules containing a wide variety of organic functional groups was used for a training set. The dielectric constant for the soil phase was optimized by trial and error to a value of $\varepsilon = 15$. While so high a dielectric constant may seem unusual for a solid organic phase, it should be kept in mind that the soil used in the experiment is "wet", i.e., it is in contact with the aqueous phase and swelled by it. The optimized dielectric constant was used for the computation of electrostatics at the SM5.42/HF/MIDI!//HF/MIDI! and SM5.42/AM1//AM1 levels.

Errors in residual G_{CDS}^{o} values were minimized by fitting the n, γ, α, and β solvent parameters as described above. For the SM5.42/HF/MIDI! level, these parameters took on values of 1.311, 45.3, 0.56, and 0.60; such values are consistent with the functionalities expected for the organic component of soil (which include aromatics, phenols, quinones, and carboxylic acids). The resulting soil partitioning model had a MUE over the training set of 0.98 kcal/mol.

A point that merits emphasis is that application of the SM*x* models to predict solvation or partitioning behavior can certainly be useful as a predictive exercise, but a potentially still more important feature of the models is that the free energy associated with the particular phase transfer can be decomposed into atomic or group contributions, and thus interpreted in a fragment basis [163,164]. The general form of the GB and surface tension equations should make this point clear: $\Delta G^{\circ}_{\text{ENP}}$ is computed as a sum of atomic-charge dependent terms, and G°_{CDS} is computed as a sum over atomic solvent-exposed surface areas.

For instance, Figure 3 illustrates the correlation between experimental and predicted $\log K_{\text{OC}}$ values for polychlorinated biphenyls (PCBs). There is a systematic error of about 1 log unit in the absolute accuracy of the predictions, but otherwise a fairly good correlation ($R^2 = 0.86$). Analysis of atomic contributions indicates that chlorine atoms near the biphenyl ipso carbons are less hydrophilic (owing to decreased exposure to the solvent) than chlorine atoms substituted at other positions. Such fragment analysis can be useful in drug design, where particular ranges of partitioning constants may enhance drug bioavailability.

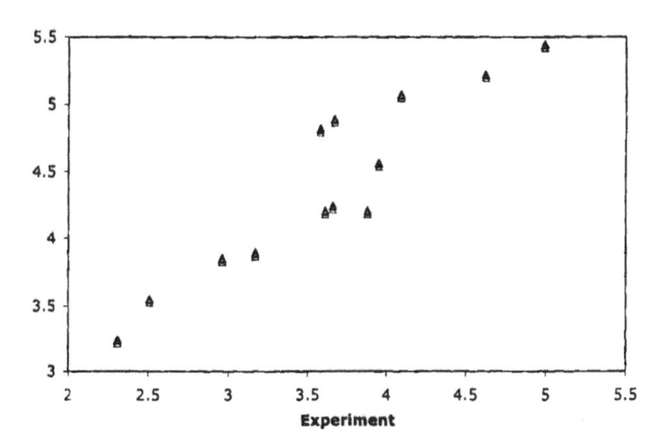

Figure 3: Predicted vs. experimental soil/water partition coefficients for PCBs ($\log_{10}$ units).

A final example of a novel phase to which the SM*x* models have been extended is the surface layer of an aqueous solution that is in contact with air, i.e., the air-water interface. This is a phase that is formally two-dimensional, although at the molecular level it has some non-zero thickness. In this case the measured equilibrium constant of interest is

$$K_{\text{i/a}} = \frac{\Gamma_i}{C_{\text{a}}} \tag{26}$$

where Γ_{i} is the concentration of the adsorbed solute at the air-water interface in mol/m^2, and C_{a} is the equilibrium vapor concentration in mol/m^3 of the solute. The adsorption coefficient for a molecule transferring from the gas phase to a air-water interface is related to the standard-state free energy of transfer by:

$$\ln K_{\text{i/a}} = -\frac{\Delta G^{*}_{\text{i/a}}}{RT} \tag{27}$$

However, the free energy in eq. (27) differs from all of the other free energies that have thus far been described in the sense that it contains two distinct components. The first component is the coupling free energy associated with the new interactions that a molecule in a condensed phase has in comparison to the gas phase—this is the component that the SM*x* models are designed to predict. In the case of $\Delta G^{*}_{\text{i/a}}$, however, there is also a term associated with the loss of entropy associated with moving from a phase of 3 dimensions to one of 2 dimensions. That is

$$\Delta G^{\circ}_{i/a} = \Delta G^{\circ}_{\text{coup(a}\rightarrow\text{i)}} + \Delta G^{\circ}_{3\text{D}\rightarrow 2\text{D}} \tag{28}$$

The coupling term may be computed directly, however, and depends only on the molecular weight. Thus, a particle of mass m at temperature T has a de Broglie wavelength equal to [165]

$$\Lambda = \sqrt{\frac{h^2}{2\pi mkT}} \tag{29}$$

where k is Boltzmann's constant, and h is Planck's constant. For an ideal gas molecule in M dimensions, the molecular translational partition function can be written as [165]

$$q = \left(\frac{L^{\circ}}{\Lambda}\right)^{M} \tag{30}$$

where L° is the standard-state unit of length. The molar translational partition function is then

$$Z = \frac{1}{N_A!}\left(\frac{L^{\circ}}{\Lambda}\right)^{MN_A} \tag{31}$$

where N_A is Avagadro's number. With this partition function and taking standard thermodynamic relationships for internal energy U, enthalpy H, and entropy S as a function of Z [165], we may compute

$$U^{\circ} = \frac{M}{2}RT \qquad H^{\circ} = U^{\circ} + PV \qquad S^{\circ} = R\ln\left[\frac{e^{(M+2)/2}}{N}\left(\frac{L^{\circ}}{\Lambda}\right)^{M}\right] \tag{32}$$

and thus, from

$$\Delta G^{\circ}_{3\text{D}\rightarrow 2\text{D}} = \Delta H^{\circ}_{3\text{D}\rightarrow 2\text{D}} - T\Delta S^{\circ}_{3\text{D}\rightarrow 2\text{D}} \tag{33}$$

we derive

$$\Delta G^{\circ}_{3D\rightarrow 2D} = \Delta(PV)_{3\text{D}\rightarrow 2\text{D}} - RT\ln\left(\frac{\Lambda}{L^{\circ}}\right) \tag{34}$$

The PV terms may be calculated from the microcanonical relationship [165]

$$\frac{P}{T} = \left\{\frac{\partial S}{\partial\left[\left(L^{\circ}\right)^{M}\right]}\right\}_{N,U} \tag{35}$$

Using S° as defined in eq. (32) gives

$$\frac{P}{T} = \frac{R}{\left(L^{\circ}\right)^{M}} \tag{36}$$

which is the ideal gas law $PV = RT$ but expressed for volume in an arbitrary number of dimensions. Thus, DPV is zero for the dimensionality change and we have

$$\Delta G^{\circ}_{3D\rightarrow 2D} = -RT\ \ln\left(\frac{\Lambda}{L^{\circ}}\right) \tag{37}$$

When this component is removed from measured free energies of adsorption, the remaining free energies of coupling can be used as a training set for the determination of an air-water interface SMx model in the usual way.

In this case, however, there is some ambiguity associated with computing the electrostatic component of solvation. One approximation might be to assume that a molecule at the interface has half the value of $\Delta G^{\mathrm{o}}_{\mathrm{ENP}}$ that would be computed in the bulk. A more accurate model would account for preferential orientation of surface molecules. In previous work, however, we instead investigated whether an SM5.0 model (i.e., one that ignores electrostatics) would be successful in making predictions. For a training set of 85 solutes, we found that a best fit of the "solvent" descriptors for the air-water interface involved taking a, b, and g to be 1.11, 0.59, and −144.6, respectively [166].

These values appear to be quite physical. Various prior simulations and experimental studies [167-170] have suggested that dangling hydrogen bond sites at the air-water interface make surface water more active as a hydrogen bond donor and acceptor than bulk water (which has a and b values of 0.82 and 0.35, respectively [171]). The negative value for the surface tension is also reasonable, since for the adsorption process no free energy is being expended in order to cavitate, but instead new dispersion interactions are being added by the solute "sticking" to the surface that were not present previously. The SM5.0surf model has a MUE (log units) of 0.47 for the 85 molecules in the training set [166], many of which are functionally rich pesticides and other environmental contaminants.

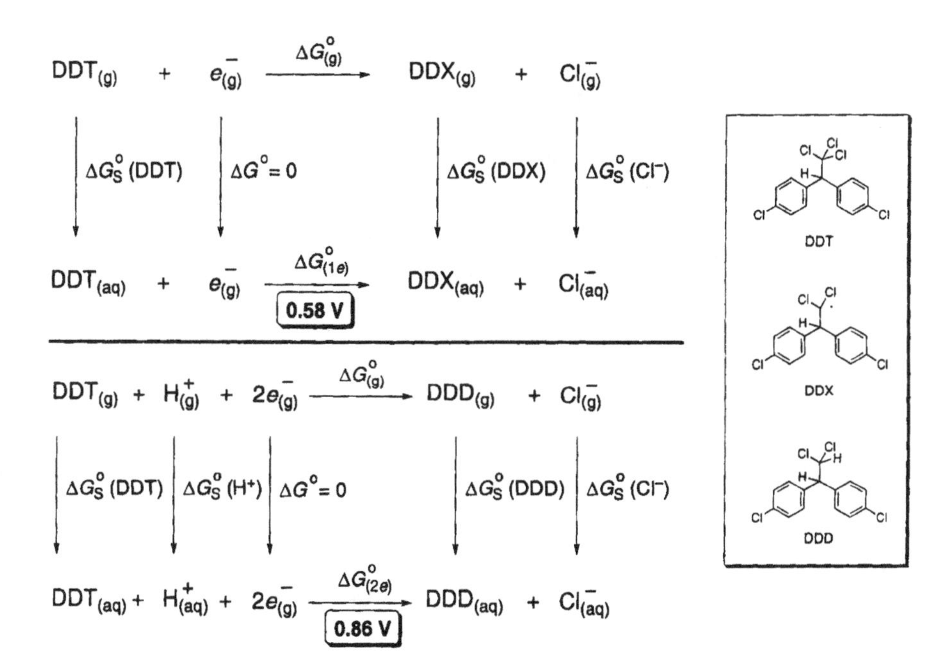

Figure 4: One- and two-electron reduction half reactions for DDT.

4. Selected Applications in Environmental Chemistry

Electron-transfer reactions play important roles in many pathways that degrade contaminants in the environment. Such electron transfers are often described with electrochemical half reactions, like those in Figure 4 for the one- and two-electron reductions of the pesticide DDT.

In order to compute the lower leg of each thermodynamic cycle, it is sufficient to compute the upper leg, which is a gas phase process, at some suitable level of electronic structure theory, and then to add the vertical legs, which quantify the differences between the solvation free energies of the various reagents. In the case of the reactions in Figure 4, calculations were carried out at the B3LYP/6-311+G(d)//B3LYP/6-31G(d) level for gas-phase energetics, the solvation free energies of the proton and chloride anion were taken from experiment [172], and the remaining solvation free energies were computed at the SM5.42/BPW91/6-31G(d)//B3LYP/6-31G(d) level [173].

The resulting free energy of reaction for a solution process $\Delta G^{\circ}_{\text{rxn}}$ may be converted to a potential relative to the normal hydrogen electrode (NHE) by

$$E^{\circ} = \frac{\Delta G^{\circ}_{\text{rxn}} - \Delta G^{\circ}_{\text{NHE}}}{nF} \tag{38}$$

where n is the number of electrons transferred, F is the Faraday constant in appropriate units, and $\Delta G^{\circ}_{\text{NHE}}$ is the free energy of reaction for the reference half reaction

$$\frac{1}{2}\text{H}_{2(\text{g})} \rightarrow \text{H}^{+}_{(\text{aq})} + e^{-}_{(\text{g})} \tag{39}$$

which is 4.28 eV [173,174]. Note that this value, and the values listed in Figure 4, differ from those reported in the original reference [173] because of (i) a standard-state error made in the original reference for the solvation free energy of the proton, as explained in subsequent work [174], an

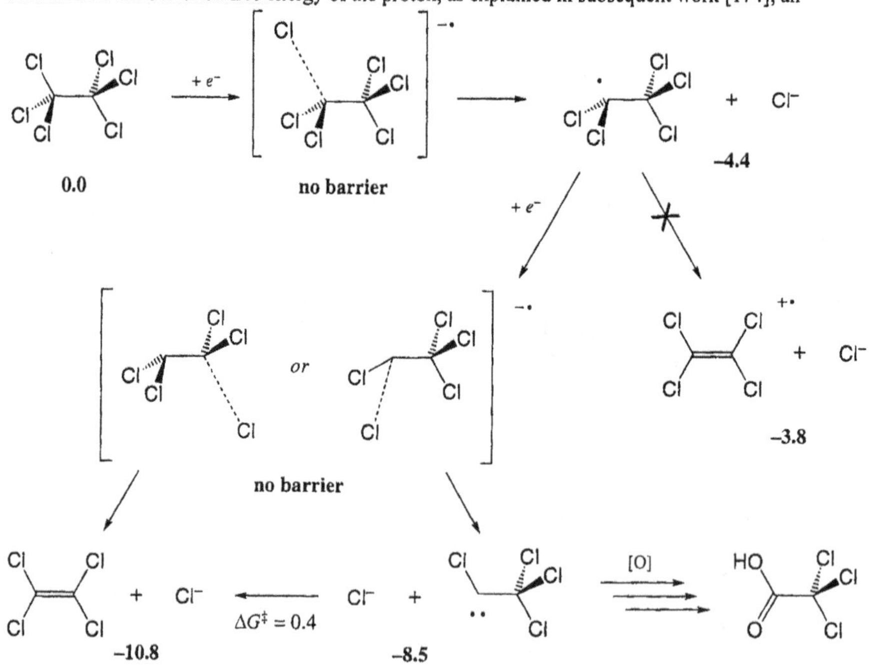

Figure 5: Steps in the reductive dechlorination of hexachloroethane in aqueous solution. Relative energies of stoichiometrically equivalent species are given in eV.

error in the 298 K free energy of the chloride anion (which should have a tabulated value of -460.31875 E_{h} in the original reference [173]), and (iii) the use here of the experimental solvation free energy for the chloride anion instead of a computed solvation free energy (a difference of 0.1 eV).

Reduction (and oxidation) potentials are useful quantities for analyzing environmental persistence in phases having either oxidizing or reducing character. Indeed, taken together with soil/water partition coefficients, which may also be computed using SM*x* methods, they are key components in the prediction of environmental fate constants for organic contaminants.

Solvation free energies can also be added to reagents in reactions that do not involve electron transfer. Accounting for and predicting changes in reaction pathways induced by solvation is a key application of solvation models. One interesting mechanism that we have studied [175], which involves both electron transfer steps and steps in which electrons do not appear as individual species, is the reductive dechlorination of hexachloroethane (Figure 5).

In the initial reductive dechlorination, loss of the first chloride anion is predicted to proceed without barrier subsequent to electron transfer in aqueous solution. One proposal that had been made in the literature was that a heterolytic ionization of the resulting radical might precede a second electron transfer, however this step is predicted at the SM5.42/BPW91/aug-cc-pVDZ//BPW91/aug-cc-pVDZ level of theory to be too endergonic to be important. Interestingly, chloride elimination is again predicted to proceed without barrier following the second electron transfer. However, that prediction holds true for both sets of stereochemically distinct chlorine atoms, those leading to the much more stable tetrachloroethene and those leading to the less stable chlorotrichloromethylcarbene. Rearrangement of the carbene is predicted to have a sufficiently high barrier that diffusion-controlled bimolecular processes might compete with rearrangement of the singlet carbene to the alkene. One such process involves attack of water on the carbene to generate an oxonium ylide that, upon rearrangement would generate a chlorohydrin that could subsequently be easily oxidized to trichloromethylacetic acid. The latter species is detected in small amounts in the reductive

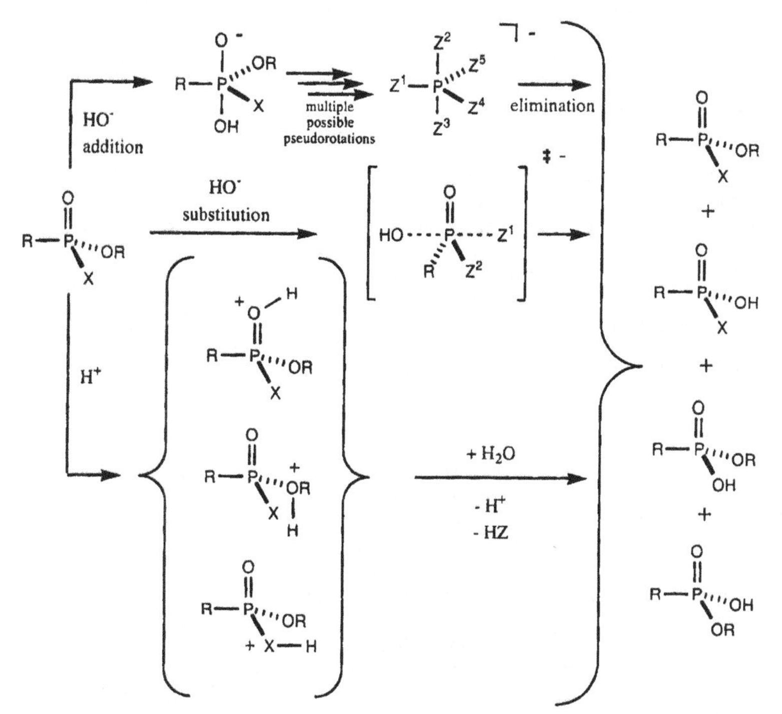

Figure 6: Possible pathways for aqueous hydrolysis of a generic phosphonate derivative.

dechlorination process, and the mechanism of its formation had not been well understood prior to our modeling efforts [175].

The prediction of pK_a values for environmental contaminants can also be particularly important because of the potentially enhanced (or retarded) activity of molecules in their conjugate acid or base forms. In the case of phosphorus-based chemical weapons agents, for example, aqueous decontamination can follow different reaction coordinates depending on aqueous pH (Figure 6). One mechanistic pathways is the reaction of the weapons agent with hydroxide acting as a nucleophile. In the upper pathway, the addition of hydroxide to phosphorus leads to a phosphorane anion intermediate. In general, the elimination of ligands from phosphoranes is more facile from apical positions than from equatorial. Stereochemistry in the phosphorane, i.e., the energetic preference for location of individual groups in apical or equatorial positions, is controlled by a number of factors, including apicophilicity and hyperconjugative interactions between the ligands [176] and solvation effects. Interconversion between different trigonal bipyramids requires a pseudorotation, and the barriers to pseudorotation in such complex phosphoranes are not well understood and may also be subject to solvation effects.

An alternative pathway (the central one in Figure 6) is a one-step bimolecular substitution. Depending on the substituents of the original phosphonate, either or both of these two pathways may be operating. In general, the direct substitution pathway becomes more favorable when the leaving group is the conjugate base of a strong acid.

Finally, under neutral or acidic conditions, the nucleophile may not be hydroxide but instead a water molecule (the lower path in the figure). Under these conditions, there are again two paths, addition/elimination and direct substitution, either or both of which are potentially operative.

The SM6 aqueous solvation model has recently been demonstrated to be particularly effective for the computation of pK_a values [177] (although they are still subject to errors of 1 to 2 pK units in unfavorable cases). Other SM*x* models have also been decisive in explaining key features in the hydrolytic detoxification of chemical weapons agents. For example, Seckute et al. [178] were able to explain the anomalous reactivity of VX-like compounds (thiophosphonates having potential neurotoxic properties) with aqueous hydroperoxide vs hydroxide, a case where solvation effects played a key role in favoring a beneficial reaction path (formation of an innocuous product) over a less useful one (formation of a long-lived toxic product).

5. Conclusions

Continuum solvation models are a critical element in the computational chemist's toolbox because of their utility for describing (i) condensed-phase effects on molecular structure and properties and (ii) the connection between gas-phase and condensed-phase potential energy surfaces. The parameterization of the SM*x* models against a large and diverse set of data, the broad variety of electronic structure levels with which they are compatible, e.g., semiempirical theory, ab initio Hartree-Fock theory, and density functional theory, and their incorporation into several general-purpose electronic structure packages makes them particularly attractive for modeling. Future developments designed to include temperature dependence, improve accuracy, and extend the range of condensed phases to which they may be applied, may be expected to further increase their general utility.

Acknowledgments

The authors are grateful to Daniel Liotard, David Giesen, Gregory Hawkins, Joey Storer, Candee Chambers, Jiabo Li, Tony Zhu, Paul Winget, Jason Thompson, James Xidos, Casey Kelly, and Adam Chamberlin for extensive contributions to the development of the SM*x* models and to Eric Weber, Eric Patterson, and Bill Arnold for collaboration on their application to environmental problems. We are also grateful to our many co-workers on the SM*x* models over the years, whose names grace the publications resulting from their efforts. This work was supported by the NIH training grant for Neuro-physical-computational Sciences, by the U.S. Army Research Office under Multidisciplinary Research Program of the University Research Initiative (MURI) through grant number DAAD19-02-1-0176, by the Minnesota Partnership for Biotechnology and Medical Genomics, by the National Science Foundation (CHE02-03446 and CHE03-49122), and by the Office of Naval Research under grant no. N 00014-05-01-0538.

References

[1] J. Tomasi and M. Persico, Molecular interactions in solution: An overview of methods based on continuous distributions of the solvent, *Chemical Reviews* **94** 2027-2094(1994).

[2] C.J. Cramer and D.G. Truhlar, Development and biological applications of quantum mechanical continuum solvation models, *Quantitative Treatments of Solute/Solvent Interactions*, P. Politzer and J.S. Murray (Eds.), Elsevier, Amsterdam, 1994, pp. 9-54.

[3] C.J. Cramer and D.G. Truhlar, Continuum solvation models: Classical and quantum mechanical implementations, *Reviews in Computational Chemistry*, K.B. Lipkowitz and D.B. Boyd (Eds.), VCH, New York, 1995, pp. 1-72.

[4] J.-L. Rivail and D. Rinaldi, Liquid state quantum chemistry: Computational applications of the polarizable continuum models, *Computational Chemistry, Review of Current Trends*, J. Leszczynski (Ed.), World Scientific, New York, 1996, pp. 139-174.

[5] C.J. Cramer and D.G. Truhlar, Continuum solvation models, *Solvent Effects and Chemical Reactivity*, O. Tapia and J. Bertran (Eds.), Kluwer, Dordrecht, 1996, pp. 1-81.

[6] C.J. Cramer and D.G. Truhlar, Implicit solvation models: Equilibria, structure, spectra, and dynamics, *Chemical Reviews* **99** 2161-2200(1999).

[7] M. Orozco and F.J. Luque, Theoretical methods for the description of the solvent effect on biomolecular systems, *Chemical Reviews* **100** 4187-4225(2000).

[8] D. Bashford and D.A. Case, Generalized Born models of macromolecular solvation effects, *Annual Review of Physical Chemistry* **51** 129-152(2000).

[9] C.J. Cramer and D.G. Truhlar, Thermodynamics of solvation and the treatment of equilibrium and nonequilibrium solvation effects by models based on collective solvent coordinates, *Free Energy Calculations in Rational Drug Design*, M.R. Reddy and M.D. Erion (Eds.), Kluwer Academic/Plenum, New York, 2001, pp. 63-95.

[10] C.J. Cramer, *Essentials of Computational Chemistry: Theories and Models*, John Wiley & Sons, Chichester, 2004.

[11] J. Tomasi, Thirty years of continuum solvation chemistry: A review, and prospects for the near future, *Theoretical Chemistry Accounts* **112** 184-203(2004).

[12] M. Feig and C.L. Brooks, Recent advances in the development and application of implicit solvent models in biomolecule simulations, *Current Opinion in Structural Biology* **14** 217-224(2004).

[13] J. Tomasi, B. Mennucci, and R. Cammi, Quantum mechanical continuum solvation models, *Chemical Reviews* **105** 2999-3093(2005).

[14] J.W. Storer, D.J. Giesen, G.D. Hawkins, G.C. Lynch, C.J. Cramer, D.G. Truhlar, and D.A. Liotard, Solvation modeling in aqueous and nonaqueous solvents: New techniques and a re-examination of the claisen rearrangement, *Structure and Reactivity in Aqueous Solution*, C.J. Cramer and D.G. Truhlar (Eds.), American Chemical Society, Washington, DC, 1994, pp. 24-49.

[15] D.J. Giesen, C.C. Chambers, G.D. Hawkins, C.J. Cramer, and D.G. Truhlar, Modeling free energies of solvation and transfer, *Computational Thermochemistry: Prediction and Estimation of Molecular Thermodynamics*, K. Irikura and D.J. Frurip (Eds.), American Chemical Society, Washington, DC, 1998, pp. 285-300.

[16] G.D. Hawkins, J. Li, T. Zhu, C.C. Chambers, D.J. Giesen, D.A. Liotard, C.J. Cramer, and D.G. Truhlar, Universal solvation models, *Combined Quantum Mechanical and Molecular Mechanical Methods*, J. Gao and M.A. Thompson (Eds.), American Chemical Society, Washington, DC, 1998, pp. 201-219.

[17] C.C. Chambers, D.J. Giesen, G.D. Hawkins, W.H.J. Vaes, C.J. Cramer, and D.G. Truhlar, Modeling the effect of solvation on structure, reactivity, and partitioning of organic solutes: Utility in drug design, *Rational Drug Design*, D.G. Truhlar, W.J. Howe, A.J. Hopfinger, J.M. Blaney, and R.A. Dammkoehler (Eds.), Springer, New York, 1999, pp. 51-72.

[18] G.D. Hawkins, J. Li, T. Zhu, C.C. Chambers, D.J. Giesen, D.A. Liotard, C.J. Cramer, and D.G. Truhlar, New tools for rational drug design, *Rational Drug Design*, A.L. Parrill and M.R. Reddy (Eds.), American Chemical Society, Washington, DC, 1999, pp. 120-140.

[19] C.J. Cramer and D.G. Truhlar, General parameterized scf model for free energies of solvation in aqueous solution, *Journal of the American Chemical Society* **113** 8305-8311(1991).

[20] G.J. Hoijtink, E. de Boer, P.H. Van der Meij, and W.P. Weijland, Potentials of various aromatic hydrocarbons, *Recueil de Travail Chimique des Pays-Bas* **75** 487-503(1956).

[21] F. Peradejordi, On the Pariser and Parr semiempirical method for computing molecular wave functions. The basic strength of N-heteroatomic compounds and their monoamines, *Cahiers Physique* **17** 393-447(1963).

[22] I. Jano, Sur l'ιnergie de solvatation, *Comptes Rendu Academie de Sciences, Paris* **261** 103-105(1965).

[23] O. Tapia, Local field representation of surrounding medium effects. From liquid solvent to protein core effects, *Quantum Theory of Chemical Reactions*, R. Daudel, A. Pullman, L. Salem, and A. Viellard (Eds.), Reidel, Dordrecht, 1980, pp. 25-72.

[24] S.C. Tucker and D.G. Truhlar, Generalized Born fragment charge model for solvation effects as a function of reaction coordinate, *Chemical Physics Letters* **157** 164-170(1989).

[25] W.C. Still, A. Tempczyk, R.C. Hawley, and T. Hendrickson, Semianalytical treatment of solvation for molecular mechanics and dynamics, *Journal of the American Chemical Society* **112** 6127-6129(1990).

[26] C.J. Cramer and D.G. Truhlar, AM1-SM2 and PM3-SM3 parameterized scf solvation models for free energies in aqueous solution, *Journal of Computer-Aided Molecular Design* **6** 629-666(1992).

[27] O. Kikuchi, T. Matsuoka, H. Sawahata, and O. Takahashi, Ab initio molecular orbital calculations including solvent effects by generalized Born formula. Conformation of zwitterionic forms of glycine, alanine, and serine in water, *Journal of Molecular Structure (Theochem)* **305** 79-87(1994).

[28] D.A. Liotard, G.D. Hawkins, G.C. Lynch, C.J. Cramer, and D.G. Truhlar, Improved methods for semiempirical solvation models, *Journal of Computational Chemistry* **16** 422-440(1995).

[29] S.L. Chan and C. Lim, Reducing the error due to the uncertainty in the Born radius in continuum dielectric calculations, *Journal of Physical Chemistry* **98** 692-695(1994).

[30] G.D. Hawkins, C.J. Cramer, and D.G. Truhlar, Pairwise solute screening of solute charges from a dielectric medium, *Chemical Physics Letters* **246** 122-129(1995).

[31] D. Qiu, P.S. Shenkin, F.P. Hollinger, and W.C. Still, The GB/SA continuum model for solvation. A fast analytical method for the calculation of approximate Born radii, *Journal of Physical Chemistry A* **101** 3005-3014(1997).

[32] B. Jayaram, Y. Liu, and D.L. Beveridge, A modification of the generalized Born theory for improved estimates of solvation energies and pK shifts, *Journal of Chemical Physics* **109** 1465-1471(1998).

[33] C.S. Babu and C. Lim, A new interpretation of the effective Born radius from simulation and experiment, *Chemical Physics Letters* **310** 225-228(1999).

[34] C.S. Babu and C. Lim, Incorporating nonlinear solvent response in continuum dielectric models using a two-sphere description of the Born radius, *Journal of Physical Chemistry* **105** 5030-5036(2001).

[35] A. Onufriev, D.A. Case, and D. Bashford, Effective Born radii in the generalized Born approximation: The importance of being perfect, *Journal of Computational Chemistry* **23** 1297-1304(2002).

[36] M.S. Lee, F.R. Salsbury, and C.L. Brooks, Novel generalized Born methods, *Journal of Chemical Physics* **116** 10606-10614(2002).

[37] B. Lee and F.M. Richards, The interpretation of protein structure: Estimation of static accessibility, *Journal of Molecular Biology* **55** 379-400(1971).

[38] J.L. Pascual-Ahuir and E. Silla, GEPOL: An improved description of molecular surfaces. I. Building the spherical surface set, *Journal of Computational Chemistry* **11** 1047-1059(1990).

[39] M.J.S. Dewar, E.G. Zoebisch, E.F. Healy, and J.J.P. Stewart, AM1: A new general purpose quantum mechanical molecular model, *Journal of the American Chemical Society* **107** 3902-3909(1985).

[40] R.S. Mulliken, Electronic population analysis on LCAO-MO molecular wave functions. I., *Journal of Chemical Physics* **23** 1833-1840(1955).

[41] C.J. Cramer and D.G. Truhlar, An SCF solvation model for the hydrophobic effect and absolute free energies of aqueous solvation including specific water interactions, *Science (Washington DC)* **256** 213-217(1992).

[42] C.J. Cramer and D.G. Truhlar, PM3-SM3: A general parameterization for including aqueous solvation effects in the PM3 molecular orbital model, *Journal of Computational Chemistry* **13** 1089-1097(1992).

[43] J.J.P. Stewart, Optimization of parameters for semiempirical methods. 1. Method, *Journal of Computational Chemistry* **10** 209-220(1989).

[44] D.R. Armstrong, R. Fortune, P.G. Perkins, and J.J.P. Stewart, Molecular orbital theory for the excited states of transition metal complexes, *Journal of the Chemical Society, Faraday Transactions 2* **68** 1839-1846(1972).

[45] D.R. Armstrong, P.G. Perkins, and J.J.P. Stewart, Bond indices and valency, *Journal of the Chemical Society, Dalton Transactions* 838-840(1973).

[46] G.D. Hawkins, D.J. Giesen, G.C. Lynch, C.C. Chambers, I. Rossi, J.W. Storer, J. Li, J.D. Thompson, P. Winget, B.J. Lynch, D. Rinaldi, D.A. Liotard, C.J. Cramer, and D.G. Truhlar, *Amsol--version 7.1*, University of Minnesota, Minneapolis, MN, 2003.

[47] *Spartan version 5*, Wavefunction, Inc., Irvine, CA, 1998.

[48] A. Acker, H.J. Hoffmann, and R. Cimiraglia, On the tautomerism of maleimide and phthalimide derivatives, *Journal of Molecular Structure (Theochem)* **121** 43-51(1994).

[49] B. Hernández, M. Orozco, and F.J. Luque, Tautomerism of xanthine and alloxanthine: A model for substrate recognition by xanthine oxidase, *Journal of Computer-Aided Molecular Design* **10** 535-544(1996).

[50] F. Lombardo, J.F. Blake, and W.J. Curatolo, Computation of brain-blood partitioning of organic solutes via free energy calculations, *Journal of Medicinal Chemistry* **39** 4750-4755(1996).

[51] J. Wang and R.J. Boyd, A theoretical study of proton transfers in aqueous *para-*, *ortho-*hydroxypyridine and *para-*, *ortho-*hydroxyquinoline, *Chemical Physics Letters* **259** 647-653(1996).

[52] G.D. Hawkins, C.J. Cramer, and D.G. Truhlar, New methods for potential functions for simulating biological molecules, *Journal de Chimie Physique* **94** 1448-1481(1997).

[53] I. Lee, C.K. Kim, B.-S. Lee, C.K. Kim, H.W. Lee, and I.S. Han, Theoretical studies on the transition-state imbalance in malononitrile anion-forming reactions in the gas phase and in water, *Journal of Physical Organic Chemistry* **10** 908-916(1997).

[54] F.C. Lightstone and T.C. Bruice, Separation of ground state and transition state effects in intramolecular and enzymatic reactions. 2. A theoretical study of the formation of transition states in cyclic anhydride formation, *Journal of the American Chemical Society* **119** 9103-9113(1997).

[55] H. Adalsteinsson and T.C. Bruice, What is the mechanism of catalysis of ester aminolysis by weak amine bases? Comparison of experimental studies and theoretical investigation of the aminolysis of substituted phenyl esters of quinoline-6- and -8-carboxylic acids, *Journal of the American Chemical Society* **120** 3440-3447(1998).

[56] M. Agback, S. Lunell, A. Hussenius, and O. Matsson, Theoretical studies of proton transfer reactions in 1-methylindene, *Acta Chemica Scandanivica* **52** 541-548(1998).

[57] C. Aleman, H.M. Ishiki, E.A. Armelin, O. Abrahao, and S.E. Galembeck, Free energies of solvation for peptides and polypeptides using scrf methods, *Chemical Physics* **233** 85-96(1998).

[58] J.R. Alvarez-Idaboya and E.S. Kryachko, Nonrigidity of molecules in solvent and its impact on infrared spectrum: Substituted phenols and trimethylamine *N*-oxide. I., *Journal of Molecular Structure (Theochem)* **433** 263-278(1998).

[59] A. Bagno, E. Menna, E. Mezzina, G. Scorrano, and D. Spinelli, Site of protonation of alkyl- and arylhydrazines probed by ^{14}N, ^{15}N, and ^{13}C NMR relaxation and quantum chemical calculations, *Journal of Physical Chemistry A* **102** 2888-2892(1998).

[60] M. Bräuer, M. Mosquera, J.L. Pérez-Lustres, and F. Rodríguez-Prieto, Ground-state tautomerism and excited-state proton transfer processes in 4,5-dimethyl-2-(2′-hydroxyphenyl)imidazole in solution: Fluorescence spectroscopy and quantum mechanical calculations, *Journal of Physical Chemistry A* **102** 10736-10745(1998).

[61] F. Döring, J. Will, S. Amasheh, W. Clauss, H. Ahlbrecht, and H. Daniel, Minimal molecular determinants of substrates for recognition by the intestinal peptide transporter, *Journal of Biological Chemistry* **273** 23211-23218(1998).

[62] J. Köhler, M. Hohla, R. Sollner, and M. Amann, The difference between cholesterol- and glycyrrhizin-γ-cyclodextrin complexes - an analysis by MD simulations in vacuo and in aquo and the calculation of solvation free energies with AMSOL, *Supramolecular Science* **5** 117-137(1998).

[63] J. Köhler, M. Hohla, R. Sollner, and H.J. Eberle, Cyclohexadecanone derivative gamma-cyclodextrin complexes md simulations and amsol calculations in vacuo and in aquo compared with experimental findings, *Supramolecular Science* **5** 101-116(1998).

[64] E.S. Kryachko and G. Zundel, Quantum chemical study of 1-methyladenine and its spectra in gas phase and in solvent. I., *Journal of Molecular Structure (Theochem)* **446** 41-54(1998).

[65] S.F. Lieske, B. Yang, M.E. Eldefrawi, A.D. MacKerell, and J. Wright, (–)3β-substituted ecgonine methyl esters as inhibitors for cocaine binding and dopamine uptake, *Journal of Medicinal Chemistry* **41** 864-876(1998).

[66] D.J. Price, J.D. Roberts, and W.L. Jorgensen, Conformational complexity of succinic acid and its monoanion in the gas phase and in solution: Ab initio calculations and monte carlo simulations, *Journal of the American Chemical Society* **120** 9672-9679(1998).

[67] G. Reinwald and I. Zimmermann, A combined calorimetric and semiempirical quantum chemical approach to describe the solution thermodynamics of drugs, *Journal of Pharmaceutical Science* **87** 745-750(1998).

[68] B. Schiψtt, Y.-J. Zheng, and T.C. Bruice, Theoretical investigation of the hydride transfer from formate to NAD^+ and the implications for the catalytic mechanism of formate dehydrogenase, *Journal of the American Chemical Society* **120** 7192-7200(1998).

[69] G. Schóórmann, Quantum chemical analysis of the energy of proton transfer from phenol and chlorophenols to H_2O in the gas phase and in aqueous solution, *Journal of Chemical Physics* **109** 9523-9528(1998).

[70] M. Hoffmann, J. Rychlewski, and U. Rychlewska, Effects of substitution of OH group by F atom for conformational preferences of fluorine-substituted analogues of (*RR*)-tartaric acid, its dimethyl diester, diamide, and *N,N,N',N'*-tetramethyl diamide. Ab initio conformational analysis, *Journal of the American Chemical Society* **121** 1912-1921(1999).

[71] P.E. Just, K.I. Chane-Ching, J.C. Lacroix, and P.C. Lacaze, Anodic oxidation of dipyrrolyls linked with conjugated spacers: Study of electronic interactions between the polypyrrole chain and the spacers, *Journal of Electroanalytical Chemistry* **479** 3-11(1999).

[72] O. Kwon, M.L. McKee, and R.M. Metzger, Theoretical calculations of methylquinolinium tricyanoquinodimethanide (CH3Q-3CNQ) using a solvation model, *Chemical Physics Letters* **313** 321-331(1999).

[73] P.-T. Chou, G.-R. Wu, C.-Y. Wei, C.-C. Cheng, C.-P. Chang, and F.-T. Hung, Photoinduced double proton tautomerism in 4-azabenzimidazole, *Journal of Physical Chemistry B* **103** 10042-10052(1999).

[74] S.D. Rychnovsky, R. Vaidyanathan, T. Beauchamp, R. Lin, and P.J. Farmer, AM1-SM2 calculations model the redox potential of nitroxyl radicals such as TEMPO, *Journal of Organic Chemistry* **64** 6745 -6749(1999).

[75] A.G. Turjanski, R.E. Rosenstein, and D.A. Estrin, Solvation and conformational properties of melatonin: A computational study, *Journal of Molecular Modeling* **5** 271-280(1999).

[76] I.J. Chen and A.D. MacKerell, Computation of the influence of chemical substitution on the pK_a of pyridine using semiempirical and ab initio methods, *Theoretical Chemistry Accounts* **103** 483-494(2000).

[77] F. D'Souza, M.E. Zandler, G.R. Deviprasad, and W. Kutner, Acid-base properties of fulleropyrrolidines: Experimental and theoretical investigations, *Journal of Physical Chemistry A* **104** 6887-6893(2000).

[78] D.E. Elmore and D.A. Dougherty, A computational study of nicotine conformations in the gas phase and in water, *Journal of Organic Chemistry* **65** 742-747(2000).

[79] F. Iribarne, M. Paulino, and O. Tapia, Hydride-transfer transition structure as a possible unifying redox step for describing the branched mechanism of glutathione reductase. Molecular-electronic antecedents, *Theoretical Chemistry Accounts* **103** 451-462(2000).

[80] A. Rastelli, R. Gandolfi, and M.S. Amade, Regioselectivity and diastereoselectivity in the 1,3-dipolar cycloadditions of nitrones with acrylonitrile and maleonitrile. The origin of endo/exo selectivity, *Advances in Quantum Chemistry* **36** 151-167(2000).

[81] M.M. Wang, B. Cornett, J. Nettles, D.C. Liotta, and J.P. Snyder, The oxetane ring in taxol, *Journal of Organic Chemistry* **65** 1059-1068(2000).

[82] C.S. Callam, S.J. Singer, T.L. Lowary, and C.M. Hadad, Computational analysis of the potential energy surfaces of glycerol in the gas and aqueous phases: Effects of level of theory, basis set, and solvation on strongly intramolecularly hydrogen-bonded systems, *Journal of the American Chemical Society* **123** 11743-11754(2001).

[83] E. Lodyga-Chruscinska, G. Micera, D. Sanna, J. Olczak, and J. Zabrocki, A new class of peptide chelating agents towards copper(II) ions, *Polyhedron* **20** 1915-1923(2001).

[84] H. Matter, K.H. Baringhaus, T. Naumann, T. Klabunde, and B. Pirard, Computational approaches towards the rational design of drug-like compound libraries, *Combinatorial Chemistry & High Throughput Screening* **4** 453-475(2001).

[85] G. Tresadern, J. Willis, I.H. Hillier, and C.I.F. Watt, Hydride shift in substituted phenyl glyoxals: Interpretation of experimental rate data using electronic structure and variational transition state theory calculations, *Physical Chemistry Chemical Physics* **3** 3967-3972(2001).

[86] P. Benedetti, R. Mannhold, G. Cruciani, and M. Pastor, GBRcompounds and mepyramines as cocaine abuse therapeutics: Chemometric studies on selectivity using grid independent descriptors (grind), *Journal of Medicinal Chemistry* **45** 1577-1584(2002).

[87] E.J. Delgado and J. Alderete, On the calculation of Henry's law constants of chlorinated benzenes in water from semiempirical quantum chemical methods, *Journal of Chemical Information and Computer Science* **42** 559-563(2002).

[88] M. Filizola and D.L. Harris, Molecular determinants of recognition and activation at GABA(a)/benzodiazepine receptors, *International Journal of Quantum Chemistry* **88** 56-64(2002).

[89] C. Fontanesi, R. Benassi, R. Giovanardi, M. Marcaccio, F. Paolucci, and S. Roffia, Computational electrochemistry. Ab initio calculation of solvent effect in the multiple electroreduction of polypyridinic compounds, *Journal of Molecular Structure* **612** 277-286(2002).

[90] B.J. McConkey, V. Sobolev, and M. Edelman, The performance of current methods in ligand-protein docking, *Current Science* **83** 845-856(2002).

[91] E.J. Delgado and J.B. Alderete, Prediction of Henry's law constants of triazine derived herbicides from quantum chemical continuum solvation models, *Journal of Chemical Information and Computer Sciences* **43** 1226-1230(2003).

[92] D. Ivanov and M. Constantinescu, Computational study of maleamic acid cyclodehydration, *Journal of Physical Organic Chemistry* **16** 348-354(2003).

[93] J.W. Storer, D.J. Giesen, C.J. Cramer, and D.G. Truhlar, Class IV charge models: A new semiempirical approach in quantum chemistry, *Journal of Computer-Aided Molecular Design* **9** 87-110(1995).

[94] D.J. Giesen, J.W. Storer, C.J. Cramer, and D.G. Truhlar, A general semiempirical quantum mechanical solvation model for nonpolar solvation free energies. *n*-hexadecane, *Journal of the American Chemical Society* **117** 1057-1068(1995).

[95] D.J. Giesen, C.J. Cramer, and D.G. Truhlar, A semiempirical quantum mechanical solvation model for solvation free energies in all alkane solvents, *Journal of Physical Chemistry* **99** 7137-7146(1995).

[96] S.E. Barrows, F.J. Dulles, C.J. Cramer, D.G. Truhlar, and A.D. French, Factors controlling the relative stability of alternative chair forms and hydroxymethyl conformations of D-glucopyranose, *Carbohydrate Research* **276** 219-251(1995).

[97] A. Bondi, van der Waals volumes and radii, *Journal of Physical Chemistry* **68** 441-451(1964).

[98] G.D. Hawkins, C.J. Cramer, and D.G. Truhlar, Parameterized models of aqueous free energies of solvation based on pairwise descreening of solute atomic charges from a dielectric medium, *Journal of Physical Chemistry* **100** 19824-19839(1996).

[99] J. Srinivasan, M.W. Trevathan, P. Beroza, and D.A. Case, Application of a pairwise generalized Born model to proteins and nucleic acids: Inclusion of salt effects, *Theoretical Chemistry Accounts* **101** 426-434(1999).

[100] C.P. Sosa, T. Hewitt, M.R. Lee, and D.A. Case, Vectorization of the generalized Born model for molecular dynamics on shared-memory computers, *Journal of Molecular Structure (Theochem)* **549** 193-201(2001).

[101] C.C. Chambers, G.D. Hawkins, C.J. Cramer, and D.G. Truhlar, Model for aqueous solvation based on class IV atomic charges and first solvation shell effects, *Journal of Physical Chemistry* **100** 16385-16398(1996).

[102] D.J. Giesen, M.Z. Gu, C.J. Cramer, and D.G. Truhlar, A universal organic solvation model, *Journal of Organic Chemistry* **61** 8720-8721 (1996).

[103] D.J. Giesen, G.D. Hawkins, D.A. Liotard, C.J. Cramer, and D.G. Truhlar, A universal model for the quantum mechanical calculation of free energies of solvation in non-aqueous solvents, *Theoretical Chemistry Accounts* **98** 85-109(1997).

[104] M.H. Abraham, Scales of solute hydrogen-bonding: Their construction and application to physicochemical and biochemical processes, *Chemical Society Reviews* 73-83(1993).

[105] D.J. Giesen, C.C. Chambers, C.J. Cramer, and D.G. Truhlar, Solvation model for chloroform based on class IV atomic charges, *Journal of Physical Chemistry B* **101** 2061-2069(1997).

[106] M.J.S. Dewar and W. Thiel, Ground states of molecules. 38. The MNDO method. Approximations and parameters, *Journal of the American Chemical Society* **99** 4899-4907(1977).

[107] W. Thiel and A.A. Voityuk, Extension of MNDO to d orbitals: Parameters and results for the halogens, *International Journal of Quantum Chemistry* **44** 807-829(1992).

[108] W. Thiel and A.A. Voityuk, Extension of MNDO to d orbitals: Integral approximations and preliminary numerical results, *Theoretica Chimica Acta* **81** 391-404(1992).

[109] W. Thiel and A.A. Voityuk, Extension of MNDO to d orbitals: Parameters and results for the second-row elements and for the zinc group, *Journal of Physical Chemistry* **100** 616-626(1996).

[110] G.D. Hawkins, C.J. Cramer, and D.G. Truhlar, Parameterized model for aqueous free energies of solvation using geometry-dependent atomic surface tensions with implicit electrostatics, *Journal of Physical Chemistry B* **101** 7147-7157(1997).

[111] G.D. Hawkins, D.A. Liotard, C.J. Cramer, and D.G. Truhlar, OMNISOL: Fast prediction of free energies of solvation and partition coefficients, *Journal of Organic Chemistry* **63** 4305-4313(1998).

[112] G.D. Hawkins, B.J. Lynch, C.P. Kelly, D.A. Liotard, C.J. Cramer, and D.G. Truhlar, *OMNISOL–version 2.0*, University of Minnesota, Minneapolis, 2005.

[113] D.M. Dolney, G.D. Hawkins, P. Winget, D.A. Liotard, C.J. Cramer, and D.G. Truhlar, A universal solvation model based on the conductor-like screening model, *Journal of Computational Chemistry* **21** 340-366(2000).

[114] A. Klamt and G. Schóórmann, COSMO: A new approach to dielectric screening in solvents with explicit expressions for the screening energy and its gradient, *Journal of the Chemical Society, Perkin Transactions II* 799-805(1993).

[115] J. Li, T. Zhu, C.J. Cramer, and D.G. Truhlar, A new class IV charge model for extracting accurate partial charges from wave functions, *Journal of Physical Chemistry A* **102** 1820-1831(1998).

[116] J. Li, J. Xing, C.J. Cramer, and D.G. Truhlar, Accurate dipole moments from Hartree–Fock calculations by means of class IV charges, *Journal of Chemical Physics* **111** 885-892(1999).

[117] P.-O. Lφwdin, On the non-orthogonality problem connected with the use of atomic wave functions in the theory of molecules and crystals, *Journal of Chemical Physics* **18** 365-375(1950).

[118] I. Mayer, Charge, bond order and valence in the ab initio SCF theory, *Chemical Physics Letters* **97** 270-277(1983).

[119] I. Mayer, Comments on the quantum theory of valence and bonding: Choosing between alternative definitions, *Chemical Physics Letters* **110** 440(1984).

[120] J. Li, G.D. Hawkins, C.J. Cramer, and D.G. Truhlar, Universal reaction field model based on ab initio Hartree-Fock theory, *Chemical Physics Letters* **288** 293-298(1998).

[121] R.E. Easton, D.J. Giesen, A. Welch, C.J. Cramer, and D.G. Truhlar, The MIDI! Basis set for quantum mechanical calculations of molecular geometries and partial charges, *Theoretica Chimica Acta* **93** 281-301(1996).

[122] J. Li, C.J. Cramer, and D.G. Truhlar, MIDI! Basis set for silicon, bromine, and iodine, *Theoretical Chemistry Accounts* **99** 192-196(1998).

[123] T. Zhu, J. Li, G.D. Hawkins, C.J. Cramer, and D.G. Truhlar, Density-functional solvation model based on CM2 atomic charges, *Journal of Chemical Physics* **109** 9117-9133(1998).

[124] J. Li, T. Zhu, G.D. Hawkins, P. Winget, D.A. Liotard, C.J. Cramer, and D.G. Truhlar, Extension of the platform of applicability of the SM5.42R universal solvation model, *Theoretical Chemistry Accounts* **103** 9-63(1999).

[125] P. Winget, J.D. Thompson, C.J. Cramer, and D.G. Truhlar, Parameterization of universal solvation model for molecules containing silicon, *Journal of Physical Chemistry A* **106** 5160-5168(2002).

[126] J. Li, B. Williams, C.J. Cramer, and D.G. Truhlar, A class IV charge model for molecular excited states, *Journal of Chemical Physics* **110** 724-733(1999).

[127] M.C. Zerner, Semiempirical molecular orbital methods, *Reviews in Computational Chemistry*, K.B. Lipkowitz and D.B. Boyd (Eds.), VCH, New York, 1991, pp. 313-365.

[128] J.E. Ridley and M.C. Zerner, An intermediate neglect of differential overlap technique for spectroscopy: Pyrrole and the azines, *Theoretica Chimica Acta* **32** 111(1973).

[129] J. Li, T. Zhu, C.J. Cramer, and D.G. Truhlar, A universal solvation model based on class IV charges and the intermediate neglect of differential overlap for spectroscopy molecular orbital method, *Journal of Physical Chemistry B* **104** 2178-2182(2000).

[130] J. Li, C.J. Cramer, and D.G. Truhlar, A two-response-time model based on CM2/INDO/S2 electrostatic potentials for the dielectric polarization component of solvatochromic shifts on vertical excitation energies, *International Journal of Quantum Chemistry* **77** 264-280(2000).

[131] T. Zhu, J. Li, D.A. Liotard, C.J. Cramer, and D.G. Truhlar, Analytical gradients of a self-consistent reaction-field solvation model based on CM2 atomic charges, *Journal of Chemical Physics* **110** 5503-5513(1999).

[132] P. Winget, J.D. Thompson, J.D. Xidos, C.J. Cramer, and D.G. Truhlar, Charge Model 3: A class IV charge model based on hybrid density functional theory with variable exchange., *Journal of Physical Chemistry A* **106** 10707-10717(2002).

[133] J.D. Thompson, C.J. Cramer, and D.G. Truhlar, Parameterization of Charge Model 3 for AM1, PM3, BLYP, and B3LYP, *Journal of Computational Chemistry* **24** 1291-1304(2003).

[134] J.M. Brom, B.J. Schmitz, J.D. Thompson, C.J. Cramer, and D.G. Truhlar, A class IV charge model for boron based on hybrid density functional theory, *Journal of Physical Chemistry A* **107** 6483-6488(2003).

[135] J.A. Kalinowski, B. Lesyng, J.D. Thompson, C.J. Cramer, and D.G. Truhlar, Class iv charge model for the self-consistent charge density-functional tight-binding method, *Journal of Physical Chemistry A* **108** 2545-2549(2004).

[136] J.D. Thompson, J.D. Xidos, T.M. Sonbuchner, C.J. Cramer, and D.G. Truhlar, More reliable partial atomic charges when using diffuse basis sets, *PhysChemComm* **5** 117-134(2002).

[137] J.D. Thompson, C.J. Cramer, and D.G. Truhlar, New universal solvation model and comparison of the accuracy of the SM5.42R, SM5.43R, C-PCM, D-PCM, and IEF-PCM continuum solvation models for aqueous and organic solvation free energies and for vapor pressures, *Journal of Physical Chemistry A* **108** 6532-6542(2004).

[138] J.D. Thompson, C.J. Cramer, and D.G. Truhlar, Density-functional theory and hybrid density-functional theory continuum solvation models for aqueous and organic solvents: Universal SM5.43 and SM5.43R solvation models for any fraction of Hartree-Fock exchange, *Theoretical Chemistry Accounts* **113** 107-131(2005).

[139] A.D. Becke, Density functional exchange energy approximation with correct asymptotic behavior, *Physical Review A* **38** 3098-3100(1988).

[140] C. Lee, W. Yang, and R.G. Parr, Development of the Colle-Salvetti correlation-energy formula into a functional of the electron density, *Physical Review B* **37** 785-789(1988).

[141] A.D. Becke, Density-functional thermochemistry. III. The role of exact exchange, *Journal of Chemical Physics* **98** 5648-5652(1993).

[142] P.J. Stephens, F.J. Devlin, C.F. Chabalowski, and M.J. Frisch, Ab initio calculation of vibrational absorption and circular dichroism spectra using density functional force fields, *Journal of Physical Chemistry* **98** 11623-11627(1994).

[143] W.J. Hehre, L. Radom, P.v.R. Schleyer, and J.A. Pople, *Ab Initio Molecular Orbital Theory*, Wiley, New York, 1986.

[144] J. Perdew and Y. Wang, Accurate and simple analytic representation of the electron-gas correlation energy, *Physical Review B* **45** 13244-13249(1992).

[145] J.P. Perdew, Unified theory of exchange and correlation beyond the local density approximation, *Electronic structure of solids '91*, P. Ziesche and H. Eschrig (Eds.), Akademie Verlag, Berlin, 1991, pp. 11-20.

[146] C. Adamo and V. Barone, Exchange functionals with improved long-range behavior and adiabatic connection methods without adjustable parameters: The *m*PW and *m*PW1PW models, *Journal of Chemical Physics* **108** 664-675(1998).

[147] J. Pu, C.P. Kelly, J.D. Thompson, J.D. Xidos, J. Li, T. Zhu, G.D. Hawkins, Y.-Y. Chuang, P.L. Fast, B.J. Lynch, D.A. Liotard, D. Rinaldi, J. Gao, C.J. Cramer, and D.G. Truhlar, *GAMESSPLUS version 4.7*, University of Minnesota, Minneapolis, 2005.

[148] H. Nakamura, A.C. Chamberlin, C.P. Kelly, J.D. Xidos, J.D. Thompson, J. Li, G.D. Hawkins, T. Zhu, B.J. Lynch, Y. Volobuev, D. Rinaldi, D.A. Liotard, C.J. Cramer, and D.G. Truhlar, *HONDOPLUS-v.5.0*, University of Minnesota, Minneapolis, MN, 2006.

[149] A.C. Chamberlin, C.P. Kelly, J.D. Thompson, B.J. Lynch, J.D. Xidos, G.D. Hawkins, J. Li, T. Zhu, Y. Volobuev, D. Rinaldi, D.A. Liotard, C.J. Cramer, and D.G. Truhlar, *SMXGAUSS–version 3.4*, University of Minnesota, Minneapolis, 2006.

[150] S. Miertus, E. Scrocco, and J. Tomasi, Electrostatic interaction of a solute with a continuum. A direct utilization of ab initio molecular potentials for the prevision of solvent effects, *Chemical Physics* **55** 117-129(1981).
[151] M. Cossi, V. Barone, R. Cammi, and J. Tomasi, Ab initio study of solvated molecules: A new implementation of the polarizable continuum model, *Chemical Physics Letters* **255** 327-335(1996).
[152] E. Canc θs, B. Mennucci, and J. Tomasi, A new integral equation formalism for the polarizable continuum model: Theoretical background and applications to isotropic and anisotropic dielectrics, *Journal of Chemical Physics* **107** 3032-3041(1997).
[153] M. Cossi, V. Barone, B. Mennucci, and J. Tomasi, Ab initio study of ionic solutions by a polarizable continuum dielectric model, *Chemical Physics Letters* **286** 253-260(1998).
[154] V. Barone and M. Cossi, Quantum calculation of molecular energies and energy gradients in solution by a conductor solvent model, *Journal of Physical Chemistry A* **102** 1995-2001(1998).
[155] M. Cossi, N. Rega, G. Scalmani, and V. Barone, Energies, structures, and electronic properties of molecules in solution with the C-PCM solvation model, *Journal of Computational Chemistry* **24** 669-681(2003).
[156] C.P. Kelly, C.J. Cramer, and D.G. Truhlar, SM6: A density functional theory continuum solvation model for calculating aqueous solvation free energies of neutrals, ions, and solute-water clusters, *Journal of Chemical Theory and Computation* **1** 1133-1152(2005).
[157] W.L. Jorgensen, D.S. Maxwell, and J. Tirado-Rives, Development and testing of the OPLS all-atom force field on conformational energetics and properties of organic liquids, *Journal of the American Chemical Society* **118** 11225-11236(1996).
[158] A.C. Chamberlin, C.J. Cramer, and D.G. Truhlar, Predicting aqueous free energies of solvation as functions of temperature, *Journal of Physical Chemistry B* **110** 5665-5675(2006).
[159] P. Winget, G.D. Hawkins, C.J. Cramer, and D.G. Truhlar, Prediction of vapor pressures from self-solvation free energies calculated by the SM5 series of universal solvation models, *Journal of Physical Chemistry B* **104** 4726-4734(2000).
[160] J.D. Thompson, C.J. Cramer, and D.G. Truhlar, Predicting aqueous solubilities from aqueous free energies of solvation and experimental or calculated vapor pressures of pure substances, *Journal of Chemical Physics* **119** 1661-1670(2003).
[161] P. Winget, C.J. Cramer, and D.G. Truhlar, Prediction of soil sorption coefficients using a universal solvation model, *Environmental Science and Technology* **34** 4733-4740(2000).
[162] W.J. Lyman, Adsorption coefficient for soils and sediments, *Handbook of Chemical Property Estimation Methods*, W.J. Lyman, W.F. Reehl, and D.H. Rosenblatt (Eds.), American Chemical Society, Washington, DC, 1990, pp. Chapter 4.
[163] C.J. Cramer and D.G. Truhlar, Polarization of the nucleic acid bases in aqueous solution, *Chemical Physics Letters* **198** 74-80(1992).
[164] D.J. Giesen, C.C. Chambers, C.J. Cramer, and D.G. Truhlar, What controls partitioning of the nucleic acid bases between chloroform and water? *Journal of Physical Chemistry B* **101** 5084-5088(1997).
[165] D.A. McQuarrie, *Statistical Mechanics*, Harper & Row, New York, 1976.
[166] C.P. Kelly, C.J. Cramer, and D.G. Truhlar, Predicting adsorption coefficients at air-water interfaces using universal solvation and surface area models, *Journal of Physical Chemistry B* **108** 12882-12897(2004).
[167] Q. Du, R. Superfine, E. Freysz, and Y.R. Shen, Vibrational spectroscopy of water at the vapor/water interface, *Physical Review Letters* **70** 2313(1993).
[168] Q. Du, E. Freysz, and Y.R. Shen, Surface vibrational spectroscopic studies of hydrogen-bonding and hydrophobicity, *Science (Washington DC)* **264** 826-828(1994).
[169] D.E. Gragson and G.L. Richmond, Investigations of the structure and hydrogen bonding of water molecules at liquid surfaces by vibrational sum frequency spectroscopy, *Journal of Physical Chemistry B* **102** 3847-3861(1998).
[170] I.-F.W. Kuo and C.J. Mundy, An ab initio molecular dynamics study of the aqueous liquid-vapor interface, *Science (Washington DC)* **303** 658-660(2004).
[171] M.H. Abraham, J. Andonian-Haftvan, G.S. Whiting, A. Leo, and R.S. Taft, Hydrogen bonding. Part 34. The factors that influence the solubility of gases and vapours in water at 298 K, and a new method for its determination, *Journal of the Chemical Society, Perkin Transactions II* 1777-1791(1994).

[172] M.D. Tissandier, K.A. Cowen, W.Y. Feng, E. Gundlach, M.H. Cohen, A.D. Earhart, J.V. Coe, and T.R. Tuttle, The proton's absolute aqueous enthalpy and Gibbs free energy of solvation from cluster-ion solvation data, *Journal of Physical Chemistry A* **102** 7787-7794(1998).

[173] A. Lewis, J.A. Bumpus, D.G. Truhlar, and C.J. Cramer, Molecular modeling of environmentally important processes: Reduction potentials, *Journal of Chemical Education* **81** 596-604(2004).

[174] C.P. Kelly, C.J. Cramer, and D.G. Truhlar, Aqueous solvation free energies of ions and ion-water clusters based on an accurate value for the absolute aqueous solvation free energy of the proton, *Journal of Physical Chemistry B* **110** 16066-16081 (2006).

[175] E.V. Patterson, C.J. Cramer, and D.G. Truhlar, Reductive dechlorination of hexachloroethane in the environment. Mechanistic studies via computational electrochemistry, *Journal of the American Chemical Society* **123** 2025-2031(2001).

[176] C.J. Cramer and S.M. Gustafson, Hyperconjugation vs. apicophilicity in trigonal bipyramidal phosphorus species, *Journal of the American Chemical Society* **115** 9315-9316(1993).

[177] C.P. Kelly, C.J. Cramer, and D.G. Truhlar, Adding explicit solvent molecules to continuum solvent calculations for the calculation of aqueous acid dissociation constants, *Journal of Physical Chemistry A* **110** 2493-2499(2006).

[178] J. Seckute, J.L. Menke, R.J. Emnett, E.V. Patterson, and C.J. Cramer, Ab initio molecular orbital and density functional studies on the solvolysis of sarin and *O,S*-dimethylmethylphosphonothioate, a VX-like compound, *Journal of Organic Chemistry* **70** 8649-8660(2005).

Brill Academic Publishers
P.O. Box 9000, 2300 PA Leiden
The Netherlands

Lecture Series on Computer and Computational Sciences
Volume 6 , 2006, pp 140-154

Probing actinide electronic structure using fluorescence and multi-photon ionization spectroscopy

Michael C. Heaven[1]

Department of Chemistry
Emory University
Atlanta, GA 30322, USA

Received 13 July, 2006; accepted 20 July, 2006

Abstract: High-level theoretical models of the electronic structures and properties of actinide compounds are being developed by several research groups. These models must be tested and evaluated through comparisons with experimental results. Gas phase data are most suitable for this purpose, but there have been very few gas phase studies of actinide compounds. We are addressing this issue by carrying out spectroscopic studies of simple uranium and thorium oxides and halides. Multiple resonance and jet cooling techniques are being used to unravel the complex electronic spectra of these compounds. Zero kinetic energy – pulsed field ionization measurements are being used to examine the cations. Recent results for the oxides will be discussed. Systematic errors in the accepted values for the ionization energies have been discovered, and the patterns of electronic states for these molecules provide information concerning the occupation of the $5f$ orbitals and their participation in bond formation.

Keywords: Actinide electronic structure, Relativistic effects, Spectroscopy, Ionization energies

Mathematics Subject Classification: 81V

PACS: 31.10.+z, 33.15.Ry, 33.15.Mt, 33.20.t, 33.80.Rv

1. Introduction

Computational electronic structure studies of the actinides are challenging because of the need to include relativistic effects (potentially requiring expensive four-component spin-orbit CI treatments), the fact that numerous low-lying electronic states must be considered, the common occurrence of open-shell electrons, and the observation that many of the nominally closed-shell electrons must be modeled to obtain good agreement with experiment. Large CI expansions are required that push the limits of computational resources. New computational strategies, basis sets, exchange-correlation functionals for density functional theory (DFT), and core potentials are actively being developed [1-15]. As the accuracy of theoretical calculations improves, it is increasingly important to have reliable experimental data against which the methods may be tested and refined.

For many years, electronic structure models for the actinides have been derived from and evaluated against spectroscopic data. The greater majority of the experimental data have been obtained under solution or condensed phase conditions. In modeling these data the effects of surrounding counter ions and/or solvent molecules are often neglected or approximated using perturbative models. A great deal of progress has been made using these approaches, but the new generation of high-level calculations will be compromised if they do not account for counter-ion and solvent interactions by more rigorous means. It will be important to establish reliable methods for treating such interactions. However, at this point, where the focus is on development of relativistic electronic structure methods and basis sets, the presence of these perturbations in the test data is an unwanted complication. Gas phase spectroscopic data for small actinide-containing molecules are free from such effects, so they are ideal for testing theoretical calculations at the present stage of development. To date, there have been few gas phase studies [16-31].

[1] Corresponding author. E-mail: mheaven@emory.edu

A number of technical difficulties are responsible for the lack of gas phase data for actinide compounds. When partially filled, the *f* orbitals give rise to dense manifolds of electronic states. As these orbitals are nominally non-bonding, the states belonging to a given electronic configuration have very similar vibrational and rotational constants. This situation results in congested electronic spectra that contain numerous overlapping band systems. In traditional spectroscopic experiments the congestion problems are exacerbated by the means used to obtain gas phase samples. Simple actinide compounds such as oxides and halides are refractory solids with negligible vapor pressures at room temperature. Knudsen cells, furnaces, laser ablation and electron beams have been used to vaporize these materials. These high temperature environments populate many of the low-lying electronic states, leading to additional spectral congestion. The problem can be so severe that it can prevent unambiguous identification of the ground state. Furthermore, the vapors produced by heating usually contain several different species in equilibrium, and the probability that spectra for these closely related species will occur in the same spectral regions is high.

Thermal population of electronically excited states and the presence of multiple species in vapor phase samples also complicates the determination of accurate ionization energies (IE) for actinide compounds (see, for example, ref. [32]). The IE's are of interest as they provide benchmark data and may be used to determine bond dissociation energies when combined with other spectroscopic and thermodynamic data. Most IE's for actinide compounds have been determined by electron impact ionization with mass spectrometric ion detection [32-35]. Although the detection method is species-specific, this technique does not fully resolve the problems associated with samples that contain multiple species as electron impact often causes fragmentation. Thermal population of excited levels degrades the accuracy of electron impact IE measurements as it blurs observation of the appearance potential. Using charge exchange reactions in an ion cyclotron mass spectrometer, Gibson and co-workers have shown that the IE's for several actinide oxides were substantially underestimated by electron impact measurements [36, 37]. The largest and most surprising discrepancy was found for PuO_2 where the electron impact measurement *overestimated* the IE by 3 eV [38-40]

In the present work we describe new gas phase spectroscopic studies of thorium and uranium oxides. Low-temperature vapor phase samples of the oxides were obtained using laser ablation in combination with jet expansion cooling. High resolution electronic and photoelectron spectra were obtained using multi-photon ionization techniques. The relevant experimental techniques are briefly outlined in the following section. This is followed by a description of the experimental results and comparisons with the predictions of recent high-level calculations. The potential value of a simple ligand field approach as a means to understand the patterns of low-lying electronic states for ionic actinide compounds is also explored.

2. Experimental techniques

Cooling of molecules in jet-expansions is widely used to simplify complex spectra. Refractory species may be entrained in jets by pulsed-laser ablation of solid samples. Cooling to temperatures below 150 K is easily achieved. In addition to reducing the spectral congestion, cooling also facilitates identification of the electronic ground state. As the number densities in expanded jets are relatively low, sensitive methods are needed to observe the spectra of entrained species. Multi-pass absorption, laser induced fluorescence (LIF) and resonance-enhanced multi-photon ionization (REMPI) techniques [41] have all been applied successfully. REMPI offers the unique advantage that it may be combined with time-of-flight mass spectrometry (TOFMS), and thereby provide a species-specific detection method.

The combination of laser ablation and jet-cooling can also improve the accuracy of IE measurements by greatly diminishing the populations in excited states. Beyond the advantages of simple cooling, there have been dramatic improvements in the techniques used to measure IE's. One of the most powerful methods is pulsed field ionization zero kinetic energy (PFI-ZEKE) photoelectron spectroscopy [42, 43]. The resolution of this technique is orders of magnitude better than can be achieved using electron impact or conventional photoelecton spectroscopy.

As REMPI, mass analyzed threshold ionization (MATI) and ZEKE techniques were used to obtain many of the results reported here, a brief description of the underlying principles of these measurements will be helpful. Pulsed lasers were used for excitation and ionization as ion detection was accomplished using a time-of-flight mass spectrometer. A schematic diagram of the apparatus used at Emory to obtain REMPI, PFI-ZEKE and MATI spectra for actinides is shown in Fig. 1. The energy level scheme used for REMPI is shown in Fig. 2a, for the case where the energy supplied by two photons is sufficient to cause ionization. In a typical REMPI experiment the first photon is tuned through resonant absorption features.

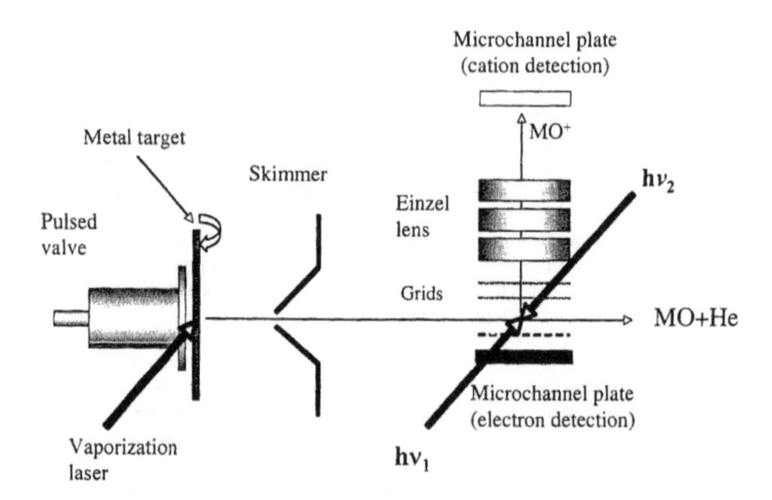

Figure 1: Schematic diagram of the apparatus used to record REMPI, PIE, MATI and PFI-ZEKE spectra for actinide oxides.

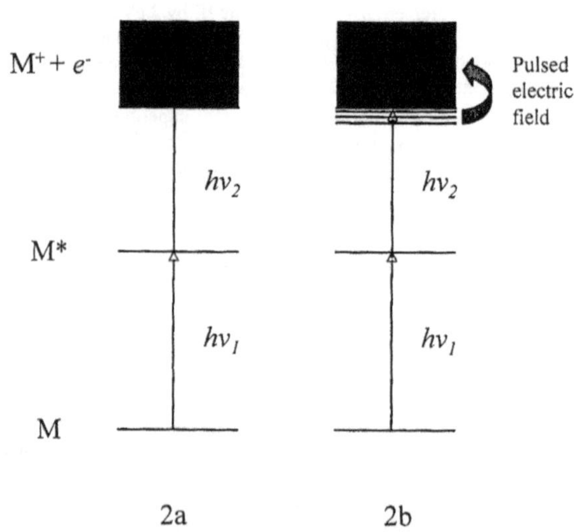

Figure 2: Energy level diagrams for two-photon excitation / ionization processes. 2a shows a typical two-photon REMPI scheme. Tuning the energy of $h\nu_1$ yields spectra for the neutral molecule excited states (M*). Tuning $h\nu_2$ can be used to locate ionization thresholds from photionization efficiency curves. 2b shows the excitation sequence used to record MATI and PFI-ZEKE spectra. $h\nu_1$ is fixed on a resonant transition while is $h\nu_2$ is used to excite long-lived high-n Rydberg levels. A pulsed electric field is used to ionize the Rydberg molecules a few microseconds after the second laser pulse.

When excitation occurs the second photon ionizes the molecule and the resulting ion current is detected. Photoionization thresholds can be measured by tuning the photon energy through the threshold, while monitoring the ion current (this type of scan is often referred to as a photo-ionization efficiency (PIE) curve). This can be done as a single photon experiment, but all significantly populated levels of the molecule will contribute to the observed ion signal. This has the effect of blurring the threshold and thereby reduces the accuracy of the measurement. Two-color ionization can be used to mitigate this problem. If the first laser can be used to move population to a single level of the intermediate excited state (or a small group of rotational levels) the threshold observed by tuning the second laser will be significantly sharper. This approach can yield vertical IE's with error limits around 10 cm^{-1}.

More accurate IE's, and spectroscopic data for the molecular cation, can be obtained from MATI and ZEKE measurements. These techniques rely on the fact that the high-*n* Rydberg levels lying just below the ionization limit are metastable [42, 43]. In the ZEKE experiment a pulsed laser is used to populate these levels while the molecules are in an environment that is free from electric fields, to the extent that this can be achieved in practice (Fig. 2b). A pulsed electric field is then used to ionize the excited molecules and accelerate the electrons towards a detector. Application of the pulsed field is delayed by a few microseconds to permit the dispersion of promptly formed electrons (e.g., electrons produced by auto-ionization processes). The MATI experiment is based on the same idea, but with mass selected detection of the molecular cations. Due to their greater mass, promptly formed cations will not diffuse out of the ion source rapidly enough under nominally field-free conditions. Consequently, a small reverse bias is supplied to the grid plates of the mass spectrometer during the excitation process. This is used to spatially separate the promptly formed ions from the Rydberg excited molecules. After a few microseconds delay the full draw-out field is applied to the grid plates to ionize and accelerate the cations. Promptly formed ions and those produced by pulsed field ionization can be detected separately as they have different flight times in the mass spectrometer. Typically, higher pulsed fields have to be applied for MATI as compared to PFI-ZEKE. As a consequence, the resolution of MATI is not quite as good.

3. Electronic spectra and theoretical calculations for ThO^+

The ThO^+ cation is of particular interest from a theoretical perspective due to the anticipated simplicity of its electronic structure. Formally, the lowest energy configuration is expected to be $Th^{3+}(7s)O^{2-}$, giving rise to a $X^2\Sigma^+$ ground state. The low-lying electronically excited states should be $^2\Delta$, $^2\Pi$ and $^2\Sigma$ derived from the $Th^{3+}(6d)O^{2-}$ configuration. We began our studies of ThO^+ by first making an accurate measurement of the IE for ThO. PIE curves and MATI spectra for ThO were recorded for this purpose. Typical data are shown in Fig. 3. The IE obtained, 6.6038(12) eV, was 0.5 eV higher than the value reported previously using electron impact ionization (6.1(1) eV) [35]. The reason for this large discrepancy was easily identified. In the electron impact study, vapor phase ThO was obtained by heating a sample of solid thorium dioxide to 2000 K. At this temperature many vibrational levels of ThO(X) would be populated, along with the low-lying states at 5317 and 6128 cm^{-1}. Hence the ThO^+

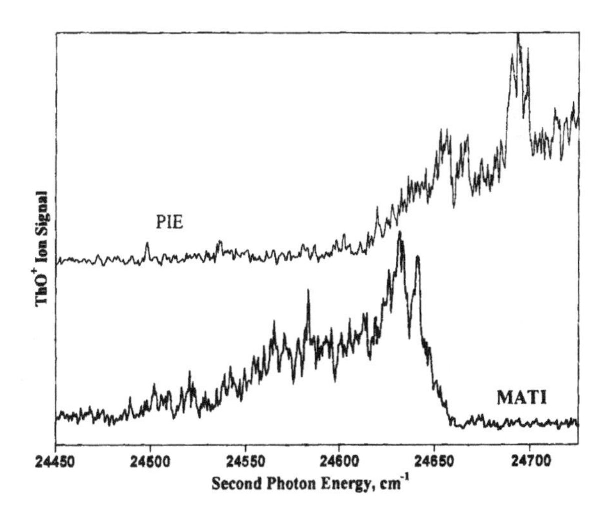

Figure 3: Photo-ionization efficiency (PIE) and mass-analyzed threshold ionization (MATI) spectra for ThO. These traces were recorded with the first laser tuned to the Q-branch feature of the $C'(1)$ - $X(0^+)$ 0-0 band at 28578.7(3) cm^{-1}. The structure in the high frequency region of the PIE curve is due to one-color three-photon resonances.

appearance potential most probably corresponded to the ionization of thermally excited ThO. This problem was avoided in the multi-photon ionization measurements as the first laser was tuned to a transition that was known to originate from the electronic ground state. Hence the true IE was obtained, regardless of the presence of excited molecules.

When combined with the IE for atomic Th, the IE for ThO yields the difference between the dissociation energies of ThO and ThO^+ (IE(ThO)-IE(Th)=D_0(ThO)-D_0(ThO^+)). With the corrected IE for ThO it was apparent that the neutral molecule is 0.3 eV more tightly bound than the cation. This was a surprising result as ionization was expected to involve removal of one of the non-bonding 7*s* electrons, thereby reducing the screening of the Th^{3+} ion core. Spectroscopic studies of ThO^+ provided crucial insights concerning this unexpected behavior. Rotationally resolved spectra were obtained using the PFI-ZEKE technique [22]. As an example, Fig. 4 shows the rotational structure of the $X\,^2\Sigma^+$, v=0 level. These data were obtained using two-color excitation, with the first laser set to excite a single rotational level of the intermediate electronically excited state. The "lines" in the PFI-ZEKE spectra actually correspond to unresolved groups of high-*n* Rydberg levels that converge on a specific rotational level of the ion. The vertical broken lines in Fig. 4 show the zero-field rotational energies that these series are converging to. Note that the range of rotational states observed in each trace increased with increasing angular momentum of the intermediate electronically excited state. This effect, caused by coupling between rotational channels, was observed in all of the PFI-ZEKE spectra for ThO^+. Spectra such as those shown in Fig. 4 supply four valuable pieces of information. These are the term energy for the state, the rotational constant, the Ω value (where Ω is the unsigned projection of the electronic angular momentum on the diatomic axis), and an indication of the angular momentum coupling case [44]. The latter is obtained by fitting energy level expressions to the rotational term series. Levels characterized by integer quantum numbers correspond to Hund's case *b* coupling while half-integer quantum numbers indicate coupling cases *a* or *c* (intermediate coupling cases may also be recognized by identifying the appropriate energy level expression) [44]. The data shown in Fig. 4 confirmed that the ground state of ThO^+ is $X\,^2\Sigma^+$. Fig. 5 shows PFI-ZEKE spectra for the second electronically excited state of ThO^+. An analysis of these data shows that this state has half-integer rotational quantum numbers and Ω=5/2, as expected for $Th^{3+}(6d)O^{2-}$, $1^2\Delta(\Omega=5/2)$. The rotational constant was smaller than that of the ground state (0.3410 vs. 0.3450 cm^{-1}), consistent with promotion of the unpaired electron to a less polarizable orbital. Vibrationally excited levels were readily observed. For example, ground state levels ranging from v=0 to 7 have been characterized. In total sixteen vibronic states of ThO^+ have been analyzed. The states located above $1^2\Delta(\Omega=3/2)$ (2934 cm^{-1}) were identified as $1^2\Delta(\Omega=5/2)$ (5814 cm^{-1}) and $1^2\Pi(\Omega=1/2)$ (7404 cm^{-1}).

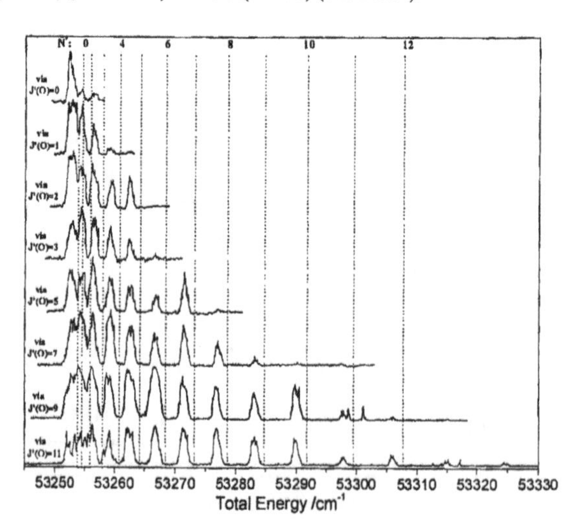

Figure 4: PFI-ZEKE spectra for ThO recorded via specific rotational levels (J' = 0, 1, 2, 3, 5, 7, 9, 11) of the intermediate state *O*. A pulsed electric field with an amplitude of 0.286 V/cm was applied to record these data. The spectra are plotted against total energy of the transition from the ThO(X) v=0, J=0 state.

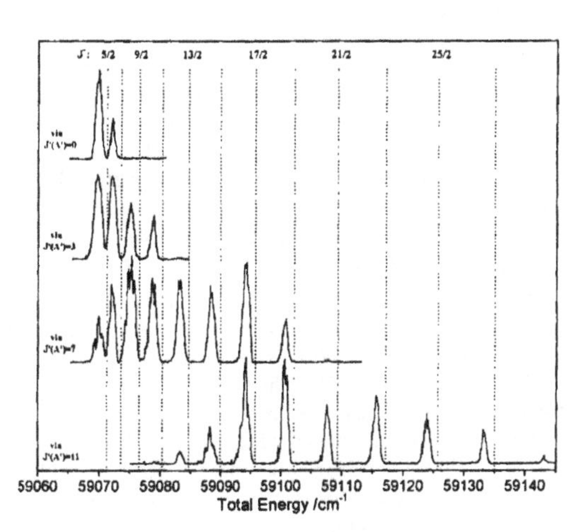

Figure 5: PFI-ZEKE spectra for ThO recorded via specific rotational levels (J' = 0, 3, 7, 11) of the intermediate state A' using a pulsed field with an amplitude of 0.286 V/cm. These data shows the rotational manifold of the state $1^2\Delta_{5/2}$ of ThO^+.

The IE measurements had shown that ThO is more strongly bound than ThO^+. In contrast, the molecular constants seemed to show the opposite trend. The bond length for $ThO^+(X)$ was shorter (R_e=1.807 vs. 1.840 Å) and the vibration frequency was higher (ω_e=955 vs. 896 cm^{-1}). Goncharov et al. [22] explored this apparent paradox using DFT calculations with relativistic core potentials. These calculations were successful in reproducing the molecular constants for ThO and ThO^+. Spin-density calculations for ThO^+ were also consistent with the formal $Th^{3+}O^{2-}$ charge separation. Scans of the potential energy curves provided insights concerning the lower dissociation energy for ThO^+. While the structure near the equilibrium distance corresponds to tightly bound $Th^{3+}O^{2-}$, the ground state must correlate with the $Th^+(7s6d^2, {}^4F)+O({}^3P)$ dissociation asymptote. Therefore, adiabatic dissociation of the molecule must involve avoided crossings with states that correlate with the $Th^{2+}+O^-$ and Th^++O dissociation asymptotes. Due to these avoided crossings the dissociation energy of ThO^+, relative to that of ThO, is influenced by the magnitude of the IE for Th^{2+}.

Relativistic ab initio calculations for ThO^+ have been reported by Tyagi [14]. In this work the MCSCF method was used with a 68 electron core potential for Th. Both spin free and spin-orbit coupled results were obtained. These calculations supported the simple picture of the electronic structure presented above. The ground state wavefunction was found to have 90% $7s$ character. The theoretical energies for the two components of the $^2\Delta$ state were in good agreement with the measured values (errors of 332 and 38 cm^{-1}). The energy for $^2\Pi_{1/2}$ was greater than the measured value by 1763 cm^{-1}, but the correlation between the observed and calculated energy level patterns was unambiguous. The wavefunctions for the $^2\Delta$ and $^2\Pi$ states all possessed greater than 85% $6d$ character.

Tyagi [14] also calculated the IE for ThO. The results were found to be very sensitive to the method used to treat dynamical correlation. The highest-level method employed, multi-reference configuration interaction with single and double excitations (MRCISD), yielded a vertical IE of 6.45 eV, just 0.15 eV below the experimental value.

Clearly, further experimental and theoretical studies of ThO^+ would be of value. The low-lying states that are within 12000 cm^{-1} of the ground state have been surveyed. Beyond this energy range the states arising from $Th^{3+}(5f)O^{2-}$ are expected, and it will be of interest to locate these states and determine their characteristic molecular constants. Similarly the theoretical calculations should be extended to higher energies and the benefits of using higher-level methods to treat dynamical correlation should be explored.

4. Electronic spectra and theoretical calculations for UO and UO^+

Studies of the electronic transitions of UO have been carried out using absorption and emission spectroscopy [30, 31], REMPI of jet-cooled samples [28], and the application of WSFES techniques to UO that was vaporized in a high temperature furnace (T=2500 K) [26, 27]. Thirty-three electronic transitions have been examined at high-resolution. Energy linkages between all of the upper and lower states sampled by these transitions were established. The ground state of UO was found to be an Ω=4 component of the $U^{2+}(5f^37s, {}^5I_4)O^{2-}$ configuration. Seven other states belonging to this configuration were identified. The lowest energy excited state, also Ω=4, was located just 294.1 cm^{-1} above the ground state. This state did not fit as a member of the $5f^37s$ group, and its molecular constants were consistent with the configuration $U^{2+}(5f^27s^2, {}^3H_4)O^{2-}$. The interactions between the three lowest energy Ω=4 states were large enough to cause a significant perturbation of the first vibrational interval for the ground state [27, 29]. A deperturbation analysis was carried out for these states (separated by 294 and 1280 cm^{-1}) [27]. The deperturbed ground state vibrational interval of $\Delta G_{1/2}$=841.9 was substantially smaller than the observed interval of 882.4 cm^{-1}. This provides an example of one of the complications encountered when comparing spectroscopic data with theoretical predictions. As calculated vibrational constants are usually derived from the second derivative of the potential energy curve, they do not include the effects of vibronic perturbations. Hence the calculated vibrational constant for UO reported by Krauss and Stevens [45] (ω_e=845 cm^{-1}) did not seem consistent with the measured gas phase vibrational interval, but it was in good agreement with the deperturbed vibrational constant (ω_e=846.5 cm^{-1}). UO also illustrates the problems that can be encountered with spectra recorded in cryogenic rare gas matrices. The fundamental vibrational transition was observed at 889.5 cm^{-1} in solid Ne [46] and 819.8 cm^{-1} in solid Ar [47, 48]. These anomalously large matrix effects are probably caused by differences in the guest-host interactions that change the energy intervals between the $5f^37s$ and $5f^27s^2$ states.

Kaledin et al. [27] performed LFT calculations for neutral UO in an attempt to provide configurational assignments for the electronic transitions they observed. They found that a LFT model based on fixed interaction parameters, derived from consideration of the effect of the O^{2-} ligand on the energy levels of U^{2+} ion, was not in agreement with the experimental results (an Ω=1 ground state was predicted, with Ω=2 for the first excited state). It was speculated that the neglect of covalent interactions and/or the inaccuracy of the atomic ion wavefunctions used in evaluating the electrostatic interaction integrals was responsible for the failure. The low-lying energy levels of UO were successfully fitted using a semi-empirical LFT method where selected interaction parameters were treated as variables. Assignments for a sub-set of the observed states were proposed using the results from the LFT analysis. Eight states were assigned to $5f^37s$ and a further five were assigned to $5f^27s^2$.

Electron impact measurements of the IE for UO provided a consistent value for the IE of 5.6(1) eV [32, 35]. Allen et al. [49] obtained a low-resolution photoelectron spectrum for UO that exhibited a broad feature (corresponding to ionization energies of 5.8-7.6 eV) where the low-energy threshold appeared to be consistent with the electron impact IE. Re-examination of the IE of UO using resonantly enhanced multi-photon ionization yielded a significantly higher value for the IE (6.0313 eV) [24, 25]. As for ThO, the low IE obtained in the earlier measurements was attributed to the ionization of thermally excited molecules. The difference between the IE's for UO and U show that UO^+ is 0.163 eV more tightly bound than neutral UO. Theoretical calculations of the IE for UO have predicted values of 6.17 (Mali [50]), 5.71 (Boudreaux and Baxter [51]), 6.05 (Paulovič et al. [7]), and 5.59 eV (Tyagi [14]). It is noteworthy that the complete active space state interaction – spin orbit coupling (CASSI-SOC) calculations of Paulovič et al. [7] yielded reliable values for the IE's of both UO and U atoms (IE(U)=6.20 (calc) vs. 6.194 (obs) eV).

Spectroscopic data for UO^+ are of interest from both practical and theoretical perspectives. The practical interest in UO^+ derives from the fact that the ion can be formed by the associative ionization reaction [7, 52] $U+O\rightarrow UO^+ + e^-$. As a consequence, spectroscopic data for UO^+ have been sought for inclusion in atmospheric radiance models used to predict phenomena associated with nuclear explosions [53]. UO^+ is of theoretical interest as it is small enough to be amenable to high-level treatments, but challenging as it is expected to possess a large number of low-lying excited states. Several high-level relativistic calculations had been carried out to predict the electronic structure and properties of UO^+ (references [7, 14, 45, 49, 53]). Krauss and Stevens [45] published one of the earliest theoretical studies. They concluded that lowest energy configuration was $U^{3+}(5f^3({}^4I))O^{2-}$, giving rise to an Ω=4.5 ground state. They also predicted the energies of all 26 states derived from the ^4I configuration (with Ω values ranging from 0.5 to 7.5). Kaledin et al. [27] used their empirically adjusted LFT model for UO to predict the states of UO^+ associated with the $U^{3+}(5f^3(^4I))$ ion core, and obtained results that were qualitatively in agreement with the calculations of Krauss and Stevens [45]. One of the motivations for spectroscopic studies of UO^+ was that an indication of the value of the LFT approach for the early actinides could be obtained by comparing the model predictions with the observed energies. Clearly, if the model for UO failed to predict the structure for UO^+, the transferability of the parameters for the treatment of other uranium compounds would be suspect.

Goncharov et al. [54] recorded rotationally resolved ZEKE spectra for thirty-three vibronic bands of UO^+. Transitions to the ground state and nine electronically excited states were characterized within the energy range from 0 to 5200 cm^{-1} (relative to $UO^+(X, v=0, J=0)$). In accordance with theoretical predictions, the ground state was Ω=4.5. Figure 6 shows the rotational structure for the zero-point level. As for ThO^+, the PFI-ZEKE spectra for UO^+ showed extensive vibrational progressions for each electronic state. Another advantage of PFI-ZEKE spectroscopy is that it is not constrained by the usual optical selection rules (e.g., $\Delta\Omega=0,\pm1$ for a single photon transition). The UO^+ spectra included transitions to states with all Ω values in the range from Ω=0.5 to 5.5, using just the [19453]Ω=3 intermediate state of UO. Molecular constants for the ground state of UO^+ (ω_e=911.9(2), B_0=0.3467(7) cm^{-1}) were larger than those of neutral UO, consistent with ionization by removal of the non-bonding 7*s* electron. The calculations of Krauss and Stevens [45] yielded reasonably good estimates for the ground state constants (ω_e=925(30), R_e=1.842 Å (R_e(exp)=1.801(5) Å)), while the CASSI-SOC calculations of Paulovič et al. [7] predicted constants that were within the experimental error limits (ω_e=912, R_e=1.802 Å).

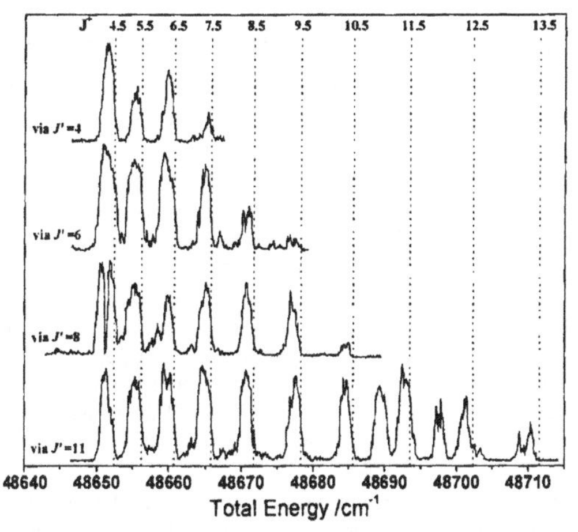

Figure 6: PFI-ZEKE spectra for UO recorded via specific rotational levels (J'=4, 6, 8, 11) of the intermediate state [19.453]3. The spectra show the rotational structure for the UO^+ $X(1)4.5$ v=0 level. These data were obtained using a pulsed electric field of 283 mV/cm applied 3 μs after photo-excitation. The spectra are plotted against total energy of the transition from the UO $X(1)4$ v=0, J=0 state.

The pattern of electronic states observed for UO^+ was readily understood using LFT. The low-lying states correlate with the $^4I_{4.5}$ and $^4I_{5.5}$ spin-orbit levels of the $U^{3+}(5f^3)$ ion. The atomic ion spin-orbit coupling strength was preserved in UO^+ to the extent that the atomic spin-orbit interval was recurrent in the energy level structure. For a given value of the atomic ion core angular momentum vector ($\boldsymbol{J_a}$) the lowest energy state corresponds to the maximum projection of $\boldsymbol{J_a}$ on the diatomic axis for the first and third quarters of the nf^N shell [55] (e.g., X(1)4.5 from $5f^3$ ($^4I_{4.5}$)). The energies of the states increase as the vector is tipped away from the molecular axis. Hence, the atomic ion $^4I_{4.5}$ core gives rise to states with Ω=4.5, 3.5, 2.5, 1.5, and 0.5 in ascending energy order. Similarly, $^4I_{5.5}$ gives rise to states with Ω from 5.5 to 0.5. These patterns were apparent in the spectrum of UO^+. All five of the states from $^4I_{4.5}$ were observed, along with the four lowest energy states of $^4I_{5.5}$. The right hand side of Fig. 7 shows the energy levels of UO^+, arranged in stacks that belong to a specific Ω value. The spin-orbit interval for the free $U^{3+}(5f^3)$ ion is indicated on the left. The $^4I_{5.5}$ -$^4I_{4.5}$ interval for $U^{3+}(5f^3)$

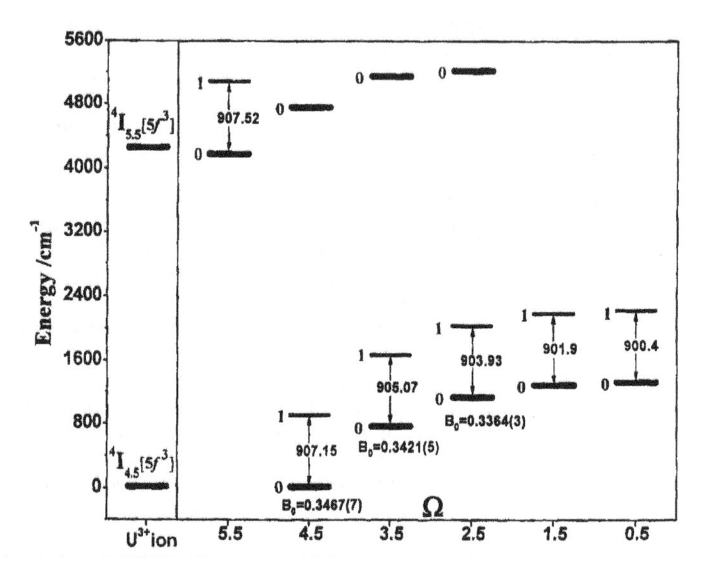

Figure 7: Observed low-lying electronic states (v=0 and 1 only) for the $U^{3+}(5f^3)O^{2-}$ configuration of UO^+, arranged according to energy and Ω. For comparison, the spin-orbit splitting for $U^{3+}(5f^3, {}^4I)$ is shown on the left hand side of the diagram.

is estimated to be 4265 cm^{-1} (ref. [56]), while the corresponding Ω=5.5-4.5 interval for UO^+ is 4178 cm^{-1}. To a first approximation it is expected that the states arising from a particular atomic ion configuration will have very similar vibrational and rotational constants [57]. The vibrational constants indicated in Fig. 7 fit this expectation, as did the rotational constants. For the $^4I_{4.5}$ states it can be seen that there was a slight, systematic decrease in the vibrational constant as the projection of J_a along the diatomic axis decreased. This trend is reasonable as the rotation of J_a away from the bond axis rotates the orbitals in a way that increases the repulsive interaction between the 5f electrons and the O^{2-} ligand (the same effect that results in the observed energy ordering of the states). Note that the Ω=3.5 state at 4982.4 cm^{-1} did not belong to the ^4I group. This particular state was identified as the lowest energy state of the $U^{3+}(5f^27s, {}^4H_{3.5})O^{2-}$ group [54].

A quantitative comparison between the observed states of UO^+ and the LFT predictions of Kaledin et al. [27] reveals specific shortcomings of the model. The observed and calculated energies are presented in Table 1, where the first five states correlate with $^4I_{4.5}$, and the next four with $^4I_{5.5}$. The systematic differences between the original LFT energies (column five of Table 1) and the experimental results indicates that the strength of the ligand field had been underestimated. The model correctly predicted that the ground state is Ω=4.5, but the ordering of the first few excited states is in error. As might be expected for a weak ligand field, the atomic spin-orbit splitting was almost preserved, yielding an Ω=4.5 to 5.5 energy spacing that was close to the estimated value of 4265 cm^{-1}. Re-optimization of the LFT model yielded better agreement with the experimental data. Given that there are appreciable correlations between the LFT parameters, it is likely that further improvements can be made by a simultaneous fit to the data for both UO and UO^+. At this stage the primary insight derived from the LFT model is that the $U^{3+}(5f^3)$ ion core retains its atomic character in UO^+. This provides direct evidence that the participation of the 5f orbitals in bond formation is negligible.

Table 1 also lists the results from ab initio calculations for the excited states of UO^+. The third and fourth columns list the energies of Krauss and Stevens [45] and Tyagi [14], respectively. In comparing with the experimental data, it can be seen that the calculations of Krauss and Stevens [45] correctly predicted the energy ordering of the $^4I_{4.5}$ states, but the energy intervals between the states were overestimated by a factor of almost two. This suggests that the electrostatic perturbation of the U^{3+} ion had been overestimated. In contrast, the interval between the Ω=4.5 and 5.5 states was underestimated, which suggests that the spin-orbit coupling of the ion core was partially quenched. This may also be a consequence of the overestimation of the strength of the ligand field. These

Table 1. Experimental and theoretical term energies (cm^{-1}) for the low energy states of UO^+.

State (*Config.*)	Measured term energy	MCSCF/VCI, 1983 [45]	MCSCF/CI, 2005 [14]	LFT, 1994 [27]	LFT*, (re-optimized [54])
X(1)4.5 ($5f^3$)	0	0	0	0	0/0
(1)3.5 ($5f^3$)	764.93(20)	1319	582	633	767
(1)2.5 ($5f^3$)	1132.42(20)	1895	856	696	1131
(1)1.5 ($5f^3$)	1284.5(3)	2094	1076	580	1270
(1)0.5 ($5f^3$)	1324.9(3)	3296	–	695	1309
(1)5.5 ($5f^3$)	4177.83(20)	2563	3744	3991	4150
(2)4.5 ($5f^3$)	4758.46(20)	3599	4180	4601	4767
(2)3.5 ($5f^3$)	5161.96(20)	4045	–	4770	5110
(3)2.5 ($5f^3$)	5219.37(20)	–	–	4744	5293
(3)3.5 ($5f^27s$)	4982.44(20)	–	4287	–	–

* - interaction parameters were treated as variables and fit to the experimental data in this semi-empirical LFT model.

opposing trends resulted in a slight overlap of the $^4I_{4.5}$ and $^4I_{5.5}$ manifolds, such that the ordering of the Ω=0.5 and 5.5 states was reversed relative to the observed ordering. The more recent calculations of Tyagi [14] were in better quantitative agreement with the experimental data, but there were a few puzzling discrepancies. States corresponding to (1)0.5, (2)3.5 and (3)2.5 were not predicted in the 0-8800 cm^{-1} energy range. With the newly available experimental data for UO^+ it is anticipated that there will be further theoretical studies in the near future.

5. Experimental and theoretical studies of UO_2

Moving up in complexity, UO_2 is a suitable prototype for studies of a polyatomic actinide compound. Ab initio calculations for UO_2 carried out in the 1980's predicted a symmetric linear equilibrium structure [1, 49, 58, 59]. The highest occupied molecular orbitals (HOMO's) were found to be the metal-centered 5*f* orbitals. Calculated energies for the $5f\phi_u$ and $5f\delta_u$ orbitals were close enough to complicate the task of identifying the ground state configuration. Wood et al. [59] and Michels and Hobbs [58] found that $(5f\phi_u)^2$ gave the lowest energy, which would produce a $^3\Sigma_g^-$ ground state. Alternatively, the calculations of Allen et al.[49] indicated that either $(5f\phi_u)(5f\delta_u)$ or $(5f\delta_u)^2$ would be lowest in energy, yielding either a ^3H_g or $^3\Sigma_g^-$ ground state. More recent investigations (ab initio and DFT) yielded a $^3\Phi_u$ ground state derived from the $(5f\phi_u)(7s\sigma_g)$ configuration [3, 14, 46, 60-62]. All of the published studies predicted partially occupied 5*f* orbitals that give rise to manifolds of low-lying electronic states, which was consistent with the large heat capacity of vapor phase UO_2.

Experimental confirmation of the linear structure was provided by molecular beam - electric field focusing measurements [63] and IR spectra for matrix isolated UO_2 (refs. [46-48]). While the IR spectra were clearly consistent with a linear structure, they were somewhat anomalous in that they exhibited a large change in the vibrational frequency for the asymmetric stretch when the matrix host was changed from Ar to Ne (776 vs. 914 cm^{-1})[46]. Prior to the work of Han et al. [24, 25] there were no gas phase spectra that could be used to determine the electronic configuration of the ground state. Allen et al. [49] recorded the He I photoelectron spectroscopy for a mixture of UO and UO_2. They observed structure corresponding to excited states of UO_2^+, but were unable to determine the IE as the relevant region of the spectrum was overlapped by a broad feature attributed to UO. Electron impact measurements yielded an IE value of 5.4(1) eV [32, 35]. It was this particular result that generated a clearly defined conflict between theory and experiment. The calculations of Zhou et al. [46] and Gagliardi et al. [61] yielded IE's of 6.27 and 6.17 eV. Gagliardi et al. [61] argued that their calculations should not be in error by such a large margin and called the experimental results into question.

The first electronic spectra for gas phase UO_2 were reported by Han et al. [24, 25] Twenty-two vibronic bands of neutral UO_2 were observed in the range from 17400 – 32000 cm^{-1}. PIE curves were used to determine the energies of the lower levels for these transitions, relative to the first ionization limit. Due to the low frequency of the ground state bending vibration (120 cm^{-1}) levels with v_b=0-2 were significantly populated. In addition, transitions were observed that originated from an electronically excited state that was approximately 360 cm^{-1} above the ground state. The presence of this low-lying state provided the key to determining the lowest energy electronic configuration. The ab initio calculations of Chang [62], Tyagi [14], and Gagliardi et al. [61] all indicated that the angular momentum coupling for the low-lying states of UO_2 is intermediate between *LS* and *Jj*. Viewed from the latter perspective, the states of $U(5f\phi_u 7s\sigma_g)O_2$ are built on the ^2F term arising from the $5f$ electron. The J_f=5/2 and 7/2 components of this term are widely separated by the spin-orbit interaction. The lower energy J_f=5/2 component is split into $J_a=3_u$ and 2_u terms by the much weaker interaction between J_f and the spin of the $7s$ electron. The lowest two energy states of UO_2 are produced when the electronic angular momentum vectors have their maximum projections along the molecular axis. The $J_a=3_u$, $\Omega=3_u$ and $J_a=2_u$, $\Omega=2_u$ components correlate with the *LS* term symbols $^3\Phi(3_u)$ and $^3\Phi(2_u)$. High-level theoretical calculations [3, 14, 60-62] predict that these states are separated by energies in the range of 378 - 439 cm^{-1}. The next pair of states correspond to $\Omega=J_a-1$ (2_u and 1_u) and they are approximately 1000 cm^{-1} above the ground state ($\tilde{X}\,^3\Phi(2_u)$). The 4_u component of the $^3\Phi$ state correlates with the $^2F_{7/2}$ core and is situated approximately 5000 cm^{-1} above $\tilde{X}\,^3\Phi(2_u)$. The pattern of low-lying states originating from $U(5f\phi_u 5f\delta_u)O_2$ is markedly different. The two lowest energy states for this configuration are $^3H(4_g)$ and $^3H(5_g)$, and they are separated by a relatively strong spin-orbit interaction. Chang's [62] calculations predict that this interval is in excess of 3800 cm^{-1}. Hence, the observation of a low-lying electronic state of UO_2 confirmed that the ground state is derived from the $5f\phi_u 7s\sigma_g$ configuration. Excited states of UO_2, along with oscillator strengths for transitions originating from $\tilde{X}\,^3\Phi(2_u)$ were calculated by Chang [62]. Han et al. [24] used these data to propose assignments for states in the 17800-31900 cm^{-1} energy range. As the oscillator strengths were not available for transitions from $^3\Phi(3_u)$, assignments for the bands associated with these states were proposed based on the application of the $\Delta\Omega=0, \pm1$ selection rule. It was evident that the stronger bands in this energy range were formally metal-centered $5f7p \leftarrow 5f7s$ transitions. A pair of transitions were identified that appeared to originate from the $\tilde{X}\,^3\Phi(2_u)$ and $^3\Phi(3_u)$ states, and terminate on a common upper level. The interval between these bands was used to refine the estimate for the $\tilde{X}\,^3\Phi(2_u)$ - $^3\Phi(3_u)$ energy difference. Subsequent high-level calculations have shown that revisions of the proposed assignments are needed [60]. Calculations of the oscillator strengths for transitions from both $\tilde{X}\,^3\Phi(2_u)$ and $^3\Phi(3_u)$ revealed an unexpected pattern. There were no occurrences of excited state levels that had good oscillator strengths for transitions from both lower states. A transition from one of the lower levels was always favored by a factor of at least 100 as compared to the transition from the other level. This implied that the transitions ascribed to a common upper state actually terminated on two nearby states. Based on the PIE curve data, which defined the $\tilde{X}\,^3\Phi(2_u)$ - $^3\Phi(3_u)$ energy difference to within ±10 cm^{-1}, the two upper levels should be separated by 10 cm^{-1} or less. Consistent with this interpretation, the calculations of Gagliardi et al. [60] yield a pair of upper states with suitable oscillator strengths that are separated by 13 cm^{-1} (c.f. Table 2). The system of assignments for the excited states of UO_2, based on the calculations by Gagliardi et al. [60], are presented in Table 2. Further experimental and theoretical work is needed to test the validity of these assignments.

The IE for UO_2 was determined from PIE curve measurements. The result obtained, 49425±20 cm^{-1} (6.128(3) eV), was 0.7 eV higher than the generally accepted literature value, but in excellent agreement with the theoretical predictions [46, 61]. The fact that theory arrived at the correct IE in advance of the experimental results seems to provide a strong endorsement for the current generation of relativistic electronic structure calculations.

The anomalous vibrational frequency difference observed for UO_2 in Ne and Ar matrices was explored by Zhou et al. [46] using DFT methods. Their spin-free calculations were the first to predict the $5f7s$ $^3\Phi_u$ ground state. They also found that the first excited state derived from the $5f\phi 5f\delta$ configuration, ^3H_g, was 1930 cm^{-1} above the ground state. The frequencies predicted for the asymmetric stretch vibration for these two states, 931 and 814 cm^{-1}, were in approximate agreement with the frequencies measured

Table 2. Tentative assignments of the electronic spectrum of UO_2 by Gagliardi et al. [60]

State	Energy obs.	Energy calc.	$f(2_u)^a$	$f(3_u)^a$	Wavefunction composition (weight in %)
2_u	0	0			$^3\Phi_u(91)$, $^3\Delta_u(05)$, $^1\Delta_u(03)$
3_u	360	378	0.0000		$^3\Phi_u(53)$, $^1\Phi_u(37)$, $^3\Delta_u(10)$
1_u	1094	2567	0.0000	0.0000	$^3\Delta_u(96)$, $^3\Pi_u(2)$, $^1\Pi_u(2)$
2_u	1401	2908	0.0000	0.0000	$^3\Delta_u(54)$, $^1\Delta_u(42)$, $^3\Pi_u(4)$
2_g	17859	15452	0.0000	0.0681	$^3\Delta_g(49)$, $^1\Delta_g(35)$, $^3\Pi_g(13)$
1_g	18159	16725	0.0777	0.0000	$^3\Delta_g(75)$, $^1\Pi_g(14)$, $^3\Sigma_g^-(08)$, $^1\Pi_g(02)$
4_g	18587	17274	0.0000	0.0466	$^3\Gamma_g(87)$, $^1\Gamma_g(08)$, $^1\Gamma_g(02)$
1_g	18423	17645	0.0102	0.0000	$^3\Sigma_g^-(55)$, $3\Pi_g(20)$, $^1\Pi_g(07)$, $^3\Delta_g(06)$
1_g	27259	26349	0.1026	0.0000	$^3\Delta_g(70)$, $^1\Pi_g(16)$, $^3\Pi_g(07)$
2_g	28700	26617	0.0000	0.0762	$^3\Delta_g(44)$, $^1\Delta_g(23)$, $^3\Pi_g(22)$
1_g	29700	28124	0.0108	0.0000	$^1\Pi_g(40)$, $^3\Pi_g(33)$, $^3\Delta_g(21)$
3_g	31838	36770	0.0009	0.1676	$^3\Phi_g(48)$, $^1\Phi_g(41)$, $^3\Delta_g(10)$
2_g	31838	36783	0.1816	0.0014	$^3\Phi_g(90)$, $^3\Delta_g(09)$

[a] Oscillator strength: only states with an oscillator strength larger than 0.01 from either the $^3\Phi_{2u}$ or the $^3\Phi_{3u}$ state, and energies <37000 cm^{-1} are included.

in Ne and Ar matrices, respectively. Consequently Zhou et al. [46] suggested that the guest-host interaction in Ar reversed the energy ordering of the $^3\Phi_u$ and ^3H_g states. Note that the vibrational frequency for the $5f\phi5f\delta$ configuration is lower because the $5f$ orbital is less polarizable than the $7s$, and therefore provides greater shielding of the charge on the metal. Anomalous differential matrix shifts have also been reported for CUO in Ne and Ar, and it has been proposed that the interaction between CUO and Ar is large enough to re-order the analogous low-lying electronic states[64, 65]. Based on the results from DFT calculations Andrews et al. [66] concluded that the rare gas atoms Ar, Kr and Xe form weak chemical bonds with CUO. However, some controversy remains concerning the re-ordering of the states by these interactions [67].

To further investigate the anomalous matrix shifts, Lue et al. [68] examined the electronic absorption and emission spectra of UO_2 in solid Ar. Laser excitation and dispersed fluorescence measurements yielded results that were in good agreement with the data for gas phase UO_2. This was in direct conflict with the suggestion that a reordering of states occurs in Ar. If reordering occurred the ground state would change from *u* to *g* symmetry, and, due to the optical selection rules, the absorption spectrum for UO_2 in Ar would not be expected to correlate with that of UO_2 in the gas phase. The dispersed fluorescence spectra for UO_2 in Ar revealed a pattern of low-lying states that was consistent with the preservation of $5f7s$ as the lowest energy configuration. The most prominent and best-resolved emission features of the matrix spectrum were assigned on the basis of the calculations of Chang [62]. Subsequent theoretical studies by Gagliardi et al. [60] and Fleig et al. [3] support the assignments for the transitions to the two lowest energy states, $\tilde{X}^3\Phi(2_u)$ and $^3\Phi(3_u)$. However, they raise doubts concerning the assignments for the next two excited states (1_u and 2_u). The assignments of the fluorescence spectrum presented in ref. [68] yield term energies for 1_u and 2_u of 1094 and 1401 cm^{-1}, respectively. Gagliardi et al.'s [60] recent calculations gave energies of 2567 and 2908 cm^{-1}, and a large scale calculation by Fleig et al. [3] predicted an energy of 2042 cm^{-1} for the 2_u state. It is possible that the matrix emission bands assigned to the 1_u and 2_u states are really vibrationally excited levels of the ground and 3_u states. Additional experimental work on the spectrum of gas phase UO_2 will be needed to resolve this question.

Theoretical calculations have also been used to examine the question of electronic state reordering for Ar matrix isolated UO_2. Li et al. [69] preformed spin-free calculations for UO_2 interacting with five Ar atoms. They found that the energy separation between the $^3\Phi$ and ^3H states was reduced from 4759 cm^{-1} for UO_2 to 734 cm^{-1} for UO_2Ar_5, and suggested that ^3H would drop below $^3\Phi$ when spin-orbit coupling was included. There have been calculations of the UO_2 $\tilde{X}^3\Phi(2_u) - {}^3H(4_g)$ energy interval with explicit treatment of the spin-orbit coupling. Energies of 3330 cm^{-1} (Gagliardi et al. [60]), 3974 cm^{-1} (Tyagi [14]) and 3102 cm^{-1} (Fleig et al. [3]) have been reported. If these estimates are reliable, they cast doubt on the proposed state reordering in solid Ar. Although it is not out of the question, it would be surprising to find that the interaction between $U(5f^2)O_2$ and a surrounding Ar lattice could stabilize this state by more than 3000 cm^{-1} relative to $U(5f7s)O_2$.

6. Summary and conclusions

Techniques suitable for obtaining well-resolved spectroscopic data for gas phase transition metal compounds have been available for many years, but there had been few attempts to apply these methods to actinide compounds. In part, these experiments were not carried out as there were concerns that the data might prove to be unintelligible. Given the recent computational advances this is no longer the case, and we are poised to advance our understanding of the structure and bonding of actinide compounds through the interplay of spectroscopic measurements and high-level theoretical calculations. The studies described here show that conventional laser ablation-jet cooling techniques can provide interpretable spectra for simple di- and triatomic actinide compounds. Measurements that utilize resonance enhanced multi-photon ionization are particularly valuable as they can provide accurate ionization energies and spectroscopic data for ions.

Gas phase spectra for ThO^+, UO, UO^+ and UO_2 have been used in the evaluation of semi-empirical and ab initio theoretical models. Ligand field theory has the potential to provide a simple model for the low-lying states of actinide compounds, but this picture is compromised if the metal-centered $5f$ orbitals participate in covalent bond formation. The data for the oxides indicate that the $5f$ orbitals retain their atomic character, which implies that LFT models do provide meaningful insights. Recent relativistic ab initio calculations have achieved impressive results. For example, the IE for UO_2 and the molecular constants for UO^+ were predicted accurately, before reliable experimental data were available. However, it is evident that further development of relativistic theoretical methods is needed. Even for simple di- and triatomic molecules it is still difficult and expensive to make reliable calculations of basic physical properties.

Looking beyond the characterization of gas phase actinide molecules, the binding of solvent molecules and ligands will be addressed in future studies. It is evident from comparisons of gas phase and matrix spectra that the physical interactions of the actinides are unusual. Experimental and theoretical studies of these intermolecular forces are of fundamental interest, but they may also have significant practical value. It is hoped that a deeper understanding of these interactions will facilitate the design of selective chelating reagents for waste separation applications.

Acknowledgments

Mr. Vasiliy Goncharov and Dr. Leonid A. Kaledin carried out most of the experimental work described in this article. Several people are acknowledged for their helpful advice concerning the implementation of MATI and PFI-ZEKE techniques. Thanks go to Profs. M. A. Duncan (University of Georgia, Athens), D.-S. Yang (University of Kentucky), M. D. Morse (University of Utah), and F. Merkt (ETH Zurich). Profs. B. O. Roos (Lund), L. Gagliardi (University of Geneva), R. M. Pitzer (Ohio State Uinversity) and Dr. R. Tyagi (OSU) are thanked for many helpful discussions concerning the electronic structures of actinide oxides, and for sharing the results of their calculations prior to publication. This work was supported by the US Department of Energy under grant DE-FG02-01ER15153-A003.

References

[1] M. Pepper, B. E. Bursten: Chem. Rev. 91 (1991) 719.
[2] M. Dolg, X. Cao: Recent Advances in Computational Chemistry 5 (2004) 1.
[3] T. Fleig, H. J. A. Jensen, J. Olsen, L. Visscher: J. Chem. Phys. 124 (2006) 104106/1.
[4] T. Fleig, J. Olsen, L. Visscher: J. Chem. Phys. 119 (2003) 2963.
[5] L. Gagliardi, B. O. Roos: Nature 433 (2005) 848.
[6] W. Kutzelnigg, W. Liu: J. Chem. Phys. 123 (2005) 241102/1.
[7] J. Paulovic, L. Gagliardi, J. M. Dyke, K. Hirao: J. Chem. Phys. 122 (2005) 144317/1.
[8] A. N. Petrov, N. S. Mosyagin, A. V. Titov, I. I. Tupitsyn: J. Phys. B: At., Mol. Opt. Phys. 37 (2004) 4621.
[9] B. O. Roos, L. Gagliardi: Inorg. Chem. 45 (2006) 803.
[10] B. O. Roos, R. Lindh, P.-A. Malmqvist, V. Veryazov, P.-O. Widmark: Chem. Phys. Lett. 409 (2005) 295.
[11] B. O. Roos, P.-A. Malmqvist: Phys. Chem. Chem. Phys. 6 (2004) 2919.
[12] O. Schwerdtfegerm: Relativistic Electronic Structure Theory, Part 2: Applications. (Dedicated to Prof. Pekka Pyykko on the Occasion of his 60th Birthday.) [In: Theor. Comput. Chem., 2004; 14], 2004.
[13] M. Krauss, W. J. Stevens: Mol. Phys. 101 (2003) 125.
[14] R. Tyagi: Ab initio studies of systems containing actinides using relativistic effective core potentials, Ph.D. Thesis, Department of Chemistry, Ohio State University, 2005. (Advisor, R. M. Pitzer).
[15] J. Paulovic, T. Nakajima, K. Hirao, R. Lindh, P. A. Malmqvist: J. Chem. Phys. 119 (2003) 798.
[16] G. Edvinsson, A. Lagerqvist: Phys. Scr. 30 (1984) 309.
[17] G. Edvinsson, A. Lagerqvist: Phys. Scr. 32 (1985) 602.
[18] G. Edvinsson, A. Lagerqvist: J. Mol. Spectrosc. 113 (1985) 93.
[19] G. Edvinsson, A. Lagerqvist: J. Mol. Spectrosc. 122 (1987) 428.
[20] G. Edvinsson, A. Lagerqvist: J. Mol. Spectrosc. 128 (1988) 117.
[21] G. Edvinsson, A. Lagerqvist: Phys. Scr. 41 (1990) 316.
[22] V. Goncharov, M. C. Heaven: J. Chem. Phys. 124 (2006) 64312.
[23] V. Goncharov, J. Han, A. Kaledin Leonid, M. C. Heaven: J. Chem. Phys.122 (2005) 204311.
[24] J. Han, V. Goncharov, L. A. Kaledin, A. V. Komissarov, M. C. Heaven: J. Chem. Phys. 120 (2004) 5155.
[25] J. Han, L. A. Kaledin, V. Goncharov, A. V. Komissarov, M. C. Heaven: J. Am. Chem. Soc. 125 (2003) 7176.
[26] L. A. Kaledin, M. C. Heaven: J. Mol. Spectrosc. 185 (1997) 1.
[27] L. A. Kaledin, J. E. McCord, M. C. Heaven: J. Mol. Spectrosc. 164 (1994) 27.
[28] M. C. Heaven, J. P. Nicolai, S. J. Riley, E. K. Parks: Chem. Phys. Lett. 119 (1985) 229.
[29] L. A. Kaledin, A. N. Kulikov: Zh. Fiz. Khim. 63 (1989) 1697.
[30] L. A. Kaledin, A. N. Kulikov, L. V. Gurvich: Zh. Fiz. Khim. 63 (1989) 801.
[31] L. A. Kaledin, E. A. Shenyavskaya, L. V. Gurvich: Zh. Fiz. Khim. 60 (1986) 1049.
[32] F. Capone, Y. Colle, J. P. Hiernaut, C. Ronchi: J. Phys. Chem. A 103 (1999) 10899.
[33] D. L. Hildenbrand, L. V. Gurvich, V. S. Yungman, The Chemical Thermodynamics of Actinide Elements and Compounds. International Atomic Energy Agency, 1985, p. 5.
[34] E. G. Rauh, R. J. Ackermann: J. Chem. Phys. 62 (1975) 1584.
[35] E. G. Rauh, R. J. Ackermann: J. Chem. Phys. 60 (1974) 1396.
[36] J. K. Gibson, J. Marcalo: Coord. Chem. Rev. 250 (2006) 776.
[37] J. K. Gibson, R. G. Haire, J. Marcalo, M. Santos, A. Pires de Matos, J. P. Leal: J. Nucl. Mater. 344 (2005) 24.
[38] K. Gibson John, M. Santos, J. Marcalo, P. Leal Joao, A. Pires de Matos, G. Haire Richard: J. Phys. Chem. A 110 (2006) 4131.
[39] J. K. Gibson, M. Santos, J. Marcalo, J. P. Leal, A. Pires de Matos, R. G. Haire: J. Phys. Chem. A 110 (2006) 4131.
[40] M. Santos, J. Marcalo, A. Pires de Matos, J. K. Gibson, R. G. Haire: J. Phys. Chem. A 106 (2002) 7190.
[41] W. Demtroder: Laser Spectroscopy, Springer-Verlag, 2002.

[42] E. W. Schlag: ZEKE Spectroscopy, Cambridge University Press, Cambridge, England 1998.
[43] T. P. Softley: Int. Rev. Phys. Chem. 23 (2004) 1.
[44] J. M. Brown, A. Carrington: Rotational spectroscopy of diatomic molecules, Cambridge Univesity Press, Cambridge, 2003.
[45] M. Krauss, W. J. Stevens: Chem. Phys. Lett. 99 (1983) 417.
[46] M. Zhou, L. Andrews, N. Ismail, C. Marsden: J. Phys. Chem. A 104 (2000) 5495.
[47] R. D. Hunt, L. Andrews: J. Chem. Phys. 98 (1993) 3690.
[48] S. D. Gabelnick, G. T. Reedy, M. G. Chasanov: J. Chem. Phys. 58 (1973) 4468.
[49] G. C. Allen, E. J. Baerends, P. Vernooijs, J. M. Dyke, A. M. Ellis, M. Feher, A. Morris: J. Chem. Phys. 89 (1988) 5363.
[50] G. L. Malli, in D.R. Salahub, M.C. Zerner (Eds.), The challenge of d and f electrons, theory and computations. ACS Symposium series, 1989, p. 291.
[51] E. A. Boudreaux, E. Baxter: Int. J. Quant. Chem. 90 (2002) 629.
[52] W. L. Fite, H. H. Lo, P. Irving: J. Chem. Phys. 60 (1974) 1236.
[53] H. H. Michels, Radiative properties of uranium oxide positive ion (UO+). United Technol. Res. Cent.,East Hartford,CT,USA., 1989, p. 25 pp.
[54] V. Goncharov, L. A. Kaledin, M. C. Heaven: J. Chem. Phys. accepted, May (2006).
[55] L. A. Kaledin, C. Linton, T. E. Clarke, R. W. Field: J. Mol. Spectrosc. 154 (1992) 417.
[56] W. T. Carnall, H. M. Crosswhite. ANL-84-90, Argonne National Laboratory, 1984.
[57] R. W. Field: Ber. Bunsenges. Phys. Chem. 86 (1982) 771.
[58] H. H. Michels, R. H. Hobbs, Theoretical study of the radiative and kinetic properties of selected metal oxides and air molecules. United Technol. Res. Cent.,East Hartford,CT,USA., 1983, p. 91 pp.
[59] J. H. Wood, M. Boring, S. B. Woodruff: J. Chem. Phys. 74 (1981) 5225.
[60] L. Gagliardi, M. C. Heaven, J. W. Krogh, B. O. Roos: J. Am. Chem. Soc. 127 (2005) 86.
[61] L. Gagliardi, B. O. Roos, P. Malmqvist, J. M. Dyke: J. Phys. Chem. A 105 (2001) 10602.
[62] Q. Chang: Ab Initio Calculations on UO_2, M.S. Thesis, Department of Chemistry, Ohio State University, 2002. (Advisor, R. M. Pitzer)
[63] M. Kaufman, J. Muenter, W. Klemperer: J. Chem. Phys. 47 (1967) 3365.
[64] L. Andrews, B. Liang, J. Li, B. Bursten: Angew. Chem. Int. Ed. 39 (2000) 4565.
[65] B. Bursten, D. Drummond, J. Li: Faraday Discuss. 124 (2003) 1.
[66] L. Andrews, B. Liang, J. Li, B. E. Bursten: J. Am. Chem. Soc. 125 (2003) 3126.
[67] B. O. Roos, P.-O. Widmark, L. Gagliardi: Faraday Discussions 124 (2003) 57.
[68] C. J. Lue, J. Jin, M. J. Ortiz, J. C. Rienstra-Kiracofe, M. C. Heaven: J. Am. Chem. Soc. 126 (2004) 1812.
[69] J. Li, B. E. Bursten, L. Andrews, C. J. Marsden: J. Am. Chem. Soc. 126 (2004) 3424.

Brill Academic Publishers
P.O. Box 9000, 2300 PA Leiden
The Netherlands

Lecture Series on Computer and Computational Sciences
Volume 1, 2006, pp. 155-164

A new hybrid DFT functional - Accurate description of response properties and van der Waals interactions

Takao Tsuneda
Department of Quantum Engineering and Systems Science, School of Engineering,
University of Tokyo, Tokyo, 113-8656, Japan

and

Kimihiko Hirao[1]
Department of Applied Chemistry, School of Engineering,
University of Tokyo, Tokyo, 113-8656, Japan

Received 31 July, 2006; accepted 2 August, 2006

Abstract: Hybrid functionals can be improved through the introduction of an Ewald partitioning of $1/r_{12}$. The new hybrid GGA functional with correct long-range electron-electron interactions has good energetics, good Rydberg behavior and good charge transfer predictions. It also describes the long-range exchange repulsion accurately, Thus, the present scheme combined with the Andersson-Langreth-Lundqvist functional describes successfully the weak van der Waals interactions.

Keywords: DFT, hybrid functional, long range electron-electron correlation, Rydberg excitation, charge transfer excitation, van der Waals interactions

1. Introduction

Density functional theory (DFT) has advanced to one of the most popular theoretical approaches to calculate molecular properties. The first-order molecular properties (energies, geometries, frequencies, dipole moments, etc) are well predicted by local generalized gradient approximation (GGA) functionals. However, DFT fails to describe induced or response properties. This failure has been attributed to the wrong long-range behavior of the standard exchange-correlation functionals. The conventional DFT models use local functional $E_{xc}(\rho)$ dependent only on ρ, or semilocal functions $E_{xc}(\rho,\nabla\rho)$ dependent on ρ and its gradient $\nabla\rho$. The most exchange-correlation functionals do not satisfy the correct asymptotic behavior

$$\lim_{R\to\infty} v_{xc}^{\sigma}(R) = -\frac{1}{R}, \quad v_{xc}^{\sigma} = \frac{\delta E_{xc}}{\delta\rho_{\sigma}} \tag{1}$$

Due to a (semi)local character of these functionals, one cannot expect that the long range electron-electron interaction will be described by current DFT models adequately.

The failure arises from the wrong long range behavior due to the local character of the approximate exchange-correlation functionals. By splitting the Coulomb interaction into short-range and long-range components, the exchange energy can be decomposed into short-range and long-range contributions. The key is a partitioning of $1/r_{12}$ for the exchange contribution. Hybrid functionals can be improved through the introduction of an Ewald partitioning of $1/r_{12}$. Hybrid GGA functional with correct long-range electron-electron interactions has good energetics, good Rydberg behavior and good CT predictions.

[1] Corresponding author. E-mail: hirao@qcl.t.u-tokyo.ac.jp

2. New Hybrid Exchange Functional

Pure DFT exchange-correlation functionals have been represented by using only local quantities at a reference point: *e.g.* electron density, gradient of density, and etc. It is, therefore, presumed that pure functionals overestimate local contributions and underestimate nonlocal contributions. The most significant nonlocal contribution neglected in pure functionals may be the long-range electron-electron exchange interaction, because it may be impossible to represent this interaction as a functional of a one-electron quantity.

In 1996, Savin suggested a long-range exchange correction scheme for LDA functional [1]. In this scheme, the two-electron operator, $1/r_{12}$, is separated into the short-range and long-range parts naturally by using the standard error function *erf* such that

$$\frac{1}{r_{12}}=\frac{1-erf(\mu r_{12})}{r_{12}}+\frac{erf(\mu r_{12})}{r_{12}}, \tag{2}$$

where $r_{12}=|\mathbf{r}_1-\mathbf{r}_2|$ for coordinate vectors of electrons, $\mathbf{r}_1$ and $\mathbf{r}_2$, and μ is a parameter that determines the ratio of these parts. Based on Eq. (2), the long-range exchange interaction is described by the HF exchange integral,

$$E_x^{lr}=-\frac{1}{2}\sum_{\sigma}\sum_{i}^{occ}\sum_{j}^{occ}\iint\psi_{i\sigma}^{*}(\mathbf{r}_1)\psi_{j\sigma}^{*}(\mathbf{r}_2)\frac{erf(\mu r_{12})}{r_{12}}\psi_{j\sigma}(\mathbf{r}_1)\psi_{i\sigma}(\mathbf{r}_2)d^3\mathbf{r}_1d^3\mathbf{r}_2, \tag{3}$$

where $\psi_{i\sigma}$ is the *i*th σ-spin orthonormal molecular orbital. The LDA exchange functional is applied to the short-range exchange interaction such that

$$\begin{aligned}E_x^{sr}=&-\frac{3}{2}\left(\frac{3}{4\pi}\right)^{1/3}\sum_{\sigma}\int\rho_\sigma^{4/3}\Bigg\{1-\frac{8}{3}a_\sigma\\&\times\left[\sqrt{\pi}erf\left(\frac{1}{2a_\sigma}\right)+(2a_\sigma-4a_\sigma^3)\exp\left(-\frac{1}{4a_\sigma}\right)-3a_\sigma+4a_\sigma^3\right]\Bigg\}d^3\mathbf{R},\end{aligned} \tag{4}$$

where $a_\sigma=\mu/(2k_\sigma)$. The averaged relative momentum k_σ is written for LDA as the Fermi momentum, *i.e.* $k_{F\sigma}=(6\pi^2\rho_\sigma)^{1/3}$. Equation (4) is derived by using the density matrix form corresponding to the LDA exchange functional,

$$P_{1\sigma}^{LDA}\left(\mathbf{R}+\frac{\mathbf{r}}{2},\mathbf{R}-\frac{\mathbf{r}}{2}\right)=3\frac{j_1(k_{F\sigma}r)}{k_{F\sigma}r}\rho_\sigma(\mathbf{R}), \tag{5}$$

where j_1 is the first-order spherical Bessel function.

However, Savin's scheme is inapplicable to conventional GGA exchange functionals, because GGA functionals usually have no corresponding density matrices unlike LDA. In 2001, we solved this problem by pushing gradient terms of GGA functionals into the momentum k_σ [2,3]. That is, the corresponding density matrix is determined for any GGA exchange functional by substituting $k_{F\sigma}$ in Eq. (5) with

$$k_\sigma^{GGA}=\left(\frac{9\pi}{K_\sigma^{GGA}}\right)^{1/2}\rho_\sigma^{1/3}, \tag{6}$$

where K_σ^{GGA} is defined in an exchange functional used: $E_x^{GGA}=\int\rho_\sigma^{4/3}K_\sigma^{GGA}d^3\mathbf{R}$. Equation (6) correctly reproduces the Fermi momentum $k_{F\sigma}$ for K_σ^{LDA}. By using k_σ^{GGA}, the short-range exchange energy in Eq. (4) is substituted by

$$E_x^{sr} = -\frac{3}{2}\left(\frac{3}{4\pi}\right)^{1/3} \sum_{\sigma} \int \rho_{\sigma}^{4/3} K_{\sigma}^{GGA} \left\{ 1 - \frac{8}{3} a_{\sigma} \times \left[\sqrt{\pi}\, erf\left(\frac{1}{2a_{\sigma}}\right) + (2a_{\sigma} - 4a_{\sigma}^3)\exp\left(-\frac{1}{4a_{\sigma}}\right) - 3a_{\sigma} + 4a_{\sigma}^3 \right] \right\} d^3\mathbf{R}. \tag{7}$$

It is easily confirmed that Eq. (7) reproduces the original GGA exchange functional for μ =0. Parameter μ is determined to optimize bond distances of homonuclear diatomic molecules up to the third period as μ =0.33. This scheme is called "long-range correction (LC) scheme". The present LC-GGA is compared with the conventional B3LYP functional in Fig.1.

One of the most critical DFT problems is the poor reproducibility of van der Waals (vdW) bondings. Several DFT studies have been made on vdW calculations by using, *e.g.*, a perturbation theory based on DFT. The most effective and general way may be the use of a vdW functional. Various types of vdW functionals have been suggested and some of these functionals reproduce accurate vdW C_6 coefficient comparable to the results of high-level *ab initio* methods. However, these functionals give poor vdW bondings of, *e.g.*, rare gas dimmers by simply combining with a conventional exchange-correlation functional in DFT calculations. It is due to the lack of long-range interactions in exchange functionals, because vdW bondings are supposed to be in the balance between attractive dispersion and long-range Pauli repulsion. To supplement dispersion interactions into DFT, we [4,5] employed the ALL vdW correlation functional [6]. This functional was developed to reproduce the exact vdW interaction for both separated uniform electron-gas regions and separated atoms. The functional is given by

$$E_{vdw}^{ALL} = -\frac{6}{(4\pi)^{3/2}} \sum_{A}^{N_{atom}} \sum_{B}^{N_{atom}} \int_{V_A} \int_{V_B} d^3\mathbf{r}_1 d^3\mathbf{r}_2 \frac{\rho^{1/2}(\mathbf{r}_1)\rho^{1/2}(\mathbf{r}_2)}{\rho^{1/2}(\mathbf{r}_1)+\rho^{1/2}(\mathbf{r}_2)} \frac{1}{r_{12}^{\,6}}, \tag{8}$$

where $\rho(r)$ is the total electron density at position $\mathbf{r}$, and N_{atom} is the number of constituent atoms. Position vectors, $\mathbf{r}_1$ and $\mathbf{r}_2$, run over the volumes V_A and V_B, which qualitatively correspond to the volume of each pair of constituent A and B (A $\neq$ B) atoms. These volumes are defined without ambiguity through the following cut-off criteria. According to a zero wave vector theory, a cut-off criterion is employed to skip the regions where the length of scale for the change of local Fermi wave vector is smaller than the electron screening length such that,

$$\frac{|\nabla\rho|}{6\rho} \geq \frac{\omega_p}{v_F}, \tag{9}$$

where $\omega_p = (4\pi\rho)^{1/2}$ is the local plasma frequency and $v_F = (3\pi^2\rho)^{1/3}$ is the local Fermi velocity.

In practical calculations, equation (8) is usually evaluated by using a numerical quadrature. Since equation (8) has a grid–grid pairwise formula, it diverges as r_{12} goes to zero due to the $1/r_{12}^{\,6}$ term. To avoid this divergence, we introduced the grid–grid pairwise damping function, $f_{damp}(r_{12})$, which provides zero for small r_{12} and gives one for sufficiently large r_{12}. For this purpose, we adapted an exponential form,

$$f_{damp}(r_{12}) = \exp[-(\frac{\alpha}{r12})^6], \tag{12}$$

where α is the parameter which determines the size of the cut-off region. Since the parameter α has the dimension of length, α should be determined by taking the effective size of pairwise regions into consideration. By using the damping function in equation (10), the grid–grid pairwise formula of the ALL vdW energy in equation (10) is expressed as

$$E_{vdw}^{ALL} \cong -\frac{6}{(4\pi)^{3/2}} \sum_{\mathbf{r}_1 \in V_1} \sum_{\mathbf{r}_2 \in V_2} w(\mathbf{r}_1)w(\mathbf{r}_2) \frac{\rho^{1/2}(\mathbf{r}_1)\rho^{1/2}(\mathbf{r}_2)}{\rho^{1/2}(\mathbf{r}_1)+\rho^{1/2}(\mathbf{r}_2)} \frac{1}{r_{12}^{\,6}} f_{damp}(r_{12}), \tag{13}$$

where w is the weight of grids at reference points.

In usual DFT calculations of molecules, numerical grids are spherically distributed around constituent atoms. It is, therefore, useful to write equation (13) as a summation of atomic pair contributions,

$$E_{vdw}^{ALL} = \sum_{A \in V_1} \sum_{B \in V_2} E_{AB}, \tag{14}$$

where

$$E_{AB} = -\frac{6}{(4\pi)^{3/2}} \sum_{\mathbf{r}_1 \in A} \sum_{\mathbf{r}_2 \in B} w(\mathbf{r}_1) w(\mathbf{r}_2) \frac{\rho^{1/2}(\mathbf{r}_1)\rho^{1/2}(\mathbf{r}_2)}{\rho^{1/2}(\mathbf{r}_1)+\rho^{1/2}(\mathbf{r}_2)} \frac{1}{r_{12}{}^6} f_{damp}(r_{12}), \tag{15}$$

in which A and B are the labels of constituent atoms in monomers 1 and 2, respectively, and w is the weight of grids at reference points. Position vectors, $\mathbf{r}_1$ and $\mathbf{r}_2$, run over the grid points distributed around atoms A and B, respectively. For the damping function, $f_{damp}(r_{12})$, we adopted the exponential form,

$$f_{damp}(r_{12}) = \exp[-(\frac{\alpha_{AB}}{r12})^6], \tag{16}$$

where α_{AB} is the parameter that determines the size of the cutoff region for participating atomic pairs. Considering the effective size of atoms, we defined this parameter as a linear formula,

$$\alpha_{AB} = C_1 R_m + C_2, \tag{17}$$

where R_m is the sum of the Bondi's vdW radii[45] of A and B atoms, and coefficients C_1 and C_2 are determined to reproduce accurate potential energy curves of rare-gas dimers as C_1=0.4290 and C_2=1.8949 in atomic units.

3. Application Calculations

The LC scheme has been examined by illustrating its applicabilities to some problems that have never been solved: 1. underestimations of Rydberg excitation energies, oscillator strengths, and charge transfer excitation energies in time-dependent density functional calculations, 2. systematic underestimation of reaction barrier heights, and 3. poor reproducibilities of van der Waals sinteractions.

Table 1 summarizes mean absolute errors in calculated excitation energies of five typical molecules by TDDFT. The table also displays calculated results of asymptotically-corrected AC (AC-BOP and LBOP) and hybrid B3LYP functionals, which are mentioned in the former section. The *ab initio* SAC-CI results are also shown to confirm the accuracies. The 6-311G++(2d,2p) basis set was used in TDDFT calculations [3]. As the table indicates, the LC scheme clearly improves Rydberg excitation energies that are underestimated for pure BOP functional, at the same (or better) level as the AC scheme does. It should be noted that LC and AC schemes also provide improvements on valence excitation energies for all molecules. LC and AC results are comparable to SAC-CI results. The LB scheme clearly modifies Rydberg excitation energies, and however brings underestimations of valence excitation energies. B3LYP results are obviously worse than LC and AC results for both valence and Rydberg excitation energies. We have calculated excitation energies of systems including benzene, five-membered ring compounds, free- and metal porphyrins, etc. and estimated the deviation from the experiments. TDDFT results with pure functional such as BLYP underestimated the Rydberg excitation energies by 1.54 eV. B3LYP improves the underestimation but the deviation is still 0.89 eV. The present LC-BLYP gives the mean absolute deviation of 0.35 eV for Rydberg excitation energy.

Calculated oscillator strengths of excited states by TDDFT are shown in Table 2. LC scheme drastically improves oscillator strengths, which are underestimated for BOP as second to hundredth part of experimental values, to the same digit. Although AC-BOP, LBOP, and B3LYP also provide closer oscillator strengths to the experimental values than BOP do, the accuracies are unsatisfactory in comparison with LC-BOP ones. It is, therefore, concluded that the lack of long-range interactions in exchange functional may also cause the underestimations of oscillator strengths in TDDFT calculations.

As an example of the charge transfer excitation, transition energies of ethylene-tetrafluoroethylene dimmer were calculated. Dreuw *et al.* recently suggested that poor charge transfer excitation energies of far-aparted molecules may be one of the main problems of TDDFT [7]. They pointed out that intermolecular charge transfer excitation energies of far-aparted molecules should have the correct asymptotic behavior for long intermolecular distance. That is, for long molecular-molecular distances R and R_0 ($R > R_0$), charge transfer energy ω_{CT} should satisfy

$$\omega_{CT}(R) - \omega_{CT}(R_0) \geq -\frac{1}{R} + \frac{1}{R_0}. \tag{18}$$

Fig.2 shows that LC-BOP gives the correct asymptotic behavior as is different from AC-BOP and LBOP do [4]. Although B3LYP recovers a part of this behavior, the degree is in proportion to the mixing rate of the HF exchange integral. Hence, this result may also indicate that problems in conventional TDDFT calculations come from the lack of long-range exchange interactions in exchange functionals rather than the poor far-nucleus asymptotic behavior of exchange functionals.

We also calculated the barrier heights and heats of reaction of more than 100 chemical reactions. The mean absolute error is 2.7 kcal/mol while that of B3LYP is 4.9 kcal/mol. LC scheme remedies the underestimation of the barrier heights.

The simplest LDA predicts the binding character of the vdW interactions. However, LDA severely overestimates the binding energy and yields a too short vdW bond. The GGA predicts purely repulsive vdW interactions. Thus, none of the conventional functionals account successfully for the vdW interaction. MP2 significantly overestimates the binding energies and MP2 results have strong basis set dependence. Only CCSD(T) with a large basis set gives the accurate estimation. The potential energy curves of naphthalene dimmer in several configurations (Fig. 3) are calculated [5]. From previous high-level *ab initio* MO calculations, it was found that PD configuration was energetically favored than T configuration in contrast that these configurations were almost isoenergetic in benzene dimer. We adopted BOP exchange-correlation functional and 6-31+G** basis set. Calculated potential energy curves are shown in Fig.4, and equilibrium intermolecular distances and binding energies are presented in Table 3. For PD configuration, we first investigated the minimum structure by varying three vertical distances (R_1, R_2, and R_3), and drew the potential curve across the fixed orientation to the minimum. As can be seen from the figure and table, LC-DFT+ALL method represents quite accurate potential for planner structure (P, C, and PD configuration), and reproduced the correct order of stability in PD and T configurations. However LC-DFT+ALL gives substantially (approximately 1.5 kcal/mol) deeper well for T configuration compared with CCSD(T) results. This trend was also seen for benzene dimer, and it seems that LC-DFT+ALL method somewhat overestimates the CH-π interaction. But these overestimations were always smaller than those by MP2 for naphthalene dimer.

4. Summary

The hybrid functionals can be improved through the introduction of an Ewald partitioning of $1/r_{12}$. The new LC-GGA functional with correct long-range electron-electron interactions has good energetics, good Rydberg behavior and good charge transfer predictions. It also describes the long-range exchange repulsion accurately and the LC scheme combined with the Andersson-Langreth-Lundqvist functional is proved to give the van der Waals interactions successfully.

The LC scheme gives better results for various chemical properties than those with conventional corrections including hybrid functionals. For some properties, the LC scheme provides equivalent improvements in comparison with B3LYP. This may indicate that accurate B3LYP results may be due to the equivalency in the mixed HF exchange energy to the LC scheme, rather than the validity of the constant weight hybridization of the HF exchange. This argument may require further examination of the LC scheme.

Table 1: Calculated vertical excitation energies of furan molecule in eV.

Excitation	LC-BOP	LC-BLYP	B3LYP	BOP	BLYP	SAC-CI[a]	MRMP[b]	CASPT2[c]	Expt
1 $^1A_2$ ($1a_2 \rightarrow 3s$)	6.15	6.05	5.44	5.09	5.03	5.99	5.84	5.92	5.91[d,e]
1 $^1B_2$ ($\pi \rightarrow \pi^*$)	6.23	6.20	5.97	5.72	5.70	6.40	5.95	6.04	6.06[f,i]
1 $^1A_1^-$ ($\pi \rightarrow \pi^*$)	7.12	7.10	6.74	6.38	6.36	6.79	6.16	6.16	
1 $^1B_1$ ($1a_2 \rightarrow 3pb_2$)	6.68	6.60	5.92	5.56	5.50	6.45	6.40	6.46	6.48[f,j,k]
2 $^1B_2$ ($1a_2 \rightarrow 3pb_1$)	6.93	6.91	6.34	5.99	6.01	6.82	6.50	6.48	6.48[f,j,k]
2 $^1A_2$ ($1a_2 \rightarrow 3pa_1$)	6.89	6.82	6.07	5.69	5.64	6.66	6.53	6.59	6.61[f,h]
3 $^1A_2$ ($1a_2 \rightarrow 3da_1$)	7.14	7.07	6.42	5.95	5.97	7.04	6.98	7.00	
2 $^1B_1$ ($1a_2 \rightarrow 3db_2$)	7.36	7.32	6.44	5.95	5.96	7.14	7.10	7.15	
4 $^1A_2$ ($1a_2 \rightarrow 3da_1$)	7.41	7.38	6.56	6.04	6.04	7.27	7.18	7.22	
3 $^1B_2$ ($1a_2 \rightarrow 3db_1$)	7.68	7.72	6.76	6.33	6.36	7.51	7.18	7.13	
2 $^1A_1$ ($1a_2 \rightarrow 3da_2$)	7.55	7.58	6.59	6.03	6.09	7.36	7.26	7.31	7.28[j]
4 $^1B_1$ ($2b_1 \rightarrow 3s$)	7.65	7.56	6.83	6.32	6.26	7.45	7.31	7.21	7.38[f,j,l]
4 $^1A_1^+$ ($\pi \rightarrow \pi^*$)	8.36	8.34	8.36	8.15	8.13	8.34	7.69	7.74	7.82[i]

Table 2: Mean absolute errors in calculated excitation energies of five typical molecules by TDDFT in eV.

Molecule		LC-BOP	BOP	AC-BOP	LBOP	B3LYP	SAC-CI
N_2	Valence	0.36	0.40	0.27	1.48	0.54	0.33
	Rydberg	0.90	2.37	0.84	0.43	1.30	0.25
	Total	0.54	1.06	0.46	1.13	0.79	0.30
CO	Valence	0.19	0.28	0.17	1.02	0.36	0.26
	Rydberg	0.75	2.06	0.79	0.42	1.16	0.27
	Total	0.47	1.17	0.48	0.72	0.76	0.27
H_2CO	Valence	0.25	0.59	0.24	0.52	0.26	0.45
	Rydberg	0.47	1.66	0.59	0.07	0.84	0.13
	Total	0.40	1.30	0.47	0.22	0.64	0.24
C_2H_4	Valence	0.30	0.47	0.24	1.52	0.47	0.11
	Rydberg	0.18	1.41	0.58	0.69	0.92	0.17
	Total	0.20	1.28	0.53	0.80	0.85	0.16
C_6H_6	Valence	0.21	0.28	0.24	0.84	0.26	0.35
	Rydberg	0.24	1.01	0.88	0.35	0.56	0.15
	Total	0.23	0.74	0.64	0.53	0.44	0.22

Table 3: Calculated equilibrium intermolecular distances in Å and binding energies in kcal/mol for P, C, T, and SP configurations of naphthalen dimer.

	P	C	T	PD			
Intermolecular distance(Å)	R	R	R	R_1	R_2	R_3	R
LC-BOP+ALL/6-31+G**[a]	3.9	3.6	5.1	3.5	1.4	1.3	4.0
estd. CBS CCSD(T)[b]	3.8	3.6	5.0	3.5	1.4	1.0	3.9
Binding energy (kcal/mol)							
LC-BOP+ALL/6-31+G**[a]	4.17	5.63	5.83	6.16			
estd. CBS MP2[b]	7.49	9.66	6.56	10.3			
estd. CBS CCSD(T)[b]	3.78	5.28	4.34	5.73			

[a]The present scheme
[b] S. Tsuzuki, K. Honda, T. Uchimaru, and M. Mikami, *J. Chem. Phys.* **120**, 647 (2004)

Figure 1: Comparison of B3LYP and LC-GGA exchange functionals

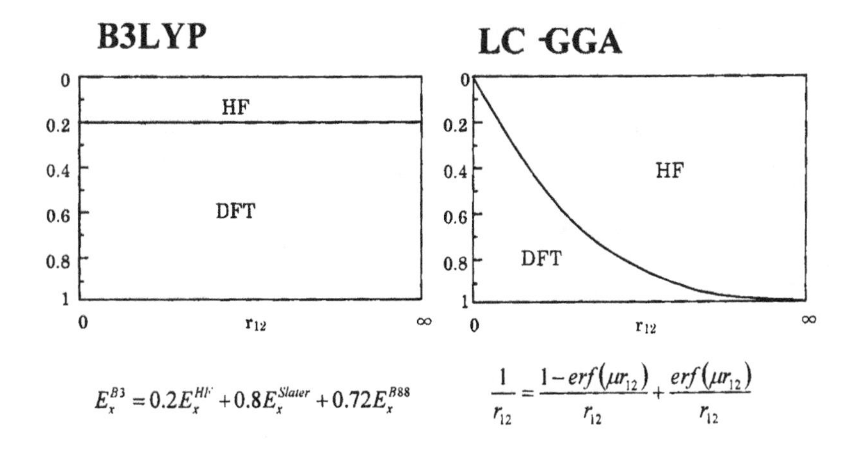

Figure 2: Comparison of the long-range behavior of the lowest CT state of the ethylene-tetrafluoroethylene dimer along the internuclear distance coordinate.

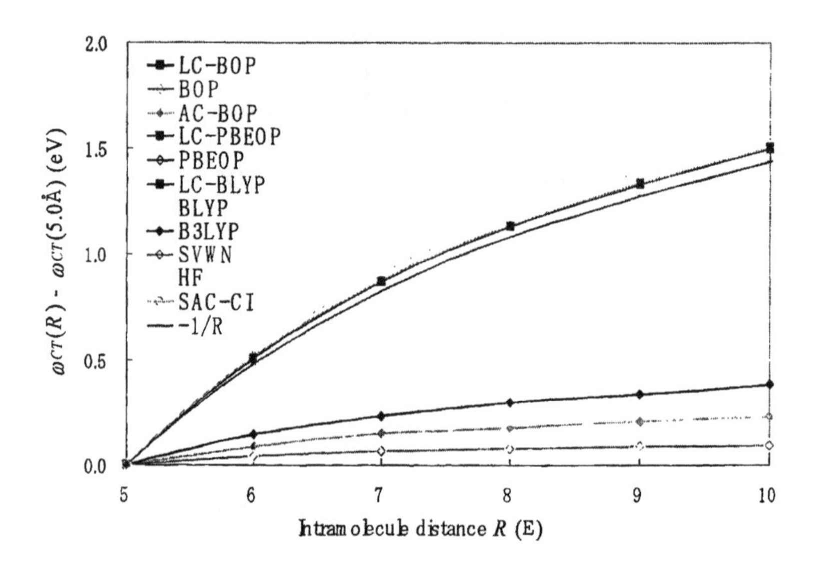

Figure 3: P, C, T, and PD configurations of naphthalene dimer.

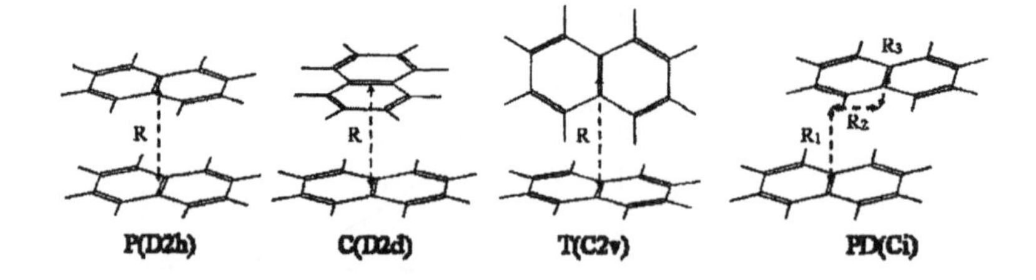

Figure 4: Calculated potential energy curves of naphthalene dimer in P, C, T, and PD configurations. The equilibrium intermolecular distances and interaction energies by estimated CBS CCSD(T) (Ref. 58) are also plotted for comparison.

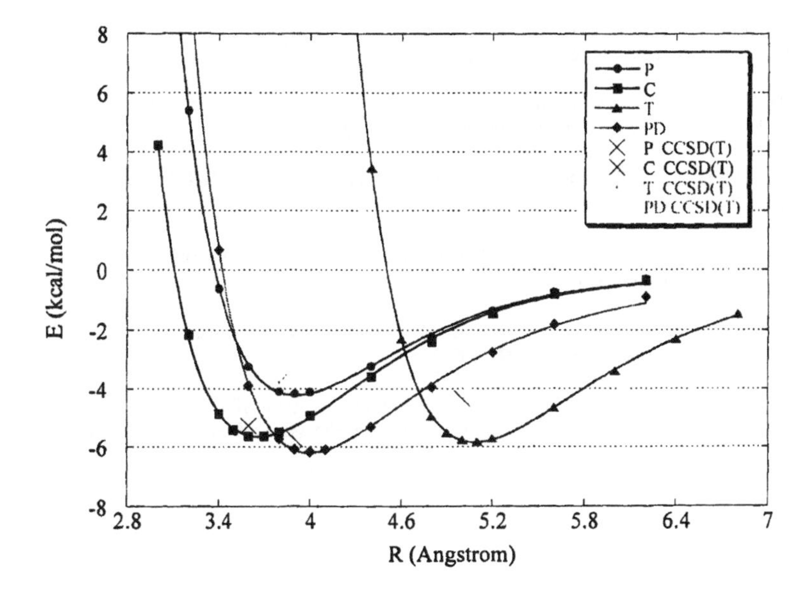

Acknowledgments

This research was supported in part by a grant-in-aid for Scientific Research in Specially Promoted Research "Simulations and Dynamics for Real Systems" and the Grant for 21st Century COE Program "Human-Friendly Materials based on Chemistry" from the Ministry of Education, Science, Culture, and Sports of Japan.

References

[1] A. Savin, *Recent Developments and Applications of Modern Density Functional Theory*, (Ed. J. M. Seminario) Elsevier, Amsterdam, 327 (1996).

[2] H. Iikura, T. Tsuneda, T. Yanai, and K. Hirao, A long-range correction scheme for generalized-gradient-approximation exchange functionals, *J. Chem. Phys.* **115**, 3540-3544 (2001)

[3] Y. Tawada, T. Tsuneda, S. Yanagisawa, T. Yanai, and K. Hirao, A long-range-corrected time-dependent density functional theory, *J.Chem.Phys.*, **120**, 8425-8433 (2004).

[4] M.Kamiya, T.Tsuneda, and K.Hirao, A density functional study of van der Waals interactions *J.Chem.Phys.* **117**, 6010 (2002).

[5] T.Sato, T.Tsuneda, and K.Hirao, A density functional study on pi-arpmotic interactions: Benzene dimer and naphthalene dimmer, *J.Chem.Phys.* **123**, 104307 (2005).

[6] Y. Andersson, D. C. Langreth and B. I. Lundqvist, Density Functional for van der Waals Forces at Surfaces, *Phys. Rev. Lett.* **76**, 102 (1996).

[7] A. Dreuw, J. L. Weisman, and M. Head-Gordon, Long-range charge-transfer excited states in time-dependent density functional theory require non-local exchange, *J.Chem.Phys.***119**, 2943 (2003).

Brill Academic Publishers
P.O. Box 9000, 2300 PA Leiden
The Netherlands

Lecture Series on Computer and Computational Sciences
Volume 6, 2006, pp. 165-176

Site-Specific Polarizabilities: Probing the Atomic Response of Silicon Clusters to an External Electric Field

K. Jackson[1] and M. Yang
Physics Department, Central Michigan University, Mt. Pleasant, Michigan 48859, USA

J. Jellinek[2]
Chemistry Division, Argonne National Laboratory, Argonne, Illinois 60439, USA

Received 14 July, 2006; accepted 25 July, 2006

Abstract: A new method for computation of local, site-specific polarizabilities in atomic clusters is presented and applied to silicon clusters, Si_N. The procedure is based on partitioning the cluster volume into atomic volumes defined as Voronoi cells associated with the atomic positions. Local contributions to the overall cluster polarizability are computed from changes in the charge density in each atomic volume caused by the applied electric field. The local polarizabilities are further decomposed into dipole and charge transfer terms. It is shown that the atomic, or site specific, polarizabilities vary substantially from site to site of a Si_N cluster, with more peripheral atoms having larger values. The charge transfer component accounts for 70% of the total polarizability in the case of the two isomers of Si_{26} studied.

Keywords: Clusters, polarizability, electric dipoles, size-dependence of properties, size-induced transition to metallicity

PACS: 36.40.-c, 32.10.Dk, 36.40.Cq

1. Introduction

Nanoscale atomic clusters exhibit unusual bonding arrangements that are different from those in bulk materials and that can change significantly with the number of atoms in the cluster. As a consequence, the physical and chemical properties of clusters are also different from those of their bulk counterparts. An example is the electric dipole polarizability, which is a measure of the response of a system to an external electric field. The polarizabilties of clusters are very different from the known polarizabilities of bulk materials and they exhibit strong and nonmonotonic size-variations. Cluster polarizabilities have been studied both experimentally and theoretically using beam-deflection techniques [1-6] and quantum chemical and density functional theory (DFT) computations [7-10].

In this paper, we formulate a new scheme that allows for partitioning of the total polarizability of a cluster into nonoverlapping contributions of its constituent atoms. Our motivation is to look beyond the global response, as captured by the total cluster polarizability, and analyze the local, atomic responses. The intent is similar to that of the "atoms in molecules" approach, [11-16] with an emphasis on the site-specific effects. For example, we are interested in the differences between the responses to an external electric field shown by atoms occupying more peripheral and less peripheral sites of a cluster.

The utility of the scheme is illustrated through applications to Si_N clusters. These have been studied extensively both experimentally [17-20] and computationally [21-25]. Among their interesting characteristics is the shape transition they undergo as they grow in size. Starting at N=10, where the lowest energy structure is a compact configuration, the energetically preferred conformations of Si_N become ever more elongated with a prolate shape. This prolate growth pattern persists up to N=28. At N=20, however, a new, energetically competitive structural pattern emerges that is represented by near-spherical, compact structures. These latter become the single "winners" for cluster sizes larger than N=28. The "coexistence" of the prolate and compact structures in the size range of N=20-28 has been

[1] Corresponding author; E-mail: jackson@phy.cmich.edu
[2] Corresponding author; E-mail: jellinek@anl.gov

demonstrated experimentally [20]. Recent computations [24] identified the likely atomic structures of intermediate size Si_N clusters.

A natural question one can ask is how the size-driven changes in the preferred structure and shape of Si_N are reflected in their polarizabilities? Earlier computational studies [26-33] have indicated that silicon clusters are more polarizable than bulk silicon. The issue of the correlation between the shape of the cluster and its polarizability has also been addressed [27,28,33]. Recently it has been shown [33] that the size-dependent trends in the polarizability of Si_N as obtained in DFT computations can be fitted to the predictions of jellium-based models [34,35]. This has been interpreted as an indication of the metallic nature of silicon clusters. Below we use the site-specific polarizabilities to further comment on this issue.

The paper is organized as follows. The new scheme for partitioning the total polarizability into site-specific contributions is formulated in the next section. Section 3 outlines the details of the computational methodology. The results are presented and discussed in Section 4. A brief summary and outlook are given in Section 5.

2. Partitioning of the Polarizability into Site-Specific Contributions

The energy E of a cluster in a weak external electric field **F** can be expressed as

$$E(\mathbf{F}) = E_0 - \boldsymbol{\mu}\cdot\mathbf{F} - \frac{1}{2}\boldsymbol{\alpha}\cdot\mathbf{F}\cdot\mathbf{F}, \tag{1}$$

where $\boldsymbol{\mu}$ is the electric dipole moment, and $\boldsymbol{\alpha}$ is the polarizability tensor. From this relationship, the dipole moment and the polarizability can be defined as derivatives of the total energy with respect to F:

$$\mu_i = \left.\frac{dE}{dF_i}\right|_{F=0}, \tag{2a}$$

$$\alpha_{ij} = \left.\frac{d^2 E}{dF_i dF_j}\right|_{F=0} = \left.\frac{d\mu_i}{dF_j}\right|_{F=0}. \tag{2b}$$

The electric dipole moment can also be expressed as

$$\mu_i = \int r_i \rho(\mathbf{r}) d^3\mathbf{r}, \tag{3}$$

where r_i is the is i-th component of the vector **r** (i=x, y, z), ρ is the cluster charge density, and the integration extends over the entire space. The charge density ρ incorporates contributions from both the electrons and the nuclei.

The site-specific contributions to the dipole moment can be obtained by decomposing space into atomic regions, where Ω_A, the volume associated with the atom *A*, is defined as the locus of points that are closer to the nucleus of atom *A* than to any other nucleus. The Ω_As are thus the Voronoi cells, or, in the nomenclature of solid state physics, the Wigner-Seitz cells. Partitioning the integration over the entire space, Eq. (3), into a sum of integrals over the atomic regions Ω_A, one arrives at partitioning of the total dipole moment μ_i into atomic contributions μ_i^A,

$$\mu_i = \sum_A \mu_i^A, \tag{4}$$

where

$$\mu_i^A = \int_{\Omega_A} r_i \rho(\mathbf{r}) d^3\mathbf{r}. \tag{5}$$

One can easily show that the electric dipole moment for a given charge distribution is unique and independent of the location of the origin of the system of coordinates, provided the distribution integrates to zero net charge. Thus, for a neutral cluster the sum in Eq. (4) is defined uniquely. This is not necessarily the case for the individual atomic dipole moments defined by Eq. (5), as the net charge in a given atomic region Ω_A may not be zero. The general origin dependence of μ^A can be made explicit by rewriting Eq. (5) as

$$\mu_i^A = \mu_i^{A,p} + \mu_i^{A,q} , \tag{6a}$$

where

$$\mu_i^{A,p} = \int_{\Omega_A} (r_i - R_i^A)\rho(\mathbf{r})d^3\mathbf{r} , \tag{6b}$$

$$\mu_i^{A,q} = \int_{\Omega_A} R_i^A \rho(\mathbf{r})d^3\mathbf{r} = R_i^A \int_{\Omega_A} \rho(\mathbf{r})d^3\mathbf{r} , \tag{6c}$$

and $\mathbf{R}^A$ defines the position of the nucleus of atom *A*. The term $\mu_i^{A,p}$ defined by Eq. (6b) represents the local dipole moment of the charge in Ω_A with respect to $\mathbf{R}^A$, and it is clearly origin independent. The term $\mu_i^{A,q}$ defined by Eq. (6c) can also be written as

$$\mu_i^{A,q} = q^A R_i^A , \tag{7}$$

where q^A is the net charge in the volume Ω_A (i.e., the charge of atom A), and it does depend on the choice of the origin of the system coordinates through $\mathbf{R}^A$. Clearly, $\mu_i^{A,q}$ is an effective dipole moment of atom A with respect to the origin, when q^A is viewed as a point charge placed at $\mathbf{R}^A$.

Using Eq. (6a), one can rewrite Eq. (2b) in the form

$$\alpha_{ij}^A = \alpha_{ij}^{A,p} + \alpha_{ij}^{A,q} , \tag{8a}$$

where

$$\alpha_{ij}^{A,p} = \left. \frac{d\mu_i^{A,p}}{dF_j} \right|_{F=0} \tag{8b}$$

and

$$\alpha_{ij}^{A,q} = \left. \frac{d\mu_i^{A,q}}{dF_j} \right|_{F=0} . \tag{8c}$$

The $\alpha_{ij}^{A,p}$ component of the atomic polarizability α_{ij}^A characterizes the change in the local dipole moment of the atomic volume Ω_A triggered by an external field. It can be viewed as a measure of a dielectric (i.e., charge shift) type of a response. The $\alpha_{ij}^{A,q}$ component accounts for the effect of the change in the value of the atomic charge q^A caused by the field. It serves as a measure of a charge transfer type of a response. Whereas $\alpha_{ij}^{A,p}$ is defined uniquely, $\alpha_{ij}^{A,q}$ depends on the choice of the origin of the system of coordinates. This is transparent when one rewrites Eq. (8c) as

$$\alpha_{ij}^{A,q} = R_i^A \left. \frac{dq^A}{dF_j} \right|_{F=0} . \tag{9}$$

In effect, the partitioning scheme outlined above represents the true cluster charge distribution by point charges and point dipole moments associated with the individual Ω_As or, alternatively, atoms. The changes in these charges and dipole moments caused by vanishingly weak external fields determine the atomic polarizabilities and, consequently, the cluster total polarizability. The latter can be written as

$$\alpha_{ij} = \alpha_{ij}^p + \alpha_{ij}^q , \tag{10}$$

where α_{ij}^{p} and α_{ij}^{q} are sums over all the atoms of the $\alpha_{ij}^{A,p}$ and $\alpha_{ij}^{A,q}$ terms, respectively. Equation (10) shows that the polarizability of a cluster can be partitioned into "dielectric" (α_{ij}^{p}) and "charge transfer" (α_{ij}^{q}) components. For a neutral cluster, both are origin independent (i.e., both are defined uniquely).

3. Computational Methodology

We applied the scheme outlined in the preceding section to Si_N clusters. The computations were performed using the gradient-corrected version of DFT with the Perdew, Burke and Ernzerhof (PBE) [36] exchange-correlation functional as implemented in the NRLMOL code [37,38]. The all-electron basis set for the Si atoms was constructed from 16 primitive Gaussian functions through contraction into 6 s-type, 5 p-type, and 4 d-type orbitals.

NRLMOL employs a numerical integration procedure [37] to compute the matrix elements and the total energy. The integration grid points and the associated volume elements are constructed to accurately evaluate three-dimensional integrands that are strongly peaked near the nuclear positions and decay slowly with the distance from the nuclei. The mesh generation algorithm is based on dividing the space into a large number of non-overlapping regions and using adaptive quadrature techniques to define region-specific grid points and their volume elements. The choice of these is guided by user-specified accuracy criteria that are to be met in integration of test integrands that mimic the spatial variation of the integrands encountered in DFT computations. For further details we refer the reader to Ref. 37. The resulting grids are relatively dense near the atomic positions and are coarser in the interstitial and remote regions of a cluster. The standard settings in the code typically yield an integrated total cluster charge that is accurate to better than 1.0E-05 electron, and an integrated total cluster dipole moment that agrees with the value obtained by differentiating the total energy with respect to the applied field [cf. Eq. (2a)] to 1.0E-04 atomic units.

The discretized version of the Eqs. (6a)-(6c) is

$$\mu_i^A = \sum_{\mathbf{r}_s \in \Omega_A} (r_{si} - R_i^A)\rho(\mathbf{r}_s)w_s + \sum_{\mathbf{r}_s \in \Omega_A} R_i^A \rho(\mathbf{r}_s)w_s, \tag{11}$$

where w_s is the volume element associated with the integration grid point r_s, and the sums extend over all the grid points lying in the volume Ω_A. Since each grid volume element is necessarily finite, the integration volume for a given atom is an approximation to a Voronoi cell. If a grid point lies near the boundary between two cells, its associated volume element may straddle the boundary, effectively transferring a portion of the volume of one cell into the other. The effect of this blurring of the boundaries on the computed values of α_{ij}^{p} and α_{ij}^{q} is small (see below).

The derivatives in Eqs. (8b) and (8c) are computed using the finite difference method. An extra term, $-\mathbf{F}\cdot\mathbf{r}$, representing the potential energy of an electron in an external electric field $\mathbf{F}$ is added to the potential in the Kohn-Sham equation. The modified equation is then solved self-consistently to yield the charge density in the presence of the field. The computations are performed for fields oriented along the "+" and "-" directions of the *x*, *y*, and *z* axes. The polarizability components are then evaluated as

$$\alpha_{ij}^{A,p} = \frac{\mu_i^{A,p}(+F_j) - \mu_i^{A,p}(-F_j)}{2F_j}, \tag{12}$$

where F_j is the magnitude of the external field applied in the j-th coordinate direction. The value of 0.005 a.u. (1 a.u. = 27.21 V/bohr) is used for F_j. It has been found to yield well converged results [27]. In the analysis presented in the next section we use the averaged quantities

$$<\alpha> = \frac{1}{3}\sum_i \alpha_{ii}. \tag{13}$$

$$<\frac{dq^A}{dF}>=\sqrt{\left(\frac{dq^A}{dF_x}\right)^2+\left(\frac{dq^A}{dF_y}\right)^2+\left(\frac{dq^A}{dF_z}\right)^2}\ . \tag{14}$$

4. Results and Discussion

A. Si_3: An illustrative example

As an introductory example, we first consider the application of the partition scheme introduced in Section 2 to Si_3. As shown in panel *a* of Fig. 1, the lowest energy configuration of this cluster is an isosceles triangle. The two shorter interatomic distances are 2.20 E whereas the longer one is 2.84 E. The cluster has a HOMO-LUMO gap of 1.02 eV. Panel *a* of Fig. 1 depicts the cluster in zero external field. The value of the local charge q^A and the magnitude of the local dipole moment μ^A in each region Ω_A are also given. The arrows indicate the direction of the atomic dipoles. The charge and the dipole moment of the apex atom differ from those of the other two atoms because of the difference in the respective bonding environments. The three local dipoles point in different directions and nearly cancel each other. The net dipole for the cluster has a value of 0.112 a.u. and points downward.

To illustrate the origin of the local charges and dipoles, the deformation charge density defined as

$$\rho'(\mathbf{r})=\rho(\mathbf{r})-\sum_A \rho_0(\mathbf{r}-\mathbf{R}^A) \tag{15}$$

is shown in panel *b* of Fig. 1. Here $\rho_0(\mathbf{r})$ is the self-consistent charge density of the free atom, computed with zero net spin. The deformation charge reflects how bonding in the cluster rearranges the overlapping free atom charge densities. A build-up of density is clearly visible in panel *b* between the atoms on the shorter sides of the triangle and at the center of the triangle. This suggests a mixture of covalent and delocalized bonding in the cluster. The dotted lines in the figure indicate where the boundaries between the atomic regions intersect the plane of the atoms. These lines make it clear that the bonding charge is distributed unevenly among the atomic regions. More charge falls into Ω_1

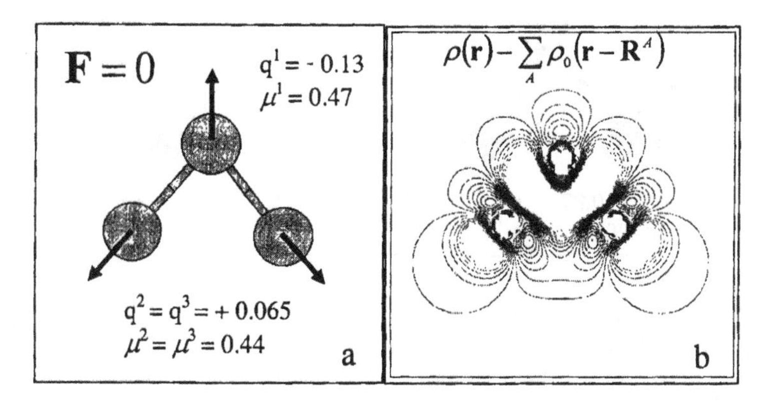

Figure 1: a) The Si_3 cluster in the absence of an external field. The computed values of the local charges (in electrons) and dipole moments (in electron·bohrs) are indicated for each atomic volume. The arrows show the direction of the local dipoles; b) Contour plot of the deformation charge density for Si_3 (see the text for details). A build-up of electronic charge can be seen between atoms on the shorter sides of the triangular framework and in the center of the triangle. The dotted lines indicate the boundaries between the atomic volumes.

making q^1 slightly negative and $q^2=q^3$ slightly positive. Likewise, the local dipoles in the individual regions are nonzero, and they point in directions defined by the positions of the nuclei and the centroids of the negative charges of the atoms.

The effect of an applied electric field on the charge distribution in Si_3 is shown in Fig. 2. Panels *a* and *b* of the figure show, respectively, $\delta\rho = \rho(\mathbf{F}) - \rho(\mathbf{0})$, the difference between the self-consistent cluster charge density computed with the indicated fields and the zero field density. The blue contours indicate a depletion of electron charge, whereas the red contours reflect its accumulation. The electronic charge is clearly driven in a direction parallel to the applied field, and the largest charge transfer occurs in the cluster exterior.

The differences in the local dipoles, $\delta\mu^A = \mu^A(\mathbf{F}) - \mu^A(\mathbf{0})$, are shown as vectors in panels *c* and *d* of Fig. 2. The magnitudes of these differences and the changes in the local charges, $\delta q^A = q^A(\mathbf{F}) - q^A(\mathbf{0})$, are also indicated. The changes in the local dipoles are largely in the direction of the applied field. The magnitudes of the changes depend on the direction of the field and on the atomic site within the cluster.

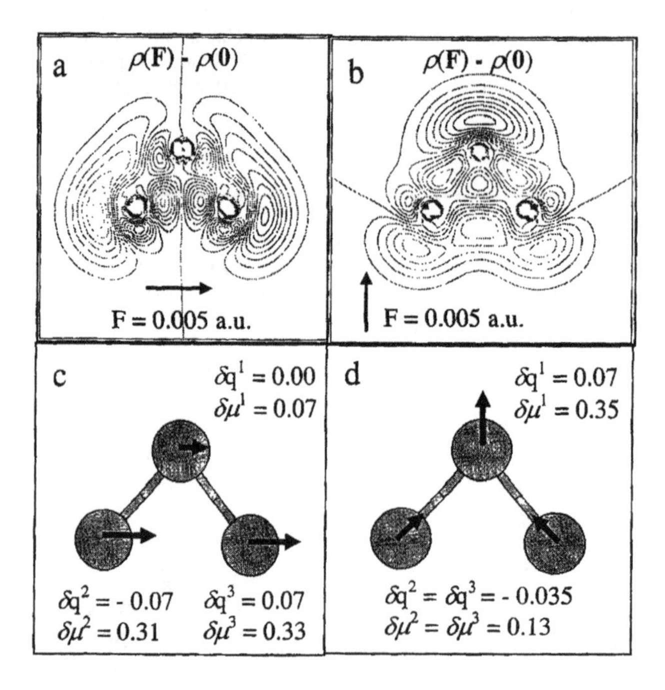

Figure 2: a) and b): Contour plots of the difference charge density ($\rho(\mathbf{F}) - \rho(\mathbf{0})$) for Si_3 computed for an external field $\mathbf{F}$ of magnitude 0.005 a.u. and oriented as shown; c) and d): Computed values of $\delta q^A = q^A(\mathbf{F}) - q^A(\mathbf{0})$ (in electrons) and $\delta\mu^A = |\mu^A(\mathbf{F}) - \mu^A(\mathbf{0})|$ (in electron·bohrs) for the same fields as in panels a and b, respectively. The arrows indicate the directions of $\delta\mu^A$.

The mesh in NRLMOL can be adjusted to create different number of integration grid points. As mentioned above, the effective location of the boundaries between the Ω_A regions depends on the location of the grid points. It is, therefore, important to evaluate the effect of the choice of the grid on the computed site-specific quantities. In Table 1 we list values of various quantities computed with different number of grid points. A measure of the overall performance of a mesh is the total electron charge Q of the cluster. Under the standard settings in NRLMOL, the mesh contains 27 000 grid points and, as mentioned above, it yields values of Q accurate to approximately 1.0E-5 electron.

Inspection of Table 1 shows that the net local charges q^A vary by as much as 0.07 electron over the range of meshes considered. Although this value is only a small fraction of the total electron charge of approximately 14 electrons in each Ω_A, it is a significant portion of the net local charge. The local

Table 1: Values of the total electron charge Q, net local charge q^A, local dipole moment μ^A (in a.u.: electron·bohrs), average local polarizability $\langle\alpha^A\rangle$ (in a.u.: bohr^3), average dipole component $\langle\alpha^{A,p}\rangle$ of the local polarizability (in bohr^3), and $\langle dq^A/dF\rangle$ (in a.u.: bohr^2) for Si_3 obtained using different number of grid points in the integration mesh. The numbering of the atoms refers to that in Fig. 1.

# of points	Q	Atom	q^A	μ^A	$\langle\alpha^A\rangle$	$\langle\alpha^{A,p}\rangle$	$\langle dq^A/dF\rangle$
13 000	42.001414	1	-0.096	0.304	35.93	21.02	14.20
		2	0.047	0.534	35.91	22.55	16.50
27 000	42.000012	1	-0.059	0.290	35.53	20.71	14.13
		2	0.029	0.503	36.18	22.72	16.59
38 000	41.999997	1	-0.061	0.280	35.28	20.52	14.07
		2	0.030	0.509	36.30	22.81	16.62
54 000	42.000000	1	-0.103	0.265	35.63	20.83	14.10
		2	0.052	0.544	36.12	22.70	16.56
76 000	42.000000	1	-0.130	0.228	35.40	20.65	14.04
		2	0.065	0.570	36.24	22.79	16.59

dipoles μ^A vary by a similarly large amount of 0.08 a.u. While the local charges and dipoles show significant variation, the changes δq^A and $\delta\mu^A$ due to an applied field are less sensitive to the grid, as long as the computations for the zero field and nonzero field cases are performed with the same mesh. An indication for this is given in the table in terms of dq^1/dF. Over the range of meshes considered, its values can be represented as 14.12+/-0.08 electron/a.u.. For an applied field of 0.005 a.u. this translates into a charge difference δq^1 of 0.0706+/-0.0004 electron. Since $\alpha^{A,p}$ and $\alpha^{A,q}$ are defined by $\delta\mu^A$ (or δq^A), Eqs. (8b), (8c), and (9), they are also only weakly sensitive to the choice of the grid. The data in the table indicate that the values of the local polarizabilities $\langle\alpha^A\rangle$ and $\langle\alpha^{A,p}\rangle$ vary by less than 2%.

B. Si_{26}: Response in prolate and compact clusters

The lowest energy prolate and compact structures of Si_{26} found in extensive unbiased structural searches [24] are shown in Fig. 3. The two configurations are very different, yet their cohesive

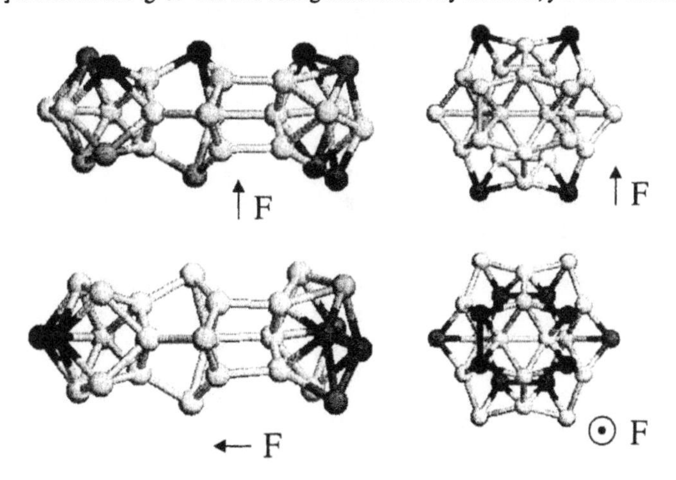

Figure 3: Charge changes δq^A obtained from the self-consistent charge distributions computed with and without the indicated applied fields for the most stable prolate and compact isomers of Si_{26}. Blue represents the largest positive change in the net local charge, whereas red indicates the largest negative change.

energies of 3.928 eV and 3.929 eV, respectively, as defined within the DFT/PBE computational framework, are essentially degenerate. The structural differences give rise to significant differences in other properties. For example, the HOMO-LUMO gap of the prolate isomer is 1.32 eV, whereas that of the compact structure is 0.42 eV. Both isomers present a diversity of local bonding arrangements. The number of bonds associated with the individual atoms ranges from 3 to 6, and the bond angles span a correspondingly broad range. Interestingly, none of the atoms in either structure adopts the tetrahedral bonding found in bulk Si.

Figure 3 also depicts the changes δq^A in the local atomic charges caused by various mutually orthogonal external electric fields. The pattern of the field-induced charge transfer across the cluster strongly correlates with the direction of the field. Atoms at the extremities of the isomers along the direction of the field experience the largest changes, whereas the most interior atoms are affected the least.

Table 2 lists the values of $\langle \alpha^{A,P} \rangle$ and $\langle dq^A/dF \rangle$ computed for each atom in the two isomers of Si_{26}. As discussed earlier, these quantities do not depend on the location of the origin of the system of coordinates and are defined uniquely. As seen in the table, their values very considerably from one atomic site to another. For example, in the prolate isomer the smallest value of $\langle \alpha^{A,P} \rangle$, 1.93 a.u., is nearly an order of magnitude smaller than the largest value, 18.20 a.u. An even larger variation in $\langle \alpha^{A,P} \rangle$ is exhibited by the compact structure of Si_{26}. The trends in the relative values of $\langle dq^A/dF \rangle$ mirror those in the values of $\langle \alpha^{A,P} \rangle$. For an applied field of 0.005 a.u., magnitudes of $\langle dq^A/dF \rangle$ of the order of 10 a.u. (cf. the values of $\langle dq^A/dF \rangle$ in Table 2 averaged over all the atoms of Si_{26}) imply charge change magnitudes $|\delta q^A|$ of the order of 0.05 electron. These are comparable to the magnitudes of the changes in the net atomic charges of Si_3 under the same field listed in panels *c* and *d* of Fig. 2.

Table 2: Values (row data) of the average local polarizability $\langle \alpha^{A,P} \rangle$ (in bohr^3) and of $\langle dq^A/dF \rangle$ (in bohr^2) computed for the prolate and compact isomers of Si_{26} shown in Fig. 3. The numbering of the atoms in the table corresponds to the left-to-right ordering of the atoms in the figure.

Atom number	Prolate		Compact	
	$\langle \alpha^{A,P} \rangle$	$\langle dq^A/dF \rangle$	$\langle \alpha^{A,P} \rangle$	$\langle dq^A/dF \rangle$
1	18.20	18.15	18.73	17.90
2	14.74	12.93	5.77	6.24
3	13.11	11.97	5.83	6.27
4	13.21	12.06	18.83	18.15
5	14.41	12.69	12.99	12.58
6	11.81	9.99	3.47	3.64
7	14.49	12.78	12.86	12.46
8	1.93	1.65	18.85	18.14
9	2.10	1.77	8.27	8.18
10	2.26	2.34	-0.12	0.84
11	14.03	10.77	8.28	8.18
12	14.55	11.43	5.57	5.09
13	15.27	12.24	4.55	4.93
14	4.42	4.05	4.55	4.92
15	4.86	4.38	5.58	5.12
16	4.90	4.26	8.31	8.22
17	3.04	3.12	-0.12	0.85
18	3.00	3.15	8.36	8.25
19	2.09	2.40	18.83	18.15
20	13.43	11.61	12.90	12.48
21	12.74	11.04	3.45	3.62
22	13.04	11.34	12.87	12.47
23	16.09	15.57	18.80	18.14
24	15.37	14.34	5.81	6.25
25	15.46	14.34	5.78	6.24
26	14.86	13.17	18.66	17.83
Average	10.52	9.36	9.53	9.42

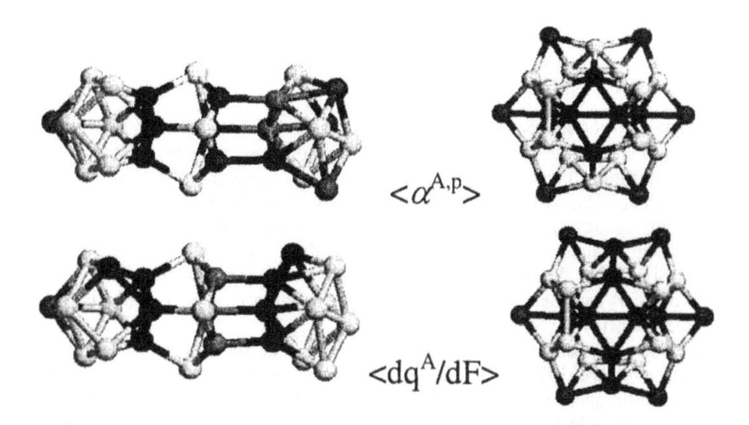

Figure 4: Pictorial representation of the values of $<\alpha^{A,p}>$ and $<dq^A/dF>$ listed in Table 2. The color change from blue to red indicates increase in value.

The values of $<\alpha^{A,p}>$ and $<dq^A/dF>$ listed in Table 2 are shown pictorially in Fig. 4. The change from blue color to red color indicates increase in value. For both isomers, the atomic polarizabilities depend on the site location within the cluster. The more peripheral is an atom, the larger its averaged polarizability. For the prolate isomer this means that atoms near the ends of the prolate axis have the largest values; for the compact isomer, atoms in the outermost ring have the largest polarizabilities.

Despite the substantial differences in shape, the prolate and compact isomers of Si_{26} exhibit remarkably similar total polarizabilities. For the prolate isomer, the values of $<\alpha>/N$, $<\alpha^p>/N$ and $<\alpha^q>/N$, are 34.37, 10.52, and 23.85 a.u., resprctively. The corresponding values for the compact isomer are 31.80, 9.53, and 22.27 a.u. For both isomers, the charge transfer component constitutes approximately 70% of the total polarizability. In Si_3, $<\alpha>/N$ is larger (35.96 a.u.), and the relative weight of the dipole and charge transfer components is reversed: The values of $<\alpha^p>/N$ and $<\alpha^q>/N$ are 22.05 and 13.91 a.u., respectively (i.e., the charge transfer component accounts for only about 39% of the total polarizability).

The compact isomer of the Si_{26} cluster has two interior atoms, labeled in Table 2 as atoms 10 and 17, which are entirely encapsulated by surface atoms. As indicated in the table, the values of $<\alpha^{A,p}>$ and $<dq^A/dF>$ for these interior atoms are -0.12 a.u. and 0.85 a.u., respectively. For an external field of 0.005 a.u., these values translate into changes of 6.0E-4 a.u. in the local dipole moment and of 4.0E-3 electron in the net local charge. In contrast, the values of $<\alpha^{A,p}>$ and $<dq^A/dF>$ averaged over the adjacent surface atoms are approximately 16 a.u. for each quantity. These translate into local dipole moment changes of the order of 0.1 a.u. and local charge changes of the order of 0.1 electron. The effect of the external field on the surface atoms is considerably larger than on the internal atoms. The significant screening of the interior of the cluster by its surface can be viewed as a metal-like attribute.

The partition scheme described in Section 2 decomposes the total cluster polarizability into local contributions from non-overlapping atomic volumes Ω_A defined as Voronoi cells. These local contributions should be distinguished from alternative definitions of atomic, or molecular, polarizabilities used in the literature [16]. As specified by Eqs. (8b), (8c), and 9, $\boldsymbol{\alpha}^{A,p}$ and $\boldsymbol{\alpha}^{A,q}$ are derivatives of the local dipole moments (or charges) with respect to the *external* field. The alternative definitions of atomic or molecular polarizabilities use derivatives of the local dipole moments with respect to the *local* field. The latter is specified as the superposition of the external field and the internal field due to the multipoles induced elsewhere in the cluster [16]. These alternative definitions of local polarizability are expected to be transferable between systems with similar bonding environments and they should, therefore, be useful in empirical molecular modeling schemes. The significance of the quantities $\alpha^{A,p}$ and $\alpha^{A,q}$ introduced here is that they represent a further partitioning of the local polarizabilities into dielectric and charge transfer components. In principle, they also can be defined as a response to the local field. Initial tests indicate that multipoles of higher order will have to be taken into account to arrive at converged values of the local fields in silicon clusters.

5. Summary and Outlook

In this paper we introduced a new scheme for decomposition of the total polarizability of an atomic cluster into local contributions from nonoverlapping atomic volumes. These local contributions are then further partitioned into dipole and charge transfer components. The utility of the scheme is illustrated through applications to silicon clusters. The local, atomic polarizabilities exhibit strong site-specificity: Atoms occupying more peripheral sites have larger polarizabilities. The charge transfer components add up to about 70% of the total polarizabilities of the most stable prolate and compact isomers of the Si_{26} cluster. In contrast, the contribution of the charge transfer components to the total polarizability of Si_3 is less than 40%. The interior of the compact isomer of Si_{26} is strongly shielded from an external electric field by the cluster surface.

We will use the new scheme as a tool for atomic-level understanding and description of the effects of composition, structure/shape, and size on the response properties of a broad variety of cluster systems. The specific questions we will address include the following: How is an anisotropy in the structure or shape of a cluster reflected in its polarizability characteristics? How does the branching ratio between the dipole and the charge transfer components of the polarizability change with the cluster size? How does this size-evolution of the branching ratio depend on the cluster material? Answers to these questions will help to enhance our understanding of different cluster phenomena, including the size-driven transition to metallicity. They will provide a characterization of this transition complementary to that based on the size-evolution of the gap between the energies of the frontier orbitals (see, e.g., Refs. 40-43 and citations therein).

Acknowledgments

This work was supported by the Office of Basic Energy Sciences, Division of Chemical Sciences, Geosciences, and Biosciences, U. S. Department of Energy under grant number DE-FGO2-03ER15489 (KAJ and MY) and contract number W-31-109-Eng-38 (JJ) and by Research Excellence Funds from the State of Michigan (KAJ).

References

[1] W. D. Knight, K. Clemenger, W. A. de Heer, and W. A. Saunders, Polarizability of alkali clusters, *Phys. Rev. B* **31**, 2539 (1985).

[2] W. A. de Heer, The physics of simple metal clusters: experimental aspects and simple models, *Rev. Mod. Phys.* **65**, 611 (1993).

[3] M. B. Knickelbein, Electric dipole polarizabilities of copper clusters, *J. Chem. Phys.* **120**, 10450 (2004).

[4] M. B. Knickelbein, Electric dipole polarizabilities of Nb_{2-27}, *J. Chem. Phys.* **118**, 6230 (2003).

[5] R. Moro, X. S. Xu, S. Y. Yin and W. A. de Heer, Ferroelectricity in free niobium clusters, *Science* **300**, 1265 (2003).

[6] M. Abd El Rahim et al., Position sensitive detection coupled to high-resolution time-of-flight mass spectrometry: Imaging for molecular beam deflection experiments, *Rev. Sci. Instr.* **75**, 5221 (2004).

[7] A. A. Quong and M. R. Pederson, Density-functional-based linear and nonlinear polarizabilities of fullerene and benzene molecules, *Phys. Rev. B* **46**, 12906 (1992).

[8] I. Vasiliev, S. Ogut and J. R. Chelikowsky, Ab initio calculations for the polarizabilities of small semiconductor clusters, *Phys. Rev. Lett.* **78**, 4805 (1997).

[9] L. Kronik, I. Vasiliev, M. Jain and J. R. Chelikowsky, Ab initio structures and polarizabilities of sodium clusters, *J. Chem. Phys.* **115**, 4322 (2001).

[10] M. Yang and K. Jackson, First-principles investigations of the polarizability of small and intermediate-sized Cu clusters, *J. Chem. Phys.* **122**, 184317 (2005).

[11] R. F. W. Bader, "Atoms in Molecules: A quantum theory", Clarendon Press: Oxford, U.K., 1990.

[12] E. R. Batista, S. S. Xantheas, and H. Jonsson, Multipole moments of water molecules in clusters and ice Ih from first principles calculations, *J. Chem. Phys.* **111**, 6011 (1999).
[13] M. in het Panhuis, P. L. A. Popeliern, R. W. Munn and J. G. Angyan, Distributed polarizability of the water dimer: Field-induced charge transfer along the hydrogen bond, *J. Chem. Phys.* **114**, 7951 (2001).
[14] F. De Proft, C. Van Alsenoy, A. Peeters, W. Langenaeker, and P. Geerlings, Atomic Charges, Dipole Moments, and Fukui Functions Using the Hirshfeld Partitioning of the Electron Density, *J. Comp. Chem.* **23**, 1198 (2002).
[15] C. F. Guerra, J. Handgraff, E. J. Baerends and F. M. Bickelhaupt, Voronoi Deformation Density (VDD) Charges: Assessment of the Mulliken, Bader, Hirschfeld, Weinhold, and VDD methods for charge analysis, *J. Comp. Chem.* **25**, 189 (2003).
[16] M. Yang, P. Senet and C. Van Alsenoy, DFT study of polarizabilities and dipole moments of water clusters, *Int. J. Quant. Chem.* **101**, 535 (2005).
[17] M. F. Jarrold, Nanosurface chemistry on size-selected silicon clusters, *Science* **252**, 1085 (1991).
[18] M. F. Jarrold and E. C. Honea, Dissociation of large silicon clusters - the approach to bulk behavior, *J. Phys. Chem.* **95**, 9181 (1991).
[19] E. C. Honea, A. Ogura, C. A. Murray, K. Raghavachari, W. O. Sprenger, M. F. Jarrold, and W. L. Brown, Raman spectra of size-selected silicon clusters and comparison with calculated structures, *Nature* **366**, 42 (1993).
[20] R. R. Hudgins, M. Imai, M. F. Jarrold, and P. Dugourd, High-resolution ion mobility measurements for silicon cluster anions and cations, *J. Chem. Phys.* **111**, 7865 (1999).
[21] E. Kaxiras and K. A. Jackson, Shape of small silicon clusters, *Phys. Rev. Lett.* **71**, 727 (1993).
[22] K. M. Ho, A. A. Shvartsburg, B. C. Pan, Z. Y. Lu, C. Z. Wang, J. G. Wacker, J. L. Fye and M. F. Jarrold, Structures of medium-sized silicon clusters, *Nature* **392**, 582 (1998).
[23] I. Rata, A. A. Shvartsburg, M. Horoi, T. Frauenheim, K. W. M. Siu, and K. A. Jackson, Single-parent evolution algorithm and the optimization of Si clusters, *Phys. Rev. Lett.* **85**, 546 (2000).
[24] K. A. Jackson, M. Horoi, I. Chaudhuri, and T. Frauenheim, Unraveling the shape transition in silicon clusters, *Phys. Rev. Lett.* **93**, 13401 (2004).
[25] S. Yoo and X. C. Zeng, Structures and relative stability of medium-sized silicon clusters. IV. Motif-based low-lying clusters Si_{21}-Si_{30}, *J. Chem. Phys.* **124**, 054304 (2006).
[26] R. Schafer, S. Schlecht, J. Woenckhaus, and J. A. Becker, Polarizabilities of isolated semiconductor clusters, *Phys. Rev. Lett.* **76**, 471 (1996).
[27] K. A. Jackson, M. R. Pederson, K.-M. Ho and C.-Z. Wang, Calculated polarizabilities of Si clusters, *Phys. Rev. A* **59**, 3685 (1999).
[28] K. Deng, J. L. Yang, C. T. Chan, Calculated polarizabilities of small Si clusters, *Phys. Rev. A* **61**, 025201 (2000).
[29] V. E. Bazterra, M. C. Caputo, M. B. Ferraro, and P. Fuentealba, On the theoretical determination of the static dipole polarizability of intermediate size silicon clusters, *J. Chem. Phys.* **117**, 11158 (2002).
[30] G. Maroulis, D. Begut, and C. Pouchan, Accurate dipole polarizabilities of small silicon clusters from ab initio and density functional theory calculations, *J. Chem. Phys.* **119**, 794 (2003).
[31] C. Pouchan, D. Begut and D. Y. Zhang, Between geometry, stability, and polarizability: Density functional theory studies of silicon clusters Si_n (n=3-10), *J. Chem. Phys.* **121**, 4628 (2004).
[32] F. Torrens, Molecular polarizability of Si/Ge/GaAs semiconductor clusters, *J. Comp. Meth. Sci. Eng.* **4**, 439 (2004).
[33] K. A. Jackson, M. Yang, I. Chaudhuri and T. Frauenheim, Shape, polarizability and metallicity in silicon clusters, *Phys. Rev. A* **71**, 033205 (2005).
[34] S. P. Apell, J. R. Sabin, S. B. Trickey, and J. Oddershede, Shape-dependent molecular polarizabilities, *Int. J. Quant. Chem.* **86**, 35 (2002).
[35] J. G. Guan, M. E. Casida, A. M. Koster, and D. R. Salahub, All-electron local and gradient-corrected density functional calculations of Na_N dipole polarizabilities for N = 1-6, *Phys. Rev. B* **52**, 2184 (1995).
[36] J. P. Perdew, K. Burke, and M. Ernzerhof, Generalized gradient approximation made simple, *Phys. Rev. Lett.* **77**, 3865 (1996).
[37] M. R. Pederson and K. A. Jackson, A variational mesh for quantum-mechanical simulations, *Phys. Rev. B* **41**, 7453 (1990).

[38] K. A. Jackson and M. R. Pederson, Accurate forces in a local orbital approach to the local density approximation, *Phys. Rev. B* **42**, 3276 (1990).
[39] F. L. Hirschfeld, Bonded-atom fragments for describing molecular charge-densities, *Theor. Chim. Acta* **44**, 129 (1977).
[40] P. H. Acioli and J. Jellinek, Electron binding energies of anionic magnesium clusters and the nonmetal-to-metal transition, *Phys. Rev. Lett.* **89**, 213402 (2002).
[41] O. C. Thomas, W. Zheng, S. Xu, and K. H. Bowen, Jr., Onset of metallic behavior in magnesium clusters, *Phys. Rev. Lett.* **89**, 213403 (2002).
[42] J. Jellinek and P. H. Acioli, Magnesium clusters: structural and electronic properties and the size-induced nonmetal-to-metal transition, *J. Phys. Chem. A* **106**, 10919 (2002); **107** 1610 (2003).
[43] B. von Issendorf and O. Cheshnovsky, Metal to insulator transitions in clusters, *Ann. Rev. Phys. Chem.* **56**, 549 (2005).

Brill Academic Publishers
P.O. Box 9000, 2300 PA Leiden,
The Netherlands

Lecture Series on Computer and Computational Sciences
Volume 6, 2006, pp. 177-189

Towards black-box linear scaling optimization in Hartree-Fock and Kohn-Sham theories

Stinne Høst, Jeppe Olsen, Branislav Jansik, Poul Jørgensen[1]

Center for Theoretical Chemistry,
Department of Chemistry,
University of Aarhus,
DK-8000 Århus C, Denmark

Simen Reine, Trygve Helgaker

Department of Chemistry,
University of Oslo,
P. O. Box 1033 Blindern, N-0315 Norway

Paweł Sałek

Laboratory of Theoretical Chemistry,
The Royal Institute of Technology,
Teknikringen 30, Stockholm SE-10044, Sweden

Sonia Coriani

Dipartimento di Scienze Chimiche, Università degli Studi di Trieste, Via Licio Giorgieri 1, I-34127 Trieste, Italy

Received 1 August, 2006; accepted 5 August, 2006

Abstract: A linear scaling implementation of the trust-region self-consistent field (LS-TRSCF) method for the Hartree-Fock and Kohn-Sham calculations is described. The convergence of the method is smooth and robust and of equal quality for small and large systems. The LS-TRSCF calculations converge in several cases where conventional DIIS calculations diverge. The LS-TRSCF method may be recommended as the standard method for both small and large molecular systems.

Keywords: Linear scaling SCF, Hartree-Fock optimization, Kohn-Sham optimization, trust-region method

Mathematics Subject Classification: 31.15-p

1 Introduction

In Hartree-Fock (HF) and Kohn-Sham (KS) density functional theory (DFT), the electronic energy E_{SCF} is minimized with respect to the density matrix of a single-determinantal wave function. In its original formulation, the minimization was carried out using the self-consistent field (SCF) method consisting of a sequence of Roothaan-Hall iterations. At each iteration, the Fock/KS matrix $\mathbf{F}$ is constructed from the current atomic-orbital (AO) density matrix $\mathbf{D}$; next, the Fock/KS matrix is diagonalized and finally an improved AO density matrix is determined from the molecular orbitals

[1] Corresponding author. E-mail: pou@chem.au.dk

(MOs) obtained by this diagonalization. Unfortunately, this simple SCF scheme converges only in simple cases.

To improve upon the convergence, the optimization is modified by constructing the Fock/KS matrix not directly from the AO density matrix of the last iteration, but rather from an averaged density matrix, obtained as a linear combination of the density matrices of the current and previous iterations. Typically, the averaged density matrix is obtained using the DIIS method of Pulay [1], by minimizing the norm of the linear combination of the gradients. The SCF/DIIS method has been implemented in most electronic-structure programs and has been successfully used to obtain optimized HF/KS energies. However, in some cases the DIIS procedure fails to converge.

During the last decade, much effort has been directed towards developing linear scaling SCF methods. In particular, the computational scaling for the evaluation of the Fock/KS matrix has been successfully reduced by use of the fast multipole method (FMM) for the Coulomb contribution [2]–[6], the order-N exchange (ONX) method and the linear exchange K (LinK) method for the exact (Hartree–Fock) exchange contribution [7]–[12], and efficient numerical quadrature methods for the exchange–correlation (XC) contribution [13]–[15]. Our SCF code uses FMM combined with density fitting for the Coulomb contribution, LinK for the exact exchange contribution, and linear-scaling numerical quadrature for the XC contribution.

In the optimization of the SCF energy, the diagonalization of the Fock/KS matrix, which scales cubically with the system size (N^3), may therefore become the time dominating step for large molecules. To remove the diagonalization bottleneck, many methods have been suggested. In the canonical-purification (CP) method of Palser and Manolopoulos [16], the density matrix $\mathbf{D}$ is generated iteratively by a scheme similar to that of the McWeeny purification [17], with the additional requirement that the purified density matrix (with eigenvalues one and zero) commutes with the Fock/KS matrix $\mathbf{F}$. Another strategy uses the fact that the density matrix obtained from a Fock/KS matrix diagonalization represents the global minimum of the Roothaan–Hall energy function $E^{\mathrm{RH}} = \mathrm{Tr}\,\mathbf{DF}$ (with $\mathbf{F}$ fixed) [18, 19], thereby replacing the diagonalization by a minimization, suitably constrained so as to satisfy the idempotency condition $\mathbf{DSD} = \mathbf{D}$. Li, Nunes and Vanderbilt proposed to deal with the idempotency constraint by replacing the density matrix in the optimization by its McWeeny-purified counterpart, noting that the variations are then idempotent to first order [20]; their approach was further developed by Millam and Scuseria [21] and by Challacombe [22]. Alternatively, the idempotency condition may be incorporated into the parametrization of the density matrix $\mathbf{D}(\mathbf{X}) = \exp(-\mathbf{XS})\,\mathbf{D}\exp(\mathbf{SX})$ with $\mathbf{X}$ antisymmetric, as described by Helgaker *et al.* [19, 23, 24]. The first attempts to use this parametrization to minimize E^{RH} employed a sequence of Newton iterations but encountered difficulties in the solution of the Newton linear equations [24]. Subsequently, these difficulties were solved by Shao *et al.* [25], in their curvy-step method, by transforming the Newton equations to the Cholesky basis, where the Hessian matrix has a lower conditioning number and is more diagonally dominant than in the AO basis. We further develop the strategy of minimizing the Roothaan-Hall energy for obtaining efficient and robust convergence in Hartree-Fock and Kohn-Sham calculations using a method that scales linearly.

In the SCF/DIIS method, the minimization of the energy is carried out in two separate steps: the diagonalization of the Fock/KS matrix and the averaging of the density matrix. In neither step an energy lowering is enforced on E_{SCF}. It is simply hoped that, at the end of the SCF iterations, an optimized state is determined. We discuss improvements to both the diagonalization and the density matrix averaging, where a lowering of the energy E_{SCF} is enforced at each iteration. For both steps, we construct a local energy model to E_{SCF} with the current density matrix as the expansion point. At the expansion point, these models have the true gradient, but only an approximate Hessian. They are therefore valid only in a restricted region about the expansion point - the trust region. When these local models are used, it is essential that steps are only generated within the trust region, as otherwise no energy lowering is guaranteed.

As discussed above the diagonalization of the Fock/KS matrix may be avoided by recognizing that the density matrix obtained by diagonalizing the Fock/KS matrix represents the global minimum of the Roothaan-Hall energy function E^{RH} The diagonalization may therefore be replaced by a minimization of E^{RH}. However, since E^{RH} is only a crude model of the true energy E_{SCF}, a complete minimization of E^{RH} (as obtained for example by diagonalization) may give steps that are too long to be trusted. When minimizing E^{RH}, we require the steps to be inside the trust region, solving a set of level shifted Newton equations where the level shift controls the size of the steps. The level shifted Newton equations may be solved using iterative algorithms where the time-dominating step is the multiplication of the Hessian by trialvectors. Linear complexity therefore may be obtained by using sparse matrix algebra. The obtained algorithm will be denoted the linear scaling trust-region Roothaan-Hall (LS-TRRH) method.

To improve on the DIIS scheme we construct an energy function where the expansion coefficients of the averaged density matrix are the variational parameters. Carrying out a second-order expansion of this energy, using the quasi-Newton condition and neglecting terms that require evaluation of new Fock/KS matrices, we arrive at the density subspace minimization (DSM) approximation to the energy E^{DSM} [26, 27]. At the expansion point, E^{DSM} has the same gradient as E_{SCF} and a good approximation to the Hessian. Again, trust-region optimization may be used to determine the optimal expansion coefficients, ensuring also an energy lowering at this step of the iterative procedure. The obtained algorithm is denoted the trust-region density subspace minimization (TRDSM) method. Combining the LS-TRRH and TRDSM amethods we obtain the LS-TRSCF method.

In the next section we describe the LS-TRRH algorithm while in section 3 the TRDSM algorithm is discussed. Section 4 contains numerical results which demonstrate the convergence of LS-TRSCF calculations and that linear scaling is obtained. The last section contains some concluding remarks.

2 Optimization of the Roothaan–Hall energy

2.1 Parametrization of the density matrix

Let $\mathbf{D}$ be a valid Kohn–Sham density matrix of an N-electron system, which together with the AO overlap matrix $\mathbf{S}$ satisfies the symmetry, trace and idempotency relations:

$$\mathbf{D}^{\mathrm{T}} = \mathbf{D} \tag{1}$$

$$\mathrm{Tr}\,\mathbf{DS} = N \tag{2}$$

$$\mathbf{DSD} = \mathbf{D} \tag{3}$$

Introducing the projectors $\mathbf{P}_{\mathrm{o}}$ and $\mathbf{P}_{\mathrm{v}}$ on the occupied and virtual spaces

$$\mathbf{P}_{\mathrm{o}} = \mathbf{DS} \tag{4}$$

$$\mathbf{P}_{\mathrm{v}} = \mathbf{I} - \mathbf{DS} \tag{5}$$

we may, from the reference density matrix $\mathbf{D}$, generate any other valid density matrix by the transformation [19, 23, 24]

$$\mathbf{D}(\mathbf{X}) = \exp[-\mathcal{P}(\mathbf{X})\mathbf{S}]\,\mathbf{D}\exp[\mathbf{S}\mathcal{P}(\mathbf{X})] \tag{6}$$

where $\mathbf{X}$ is an anti-Hermitian matrix and where

$$\mathcal{P}(\mathbf{X}) = \mathbf{P}_{\mathrm{o}}\mathbf{X}\mathbf{P}_{\mathrm{v}}^{\mathrm{T}} + \mathbf{P}_{\mathrm{v}}\mathbf{X}\mathbf{P}_{\mathrm{o}}^{\mathrm{T}} \tag{7}$$

projects out the redundant occupied-occupied and virtual-virtual components of $\mathbf{X}$.

The density matrix $\mathbf{D}(\mathbf{X})$ may be expanded in orders of $\mathbf{X}$ as

$$\mathbf{D}(\mathbf{X}) = \mathbf{D} + [\mathbf{D}, \mathcal{P}(\mathbf{X})]_S + \tfrac{1}{2}\left[[\mathbf{D}, \mathcal{P}(\mathbf{X})]_S, \mathcal{P}(\mathbf{X})\right]_S + \cdots \quad (8)$$

where we have introduced the S commutator

$$[\mathbf{A}, \mathbf{B}]_S = \mathbf{ASB} - \mathbf{BSA} \quad (9)$$

2.2 The Roothaan–Hall Newton equations in the AO basis

In an SCF optimization, the diagonalization of the Fock/KS matrix $\mathbf{F}$ is equivalent to the minimization of the Roothaan-Hall energy [18, 19]

$$E^{\mathrm{RH}}(\mathbf{X}) = \mathrm{Tr}\,[\mathbf{FD}(\mathbf{X})] \quad (10)$$

in the sense that both approaches yield the same density matrix. Inserting the S-commutator expansion of the density matrix $\mathbf{D}(\mathbf{X})$, we obtain

$$\begin{aligned}\mathrm{Tr}\,[\mathbf{FD}(\mathbf{X})] = \mathrm{Tr}\,(\mathbf{FD}) &+ \mathrm{Tr}\,(\mathbf{F}^{\mathrm{vo}}\mathbf{X} - \mathbf{F}^{\mathrm{ov}}\mathbf{X}) \\ &+ \mathrm{Tr}\,(\mathbf{F}^{\mathrm{oo}}\mathbf{X}\mathbf{S}^{\mathrm{vv}}\mathbf{X} - \mathbf{F}^{\mathrm{vv}}\mathbf{X}\mathbf{S}^{\mathrm{oo}}\mathbf{X}) + \cdots\end{aligned} \quad (11)$$

where we have made repeated use of the idempotency relations $\mathbf{P}_{\mathrm{o}}^2 = \mathbf{P}_{\mathrm{o}}$ and $\mathbf{P}_{\mathrm{v}}^2 = \mathbf{P}_{\mathrm{v}}$ and of the orthogonality relations $\mathbf{P}_{\mathrm{o}}\mathbf{P}_{\mathrm{v}} = \mathbf{P}_{\mathrm{v}}\mathbf{P}_{\mathrm{o}} = 0$ and $\mathbf{P}_{\mathrm{o}}^{T}\mathbf{S}\mathbf{P}_{\mathrm{v}} = \mathbf{P}_{\mathrm{v}}^{T}\mathbf{S}\mathbf{P}_{\mathrm{o}} = 0$ and introduced the short-hand notation

$$\mathbf{F}^{\mathrm{ab}} = \mathbf{P}_{\mathrm{a}}^{T}\mathbf{F}\mathbf{P}_{\mathrm{b}} \quad (12)$$

Note that, whereas the off-diagonal projections $\mathbf{F}^{\mathrm{ov}}$ and $\mathbf{F}^{\mathrm{vo}}$ of $\mathbf{F}$ contribute to the terms linear in $\mathbf{X}$, the diagonal projections $\mathbf{F}^{\mathrm{oo}}$ and $\mathbf{F}^{\mathrm{vv}}$ contribute to the quadratic terms.

The Roothaan-Hall energy E^{RH} is only a crude model of the true HF/KS energy E_{SCF}, having the correct gradient but an approximate Hessian at the point of expansion; this can be understood from the observation that E^{RH} depends linearly on $\mathbf{D}(\mathbf{X})$, whereas the true energy E_{SCF} depends quadratically on $\mathbf{D}(\mathbf{X})$. Therefore, a complete minimization of E^{RH} (as achieved, for example, by diagonalization of the Fock/KS matrix), may give steps that are too long to be trusted. Such steps may, for example, increase rather than decrease the total SCF energy. We therefore impose on the energy minimization the constraint that the new occupied space does not differ appreciably from the old occupied space. The step must therefore be inside or on the boundary of the trust region of E^{RH}, which we define as a hypersphere with radius h around the density at the current expansion point. In the S metric norm, the length of the step

$$||\mathcal{P}(\mathbf{X})||_S^2 = \mathrm{Tr}[\mathcal{P}(\mathbf{X})\mathbf{S}\mathcal{P}(\mathbf{X})\mathbf{S}] \quad (13)$$

is thus restricted to h^2. To satisfy this constraint, we introduce an undetermined multiplier μ and set up the Lagrangian

$$L^{\mathrm{RH}}(\mathbf{X}) = \mathrm{Tr}\,[\mathbf{FD}(\mathbf{X})] - \frac{1}{2}\mu\left(\mathrm{Tr}[\mathcal{P}(\mathbf{X})\mathbf{SXS}] - h^2\right) \quad (14)$$

Expanding this Lagrangian to second order in $\mathbf{X}$ using Eq. (11), we obtain

$$\begin{aligned}L^{\mathrm{RH}}(\mathbf{X}) = &\,\mathrm{Tr}\,(\mathbf{FD}) + \mathrm{Tr}\,(\mathbf{F}^{\mathrm{vo}}\mathbf{X} - \mathbf{F}^{\mathrm{ov}}\mathbf{X}) \\ &+ \mathrm{Tr}\,(\mathbf{F}^{\mathrm{oo}}\mathbf{X}\mathbf{S}^{\mathrm{vv}}\mathbf{X} - \mathbf{F}^{\mathrm{vv}}\mathbf{X}\mathbf{S}^{\mathrm{oo}}\mathbf{X}) + \mu\left[\mathrm{Tr}\,(\mathbf{S}^{\mathrm{oo}}\mathbf{X}\mathbf{S}^{\mathrm{vv}}\mathbf{X}) - \frac{1}{2}h^2\right]\cdots\end{aligned} \quad (15)$$

Differentiating this function with respect to the elements of $\mathbf{X}$, we obtain

$$\frac{\partial L^{\mathrm{RH}}(\mathbf{X})}{\partial \mathbf{X}} = \mathbf{F}^{\mathrm{ov}} - \mathbf{F}^{\mathrm{vo}} - \mathbf{S}^{\mathrm{vv}}\mathbf{X}\mathbf{F}^{\mathrm{oo}} - \mathbf{F}^{\mathrm{oo}}\mathbf{X}\mathbf{S}^{\mathrm{vv}} + \mathbf{F}^{\mathrm{vv}}\mathbf{X}\mathbf{S}^{\mathrm{oo}} + \mathbf{S}^{\mathrm{oo}}\mathbf{X}\mathbf{F}^{\mathrm{vv}} - \mu\left(\mathbf{S}^{\mathrm{vv}}\mathbf{X}\mathbf{S}^{\mathrm{oo}} + \mathbf{S}^{\mathrm{oo}}\mathbf{X}\mathbf{S}^{\mathrm{vv}}\right) + \cdots \tag{16}$$

where we have used the relation

$$\frac{\partial \,\mathrm{Tr}(\mathbf{AX})}{\partial \mathbf{X}} = \mathbf{A}^{\mathrm{T}} \tag{17}$$

Finally, setting the right-hand side equal to zero and ignoring higher-order contributions, we obtain the matrix equation

$$\mathbf{F}^{\mathrm{vv}}\mathbf{X}\mathbf{S}^{\mathrm{oo}} - \mathbf{F}^{\mathrm{oo}}\mathbf{X}\mathbf{S}^{\mathrm{vv}} + \mathbf{S}^{\mathrm{oo}}\mathbf{X}\mathbf{F}^{\mathrm{vv}} - \mathbf{S}^{\mathrm{vv}}\mathbf{X}\mathbf{F}^{\mathrm{oo}} - \mu\left(\mathbf{S}^{\mathrm{vv}}\mathbf{X}\mathbf{S}^{\mathrm{oo}} + \mathbf{S}^{\mathrm{oo}}\mathbf{X}\mathbf{S}^{\mathrm{vv}}\right) = \mathbf{F}^{\mathrm{vo}} - \mathbf{F}^{\mathrm{ov}} \tag{18}$$

for the stationary points on the trust sphere of the Roothaan–Hall energy function.

Eq. (18) is equivalent to a level shifted set of Newton equations

$$(\mathbf{H} - \mu\mathbf{M})\mathbf{x} = \mathbf{G} \tag{19}$$

where

$$\mathbf{H} = \mathbf{F}^{\mathrm{vv}} \otimes \mathbf{S}^{\mathrm{oo}} - \mathbf{F}^{\mathrm{oo}} \otimes \mathbf{S}^{\mathrm{vv}} + \mathbf{S}^{\mathrm{oo}} \otimes \mathbf{F}^{\mathrm{vv}} - \mathbf{S}^{\mathrm{vv}} \otimes \mathbf{F}^{\mathrm{oo}} \tag{20}$$

$$\mathbf{M} = \mathbf{S}^{\mathrm{vv}} \otimes \mathbf{S}^{\mathrm{oo}} - \mathbf{S}^{\mathrm{oo}} \otimes \mathbf{S}^{\mathrm{vv}} \tag{21}$$

$$\mathbf{G} = \mathrm{Vec}(\mathbf{F}^{\mathrm{vo}} - \mathbf{F}^{\mathrm{ov}}) \tag{22}$$

$$\mathbf{x} = \mathrm{Vec}\mathbf{X} \tag{23}$$

2.3 The Roothaan–Hall Newton equations in an orthonormal basis

The conditioning number of the level shifted Hessian matrix in Eq. (19) is greatly reduced by transforming the equation to an orthogonal basis. We consider transformations based on the factorization of the overlap in the form

$$\mathbf{S} = \mathbf{V}^{\mathrm{T}}\mathbf{V} \tag{24}$$

Such a factorization may be accomplished in infinitely many ways – for example, by introducing a Cholesky factor or the principal square root

$$\mathbf{V}_{\mathrm{c}} = \mathbf{U} \tag{25}$$

$$\mathbf{V}_{\mathrm{s}} = \mathbf{S}^{1/2} \tag{26}$$

where $\mathbf{U}$ is an nonsingular upper triangular matrix and where $\mathbf{S}^{1/2}$ is a positive-definite symmetric matrix. In the chosen orthonormal basis, the Roothaan–Hall Newton equations Eq. (18) take the form

$$(\mathbf{F}_{\mathrm{V}}^{\mathrm{vv}} - \mathbf{F}_{\mathrm{V}}^{\mathrm{oo}} - \mu\mathbf{I})\mathbf{X}^{\mathrm{V}} + \mathbf{X}^{\mathrm{V}}(\mathbf{F}_{\mathrm{V}}^{\mathrm{vv}} - \mathbf{F}_{\mathrm{V}}^{\mathrm{oo}} - \mu\mathbf{I}) = \mathbf{F}_{\mathrm{V}}^{\mathrm{vo}} - \mathbf{F}_{\mathrm{V}}^{\mathrm{ov}} \tag{27}$$

where we have introduced the notation

$$\mathbf{A}_{\mathrm{V}} = \mathbf{V}^{-\mathrm{T}}\mathbf{A}\mathbf{V}^{-1} \tag{28}$$

$$\mathbf{A}^{\mathrm{V}} = \mathbf{V}\mathbf{A}\mathbf{V}^{\mathrm{T}} \tag{29}$$

and where we have further assumed that $\mathbf{X}^{\mathrm{V}}$ contains only non-redundant components.

Eq. (27) represents the solution of a level shifted set of Newton equations

$$(\mathbf{H}_{\mathrm{V}} - \mu\mathbf{I})\mathbf{x}^{\mathrm{V}} = \mathbf{G}_{\mathrm{V}} \tag{30}$$

where

$$\mathbf{H}_{\mathrm{V}} = (\mathbf{F}_{\mathrm{V}}^{\mathrm{vv}} - \mathbf{F}_{\mathrm{V}}^{\mathrm{oo}}) \otimes \mathbf{I} + \mathbf{I} \otimes (\mathbf{F}_{\mathrm{V}}^{\mathrm{vv}} - \mathbf{F}_{\mathrm{V}}^{\mathrm{oo}}) \tag{31}$$

$$\mathbf{x}^{\mathrm{V}} = \mathrm{Vec}\mathbf{X}^{\mathrm{V}} \tag{32}$$

$$\mathbf{G}_{\mathrm{V}} = \mathrm{Vec}(\mathbf{F}_{\mathrm{V}}^{\mathrm{vo}} - \mathbf{F}_{\mathrm{V}}^{\mathrm{ov}}) \tag{33}$$

When solving Eq. (30) by the conjugate gradient method, it is advantageous to use a diagonal preconditioner.

In the global region of an SCF optimization, the boundary of the trust region is represented by $X_{\mathrm{V}}^{\max} = k$, where $X_{\mathrm{V}}^{\max}$ is the largest component of $\mathbf{X}^{\mathrm{V}}$ and k is 0.35. Unlike $||\mathbf{X}^{\mathrm{V}}||_S$, $X_{\mathrm{V}}^{\max}$ is size-intensive. To ensure that the minimum is determined on the boundary of the trust region, the level shift must be restricted to the interval $-\infty < \mu < \epsilon_{\min}$ where $\epsilon_{\min}$ is the lowest eigenvalue of the Hessian Eq. (31). In principle, the lowest Hessian eigenvalue should therefore be determined and a line search carried out in the interval $-\infty < \mu < \epsilon_{\min}$ to find the level shift μ with $X_{\mathrm{V}}^{\max} = 0.35$. However, a simpler strategy is obtained by recognizing that the solution of the level shifted Newton equations can be determined from the eigenvectors of the augmented Hessian eigenvalue equation [28, 29, 30]. If the solution with the lowest eigenvalue is determined, the level shift is restricted to the interval $-\infty < \mu < \epsilon_{\min}$.

The level shifted Newton equations may be solved using an iterative procedure where the reduced space Hessians and gradients are set up in each iteration. At each iteration, the augmented Hessian may therefore also be set up in the subspace at essentially no cost and the lowest eigenvalue determined. Consequently the level shift may be updated by solving the reduced space augmented Hessian eigenvalue problem at no extra cost. With the updated level shift, a new Newton iteration may be carried out and the iterations continued until convergence is obtained with respect to level shift and the residual of the Newton equations (see Ref. [31]).

When the level shifted Newton equations are solved using iterative algorithms, the time consuming step is the linear transformation of the Hessian matrix on trial vectors. Using sparse matrix algebra, linear scaling may be obtained in these linear transformations.

3 The density subspace minimization (DSM) algorithm

After a sequence of Roothaan-Hall iterations, we have determined a set of density matrices $\mathbf{D}_i$ and a corresponding set of Fock/KS matrices $\mathbf{F}_i = \mathbf{F}(\mathbf{D}_i)$. We now discuss how to make the best use of the information contained in these matrices.

3.1 Parametrization of the DSM density matrix

Using $\mathbf{D}_0$ as the reference density matrix, the improved density matrix may be expressed as a linear combination of the current and previous density matrices [26, 27]

$$\overline{\mathbf{D}} = \mathbf{D}_0 + \sum_{i=0}^{n} c_i \mathbf{D}_i. \tag{34}$$

Ideally $\overline{\mathbf{D}}$ should satisfy the symmetry, trace and idempotency conditions Eqs. (1-3). The symmetry condition Eq. (1) is trivially satisfied while the trace condition Eq. (2) holds only if

$$c_0 = -\sum_{i=1}^{n} c_i. \tag{35}$$

Using c_i with $1 \leq i \leq n$ as independent parameters the density matrix $\overline{\mathbf{D}}$ may be expressed as

$$\overline{\mathbf{D}} = \mathbf{D}_0 + \mathbf{D}_+, \tag{36}$$

where we have introduced the notation

$$\mathbf{D}_+ = \sum_{i=1}^{n} c_i \mathbf{D}_{i0}, \tag{37a}$$

$$\mathbf{D}_{i0} = \mathbf{D}_i - \mathbf{D}_0. \tag{37b}$$

While $\overline{\mathbf{D}}$ satisfies the symmetry and trace conditions Eqs. (1) and (2), the idempotency condition Eq. (3) is not fulfilled. A smaller idempotency error may be obtained using the purified density matrix of McWeeny [17, 32]

$$\tilde{\mathbf{D}} = 3\overline{\mathbf{D}}\mathbf{S}\overline{\mathbf{D}} - 2\overline{\mathbf{D}}\mathbf{S}\overline{\mathbf{D}}\mathbf{S}\overline{\mathbf{D}}. \tag{38}$$

Emphasizing that $\mathbf{D}_0$ is the reference density matrix, the first-order purified density matrix may be expressed as

$$\tilde{\mathbf{D}} = \mathbf{D}_0 + \mathbf{D}_+ + \mathbf{D}_\delta. \tag{39}$$

where we have introduced the idempotency correction

$$\mathbf{D}_\delta = \tilde{\mathbf{D}} - \overline{\mathbf{D}}. \tag{40}$$

3.2 Construction of the DSM energy function

Expanding the energy for the purified averaged density matrix, Eq. (39), around the reference density matrix $\mathbf{D}_0$, we obtain to second order

$$E(\tilde{\mathbf{D}}) = E(\mathbf{D}_0) + (\mathbf{D}_+ + \mathbf{D}_\delta)^{\mathrm{T}} \mathbf{E}_0^{(1)} + \frac{1}{2} (\mathbf{D}_+ + \mathbf{D}_\delta)^{\mathrm{T}} \mathbf{E}_0^{(2)} (\mathbf{D}_+ + \mathbf{D}_\delta) \tag{41}$$

To evaluate the terms containing $\mathbf{E}_0^{(1)}$ and $\mathbf{E}_0^{(2)}$, we first recall that the Fock/KS matrix is defined as

$$\mathbf{E}_0^{(1)} = 2\mathbf{F}_0 \tag{42}$$

Next we carry out an expansion of $\mathbf{E}_i^{(1)}$ with $\mathbf{D}_0$ as expansion point

$$\mathbf{E}_i^{(1)} = \mathbf{E}_0^{(1)} + \mathbf{E}_0^{(2)}(\mathbf{D}_i - \mathbf{D}_0) + \mathcal{O}(\mathbf{D}_i - \mathbf{D}_0)^2 \tag{43}$$

Neglecting terms of order $\mathcal{O}(\mathbf{D}_i - \mathbf{D}_0)^2$ we obtain the quasi-Newton condition

$$\mathbf{E}_0^{(2)}(\mathbf{D}_i - \mathbf{D}_0) = 2\mathbf{F}_i - 2\mathbf{F}_0 = 2\mathbf{F}_{i0} \tag{44}$$

which may be used to obtain

$$\mathbf{E}_0^{(2)}\mathbf{D}_+ = 2\mathbf{F}_+ + \mathcal{O}(\mathbf{D}_+^2), \tag{45}$$

where we have generalized the notation Eq. (37a) to the Fock/KS matrix

$$\mathbf{F}_+ = \sum_{i=1}^{n} c_i \mathbf{F}_{i0} \tag{46}$$

Using Eq. (42) and Eq. (45) and ignoring the terms quadratic in $\mathbf{D}_\delta$ in Eq. (41) and quadratic in $\mathbf{D}_+$ in Eq. (45), we then obtain the DSM energy

$$E^{\mathrm{DSM}}(\mathbf{c}) = E(\mathbf{D}_0) + 2\,\mathrm{Tr}\,\mathbf{D}_+\mathbf{F}_0 + \mathrm{Tr}\,\mathbf{D}_+\mathbf{F}_+ + 2\,\mathrm{Tr}\,\mathbf{D}_\delta\mathbf{F}_0 + 2\,\mathrm{Tr}\,\mathbf{D}_\delta\mathbf{F}_+. \tag{47}$$

Note that $E^{\mathrm{DSM}}(\mathbf{c})$ is expressed solely in terms of the density and Fock/KS matrices of the current and previous iterations. For a more compact notation, we introduce the weighted Fock/KS matrix

$$\overline{\mathbf{F}} = \mathbf{F}_0 + \mathbf{F}_+ = \mathbf{F}_0 + \sum_{i=1}^{n} c_i\mathbf{F}_{i0} \tag{48}$$

and find that the DSM energy may be written in the form

$$E^{\mathrm{DSM}}(\mathbf{c}) = E(\overline{\mathbf{D}}) + 2\,\mathrm{Tr}\,\mathbf{D}_\delta\overline{\mathbf{F}}, \tag{49}$$

where the first term is quadratic in the expansion coefficients c_i

$$E(\overline{\mathbf{D}}) = E(\mathbf{D}_0) + 2\,\mathrm{Tr}\,\mathbf{D}_+\mathbf{F}_0 + \mathrm{Tr}\,\mathbf{D}_+\mathbf{F}_+, \tag{50}$$

and the second (idempotency correction) term is quartic in these coefficients:

$$2\,\mathrm{Tr}\,\mathbf{D}_\delta\overline{\mathbf{F}} = \mathrm{Tr}(6\overline{\mathbf{D}}\mathbf{S}\overline{\mathbf{D}} - 4\overline{\mathbf{D}}\mathbf{S}\overline{\mathbf{D}}\mathbf{S}\overline{\mathbf{D}} - 2\overline{\mathbf{D}})\overline{\mathbf{F}}. \tag{51}$$

The derivatives of $E^{\mathrm{DSM}}(\mathbf{c})$ are straightforwardly obtained by inserting the expansions of $\overline{\mathbf{F}}$ and $\overline{\mathbf{D}}$, using the independent parameter representation and the minimization of $E^{\mathrm{DSM}}(\mathbf{c})$ may straightforwardly be carried out using the trust-region method.

4 Numerical Illustrations

4.1 Convergence of test calculations

We now describe the convergence of test calculations for Hartree-Fock and DFT LDA using the LS-TRSCF algorithm where the level shifted Newton equations are solved in the basis defined by the principal square root in Eq. (26). For comparison, the convergence of the standard SCF/DIIS calculations (diagonalization + DIIS, no level shift) will also be reported. In both DIIS and TRDSM a maximum of eight densities and Fock/KS matrices are stored.

In Fig. 1 we display the convergence (the difference between the energy of a given iteration and the converged energy) of Hartree-Fock calculations using LS-TRSCF (left panel) and SCF/DIIS (right panel) on six molecules representing different types of chemical compounds: 1). Water, stretched: H_2O where the O–H bond is twice its equilibrium value (d-aug-pVTZ basis). 2). Rh complex: The rhodium complex of Ref. [26] (AhlrichsVDZ basis [33], STO-3G on Rh). 3). Cd complex: The cadmium-imidazole complex of Ref. [27] (3-21G basis). 4). Zn complex: The zinc-EDDS complex of Ref. [27] (6-31G basis). 5). Polysaccharide: A polysaccharide containing 438 atoms (6-31G basis). 6). Polyalanine, 24 units: A polypeptide containing 24 alanine residues (6-31G basis). As initial guess we have used H1 core for molecules 1–3 and Hückel for molecules 4–6. Smooth convergence to 10^{-8} a.u. is obtained in all LS-TRSCF calculations. Convergence is obtained in 12-30 iterations. The convergence is very similar for the SCF/DIIS and the LS-TRSCF calculations except for the rhodium complex, where the SCF/DIIS calculation diverges while smooth convergence is obtained using the LS-TRSCF algorithm. The local convergence is very similar for SCF/DIIS and LS-TRSCF reflecting that in both DIIS and DSM, the local convergence is determined by the fact that the quasi-Newton condition is satisfied [27]. In Fig. 2, we report calculations similar to those in Fig. 1 but where the Hartree-Fock model is replaced

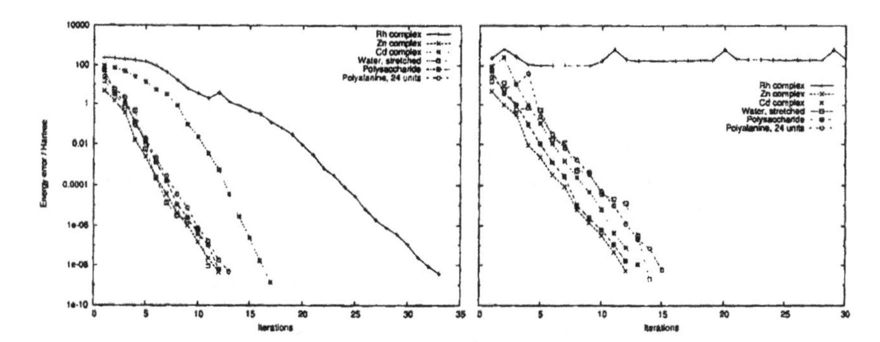

Figure 1: Hartree-Fock: Convergence of the LS-TRSCF (left panel) and SCF/DIIS (right panel) calculations for the rhodium complex, the zinc complex, the cadmium complex, the stretched water, the polysaccharide and the polyalanine. The energy error (a.u.) in each iteration is plotted versus number of iterations.

by LDA. The convergence of the LS-TRSCF Hartree-Fock and LDA calculations is very similar with the exception of the Rh complex, where the LDA calculation has a rather erratic behaviour from about iteration 20 to 80 after which fast convergence is obtained. The SCF/DIIS LDA calculations in the right panel in Fig. 2 show a rather erratic convergence behaviour, in particular for the Cd complex and polyalanine where the calculations diverge, and for the polysaccharide calculation in the first 25 iterations. The erratic behaviour which in general is observed in the initial iterations of SCF/DIIS calculations reflects that energy lowering is not an issue in the SCF/DIIS scheme. Surprisingly, the SCF/DIIS LDA calculation on the rhodium complex converges, while the corresponding Hartree-Fock calculation diverges.

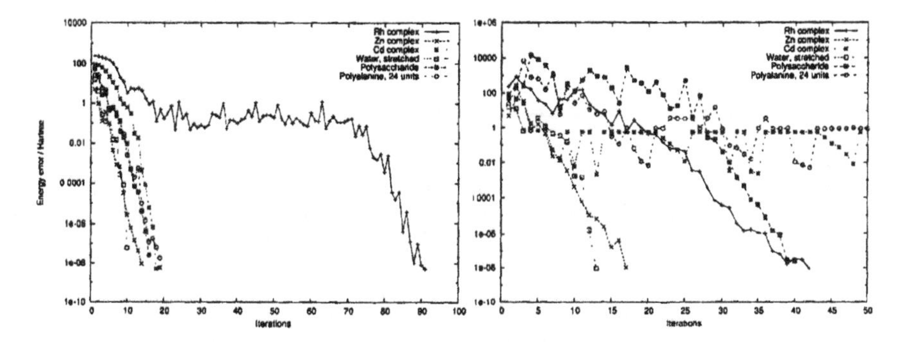

Figure 2: LDA: Convergence of the LS-TRSCF (left panel) and SCF/DIIS (right panel) calculations for the rhodium complex, the zinc complex, the cadmium complex, the stretched water, the polysaccharide and the polyalanine. The energy error (a.u.) in each iteration is plotted versus number of iterations.

To sum it up, similar convergence is seen in Hartree-Fock SCF/DIIS and LS-TRSCF calculations, whereas for LDA, a much more smooth and robust convergence is obtained by using the LS-TRSCF scheme. Particularly in the initial iterations, a more erratic behaviour is seen with the

SCF/DIIS algorithm. In several cases the LS-TRSCF calculations converge, where the SCF/DIIS calculations diverge.

4.2 Linear scaling using the LS-TRSCF algorithm

In this subsection, we will illustrate that linear scaling is obtained using the LS-TRSCF algorithm. We consider calculations on a polyalanine peptide where we extend the number of alanine residues. We consider both Hartree-Fock and B3LYP calculations in the 6-31G basis. The largest alanine peptide contains 119 alanine residues (a total of 1192 atoms). The convergence of the alanines is similar to the one for the 24 residue peptide given in Figs. 1-2.

In Fig. 3 we have shown the CPU time used in the different parts of the LS-TRSCF algorithm for the Hartree-Fock calculations using sparse matrix algebra. In all figures, the timings are for the first iteration in the local region, except for the DSM time, which is dependent on the number of previous densities. Therefore, the DSM time is always given for iteration 8, where we have the maximum number of previous densities involved. The timings are given for the evaluation of the Coulomb (Fock J) and exchange (Fock X) parts of the Fock matrix, respectively, and for the LS-TRRH step and for the TRDSM step. The curve for the most expensive step – the exchange part of the Fock matrix – has a bend, probably due to an N^2 scaling sorting routine. For both the LS-TRRH and TRDSM steps, the time consuming part consists of matrix multiplications. Both LS-TRRH and TRDSM scale linearly with system size in the calculations in Fig. 3, showing that sparsity is exploited efficiently in the matrix multiplications. The benefits from exploiting the sparsity of the

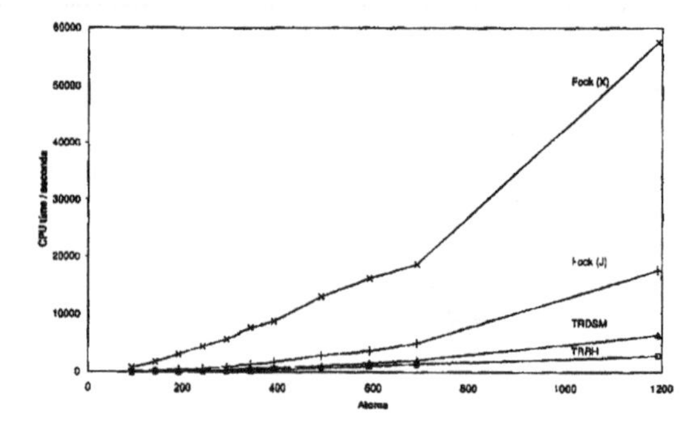

Figure 3: CPU timings for one iteration of a Hartree-Fock calculation using a 6-31G basis plotted as a function of the number of atoms in a polyalanine peptide. The considered contributions are the exchange (X) and Coulomb (J) contributions to Fock matrix in addition to the LS-TRRH and TRDSM optimization steps where sparse matrix algebra is used.

involved matrices are evident from Fig. 4, where we have plotted the CPU times for the LS-TRRH and TRDSM steps from Fig. 3 in combination with timings for calculations where the matrix multiplications involve full (dense) matrices. The timings for full matrix representations increase with system size in accordance with cubic scaling, but become linear when sparsity is exploited. As seen on the figure, the advantage of going to the sparse matrix representation has an earlier onset for TRDSM than for LS-TRRH, because TRDSM contains more matrix multiplications than LS-TRRH. Fig. 5 shows the CPU timings for the B3LYP calculations in the sparse matrix representation. The timings shown are the same as in Fig. 3, with the addition of the timing for the exchange-correlation (Kohn-Sham XC) contribution. Like the other contributions to the KS matrix (Coulomb and exchange), the exchange-correlation contribution has reached the linear

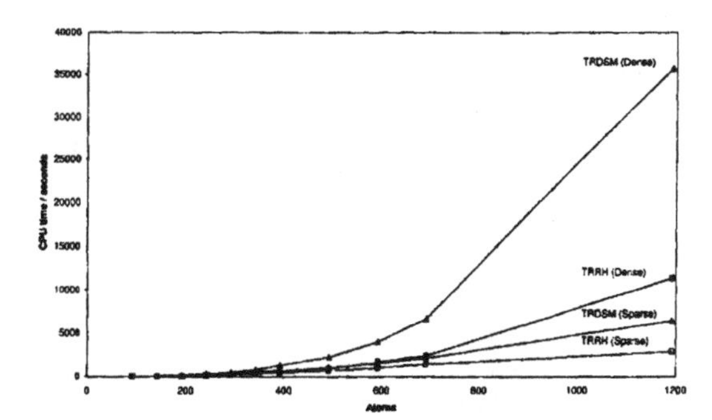

Figure 4: CPU timings for one iteration of a Hartree-Fock calculation using a 6-31G basis for the LS-TRRH and TRDSM steps for sparse and dense matrices plotted as a function of the number of atoms in a polyalanine peptide.

scaling regime. In general, the behaviour of the B3LYP curves is similar to the one observed for the Hartree-Fock curves.

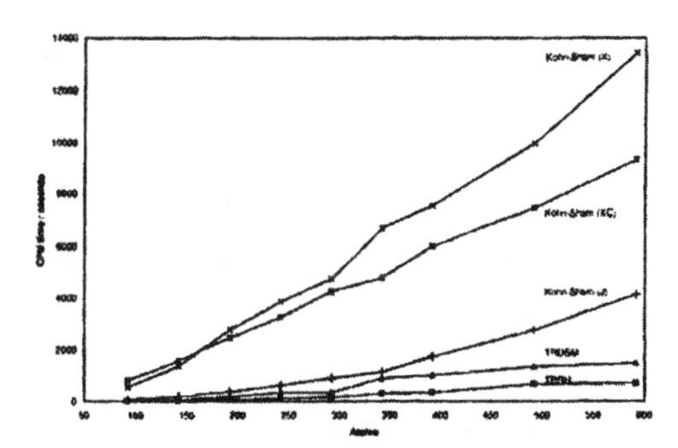

Figure 5: CPU timings for one iteration of a B3LYP calculation using a 6-31G basis plotted as a function of the number of atoms in a polyalanine peptide. The same contributions as in Fig. 3 are considered, in addition to the exchange correlation (XC) contribution.

5 Conclusion

We have described a linear scaling implementation of the trust-region self-consistent field (LS-TRSCF) method. In the LS-TRSCF method, each iteration consists of a incomplete optimization of the Roothaan-Hall energy giving a new density matrix (see Section 2.3) followed by the determination of an improved density matrix in the subspace containing the current and previous density matrices. A linear scaling algorithm is obtained using iterative methods to solve the level shifted Newton equations and sparse matrix algebra.

The convergence of the LS-TRSCF method is examined, and for comparison the convergence of conventional SCF/DIIS calculations have been reported. The LS-TRSCF calculations show smooth and robust convergence, and in several cases the LS-TRSCF calculations converge where the SCF/DIIS calculations diverge. The convergence of the LS-TRSCF method is in general equally good for small and large systems. For small systems, a TRSCF implementation based on an explicit diagonalization of the Fock/KS matrix may be more efficient. However, for small systems the computational time for optimizing the density matrix is insignificant compared to the computational time for setting up the Fock/KS matrix. Consequently we recommend using LS-TRSCF as standard method for calculations on both small and large systems.

Acknowledgments

This work has been supported by the Danish Natural Research Council and the Norwegian Research Council. We also acknowledge support from the Danish Center for Scientific Computing (DCSC) and the European Research and Training Network NANOQUANT, Understanding Nanomaterials from the Quantum Perspective, contract No. MRTN-CT-2003-506842.

References

[1] P. Pulay, *Chem. Phys. Lett.* **73** 393 (1980).

[2] C. A. White, B. G. Johnson, P. M. W. Gill and M. Head-Gordon, *Chem. Phys. Lett.* **230**, 8 (1994).

[3] C. A. White,B. G. Johnson, P. M. W. Gill and M. Head-Gordon, *Chem. Phys. Lett.* **253**, 268 (1996).

[4] M. C. Strain, G. E. Scuseria and M. J. Frisch, *Science* **271**, 51 (1996).

[5] M. Challacombe and E. Schwegler, *J. Chem. Phys.* **106**, 5526 (1997).

[6] Y. Shao, and M. Head-Gordon, *Chem. Phys. Lett.* **323**, 425 (2000).

[7] E. Schwegler and M. Challacombe , *J. Chem. Phys.* **105**, 2726 (1996).

[8] E. Schwegler, M. Challacombe and M. Head-Gordon , *J. Chem. Phys.* **106**, 9708 (1997).

[9] E. Schwegler and M. Challacombe , *J. Chem. Phys.* **111**, 6223 (1999).

[10] E. Schwegler and M. Challacombe , *Theor. Chem. Acc.* **104**, 344 (2000).

[11] C. Ochsenfeld, C. A. White and and M. Head-Gordon , *J. Chem. Phys.* **109**, 1663 (1998).

[12] J. C. Burant, G. E. Scuseria and M. J. Frisch, *J. Chem. Phys.* **105**, 8969 (1996).

[13] J. M. Pérez-Jordá and W. Yang, *Chem. Phys. Lett.* **241**, 469 (1995).

[14] B. G. Johnson, C. A. White, Q. Zang, B. Chen, R. L. Graham, P. M. W. Gill and M. Head-Gordon, in *Recent Developments in Density Functional Theory.* Edited by J. M. Seminario (Elsevier Science, Amsterdam, 1996), Vol. 4

[15] R. E. Stratman, G. E. Scuseria and M. J. Frisch, *Chem. Phys. Lett.* **257**, 213 (1996).

[16] A. H. R. Palser and D. E. Manolopoulos, Phys. Rev. B **58**, 12704 (1998).

[17] R. McWeeny, *Rev. Mod. Phys.* **32**, 325 (1960).

[18] R. McWeeny. *Methods of Molecular Quantum Mechanics*, 2nd Edition. Academic Press, 1992.

[19] T. Helgaker, P. Jørgensen and J. Olsen. *Molecular Electronic-Structure Theory*. Wiley, New York, 2000.

[20] X. P. Li, R. W. Nunes and D. Vanderbilt, Phys. Rev. B **47**, 10891 (1993).

[21] J. M. Millam and G. E. Scuseria, J. Chem. Phys. **106**, 5569 (1997).

[22] M. Challacombe, J. Chem. Phys. **110**, 2332 (1999).

[23] T. Helgaker, H. Larsen, J. Olsen and P. Jørgensen, Chem. Phys. Lett. **327**, 379 (2000).

[24] H. Larsen, J. Olsen and P. Jørgensen and T. Helgaker, J. Chem. Phys. **115**, 9685 (2001).

[25] Y. Shao, C. Saravanan, M. Head-Gordon and C. A. White, *J. Chem. Phys.* **118**, 6144 (2003).

[26] L. Thøgersen, J. Olsen, D. Yeager, P. Jørgensen, P. Sałek, T. Helgaker, *J. Chem. Phys.* **121** 16 (2004).

[27] L. Thøgersen, J. Olsen, A. Köhn, P. Jørgensen, P. Sałek, T. Helgaker, *J. Chem. Phys.* **123** 074103 (2005).

[28] H. J. Aa. Jensen and P. Jørgensen, *J. Chem. Phys.* **80**, 1204 (1984).

[29] B. Lengsfield III, *J. Chem. Phys.* **73**, 382 (1980).

[30] R. Shepard, I. Shavitt and J. Simons, *J. Chem. Phys.* **76**, 543 (1982).

[31] P. Sałek *et al.*, *J. Chem. Phys.*, to be submitted.

[32] R. W. Nunes and D. Vanderbilt, *Phys. Rev. B* **50**, 17611 (1994).

[33] A. Schafer, H. Horn and R. Ahlrichs, *J. Chem. Phys.* **97**, 2571 (1992).

Brill Academic Publishers
P.O. Box 9000, 2300 PA Leiden
The Netherlands

Lecture Series on Computer and Computational Sciences
Volume 6, 2006, pp. 190-195

Problems in the Density Functional Method with the Total Spin and Space Degeneracy

G. Kaplan[1]

Instituto de Investigaciones en Materiales, UNAM, Apdo. Postal 70-360, 04510 Mιxico D.F. Mιxico.

Received 28 July, 2006; accepted 2 August, 2006

Abstract: The problems in the density functional theory arising when it is applied to the spin- and space-degenerate states are discussed. It is rigorously proved that the electron density of an arbitrary N-electron system cannot, in principle, depend upon the total spin S and for all values of S has the same form as it has for a single-determinantal wave function. It is also proved that the diagonal element of the density matrix is invariant in respect to the symmetry of the state and in the frame of density matrix description there is no difference between degenerate and nondegenerate states. From this follows that the problems in DFT connected with the total spin and degenerate states cannot be solved within the framework of the density matrix formalism.

Keywords: Density Functional Theory; Spin: Degenerate states.

1. Introduction

In the last two decades, the density functional theory (DFT) [1-3] has been widely used for solution of different problems in atomic, molecular and solid-state physics. The numerous results of the DFT calculations performed in this period [4-7] demonstrated that its gradient corrected versions yield quite reliable results in the prediction of the global minima on the potential energy surfaces and give reasonable binding energies for different classes of atomic and molecular complexes in the ground states comparable with the *ab initio* calculations by the Mψller-Plesset perturbation theory in the second order (MP2). The great advantage of the DFT methods is in its applicability to calculation of large systems for which *ab initio* methods are very expensive or limited by currently available computing power.

In spite of great success in the application of the DFT method to the ground-state properties, it was long ago recognized [8] that DFT cannot be directly applied to calculation of the spin and space multiplet structure. The special procedures developed for overcoming these difficulties [8-10], strictly speaking, are all beyond the DFT scope, see also the discussion in Ref. [11]. For instance, the so-called multiplet-sum method (MSM) [8, 10], in which the linear combinations of single-determinantal DFT energies are constructed in correspondence with the appropriate linear combinations of the Slater determinants for the state with the definite value of the total spin, does not follow from the DFT formalism; the same concerns to the restricted open-shell Kohn-Sham method (ROKS) [9], see discussion in next section.

The Kohn-Sham (KS) equations [12]

$$h_i^{KS}\left[\rho(\mathbf{r}_i)\right]\varphi^{KS}(\mathbf{r}_i)=\varepsilon_n\,\varphi^{KS}(\mathbf{r}_i)\qquad n=1,2,\dots,\frac{N}{2}\tag{1}$$

on which the DFT methods are based, depend upon the ground-state electron density $\rho(\mathbf{r})$ that is the diagonal element of the spinless one-electron reduced density matrix [13]

[1] *E-mail: kaplan@iim.unam.mx; fax: 5255-5616-1251.*

$$\rho(\mathbf{r}_1) = N \sum_{\sigma_1,..,\sigma_N} \int |\Psi(\mathbf{r}_1\sigma_1,...\mathbf{r}_N\sigma_N)|^2 \, dV^{(1)} \tag{2}$$

where sum is taken over the whole spin space and integration is performed over the configuration space of all electrons except the first. For a single-determinantal double-occupied wave function

$$\rho^{KS}(\mathbf{r}_i) = 2\sum_n \left[\varphi_n^{KS}(\mathbf{r}_i)\right]^2 \tag{3}$$

The Kohn-Sham equations (as the Hartree-Fock equations) correspond to the independent particle approximation. The Hohenberg-Kohn theorem [14] and based on it the KS equations were formulated for nondegenerate ground state described by a single-determinantal wave function. This limitation, as was accepted by most of the DFT users, was removed by the Levi-Lieb constraint search procedure [15,16]. In this procedure one searches a set of antisymmetric wave functions that leads to the same electronic density and then constructs a combination of these functions that minimizes the expectation value of the energy. As was demonstrated by McWeeny [17] and Bersuker [18] and follows from our results, in reality the constraint search cannot solve the problems arising in DFT in the case of spin and space degenerate states.

The spin-dependent DFT was developed first by von Barth and Hedin [19] who formulated the KS equations in so-called local spin density approximation (LSDA). The latter operates with different electron densities for different spin projections (ρ_α and ρ_β). The LSDA method, as the unrestricted Hartree-Fock method, corresponds to a state with a definite value of the spin projection S_z, but does not correspond to the state with a definite value of the total spin S.

$$\begin{gathered} \hat{S}^2\,\Psi^{LSDA} \neq S(S+1)\Psi^{LSDA} \\ \hat{S}_z\,\Psi^{LSDA} = S_z\,\Psi^{LSDA} \end{gathered} \tag{4}$$

Its solution includes all possible values of $S \geq S_z$ that may exist in the studied N-electron system, in other words, it is spin-contaminated. The proper S is extracted from unrestricted KS calculations by some projection procedure that does not always lead to correct result; Note, the generalization of DFT by the pair density formulation [20] did not reveal new possibilities of treating states with definite S.

As was mentioned by McWeeny [17]: "electron spin is in a certain sense extraneous to the DFT". On the example of the two-electron system in the singlet and triplet spin states, McWeeny [17] showed that knowing only the electron density, one cannot identify the spin sate.

In this lecture we consider the general case on N-electron system in a state with an arbitrary total spin S and prove the invariance of the electron density in respect to S. As we discuss below, the problems in DFT with description of spin- and space-degenerate states are connected with an inherent property of the electron density. Namely, the electron density of a quantum state does not depend upon the permutation and space symmetry of the state. We begin with consideration of states with a definite value of the total spin.

2. Total spin and DFT

As is well known, the value of the total spin S of an arbitrary N-electron system is uniquely connected with the permutation symmetry of the spin wave function characterized by the Young diagram $[\tilde{\lambda}]$ with N boxes [21,22]. The total wave function, corresponding to the spin S and satisfying the Pauli principle, can be constructed as a bilinear combination of the coordinate, $\Phi_{rt}^{[\lambda]}$, and spin, $\Omega_{\tilde{r}}^{[\tilde{\lambda}]}$, wave functions symmetrized according to the conjugate representations $\Gamma^{[\lambda]}$ and $\Gamma^{[\tilde{\lambda}]}$ of the permutation group with the dual Young diagrams [22,23],

$$\Psi_t^{[\lambda]} = \frac{1}{\sqrt{f_\lambda}} \sum_r \Phi_{rt}^{[\lambda]} \Omega_{\tilde{r}}^{[\tilde{\lambda}]} \tag{5}$$

where index t enumerates the different bases that can be constructed for $\Gamma^{[\lambda]}$ and f_λ is the dimension of irreducible representations $\Gamma^{[\lambda]}$ and $\Gamma^{[\tilde{\lambda}]}$.

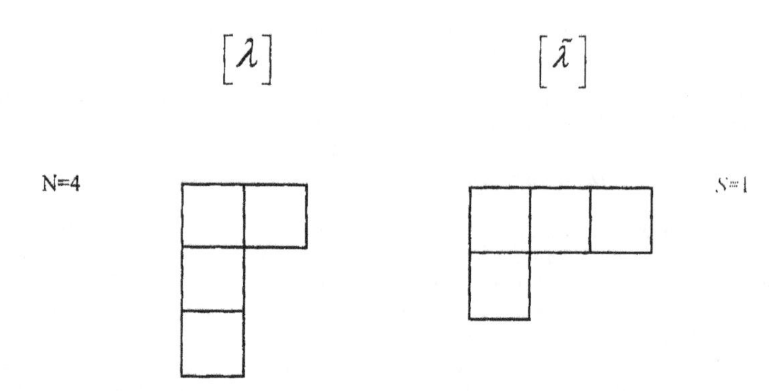

In nonrelativistic quantum mechanics, all properties of the studied system are completely determined by the coordinate wave function $\Phi_{rt}^{[\lambda]}$. The latter can be built on some nonsymmetric product of orthonormal orbitals (for simplicity we consider the single-occupied orbital configuration)

$$\Phi_0 = \varphi_1(\mathbf{r}_1)\varphi_2(\mathbf{r}_2)...\varphi_N(\mathbf{r}_N), \tag{6}$$

$$\int \varphi_n^*(r)\varphi_m(r)d^3r = \delta_{nm}, \tag{7}$$

in the form [22,23]

$$\Phi_{rt}^{[\lambda]} = \sqrt{\frac{f_\lambda}{N!}} \sum_P \Gamma_{rt}^{[\lambda]}(P) P\Phi_0 \tag{8}$$

where $\Gamma_{rt}^{[\lambda]}(P)$ are the matrix elements of representation $\Gamma^{[\lambda]}$ and P runs over all $N!$ permutations of the permutation group π_N. The Young diagram $[\lambda]$ is uniquely connected with the value of spin S. Thus, each value of S corresponds to a definite expectation value of energy due to the permutation symmetry of the corresponding coordinate wave function.

As we prove in Theorem 1, it is not true for the energy functional expressed via the electron density because of some specific properties of the latter. Note that for two-electron system, it was long ago known that the electron density is the same for the singlet and triplet states. But it is valid in the general case.

Theorem 1

The electron density of an arbitrary N-electron system does not depend upon the total spin S of the state and always preserves the same form as it is for a single-determinantal wave function.

The proof is presented in Ref. [24]. From it follows that

$$\rho_t^{[\lambda]}(\mathbf{r}) = \sum_{n=1}^{N} |\varphi_n|^2. \tag{9}$$

It is the well-known expression of the electron density for the state described by the single-occupied determinantal function. In the same manner it is easy to show that in the case of orbital configuration with arbitrary occupation numbers, the final expression will also correspond to the electron density for the one-determinantal function but with arbitrary occupation numbers.

Thus, regardless of the permutation symmetry of the coordinate wave function, which is uniquely connected with the value of the total spin S, the electron density for all S, describing by different multi-determinantal wave functions, has the same form as for a single Slater determinant. This result is rather surprising: different linear combinations of determinants correspond to the same expression as it is obtained with one determinant. From this follows that multi-determinantal versions of DFT cannot resolve the problem with the total spin. Due to the independence of the electron density on the total spin S, the density functionals, and, consequently, the KS equations will be the same for all multi-determinantal wave functions corresponding to different S.

The mentioned above MSM and ROKS procedures, elaborated for the study of states with definite S, are based on the wave-function formalism, hence they are both beyond the DFT approach. In MSM [8,10], the DFT energies are summed in the same way as the Slater determinants describing the state with given S. The functionals arising in this procedure may not be considered as a density functionals corresponding to a given value of the total spin, because, according to Theorem 1, such functionals do not exist. The MSM procedure is, in fact, a some practical recipe *ad hoc* that allows to calculate approximately the multiplet structure. It is worth-while to stress an approximate nature of MSM, it includes only the first order electrostatic interactions, ignoring the second order (correlation) effects.

In the ROKS procedure [9], the Hamiltonian for the spin-restricted open-shell calculations is constructed similar to the Hamiltonian of the restricted open-shell Hartree-Fock (ROHF) method with an exception of the exchange part, which is replaced by the exchange-correlation functional. Evidently, it leads to errors since in the states with a definite total spin S, the coefficients in the exchange and correlations parts must be different. Thus, in both procedures MSM and ROKS, the correlation functional does not correspond to the proper S.

The paradox with an independence of the electron density on the total spin stems from the general invariance of the density matrix in respect to the symmetry of the state. This inherent property of the density matrix will be discussed in the next section.

3. Symmetry properties of the density matrix

Let us consider some degenerate quantum state with symmetry of a point group **G**. The wave functions pertaining to this state must transform according to one of irreducible representations $\Gamma^{(\alpha)}$ of **G**. As shown in Refs. [21,22], they can be constructed as

$$\Psi_{ik}^{(\alpha)} = \frac{f_\alpha}{g}\sum_R \Gamma_{ik}^{(\alpha)}(R)^* R\Psi_0 \tag{10}$$

where $\Gamma_{ik}^{(\alpha)}(R)$ are the matrix elements of the representation $\Gamma^{(\alpha)}$, f_α is its dimension, R runs over all g element of the group **G**, and Ψ_0 is some arbitrary function. The set of f_α functions (10) with fixed second index k forms a basis for $\Gamma^{(\alpha)}$, index k enumerates different bases.

In a degenerate state, the system can be described with equal probability by any one of the basis vectors of this state. As a result, we can no longer select a pure state (the one that is described by a single wave function) and should regard the degenerate state as a mixed one, where each basis vector enters with the same probability. As was first shown by von Neumann [25], the diagonal element of the full density matrix for a degenerate state has the following form

(11)

The following theorem can be easily proved [24]:

Theorem 2

The diagonal element of the full density matrix is invariant respecting all operations of the group symmetry of the state, that is, it is a group invariant.

This means that

$$RD_k^{(\alpha)} = \frac{1}{f_\alpha}\sum_{l=1}^{f_\alpha}\left|\Psi_{lk}^{(\alpha)}\right|^2 = D_k^{(\alpha)}.$$
$$D_k^{(\alpha)} = \frac{1}{f_\alpha}\sum_{l=1}^{f_\alpha}\left|\Psi_{ik}^{(\alpha)}\right|^2. \tag{12}$$

For all irreducible representations, characterizing the quantum state, the diagonal element of the density matrix transforms according to the totally symmetric one-dimensional representation A_1 of **G** and in this respect one cannot distinguish between degenerate and nondegenerate states. Thus, the diagonal element of the density matrix is a group invariant.

The invariance of the diagonal elements of the density matrix following from the Theorem 2 could be expected. For the permutation group, this result was already used in Ref. [26]. Nevertheless, to the best of our knowledge, it was not discussed in literature. Even in the specialized monograph by Davidson [13], the total symmetry of the reduced density matrix is attributed only to nondegenerate states, but the latter is trivial.

The pure electronic degenerate states may not be stationary. In the ordinary stationary case, e.g., for degenerate ground states, the Born-Oppenheimer adiabatic approximation fails and the nonadiabatic approach with the vibronic interaction, mixing electronic states, has to be applied [27].

The applicability of the DFT approaches to degenerate states was analyzed in detail by Bersuker [18] who showed that the DFT method cannot, in principle, be applied to degenerate and psedodegenerate states. On one hand, because in these states the electronic and nuclear motions are nonseparable and cannot be described by appropriate densities. On the other hand, the important in the Jahn-Teller systems Berry phase problem [28, 29]: the strong dependence of the resulting energy spectrum and wave functions of the degenerate term of the phase properties of the electronic wave function, is beyond the DFT method. The latter is evident, since in the density formalism, operating with the square modulus of the wave function, the phase vanishes.

In Refs. [30, 31] the authors claimed that they formulated the non-Born-Oppenheimer DFT. The simple analysis shows that both formulations [30, 31] must be attributed to the Born-Oppenheimer (BO) approximation. The approach developed by Capitani et al. [30] corresponds to the BO approximation in its crude form (the so-called Condon adiabatic approximation), in which the electronic wave function does not depend upon the nuclear coordinates. Although in the approach by Kryachko et al. [31], the electronic density contains the nuclear coordinates, it also corresponds to the BO approximation, since the authors proceed from the multiplicative form of electronic and nuclear densities that is valid only for the nondegenerate states.

4. Conclusions

From the presented theorems and discussion follows that the problems in DFT with the total spin and, in general, with degenerate states cannot be solved within the framework of the density matrix formalism, on which DFT is based.

The insertion inside DFT the wave function approaches (e.g., the expressions of the Hartree-Fock method) can help to solve some of the discussed problems, at least, the problem with the total spin, but only for the exchange energy. The construction of the correlation functional depending on the total spin is still unsolved problem.

Acknowledgments

I would like to acknowledge Isaak Bersuker, Andreas Koster, Brian Sutcliffe, Andrei Tchougreeff, and Samuel Trickey for sending important references and numerous useful discussions. The study was partly supported by projects PAPIIT UNAM, 1N107305 and CONACYT (Mιxico) No. 46770.

References

[1] R.G. Parr and W. Yang, *Density-Functional Theory of Atoms and Molecules*, Oxford University Press, New York, 1989.

[2] E.S. Kryachko and E.V. Ludena, *Energy Density Functional Theory of Many-Electron Systems*, Kluwer Academic Press, Dordrecht, 1990.

[3] N.H. March, *Electron Density Theory of Atoms and Molecules*, Academic Press, London, 1992.

[4] T. Ziegler, Chem. Rev. 91 (1991) 651.

[5] E.K.U. Gross and R.M. Dreizler (Eds.), *Density Functional Theory*, Plenum, New York, 1995.

[6] H. Guo, S. Sirois, E.I. Proyanov, and D.R. Salahub, in *Theoretical Treatments of Hydrogen Bonding*, ed. D. Had☐i, Wiley & Sons, Chichester, 1997, pp 49-74.
[7] H.L. Schmider and A.D. Becke, J. Chem. Phys. 108 (1998) 9624.
[8] T. Ziegler, A. Rauk, and E. Baerends, Theor. Chim. Acta 43 (1977) 261..
[9] T.V. Russo, R.L. Martin, and P.J. Hay, J. Chem. Phys. 101 (1994) 7729.
[10] T. Mineva, A. Goursot, and C. Daul, Chem. Phys. Lett. 350 (2001) 147.
[11] I.G. Kaplan, *Intermolecular Interactions: Physical Picture, Computational Methods and Model Potentials*, Wiley & Sons, Chichester, 2006.
[12] W. Kohn and L. J. Sham, Phys. Rev. 140 (1965) A1133.
[13] E.R. Davidson, Reduced Density Matrices in Quantum Chemistry, Academic Press, New York, 1976.
[14] P. Hohenberg and W. Kohn, Phys. Rev. 136 (1964) B864.
[15] M. Levi, Proc. Natt. Acad. Sci. USA 76 (1979) 6062.
[16] E. Lieb, Int. J. Quant. Chem. 24 (1983) 243.
[17] R. McWeeny, Phil. Mag. B 69 (1994) 727.
[18] I.B. Bersnker, J. Comput. Chem 18 (1997) 260.
[19] U. von Barth and L. Hedin, J. Phys. C 5 (1972) 1629.
[20] P.Ziesche, Int. J. Quant. Chem. Symp. 30 (1996) 1361.
[21] L.D. Landau and E.M. Lifschitz, *Quantum Mechanics (Non-relativistic Theory)*, Third Edition, Pergamon Press, Oxford, 1977.
[22] I.G. Kaplan, *Symmetry of Many-Electron Systems*, Academic Press, New York, 1975.
[23] I.G. Kaplan, in *Handbook of Molecular Physics and Quantum Chemistry*, Vol. 2, ed. S. Wilson, Wiley & Sons, Chichester, 2003, p.15.
[24] I.G. Kaplan, J. Mol. Struct. (in Press).
[25] J.V.von Neumann, Mathematische Grundlagen der Quantenmechanik, Springer-Verlag, Berlin, 1932.
[26] I.G. Kaplan, Int. J. Quant. Chem. 89 (2002) 268.
[27] I.B. Bersuker, *The Jahn-Teller Effect*, Cambridge University Press, Cambridge, 2006.
[28] M.V. Berry, Proc. Roy. Soc. A 392 (1984) 45.
[29] C.A. Mead, Rev. Mod. Phys. 64 (1992) 51.
[30] J.F. Capitani, R.F. Nalewajski, and R.G. Parr, J. Chem. Phys., 76 (1982) 568.
[31] E.S. Kryachko, E.V. Ludena, and V. Mujica, Int. J. Quantum Chem., 40 (1997) 260.

Brill Academic Publishers
P.O. Box 9000, 2300 PA Leiden
The Netherlands

Lecture Series on Computer and Computational Sciences
Volume 6, 2006, pp. 196-208

Polarizability anisotropy dynamics in one- and two-component aromatic liquids

B.M.Ladanyi[1] and M.D. Elola

Department of Chemistry,
Colorado State University,
Fort Collins, CO 80523-1872, U.S.A.

Received 30 July, 2006; accepted August 2, 2006

Abstract: In this lecture will be presented the results of our molecular dynamics (MD) computer simulation studies of the structure and dynamics of several aromatic liquids and their mixtures. Benzene (Bz), hexaflourobenzene (HFB), and 1,3,5-trifluorobenzene (TFB) are molecules of very similar shapes, but different charge distributions, in view of the fact that C-H and C-F bonds have opposite polarities. They are therefore an interesting set of species to study in order to determine the balance of electrostatic and steric effects on the local structure and collective dynamics in the liquid state. Using MD simulation, we have studied Bz, HFB, and TFB neat liquids as well as mixtures of Bz and HFB. The main focus of our work has been on the relaxation of the collective polarizability anisotropy, which can be measured in optical Kerr effect (OKE) and depolarized light scattering experiments. We have calculated the polarizability response arising from intermolecular interactions by using the first-order dipole-induced-dipole model with the polarizability distributed over the carbon sites of each molecule. In the case of Bz-HFB mixtures, we have collaborated with John Fourkas and his group to carry out a parallel experimental-simulation study of the effects of interspecies orientational correlations on the polarizability dynamics. We found good qualitative agreement between experiments and simulations in the features exhibited by the OKE nuclear response function for pure liquids and mixtures. Through analysis of MD simulation data, we have determined how the local intermolecular structure and dynamical orientational correlations influence the OKE response. We have also examined two different schemes, which have been proposed to separate the OKE response into orientational relaxation and other dynamical processes, and found them to lead to quite different results. The implications of these results will be discussed.

Keywords: Collective polarizability anisotropy, aromatic liquids, orientational relaxation, molecular dynamics simulation.

PACS: 61.20.Ja, 61.25.Em, 42.65.Hw, 78.30.Cp.

1. Introduction

Dynamics and structure of aromatic liquids have been the subject of intense investigation for many years. Experimental measurements using neutron and x-ray diffraction techniques show that the average molecular arrangement in liquid benzene and hexafluorobenzene is similar to that in the solid, namely, neighboring molecules are oriented preferentially perpendicular to each other [1]. This feature has also been confirmed by molecular dynamics (MD) simulations [2-4]. Analysis of the local order from the radial-angular pair correlation functions obtained from simulations shows that parallel and perpendicular arrangements coexist between pairs of molecules within the first coordination shell in both liquids. However, while perpendicular arrangements are preferred in benzene, parallel configurations are slightly favored at very short distances in hexafluorobenzene.

In addition to similar shapes, benzene (Bz) and hexafluorobenzene (HFB) molecules have nearly the same polarizabilities and polarizability anisotropies (see Table I of Ref. [5]). However, there are significant differences in the molecular charge distributions due to the fact that the C-H and C-F bonds have opposite polarities A consequence of this fact is that the electric quadrupole moments of Bz (-

[1] Corresponding author. E-mail: bl@lamar.colostate.edu

2.90×10^{-39} C m²) and HFB (3.17×10^{-39} C m²) are roughly equal in magnitude but opposite in sign [6]. The quadrupole-quadrupole interaction favors a perpendicular arrangement of neighboring molecules while the dispersion interactions tend to favor parallel stacking. Thus both arrangements are possible for neighboring molecules in the liquid phase, as indicated above. In the case of 1,3,5-trifluorobenzene (TFB), alternating C-H and C-F bonds give rise to a much smaller quadrupole [7], 0.31×10^{-39} C m², than those of Bz and HFB. Thus the contributions of higher multipole moments will not be negligible and it is reasonable to expect that the structure of liquid TFB will be different from that of the other two liquids. Indeed, neutron diffraction experiments and molecular dynamics simulations [2,8] show that the short-range local order in TFB involves parallel stacked configuration with nearest-neighbor separation of less than 4 Å. This is consistent with the expectation that electrostatic interactions will favor y parallel alignment of C-H and C-F bonds of neighboring molecules. Therefore, Bz, TFB and HFB form an interesting set of molecules in which the effects of local order on intermolecular dynamics can be investigated.

Opposite polarities of C-H and C-F bonds also lead to a strong attractive interaction between Bz and HFB molecules. Mixtures of these species are therefore quite nonideal. The equimolar mixture freezes, forming a structure of alternating parallel stacks of Bz and HFB molecules, at 23.4°C, a temperature that is about 18°C higher than the freezing temperatures of either species.[9] Diffraction experiments and MD simulations have shown that alternate parallel stacking also dominates the short-range order in the liquid phase [10,11]. Our MD studies [3] have further revealed that the heterdimers are long-lived at room temperature, attaining their longest lifetimes at equimolar concentration.

The different structural features of different aromatic liquids and their mixtures give rise to different dynamics. Information about the dynamics of intermolecular modes in a liquid can be obtained experimentally from depolarized Rayleigh scattering (DRS) and nonlinear optical techniques, such as femtosecond optical heterodyne-detected Raman-induced Kerr effect spectroscopy [12]. This technique has been used on a number of occasions to explore the collective dynamics in aromatic liquids [13-15]. Several MD simulation studies have been carrWe have used MD simulation to investigate the connections the The time-dependent signal carried out to investigate the polarizability anisotropy relaxation of aromatic liquids in greater detail in order to gain further insight into the connections between these dynamics and local structure [4,5,15-17]. In this lecture, the focus will be on our work in which MD studies have been used to connect the local structure of Bz, HFB, TCF, and Bz-HFB mixtures to the polarizability anisotropy dynamics. This work has been reported in several papers [3,5,15]. Some of the highlights will be described here.

In the next section, we outline the theoretical background associated with the modeling of the collective polarizability and the determination of the dynamical origin of the polarizability anisotropy relaxation. The details of our MD simulation methods are given in Sec. 3. The results for neat liquids, Bz, HFB, and TFB, are given in Sec. 4 and for the Bz-HFB mixtures in Sec. 5. Our main findings are summarized in Sec. 6.

2. Theoretical Background: Polarizability Anisotropy Time Correlation

In a condensed-phase system, the collective polarizability $\mathbf{\Pi}$ of an N-molecule system is given by

$$\mathbf{\Pi} = \mathbf{\Pi}^{\mathrm{M}} + \mathbf{\Pi}^{\mathrm{I}}, \tag{1}$$

where

$$\mathbf{\Pi}^{\mathrm{M}} = \sum_{i=1}^{N} \boldsymbol{\alpha}_i, \tag{2}$$

is the 'molecular' component of the collective polarizability. It corresponds to a sum of isolated-molecule polarizabilities, $\boldsymbol{\alpha}_i$ and $\mathbf{\Pi}^{\mathrm{I}}$ is the polarizability arising from interactions of molecular induced moments. In its simplest form it is given in terms of ideal induced-dipole interactions between molecular centers of mass. The leading term is then

$$\mathbf{\Pi}^{\mathrm{I}} \cong \sum_{i=1}^{N}\sum_{j\neq i}^{N} \boldsymbol{\alpha}_i \cdot \mathbf{T}(\mathbf{r}_{ij}) \cdot \boldsymbol{\alpha}_j \tag{3}$$

where

$$\mathbf{T}(\mathbf{r}) = \frac{3\hat{\mathbf{r}}\hat{\mathbf{r}} - \mathbf{1}}{r^3} \quad (4)$$

is the dipole tensor.

For polyatomics, this form of $\mathbf{\Pi}^{\mathrm{I}}$ is often inaccurate at short intermolecular separations, given that intra- and intermolecular distances are then comparable in magnitude and the distribution of polarizability within the molecule becomes important. In these cases, it is more reasonable to use a site-site interaction model. In the case of Bz and its fluorinated analogs, we are using a model in which the polarizability is assigned to six sites, centered on the carbon atoms [18]. This leads to the following form of the site-site dipole-induced dipole (ss-DID) model

$$\mathbf{\Pi}^{\mathrm{I}} \cong \sum_{i=1}^{N}\sum_{j\neq i}^{N}\sum_{m=1}^{6}\sum_{n=1}^{6} \boldsymbol{\alpha}_{im} \cdot \mathbf{T}(\mathbf{r}_{im,jn}) \cdot \boldsymbol{\alpha}_{jn} \,. \quad (5)$$

Our work indicates [15] that this model is more accurate than Eq. (3) when compared to OKE results for Bz and HFB.

In DRS and optical Kerr effect (OKE), the measured response is related to the time correlation function (TCF) of an anisotropic component of the collective polarizability. To calculate this response, one may choose any off-diagonal component or the traceless part of a diagonal component. Statistics can be improved by employing all the equivalent components in evaluating the TCF from simulation data. We choose here Π_{xz} and then the relevant TCF is

$$\Psi_{xz}(t) = \frac{\langle \Pi_{xz}(0)\Pi_{xz}(t)\rangle}{\Gamma^2} \quad (6)$$

where $\Gamma^2 = N\gamma_0^2/15$ is the ideal-gas DRS intensity and γ_0 the isolated-molecule polarizability anisotropy. For a two-component system, this definition generalizes to $\Gamma^2 = N(x_A\gamma_{0A}^2 + x_B\gamma_{0B}^2)/15$, where x_p and γ_{0p} are the mole fraction and the molecular polarizability anisotropy for component p.

The DRS spectrum corresponds to the Fourier transform of the above TCF, while the nuclear portion of the OKE response is proportional to its time derivative

$$R_n(t) \propto -\frac{1}{k_B T}\frac{\partial \Psi_{xz}(t)}{\partial t} \,. \quad (7)$$

There has been considerable interest in finding out the dynamical origin of the collective polarizability response. The substitution of Eq. (1) into Eq. (6) leads to

$$\Psi_{xz}(t) = \Psi_{xz}^{\mathrm{MM}}(t) + \Psi_{xz}^{\mathrm{MI}}(t) + \Psi_{xz}^{\mathrm{II}}(t) \,. \quad (8)$$

As can be seen from Eqs. (2) and (5), for molecules lacking torsional degrees of freedom, Π_{xz}^{M} depends on molecular orientational coordinates, while Π_{xz}^{I} depends both on molecular orientations and intermolecular distances. Thus $\Psi_{xz}^{\mathrm{MM}}(t)$ relaxes solely through collective reorientation, $\Psi_{xz}^{\mathrm{MI}}(t)$ and $\Psi_{xz}^{\mathrm{II}}(t)$ relax through both rotational and translational dynamics. Several strategies have been used to analyze the dynamics contributing to the interaction-induced portions of the polarizability anisotropy TCF [16,19,20].

The portion of Π_{xz}^{I} that relaxes at the same rate as Π_{xz}^{M} can be identified by projecting it along this quantity [19,21]. In this projection scheme, the part of the polarizability Π_{xz}^{R} that relaxes through collective reorientation is given by

$$\Pi_{xz}^{\mathrm{R}} = (1 + f_{xz})\Pi_{xz}^{\mathrm{M}} \quad (9)$$

where,

$$f_{xz} = \langle \Pi_{xz}^{\mathrm{I}}\Pi_{xz}^{\mathrm{M}}\rangle / \langle \Pi_{xz}^{\mathrm{M}}\Pi_{xz}^{\mathrm{M}}\rangle \,. \quad (10)$$

The remaining portion of the interaction-induced polarizability,

$$\Delta\Pi_{xz} = \Pi^{\mathrm{I}}_{xz} - f_{xz}\Pi^{\mathrm{M}}_{xz}, \tag{11}$$

sometimes called the 'collision-induced' polarizability [19], relaxes through fast dynamics, much of which corresponds to intermolecular translation.

The polarizability anisotropy TCF can now be rewritten as

$$\Psi_{xz}(t) = \Psi^{\mathrm{RR}}_{xz}(t) + \Psi^{\mathrm{R\Delta}}_{xz}(t) + \Psi^{\Delta\Delta}_{xz}(t), \tag{12}$$

where

$$\Psi^{\mathrm{RR}}_{xz}(t) = (1+f_{xz})^2\Psi^{\mathrm{MM}}_{xz}(t). \tag{13}$$

Thus $(1+f_{xz})^2$ serves as a local field factor which modifies the intensity of the collective reorientation component, without altering its time-dependence.

As will be illustrated in the case of aromatics, the above projection scheme is successful in identifying the slowly relaxing portion of $\Psi_{xz}(t)$ as being due to collective reorientation, but somewhat less in unraveling the fast dynamics contributing to polarizability anisotropy relaxation.

A different method of analysis can be employed in the case of the nuclear respo(14)nse function $R_n(t)$, measured in OKE. This analysis, proposed by Steele [22], is based on a dynamical separation of the time derivative of the collective variable of interest, in the present case, of Π_{xz}. Thus, by using the chain rule of differentiation,

$$\dot{\Pi}_{xz} = \sum_{i=1}^{N}\sum_{\mu=1}^{6}\frac{\partial\Pi_{xz}}{\partial r_{i\mu}}\dot{r}_{i\mu} = \dot{\Pi}^{Tr}_{xz} + \dot{\Pi}^{Rot}_{xz}, \tag{14}$$

where $\dot{\Pi}^{Tr}_{xz}$ and $\dot{\Pi}^{Rot}_{xz}$ are the polarizability anisotropy time derivatives for which $\dot{r}_{i\mu}$ are, respectively, center-of-mass and angular velocities. Rotational and translational components of instantaneous normal mode influence spectra [16,20] are based in this same separation. Eq. (14) can be used to construct the 'polarizability velocity' TCF

$$\begin{aligned} G_{xz}(t) &= \langle\dot{\Pi}_{xz}(0)\dot{\Pi}_{xz}(t)\rangle/\Gamma^2 = -\ddot{\Psi}_{xz}(t) \\ &= G^{Rot}_{xz}(t) + G^{RotTr}_{xz}(t) + G^{Tr}_{xz}(t). \end{aligned} \tag{15}$$

The OKE response can be obtained from a time integral of $G_{xz}(t)$

$$R_n(t) \propto \int_0^t G_{xz}(\tau)d\tau \tag{16}$$

and can be separated into the same set of rotational (Rot), translational (Tr) and their cross-correlation (RotTr) components as $G_{xz}(t)$.

3. Intermolecular Potentials and Simulation details

Our MD simulations of aromatic liquids were carried out in the microcanonical ensemble for systems of 256 molecules, the average temperature of 298 K, and at densities appropriate for the liquids at this temperature and ambient pressure. The molecules were assumed to be rigid with bond lengths and angles corresponding to equilibrium geometry. Intermolecular interactions were calculated using exponential-6 + Coulomb potentials. In the case of neat liquids, the parameters were taken from the work of Cabaço et al. [2]. In the case of Bz-HFB mixtures, most of the potential parameters were taken from the work of Williams and coworkers [23]. The Exp-6 parameters were modified [3] in order to properly take into account the fact that H-F interactions are less attractive than one would predict using the conventional combining rules for interactions between unlike sites [24].

The simulations were carried out using the Verlet algorithm [25] and the SHAKE algorithm [26]was used to maintain the fixed molecular geometry. Ewald sums with conducting boundary conditions [27] were applied to Coulomb interactions and the Exp-6 interactions were cut off at half the box length. After equilibration, trajectories of at least 2 ns were used for further data analysis.

4. Local Structure and Polarizability Anisotropy Dynamics in Neat Liquids

The molecular shapes of Bz, HFB and TFB are quite similar, but different attractive interactions give rise to different local structures in the liquid phase [2,5]. This is illustrated in Figure 1 in which the pair angular-radial distributions $g(r,\theta)$, where r is the center-of-mass intermolecular distance and θ the angle between the highest symmetry axes of a pair of molecules, are displayed.

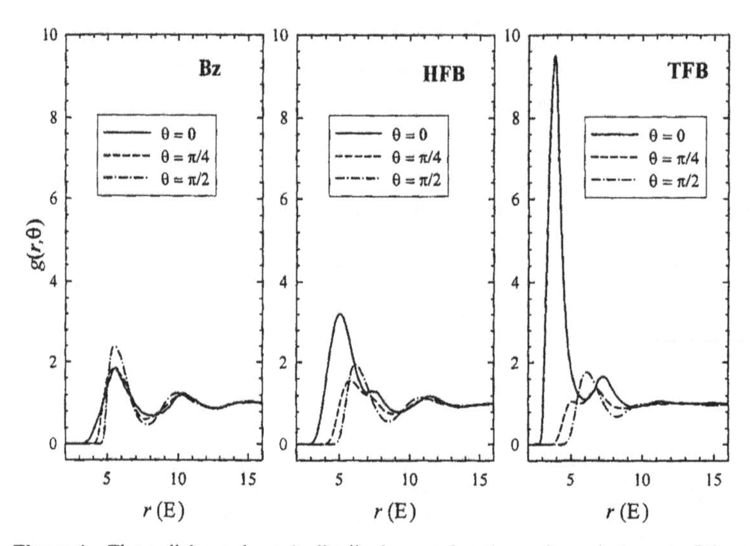

Figure 1. The radial-angular pair distributions as functions of r and at a set of three angles, θ, between molecular axes of highest symmetry. The left, middle and right panels depict the results for Bz, HFB and TFB, respectively.

The $g(r,\theta)$ results show that in benzene there is a slight preference of T-shaped ($\theta = \pi/2$) over parallel ($\theta = 0$) and tilted ($\theta = \pi/4$) orientations, while in HFB the parallel structures are favored somewhat more strongly than the other two relative orientations. The TFB case is dramatically different, with a strong preference of the parallel over the other two structures. In this case, the alternating polarities of C-H and C-F bonds favor parallel stacking, which is reflected in a very prominent first peak in $g(r,0)$ and a quite prominent second peak.

The polarizability anisotropy dynamics is sensitive to intermolecular orientational correlations, especially through the $\Psi_{xz}^{MM}(t)$ term. How much this term contributes is to some extent governed by the magnitudes of the molecular polarizability components. The molecules under consideration here are axially symmetric and thus have two independent polarizability components, $\alpha_{\parallel}$ and $\alpha_{\perp}$, in the molecular frame, leading to isotropic and anisotropic molecular polarizabilities

$$\bar{\alpha} = (\alpha_{\sim} + 2\alpha\ );\ \gamma_0 = \alpha_{\sim} - \alpha \qquad (17)$$

Table 1: Molecular polarizability components for benzene [28], hexafluorobenzene [29], and 1,3,5-trifluorobenzene [29].

molecule	$\alpha_{\sim}$ (Å^3)	$\alpha_{\perp}$ (Å^3)	$\gamma_0 / \bar{\alpha}$
Bz	6.54	11.73	-0.519
HFB	4.58	12.08	-0.783
TFB	6.14	12.23	-0.676

As can be seen from Table 1, the polarizability components of the three types of isolated molecules do not differ greatly. Benzene has the smallest and HFB the largest relative polarizability anisotropy, $\gamma_0 / \bar{\alpha}$, which governs the relative contributions of interaction-induced and molecular polarizabilities to $\Psi_{xz}(t)$ [16]. Figure 2, in which the components $\Psi_{xz}^{MM}(t)$, $\Psi_{xz}^{MI}(t)$,and $\Psi_{xz}^{II}(t)$ of the polarizability anisotropy TCF of the three liquids are shown, illustrates this trend.

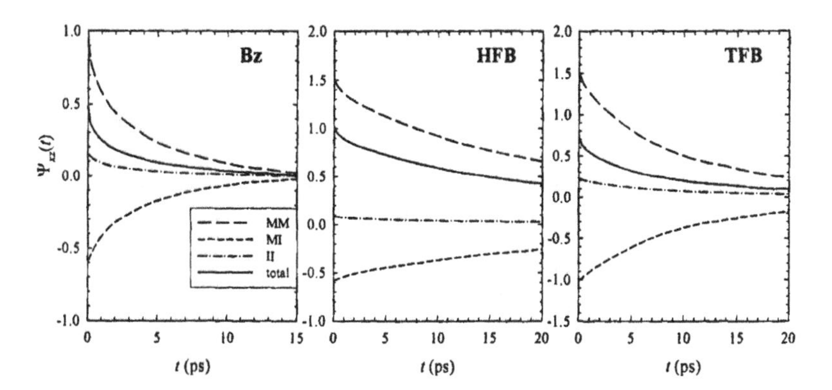

Figure 2. The polarizability anisotropy time correlation, $\Psi_{xz}(t)$, and its components $\Psi_{xz}^{MM}(t)$, $\Psi_{xz}^{MI}(t)$,and $\Psi_{xz}^{II}(t)$ for the three liquids: Benzene (left), HFB (middle), and TFB (right).

The interaction-induced contributions to $\Psi_{xz}(t)$ are quite large for all three liquids. A comparison of $\Psi_{xz}^{MM}(t)$ and $\Psi_{xz}^{MI}(t)$ shows that the relative magnitude of the cross term is largest in Bz, as expected, and smallest in HFB. The relaxation rates of $\Psi_{xz}^{MM}(t)$ and $\Psi_{xz}^{MI}(t)$ are quite similar. The Keyes-Kivelson projection scheme [19,21], defined by Eqs. (9) - (12), makes it possible to separate out the portions of $\Psi_{xz}^{MI}(t)$,and $\Psi_{xz}^{II}(t)$ that relax at the same rate as $\Psi_{xz}^{MM}(t)$. Its effects are illustrated in Figure 3, which shows that most of the longer-time dynamics contributing to $\Psi_{xz}(t)$ is due to collective reorientation.

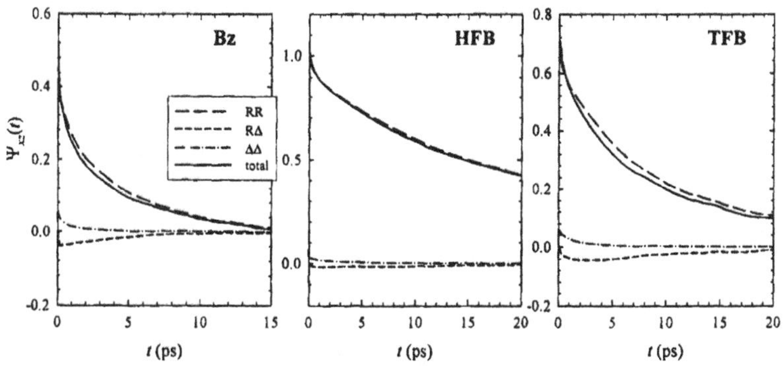

Figure 3. The polarizability anisotropy time correlation, $\Psi_{xz}(t)$, and its components in the projection scheme, $\Psi_{xz}^{RR}(t)$, $\Psi_{xz}^{R\Delta}(t)$, and $\Psi_{xz}^{\Delta\Delta}(t)$, for the three liquids: Benzene (left), HFB (middle), and TFB (right).

The results shown in Figure 3 clearly indicate that, for the aromatic liquids considered, the dynamics observed in OKE beyond about 5 ps is collective reorientation. On this time scale, the interaction-induced polarizability modifies the intensity of the response, but not its time-evolution.

Since the reorientation that contributes to OKE and light scattering is collective, intermolecular correlations can play an important role. As was shown in Figure 1, static orientational pair correlations are strongest in TFE. The contributions of the dynamical pair correlations to $\Psi_{xz}^{\mathrm{MM}}(t)$ can be determined by separating it into single-molecule, $\Psi_{xz}^{\mathrm{MMs}}(t)$, and pair, $\Psi_{xz}^{\mathrm{MMp}}(t)$, contributions:

$$\begin{aligned}\Psi_{xz}^{\mathrm{MM}}(t) &= \frac{1}{N}\sum_{i=1}^{N}\sum_{j=1}^{N} P_2\left[\hat{\mathbf{u}}_i(0)\cdot\hat{\mathbf{u}}_j(t)\right] \\ &= \frac{1}{N}\sum_{i=1}^{N} P_2\left[\hat{\mathbf{u}}_i(0)\cdot\hat{\mathbf{u}}_i(t)\right] + \frac{1}{N}\sum_{i=1}^{N}\sum_{j\neq i} P_2\left[\hat{\mathbf{u}}_i(0)\cdot\hat{\mathbf{u}}_j(t)\right] = \Psi_{xz}^{\mathrm{MMs}}(t) + \Psi_{xz}^{\mathrm{MMp}}(t)\end{aligned} \tag{18}$$

where $\hat{\mathbf{u}}_i$ is a unit vector along the molecular symmetry axis and P_2 is the second-rank Legendre polynomial.

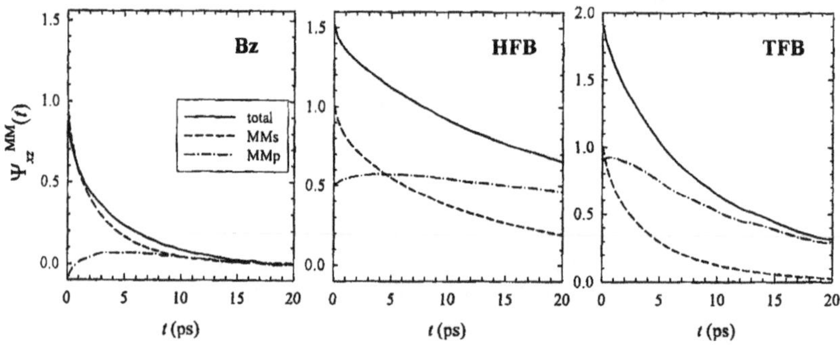

Figure 4. The separation of the molecular component, $\Psi_{xz}^{\mathrm{MM}}(t)$, of the polarizability anisotropy TCF into single-molecule and pair contributions for the three liquids: Benzene (left), HFB (middle), and TFB (right).

Figure 4 illustrates the contributions of orientational pair correlations to the relaxation of the molecular component of the collective polarizability. As anticipated on the basis of static pair correlations, the relative importance of $\Psi_{xz}^{\mathrm{MMp}}(t)$ is largest for TFB and smallest for benzene. In all cases, the pair term is more slowly relaxing than the single-molecule component and leads to slower relaxation of $\Psi_{xz}^{\mathrm{MM}}(t)$ when compared with $\Psi_{xz}^{\mathrm{MMs}}(t)$. The values of $\Psi_{xz}^{\mathrm{MMp}}(0)$, which is negative for benzene and positive for HFB and TFB reflect the balance between parallel and T-shaped arrangements of near neighbors, as we have shown through further analysis [5].

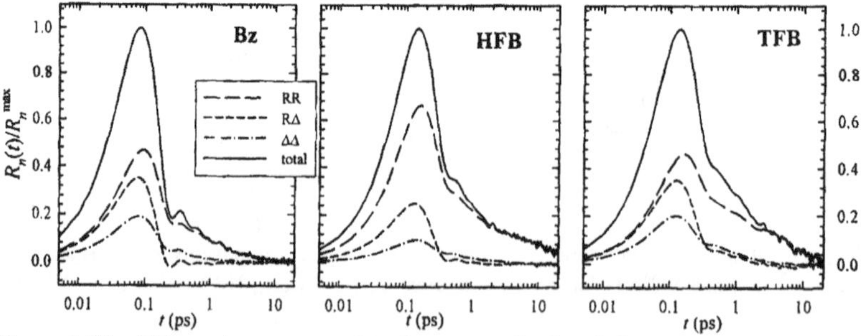

Figure 5. The OKE nuclear response functions, normalized to their maximum values, vs. time on the logarithmic scale. Depicted is the total $R_n(t)$, Eq. (7), and its components in the projected reprsentation for the three liquids, benzene (left), HFB (middle) and TFB (right).

While the longer-time dynamics are dominated by collective reorientation through $\Psi_{xz}^{\mathrm{RR}}(t)$ (which differs from $\Psi_{xz}^{\mathrm{MM}}(t)$ by a constant local-field factor), the origin of the short-time portion of the OKE response is more complicated, given that $\Psi_{xz}^{R\Delta}(t)$, and $\Psi_{xz}^{\Delta\Delta}(t)$ also make a significant contribution. This is illustrated in Figure 5, which depicts the contributions to $R_n(t)$ for the three liquids.

The results shown in the figure indicate that near the peak of the nuclear response function, $R_n(t)$, the three components, $R_n^{RR}(t)$, $R_n^{R\Delta}(t)$ and $R_n^{\Delta\Delta}(t)$ have similar magnitudes. While one can tentatively correlate $R_n^{R\Delta}(t)$ and $R_n^{\Delta\Delta}(t)$, respectively, with rotation-translation coupling and intermolecular translation, these assignments are not clear-cut, since rotational and translational dynamics contribute to both quantities. We have therefore also examined a different approach to determining the dynamical origin of the contributions to $R_n(t)$. This approach is due to Steele and identifies the dynamical origin by following the corresponding velocity components. The corresponding expressions are given by Eqs. (15) and (16).

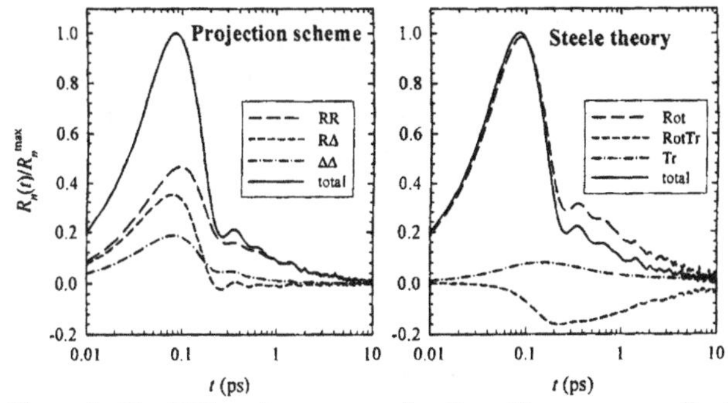

Figure 6. The OKE nuclear response function of benzene, normalized to its maximum value, vs. time on a logarithmic scale. The results of two methods of analysis of $R_n(t)$ are compared. The left panel depicts the components obtained in the projection scheme and the right panel in the Steele theory.

A comparison of the results of the two analysis methods for $R_n(t)$ of benzene is shown in Figure 6. As can be seen from this figure, three components into which $R_n(t)$ is divided in the projection and Steele analysis schemes are definitely not equivalent. For example, the rotational component in the Steele theory, $R_n^{Rot}(t)$, clearly does not correspond to collective reorientation, represented by $R_n^{RR}(t)$. Steele theory is most useful at short times at which the dynamics that is observed is associated with velocity relaxation, which usually occurs at a faster rate than the relaxation of the generalized forces (coordinate derivatives of the variables of interest, here Π_{xz}). Thus the Steele-theory analysis shows that the short-time portion of $R_n(t)$, i.e., its rise and decay close to the peak, occur predominantly through angular velocity relaxation. However the Steele theory does not the correctly identify the long-time decay with diffusive reorientation, while the projection scheme does.

5. Polarizability anisotropy dynamics in mixtures of benzene and hexafluorobenzene

As noted in the introduction, the quadrupole moments of benzene and HFB are of similar magnitude, but of opposite sign. This leads to strong Bz-HFB electrostatic attraction. The binding energy for the heterodimer is about twice as attractive as for HFB-HFB homodimers and almost three times as attractive as for Bz-Bz homodimers [3]. This strong attraction between unlike species leads to interesting structural and dynamical properties of Bz-HFB liquid mixtures [3,11,14,15].

The dynamics are characterized by the existence of long-lived parallel stacks of alternating molecular species. We have used MD simulation to investigate heterodimer lifetimes [3], using a geometrical definition, based on the location of the first peak of $g_{BzHFB}(r,0)$, to set the limits on the nearest-neighbor heterodimer center-of-mass separations. A method that we used involved calculating the heterodimer lifetime as the time-integral of the time correlation

$$C(t) = \frac{\left\langle \sum_k h_k(0) h_k(t) \right\rangle}{\left\langle N_{bonds} \right\rangle} \tag{19}$$

where, h_k is the occupation number for the kth Bz-HFB pair, r_k is the center-of-mass separation for the pair and r_1 the location of the minimum after the first peak ofn $g_{BzHFB}(r,0)$:

$$h_k = \begin{cases} 1 \text{ if } r_k \le r_1 \\ 0 \text{ if } r_k > r_1 \end{cases}. \tag{20}$$

The lifetime is given by

$$\tau = \int_0^\infty C(t)dt\,. \tag{21}$$

Different values are obtained if continuous (i.e., $h_k(t) = 0$ for all times after the bond is broken) or intermittent occupation are considered. Table 2 displays the two sets of values, $\tau_{cont.}$ and $\tau_{int.}$. As expected, $\tau_{int.} > \tau_{cont.}$, and both have their largest values for the equimolar mixtures. $\tau_{cont.}$ comparable to the diffusive relaxation times of $\Psi_{xz}^{MM}(t)$ for the pure components, (cf. Figure 4), while $\tau_{int.}$ is considerably longer. Thus we expect interspecies orientational correlations to play an important role in the relaxation of the polarizability anisotropy relaxation of the mixtures.

Table 2. Heterodimer lifetimes

x_{Bz}	τ_{cont} (ps)	$\tau_{int.}$ (ps)
0.8	19.20	60.11
0.5	30.29	120.79
0.2	24.91	110.75

We have investigated the polarizability anisotropy dynamics of the Bz-HFB mixtures in a parallel simulation – experimental MD – OKE study [15]. Figure 7 summarizes the results of the direct comparison.

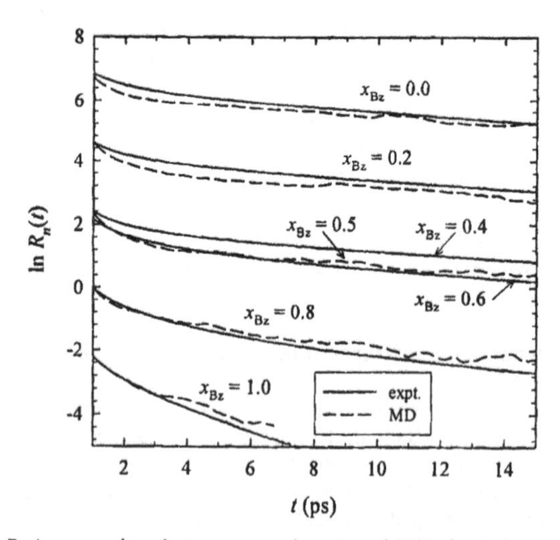

Figure 7. A comparison between experiments and MD simulation results for the long-time OKE response of benzene-hexafluorobenzene mixtures [15]. Data for different benzene mole fractions are offest for clarity. Note that the MD results for x_{Bz} = 0.5 are compared to the experimental data for x_{Bz} = 0.4 and 0.6.

As can be seen from this figure, the agreement between simulation and experiment is quite good, which allows us to use the analysis of the MD data in order to explore further the origin of the behavior observed in the experiment. One of the interesting features is the observation that the long-time decay of

$R_n(t)$, which, as we have seen in the prrevious section, is due to the relaxation of the molecular component, can be fit to a sum of two exponentials. One of the corresponding relaxation times peaks at around $x_{Bz} = 0.5$, while the other increases steadily with decreasing x_{Bz}. One might naively assign the two times to the two mixture components, but analysis of $\Psi_{xz}^{MM}(t)$ indicates that this interpretation is incorrect in view of the existence of long-lived interspecies pairs. Their contribution to $\Psi_{xz}^{MM}(t)$ can be studied by decomposing this TCF into contributions from each component and their cross-correlation

$$\Psi_{xz}^{MM}(t) = \Psi_{xz,Bz\text{-}Bz}^{MM}(t) + \Psi_{xz,Bz\text{-}HFB}^{MM}(t) + \Psi_{xz,HFB\text{-}HFB}^{MM}(t)\,. \quad (22)$$

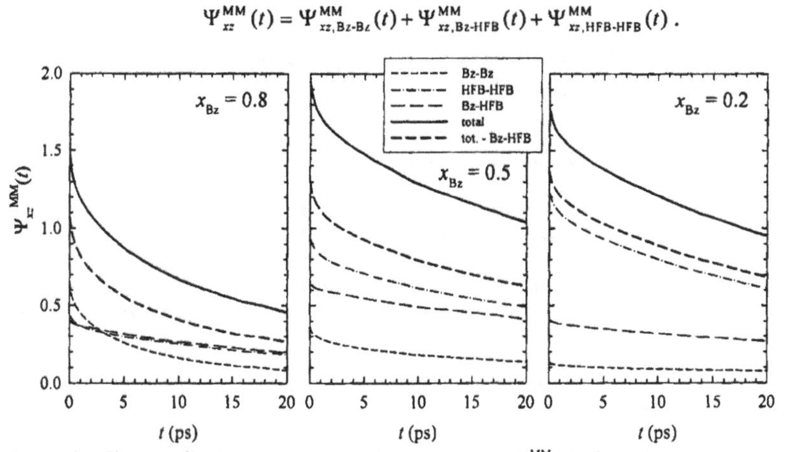

Figure 8. The contributions to the molecular component, $\Psi_{xz}^{MM}(t)$, from the component autocorrelations and interspecies cross-correlations. $\Psi_{xz}^{MM}(t) - \Psi_{xz,Bz\text{-}HFB}^{MM}(t)$ -is also shown. The three panels are for $x_{Bz} = 0.8$ (left), $x_{Bz} = 0.5$ (middle), and $x_{Bz} = 0.2$ (right).

Figure 8 illustrates this decomposition of $\Psi_{xz}^{MM}(t)$ and also shows the contribution of just the single-species autocorrelations. As can be seen from the figure, the contribution of the positive cross-correlation, $\Psi_{xz,Bz\text{-}HFB}^{MM}(t)$, is quite large and always relaxes more slowly than the autocorrelation contributions. Thus the role of $\Psi_{xz,Bz\text{-}HFB}^{MM}(t)$ is always to slow down the relaxation $\Psi_{xz}^{MM}(t)$ and, consequently to be a major contributor to the larger of the two relaxation times.

5. Summary

We have presented an overview of our research on the polarizability anisotropy relaxation in aromatic liquids and its connection to the local structure. We have focused on molecules that have very similar shapes and different charge distributions. Thus the differences that we observe among different one-liquids and two-component liquids arise primarily from electrostatic interactions. Our results indicate that strong electrostatic interactions which favor parallel alignment, as in liquid TFB and in Bz-HFB mixtures, give rise to slower relaxation of the collective polarizability anisotropy.

In the case of the neat liquids, we investigated and compared two approaches, the Keyes-Kivelson projection scheme and the Steele theory, that have been proposed as methods to determine the dynamical origin of the collective time correlations. We found that these approaches lead to quite different identification of the contributions of translational and rotational dynamics to the OKE response, with the former more useful at long and the latter at short times.

Acknowledgments

Financial support of this work from the U.S. National Science Foundation grants CHE 9981539 and CHE 0608640 is gratefully acknowledged.

References

[1] A. H. Narten, *Diffraction Pattern And Structure Of Liquid Benzene*, J. Chem. Phys. **48**, 1630-1634 (1968); A. H. Narten, *X-Ray-Diffraction Pattern And Models Of Liquid Benzene*, J. Chem. Phys. **67**, 2102-2108 (1977); E. Bartsch, H. Bertagnolli, G. Schulz et al., *A Neutron and X-Ray-Diffraction Study of the Binary-Liquid Aromatic System Benzene-Hexafluorobenzene.1. The Pure Components*, Ber. Bunsen-Ges. Phys. Chem. **89**, 147-156 (1985).

[2] M. I. Cabaço, Y. Danten, M. Besnard et al., *Neutron diffraction and molecular dynamics study of liquid benzene and its fluorinated derivatives as a function of temperature*, J. Phys. Chem. B **101**, 6977-6987 (1997).

[3] M. Dolores Elola and Branka M. Ladanyi, *Investigation of benzene-hexafluorobenzene dynamics in liquid binary mixtures*, J. Chem. Phys. **122**, 224508 (2005).

[4] R. Chelli, G. Cardini, M. Ricci et al., *The fast dynamics of benzene in the liquid phase - Part II. A molecular dynamics simulation*, Phys. Chem. Chem. Phys. **3**, 2803-2810 (2001).

[5] M.D. Elola and B.M. Ladanyi, *Molecular Dynamics Study of Polarizability Anisotropy Relaxation in Aromatic Liquids and Its Connection with Local Structure*, J. Phys. Chem. B., http://dx.doi.org/10.1021/jp062071b (2006).

[6] M. R. Battaglia, A. D. Buckingham, and J. H. Williams, *The Electric Quadrupole-Moments Of Benzene And Hexafluorobenzene*, Chem. Phys. Lett. **78**, 420-423 (1981).

[7] J. Vrbancich and G. L. D. Ritchie, *Quadrupole-Moments of Benzene, Hexafluorobenzene and other Non-Dipolar Aromatic-Molecules*, J. Chem. Soc., Faraday Trans. II **76**, 648-659 (1980).

[8] M. I. Cabaço, Y. Danten, M. Besnard et al., *Evidence of dimer formation in neat liquid 1,3,5-trifluorobenzene*, Chem. Phys. Lett. **262**, 120-124 (1996); M. I. Cabaço, T. Tassaing, Y. Danten et al., *Evolution of the local order in 1,3,5-trifluorobenzene from the liquid state up to supercritical conditions*, J. Phys. Chem. A **104**, 10986-10993 (2000); M. I. Cabaço, T. Tassaing, Y. Danten et al., *Structural study of the 1-3-5 trifluorobenzene dimer stability: from liquid to gas densities using supercritical conditions*, Chem. Phys. Lett. **325**, 163-170 (2000).

[9] C. R. Patrick and G. S. Prosser, *Molecular Complex Of Benzene And Hexafluorobenzene*, Nature **187**, 1021-1021 (1960); W. A. Duncan and F. L. Swinton, *Thermodynamic Properties Of Binary Systems Containing Hexafluorobenzene.1. Phase Diagrams*, Transactions Of The Faraday Society **62**, 1082-& (1966); J. W. Schroer and P. A. Monson, *Understanding congruent melting in binary solids: Molecular models of benzene-hexafluorobenzene mixtures*, J. Chem. Phys. **118**, 2815-2823 (2003).

[10] E. Bartsch, H. Bertagnolli, and P. Chieux, *A Neutron And X-Ray-Diffraction Study Of The Binary-Liquid Aromatic System Benzene-Hexafluorobenzene.2. The Mixtures*, Ber. Bunsen-Ges. Phys. Chem. **90**, 34-46 (1986).

[11] M. I. Cabaço, Y. Danten, M. Besnard et al., *Structural investigations of liquid binary mixtures: Neutron diffraction and molecular dynamics studies of benzene, hexafluorobenzene, and 1,3,5-trifluorobenzene*, J. Phys. Chem. B **102**, 10712-10723 (1998).

[12] J.T. Fourkas, in *Ultrafast Infrared and Raman Spectroscopy*, edited by M. D. Fayer (Marcel Dekker, New York, N. Y, 2001); Neil A. Smith and Stephen R. Meech, *Optically-heterodyne-detected optical Kerr effect (OHD-OKE): Applications in condensed phase dynamics*, Int. Rev. Phys. Chem. **21**, 75-100 (2002).

[13] D. McMorrow and W. T. Lotshaw, *Evidence For Low-Frequency (Approximate-To-15 Cm(-1)) Collective Modes In Benzene And Pyridine Liquids*, Chem. Phys. Lett. **201**, 369-376 (1993); M. Neelakandan, D. Pant, and E. L. Quitevis, *Structure and intermolecular dynamics of liquids: Femtosecond optical Kerr effect measurements in nonpolar fluorinated benzenes*, J. Phys. Chem. A **101**, 2936-2945 (1997); M. Ricci, P. Bartolini, R. Chelli et al., *The fast dynamics of benzene*

in the liquid phase - Part I. Optical Kerr effect experimental investigation, Phys. Chem. Chem. Phys. **3**, 2795-2802 (2001); M. C. Beard, W. T. Lotshaw, T. M. Korter et al., *Comparative OHD-RIKES and THz-TDS probes of ultrafast structural dynamics in molecular liquids*, J. Phys. Chem. A **108**, 9348-9360 (2004); B. J. Loughnane, A. Scodinu, and J. T. Fourkas, *Temperature-dependent optical Kerr effect spectroscopy of aromatic liquids*, J. Phys. Chem. B **110**, 5708-5720 (2006).

[14] M. Neelakandan, D. Pant, and E. L. Quitevis, *Reorientational and intermolecular dynamics in binary liquid mixtures of hexafluorobenzene and benzene: Femtosecond optical Kerr effect measurements*, Chem. Phys. Lett. **265**, 283-292 (1997).

[15] M. D. Elola, B. M. Ladanyi, A. Scodinu et al., *Effects of molecular association on polarizability relaxation in liquid mixtures of benzene and hexafluorobenzene*, J. Phys. Chem. B **109**, 24085-24099 (2005).

[16] S. Ryu and R. M. Stratt, *A case study in the molecular interpretation of optical Kerr effect spectra: Instantaneous-normal-mode analysis of the OKE spectrum of liquid benzene*, J. Phys. Chem. B **108**, 6782-6795 (2004).

[17] G. H. Tao and R. M. Stratt, *Why does the intermolecular dynamics of liquid biphenyl so closely resemble that of liquid benzene? - Molecular dynamics simulation of the optical-Kerr-effect spectra*, J. Phys. Chem. B **110**, 976-987 (2006).

[18] S. Mossa, G. Ruocco, and M. Sampoli, *Orientational and induced contributions to the depolarized Rayleigh spectra of liquid and supercooled ortho-terphenyl*, J. Chem. Phys. **117**, 3289-3295 (2002).

[19] D. Frenkel and J. P. McTague, *Molecular-Dynamics Studies Of Orientational And Collision-Induced Light-Scattering In Molecular Fluids*, J. Chem. Phys. **72**, 2801-2818 (1980).

[20] B. M. Ladanyi and S. Klein, *Contributions of rotation and translation to polarizability anisotropy and solvation dynamics in acetonitrile*, J. Chem. Phys. **105**, 1552-1561 (1996).

[21] T. Keyes and D. Kivelson, *Depolarized Light-Scattering - Theory Of Sharp And Broad Rayleigh Lines*, J. Chem. Phys. **56**, 1057-& (1972).

[22] W. A. Steele, *A Theoretical Approach To The Calculation Of Time-Correlation Functions Of Several-Variables*, Mol. Phys. **61**, 1031-1043 (1987).

[23] D. E. Williams and S. R. Cox, *Nonbonded Potentials For Azahydrocarbons - The Importance Of The Coulombic Interaction*, Acta Crystallogr., Sect. B: Struct. Sci. **40**, 404-417 (1984); D. E. Williams and D. J. Houpt, *Fluorine Nonbonded Potential Parameters Derived From Crystalline Perfluorocarbons*, Acta Crystallogr., Sect. B: Struct. Sci. **42**, 286-295 (1986).

[24] W. Song, P. J. Rossky, and M. Maroncelli, *Modeling alkane plus perfluoroalkane interactions using all-atom potentials: Failure of the usual combining rules*, **119**, 9145-9162 (2003).

[25] L. Verlet, *Computer Experiments On Classical Fluids.I. Thermodynamical Properties Of Lennard-Jones Molecules*, Physical Review **159**, 98-103 (1967).

[26] J. P. Ryckaert, G. Ciccotti, and H. J. C. Berendsen, *Numerical-Integration Of Cartesian Equations Of Motion Of A System With Constraints - Molecular-Dynamics Of N-Alkanes*, J. Comput. Phys. **23**, 327-341 (1977).

[27] M. P. Allen and D. J. Tildesley, *Computer Simulation of Liquids*. (Oxford, New York, 1987).

[28] G. R. Alms, A. K. Burnham, and W. H. Flygare, *Measurement Of Dispersion In Polarizability Anisotropies*, J. Chem. Phys. **63**, 3321-3326 (1975).

[29] I. R. Gentle and G. L. D. Ritchie, *2nd Hyperpolarizabilities And Static And Optical-Frequency Polarizability Anisotropies Of Benzene, 1,3,5-Trifluorobenzene, And Hexafluorobenzene*, J. Phys. Chem. **93**, 7740-7744 (1989).

Brill Academic Publishers
P.O. Box 9000, 2300 PA Leiden
The Netherlands

Lecture Series on Computer and Computational Sciences
Volume 6, 2006, pp. 209-221

Guanine: Structures, Properties and Interactions – From the Isolated Ground State to Excited States in Polar Solvent

M.K. Shukla and Jerzy Leszczynski[1]

Computational Center for Molecular Structure and Interactions,
Department of Chemistry,
Jackson State University,
Jackson, MS 39217, USA

Abstract: We review the results of recent theoretical investigations on the structures, properties and interactions of purine base guanine in the ground and singlet $\pi\pi^*$ excited states. With the advancement of computer hardware and state-of-the-art algorithms, it is now possible to perform high level of theoretical calculation not only in the ground state but also for the excited states. The rigorous study of excited state structure is important since it reveals the mystery of photostability of genetic molecules. Further, it may also help designing efficient photostable materials. Change in the different electronic properties of guanine under different hydrogen bonding environments both in the ground and under electronic excitation is discussed.

Keywords: Guanine, DNA, Excited State, Polar Medium, Hydration, Theoretical Study, Deactivation

Mathematics Subject Classification: 92E10, 92-08, 92C05

PACS: 87.15.Mi, 87.64.Ni, 31.50.Df, 31.15.Ne, 87.15.Mi, Rn

1. Introduction

Computational quantum chemical techniques are fast becoming an attractive alternative to the expensive and time consuming experimental methods in determining the structures and activities of molecular systems. For many experimentalists it may appear as an exaggeration, but with the continuous development of state-of-the-art computer hardware and efficient computational algorithms, it is now a reality. Although, we would like to state that modeling of the exact experimental environment, in particular the large biological systems in vivo, is not yet possible but computational methods can still provide reliable predictions and thus can guide experimentalists. Theoretical methods are especially attractive in the area where experimental measurements are still not possible e.g. the determination of excited state geometry of complex molecular systems. For the smaller molecular species one can routinely apply high level of electron correlated methods and large basis sets, however, for larger molecular system one has to make a compromise between the level of theory and the basis set and thus with the computational accuracy. However, we believe that theoretical and experimental methods are complementary to each other, and a judicious decision is needed for their efficient application.

Guanine is one of the most important building-blocks of nucleic acids. This molecule exhibits the maximum number of tautomers in different environments among nucleic acid bases and is the most reactive site for oxidative damages. Numerous theoretical and experimental studies have been performed on this small molecule and it is still continued to be the topic of active research. Therefore, it is not surprising that the search of "guanine" on Scifinder [1] yielded over 100000 results. In this paper we intend to review the recent progress made on uncovering the structures and properties of guanine tautomers for the ground and electronic singlet excited states in the gas phase and in polar media. Interested readers are also referred to some other review articles appeared recently on different aspects of nucleic acid bases and their interaction with different environments both in the ground and excited states [2-5].

[1] Corresponding author. E-mail: jerzy@ccmsi.us

2. Tautomerism in Guanine

The knowledge of the tautomeric stability of DNA bases is important in view of mutation. A rare tautomer in DNA would produce mispairing (e.g., guanine might form hydrogen bonds with thymine; a rare form of adenine would match cytosine) and the formation of such a mispair may yield spontaneous mutation. In principal, guanine can have as many as 12 possible tautomers (excluding rotamers and zwitterionic forms), but only up to four tautomers have been shown to exist in different environments. Jet-cooled spectroscopic investigations have shown the existence of up to four tautomeric forms of guanine (keto-N9H, keto-N7H, enol-N9H, and enol-N7H; Figure 1) in the gas phase, however, it is not clear if these tautomers were equilibrated after the laser desorbtion techniques used in these experiments [6,7]. A recent study based on the infra-red (IR) spectroscopy of guanine in helium droplets and MP2 level of theoretical calculation using the 6-311++G(d,p) and aug-cc-pVDZ basis sets shows that the enol-N7H tautomer is not present, instead two rotamers of the enol-N9H tautomer (enol-N9H-cis and enol-N9H-trans) were shown to exist [8].

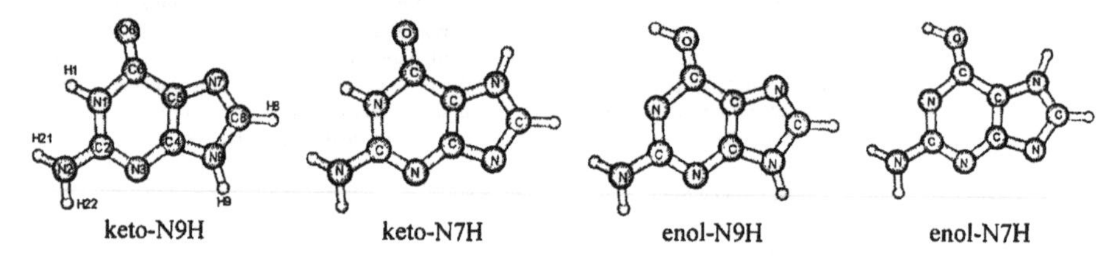

Figure 1. Structures and atomic numbering schemes of the four tautomers of guanine.

The molecular environment has been found to have a profound effect on the relative stability of different tautomers [2-4,9-11]. For example, in low temperature argon matrices, both keto-enol forms of guanine exist in equal proportions, while in a polar solvent the keto-N9H form dominates [9-11]. The accurate prediction of the relative stability of different tautomers of guanine in the gas phase at high levels of computation is compounded by the basis set dependency of the results [2,12,13]. Generally, theoretical methods predict that the keto-N7H tautomer is the most stable in the gas phase, but in water solution the tautomeric equilibrium is shifted to the keto-N9H tautomeric form [2,12-16]. Gorb et al. [14] by using the MP2 and CCSD(T) methods and several large basis sets have shown that the keto-N9H, keto-N7H and cis- and trans- rotamers of the enol-N9H tautomers of guanine are within 1 kcal/mol of energy. Hobza and coworkers [17] studied the microhydration of guanine taurtomers at the RI-MP2 level and found significant stabilization of rare tautomers under bulk solvation. The proton transfer barrier height calculation at the MP2 level suggested very high barrier, however, the inclusion of one and two water molecules in the proton transfer reaction path was found to significantly reduce the barrier height [13].

3. Amino Group Nonplanarity and Conformational Flexibility in Guanine

The amino groups of nucleic acid bases are pyramidal and this phenomenon arises due to the partial sp^3 hybridization of amino nitrogen. The interaction with the neighboring hydrogen atom also plays crucial role for the out-of-plane displacement of amino hydrogens. Guanine has the largest amino group nonplanarity among nucleic acid bases [2,18]. Experimentally, quantitative information about amino group nonplanarity is not yet known. The experimental evidence appears from the vibrational transition moment direction measurement study by Dong and Miller for adenine and cytosine [19] and that of the neutron diffraction study of adenine crystal [20]. In a recent theoretical study the consideration of nonplanar amino groups was found to improve the quality of hydrogen bonds in the DNA-protein recognition [21]. Theoretically the pyrimidine ring of DNA bases has been found to be very flexible. These results were based on the vibrational analysis of low energy vibrations of these molecules at the MP2/6-31G(d,p) level [22]. For guanine the change in the C2N1C6C5 dihedral angle by 10° only requires 0.27 kcal/mol of energy.

4. Electron Affinity, Ionization Potential, Protonation and Deprotonation Properties of Guanine

The electron (proton) affinity of a molecule is measured in terms of the amount of energy released when an electron (proton) is added to the molecule. It is computed as the energy difference between the neutral and anionic (cationic) form of the molecule. Ionization potential on the other hand is defined as the amount of energy required to remove an electron from a molecule. It is computed as the energy difference between the cationic and neutral forms of the molecule. Experiment and theory both suggest that the vertical electron affinity of nucleic acid bases are negative while according to theory the adiabatic electron affinities of pyrimidine bases are slightly positive [23,24]. In a recent theoretical study, Li et al. [24] with the help of available experimental data have estimated the value of adiabatic valence electron affinity for guanine to be -0.75 eV. Experiments and theory both agree that purines have lower and pyrimidines have higher ionization potentials and guanine has the lowest ionization potential among them and therefore is the most susceptible base for the oxidation in DNA under irradiation [25]. Experimentally the vertical and adiabatic ionization potential of guanine was determined to be at 8.24 and 7.77 eV, respectively [25-27]. Significant decrease in the ionization energy threshold of guanine in water solution has also been reported [28,29]. The protonation and deprotonation properties of guanine have been studied at the HF, MP2 and DFT levels [30,31]. Expermentally proton afinity of guanine is revealed to be 9.86 eV [32]. Theoretical studies suggest that the N7 site of the guanine is the most probable site for protonation and the computed protonation energy agrees well with the experimental data [30-32]. Recently, our group [30] have computed proton afinities of all nucleic acid bases up to the MP4(SDTQ) level and found that the computed proton affinities are very close to the experimental data; the computed error was found to be within the 2.1% . The quantitative information regarding the deprotonation energy of nucleic acid bases experimentally are not available, however, results from the Fourier transform ion cyclotron spectroscopy suggest the following acidity order among nucleic acid bases: adenine > thymine > guanine > cytosine [33]. The results of theoretical calculations suggest that the N9H site of guanine will be deprotonated first in the alkaline medium and depending upon the pH of the solution the deptrotonation of the N1H and amino group will be followed [31,34].

5. Metal Cation-Guanine Interaction

Metal cations are important for different natural biological processes and therefore are of great buiological interest. They are responsible for the stabilization of double helical structure of DNA and are located near the negatively charged phosphate group and thus responsible for the reutralization of negative charged backbone. Several metals are used in drugs. On the other hand, however, excess of metal cations or the presence of metal cations at the wrong place can cause metal toxicity. Burda et al. [35] have performed comprehensive investigation on the interaction of metal cations (M^{n+}) belonging to the IA, IB, IIA, and IIB groups with the adenine and guanine at the HF and MP2 level. In this investitaion, metal cation interactions were considered with the N7 site of adenine and the N7 and O6 sites of guanine; geometries were optimized at the HF level with planar symmetry and interaction energy were computed at the MP2 level. It was revealed that the M^{n+}–N7 distance in guanine is longer than the corresponding distance in adenine and this result was due to the fact that in case of guanine the metal cations are involved in the bidentate interaction while with adenine only unidentate binding is present. Further, in the guanine complexes, the M^{n+}–O6 distances were revealed generally shorter than the M^{n+}–N7 distances. Labiuk et al [36] have used X-ray crystallography to study the interaction of Co^{2+}, Ni^{2+} and Zn^{2+} cations with the the B-form DNA hexamer d(GGCGCC). It was found that only the N7 site of terminal guanine resudue is involved in interaction with metal cations. Thus, the metal cation bindings were not formed with non-terminal guanine sites. Moussatova et al. [37] have studied the structure and properties of complexes of Al with different tautomers of guanine at the DFT, MP2 and electron propagator methods with large basis sets. Authors have found that the most stable complex has unique structure where guanine is in the zwitterionic form (the hydrogen from the N1 site is moved to the five membered ring and therefore, both the N7 and N9 sites have hydrogen) and Al is bidentate with the N1 and O6 site of the oxygen. The compted ionization energy of this complex was found to be in good agreement with the corresponding experimental data for Al-guanine complex. Interestingly, the complex where Al is bonded at the N7 and O6 sites of guanine is about 4 kcal/mol less stable than the complex discussed earlier and the computed ionization energy of this complex was also revealed to be in good agreement with the corresponding experimental data [37]. Recently, Zhu et al. [38] performed an extensive theoretical investigation at the B3LYP/6-311++G(d,p) level on the interaction of alkali and alkaline earth metal cations with all nucleic acid bases; the studied complexes included the interaction of metal cations with the π-charge cloud as well as heteroatom of bases.

6. Excited States of Guanine: Structures, Properties and Interactions

It is well known that we are exposed to different kinds of irradiation and nucleic acid bases absorb ultra-violet (UV) radiation efficiently. However, the absorbed energy is released quickly through the ultra-fast nonradiative decay processes and therefore, making these genetic molecules very photostable. Several investigations have been performed to understand the remarkable photostability of these molecules, but exact mechanism is still ellusive [5]. However, it appears that the excited state structural nonplanarity has important role in this process [4,5]. It should be noted that the ground state geometries of all nucleic acid bases are planar except the amino group which shows the pyramidal character and such pyramidalization is largest for the guanine [2]. Geometries of bases in the electronic excited states, on the other hand, are generally significantly nonplanar [4].

The canonical form of guanine (keto-N9H) shows five electronic transitions in the UV region [4,39]. The first absorption peak of guanine is relatively weak and lies near 275 nm (4.51 eV) while the second transition is stronger than the first one and is located near 250 nm (4.96 eV) [40-42]. The third transition is not very often observed. It is very weak with oscillator strength in the range of 0.01-0.03 and is observed near 225 nm (5.51 eV) in the CD spectra [43,44], crystal spectra of guanine and 9-ethylguanine and in the protonated guanine [40]. However, it should be noted that in the CD spectra the unambiguous assignment for this transition has not been made [43]. The fourth transition is located near 204 nm (6.08 eV) and fifth transition is located near 188 nm (6.59 eV), respectively, the intensity of both peaks are strong [4,39-41]. The existence of three transitions of the nπ* type near 238, 196, and 175 nm (5.21, 6.32, and 7.08 eV) in guanine has been tentatively suggested by Clark [41]. An elegant study on the transition moment directions of guanine was also performed by Clark [41] and it was found that the precise measurement is complicated due to the crystal field effect. The 4.46, 5.08, 6.20 and 6.57 eV region peaks of guanine have been suggested to have transition moment direction of -12°, 80°, 70° and -10°, respectively [41]. Some advanced spectroscopic investigations on guanine and substituted analogs provided data for the spectral origin (0-0 transition) of different guanine tautomers, namely, the enol-N7H (32864 cm^{-1}), keto-N7H (33269 cm^{-1}), keto-N9H (33910 cm^{-1}) and enol-N9H (34755 cm^{-1}); the values in parentheses correspond to the 0-0 transition of the corresponding tautomer, in the gas phase [6]. However, the recent result by Choi and Miller [8] revealing the presence of two rotamers of the enol-N9H tautomer of guanine in the gas phase rather than that of the enol-N7H tautomer suggests the reexamination of spectral assignment by earlier researchers [6,7].

Table 1. TD-B3LYP computed ππ* singlet electronic transition energies (ΔE) and oscillator strengths (f) of guanine using the MP2/6-311G(d,p) optimized geometries and different basis sets [47].

Basis set[a]					
C	B	A	C[b]	CASPT2/CASSCF[c]	Observed[d]
ΔE (f)	ΔE (f)	ΔE (f)	ΔE (f)	$ΔE^1/ΔE^2/f$	ΔE
ππ* Transitions					
4.64 (0.0312)	4.64 (0.0314)	6.67 (0.0382)			
4.86 (0.1275)	4.86 (0.1258)	4.88 (0.1219)	4.85 (0.1408)	4.76/6.08/0.113	4.4-4.6
5.17 (0.2193)	5.17 (0.2205)	5.18 (0.2241)	5.11 (0.2487)	5.09/6.99/0.231	4.8-5.1
5.75 (0.0028)	5.81[e](0.0029)		5.59 (0.0136)	5.96/7.89/0.023	5.4-5.5
6.08 (0.0028)	6.00 (0.0047)	6.06[f](0.0053)	5.83 (0.0029)	6.65/8.60/0.161	6.0-6.3

[a]Used basis sets are: A = 6-311++G(d,p), B = 6-311(2+,2+)G(d,p), and C = 6-311(3+,3+)G(df,pd); [b]Using the planar geometry of guanine (MP2/6-311G(d,p) level); [c]$ΔE^1$:CASPT2 and $ΔE^2$:CASSCF energy (see ref. 46); [d]Observed experimental range of transition energies (see Refs. 4, 39, 46 and 47 for details); [e]Rydberg (πσ*) contamination (about 16%), [f] nπ* contamination (about 63%).

Several theoretical investigations were performed to study the electronic transitions of guanine [15,16,45-49]. The first ab initio study was carried out at the random phase approximation (RPA) and multireference configuration interaction (MRCI) levels and electronic transition energies were revealed to be higher by 1.48—1.86 eV compared to the experimental ones, and linear scaling was used for comparison with the experimental data [45]. The first CASSCF and CASPT2 level theories to compute electronic transition energies were used by Roos group [46]; the effect of the aqueous solvent on electronic transitions was considered using the self-consistent reaction field (SCRF) model. The computed accuracy of the CASPT2 transition energies were found to be within 0.3 eV compared to the experimental data. The CIS level computed transition energies of hydrated guanine after scaling (scaling factor 0.72) were also found to be in good agreement with the experimental data and CASPT2 transition energies [4]. The CIS level computed and scaled $n\pi^*$ transition energies of the hydrated guanine were also found to be in good agreement with the experimental data [4]. It should be noted that three water molecules were used in the hydration to model the water environment. Mennucci et al. [15] have studied the photophysical properties of guanine tautomers (keto-N9H and keto-N7H) including the excited state tautomerization theoretically at the DFT, CIS and multireference perturbation configuration interaction (CIPSI) methods both in the gas phase and in water solution modeled using the continuum model. The role of protonation in the excited state proton transfer from the keto-N9H to the keto-N7H form and the occurrence of fluorescence from the latter tautomer in the water solution have also been discussed. The TDDFT investigation on the electronic transitions of guanine as well as other bases have also been performed by us [47] as well as by Tsolakidis and Kaxiras [48] and Varsano et al [49]. Latter authors [49] also studied Watson-Crick hydrogen bonded and Watson-Crick stacked assemblies. We have used the B3LYP functional along with the 6-311++G(d,p), 6-311(2+,2+)G(d,p) and 6-311(3+,3+)G(df,pd) basis sets and the MP2/6-311G(d,p) optimized geometries in the TDDFT transition energy calculations; the computed $\pi\pi^*$ transition energies are shown in the Table 1 [47]. It should be noted that in the Table 1, the first transition of the nonplanar guanine is defined by the configurations HOMO → LUMO (95%) and HOMO → LUMO+1 (5%). The LUMO orbital is the mixture of both π^* character localized at the purine ring and σ^* character localized at the amino group. However, for the planar structure this transition is $\pi\sigma^*$ type defined by the HOMO → LUMO configuration and LUMO is the σ^* type being localized at the amino and N1H groups. The computed $\pi\pi^*$ transitions of guanine are generally in good agreement with the corresponding experimental data and the CASPT2 results and the transition energies are generally converged at the 6-311(2+,2+)G(d,p) basis set.

Table 2. Selected geometrical parameters of guanine tautomers and transition states in the ground and lowest singlet $\pi\pi^*$ excited state obtained at the HF/6-311G(d,p) and CIS/6-311G(d,p) levels, respectively. Corresponding ground state parameters at the B3LYP/6-311++G(d,p) level are given in parenthesis [16].

Parameters	keto-N9H		keto-N7H		enol-N9H		enol-N7H		TS-N9H		TS-N7H	
	S0	S1	S0	S1	S0	S1	S0	S1	S0	S1	S0	S1
H21N2C2	117.9 (118.2)	115.3	116.8 (117.7)	117.3	116.9 (117.3)	119.7	116.6 (117.3)	119.6	118.5 (118.5)	118.9	118.0 (118.5)	118.9
H22N2C2	113.8 (113.8)	112.7	112.5 (112.7)	114.1	116.7 (117.1)	119.1	115.8 (116.4)	119.0	116.6 (116.4)	118.1	115.7 (115.8)	117.8
H21N2H22	115.0 (114.7)	111.4	113.8 (114.0)	114.5	118.0 (118.2)	121.2	117.3 (117.8)	121.4	118.1 (117.6)	119.2	117.5 (117.5)	119.4
360-ΣHNH	13.3 (13.3)	20.6	16.9 (15.6)	14.1	8.4 (7.4)	0.0	10.3 (8.5)	0.0	6.8 (7.5)	3.8	8.8 (8.2)	3.9
N3C2N1C6	-0.6	-64.0	-0.5	-45.1	-0.7	1.9	-0.9	0.0	-0.6	28.8	-0.7	-21.9
C4N3C2N1	0.8	44.2	0.9	22.9	0.8	5.1	0.9	0.0	0.6	1.9	0.7	10.0
C4N3C2N2	-177.2	-161.4	-177.0	-165.3	-177.9	-175.6	-177.7	180.0	-178.1	-175.8	-177.9	-172.1
C5C4N3C2	-1.0	-2.4	-1.2	7.3	-0.3	-5.6	-0.2	0.0	-0.3	-30.8	-0.3	4.7
C6C5C4N3	0.9	-18.5	1.1	-14.4	-0.2	-0.2	-0.4	0.0	0.1	28.0	-0.1	-6.4
C5C6N1C2	0.3	36.2	0.3	32.6	0.1	-8.3	0.1	0.0	0.3	-29.9	0.3	18.5
N1C6C5C4	-0.4	-0.6	-0.6	-7.0	0.2	7.5	0.5	0.0	-0.1	3.5	0.0	-5.6
H21N2C2N1	30.4 (31.2)	42.3	36.6 (35.7)	32.3	17.2 (16.4)	-0.2	19.9 (18.3)	0.0	16.9 (18.0)	-11.1	19.9 (19.6)	13.0
H22N2C2N1	169.4 (170.2)	171.8	170.9 (171.5)	170.2	164.8 (165.8)	-178.8	164.1 (165.6)	180.0	167.4 (167.2)	-169.3	166.7 (167.4)	170.6

Geometries of guanine tautomers in the excited states were computed both at the CIS [15,16] and the TDDFT [50] levels. The latter study is based upon the restricted open-shell Kohn-Sham method (ROKS) where only single virtual orbital and all occupied orbitals were used in the calculation [50]. We have used the CIS/6-311G(d,p) method for the lowest singlet $\pi\pi^*$ excited state geometry optimization of four guanine tautomers [16]. For comparison, the ground state geometries were also optimized at the HF/6-311G(d,p) level, since the CIS method is the HF analog for electronic excited states [51]. Selected geometrical parameters are shown in the Table 2 while optimized geometries are shown in the Figure 2. It is clear from the Table 2 that the amino groups of tautomers are pyramidal in the ground state and the keto-N7H tautomer has the largest and enol-N9H tautomer has the smallest amino group pyramidalization. In the excited state, the keto-N9H tautomer has the largest nonplanarity of the amino group while for enol tautomers the amino groups were found to be planar. The excited state geometries of guanine tautomers are significantly nonplanar except the enol-N7H tautomer which is planar. The excited state geometrical nonplanarity is largest for the keto-N9H tautomer and smallest for the enol-N9H tautomer and is localized at the pyrimidine ring of tautomers. Mennucci et al. [15] have also optimized the lowest singlet ππ* excited state geometries of the keto-N9H and keto-N7H tautomers in the gas phase and in water solution at the CIS/cc-pVDZ level; the aqueous environment was considered using the integral equation formalism continuum model. Significant solvent effect was revealed in the excited state geometries, mainly affecting bond lengths and bond angles of the molecules.

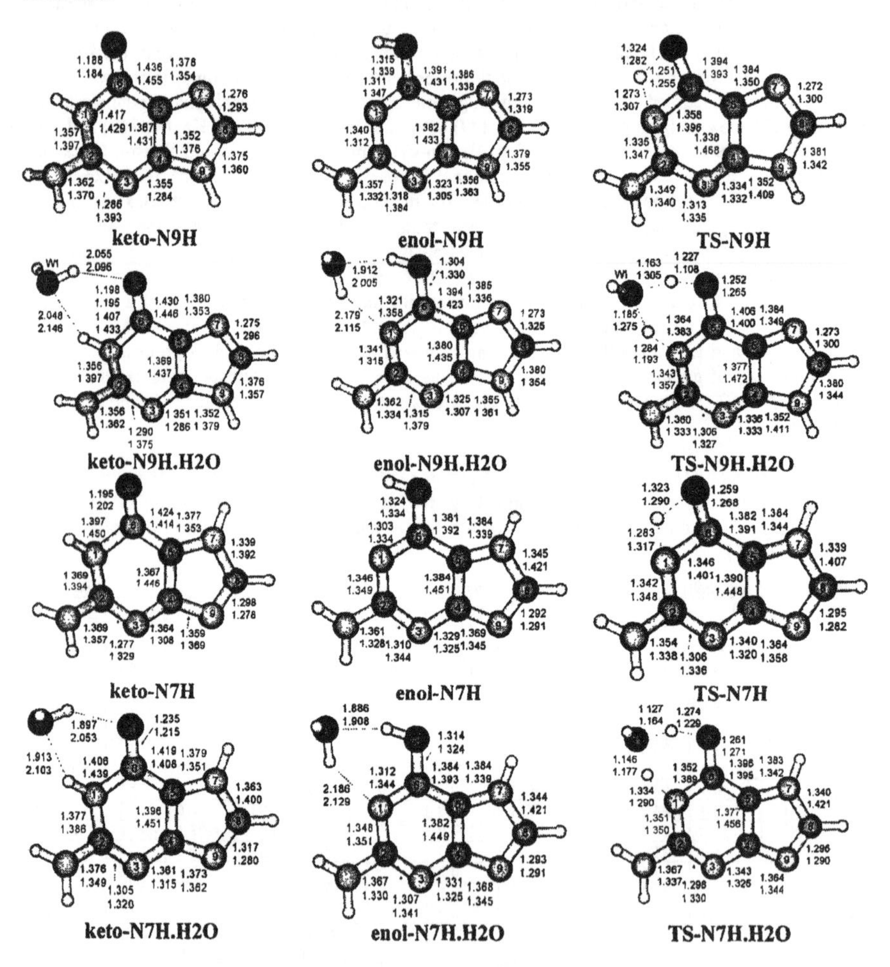

Figure 2. Lowest singlet ππ* excited state geometries of guanine tautomers and transition states corresponding to the proton transfer from the keto to the enol form in the isolated and monohydrated forms in the gas phase. The top and bottom indices correspond to the ground and excited state obtained at the HF/6-311G(d,p) and CIS/6-311G(d,p) levels, respectively.

Table 3. Computed ground and excited state barrier height (kcal/mol) for the keto-enol guanine tautomerism at the B3LYP/6-311++G(d,p) and TD-B3LYP/6-311++G(d,p)//CIS/6-311G(d,p) level, respectively in the gas phase and in water solution [16].

Species	Ground State		Excited State	
	Gas	Water	Gas	Water
keto-N9H →TS-N9H	37.5	45.2	42.9	45.7
enol-N9H →TS-N9H	36.3	38.2	36.9	40.8
keto-N7H →TS-N7H	40.6	46.5	36.8	41.4
enol-N7H →TS-N7H	35.9	38.5	36.8	39.3
keto-N9H.H2O →TS-N9H.H2O	15.9	16.7	19.8	18.3
enol-N9H.H2O →TS-N9H.H2O	12.8	10.4	12.8	13.5
keto-N7H.H2O →TS-N7H.H2O	17.3	17.1	13.9	13.5
enol-N7H.H2O →TS-N7H.H2O	12.0	10.2	13.4	11.2

We have also studied the keto-enol proton transfer phenomena in guanine in the lowest singlet $\pi\pi^*$ excited state at the TDDFT and CIS levels [16]. Ground state geometries including transition states were also optimized at the B3LYP/6-311++G(d,p) level. The excited state geometries including the transition states were optimized at the CIS/6-311G(d,p) level and vertical transition energies were computed at the TD-B3LYP/6-311++G(d,p) level using the CIS/6-311G(d,p) geometries. Barrier heights were also computed for hydrated species where a single water molecule was placed in the proton transfer reaction path. Effect of bulk water environment was considered using the PCM solvation model. The computed barrier height is shown in the Table 3 and the excited state transition state geometries are shown in the Figure 2 and selected geometrical parameters are presented in the Table 2. It is evident that the ground and excited state proton transfer barrier height is quite large both in the gas phase and in water solution. However, significant decrease in the barrier height is revealed when a water molecule was placed in the proton transfer reaction path. Surprisingly, the excited state barrier heights were generally revealed slightly larger than the corresponding ground state values. Theoretical calculation suggested that the singlet electronic excitation of guanine may not facilitate the keto-enol tautomerization both in the gas phase and in the water solution. The geometries of transitions state were also found significantly nonplanar. Geometries of the hydrated transition states in the ground and lowest singlet $\pi\pi^*$ excited states were found to be zwitterionic form in which water molecule is in the form of hydronium cation ($H3O^+$) and guanine is in the anionic form, except for the N9H form in the excited state where water molecule is in the hydroxyl anionic form (OH^-) and the guanine is in the cationic form.

Hydrogen bonding is ubiquitous. The specific patterns of hydrogen bonds in DNA are keys to genetic code. It is well known that in vivo DNA is heavily hydrated and water molecules play an important role towards the three dimensional structure of DNA. In order to uncover the interaction of DNA with water molecules under electronic excitation, it is prerequisite to understand the interaction of nucleic acid bases with water molecules in the electronic excited states. We have recently performed a systematic theoretical study of interaction of water molecules with guanine in the singlet $\pi\pi^*$ excited state [52]. In this investigation, 1,3,5,6 and 7 water molecules were considered in the first solvation shell of guanine. The ground state geometries were optimized at the HF level while excited state geometries were optimized at the CIS level; the 6-311G(d,p) basis set was used in all calculations. The optimized geometries are shown in the Figure 3, while selected parameters are shown in the Table 4. Further, selected stretching vibrational frequencies are shown in the Table 5. It was found that in the ground state the first solvation shell of guanine can accommodate up to six water molecules. Further, it is evident that water molecules when interacting with amino group of guanine (penta and hexahydrated guanine) induce planarity in the amino group both in the ground and excited states. The excited state geometries of guanine were found strongly nonplanar in all complexes, but the mode of the structural nonplanarity which is localized in the pyrimidine ring, was found to be dependent on the degree of hydration (Table 4). The excited state geometry of guanine was revealed similar for isolated, mono and trihydrated guanine but exhibited significantly different structural distortion for the penta and hexahydrated guanine. This result suggests that the degree of hydration may significantly influence the photophysical properties especially the excited state dynamics of guanine and therefore the DNA. The changes in different stretching vibrational frequencies of guanine in the isolated and hydrated complexes were also found to be in accordance with the change in the hydrogen bond distances in going from the ground state to the excited state and with increase in the number of water molecules in the solvation shell. The observed differences in the excited state geometrical distortions of guanine under different degree of hydration were explained in terms of the virtual orbital contributing to the electronic excitation.

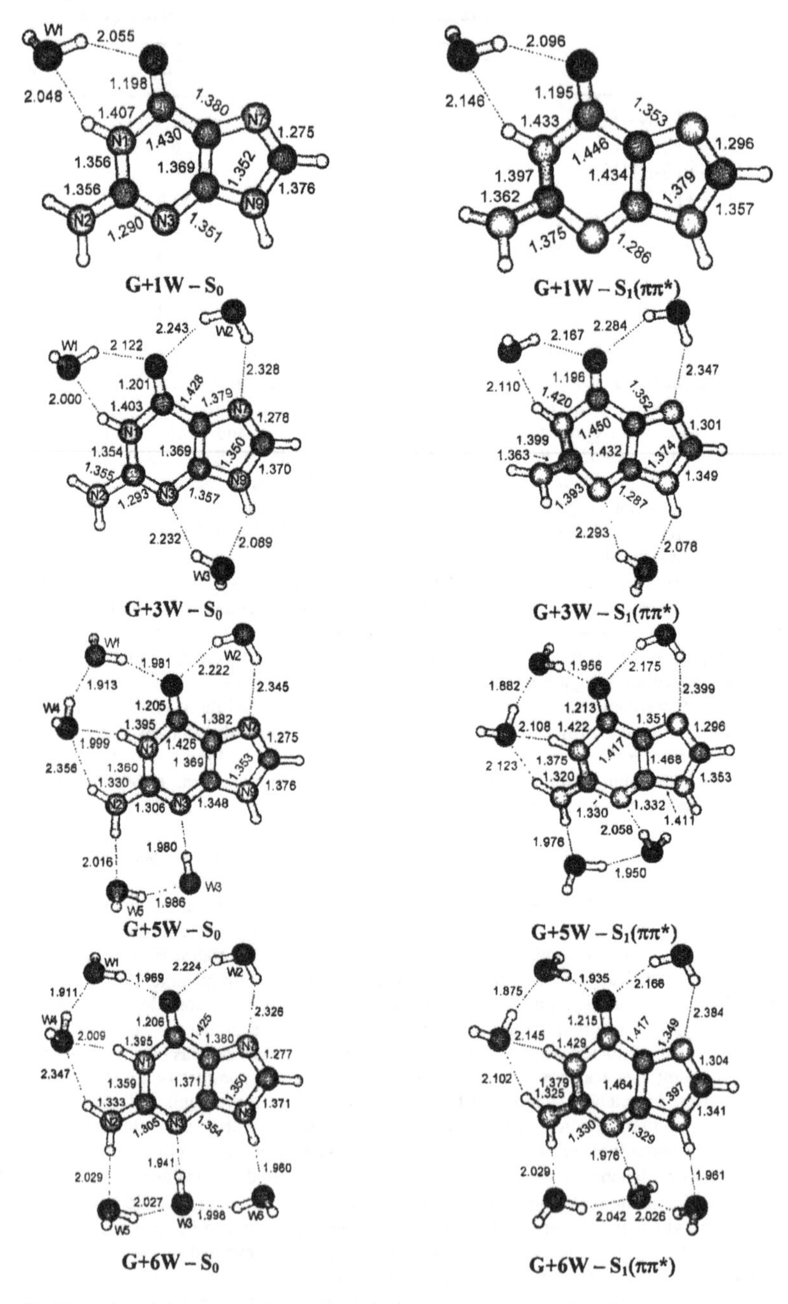

Figure 3. Ground and lowest singlet $\pi\pi^*$ excited state structures of different hydrated guanine. Distances are in Å.

Table 4. Selected dihedral angles (°) and amino group angles (°) of the guanine in different hydrated forms in the ground and lowest singlet $\pi\pi^*$ excited state obtained at the HF/6-311G(d,p) and CIS/6-311G(d,p) levels, respectively [52].

	G+1W		G+3W		G+5W		G+6W	
Parameters	S_0	S_1	S_0	S_1	S_0	S_1	S_0	S_1
H21N2C2	119.2	117.1	118.8	116.7	119.7	119.2	119.6	119.2
H22N2C2	115.2	114.3	115.7	114.4	120.1	120.4	120.4	120.4
H21N2H22	116.7	113.8	116.3	113.3	120.0	119.8	119.9	119.9
360-ΣHNH	8.9	14.8	9.2	15.6	0.2	0.6	0.1	0.5
C6N1C2N3	-0.9	-64.7	0.2	-64.8	1.3	38.1	1.3	43.0
N1C2N3C4	0.8	39.2	0.8	42.4	0.6	-1.7	0.9	-6.5
C2N3C4C5	-0.6	2.0	-1.2	0.0	-1.6	-32.2	-2.1	-29.0
N3C4C5C6	0.4	-18.1	0.5	-18.8	0.8	32.2	1.2	29.7
N1C6C5C4	-0.3	-7.3	0.5	-3.3	1.0	2.6	1.0	5.4
N2C2N3C4	-177.5	-159.7	-177.5	-160.5	-179.2	-177.3	-179.1	179.7
H21N2C2N1	22.1	31.0	23.3	34.8	6.4	10.2	5.4	10.6
H22N2C2N1	168.4	167.9	169.0	170.4	-178.6	-178.7	-176.5	-177.4

Table 5. Selected stretching vibrational frequencies (cm^{-1}) of guanine and hydrated guanine complexes in the ground and the lowest singlet $\pi\pi^*$ excited state obtained at the HF/6-311G(d,p) and CIS/6-311G(d,p) levels, respectively [52].[a]

Modes	G		G+1W		G+3W		G+5W		G+6W	
	S_0	S_1	S_0	S_1	S_0	S_1	S_0	S_1	S_0	S_1
νN1H	3844	3809	3741	3748	3719	3738	3746	3783	3754	3790
νN9H	3900	3900	3900	3903	3799	3795	3884	3864	3715	3724
ν_{asym}NH2	3919	3852	3940	3881	3937	3876	3885	3832	3890	3844
ν_{sym}NH2	3805	3633	3819	3705	3817	3682	3714	3676	3731	3706

[a]:Unscaled values

Ground and the lowest singlet $\pi\pi^*$ excited state geometries of the guanine-cytosine (GC) and guanine-guanine base pairs were also investigated at the HF and CIS levels using the 6-311G(d,p) basis set and computed results were compared with that of isolated guanine obtained at the same level of theory [53]. Two different hydrogen bonding configurations namely GG16 and GG17 of the guanine-guanine base pair were studied. In the GG16 base pair, guanine monomers form cyclic and symmetrical hydrogen bonds in between the N1H site of the first monomer and the carbonyl group of the second monomer, and vice versa. Whereas, the GG17 base pair is mainly connected by two strong hydrogen bonds; the amino group and the N1H site of the first guanine monomer acting as hydrogen bond donors (G_D) is hydrogen bonded with the carbonyl group and the N7 site of the second monomer acting as the hydrogen bond acceptor (G_A).

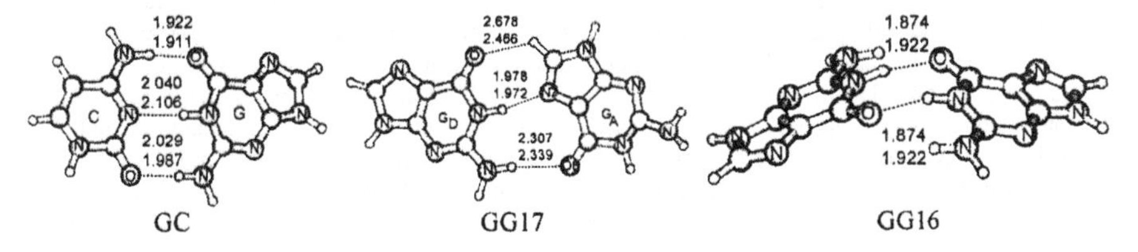

Figure 4. Optimized geometries of GC, GG17 and GG16 base pairs in the lowest singlet $\pi\pi^*$ excited state. The top indices correspond to ground state obtained at the HF/6-311G(d,p) level and bottom indices correspond to excited state obtained at the CIS/6-311G(d,p) level.

Table 6. Selected parameters of guanine monomer in the GC, GG16 and GG17 base pairs in the ground and lowest singlet $\pi\pi^*$ excited state obtained at the HF/6-311G(d,p) and CIS/6-311G(d,p) levels, respectively [53].[a]

	GC		GG16		GG17	
Parameters	S_0	$\pi\pi^*$	S_0	$\pi\pi^*$	S_0	$\pi\pi^*$
H21N2C2	122.8	122.8	120.5	120.7	118.4	116.6
H22N2C2	117.1	117.4	117.4	118.1	114.2	113.5
H21N2H22	120.1	119.7	119.3	119.9	115.3	112.2
360-ΣHNH	0.0	0.1	2.8	1.3	12.1	17.7
C6N1C2N3	0.0	36.4	0.1	3.1	-0.5	-64.7
N1C2N3C4	0.0	-3.3	-0.4	-0.7	0.7	44.4
C2N3C4C5	0.0	-28.8	0.3	-1.3	-1.1	-2.2
N3C4C5C6	0.0	29.6	0.1	1.1	1.3	-18.1
N1C6C5C4	0.0	2.6	-0.5	1.0	-0.9	-1.9
N2C2N3C4	180.0	-177.7	178.5	-179.8	-177.4	-156.9
H21N2C2N1	0.0	3.7	-10.9	8.2	29.3	41.6
H21N2C2N1	180.0	-178.5	-171.7	175.2	170.1	174.3

[a]: In the GG17 base pair, the data corresponds to the G_A guanine moiety which acts as hydrogen bond acceptor, while in the GG16 base pair, both monomers have similar geometry due to the symmetry.

The electronic excitation to the lowest singlet $\pi\pi^*$ excited state of the GC base pair was found to be dominated by orbitals mainly localized at the guanine moiety while those for the GG17 base pair were found to be localized at the G_A monomer. In the case of the GG16 base pair, the orbitals involved in the lowest singlet $\pi\pi^*$ electronic excitation were found to be delocalized [53]. Geometries of base pairs are shown in the Figure 4 and selected geometrical parameters are shown in the Table 6. It is evident from the data shown in the Table 6 that the amino groups of guanine belonging to the GC and GG16 base pairs are almost planar while that in the GG17 base pair is significantly pyramidal both in the ground state and in the lowest singlet $\pi\pi^*$ excited state. The excited state geometries of the isolated guanine and those of the GC and GG17 base pairs were found to be nonplanar and the predicted structural nonplanarity in the GC and GG17 base pairs was located at the excited guanine monomer. The remarkable difference was revealed in the mode of nonplanarity which was found to be significantly influenced by the hydrogen bonding in the base pair. The geometrical deformation of isolated guanine and the G_A monomer of the GG17 base pair in the excited state were similar but found to be significantly different than the structural deformation of guanine monomer of the GC base pair in the excited state. The geometry of the GG16 base pair was predicted to be almost planar in the excited state except both guanine monomers are folded with respect to each other. Thus, this study also suggested that the excited state dynamics of bases and DNA may be significantly influenced by the hydrogen bonding environments.

7. Non-radiative Decay in Guanine

It is now well known that the excited state life-times of nucleic acid bases are in the sub-pico second order and the absorbed energies are quickly dissipated through the ultrafast nonradiative process [5]. For nucleic acid bases the triplet quantum yield has been found to be very small at the room temperature [54]. Therefore, the intersystem crossing (ISC) would not be mechanism for ultrafast nonradiative decay in nucleic acid bases and internal conversion (IC) would account for such decay. Different mechanisms have been suggested for ultrafast nonradiative relaxation processes in nucleic acid bases through IC, the exact mechanism is not yet known. But it is becoming apparent that excited state structural nonplanarity plays an important role in facilitating such processes. A detailed and lucid analysis of nonradiative decays in nucleic acid bases has been provided by Kohler and coworkers in a recent review article [5]. Therefore here we will discuss very briefly the recent progress in the nonradiative decay of guanine. Langer and Doltsinis [55] have studied the nonradiative decay of methylated guanine upon excited to the first singlet excited state using an "on the fly" surface hopping approach based on density functional theory. If was found that the 9-methylguanine undergo large geometrical change in the S_1 state giving rise to an increased nonadiabatic transition probability and thus shorter lifetime compared to the 7-methylguanine. The global minimum structure of 9-methyl guanine shows strong out-of-plane bending in amino group, while such deformation for the 7-

methylguanine is small. Further, for the keto-9H tautomer, the methylation at the N9 site was found to have significant effect on the geometry in the S_1 excited state while for the keto-N7H tautomer, methylation at the N7 site does not induce any significant change. Therefore, it was suggested that methylation at the N9 site of guanine induces significant S_1 structural changes leading to a much reduced optical activity in the Franck-Condon region. This finding is explained as a possible explanation for the absence of a 9-methylguanine (keto form) signal in the R2PI experiment [6]. Recently, Canuel et al. [56] have studied the nonradiative dynamics of nucleic acid bases using the mass-selected femtosecond resolved pump-probe resonance ionization at the 80 fs time resolution. These authors suggested that nonradiative decays in bases are two step processes and for guanine the first step is completed in 148 fs while 36 fs is needed for the second step. Chen and Li [57] have recently investigated the deactivation pathways of guanine under electronic excitation at the CASPT2/CASSCF level. It has been found that a conical intersection between the lowest singlet $\pi\pi^*$ excited state and the ground state would facilitate the ultrafast nonradiative decay in guanine.

8. Conclusions

We have presented a brief review of the ground and excited state structures and properties of guanine in the gas phase and in water solution. It is clear that guanine is a simple molecule hiding very complex properties. It has several low energy tautomers and thus shows maximum number of rare tautomers among all nucleic acid bases in different environments, possesses the lowest ionization potential and shows negative electron affinity. It interacts strongly with metal cations through the O6 and N7 sites. The amino group of guanine exhibits the largest nonplanarity among all bases. We have investigated the electronic singlet $\pi\pi^*$ excited state structure of guanine in different environments e.g. hydration, base pairing. All these investigations have suggested strong nonplanarity in guanine in the excited state and this structural nonplanarity is localized in the six-membered ring of the molecule. The excited state structural nonplanarity is significantly influenced by the environment. It is becoming clear that excited state structural nonplanarity is involved in facilitating the ultrafast nonradiative decay in guanine as well as other nucleic acid bases. However, the surrounding environment will also have significant influence on the rate of deactivation. One of the major conclusion from this overview is that theoretical quantum chemical methods are matured enough to be applied to almost every branch of chemical sciences and reliable information can be obtained regarding photophysical properties of nucleic acid bases and base assemblies.

Acknowledgments

The authors are thankful to financial supports from NSF-CREST grant No. HRD-0318519, ONR grant No. N00034-03-1-0116, NIH-SCORE grant No. 3-S06 GM008047 31S1 and NSF-EPSCoR grant No. 02-01-0067-08/MSU. Authors are also thankful to the Mississippi Center for Supercomputing Research (MCSR) for the generous computational facility.

References

[1] SciFinder Scholar Version: 2006, 2005 American Chemical Society.

[2] J. Leszczynski, In *Advances in Molecular Structure and Research,* Vol. 6; M. Hargittai and I. Hargittai, (Eds.), JAI Press Inc., Stamford, Connecticut, 2000.

[3] P. Hobza, and Z. Havlas *Chem. Rev.* **100,** 4253(2000).

[4] M.K. Shukla and J. Leszczynski, In *Computational Chemistry: Reviews of Current Trends,* Vol. 8, J. Leszczynski (Ed.), World Scientific, Singapore, 2003, p. 249.

[5] C.E. Crespo-Hernandez, B. Cohen, P.M. Hare, and B. Kohler *Chem. Rev.* **104,** 1977(2004).

[6] M. Mons, I. Dimicoli, F. Piuzzi, B. Tardivel, and M. Elhanine *J. Phys. Chem. A* **106,** 5088(2002).

[7] E. Nir, Ch. Janzen, P. Imhof, K. Kleinermanns and M.S. de Vries *J. Chem. Phys.* **115**, 4604(2001).

[8] M.Y. Choi and R.E. Miller *J. Am. Chem. Soc.* **128**, 7320(2006).

[9] G.G. Sheina, S.G. Stepanian, E.D. Radchenko and Yu.P. Blagoi *J. Mol. Struct.* **158**, 275(1987).

[10] K. Szczepaniak, M. Szczesniak, and W.B. Person *Chem. Phys. Lett.* **153**, 39(1988).

[11] K. Szczepaniak, M. Szczesniak, W. Szajda, W.B. Person, and J. Leszczynski *Can. J. Chem.* **69**, 1705(1991).

[12] J. Leszczynski *J. Phys. Chem. A* **102**, 2357(1998).

[13] L. Gorb and J. Leszczynski *J. Am. Chem. Soc* **120**, 5024(1998).

[14] L. Gorb, A. Kaczmarek, A. Gorb, A.J. Sadlej and J. Leszczynski *J. Phys. Chem. B* **109**, 13770(2005).

[15] B. Mennucci, A. Toniolo and J. Tomasi *J. Phys. Chem. A* **105**, 7126(2001).

[16] M.K. Shukla and J. Leszczynski *J. Phys. Chem. A* **109**, 7775(2005).

[17] M. Hanus, F. Ryjacek, M. Kabelac, T. Kubar, T.V. Bogdan, S.A. Trygubenko and P. Hobza *J. Am. Chem. Soc.* **125**, 7678(2003).

[18] J. Sponer, J. Florian, P. Hobza and J. Leszczynski *J. Biomol. Struct. Dynam.* **13**, 827(1996).

[19] F. Dong and R.E. Miller *Science* **298**, 1227(2002).

[20] R.K. McMullan, P. Benci and B.M. Craven *Acta Crystal. B* **36**, 1424(1980).

[21] S. Mukherjee, S. Majumdar and D. Bhattacharyya *J. Phys. Chem. B* **109**, 10484(2005).

[22] O.V. Shishkin, L. Gorb, P. Hobza and J. Leszczynski *Int. J. Quantum Chem.* **80**, 1116(2000).

[23] K. Aflatooni, G.A. Gallup and P.D. Burrow *J. Phys. Chem. A* **102**, 6205(1998).

[24] X. Li, Z. Cai and M.D. Sevilla *J. Phys. Chem. A* **106**, 1596(2002) and references cited therein.

[25] X. Li, Z. Cai and M.D. Sevilla *J. Phys. Chem. B* **105**, 10115(2001) and references cited therein.

[26] V.M. Orlov, A.N. Smirnow and Y.M. *Tetrahedron Lett.* **48**, 4377(1976).

[27] N.S. Hush and A.S. Cheung *Chem. Phys. Lett.* **34**, 11(1975).

[28] C.E. Crespo-Hernandez, R. Arce, Y. Ishikawa, L. Gorb, J. Leszczynski and D.M. Close *J. Phys. Chem. A* **108**, 6373(2004).

[29] G.A. Papadantonakis, R. Tranter, K. Brezinsky, Y. Yang, R.B. van Breemen and P.R. LeBreton *J. Phys. Chem. B* **106**, 7704(2002).

[30] Y. Podolyan, L. Gorb and J. Leszczynski *J. Phys. Chem. A* **104**, 7346(2000).

[31] A.K. Chandra, M.T. Nguyen, T. Uchimaru and T. Zeegers-Huyskens *J. Phys. Chem. A* **103**, 8853(1999).

[32] F. Greco, A. Liguori, G. Sindona and N. Uccella *J. Am. Chem. Soc.* **112**, 9092(1990).

[33] M.T. Rodgers, S. Campbell, E.M. Marzluff and J.L. Beauchamp *Int. J. Mass Spectrom. Ion Processes* **37**, 121(1994).

[34] T.L. McConnell, C.A. Wheaton, K.C. Hunter and S.D. Wetmore *J. Phys. Chem. A* **109**, 6351(2005).

[35] J.V. Burda, J. Sponer and P. Hobza *J. Phys. Chem. A* **100**, 7250(1996).

[36] S.L. Labiuk, L.T.J. Delbaere and J.S. Lee *J. Biol. Inorg. Chem.* **8**, 715(2003).

[37] A. Moussatova, M.-V. Vazquez, A. Martinez, O. Dolgounitcheva, V.G. Zakrzewski, J.V. Ortiz, D.B. Pedersen and B. Simard *J. Phys. Chem. A* **107**, 9415(2003).

[38] W. Zhu, X. Luo, C.M. Puah, X. Tan, J. Shen. J. Gu, K. Chen and H. Jiang *J. Phys. Chem. A* **108**, 4008(2004).

[39] P.R. Callis *Ann. Rev. Phys. Chem.* **34**, 329(1983).

[40] L.B. Clark *J. Am. Chem. Soc.* **99**, 3934(1977).

[41] L.B. Clark *J. Am. Chem. Soc.* **116**, 5265(1994).

[42] C. Santhosh and P.C. Mishra *J. Mol. Struct.* **198**, 327(1989).

[43] C.A. Sprecher and W.C. Johnson Jr. *Biopolymers* **16**, 2243(1977).

[44] D.W. Miles, S.J. Hann, R.K. Robins, H. Eyring *J. Phys. Chem.* **72**, 1483(1968).

[45] J.D. Petke, G.M. Maggiora and R.F. Christoffersen *J. Am. Chem. Soc.* **112**, 5452 (1990).

[46] M.P. Fulscher, L. Serrano-Andres and B.O. Roos *J. Am. Chem. Soc.* **119**, 6168(1997).

[47] M.K. Shukla and J. Leszczynski *J. Comput. Chem.* **25**, 768(2004).

[48] A. Tsolakidis and E. Kaxiras *J. Phys. Chem. A* **109**, 2373(2005).

[49] D. Varsano, R.D. Felice, M.A.L. Marques and A. Rubio *J. Phys. Chem. B* **110**, 7129(2006).

[50] H. Langer and N.L. Doltsinis *J. Chem. Phys.* **118**, 5400(2003).

[51] J.B. Foresman, M. Head-Gordon, J.A. Pople and M.J. Frisch *J. Phys. Chem.* **96**, 135(1992).

[52] M.K. Shukla and J. Leszczynski *J. Phys. Chem. B* **109**, 17333(2005).

[53] M.K. Shukla and J. Leszczynski *Chem. Phys. Lett.* **414**, 92(2005).

[54] J. Cadet and P. Vigny, in Bioorganic Photochemistry, H. Morrison (Ed.), Vol. 1, p. 1, Wiley, 1990, New York.

[55] H. Langer and N.L. Doltsinis *Phys. Chem. Chem. Phys.* **6**, 2742(2004).

[56] C. Canuel, M. Mons, F. Piuzzi, B. Tardivel, I. Dimicoli and M. Elhanine *J. Chem. Phys.* **122**, 74316(2005).

[57] H. Chen and S. Li *J. Chem. Phys.* **124**, 154315(2006).

Brill Academic Publishers
P.O. Box 9000, 2300 PA Leiden
The Netherlands

Lecture Series on Computer and Computational Sciences
Volume 6, 2006, pp. 222-230

Computational Quantum Chemistry Design of Nanospirals and Nanoneedles

P.G. Mezey[1]

Canada Research Chair in Scientific Modeling and Simulation
Department of Chemistry and
Department of Physics and Physical Oceanography
Memorial University of Newfoundland
St. John's, NL, Canada A1B 3X7
and
Institute for Advanced Study, Collegium Budapest
Szentharomsag u. 2,
1014 Budapest, Hungary

Received 15 July, 2006; accepted 20 July, 2006

Abstract: Advanced molecular modeling methods based on quantum chemistry serve as efficient tools for designing potentially novel materials, including novel non-biological macromolecules. Two families of such macromolecules are the nanospirals and nanoneedles. Whereas the origins of nanospirals, also called nanocoils can be traced back to hexahelicene, the prototype of nanoneedles is a structure obtained by fused adamantanes. Several families of more complex structures can be obtained by combining these structural elements into novel types of macromolecules with potential applications in a wide range of nano-design. Some of the simple rules governing the main structural features of such macromolecules are discussed.

Keywords: Molecular modeling, macromolecular electron density, nanospirals, nanoneedles

Mathematics Subject Classification: 93A30, 05C10, 65D17
PACS: 31.15.Ar, 87.15.He

1. Motivation for designing spiral and needle-like nanostructures

Molecules with structures imitating the shapes of macroscopic objects have always intrigued and motivated chemists, bringing forward interesting and often useful analogies and ideas for potential applications. Whereas for quantum mechanical reasons, chemical structures, even on the macromolecular and supramolecular level possess properties seldom manifested in macroscopic objects, nevertheless, such macroscopic analogies serve a very useful role. With the rapid advances of quantum chemical modeling computations, the actual design of large molecular structures, even of macromolecular structures has become a viable approach, extending the power of quantum chemical modeling and design from the earlier size range of ten to thirty atoms to truly large systems, even to those with potential relevance to nanotechnology.

Two of the simple macroscopic structures with a wide range of applications are spirals and needles. Macroscopic spirals may serve as springs and coils, used in mechanical devices or as inductive or magnetizing electric circuits in appliances, among other applications. On the other hand, macroscopic needles also have a wide range of uses, as structural elements, as reversible fasteners for various objects, or as tools of assembly. It is not beyond the possibilities of current nanotechnology, and certainly well within the possibilities of projected developments in the near future, that similar uses can be found for nanoscale molecular objects with analogous structural features. What makes these expectations especially intriguing, it is the additional physical properties which may arise due to increased relevance of quantum mechanical phenomena on this size level of nature.

[1] Corresponding author. E-mail: pmezey@mun.ca,paul.mezey@gmail.com

In particular, one may expect that the electronic properties of molecular objects at the border region of size where classical, large-scale patterns of behavior are already emerging, but the dominance of quantum phenomena is still well preserved, could lead to useful applications. Nanospirals acting as conducting nanocoils, yet also exhibiting some of the properties of electron distributions within single molecules, may represent a potentially new class of electronic materials, especially, if by careful selection of structural elements the electronic properties become tunable.

In addition, such materials may also be applied as structural elements of larger assemblies of nanomaterials, and their mechanical properties when acing as nanosprings may become a useful feature in nanodesign. The controlled flexibility of nanodevices is an important consideration when one is interested in transferring macroscopic design principles to the nanoscale level, and nanosprings provide one potential tool for accomplishing these ideas.

On the other hand, nanoneedles are likely to serve as structural elements providing the complementing feature of structural desing: not so much the flexibility, but rigidity, when needed. Whereas there are indications based on recent studies that nanoneedles also exhibit tunable electronic properties, especially, if controlled by their local environment, nevertheless, their primary role may be linked to their structural, rather than to their electronic properties.

Thus, these two novel types of nanomaterials, nanospirals and nanoneedles are expected to have unusual electronic properties, and also represent two complementing aspects of structural nanodesign: flexibility and rigidity.

The novel nanospiral structures used as examples in this report have been designed using hexahelicene [1] as the main structural reference. With increasing interest and with the development of viable methodologies in macromolecular quantum chemistry and related modeling methods [2], it has become natural to envisage and attempt to design large-scale extensions of this molecule, and to extend the structural principle shown by hexahelicene to macromolecules and nanoscale structures. Whereas hexahelicene has served as the motivation for the novel nanospirals [3,4], on the other hand, for the novel nanoneedles [5,6] the molecule adamantane has served in a similar role. Whereas the initial studies were focused on carbon-based structures [3,5], analogous all-nitrogen and nitrogen-rich structures have also been investgated [4,6], showing additional unusual properties. All of the reported nanospiral and nanoneedle structures correspond to energy optimized geometries of energy minima (with all real vibrational frequencies), determined using GAUSSIAN program system [7] with 6-31G* and 6-31G** basis sets [8,9] at the HF and B3LYP levels.

2. Some carbon and nitrogen nanospirals

The non-planar, fused aromatic ring structure of hexahelicene was synthetized in 1956 by M.S. Newman and D. Lednicer [1], and chemists have been intrigued by this molecule ever since. Whereas there exists a conjugation throughout the ring system, the steric hindrance caused by the partially overlapping edges of the terminal rings results in a spiral, chiral structure, with very prominent optical activity. A large number of modeling studies on this and related molecules have been reported; some of these reports can be found in the literature quoted in ref. [3].

The hexahelicene structure naturally lends itself to extensions, both by simply adding rings to continue the spiral ("vertical extension"), or by replacing the hydrogen atoms by additional carbon atoms at the periphery of the molecule and constructing additional rings, extending the molecule "laterally" into larger conjugated systems. By continuing both of these processes, macromolecular, even nanoscale structures can be built, which contain a central spiral channel and otherwise it is reminiscent to graphite, where ring systems are arranged essentially in parallel sheets. However, in these structures there is just one, single conjugated sheet of aromatic rings, where the central spiral "twirl" connects the distant, nearly parallel regions of the layers on top of each other. Effectively, this structure qualifies as chiral, spiral graphite [3]. It is of some interest that according to the quantum chemical modeling studies, analogous all-nitrogen spiral structures are also stable, actual energy minima with all real vibrational frequencies [4].

In ordinary graphite, the electronic properties are determined by the dual feature of conjugation along the aromatic ring systems within each of the separate sheets and the sheet to sheet interactions perpendicular to the sheets. By contrast, in the "spiral graphite" obtained by the vertical and lateral extensions of hexahelicene, the fact that the sheet to sheet interaction is in fact a self-interaction of a conjugated supersystem, is expected to lead to unusual electronic properties. If at some distance from the spiral channel one considers two locations immediately above each other on two layers, then in a classical sense there is expected to be a competition between direct, local sheat to sheat interaction and a conjugative interaction carried along the aromatic rings, back and forth, involving the spiral channel region.

In Figures 1 and 2 the traditional structural models of the reference hexahelicene and some of the extended structures are shown where the main features leading to spiral graphite are already evident. In Figure 3, some of the molecular isodensity contours (MIDCOs) of hexahelicene are shown, corresponding to the electron density contour values of 0.05 a.u. and 0.20 a.u., where the atomic unit of electron density is used (1 a.u. = 1 e^-/bohr^3). In Figure 4, the 0.05 a.u. and 0.20 a.u MIDCOs of helical [6]-poly-methylenyl-napthalene, a laterally extended helicene, are shown.

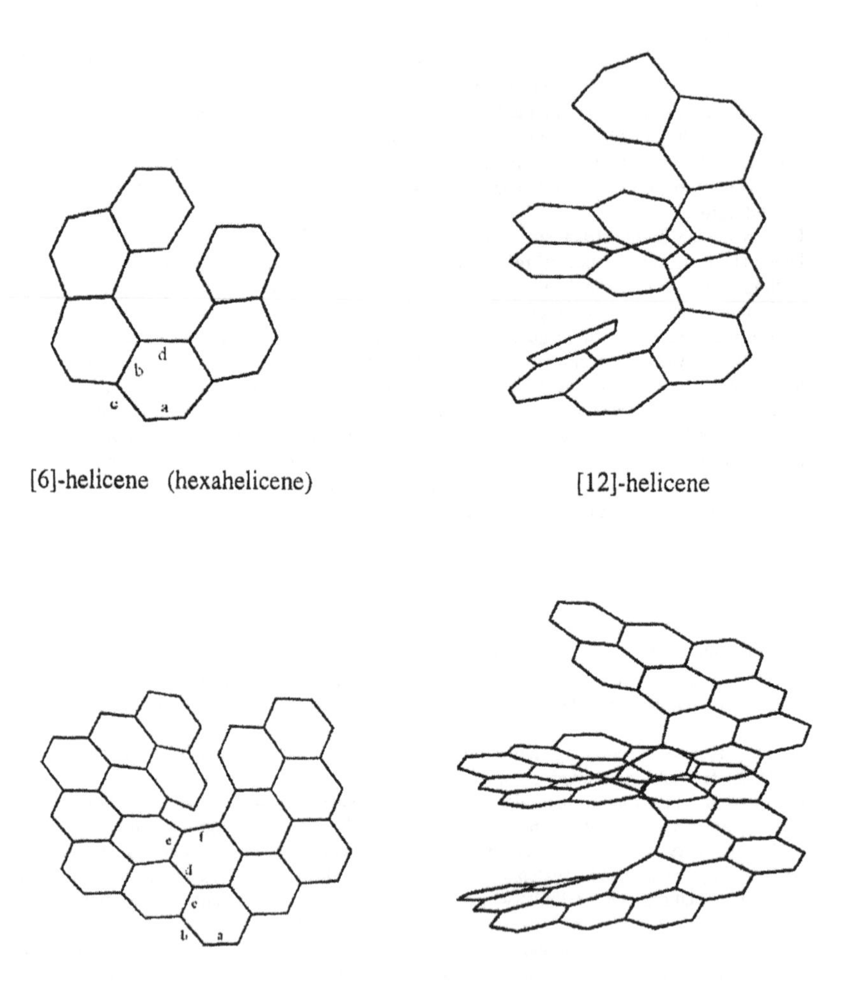

Figure 1. The geometry of bonding pattern of the carbon skeleton of hexahelicene, and some of the vertically and laterally extended helicenes

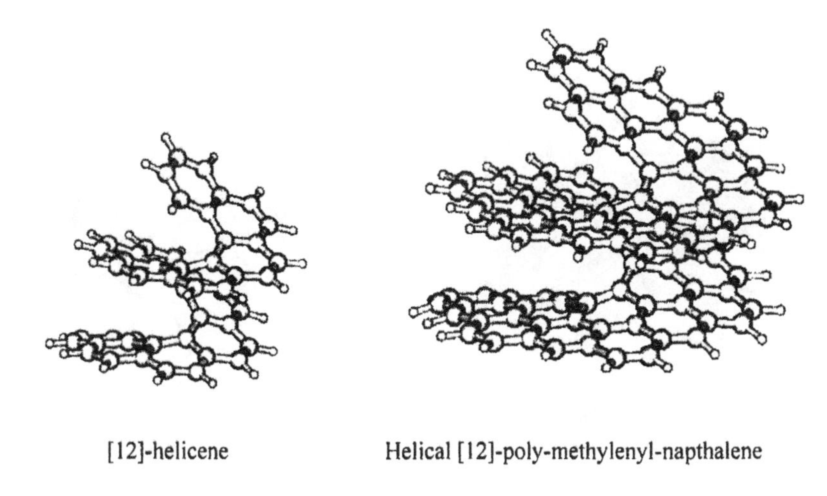

[12]-helicene Helical [12]-poly-methylenyl-napthalene

Figure 2. Ball-and-stick models of some vertically and laterally extended helicenes, showing the arrangements of the hydrogen atoms. In the larger system, the graphite-like layer-to-layer interactions are possible, however, in this case it is a self-interaction within a single spiral structure, indicating the possibility of unusual electronic properties if the system is further extended.

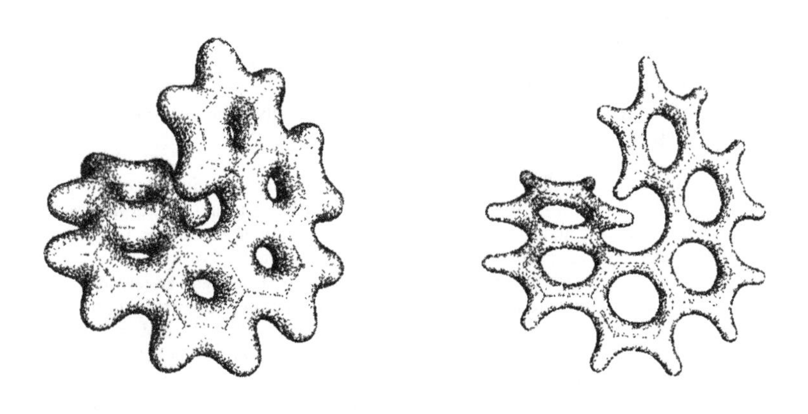

[6]-helicene (hexahelicene)

Figure 3. Molecular isodensity contours (MIDCOs) of hexahelicene at iso-values (density thresholds) a = 0.1 a.u. and a = 0.2 a.u., respectively. (1 a.u. is the atomic unit for electron density, corresponding to the charge of one electron being present in a box of volume of one cubic bohr).

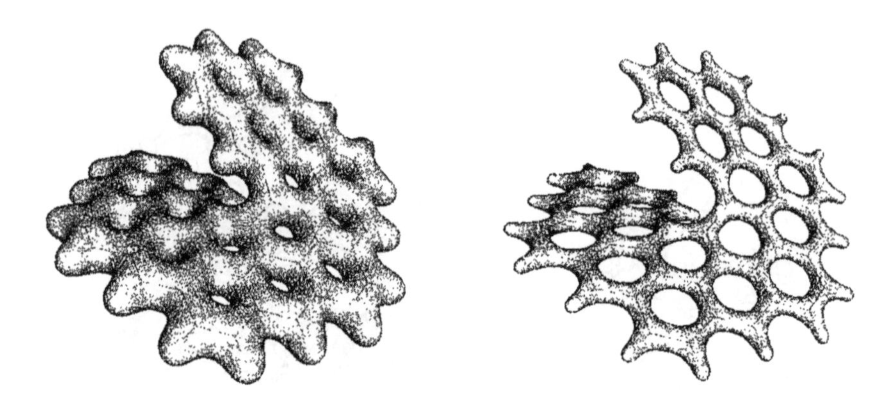

Helical [6]-poly-methylenyl-napthalene

Figure 4. Molecular isodensity contours (MIDCOs) of helical [6]-poly-methylenyl-napthalene, a laterally extended helicene, at iso-values (density thresholds) a = 0.1 a.u. and a = 0.2 a.u., respectively.

It is of some interest to investigate the possibilities of combining such structures into a single superstructure, and one may expect certain topological constraints on the generation of such superstructures.

In order to approach this question systematically, we are going to consider a large enough cube containing the extended systems, where

(a) the sheets are approximately parallel with two opposite faces of the cube,
(b) the deviations, if any, from the perfect graphite structure are approximately the greatest along lines interconnecting the centers of these two faces,
(c) the local structures at and near all other faces are essentially unperturbed by the distortions (if any) along the central lines described in (b).

On the simplest level, we are dealing with three reference structures: ordinary graphite (we shall denote this reference by G), and a left-handed version (denoted by L) and a right-handed version (denoted by R) of an extended helical sheet obtained by extending hexahelicene of L and R handedness (respectively) both vertically and laterally throughout the cube.

On the next level of complexity, and focusing on reference structures L or R, we may combine two or more such structures into topologically different structures. By removing those carbon atoms which are nearest to the central core channel (near the line specified in point (b)), this central channel effectively becomes wider, that may also allow to have greater separation between subsequent layers. This process can be repeated, by removing those carbon atoms which are nearest to the new wider channel, and by repeating this again and again, an arbitrarily wide central channel can be created, allowing arbitrarily large separation between subsequent layers. This in turn also allows two or more such spiral structures to get "screwed together", just like two or more sufficiently stretched "slinky" toys can generate double, triple or multiple helices. In such structures there are more then just one spiral sheet, and if one uses the integer n as the number of independent sheets, then the notation Ln or Rn can be used for left or right handed structures. Note that by widening the gap between layers of ordinary graphite G and placing new layers between them, one does not create any topologically different structure, instead one is merely reconstructing the original topological structure of G, hence the above treatment is of importance only for structures L and R.

Using these steps, the collection of systems now include structures G, and Rn, as well as Ln, for arbitrary finite positive n.

One may consider an additional series of constructions, leading to a different set of new structures. It is of some interest to consider, under what conditions can one have two or more helical channels within a structure with a fixed number n of independent sheets. We shall consider first the simplest case, with n = 1, resulting in structures derived from either of the three reference structures G, L=L1, and R=R1. If there are only two spirals of opposite chirality within the same structure, the structure is equivalent to generating U-cuts (cuts having the shape of letter U) above one another on each layer of ordinary graphite G, and joining the loose bottom segment of each letter U to the firm bottom segment of the hole created by the U-cut on the level intermediately above (resulting on each layer an "up rump" by one level, denoted by U1) or below (resulting on each layer a "down rump" by one level, denoted by D1). Hence, there is a direct connection between pairs of spirals of opposite chirality and the double staircase structure obtained by the U-cut method. If the up or down rumps connect sheets immediately next to one another when modifying graphite G, then for the resulting structure the notation G(U1) or G(D1) is used. However, the rumps on G may join sheets further apart, by joining the loose bottom segment of each letter U to the firm bottom segment of the hole created by the U-cut k levels above; then in the case of stepping up or down k levels, the notations G(Uk) or G(Dk) are used for the resulting structures.

Note, that the same U-cut method is equally applicable not only to the graphite reference structure G, but also to any of the n-spiral structures Rn or Ln of either chirality, for any positive integer n. This leads to structures Rn(Uk), Rn(Dk), Ln(Uk), and Ln(Dk).

One can get the U-cut and the "up rump", "down rump" method further generalized by generating more than one set of U-cuts. This way, one can generate 2 sets of cuts, or 3 sets of cuts, etc. For example, when starting with R4 and introducing a set of U-cuts completed by a set of up rump raising two levels, and another U-cut combined with a set of down rumps lowering three levels, the resulting structure has the topological notation R4(U2, D3).

For the three general types the following notations express the topological patterns:

G(Uk, Uk', ...Uk'',Dk''',...Dk''''), (1)

Ln(Uk, Uk', ...Uk'',Dk''',...Dk''''), (2)

Rn(Uk, Uk', ...Uk'',Dk''',...Dk''''). (3)

The above listed types of structures represent the main categories of arrangements derived from the structural pattern of hexahelicene. In particular, within a single structure the total number of left-handed and right handed spirals, n_L and n_R, respectively, may differ at most by one. On the simplest level, this is manifested by having only three main reference structures, G, L, and R, where

$$n_L = n_R = 0 \tag{4}$$

$$n_L = 1, \ n_R = 0, \tag{5}$$

and

$$n_L = 0, \ n_R = 1, \tag{6}$$

respectively. A more general expression valid also for more complex cases is

$$\text{abs}\,(n_L - n_R) < 2. \tag{7}$$

The above topological constraints limit the possible variations when designing new nanomaterials combining spiral graphite-like structures.

3. Carbon and nitrogen nanoneedles

The fundamental unit of adamantane is the basis for one family of nanoneedles [5,6]. By fusing adamantine skeletons, a very narrow nanotube is obtained, that does not really qualify as a tube, since the internal diameter of the structure is too small for any chemical structure to enter. Hence, the expression "nanoneedle" appears more appropriate for such structures [5,6].

The principle of fusing adamantane units is illustrated in Figure 5 using a pair of adamantane units, where a system of conjugated single and double bonds, as well as terminating oxygen atoms are shown.

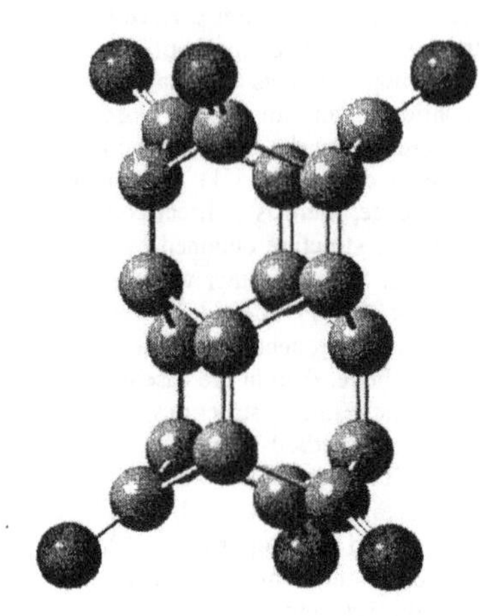

Figure 5. Two fused adamantane skeleton units with terminating oxygen atoms, sowing the principle of generating larger poly-C_6 nano-needles.

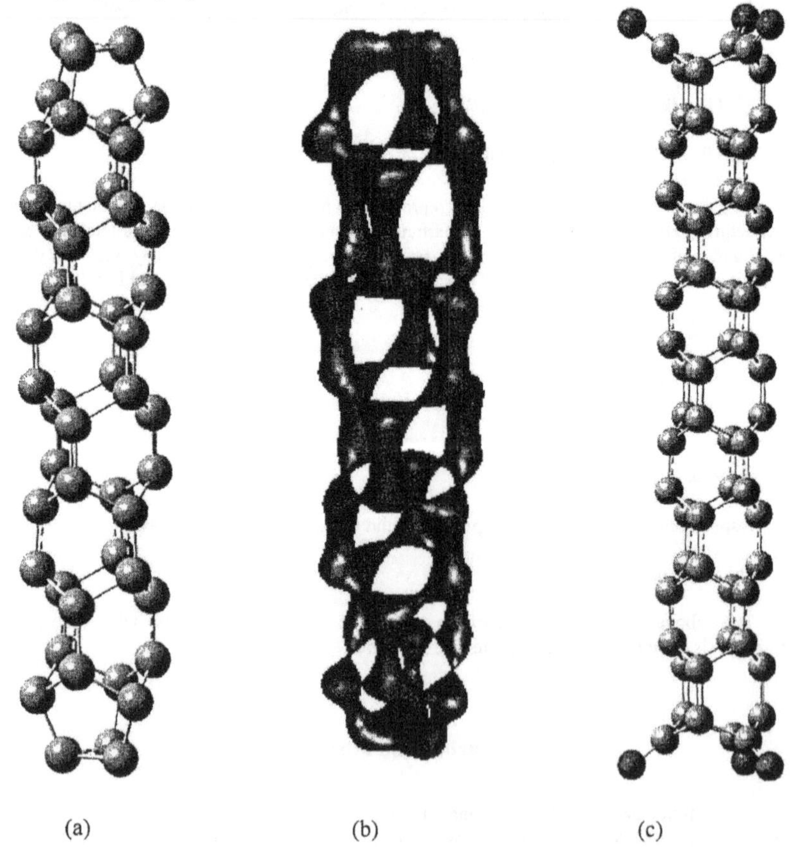

Figure 6. Various models for poly-C_6 nano-needles: (a) Ball-and-stick model, termination by rings of three carbon atoms, (b) Isodensity contour (density threshold a = 0.25 $e^-/bohr^3$), termination by rings of three carbon atoms, (c) Ball-and-stick model, termination by oxygen atoms.

Some of the models of larger needle structures are shown in Figure 6, including carbon-capped and oxygen-terminated variants. In the electron density contours, as illustrated by the MIDCO of density threshold a = 0.25 $e^-/bohr^3$, it is evident that along the body of the carbon-capped needle the conjugation is rather limited, and the carbon-carbon bonds of primarily single and double bond character are clearly distinguishable. Near the two ends of these structures the conjugation appears more pronounced, although the geometrical constraints of placing a carbon triangle on top and bottom of the needle impose a less than ideal mutual orientation for the bonds participating in the conjugative interactions.

The adamantane-based nanoneedles can be considered as diamond filaments of the smallest diameter which still preserve some of the essential structural principles of diamond itself.

Alternative nanoneedle structures can also be obtained using stacked C4 units as layers, and the corresponding columns can be terminated by a variety of atoms [5].

It is also noteworthy that according to modeling studies [6] similar adamantane skeleton and C4 skeleton nanoneedles can be constructed from nitrogen, where all the carbon atoms of some of the carbon structures are replaced by nitrogen. Whereas these latter structures are of much lower energetic stability than the carbon versions, nevertheless, all the reported N-nanoneedles are stable energy minima with all real vibrational frequencies. Apparently, ionic species of the same structural frameworks can also be stabilized by appropriate counter-ions.

Acknowledgments

This study has been supported by the Natural Sciences and Engineering Research Council of Canada, the Canada Research Chairs and Canada Foundation for Innovation projects, the e-Science RET Project, Eotvos University of Budapest, and the Institute for Advanced Study, Collegium Budapest, Hungary.

References

[1] M.S. Newman and D. Lednicer, Hexahelicene, *J. Am. Chem. Soc.* **78** (1956) 4765-4770.

[2] Zs. Szekeres, T.E. Exner, and P.G. Mezey, Fuzzy fragment selection strategies, basis set dependence, and HF – DFT comparisons in the applications of the ADMA method of macromolecular quantum chemistry, *Internat. J. Quantum Chem.*, **104** (2005) 847-860.

[3] L.J. Wang, P.L. Warburton, Zs. Szekeres, P. Surjan, and P.G. Mezey, Stability and Properties of Annelated Fused-Ring Carbon Helices: Models Towards Helical Graphites, *J. Chem. Inf. Mod.*, **45** (2005) 850-855.

[4] L.J. Wang and P.G. Mezey, Predicted High-Energy Molecules: Helical All-Nitrogen and Helical Nitrogen-Rich Ring-Clusters, *J. Phys. Chem. A*, **109** (2005) 3241-3243.

[5] J.L. Wang and P. G. Mezey, The Electronic Structures and Properties of Open-Ended and Capped Carbon Nano-Needles *J. Chem. Inf. Mod.*, **46** (2006) 801-807.

[6] J.L. Wang, G. H. Lushington, and P. G. Mezey, Stability and Electronic Properties of Nitrogen Nano-Needles and Nanotubes *J. Chem. Inf. Mod.*, in press (accepted June 28, 2006).

[7] M. J. Frisch, G. W. Trucks, H. B. Schlegel, G. E. Scuseria, M. A. Robb, J. R. Cheeseman, V. G. Zakrzewski, J. A. Montgomery, R.E. Stratmann, J. C. Burant, S. Dapprich, J. M. Millam, A. D. Daniels, K. N. Kudin, M. C. Strain, O. Farkas, J. Tomasi, V.Barone, M. Cossi, R. Cammi, B. Mennucci, C. Pomelli, C.Adamo, S. Clifford, J. Ochterski, G. A. Petersson, P. Y. Alyala, Q. Cui, K. Morokuma, D. K. Malick, A. D. Rabuck, K. Raghavachari,J. B. Foresman, J. Cioslowski, J. V. Ortiz, B. B. Stefanov,G. Liu, A. Liashenko, P. Piskorz, I. Komaromi, R. Gomperts, R.L. Martin, D. J. Fox, T. Keith, M. A. Al-Laham, C. Y. Peng, A.Nanayakkara, C. Gonzalez, M. Challacombe, P. M. W. Gill, B. G. Johnson, W. Chen, M. W. Wong, J. L. Andres, M. Head-Gordon, E. S. Replogle, J. A. Pople, GAUSSIAN 98 (Revision A.9), Gaussian, Inc., Pittsburgh, PA, 1998.

[8] A.D. Becke, *J. Chem. Phys.* **98** (1993) 5648-5652.

[9] C. Lee, W. Yang, R.G. Parr, *Phys. Rev. B* **37** (1988) 785-787.

Brill Academic Publishers
P.O. Box 9000, 2300 PA Leiden
The Netherlands

Lecture Series on Computer and Computational Sciences
Volume 6, 2006, pp. 231-240

Theoretical Study on the Second Hyperpolarizabilities of Diphenalenyl Radical Systems

[a]Masayoshi Nakano[1], [b]Takashi Kubo, [c]Kenji Kamada, [c]Koji Ohta, [d]Benoît Champagne, [b]Edith Botek and [b]Kizashi Yamaguchi

[a]Department of Materials Engineering Science, Graduate School of Science, Osaka University, Toyonaka, Osaka 560-8531, Japan,
[b]Department of Chemistry, Graduate School of Science, Osaka University, Toyonaka, Osaka 560-0043, Japan,
[c]Photonics Research Institute, National Institute of Advanced Industrial Science and Technology (AIST), Ikeda, Osaka 563-8577, Japan,
and
[d]Laboratoire de Chimie Théorique Appliquée, Facultés Universitaires Notre-Dame de la Paix, rue de Bruxelles, 61, B-5000 Namur, Belgium

Received 23 June, 2006; accepted 26 June, 2006

Abstract: The static second hyperpolarizabilities (γ) of diphenalenyl radical systems are investigated from a viewpoint of diradical character dependence of γ using a hybrid density functional theory (DFT) method. It turns out that neutral singlet diradical systems with intermediate diradical characters (y) tend to exhibit □□□□ely enhanced γ values as compared to those with small or negligible diradical characters (y~0: closed-shell systems) and pure diradical systems (y~1). The spin-state dependence of γ is investigated for these diradical systems. In the intermediate diradical systems in the singlet state, the triplet state tends to significantly reduce the γ compared to singlet state due to the Pauli principle. Some structural effects on the diradical characters and γ values are also examined. On the basis of these results, we propose a novel control scheme of γ for diradical NLO systems.

Keywords: Second hyperpolarizability, open-shell molecule, diradical, phenalenyl radical, density. functional
PACS: 31.15.Ar., 31.1.5.Ew, 33.1.5.Kr.

1. Introduction

A new class of nonlinear optical (NLO) molecules, open-shell NLO molecules [1-7], have been proposed based on our structure-property relations extracted from the theoretical and computational studies on the variation of second hyperpolarizability (γ), which is the microscopic origin of the third-order NLO property, of H_2 molecule during the bond dissociation [7,8] and *p*-quinodimethane model molecules with different diradical characters [2]. The structure-property relation states that the singlet diradical molecules with intermediate and somewhat large diradical characters tend to provide significantly enhanced γ values than closed-shell and pure diradical molecules. Such feature is understood by the fact that the intermediate bonding electrons are sensitive to the applied field, leading to large fluctuation of electrons, i.e., large polarization. This structure-property relation is substantiated by real π-conjugated molecules involving imidazole and triazole rings [6].

In this study, we investigate the γ values of thermally stable diradical molecules involving diphenalenyl radicals (DPLs). Firstly, we examine polycyclic aromatic hydrocarbons involving DPLs, *i.e.*, pentaleno[1,2,3-*cd*;4,5,6-*c'd'*]diphenalene (PDPL) and *s*-indaceno[1,2,3-*cd*;5,6,7-*c'd'*]diphenalene (IDPL) [9] in order to elucidate the diradical character dependence of static γ calculated by the hybrid density functional theory (DFT) method [4]. As a reference system, we examine a closed-shell polycyclic aromatic hydrocarbon with a similar π-conjugation size to PDPL and IDPL. Second, we

[1] Corresponding author. E-mail: mnaka@chemg.es.osaka-u.ac.jp

examine the DPLs linked with a π-conjugated bridge, e.g., 1,2-bis(phenalen-1-ylidene)ethene (BPLE1) [10], as well as pure diradical (BPLE2) and closed-shell, i.e., bis(pyren-1-yl)ethyne (BPRY2), systems [5]. The spin-state dependence of γ is investigated for these DPLs linked with acetylene bridge. Third, we examine the effect of *para-* and *ortho-* quinoid resonance forms of central indacenes in two pure hydrocarbon polycyclic DPLs, *i.e.*, IDPL and *as*-indaceno[1,2,3-*cd*;6,7,8-*c'd'*]diphenalene (*as*-IDPL), on the diradical character and the γ values [11]. As an isoelectronic molecule to *as*-IDPL, we also investigate the γ of dicyclopenta[*b*;*d*]thieno[1,2,3-*cd*;5,6,7-*c'd'*]diphenalene (TDPL) [12,13], which involves a thiophene ring instead of the central benzene ring, from the viewpoint of the influence of the aromaticity of central ring on the diradical character and on γ as well as from the viewpoint of the hypervalency of the S atom in the thiophene ring. From these results, we discuss a possibility of diradical NLO systems involving phenalenyl units and propose a control scheme of γ by tuning the diradical character and spin state.

2. Origin of the diradical character dependence of γ

The origin of the diradical character dependence of γ of neutral singlet diradical systems is elucidated based on the perturbation formula of γ using the simplest diradical molecular model with different diradical characters, i.e., H_2 under bond dissociation. In this study, we consider the longitudinal γ of symmetric diradical systems. Because the dipole moment along the longitudinal axis disappears, the γ value is described by the detailed balance between types II (negative) and III-2 (positive) [14,15] in the perturbation formula. Since the electronic states of diradical systems is characterized by three states (g, e1 and e2) composed of the HF ground, singly and doubly excited configurations concerning the highest occupied molecular orbital (HOMO) and the lowest unoccupied MO (LUMO), the dominant diradical contributions to γ is described by these three states (g, e1 and e2) with non-zero μ_{ge1} and μ_{e1e2} transition moments (μ_{ge2} = 0). Although for large-size diradical compounds there is an additional contribution to γ, which originates from the remaining (non-diradical) part of the system, we focus on the diradical contribution to γ in order to clarify the origin of the diradical character dependence of γ. In this case, the perturbation formula of γ is expressed as

$$\gamma = \gamma^{\mathrm{II}} + \gamma^{\mathrm{III-2}} \approx -4\frac{(\mu_{ge1})^4}{(E_{e1g})^3} + 4\frac{(\mu_{ge1})^2(\mu_{e1e2})^2}{(E_{e1g})^2 E_{e2g}}, \quad (1)$$

For neutral singlet diradical systems, type III-2 is a dominant process, giving a positive γ. In general, for symmetric conjugated systems, the small difference between E_{e2g} and E_{e1g} as well as the relative increase of $|\mu_{e1e2}|$ compared to $|\mu_{ge1}|$ turn out to cause a large enhancement of γ (positive) dominated by type III-2 process.

The electronic structure of the ground and excited states of singlet diradical systems is well understood from the viewpoint of covalent (or diradical) and zwitterionic contributions [16]. The variation in relative contribution of covalent (or diradical) and ionic characters to the ground (g), the first (e1) and the second (e2) excited states of singlet diradical systems are explained using a two-center two-electron model system in the MO picture [7,8]. The diradical character is determined from spin-unrestricted Hartree-Fock (UHF) calculations. The diradical character y_i related to the HOMO-i and LUMO+i is defined by the weight of the doubly-excited configuration in the MC-SCF theory and is formally expressed in the case of the spin-projected UHF (PUHF) theory as [17,18]

$$y_i = 1 - \frac{2T_i}{1+T_i^2}, \quad (2)$$

where T_i is the orbital overlap between the corresponding orbital pairs. T_i can also be represented by using the occupation numbers (n_j) of UHF natural orbitals (UNOs):

$$T_i = \frac{n_{\mathrm{HOMO}-i} - n_{\mathrm{LUMO}+i}}{2}. \quad (3)$$

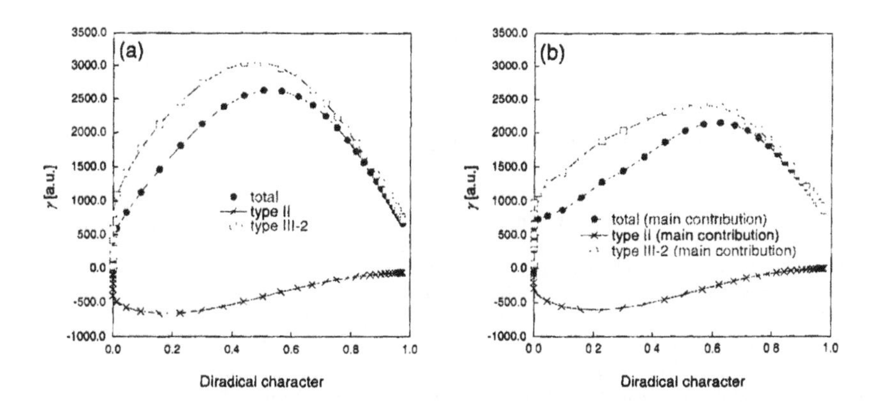

Figure 1: Variation of total, type II and type III-2 contributions to γ [a.u.] (1.0 a.u. of γ = 6.235377 × 10^{-65} C^4 m^4J^{-3} = 5.0367 × 10^{-40} esu) calculated by the all-state model (a) and the three-satte model with primary contributions at the CISD/6-31G**+*sp* level with the increase of diradical character *y*.

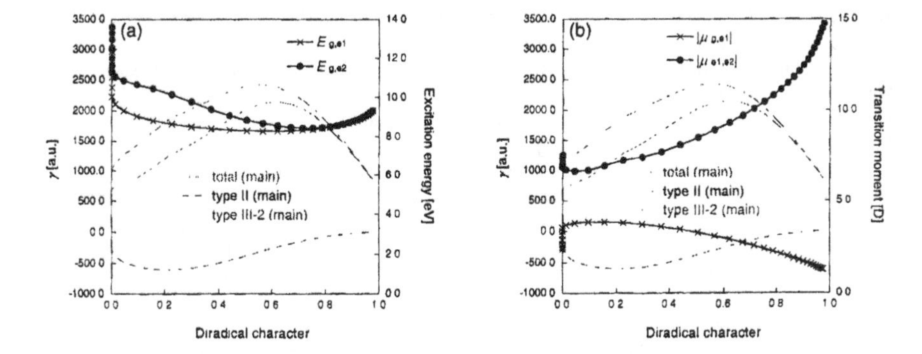

Figure 2: (a) Variation of excitation energies E_{e2g} and E_{e1g} [eV] calculated by the three-state model with primary contributions to γ at the CISD/6-31G**+*sp* level with the increase of diradical character *y*. (b) Variation in the amplitudes of transition moments $|\mu_{e1,e2}|$ and $|\mu_{ge1}|$ [D] calculated by the three-state model at the CISD/6-31G**+*sp* level with the increase of diradical character *y*. The variation of total, type II and type III-2 contributions to γ [a.u.] calculated by the three-state model at the CISD/6-31G**+*sp* level are also shown in both figures. All the calculation are performed by the GAMESS program package [19,20].

Since the PUHF diradical characters amount to 0 % and 100 % for closed-shell and pure diradical states, respectively, y_i represents the diradical character, *i.e.*, the instability of the chemical bond. Figure 1 shows the variation of total, type II and type III-2 contributions to γ calculated by the all-state model (a) and by the three-state model (b) with primary contributions at the CISD/6-31G**+*sp* level with the increase of diradical character *y* (*i*=0). The qualitative feature of all-stat model is well reproduced by the three-state model. Figure 2 shows significant increase and decrease of $|\mu_{e1e2}|$ and $|\mu_{ge1}|$, respectively, as well as a slight variation of E_{e2g}/E_{e1g} (≈ 1) with diradical character, leading to a specific behavior of γ with diradical character, i.e., an increase until a maximum is reached and then a decrease because γ is approximately proportional to $(\mu_{ge1})^2(\mu_{e1e2})^2$ (type III-2) in the case of singlet diradical systems. These trends in the amplitudes of transition moments are explained by the variation in the relative covalent (or diradical) and ionic characters in the three electronic states primarily contributing to type III-2. As seen from the configurations involved in wavefunctions, the relative increase of the diradical character in the ground state corresponds to the relative increase in the ionic character in the second excited state, whereas the first excited state is well described by the ionic configuration.

3. Polycyclic aromatic diphenalenyl radical systems

Figure 3 shows the structures of polycyclic aromatic diphenalenyl molecules, PDPL (a) and IDPL (b), and a similar size closed-shell polycyclic hydrocarbon, PY2 (c), in their singlet ground states, which are optimized at the RB3LYP level of approximation using the 6-31G** basis set. It is found that the diradical characters y of PDPL and IDPL are 0.5833 and 0.7461, respectively, which represent intermediate diradical systems, while that of PY2 is zero, which implies a closed-shell system. The fact that the y value of IDPL is larger than that of PDPL can be understood by a larger stabilization of both-end phenalenyl radicals due to the recovery of aromaticity in the central benzene ring for IDPL than PDPL. We use the hybrid DFT, i.e., spin unrestricted (U)/restricted (R) BHandHLYP, method to calculate the longitudinal γ values. Although it is well-known that the use of extended basis sets are necessary for obtaining quantitative γ values for π-conjugated systems, we use the standard basis set, 6-31G*, since the size of the systems in this study prohibits the use of such large basis sets and the γ value for IDPL using the 6-31G*+diffuse p ($\zeta = 0.0523$) is shown to be slightly (10%) enhanced as compared to that using the 6-31G* at the RHF level, which suggests that the use of 6-31G* basis set is adequate for semi-quantitative comparisons of, at least, longitudinal components of γ among the large-size systems in this study. It is shown from the previous study on the p-quinodimethane models [2] that the UBHandHLYP gives reliable γ values for diradical molecules with intermediate and large diradical characters, while the RBHandHLYP method gives reliable γ values for diradical molecules with small diradical characters or closed-shell molecules, at least for molecules of the size investigated in this study. We confine our attention to the dominant longitudinal components of static γ (γ_{xxxx}) (see Fig. 3) using the finite field approach [21], which consists in the fourth-order differentiation of energy E with respect to different amplitudes of the applied external electric field. All calculations are performed using the Gaussian 98/03 program package [22].

The UBHandHLYP/6-31G* results show that the γ_{xxxx} (γ) values for PDPL and IDPL are 1255 x 10^3 a.u. and 2383 x 10^3 a.u., respectively. This enhancement of γ for IDPL as compared to that for PDPL is predicted to be caused by the extension of π-conjugation length and the increase in the diradical character (y = 0.5833 for PDPL and y = 0.7461 for IDPL) for IDPL. PY2, which is a closed-shell system, is found to give a much smaller γ value (194 x 10^3 a.u.) than those of PDPL and IDPL (the ratio of γ : PDPL/PY2 = 6.47 and IDPL/PY2 = 12.3). This feature reconfirms that singlet diradical molecules with intermediate and somewhat large diradical characters tend to give about one-order larger γ value than closed-shell systems with similar π-conjugation size.

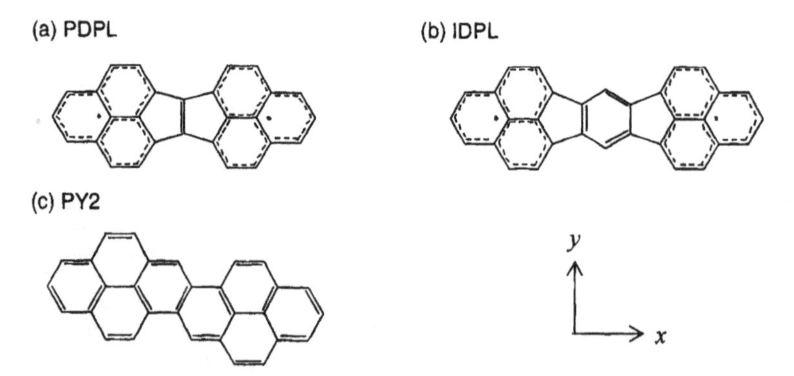

Figure 3: Molecular structures of singlet states of PDPL (a), IDPL (b) and PY2 (c). Coordinate axis is also shown. The corresponding structures are planar and those of PDPL and IDPL belong to D_{2h} while PY2 does to C_{2h} point groups. The middle carbon-carbon bond of PY2 has an angle of 30° with respect to the longitudinal (x) axis.

Table 1: Diradical character (y) and second hyperpolarizability (γ_{xxxx}) of PDPL (a), IDPL (b) and PY2 (c) shown in Fig. 3. The γ values are given in a.u. (1.0 a.u. of second hyperpolarizability = 6.235377 $\times$ 10^{-65} C^4 m^4J^{-3} = 5.0367 $\times$ 10^{-40} esu). The calculations of diradical characters are performed using the occupation numbers of UNOs and those of γ are done using the UBHandHLYP method for (a) and (b) and RBHandHLYP method for (c) with the 6-31G* basis set.

System	y	γ[x 10^3 a.u.]
(a) PDPL	0.5833	1255
(b) IDPL	0.7461	2383
(c) PY2	0.0	194

4. Diphenalenyl radical systems involving π-conjugated bridge

Figure 4 shows the structures of singlet diphenalenyl radical compounds linked with acetylene bridge (a and b) and closed-shell dipyren compounds (c). These structures are optimized at the RB3LYP level of approximation for a closed-shell system and the UB3LYP level for open-shell systems using the 6-31G** basis set. The calculations of diradical characters are performed using the occupation numbers of UNOs. As shown in Table 2, the γ value of BPLE1 with an intermediate diradical character (y = 0.652) is shown to be larger than that of a closed-shell system, BPRY2 (y = 0), and that of a pure diradical system, BPLE2 ($y \sim 1$).

We next investigate the spin state dependence of γ of BPLE1 and BPLE2. The γ value of BPLE1 in the singlet state, which has an intermediate diradical character, is found to be about 3.4 times as large as that in the triplet state (see Table 3). In contrast, the singlet BPLE2, which is a pure diradical system, shows almost the same γ value as that of triplet one. Such distinct difference in spin state dependence of γ can be understood from the difference in the degree of field-induced electron fluctuations between intermediate and pure diradical singlet states: in the singlet state, intermediate diradical state tends to be more polarized than the pure diradical state, in which the localization feature of electrons is resemble to that of triplet state due to the Pauli principle.

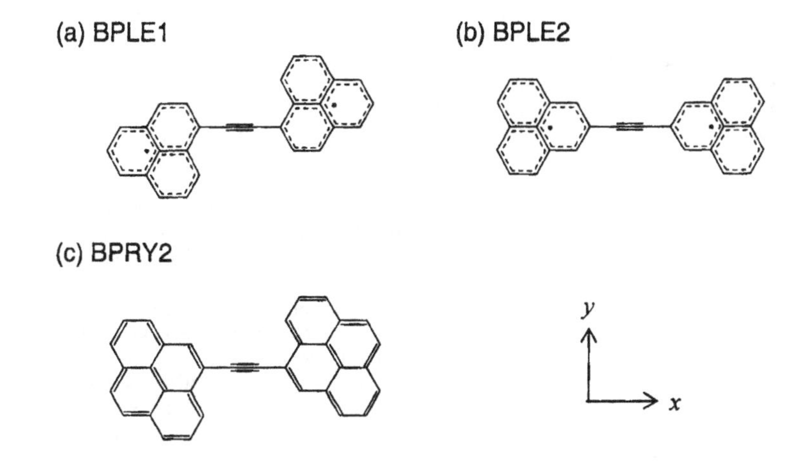

Figure 4: Molecular structures of singlet states of BPLE1 (a), BPLE2 (b) and BPRY2 (c). Coordinate axis is also shown. The optimized structures are planar and those of BPLE1 and BPRY2 belong to C_{2h} while BPLE2 does to D_{2h} point groups.

Table 2: Diradical character (y) and second hyperpolarizability (γ_{xxxx}) of BPLE1 (a), BPLE2 (b) and BPRY2 (c) shown in Fig. 4. The calculations of diradical characters are performed using the occupation numbers of UNOs and those of γ are done using the UBHandHLYP method for (a) and (b) and RBHandHLYP method for (c) with the 6-31G* basis set.

System	y	γ [x 10^3 a.u.]
(a) BPLE1	0.652	1560
(b) BPLE2	0.999	205
(c) BPRY2	0.0	366

Table 3: Calculated γ values of singlet and triplet states for BPLE1 and BPLE2

System	y (singlet)	γ [x 10^3 a.u.]	
		Singlet	Triplet
(a) BPLE2	0.652	1560	460
(b) BPLE2	0.999	205	199

5. Structural effects on γ for polycyclic diphenalenyl radical systems

In this section, we first examine the effect of *para-* and *ortho-* quinoid resonance forms of central indacenes in two pure hydrocarbon polycyclic diphenalenyl radicals, i.e., IDPL and *as*-IDPL, on the diradical character and the γ values. As an isoelectronic molecule to *as*-IDPL, we next investigate the γ of TDPL, which involves a thiophene ring instead of the central benzene ring, from the viewpoint of the influence of the aromaticity of central ring on the diradical character and on γ as well as from the viewpoint of the hypervalency of the S atom in the thiophene ring.

The geometrical structures of IDPL (a), *as*-IDPL (b) and TDPL (c) in their singlet ground state are shown in Fig. 5. These were optimized at the UB3LYP/6-31G** method. IDPL belongs to the D_{2h} point group while *as*-IDPL and TDPL to C_{2v}. The resonance structures between the diradical (I) and quinoid (II and III) forms are also shown. In the central region, two types of *para*-quinoid forms (*p*-xylylene shown in Figs. 5a II and III) contribute to IDPL, whereas *para-* and *ortho-* quinoid forms (*p*- and *o*- xylylenes shown in Figs. 5b II and III, respectively) contribute to *as*-IDPL. For TDPL, thieno-quinoid forms (*p*- and *o*-dimethylenethiophene shown in Figs. 5c II and III, respectively) describe partly the central region. The region composed of successive carbon-carbon units with the same or similar bond lengths in the central benzene ring is more extended in *as*-IDPL (C3-C1-C2-C4; *r*(C1-C2) = 1.400 Å and *r*(C2-C4) = 1.389 Å) than in IDPL (C1-C5-C2 *r*(C1-C5,C5-C2) = 1.396 Å), which suggests that the aromaticity of the central benzene ring is larger in *as*-IDPL than in IDPL. Then, the carbon-carbon bond length alternation (Δr) (described by the deviation here) in the phenalenyl rings of *as*-IDPL (Δr = 0.0122 Å) is smaller than that of IDPL (Δr = 0.0137 Å), substantiating the fact that *as*-IDPL has a larger contribution of diradical form (I) than IDPL. Both geometrical features indicate that the diradical character of *as*-IDPL is larger than that of IDPL since the increase of aromaticity in the central benzene ring is also predicted to increase the diradical character in the both-end phenalenyl rings as shown in Fig. 5. The bond length alternation in *as*-IDPL [*r*(C6-C8) – *r*(C4-C12) = 0.021 Å] is smaller than in TDPL [*r*(C5-C7) – *r*(C3-C11) = 0.046 Å], demonstrating that the thieno-quinoid (*p*-dimethylenethiophene) contribution in TDPL (Fig. 5c II) is larger than the *para*-quinoid (*p*-xylylene) contribution in *as*-IDPL (Fig. 5b II). Thus, the diradical character of TDPL is expected to be smaller than that of *as*-IDPL since the aromaticity of thiophene ring in TDPL is smaller than that of central benzene ring in *as*-IDPL. Moreover, judging from the long carbon-sulfur bond length (1.744 Å) in TDPL, the contribution of *o*-dimethylenethiophene form, which corresponds to a hypervalent structure of sulfur atom, is predicted to be negligible in the ground state.

The γ_{xxxx} (γ) values for IDPL, *as*-IDPL, and TDPL calculated by the UBHandHLYP/6-31G* are 2284 x 10^3 a.u., 472 x 10^3 a.u., and 1375 x 10^3 a.u., respectively. The significant decrease in γ for *as*-IDPL relative to IDPL is predicted to be caused by the almost pure diradical character for *as*-IDPL (y = 0.923) compared to the intermediate diradical character for IDPL (y = 0.770). On the other hand, the replacement of the central benzene ring of *as*-IDPL by a thiophene ring leads to an enhancement of γ by about a factor of 3, in agreement with the decrease of diradical character (y = 0.766). These results are in agreement with the feature of diradical characters shown in Fig. 5: *as*-IDPL has a larger diradical character than IDPL and TDPL.

In order to clarify the effect of the central *p*- and *o*- xylylene structures in IDPL and *as*-IDPL on the diradical characters and γ values, we compare their frontier orbitals with those of *p*- and *o*-xylylenes at

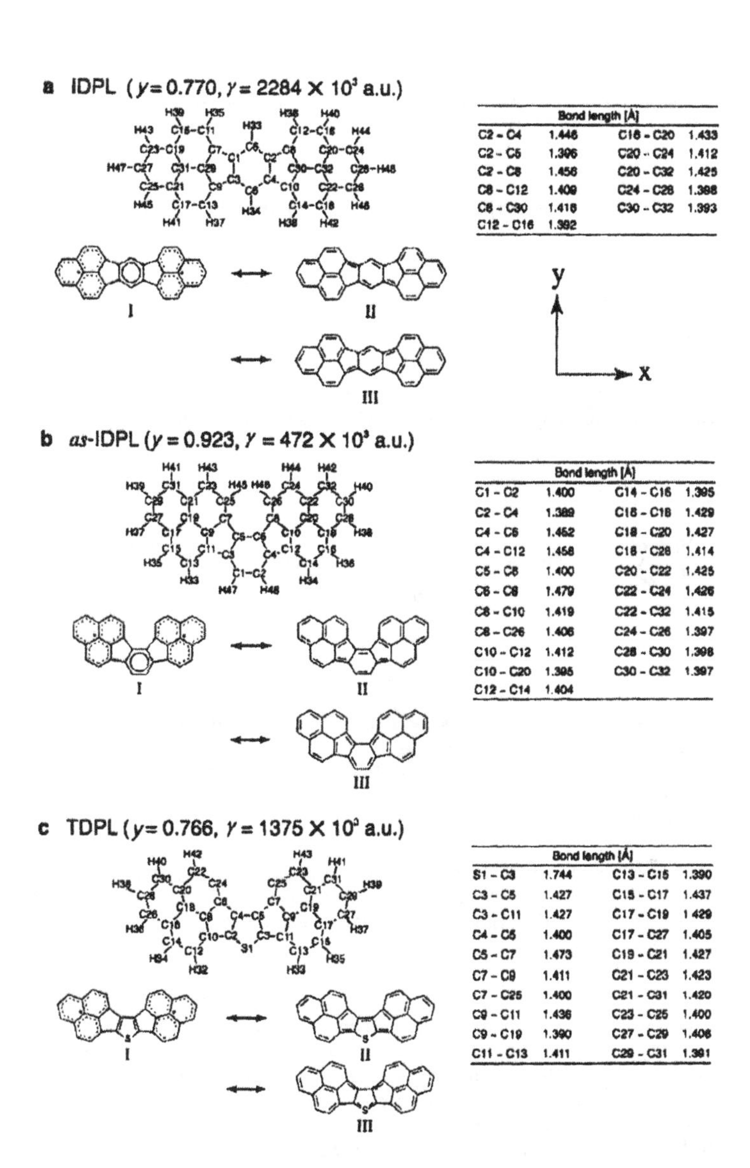

Bond length [Å]			
C2 - C4	1.446	C16 - C20	1.433
C2 - C5	1.396	C20 - C24	1.412
C2 - C8	1.456	C20 - C32	1.425
C8 - C12	1.409	C24 - C28	1.388
C8 - C30	1.416	C30 - C32	1.393
C12 - C16	1.392		

Bond length [Å]			
C1 - C2	1.400	C14 - C16	1.395
C2 - C4	1.389	C16 - C18	1.429
C4 - C6	1.452	C18 - C20	1.427
C4 - C12	1.458	C16 - C26	1.414
C5 - C8	1.400	C20 - C22	1.425
C6 - C8	1.479	C22 - C24	1.426
C8 - C10	1.419	C22 - C32	1.415
C8 - C26	1.406	C24 - C26	1.397
C10 - C12	1.412	C28 - C30	1.398
C10 - C20	1.395	C30 - C32	1.397
C12 - C14	1.404		

Bond length [Å]			
S1 - C3	1.744	C13 - C15	1.390
C3 - C5	1.427	C15 - C17	1.437
C3 - C11	1.427	C17 - C19	1.429
C4 - C6	1.400	C17 - C27	1.405
C5 - C7	1.473	C19 - C21	1.427
C7 - C9	1.411	C21 - C23	1.423
C7 - C25	1.400	C21 - C31	1.420
C9 - C11	1.436	C23 - C25	1.400
C9 - C19	1.390	C27 - C29	1.406
C11 - C13	1.411	C29 - C31	1.391

Figure 5: Geometries optimized at the UB3LYP/6-31G* level, main contributing resonance forms [diradical (I) and quinoid (II and III)], diradical characters (y), and second hyperpolarizabilities (γ_{xxxx}) of the singlet ground state of IDPL (a), *as*-IDPL (b), and TDPL (c). The γ values, evaluated at the UBHandHLYP level of approximation. See the text for more details.

the RHF/6-31G* level of approximation (Figs. 6a, b, d and e). It is found that the HOMO-LUMO energy gap of *as*-IDPL is smaller than that of IDPL, in agreement with the larger diradical character of the former. Indeed, the weight of double excitation CI related to the diradical character, gets larger for systems with smaller HOMO-LUMO energy gap. Similarly, the HOMO-LUMO energy gap is smaller in *o*- xylylene than in *p*- xylylene, demonstrating the impact of the *o*-quinoid xylylene forms in *as*-IDPL. As a result, the pure-hydrocarbon polycyclic phenalenyl radical molecules involving *ortho*-

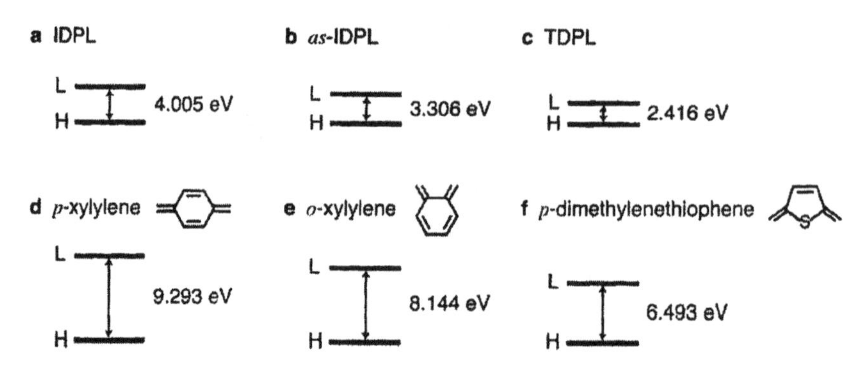

Figure 6: HOMO (H) - LUMO (L) energy gaps [eV] of IDPL (a), *as*-IDPL (b), TDPL (c), *p*-xylylene (d), *o*-xylylene (e) and *p*-dimethylenethiophene (f) calculated by the RHF/6-31G* method.

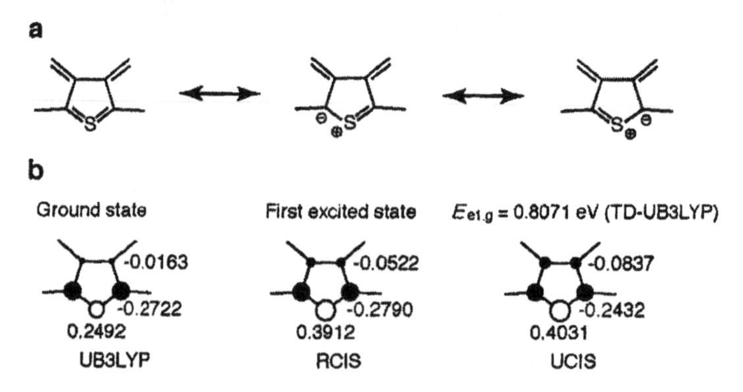

Figure 7: Hypervalent resonance forms of thiophene (a) and atomic charges for central thiophene ring in the ground (g) (UB3LYP/6-31G*) and the first excited (e1) (UCIS and RCIS/6-31G*) states of TDPL by the Mulliken population analysis (b). Excitation energy (E_{e1g}) at the TD-UB3LYP level of approximation is also shown.

quinoid resonance structure in the middle region generally tend to provide larger diradical characters than those involving *para*-quinoid resonance structure in the middle region. Although in the present case y is close to one for *as*-IDPL and thus the γ value is significantly reduced, the γ value for *ortho*-quinoid structures could be enhanced if their *para*-quinoid structures possess smaller diradical characters. We next examine the effect of the central ring modification using *as*-IDPL and TDPL. TDPL and *as*-IDPL are isoelectronic and present similar geometrical structures. In TDPL, one of the resonant (closed-shell) forms involves the *p*-dimethylenethiophene unit, which presents a small HOMO-LUMO energy gap (6.493 eV) (see Fig. 6), in agreement with its smaller aromaticity with respect to *p*-xylylene. In other words, the contribution of the quinoid resonance form is larger in *p*-dimethylenethiophene than in *p*-xylylene. Thus, a larger contribution of the quinoid structure is predicted to be at the origin of the reduction of HOMO-LUMO energy gap (2.416 eV) in TDPL and of the decrease of the diradical character of TDPL as compared to *as*-IDPL. On the other hand, the resonance structure involving *o*-dimethylenethiophene (see Fig. 5c III), which corresponds to a hypervalent structure of sulfur atom, is predicted to have a significant contribution in the dipole-allowed first excited state (e1), which is well described as a zwitterionic state [7,16], as shown in Fig. 7a. Indeed, the larger ionicity of the excited state is confirmed by using Mulliken population analysis combined with the spin-unrestricted (restricted) single-excitation CI, UCIS(RCIS)/6-31G*, method. The atomic charges on the S atom and on its neighboring C atoms amount to 0.403(0.391) and -0.243(-0.279), respectively, whereas, the ground state values determined with the UB3LYP/6-31G* method are, in the same order, 0.2492 and -0.2722. The good agreement between the UCIS and RCIS charges further confirms the zwitterionic nature of the first excited state, which can well be described by a spin-adapted singly excited determinant. Furthermore, the smaller excitation energy (E_{e1_g}) of TDPL (0.807 eV) with respect to IDPL (1.616 eV) – calculated by the time-dependent (TD)-UB3LYP/6-31G* method, which reproduces well the experimental singlet excitation energies of the phenalenyl diradical compounds [23] – is associated with an enhancement of the negative contributions to γ due to the type II process g-e1-g-e1-g. This larger negative contribution can therefore explain the smaller γ value of TDPL (1375 x 10^3 a.u.) with respect to IDPL (2284 x 10^3 a.u.) though both present similar diradical characters.

In summary, the *ortho*-quinoid structure present in *as*-IDPL turns out to increase the diradical character in comparison with the *para*-quinoid structure, and subsequently to decrease the γ value with respect to the IDPL compound displaying only *para*-quinoid structures. On the other hand, the replacement of the central benzene ring in *as*-IDPL by a thiophene ring in *as*-IDPL decreases the diradical character, which leads to an increase of the γ value for TDPL as compared to *as*-IDPL. This variation has been shown to originate from the decrease of aromaticity of the central ring, i.e., the increase in the contribution of *p*-thienoquinoid structure in the central region. Then, the γ value of IDPL is larger than in TDPL while their diradical characters are very similar. This has been explained by an enhanced negative type II contribution to γ in TDPL, which, in turn, is associated with a low-lying zwitterionic excited state originating form the hypervalency of the S atom in TDPL. These results suggest the possibility of controlling the relationships between aromaticity, diradical character, and third-order nonlinear optical properties by structural modification of phenalenyl compounds.

6. Summary

We have theoretically predicted the diradical character dependence of γ for singlet diradical systems and investigated the γ values of several real diradical systems, PDPL, IDPL, BPLE, *as*-IDPL and TDPL, involving diphenalenyl moieties. In agreement with our prediction, we have observed the significant enhancement of γ in the intermediate diradical character region as compared to the closed-shell and pure diradical systems with similar π-conjugation lengths. The spin state dependence of γ is also shown to be remarkable for the systems with intermediate diradical characters: the γ value of triplet state tends to be about three times as large as that of the singlet state. We have further found the structural dependence of diradcial character and γ of IDPL, *as*-IDPL and TDPL. This is turned out to be closely related to the degree of contribution of *para*/*ortho*-quinoid structures and of aromaticity of the central ring as well as of hypervalency of sulfur atom in TDPL. In conclusion, it is expected that the third-order NLO properties of diphenalenyl diradcial systems can be controlled by tuning the spin multiplicity and diradical character by chemical modification. Furthermore, super- and supra-molecular systems composed of these phenalenyl radical units [5] are also expected to be candidates for not only nonlinear optical property but also attractive magnetic property and electric conductivity.

Acknowledgments

This work was supported by Grant-in-Aid for Scientific Research (No. 18350007) from Japan Society for the Promotion of Science (JSPS).

References

[1] M. Nakano, T. Nitta, K. Yamaguchi, B. Champagne and E. Botek, *J. Phys. Chem. A* **108**, 4105-4111 (2004).

[2] M. Nakano, R. Kishi, T. Nitta, T. Kubo, K. Nakasuji, K. Kamada, K. Ohta, B. Champagne, E. Botek, and K. Yamaguchi, *J. Phys. Chem. A*, **109**, 5, 885 - 891 (2005).

[3] B. Champagne, E. Botek, M. Nakano, T. Nitta, and K. Yamaguchi, *J. Chem. Phys.* **122**, 114315-1-12 (2005).

[4] M. Nakano, T. Kubo, K. Kamada, K. Ohta, R. Kishi, S. Ohta, N. Nakagawa, H. Takahashi, S. Furukawa, Y. Morita, K. Nakasuji, *Chem. Phys. Lett.* **418**, 142-147 (2006).

[5] S. Ohta, M. Nakano, T. Kubo, K. Kamada, K. Ohta, R. Kishi, N. Nakagawa, B. Champagne, E. Botek, S. Umezaki, A. Takebe, H. Takahashi, S. Furukawa, Y. Morita, K. Nakasuji, K. Yamaguchi, *Chem. Phys. Lett.* **420**, 432-437 (2006).

[6] M. Nakano, R. Kishi, N. Nakagawa, S. Ohta, H. Takahashi, S. Furukawa, K. Kamada, K. Ohta, B. Champagne, E. Botek, S. Yamada and K. Yamaguchi, *J. Phys. Chem. A* **110**, 4238-4243 (2006).

[7] M. Nakano, R. Kishi, S. Ohta, A. Takebe, H. Takahashi, S. Furukawa, T. Kubo, Y. Morita, K. Nakasuji, and K. Yamaguchi, K. Kamada, K. Ohta, B. Champagne and Edith Botek, *J. Chem. Phys.* in press.

[8] M. Nakano, H. Nagao, K. Yamaguchi, *Phys. Rev. A* **55**, 1503 (1997).

[9] K. Nakasuji and T. Kubo, *Bull. Chem. Soc. Jpn.* **77**, 1791 (2004).

[10] K. Nakasuji, K. Yoshida and I. Murata, *Chem. Lett.* 969 (1982).

[11] M. Nakano, N. Nakagawa, S. Ohta, R. Kishi, T. Kubo, K. Kamada, K. Ohta, B. Champagne, E. Botek, H. Takahashi, S. Furukawa, Y. Morita, K. Nakasuji and K. Yamaguchi, submitted.

[12] T. Kubo, K. Yamamoto, K. Nakasuji and T. Takui, *Tetrahedron Lett.* **42**, 7997 (2001).

[13] T. Kubo, M. Sakamoto and K. Nakasuji, *Polyhedron* **24**, 2522 (2005).

[14] M. Nakano and K. Yamaguchi, *Chem. Phys. Lett.* **206**, 285 (1993).

[15] M.Nakano, I. Shigemoto, S. Yamada and K. Yamaguchi, *J. Chem. Phys.* **103**, 4175-4191 (1995).

[16] L. Salem and C. Rowland, *Angew. Chem., Int. Ed.* **11**, 92 (1972).

[17] K. Yamaguchi, "*Self-Consistent Field Theory and Applications*", R. Carbo and M. Klobukowski, Eds. (Elsevier: Amsterdam, 1990) 727.

[18] S. Yamanaka, M. Okumura, M. Nakano, K. Yamaguchi, *J. Mol. Structure* **310**, 205 (1994).

[19] H.D. Cohen and C.C.J. Roothaan, *J. Chem. Phys.* **43**, S34 (1965).

[20] M. S. Gordon, J. H. Jensen, S. Koseki, N. Matsunaga, K. A. Nguyen, S. J. Su, T. L. Windus, M. Dupuis and J. A. Montgomery, *J.Comput.Chem.*, 1993, 14, 1347.

[21] M. S. Gordon, M. W. Schmidt, in "*Theory and Applications of Computational Chemistry, the first forty years*", ed. C. E. Dykstra, G. Frenking, K. S. Kim, G. E. Scuseria (Elsevier, Amsterdam, 2005).

[22] (a) M.J. Frisch, et al. GAUSSIAN 98, Revision A11, Gaussian Inc., Pittsburgh, PA, 1998. (b) M. J. Frisch et al., GAUSSIAN 03, Revision B.04, Gaussian Inc., Pittsburgh, PA, 2003.

[23] T. Kubo, A. Shimizu, M. Uruichi, K. Yakushi, M. Nakano, D. Shiomi, K. Sato, T. Takui, Y. Morita and K. Nakasuji, in preparation.

Brill Academic Publishers
P.O. Box 9000, 2300 PA Leiden,
The Netherlands

Lecture Series on Computer and Computational Sciences
Volume 6, 2006, pp. 241-244

QED and the Valence Shell

P. Pyykkö[1]

Department of Chemistry,
University of Helsinki,
P O B 55 (A. I. Virtasen aukio 1),
FI-00014 Helsinki, Finland

Received 20 July, 2006; accepted 25 July, 2006

Abstract: The practical ways of including the valence-shell QED terms of lowest-order vacuum-polarization and self-energy effects in quantum chemical calculations on atoms, molecules or solids are discussed.

Keywords: Vacuum polarization, self-energy, Uehling potential.

Mathematics Subject Classification: PACS Numbers 31.10.+z, 31.30.Jv.

PACS: PACS Numbers 31.10.+z, 31.30.Jv.

1 Introduction

The Dirac-Coulomb-Breit (DCB) Hamiltonian is an excellent description of normal matter, if realistic finite-nucleus models are used. The next correction to it comes from quantum electrodynamics (QED), whose two leading terms, of opposite sign, are the self-energy (SE) and vacuum polarization (VP). These QED terms are clearly visible in comparison with experiment for both one-electron atoms (the Lamb shift), highly charged few-electron atoms, innermost shells of neutral atoms and for valence shells of *e.g.* Zn-like ions (Z = 60 - 90 range). They also are visible in the lightest neutral systems, such as H_2 or Li, which can be treated very accurately. For the valence electrons of heavier neutral, or nearly neutral systems, the precision is not yet quite high enough to see the QED terms. The bottleneck is in the handling of electron correlation, and sometimes the basis. The QED terms would be of interest for several reasons: 1) To the extent that they are small, this smallness justifies the current DCB-level calculations. 2) If the QED effects will be seen in a head-on comparison between ab initio theory and experiments, they will be interesting in their own right. After they are included, the rest should be quantum chemical technology. 3) In that sense these terms may have been 'the last train from physics to chemistry' at the level of fundamental Hamiltonians. Further physical terms, notably parity non-conservation (PNC), do exist and can be separately observed, but are 10^{-10} times smaller and hence insignificant for normal chemical energetics. 4) Finally, for magnetic dipole ($M1$) hyperfine effects of heavy s elements, such as Hg, the SE correction is of the order of -1.5 % per nucleus, which borders on other effects on nuclear spin-spin coupling, such as solvation.
A large number of careful studies of the QED effects exist for one- or few-electron atoms. Examples will be shown. The fundamental way of treating these terms is to first create a complete basis of one-particle states and to then evaluate the relevant Feynman diagrams. This can also be done

[1]E-mail: Pekka.Pyykko@helsinki.fi

for the SE of valence electrons in neutral, or nearly neutral atoms, to the extent that an effective one-electron potential is available [1]. For the lowest-order VP term, a local potential by Uehling and Serber in 1935 is available. Various approximate methods for the SE term, or the total Lamb shift, have also been tried [2].
To give an idea of the orders-of-magnitude, for heavy elements ($Z > 50$), the QED effects cancel about -1% of the Dirac-level relativistic effects [1], for the energy shifts of ns valence electrons. For the ionization potential of the gold atom this means about -0.3% of the total value. Here it should be mentioned that the very first estimates for Cs and Fr were published by Dzuba et al.[3] already in 1983.

The VP part is to lowest order a local potential, a property of space, same for all elements. Attempts also exist [4,5] to simulate the SE effects with an effective local potential for the energy and the $M1$ hyperfine integrals of s states. Note that both the relativistic and the QED effects are *smaller* for the $1s$ states than for the higher ns states. The results for the $M1$ hyperfine effects of the alkali metals Li-Fr can be compared against Sapirstein and Cheng [6].
For properties close to a threshold, the percental QED effects can become very large. An example is the electron affinity of eka-radon, E118, where a decrease of -9% is obtained [7]. The final calculated EA is -0.064(2) eV.

For the lighter elements and their compounds, a practical way to evaluate the leading QED terms could be to use scaled nuclear and electronic Darwin terms. The necessary scaling factors and a test application on ro-vibrational levels of H_2O were given in ref. [8]. Again, a slight improvement of other parts should make these contributions visible.

2 Theory and Results

The Dirac-Coulomb-Breit Hamiltonian is

$$H = \sum_i h_i + \sum_{i<j} h_{ij}. \tag{1}$$

The one-particle Hamiltonian is

$$h_{\mathrm{D}} = c\alpha \cdot \mathbf{p} + \beta c^2 + V_n, \quad \mathbf{p} = -i\nabla, \tag{2}$$

The two-particle Hamiltonian is

$$h_{ij} = 1/r_{ij}.h_{\mathrm{B}} = -\frac{1}{2r_{ij}}[\alpha_i \cdot \alpha_j \ + (\alpha_i \cdot \mathbf{r}_{ij})(\alpha_j \cdot \mathbf{r}_{ij})/r_{ij}^2]. \tag{3}$$

In correlated calculations, electron-like projection operators, P, should be added:

$$h_{ij}^{eff} = P h_{ij} P. \tag{4}$$

In the Coulomb gauge this Hamiltonian will form a legitimate starting point, valid to order α^2. The first correction terms, of order α^3, were obtained by Araki [9] and Sucher [10] for helium. The higher-order terms in this *Furry picture* can also be systematically calculated [11]. The lowest-order vacuum-polarization term can be calculated using the Uehling potential

$$V_n^{eff}(r) = -\frac{Z}{r}(1 + S(r)) = V_n + V_{Ue}, \tag{5}$$

$$S(r) = \frac{2\alpha}{3\pi} \int_1^\infty \exp(-2r\chi/\alpha)(1 + \frac{1}{\chi^2})\frac{\sqrt{\chi^2 - 1}}{\chi^2} d\chi. \tag{6}$$

For the self-energy term, we have used a number of possible approaches. In the QED sense the most strict is to actually calculate it. This can be done in the Furry picture, if the valence electron is described by an effective one-electron potential[1]. That potential can also simulate the electron correlation effects. In other words, one has strictly correct lowest-order QED, but rather approximate quantum chemistry. As mentioned, for lighter elements a renormalization of the Darwin term in the Pauli approximation could be used to model SE. This method was introduced and tested on the ro-vibrational levels of the water molecule[8]. It was concluded that a slight improvement of the quantum chemical accuracy would make the QED terms visible in comparison with experiment. An effective Gaussian SE potential was suggested by Pyykkö and Zhao[5]. Its two parameters for each element were fitted to the reference data for the 2s energy shift and M1 hyperfine shift. The other approximate methods for SE were reviewed there. A recent attempt to formulate a SE potential was reported by Flambaum and Ginges[12].

Table 1: Calculated SE-contributions or total Lamb-shifts for the valence *ns* electrons of alkali-metal atoms.

Atom	E_{SE}^{ns}/meV							
	'Correlated'			Dirac-Fock				
	[3]	$[1]^{(3)}$	[13]	$[2]^{a}$	$[1]^{(1)}$	[5]	[14]	[12]
K		$0.87(0.81^{t})$	0.75	0.51	0.49	0.60		
Rb		$2.22(1.99^{t})$	1.99	1.33	1.23	1.48	$1.43\ (1.28^{t})$	
Cs	3.5^{t}	$3.83(3.30^{t})$	$3.35(1.6\text{-}2.9^{t})$	2.22	2.15	2.36	$2.59\ (2.23^{t})$	1.92^{t}
Fr	9.5^{t}	$9.83(7.58^{t})$	$8.44(3.0\text{-}6.5^{t})$	6.18	6.03	6.46	$4.68\ (3.57^{t})$	
119				27.7	27.4	27.3	$16.33\ (10.32^{t})$	

^{a} DF $< V_{\mathrm{Ue}} > \times$ H-like SE/VP. $[1]^{(1)}$: Local V_{DF}. $[1]^{(3)}$: IP fit. ^{t} $E_{\mathrm{Lamb}}^{\mathrm{tot}}$.

3 Conclusion

It is expected that only a slight improvement of the ab initio accuracy will make valence-shell QED effects visible in a head-on comparison of theoretical and experimental ionization potentials and other properties of neutral or nearly neutral atoms and molecules.

Acknowledgment

This work belongs to the Finnish Centre of Excellence in Computational Molecular Science, CMS (2006-11). I thank Ingvar Lindgren for helpful correspondence.

References

[1] L. Labzowsky, I. Goidenko, M. Tokman, P. Pyykkö, *Phys. Rev. A*, **59** (1999) 2707.

[2] P. Pyykkö, M. Tokman, L.N. Labzowsky, *Phys. Rev. A*, **57**, (1998) R689.

[3] V. A. Dzuba, V. V. Flambaum, O. P. Sushkov, *Phys. Lett. A*, **95** (1983) 230; Novosibirsk Preprint 82-127.

[4] B. Fricke, *Lett. Nuovo Cimento*, **2** (1971) 293.

[5] P. Pyykkö, L.-B. Zhao, *J. Phys. B: At. Mol. Opt. Phys.*, **36** (2003) 1469.

[6] J. Sapirstein, K. T. Cheng, *Phys. Rev. A*, **67** (2003) 022512.

[7] I. Goidenko, L. Labzowsky, E. Eliav, U. Kaldor, P. Pyykkö, *Phys. Rev. A*, **67** (2003) 020102.

[8] P. Pyykkö, K. G. Dyall, A. G. Császár, G. Tarczay, O. L. Polyansky, J. Tennyson, *Phys. Rev. A*, **63** (2001) 024502.

[9] H. Araki, *Progr. Theor. Phys.* **17** (1957) 619.

[10] J. Sucher, *Phys. Rev.* **109** (1958) 1010.

[11] I. Lindgren, S. Salomonson, B. Åsén, *Phys. Rep.* **389** (2004) 161.

[12] V. V. Flambaum, J. S. M. Ginges, *Phys. Rev. A* **72** (2005) 052115.

[13] J. Sapirstein, K. T. Cheng, *Phys. Rev. A* **66** (2002) 042501.

[14] E. Eliav, M. J. Vilkas, Y. Ishikawa, U. Kaldor, *Chem. Phys.* **311** (2005) 163.

Brill Academic Publishers
P.O. Box 9000, 2300 PA Leiden
The Netherlands

Lceture Series on Computer
and Computational Sciences
Volume 6, 2006, pp. 245-264

Lesions in DNA Subunits: The Nucleic Acid Bases

Partha P. Bera* and Henry F. Schaefer III*
Center for Computational Chemistry, University of Georgia,
Athens, Georgia, U. S. A. 30602.

Received 1 August, 2006; accepted 3 August, 2006

Abstract: The nucleic acid bases adenine, thymine, uracil, guanine, and cytosine are among the most important building blocks of life. Molecular modifications, or lesions, in the nucleic acid bases (NABs) can give rise to profound physiological changes. The simplest such set of lesions are considered in this research: radical anions, deprotonated NABs, hydrogen abstracted radicals, hydrogen appended radicals, and hydride appended anions. A whole range of isomeric structures is possible when hydrogen atoms are added to or subtracted from the NABs. This research provides a foundation for the theoretical examination of base pairs, nucleoside pairs, nucleotide pairs, and larger DNA subunits.

Keywords: DNA, lesion, lesions, NAB, nucleic acids, purine, pyrimidine, adenine, guanine, cytosine, thymine, base pair, bases, strand breaks, radiation damage, electron affinities, energies, hydrogen abstraction.

* Corresponding author: partha@ccc.uga.edu
* Corresponding author: hfs@uga.edu

1. Introduction

The flow of genetic information is preserved by the basic building blocks of the nucleic acids DNA and RNA. These macro-molecules contain and transfer the genetic information and misinformation from generation to generation by means of chemical reactions. The stacked and coiled natures of DNA and RNA facilitate charge transfer through their backbones. The core constituents of double helical DNA and RNA are the purine and pyrimidine bases. The information stored in these highly precise structural patterns is enormous, and exceedingly dependent on the structural integrity of the sequence. Behind these complex but discernible structural patterns are the basic forces of nature. These forces encompass atom-atom attractions, dispersion effects, hydrogen bonding considerations, and the base pair and stacking interactions. The structural integrity is lost by the breaking of bonds and by fragmentation processes. Understanding the underlying chemistry and basic interactions is absolutely vital to science.

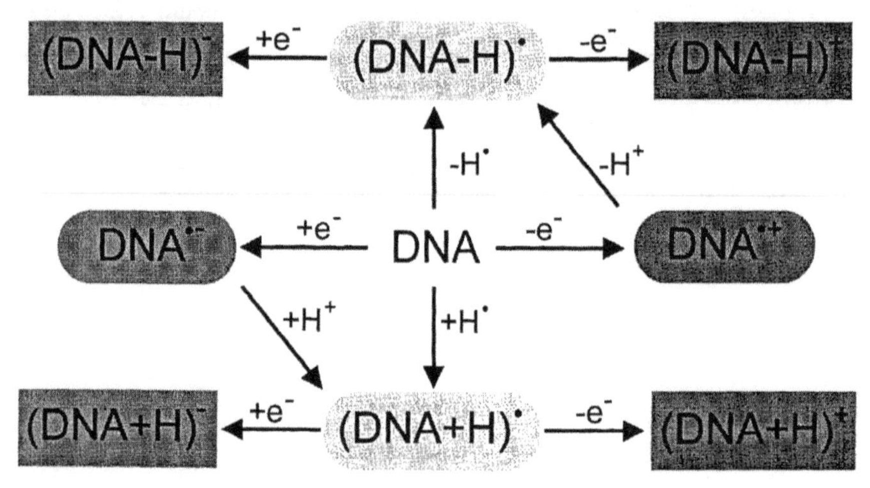

Scheme 1.

The present paper is aimed at describing research done using *ab initio* quantum chemistry, specifically density functional methods, on the bases and their fragments in completely isolated environments.[1-9]. Investigating the different interactions contributes to the chemical physics of elementary bio-molecules in the gas phase. The philosophy of our research is that the deepest and most comprehensive understanding of biochemistry will result from the study of: i) the isolated bases, base pairs and damaged fragments (cumulatively – molecules); ii) isolated molecules in micro-solvated environments (e.g., discrete numbers of water molecules); and iii) molecules in the fully solvated environment. In this paper we have summarized the understandings obtained from theory for the isolated bases and the fragments resulting from the abstraction and addition of hydrogen atom and hydride ions to the purine and pyrimidine bases of DNA.

The reaction channels leading to the production of the radicals and anions discussed in this work are represented in the cartoon above (Scheme 1).[2] In scheme 2 below we show the conceptual reactions of adenine in a pictorial manner to explain the formation of several adenine anions.[2]

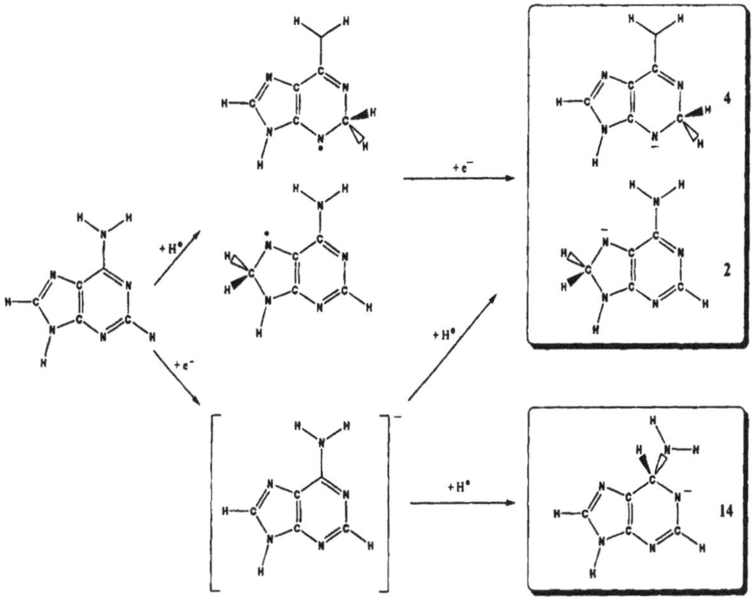

Scheme 2.

The causes of the production of radicals and anions in the DNA subunits, bases and base pairs are manifold; chemical damage, direct radiation damage and indirect damage [10-14]. Direct damage to bio-molecules like purines and pyrimidines are caused by radiation coming from electromagnetic sources including the sun. Direct bond breaking and ionization of the inner and outer shell electrons are the immediate fallout of this direct radiation damage. The fact that secondary electrons having energies as low as 3 eV, generated by ionizing radiation, plays an important part in the mechanism of indirect radiation damage by electron attachment (see Scheme 2) has been discussed at length recently and opened up new dimensions to the exploration of DNA damage[15]. Low energy electron capture and the understanding of subsequent mechanisms for the formation of the fragmented molecules represent a growing field of research[10,14] due to the alleged role in the formation of cancer cells. Ionizing radiation is one of most important causes of the primary and secondary damage to DNA [16] Hydrogen ions and radicals produced by chemical reactions inside the cell also contribute to the chemical damage done on the molecular level.[15,17-22]. Consequently the fragmentation channels of DNA damage starting from the molecular level have been the focus of intense investigation in recent years, both experimentally and theoretically.[20,21,23-26] A variety of damage occurring to nucleic acids including oxidation, reduction, abstraction and addition of electrons, protons or hydrogen atoms are being studied, and pathways of fragment (lesion) formation are being explored by many researchers [17,27-32]. There is ample experimental evidence pointing toward the formation of the radicals and anions due to irradiation of the bases.[17,22,28,30,33-35].

From an experimental point of view the group including Abdul-Carime has done extensive explorations of dissociative electron attachment (DEA) to the bases and base pairs [15,19,36]. Their research has pointed out various dissociation channels including N-H and C-H bond breaking, following electron capture by the bases. The work of the research groups of Collins [12], Sanche [12,37], Close [28,38] and Illenberger [15,19,20,34] have experimentally explored fragmentation channels, by creating and identifying the intermediates of radiation damage. Other researchers have experimentally established the energetic requirements for lesion formation in DNA subunits and subsequently for single and double strand breaks [10-12,33,38,39].

To contribute to the physical understanding of mechanisms of DNA damage, the theoretical community has also extensively studied the possible intermediates in the fragmentation channels.

Theoretical studies of the nucleic acids and the fragments generated from them provide valuable insights into the energetic (e.g., ionization potentials, electron affinities, dissociation energies, gas phase acidities, etc.) and structural (e.g,. dihedral angles and interatomic distances) properties of them. The research efforts of our group [1,2,6-9,40,41], Sevilla [32,42] Wetmore [31], Turecek [27], Chen [33,43,44], Hobza [45,46] and Lesczynski [38,46-48] among others, are particularly relevant in this respect.

2. Ab initio and density functional methods for large molecules

Studies of the physicochemical properties of bio-molecules have been attempted both experimentally and theoretically over the years. The experiments done on these systems can be difficult. Computational approaches towards predicting the fundamental properties are thus becoming ever more popular. Traditionally, wave function based quantum mechanical methods, especially 'post Hartree-Fock' correlated methods, have proved themselves to be possible alternatives to good quality experimental results for small molecules. Although modern computers are getting faster yearly and wave function based *ab initio* quantum chemistry methods are able to solve diverse chemical problems, it is still not possible to compute the physical properties of very large molecules by these methods on a regular basis. Density Functional Theory (DFT) methods prove to be very important in this context. DFT methods attempt to strike a balance between the theoretical reliability and computational requirements. One important aspect of both the correlated ab initio and DFT based methods is that they involve at most a minimal number of adjustable parameters, unlike typical semi-empirical and empirical procedures. At the same time DFT functionals give reliable structural parameters and other chemical properties of concern [40,48]. Absolute and mean errors of the theoretical parameters will be discussed in the appropriate sections below.

The DFT functionals used in the research discussed herein are well tested and broadly proven to approach chemically relevant accuracy [40]. The three density functionals discussed here are the pure functional BLYP [49,50] the hybrid density functional B3LYP [50,51] and the pure DFT functional BP86 [49,52]. Interested readers are directed to the detailed report by Papas [53] on the accuracy of the DFT functionals using different grids and various computational packages. The double zeta plus polarization and diffuse function basis set, DZP++ [54] suggested by Lee and Schaefer has proven to provide excellent geometries and some energetic properties within a few kcal/mol of the experimental values, as discussed in the review article of by Rienstra-Kiracofe et. Al [40].

All the results shown here were carried out using the Gaussian 94 suite of density functional programs.[55] Default integration grids were used for all energy and analytic gradient evaluations.

3. Numbering scheme

Throughout this review we have followed the IUPAC recommended numbering scheme for all the bases as shown below

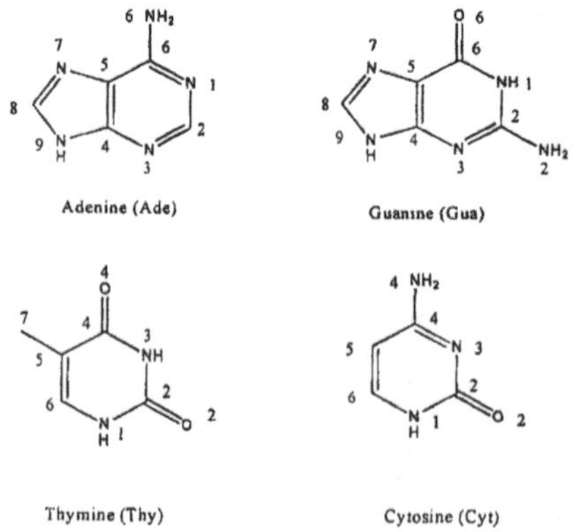

Adenine (Ade)

Guanine (Gua)

Thymine (Thy)

Cytosine (Cyt)

Figure 1.

The radicals, cations and anions arising from the above are represented using the number of the particular atom (in parentheses) from which an electron of H atom is abstracted or added. For example, the radical generated by the homolytic cleavage of the N9-H bond of Guanine (Gua) is represented as G(N9-H)˙. The closed shell anion created by adding (pedagogically) an electron to N9 of the open shell radical G(N9-H)˙ is labeled $G(N9\text{-}H)^-$, and the cation qualitatively generated by removing an electron from the same radical is represented as $G(N9\text{-}H)^+$.

All the important tautomers of the gas phase purine and pyrimidine bases are considered with respect to the addition of a hydrogen atom or a proton. The amino-oxo Watson-Crick and the amino-hydroxy tautomers of cytosine, and the subsequent hydrogen added species are shown in Figure 2. The three tautomers of cytosine considered in this work are named as follows;

Watson-Crick amino-oxo	C
Amino-hydroxy-trans	*t-amino-hyd*C
Amino-hydroxy-cis	*c-amino-hyd*C

The other purine and pyrimidine bases are named in similar fashion to maintain uniformity and simplicity.

4. Results and discussion

A. Structural

Important gas phase isolated structural parameters were predicted using the density functionals mentioned above.

Bond distances

The accuracy of predicted structural parameters from the density functional methods used to study the bio-molecules is high. The average error in the predicted bond lengths is on the order of hundredth of an angstrom. The theoretical bond distances are often comparable to those from higher accuracy MP2 or coupled cluster (CC) methods. All the structures discussed in this review are fully optimized and represent true minima on the potential energy hypersurfaces for the species of concern. The structures discussed here fall on the ground electronic state potential energy surfaces of the particular molecules, radicals or anions considered.

Figures 5 and 6 in the paper by Profeta[6] describe the bond angle and bond distance differences between neutral thymine and the analogous radical and/or anionic species. In the gas phase the bond lengths vary by as much as 0.04 angstroms for the radical T(N3-H)˙. Bond angle changes relative to neutral thymine molecule are also profound for this radical. This radical is the only one of all the thymine radicals that departs from the puckered neutral (C_1 symmetry) radical to a planar anion geometry.

Hydrogen abstraction from the isolated guanine molecule creates more structural deformations in the five-membered ring of the molecule than in its six-member ring. This indicates the substantial amount of electron delocalization in the five-member part of this purine. Nitrogen centered radicals show greater stability compared to the carbon centered structures, as explained by Luo [5]. The structural deformations in the lowest energy radical G(N2-H2)˙ (**6**) and the lowest energy anion, $G(N9\text{-}H)^-$ (**9**) are profound in the vicinity of the lesion. The largest changes in bond distances occur for the N-C bond and are more than 0.1 Å. A detailed study of all the radicals and anions is provided by Luo [5].

The purine base adenine shows similar trends to the guanine molecule. Hydrogen abstraction from the exocyclic nitrogen leads to two rotational isomers A(N6-H1)˙ and A(N6-H2)˙. These two isomers are structurally very similar, and are within one kcal/mol of each other, unlike guanine, where the analogous energy separation is slightly higher. Electron addition to these radicals requires considerable electron redistribution, and hence produces significant change in the bond lengths. Comparison of the bond distances in some of the radicals shows the C_4-N_9 and N_9-C_8 bond distances to change by almost 0.1 Å. Some of the planar radicals depart from planarity when an electron is added to generate the anion. A detailed structural analysis is also given in Table 1 in the paper of Evangelista [1].

Figure 2. Radicals Generated by the Addition of a Hydrogen Atom to Cytosine.

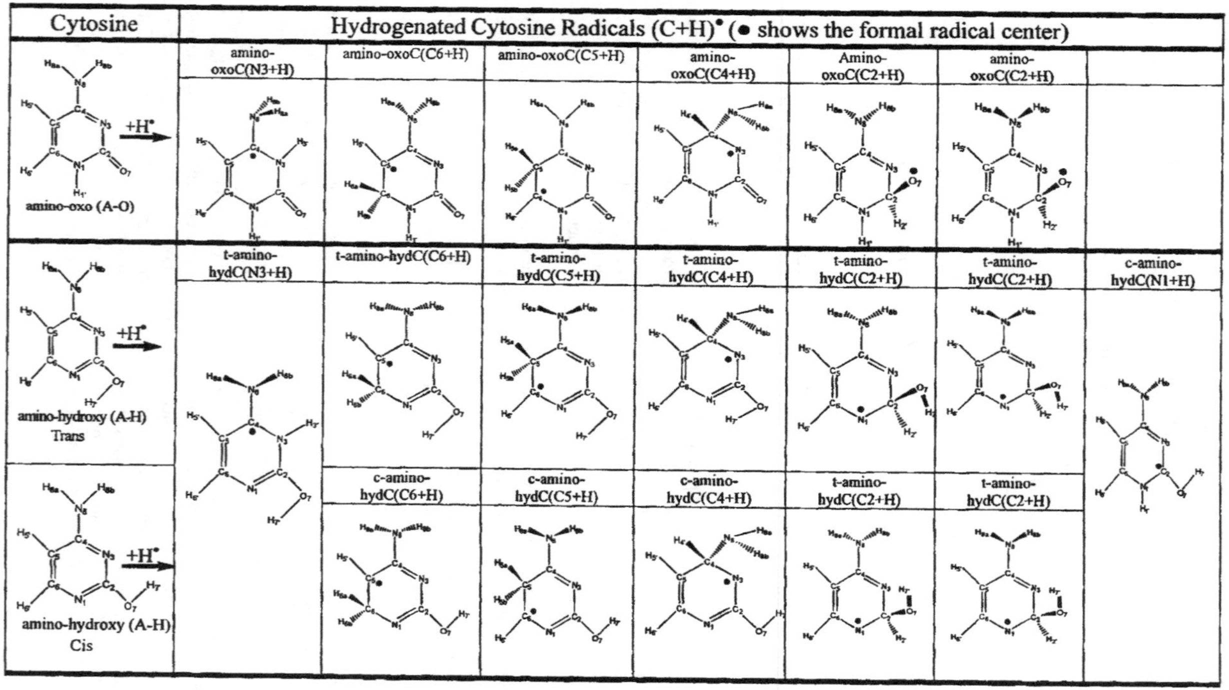

Cytosine	Hydrogenated Cytosine Radicals $(C+H)^{\bullet}$ (● shows the formal radical center)						
amino-oxo (A-O) $+H^{\bullet}$	amino-oxoC(N3+H)	amino-oxoC(C6+H)	amino-oxoC(C5+H)	amino-oxoC(C4+H)	Amino-oxoC(C2+H)	amino-oxoC(C2+H)	
amino-hydroxy (A-H) Trans $+H^{\bullet}$	t-amino-hydC(N3+H)	t-amino-hydC(C6+H)	t-amino-hydC(C5+H)	t-amino-hydC(C4+H)	t-amino-hydC(C2+H)	t-amino-hydC(C2+H)	c-amino-hydC(N1+H)
amino-hydroxy (A-H) Cis $+H^{\bullet}$		c-amino-hydC(C6+H)	c-amino-hydC(C5+H)	c-amino-hydC(C4+H)	t-amino-hydC(C2+H)	t-amino-hydC(C2+H)	

For cytosine also both radical formation and subsequent electron addition lead to structural deformations. While the radicals and anions retain their planar structures, hydrogen abstraction leads to a redistribution of electrons and resultant bond length changes, in some cases, by more than a tenth of an angstrom. Luo[4] discussed the detailed structural changes (see Figure 3) due to radical formation from cytosine. The rigid cyclic structures of the bases make the potential energy surface rather steep. The existence of a small number of conformational isomers for these molecules enables the separation of the radicals and anions by their geometrical positions.

Structural aspects of H^+, $H^\cdot$, and H^- addition to the DNA bases are addressed in the research by Zhang [8,9] Evangelista [2] and Ipek [3]. For the sake of simplicity we will restrict ourselves to the discussion of the species originating from the hydrogen addition to the Watson-Crick tautomers. For a more thorough discussion, interested readers are directed to the original papers.

Hydrogen addition to the N_7, C_8 and C_2 positions leads to the most stable guanine radical species, as explained by Zhang [8]. Hydrogen addition inevitably comes at the expense of a double bond. The single and double bonds in the vicinity shorten or elongate quite significantly. On going from the radical to the anion, again the extra electron causes the bonds to lengthen so as to nullify the effect of radical formation. Figure 2 shows the effects of H^+, $H^\cdot$ and H^- addition to the N7 nitrogen of the isolated guanine molecule.

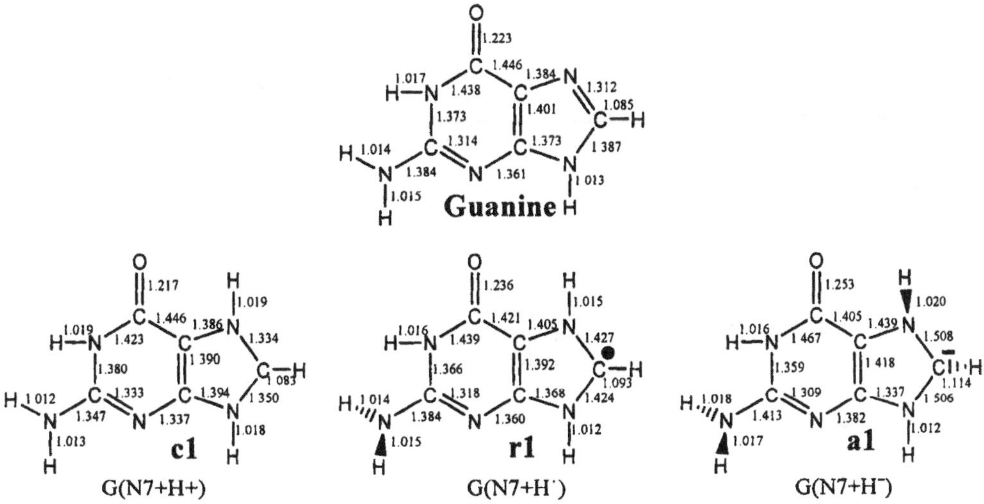

Guanine

c1 G(N7+H+)

r1 G(N7+H˙)

a1 G(N7+H⁻)

Figure 3.

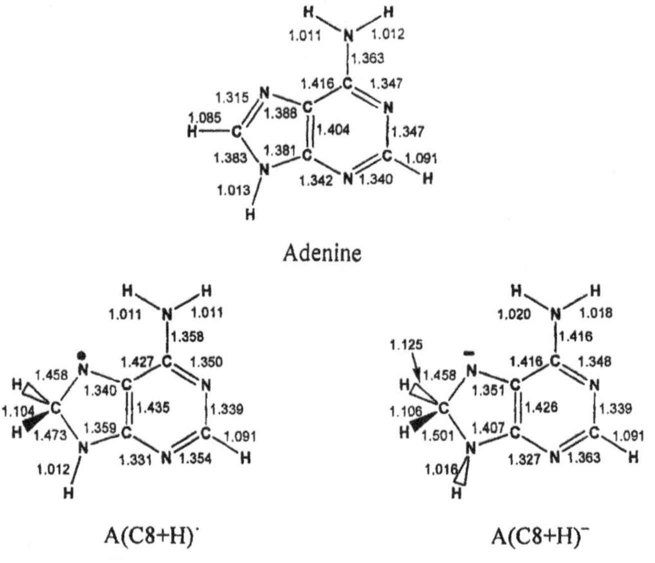

Adenine

A(C8+H)$^{\cdot}$

A(C8+H)$^{-}$

Figure 4.

Figure 4 shows the neutral adenine molecule and the hydrogen added adenine molecules. The lowest energy radical and anions are A(C8+H)$^{\cdot}$ and A(C8+H)$^{-}$. The breaking of the double bond due to the addition of a hydrogen atom is clear from the elongation of the C-N bond. The other radicals and anions show analogous bond lengthening whenever a double bond is broken. The carbon atom to which the hydrogen atom is added displays a tetrahedral geometry, and the amino group of the adenine goes out of plane for some radicals and anions.

Hydrogen atom and hydride anion addition to the double bonds of the cytosine molecule bring in geometrical changes by transforming the appropriate double bond to a single bond distance and shrinking the adjacent bonds. The lowest energy radical (shown in Figure 5) accepts an electron to become an anion (equivalent to addition of a hydride anion to cytosine) with subsequent changes in the bond lengths. Some of these radicals and anions will still be able to associate with the guanine moiety in a G-C pair. A hydrogen atom added to the C5 or C6 positions of cytosine will still be able to pair up with guanine. Of course, the energetic ordering of these radicals (C(C+H)$^{\cdot}$-G) will be different from the energetic orders of the isolated C(C+H)$^{\cdot}$ radicals [1,4,5].

Bond angles

The bond angles and the dihedral angles are important geometrical parameters because they play a vital role in keeping the overall structural features of DNA intact. Large and/or abrupt changes (Figures 3-6, reference 10) in the angles and dihedrals can cause a fatal disruption of the chain. For example, the out of plane dihedral angles D(3,4,8, 8a) and D(3,4,8,8b) of the cytosine radical **10** and anion **11** (see Figure 2 of reference 4) change by 12 and 18 degrees, respectively, due to acceptance of an extra electron. Changes in bond angles and dihedral angles are common in other bases and radicals as well.

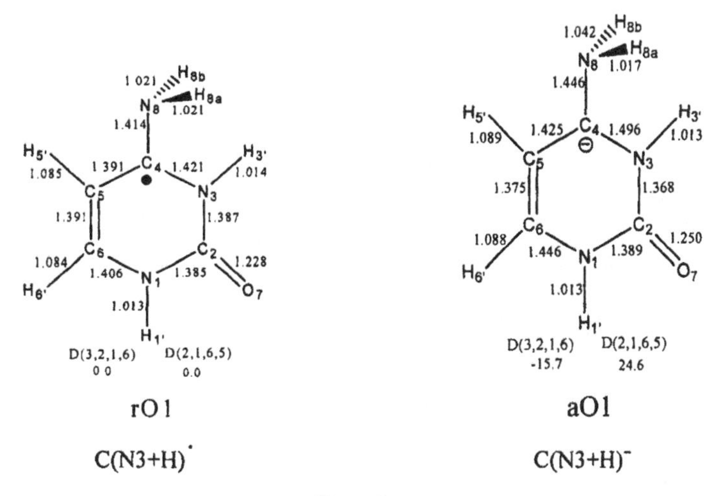

Figure 5.

B. Energetics

The expected accuracies of the absolute and relative energies associated with the different DFT functionals are discussed at length in the *Chemical Reviews* paper of Rienstra-Kiracofe. In this section we will discuss fundamental physical properties like the adiabatic electron affinities (AEA), vertical detachment energies (VDE), gas phase acidities (GPA) and association/dissociation energies of the bases and the radicals and anions treated by the various methods discussed in the introduction.

Relative stabilities/ energy ordering

The total energies of the isolated gas phase bases, radicals and anions were computed using the above discussed levels of theory. In Table 6, reported are the optimized total energies of the radicals and anions generated by the abstraction of an exo-cyclic hydrogen from the isolated bases at the B3LYP/DZP++ level of theory. The corresponding papers [1,4,5] report the total and zero point vibrationally corrected energies of the radicals and anions at other (BLYP and BP86) levels of theory.

All three functionals show the same energetic ordering of the radicals and the anions. For adenine, the energetic trend reported in Table 6 shows that the nitrogen centered radicals A(N9-H)$^{\cdot}$, A(N6-H1)$^{\cdot}$ and A(N6-H)$^{\cdot}$ are close in energy and more stable than the carbon centered radicals A(C1-H)$^{\cdot}$ and A(C2-H)$^{\cdot}$. The energetically most favorable radical is A(N9-H)$^{\cdot}$, created by removing a H atom from the N9 position of the isolated adenine molecule. The anions, like the radicals, prefer the nitrogen centers to the carbon centers for the qualitative location of the extra electron. The three nitrogen centered anions are lower in energy than the carbon centered structures with the A(N9-H)$^-$ being the lowest of the group. In adenosine, the N9 position is attached to the ribose by a carbon-nitrogen glycosidic bond; thus the formation of radical A(N9-H)$^{\cdot}$ or anion A(N9-H)$^-$ would break the glycosidic bond. The carbon centered radicals switch their energetic position in passing from the radical to the anion. The corresponding anion A(C1-H)$^-$ is much lower in energy than the other carbon centered anion, namely A(C2-H)$^-$.

For the guanine molecule, five distinct radicals can be generated by removing a hydrogen atom. It is to be remembered that three of these hydrogen atoms are H-bonded to the cytosine molecule in the GC base pair. Removing a hydrogen from the N2 position of guanine results in two rotational isomers G(N2-H2)$^{\cdot}$ and G(N2-H1)$^{\cdot}$. These radicals can be formed from the one-electron-oxidized guanine by reversible proton loss from the external nitrogen (N2). The radical G(N1-H1)$^{\cdot}$ is predicted to lie lowest in energy of the guanine radicals generated by H abstraction, while the radical at the N9 position is the second lowest in energy. Deprotonation at the N9 position produces the lowest energy anion. Due to the greater electronegativity of nitrogen, the nitrogen centered radicals and anions are lower in energy than the carbon centered structures. The carbon center radical G(C8-H)˙ and the corresponding anion are separated from the nearest nitrogen centered radicals and anions by 21 and 35 kcal/mol. The energy ordering, in going from the radicals to anions, remains the same except for the two lowest energy anions, which are switched (see Table 6). The other functionals predict same energy ordering for the radicals and anions as that predicted by B3LYP/DZP++, as shown in Table 6. Three nitrogen centered and two carbon centered anions are generated by removing the five exo-cyclic hydrogens of the cytosine molecule. In Table 6 are given the energies of the radicals and the anions produced by the abstraction of each such hydrogen at the B3LYP/DZP++ level of theory. The other density functionals (BLYP and BP86) predict the same energetic ordering of the radicals and the anions. The ZPVE corrected energies are discussed in the paper of Luo [4]. The radicals produced by hydrogen abstraction are close to each other in energy, within 15 kcal/mol. As seen for the isolated purines, the nitrogen centered radicals C(N1-H)˙, C(N2-H1)˙, and C(N8-H2)˙ are lower in energy than the carbon centered structures C(C5-H)˙ and C(C6-H)˙ for the pyrimidine also. Addition of an electron to the radicals does not alter the relative stabilities of the anions. The lowest energy cytosine radicals and anions are created by abstracting a H atom or a proton from N1 which is the sugar binding site in a cystidine residue. The rotational isomers, created by removing the hydrogen or proton from the amino group, are very close in energy to the lowest lying radical or anion. This fact indicates that the radical generated at the C8 position is very likely to be formed under suitable experimental conditions. The energetic ordering is confirmed by other levels of theory, with minor differences in the energy spacing.

Electron affinities

An electron affinity is best measured in a photodetachment experiment by knocking out the most weakly bound electron of the anion and generating the appropriate neutral species. The energy difference between the anion and the neutral gives the electron affinity (EA).

$$B + e = B^-$$

Adiabatic Electron Affinities: Each adiabatic electron affinities (AEA) is predicted from the energy difference between the neutral base or the hydrogen abstracted/added base and the corresponding anion, both at their theoretical equilibrium geometries. The anions are conceptually produced by adding an electron to the bases and re-optimizing the geometries of the species. The extra electron can go into a valence orbital or an anti-bonding orbital of the neutral species. However, the anion orbitals obtained this way are fully relaxed and hence are not identical to the corresponding orbitals of the neutral species. The predicted adiabatic electron affinities (AEA) from the DFT methods used here are typically within 0.15 eV of the experimental electron affinities. The accuracy of the different DFT functionals for the prediction of the electron affinities is described in detail in the review by Reinstra-Kiracofe [40]

The adiabatic electron affinities of the neutral nucleobases show some functional dependence, as discussed in the work of Wesolowski [26] (see Table 5 reference 33). The predicted AEA of three of the pyrimidine bases by the B3LYP, BLYP and BP86 are in good agreement with the available experimental values. The absolute values of the adiabatic electron affinities of the pyrimidine bases cytosine, thymine and uracil are greater than those for the purine bases. The final relative electron binding ability of the neutral nucleobases are summarized as follows.

$$U > T > C \sim G > A$$

The purines have five hydrogen atoms that can be abstracted due to reaction or interaction with radiation. In both purines, adenine and guanine, the radical created at the N9 position has the largest adiabatic electron affinity. The N9 position happens to be the site of the sugar-base connection. The absolute values of these dehydrogenated radical electron affinities are much higher than those of the neutral purines. The doublet radicals generated due to the loss of hydrogen are ravenous for another electron. The qualitative nitrogen sites are more favorable for the "last" electron than the carbon sites due to the greater electronegativity of nitrogen. This nitrogen bias of the "last" electron has been explained by Evangelista [1]. All the hydrogen abstracted radicals of adenine and guanine have high adiabatic electron affinities (AEA), as shown in Table 2.

Thymine has six and cytosine has five dissociable hydrogen atoms. The radicals generated by hydrogen abstraction from cytosine and thymine are discussed in the two work of Luo[4,5] and Profeta [6]. The radicals generated from either of the pyrimidines have high electron affinities (AEA), ranging from 1 eV to 3 eV. Like the purine radicals the cytosine radicals preferentially bind an extra electron on the N1 position, which is the sugar binding site in a normal strand. However, radical T(N3-H)˙ possesses the largest electron affinity of all the thymine radicals generated by hydrogen abstraction. The two carbonyl groups next to the N3 position bring in considerable electron delocalization to the anion, and lower its energy, yielding the largest AEA.

In this section we discuss the electron affinities of the species resulting from the addition to the single bases of a hydrogen atom (see cartoon, Scheme 1). In gas phase experimental studies of bio-molecules, both the Watson-Crick and non-Watson-Crick structures (tautomers) of the bases are important species, as pointed out by Bowen and coworkers [56]. All the species resulting from the addition of a hydrogen atom to the unsaturated part of the Watson-Crick amino-oxo and other tautomers of the purine and pyrimidine bases are considered. The following equations set out the electron capture processes.

$$B + H \rightarrow (B{+}H)^{\cdot} \;;\quad (B{+}H)^{\cdot} + e \rightarrow (B{+}H)^{-}$$

$$B + H^{-} \rightarrow (B{+}H)^{-}$$

The AEA values of the hydrogen-appended purine and pyrimidine bases are summarized in Table 3. The results show the familiar trend; BP86 with the highest AEA among all three functionals followed by B3LYP and BLYP. The electron affinities are much higher than those of the neutral bases but smaller compared to the hydrogen abstracted bases. Hydrogen atom addition to the double bonds was considered, leading to the formation of radicals. Most of the radicals readily accept an electron, except the radicals generated on the N1 nitrogen of adenine and cytosine. As a group, the AEAs span a wide range, from 0.01 eV to 3.0 eV. Some of the predicted AEAs are too small to be able to absolutely establish their signs, because the predicted values are smaller than the average errors of the methods employed. The AEAs for the radicals produced by adding H atom to one of the carbon atoms are higher than the AEA of the radicals produced by the addition of H to one of the N atoms.

The pure functional BLYP shows the smallest energy differences between the radical and the corresponding anion with a given basis set (DZP++), and hence predicts the smallest electron affinities. Among the three functionals used in these studies, BP86 predicts the highest electron affinities. The values for the most commonly used hybrid Hartree-Fock/density functional B3LYP method fall in between the pure (BLYP) and the hybrid (BP86) functionals.

Vertical Detachment Energies (VDE): Each vertical detachment energy is evaluated as the difference of energies between the anion and the neutral species, both at the optimized anion geometry. In a typical photodetachment experiment the vertical detachment energy of the anion is measured by knocking out one electron from the valence shell of the anion to produce the neutral species. The geometry may be assumed to be fixed in the short time period following electron removal. The anion geometry being different from that of the radical, the vertical dissociation energies are always upper bounds to the adiabatic electron affinities. This may be visualized from the cartoon (Figure 1; reference 68) presented in the review on the electron affinities of small molecules by Rienstra-Kiracofe.

From Tables 4 and 5 it may be seen that the VDEs of the anions are larger by a tenth of an eV to one eV than the AEAs of the corresponding radicals. In Table 5 are reported the VDEs of the radicals produced by the addition of H to the bases. With all the bases being closed shell species, hydrogen abstraction or addition and subsequent electron capture binds the electrons strongly. This is manifested in the relatively high VDE predictions at all levels of theory. The nitrogen centered anions are energetically more stable than the carbon centered anions. Hence, all the N-deprotonated anions have higher VDEs than the C-deprotonated structures (see Table 4). However, hydrogen (H) addition to a carbon center leaves an unpaired electron on the nitrogen (in the Lewis dot sense). So the anions for which the hydrogen is added to the carbon are more stable and have higher VDE values than the anions where the hydrogen atom is added to nitrogen.

Most of the anions, being valence bound, should be stable in the gas phase. The stabilizing effects of electron addition to either hydrogen abstracted or hydrogen added bases are manifested in their typically high electron affinities.

5. Conclusions

Results from several theoretical studies done on isolated gas phase purine and pyrimidine bases using carefully calibrated electronic structure methods are presented here. Three density functionals B3LYP, BLYP and BP86 were used in conjunction with the DZP++ basis set in all the computations. The results described above may be summarized as follows.

1. All the nitrogen centered radicals possess greater adiabatic electron affinities than the carbon centered radicals.
2. Both the dehydrogenated and hydrogenated radicals of the nucleic acid bases show typically high electron affinities. This indicates that the formation of these radicals will lead to electron transfer in or out of the bases, and eventually formation of the either the proton transferred or hydride transferred species in the gas phase.
3. Distortions of the geometries of the bases may jeopardize the integrity of the base pairing.

Acknowledgments

This research was funded by the U. S. National Science Foundation, Grant CHE-0451445. P. P. B. will like to thank Francesco Evangelista for his helpful advices and for providing some special graphics. P. P. B. will also like to thank Semran Ipek for providing access to her results prior to publication and David Zhang for some graphics. Dr. Luboš Horný and Dr. Chaitanya S. Wannere have helped a great deal by their constant encouragement throughout this work.

Table 1. Adiabatic Electron Affinities (AEA in eV) of the Four Neutral DNA Bases.

Level of theory	Adenine	Guanine	Thymine	Cytosine
B3LYP	-0.28	-0.10	0.20	0.03
BLYP	-0.19	-0.01	0.12	-0.01
BP86	-0.05	0.11	0.28	0.13
Experiment [a]	-0.045			-0.055
Experiment [b]	0.95±0.05	1.51±0.05	0.79±0.05	0.56±0.05
Experiment [c]			0.120±0.120	0.130±0.120

[a] Gas phase data[24]
[b] Scaled reduction potential data [44].
[c] Extrapolated AEAs from the photoelectron spectra of neucleobase·$(H_2O)_n$ clusters [25].

Table 2. Adiabatic Electron Affinities (in eV) of the H-Abstracted Radicals of the DNA Bases.

Structure	B3LYP	BLYP	BP86
Adenine	-0.28	-0.19	-0.05
A(N9-H)˙	3.26	3.05	3.28
A(N6-H1)˙	2.55	2.43	2.65
A(N6-H2)˙	2.58	2.46	2.68
A(C2-H)˙	0.90	0.87	0.98
A(N1-H)˙	2.38	2.27	2.42
Thymine	0.20	0.12	0.28
T(N1-H)˙	3.22	3.18	3.26
T(N3-H)˙	3.74	3.34	3.53
T(C6-H)˙	2.60	2.41	2.56
T(C7-H2)˙	1.04	1.04	1.22
T(C7-H2)˙	1.04	1.04	1.22
T(C7-H3)˙	1.04	1.04	1.22
Guanine	-0.10	-0.01	0.11
G(N1-H)˙	2.95	2.75	2.95
G(N2-H1)˙	2.75	2.57	2.77
G(N2-H2)˙	2.77	2.58	2.79
G(N9-H)˙	2.99	2.80	3.01
G(C8-H)˙	2.18	2.10	2.24
Cytosine	-0.01	0.03	0.13
C(N1-H)˙	3.00	2.84	3.05
C(N8-H1)˙	2.97	2.83	3.03
C(N8-H2)˙	2.94	2.81	3.02
C(C5-H)˙	2.22	2.05	2.22
C(C6-H)˙	2.41	2.17	2.32

Table 3. Adiabatic Electron Affinities (AEA in eV) of the Base Plus H Radicals.

Structure	B3LYP	Structure	B3LYP
Adenine	-0.28	**Guanine**	-0.10
A(C8+H)	1.22	G(N7+H)	0.01
A(C2+H)	1.38	G(C8+H)	0.12
A(N7+H)	0.27	G(C6+H)	0.58
A(N1+H)	-0.05	G(N3+H)	0.65
A(N3+H)	0.02	G(C5+H)	0.84
A(C5+H)	0.68	G(C4+H)	1.48
A(C6+H)	1.99	G(C6+H)	2.10
A(C4+H)	1.57	G(C2+H)	2.24
		G(N2+H)	3.00
Thymine	0.20	**Cytosine**	-0.01
diketo-T(C6+H)$^{\cdot}$	0.43	amino-oxoC(N3+H)	0.13
2-keto-T(N3+H)$^{\cdot}$	0.30	amino-oxoC(C6+H)	1.97
2-keto-T(C6+H)$^{\cdot}$	2.25	amino-oxoC(C5+H)	0.62
2-keto-T(C2+H)$^{\cdot}$	-0.28	amino-oxoC(C4+H)	2.59
4-keto-T(C6+H)$^{\cdot}$	1.65	amino-oxoC(C2+H)	2.34
Dienol-T(C6+H)$^{\cdot}$	0.95	amino-oxoC(C2+H)	2.38
diketo-T(C2+H)$^{\cdot}$	1.01	t-amino-hydC(N3+H)	-0.20
4keto-T(C4+H)$^{\cdot}$	0.53	t-amino-hydC(C6+H)	1.63
diketo-T(C4+H)$^{\cdot}$	2.5	t-amino-hydC(C5+H)	0.44
Dienol-T(N3+H)$^{\cdot}$	0.13	t-amino-hydC(C4+H)	2.24
4keto-T(N1+H)$^{\cdot}$	0.74	t-amino-hydC(C2+H)	2.10
Dienol-T(C4+H)$^{\cdot}$	2.06	t-amino-hydC(C2+H)	2.04
		c-amino-hydC(N1+H)	-0.16
		c-amino-hydC(C6+H)	1.67
		c-amino-hydC(C5+H)	0.45
		c-amino-hydC(C4+H)	2.22
		c-amino-hydC(C2+H)	2.09
		c-amino-hydC(C2+H)	2.05

Table 4. Vertical Detachment Energies (VDE in eV) of Deprotonated DNA Bases.

Structure	B3LYP	BLYP	BP86	Structure	B3LYP	BLYP	BP86
Adenine				**Guanine**			
A(N9-H)$^{-}$	3.61	3.35	3.58	G(N1-H)$^{-}$	3.17	2.94	3.15
A(N6-H1)$^{-}$	2.64	2.50	2.72	G(N2-H1)$^{-}$	2.86	2.66	2.87
A(N6-H2)$^{-}$	2.67	2.53	2.75	G(N2-H2)$^{-}$	2.87	2.66	2.87
A(C2-H)$^{-}$	1.73	1.66	1.78	G(N9-H)$^{-}$	3.34	3.13	3.33
A(C1-H)$^{-}$	2.89	2.73	2.90	G(C8-H)$^{-}$	2.68	2.55	2.71
Thymine				**Cytosine**			
T(N1-H)$^{-}$	3.17	3.17	3.38	C(N1-H)$^{-}$	3.12	2.96	3.17
T(N3-H)$^{-}$	4.59	4.59	4.23	C(N8-H1)$^{-}$	3.17	2.97	3.18
T(C6-H)$^{-}$	2.87	2.87	3.03	C(N8-H2)$^{-}$	3.14	2.96	3.17
T(C7-H2)$^{-}$	1.16	1.18	1.34	C(C5-H)$^{-}$	2.67	2.45	2.64
T(C7-H2)$^{-}$	1.16	1.18	1.34	C(C6-H)$^{-}$	2.94	2.60	2.77
T(C7-H3)$^{-}$	1.16	1.18	1.34				

Table 5. Vertical Detachment Energies (VDE in eV) of Hydride (H^{-}) Appended DNA Bases.

Structure	B3LYP	Structure	B3LYP
Adenine		**Guanine**	
A(C8+H)$^{-}$	1.44	G(N7+H)$^{-}$	1.06
A(C2+H)$^{-}$	1.76	G(C8+H)$^{-}$	1.21
A(N7+H)$^{-}$	1.21	G(C6+H)$^{-}$	1.20
A(N1+H)$^{-}$	0.53	G(N3+H)$^{-}$	1.25
A(N3+H)$^{-}$	0.75	G(C5+H)$^{-}$	1.23
A(C5+H)$^{-}$	0.90	G(C4+H)$^{-}$	1.83
A(C6+H)$^{-}$	2.14	G(C6+H)$^{-}$	2.72
A(C4+H)$^{-}$	1.88	G(C2+H)$^{-}$	2.30
		G(N2+H)$^{-}$	3.18
Thymine		**Cytosine**	
diketo-T(C6+H)$^{-}$	4.48	amino-oxoC(N3+H)$^{-}$	0.75
2-keto-T(N3+H)$^{-}$	5.03	amino-oxoC(C6+H)$^{-}$	2.30
2-keto-T(C6+H)$^{-}$	5.13	amino-oxoC(C5+H)$^{-}$	1.37
2-keto-T(C2+H)$^{-}$	0.21	amino-oxoC(C4+H)$^{-}$	2.94
4-keto-T(C6+H)$^{-}$	3.65	amino-oxoC(C2+H)$^{-}$	2.82
Dienol-T(C6+H)$^{-}$	1.31	amino-oxoC(C2+H)$^{-}$	2.86
diketo-T(C2+H)$^{-}$	0.52	t-amino-hydC(N3+H)$^{-}$	0.81
4keto-T(C4+H)$^{-}$	4.47	t-amino-hydC(C6+H)$^{-}$	1.90
diketo-T(C4+H)$^{-}$	4.56	t-amino-hydC(C5+H)$^{-}$	0.88
Dienol-T(N3+H)$^{-}$	-0.29	t-amino-hydC(C4+H)$^{-}$	2.43
4keto-T(N1+H)$^{-}$	2.67	t-amino-hydC(C2+H)$^{-}$	2.39
Dienol-T(C4+H)$^{-}$	3.16	t-amino-hydC(C2+H)$^{-}$	2.38
		c-amino-hydC(N1+H)$^{-}$	1.06
		c-amino-hydC(C6+H)$^{-}$	1.93
		c-amino-hydC(C5+H)$^{-}$	0.90
		c-amino-hydC(C4+H)$^{-}$	2.43
		c-amino-hydC(C2+H)$^{-}$	2.38
		c-amino-hydC(C2+H)$^{-}$	2.36

Table 6. Total Energies of the Hydrogen Abstracted Radicals and the Corresponding Anions Derived from the Bases at B3LYP/DZP++.

Radicals	B3LYP	Anion	B3LYP	Radical	B3LYP	Anion	B3LYP
Adenine				**Guanine**			
A(N9-H)$^{\cdot}$	-466.73827	A(N9-H)$^{\cdot}$	-466.85796	G(N2-H2)$^{\cdot}$	-542.01025	G(N9-H)$^{\cdot}$	-542.11524
A(N6-H1)$^{\cdot}$	-466.73409	A(N6-H1)$^{\cdot}$	-466.82778	G(N9-H)$^{\cdot}$	-542.00549	G(N2-H2)$^{\cdot}$	-542.11205
A(N6-H2)$^{\cdot}$	-466.73238	A(N6-H2)$^{\cdot}$	-466.82706	G(N1-H)$^{\cdot}$	-542.00277	G(N1-H)$^{\cdot}$	-542.11110
A(C2-H)$^{\cdot}$	-466.72310	A(N1-H)$^{\cdot}$	-466.79846	G(N2-H1)$^{\cdot}$	-542.00240	G(N2-H1)$^{\cdot}$	-542.10335
A(C1-H)$^{\cdot}$	-466.71107	A(C2-H)$^{\cdot}$	-466.75625	G(C8-H)$^{\cdot}$	-541.96769	G(C8-H)$^{\cdot}$	-542.04776
Thymine				**Cytosine**			
T(N1-H)$^{\cdot}$	-453.56938	T(N1-H)$^{-}$	-453.58310	C(N1-H)$^{\cdot}$	-394.34116	C(N1-H)$^{\cdot}$	-394.45144
T(N3-H)$^{\cdot}$	-453.53045	T(N3-H)$^{-}$	-453.56432	C(N8-H1)$^{\cdot}$	-394.33742	C(N8-H1)$^{\cdot}$	-394.44652
T(C6-H)$^{\cdot}$	-453.54340	T(C6-H)$^{-}$	-453.54539	C(N8-H2)$^{\cdot}$	-394.32879	C(N8-H2)$^{\cdot}$	-394.43700
T(C7-H1)$^{\cdot}$	-453.58282	T(C7-H1)$^{-}$	-453.58704	C(C5-H)$^{\cdot}$	-394.32658	C(C5-H)$^{\cdot}$	-394.41507
T(C7-H2)$^{\cdot}$	-453.58281	T(C7-H2)$^{-}$	-453.58704	C(C6-H)$^{\cdot}$	-394.31997	C(C6-H)$^{\cdot}$	-394.40154
T(C7-H3)$^{\cdot}$	-453.58282	T(C7-H3)$^{-}$	-453.58704				

Table 7. Total Energies (in Hartrees) of the Radicals and Anions Generated from the Addition of H to the Nucleic Acid Bases. The Abbreviation C(N3+H) Represents Amino-oxoC(N3+H) in the Watson-Crick Form.

Radicals	B3LYP	Anion	B3LYP	Radical	B3LYP	Anion	B3LYP
Adenine				**Guanine**			
A(C8+H)$^{\cdot}$	-467.96688	A(C8+H)$^{-}$	-468.01164	G(N7+H)$^{\cdot}$	-543.20691	G(N7+H)$^{-}$	-543.20667
A(C2+H)$^{\cdot}$	-467.95358	A(C6+H)$^{-}$	-468.00651	G(C8+H)$^{\cdot}$	-543.22268	G(C8+H)$^{-}$	-543.25349
A(N7+H)$^{\cdot}$	-467.94586	A(N7+H)$^{-}$	-468.00443	G(C6+H)$^{\cdot}$	-543.19648	G(C6+H)$^{-}$	-543.20080
A(N1+H)$^{\cdot}$	-467.94337	A(N1+H)$^{-}$	-467.99039	G(N3+H)$^{\cdot}$	-543.19012	G(N3+H)$^{-}$	-543.21153
A(N3+H)$^{\cdot}$	-467.94217	A(N3+H)$^{-}$	-467.96428	G(C5+H)$^{\cdot}$	-543.19773	G(C5+H)$^{-}$	-543.22176
A(C5+H)$^{\cdot}$	-467.93942	A(C5+H)$^{-}$	-467.95589	G(C4+H)$^{\cdot}$	-543.20105	G(C4+H)$^{-}$	-543.25546
A(C6+H)$^{\cdot}$	-467.93346	A(C6+H)$^{-}$	-467.94281	G(C6+H)$^{\cdot}$	-543.17382	G(C6+H)$^{-}$	-543.25617
A(C4+H)$^{\cdot}$	-467.93269	A(C4+H)$^{-}$	-467.94162	G(C2+H)$^{\cdot}$	-543.20930	G(C2+H)$^{-}$	-543.28633
				G(N2+H)$^{\cdot}$	-543.15876	G(N2+H)$^{-}$	-543.26919
Thymine				**Cytosine**			
diketo-T(C6+H)$^{\cdot}$	-454.66484	diketo-T(C6+H)$^{-}$	-454.68073	C(N3+H)$^{\cdot}$	-395.57852	C(N3+H)$^{-}$	-395.58319
2-keto-T(N3+H)$^{\cdot}$	-454.51485	2-keto-T(N3+H)$^{-}$	-454.52591	C(C6+H)$^{\cdot}$	-359.56521	C(C6+H)$^{-}$	-395.63759
2-keto-T(C6+H)$^{\cdot}$	-454.64683	2-keto-T(C6+H)$^{-}$	-454.72159	C(C5+H)$^{\cdot}$	-395. 56846	C(C5+H)$^{-}$	-395.59120
2-keto-T(C2+H)$^{\cdot}$	-454.63703	2-keto-T(C2+H)$^{-}$	-454.62667	C(C4+H)$^{\cdot}$	-395.53060	C(C4+H)$^{-}$	-395.62566
4-keto-T(C6+H)$^{\cdot}$	-454.63667	4-keto-T(C6+H)$^{-}$	-454.69760	C(C2+H)$^{\cdot}$	-395.51002	C(C2+H)$^{-}$	-395.59608
Dienol-T(C6+H)$^{\cdot}$	-454.63612	Dienol-T(C6+H)$^{-}$	-454.67101	C(C2+H$^{\cdot}$)	-395.50838	C(C2+H)$^{-}$	-395.59588
diketo-T(C2+H)$^{\cdot}$	-454.63035	diketo-T(C2+H)$^{\cdot}$	-454.67057	t-amino-hydC(N3+H)$^{\cdot}$	-395.55850	t-amino-hydC(N3+H)$^{-}$	-395.55116
4keto-T(C4+H)$^{\cdot}$	-454.62991	4keto-T(C4+H)$^{-}$	-454.64930	t-amino-hydC(C6+H)$^{\cdot}$	-395.55521	t-amino-hydC(C6+H)$^{-}$	-395.61517
diketo-T(C4+H)$^{\cdot}$	-454.49337	diketo-T(C4+H)$^{-}$	-454.58714	t-amino-hydC(C5+H)$^{\cdot}$	-395.56303	t-amino-hydC(C5+H)$^{-}$	-395.57938
Dienol-T(N3+H)$^{\cdot}$	-454.62837	Dienol-T(N3+H)$^{-}$	-454.63305	t-amino-hydC(C4+H)$^{\cdot}$	-395.53634	t-amino-hydC(C4+H)$^{-}$	-395.61862
4keto-T(N1+H)$^{\cdot}$	-454.62791	4keto-T(N1+H)$^{-}$	-454.65512	t-amino-hydC(C2+H)$^{\cdot}$	-395.53707	t-amino-hydC(C2+H)$^{-}$	-395.61355
				t-amino-	-395.53642	t-amino-	-395.61316

Dienol-T(C4+H)$^{\cdot}$	-454.60033	Dienol-T(C4+H)$^{-}$	-454.67641	hydC(C2+H)$^{\cdot}$		hydC(C2+H)$^{-}$	
				c-amino-hydC(N1+H)$^{\cdot}$	-395.56029	c-amino-hydC(N1+H)$^{-}$	-395.55441
				c-amino-hydC(C6+H)$^{\cdot}$	-395.55350	c-amino-hydC(C6+H)$^{-}$	-395.61475
				c-amino-hydC(C5+H)$^{\cdot}$	-395.56162	c-amino-hydC(C5+H)$^{-}$	-395.57821
				c-amino-hydC(C4+H)$^{\cdot}$	-395.53709	c-amino-hydC(C4+H)$^{-}$	-395.61865
				c-amino-hydC(C2+H)$^{\cdot}$	-395.53694	c-amino-hydC(C2+H)$^{-}$	-395.61323
				c-amino-hydC(C2+H)$^{\cdot}$	-395.53646	c-amino-hydC(C2+H)$^{-}$	-395.61231

References

[1] F. A. Evangelista, A. Paul, and H. F. Schaefer, J. Phys. Chem. A **108** (16), 3565 (2004).
[2] F. A. Evangelista and H. F. Schaefer, Chem. Phys. Chem. **7** (7), 1471 (2006).
[3] S. Ipek, P. P. Bera, and H. F. Schaefer, In preparation. (2006).
[4] Q. Luo, J. Li, Q. S. Li, S. Kim, S. E. Wheeler, Y. M. Xie, and H. F. Schaefer, Phys. Chem. Chem. Phys. **7** (5), 861 (2005).
[5] Q. Luo, Q. S. Li, Y. M. Xie, and H. F. Schaefer, Collect. Czech. Chem. Comm. **70** (6), 826 (2005).
[6] L. T. M. Profeta, J. D. Larkin, and H. F. Schaefer, Mol. Phys. **101** (22), 3277 (2003).
[7] N. A. Richardson, S. S. Wesolowski, and H. F. Schaefer, J. Am. Chem. Soc. **124** (34), 10163 (2002); N. A. Richardson, S. S. Wesolowski, and H. F. Schaefer, J. Phys. Chem. B **107** (3), 848 (2003).
[8] J. D. Zhang, Y. M. Xie, and H. F. Schaefer, J. Phys. Chem. A, submitted (2006).
[9] J. D. Zhang, Y. M. Xie, H. F. Schaefer, Q. Luo, and Q. Li, Mol. Phys. **104** (13-14), 2347 (2006).
[10]J. Berdys, I. Anusiewicz, P. Skurski, and J. Simons, J. Am. Chem. Soc. **126** (20), 6441 (2004).
[11]B. Brocklehurst, Radiat. Res. **155** (4), 637 (2001).
[12]G. P. Collins, Sci. Am. **289** (3), 26 (2003).
[13]S. G. Swarts, M. D. Sevilla, D. Becker, C. J. Tokar, and K. T. Wheeler, Radiat. Res. **129** (3), 333 (1992).
[14]A. Yarnell, Chem. Eng. News **81** (19), 33 (2003).
[15]H. Abdoul-Carime, S. Gohlke, and E. Illenberger, Phys. Rev. Lett. **92** (16), 168103 (2004).
[16]T. Douki, D. Angelov, and J. Cadet, J. Am. Chem. Soc. **123** (46), 11360 (2001); T. Douki, J. L.Ravanat, D. Angelov, J. R. Wagner, and J. Cadet, Top. Curr. Chem. **236**, 1 (2004); S. Steenken and L. Goldbergerova, J. Am. Chem. Soc. **120** (16), 3928 (1998).
[17]S. Steenken, Chem. Rev. **89** (3), 503 (1989).
[18]S. Steenken, S. V. Jovanovic, L. P. Candeias, and J. Reynisson, Chem. Eur. J. **7** (13), 2829 (2001); C. Chatgilialoglu, M. Guerra, and Q. G. Mulazzani, J. Am. Chem. Soc. **125** (13), 3839 (2003); W. K.Pogozelski and T. D. Tullius, Chem. Rev. **98** (3), 1089 (1998).
[19]H. Abdoul-Carime, S. Gohlke, E. Fischbach, J. Scheike, and E. Illenberger, Chem. Phys. Lett. **387** (4-6), 267 (2004).
[20]M. A. Huels, I. Hahndorf, E. Illenberger, and L. Sanche, J. Chem. Phys. **108** (4), 1309 (1998).
[21]L. E. Ramirez-Arizmendi, J. L. Heidbrink, L. P. Guler, and H. I. Kenttamaa, J. Am. Chem. Soc. **125** (8), 2272 (2003).
[22]A. J. S. C. Vieira and S. Steenken, J. Am. Chem. Soc. **112** (19), 6986 (1990).
[23]C. Desfrancois, H. Abdoul-Carime, and J. P. Schermann, J. Chem. Phys. **104** (19), 7792 (1996); M. A. Huels, B. Boudaiffa, P. Cloutier, D. Hunting, and L. Sanche, J. Am. Chem. Soc. **125** (15), 4467 (2003); J. R. Wiley, J. M. Robinson, S. Ehdaie, E. C. M. Chen, E. S. D. Chen, and W. E. Wentworth, Biochem. Biophys. Res. Comm. **180** (2), 841 (1991).
[24]V. Periquet, A. Moreau, S. Carles, J. P. Schermann, and C. Desfrancois, J. Elec. Spec. Rel. Phenom. **106** (2-3), 141 (2000).
[25]J. Schiedt, R. Weinkauf, D. M. Neumark, and E. W. Schlag, Chem. Phys. **239** (1-3), 511 (1998).
[26]S. S. Wesolowski, M. L. Leininger, P. N. Pentchev, and H. F. Schaefer, J. Am. Chem. Soc. **123** (17), 4023 (2001).
[27]X. H. Chen, E. A. Syrstad, M. T. Nguyen, P. Gerbaux, and F. Turecek, J. Phys. Chem. A **109** (36), 8121 (2005); F. Turecek and C. X. Yao, J. Phys. Chem. A **107** (43), 9221 (2003).
[28]D. M. Close, W. H. Nelson, E. Sagstuen, and E. O. Hole, Radiat. Res. **137** (3), 300 (1994); D. M. Close, Radiat. Res. **135** (1), 1 (1993).
[29]J. J. Lichter and W. Gordy, Proc. Natl. Acad. Sci. USA **60** (2), 450 (1968); J. Reynisson and S. Steenken, Phys. Chem. Chem. Phys. 7 (4), 659 (2005); J. Schmidt and D. C. Borg, Radiat. Res. **46** (1), 36 (1971); K. Sieber and J. Huttermann, Int. J. Radiat. Biol. **55** (3), 331 (1989); H. Zehner, E. Westhof, W. Flossmann, and A. Muller, Z. Naturforschung. C: Biosciences **32** (1-2), 1 (1977); M. G. Debije and W. A. Bernhard, J. Phys. Chem. A **106** (18), 4608 (2002); M. G. Debije, D. M. Close, and W. A. Bernhard, Radiat. Res. **157** (3), 235 (2002); F. Greco, A. Liguori, G. Sindona, and N. Uccella, J. Am. Chem. Soc. **112** (25), 9092 (1990); Y. H. Jang, W. A. Goddard, K. T. Noyes, L. C. Sowers, S. Hwang, and D. S. Chung, J. Phys. Chem. B **107** (1), 344 (2003); Y. Podolyan, L. Gorb, and J. Leszczynski, J. Phys. Chem. A **104** (31), 7346 (2000).

[30] W. H. Nelson, E. Sagstuen, E. O. Hole, and D. M. Close, Radiat. Res. **131** (3), 272 (1992).
[31] S. D. Wetmore, R. J. Boyd, and L. A. Eriksson, J. Phys. Chem. B **102** (51), 10602 (1998).
[32] A. O. Colson, D. Becker, I. Eliezer, and M. D. Sevilla, J. Phys. Chem. A **101** (47), 8935 (1997).
[33] E. S. D. Chen, E. C. M. Chen, and N. Sane, Biochem. Biophys. Res. Comm. **246** (1), 228 (1998).
[34] S. Gohlke, H. Abdoul-Carime, and E. Illenberger, Chem. Phys. Lett. **380** (5-6), 595 (2003).
[35] L. Kar and W. A. Bernhard, Radiat. Res. **93** (2), 232 (1983).
[36] H. Abdoul-Carime, P. Cloutier, and L. Sanche, Radiat. Res. **155** (4), 625 (2001); H. Abdoul-Carime and L. Sanche, Int. J. Radiat. Biol. **78** (2), 89 (2002).
[37] L. Sanche, Radiat. Protect. Dos. **99** (1-4), 57 (2002); L. Sanche, Physica Scripta **68** (5), C108 (2003).
[38] D. Close, G. Forde, L. Gorb, and J. Leszczynski, Struct. Chem. **14** (5), 451 (2003).
[39] J. Bertran, A. Oliva, L. Rodriguez-Santiago, and M. Sodupe, J. Am. Chem. Soc. **120** (32), 8159 (1998); M. Bixon and J. Jortner, J. Phys. Chem. A **105** (45), 10322 (2001); B. Boudaiffa, P. Cloutier, D.Hunting, M. A. Huels, and L. Sanche, M S-Medicine Sciences **16** (11), 1281 (2000); J. Cadet, T. Douki, D. Gasparutto, and J. L. Ravanat, Mutat. Res. Funda. Mol. Mech. of Mutat. **531** (1-2), 5 (2003); P. M. Cullis, M. E. Malone, and L. A. Merson-Davies, J. Am. Chem. Soc. **118** (12), 2775 (1996); D. Dee and M. E. Baur, J. Chem. Phys. **60** (2), 541 (1974); D. C. Malins, N. L. Polissar, G. K. Ostrander, and M. A. Vinson, Proc. Natl. Acad. Sci. USA **97** (23), 12442 (2000); T. Melvin, S. W. Botchway, A. W. Parker and P. ONeill, J. Am. Chem. Soc. **118** (42), 10031 (1996).
[40] J. C. Rienstra-Kiracofe, G. S. Tschumper, H. F. Schaefer, S. Nandi, and G. B. Ellison, Chem. Rev. **102** (1), 231 (2002).
[41] P. P. Bera and H. F. Schaefer, Proc. Natl. Acad. Sci. USA **102** (20), 6698 (2005).
[42] A. O. Colson, B. Besler, D. M. Close, and M. D. Sevilla, J. Phys. Chem. **96** (2), 661 (1992); A. O. Colson and M. D. Sevilla, J. Phys. Chem. A **99** (34), 13033 (1995); A. O. Colson and M. D. Sevilla, Int. J. Radiat. Biol. **67** (6), 627 (1995); A. O. Colson and M. D. Sevilla, J. Phys. Chem. **100** (11), 4420 (1996).
[43] E. C. M. Chen and E. S. Chen, Abs. of Papers Am. Chem. Soc. **220**, U497 (2000).
[44] E. C. M. Chen and E. S. Chen, J. Phys. Chem. B **104** (32), 7835 (2000).
[45] P. Hobza and J. Sponer, THEOCHEM **388**, 115 (1996); P. Hobza and J. Sponer, Chem. Phys. Lett. **288** (1), 7 (1998); P. Hobza and J. Sponer, Chem. Rev. **99** (11), 3247 (1999).
[46] P. Hobza, J. Sponer, and J. Leszczynski, J. Phys. Chem. B **101** (40), 8038 (1997).
[47] F. Hagelberg, I. Yanov, and J. Leszczynski, THEOCHEM **487** (1-2), 183 (1999); J. Leszczynski, THEOCHEM **487** (1-2), ix (1999); J. Sponer and P. Hobza, Encyc. Comput. Chem. **1**, 777 (1998); J. Sponer, P. Hobza, and J. Leszczynski, Comput. Chem. Rev. Curr. Trends **1**, (World Scientific (1996); J. Sponer, J. Leszczynski, and P. Hobza, J. Phys. Chem. **100** (5), 1965 (1996).
[48] J. Sponer, J. Leszczynski, and P. Hobza, Biopolymers **61** (1), 3 (2001).
[49] A. D. Becke, Phys. Rev. A **38** (6), 3098 (1988).
[50] C. T. Lee, W. T. Yang, and R. G. Parr, Phys. Rev. B **37** (2), 785 (1988).
[51] A. D. Becke, J. Chem. Phys. **98** (7), 5648 (1993).
[52] J.P. Perdew, Phys. Rev. B **33** (12), 8822 (1986); J. P. Perdew, Phys. Rev. B **34** (10), 7406 (1986).
[53] B. N. Papas and H. F. Schaefer, THEOCHEM, In press. (2006).
[54] T. H. Dunning, J. Chem. Phys. **53** (7), 2823 (1970); S. Huzinaga, J. Chem. Phys. **42** (4), 1293 (1965);T. J. Lee and H. F. Schaefer, J. Chem. Phys. **83** (4), 1784 (1985).
[55] M. J. Frisch, G. W. Trucks, H. B. Schlegel, G. E. Scuseria, P. M. W. Gill, B. G. Johnson, M. A. Robb, J. R. Cheeseman, T. Keith, J. A. Montgomery, R. E. Stratmann, J. C. Burant, S. Daprich, J. M. Millam, A. D. Daniels, K. N. Kudin, M. C. Strain, O. Farcas, J. Tomasi, V. Barone, M. Cossi, R. Cammi, B. Mennuchi, C. Pomelli, C. Adamo, S. Clifford, J. Ochterski, G. A. Petersson, Q. Cui, K. Morukuma, P. Salvador, J. J. Dannenberg, D. K. Malick, A. D. Rabick, A. G. Baboul, K. Raghavachari, B. B. Stefanov, G. Liu, A. Liashenko, P. Piskorz, I. Komaromi, R. Gomperts, M. A. Al-Laham, V. G. Zakrzewski, J. V. Ortiz, J. B.Foresman, J. Cioslowski, B. B. Stefanov, A. Nanayakkara, M. Challacombe, C. Y. Peng, P. Y. Ayala, W. Chen, M. W. Wong, J. L. Andres, E. S. Replogle, R. Gomperts, R. L. Martin, D. J. Fox, J. S. Binkley, D. J.
Defrees, J. Baker, J. P. Stewart, M. Head-Gordon, C. Gonzalez, and J. A. Pople, Gaussian 94, Revision E.2 (Gaussian, Inc., Pittsburgh PA., 1995).
[56] D. Radisic, K. H. Bowen, I. Dabkowska, P. Storoniak, J. Rak, and M. Gutowski, J. Am. Chem. Soc. **127** (17), 6443 (2005).

Brill Academic Publishers
P.O. Box 9000, 2300 PA Leiden,
The Netherlands

Lecture Series on Computer and Computational Sciences
Volume 265, 2006, pp. 265-285

Optimized virtual orbital space (OVOS) as a tool for more efficient correlated and relativistic calculations of molecular properties and interactions

M. Urban[1], Michal Pitoňák, Pavel Neogrády

Department of Physical and Theoretical Chemistry, Faculty of Natural Sciences, Comenius University,
SK-842 15 Bratislava, Slovakia

Received 4 July, 2006; accepted 10 July, 2006

Abstract: The performance of the CCSD(T) Coupled cluster treatment of the electron correlation with the reduced optimized virtual orbital space (OVOS) is reviewed for a series of molecular properties. Most results were obtained by using the functional which optimizes the overlap between the first order correlated wave function in the full virtual space and the reduced space, respectively. The size of OVOS is typically one half of the full virtual orbital space which leads to the computer time savings in CCSD(T) more than order of magnitude. The performance of OVOS improves with increasing the size of the basis set. When combined with the series of correlation consistent basis sets, (aug)–cc–pVXZ, X=D, T, Q, and 5, OVOS allows extrapolations to the complete basis set (CBS) limit with significantly less computational effort in CCSD(T) and other highly correlated methods than is needed with the full virtual space. The usefulness of the OVOS approach is demonstrated by calculations of reaction energies (isomerization energy of HCN – HNC; the dissociation energy of pentane to propene and ethane), spectroscopic constants of the F_2 molecule and dipole moments and dipole polarizabilities. Electric properties of carbon monoxide and thiophene, serve as a touchstone of the performance of OVOS. For all systems are OVOS results compared with the full virtual space CCSD(T) results. With the medium sized aug-cc-pVTZ basis set CCSD(T) dipole moments obtained by the OVOS treatment differ from the full virtual space results typically by 0.002–0.006 a.u. Components of the dipole polarizabilities are typically accurate to within 0.02–0.1 a.u. Hydrogen bonding or stacking interaction energies of the formamidine and the formaldehyde dimers are more accurate (by up to 10 kJ/mol) with larger basis set reduced by OVOS than with the full–space smaller basis set having equivalent number of virtuals. The OVOS technique is used for the first time in relativistic calculations. Particularly useful is OVOS when combined with the uncontracted basis sets allowing thus to circumvent problems with proper contraction needed for a specific approximate relativistic Hamiltonian. The approach is applied to dipole moments and polarizabilities of a heavy–element containing molecules like AuH, AuF, and the Au^- anion using the Douglas–Kroll–Hess spin–averaged method.

Keywords: OVOS; optimization functionals; CCSD(T); reaction energies; dipole moments; dipole polarizabilities; intermolecular interactions; relativistic effects; basis sets; F_2; CO; pentene; formamidine dimer; formaldehyde dimer; thiophene; AuH; AuF; Au^-.

Mathematics Subject Classification: 81V55

PACS: 31.15.Ne, 31.30.Jv, 33.15.Kr, 34.20.Gj, 39.30.+w, 87.15.Kg

[1]Corresponding author. E-mail: urban@fns.uniba.sk

1 Introduction

Computer modeling is presently a standard tool in most branches of chemistry, physics, biology and other research disciplines in natural sciences and in many applications. When considering calculations and predictions of molecular properties we can recognize the two mainstreams typical for present efforts in the development and applications of the theoretical and computational chemistry. Obviously, a dominating tool applicable to large molecules and large molecular clusters are methods based on the Density Functional Theory (DFT). Another large group of methods are the Wave Function (WF) methods. Results from both these methods applied mostly to individual molecules can be used in subsequent calculations in other areas like theoretical spectroscopy, statistical thermodynamics, molecular dynamics (to treat problems, e.g., in the condensed phase) and other.

¿From the point of view of the applicability the DFT methods [1] have one obvious advantage, namely they scale with the number of electrons much better than correlated WF methods. The disadvantage of DFT methods is the jungle of different functionals frequently developed to solve specific problems and particular molecular properties. Also, controlling the accuracy is difficult. Promising is recent progress in defining DFT methods by extracting a functional form from *ab initio* WF methods in an alternative way [2]. Nevertheless this approach is not yet as effective as the standard DFT methods and remains far from a routine use. For accurate calculations of molecular properties highly correlated WF methods remain important. In spite of enormous progress in theoretical formulations and computer algorithms the scaling of WF methods with respect to the number of electrons and basis sets prevents more extensive applications of such methods to large molecules and molecular clusters. One of the most efficient and sophisticated methods are Coupled cluster (CC) methods [3, 4, 5, 6, 7, 8] in which an exponential ansatz is exploited for defining the wave function $|\Psi\rangle = e^T|\Phi_0\rangle$ where T is an excitation operator and $|\Phi_0\rangle$ is the reference wave function, usually the single–determinant Hartree–Fock wave function. The most frequently used CC method for accurate calculations is CCSD[9] in which the amplitudes of the single and double excitation operators, t_i^a, t_{ij}^{ab}, are optimized iteratively, while triples are treated perturbatively [10] exploiting the t_i^a, t_{ij}^{ab} amplitudes. Most demanding steps in CCSD scale as $N_o^2 N_v^4$, $N_o^3 N_v^3$, in noniterative triples as $N_o^3 N_v^4$. N_o is the number of occupied orbitals (OO), N_v is the number of virtual orbitals (VO).

Most frequently used methods for calculating the energy and molecular properties are based on the algebraic approximation. Within the CC methods this means that one creates a one–electron reference, usually a single–determinant Hartree–Fock (HF) Self Consistent Field (SCF) reference using a suitable set of basis functions. The number of occupied molecular orbitals N_o depends on the number of electrons (n_e) and is $N_o = 1/2n_e$ for a closed shell molecule. Obviously, it is useful to create as accurate occupied orbitals as possible, i.e. SCF orbitals approaching the HF limit. Such orbitals may serve as a good starting point for subsequent correlation calculations within the CC framework or the variational methods like Configuration Interaction (CI). To approach the HF limit large basis sets are required. This, in turn, creates large space of virtual orbitals which do not affect the SCF energy. It is questionable whether so many virtuals are really needed in the electron correlation step. In fact, VOs are just a byproduct of the SCF procedure, fulfilling the only necessary condition, namely the orthogonality to the subspace of the occupied orbitals space. A routine way of reducing the computer time in the CCSD(T) step is keeping inner shell electrons uncorrelated. Depending on the selection of the valence space this means that the inner shell (core) and the core–valence correlation is neglected. Computer time saving can be significant for molecules containing a heavy element. For most practical purposes the accuracy of molecular properties is affected little, but computer time savings are small as well (may be with the exception of a heavy metal containing systems which have many inner shell electrons). In almost all accurate calculations is the number of virtual orbitals N_v much larger than N_o, depending on the basis set. Considering unfavorable scaling of the most demanding steps in CCSD(T) with N_v clearly shows that a proper target for our effort aimed at the computer time savings should be reducing N_v. Deleting virtuals which are in a way a "counterpart" of inner shell electrons is not enough. First, such orbitals are even less clearly defined than are inner shells, and, second, the number of such orbitals is too small to reduce the computer time of the large basis set calculation considerably. Satisfactory procedure should reduce the computer time by at least order of magnitude. Thus, in CCSD(T), for example, this means reducing the virtual space at least to one half (which should reduce the most demanding terms theoretically by a factor of 16). Cutting off just plain HF virtuals leads to unacceptable lowering of the accuracy of correlated CC results. More promising are techniques like OVOS proposed long ago by Adamowicz, Bartlett and Sadlej [11, 12, 13]. General aim of OVOS is the rotation of the original full set of virtuals which allows "mapping" of the information contained in the full space into the reduced OVOS space. In our recent papers we have introduced alternative optimization techniques [14] and applied these techniques to a series of properties of selected molecules. We have also shown that OVOS is useful in relativistic calculations [15]. Different contraction schemes should be used with any approximate relativistic Hamiltonian which may lead to difficulties in defining a proper contraction scheme. Alternatively, we may stay with uncontracted basis sets and relief enormous computer demands by reducing the virtual space using OVOS.

There are numerous alternative attempts to alleviate the unfavorable scaling of CC and CI methods preserving at the same time their controlled accuracy. Closely related to OVOS are other techniques which exploit rotation of molecular orbitals [16, 17, 18]. In many applications and particularly in spatially extended molecules are very useful techniques based on localized orbitals [19, 20, 21, 22, 23]. Different idea is based on the treatment of small integrals based on the Cholesky decomposition [24] and the Laplace factorization of the energy denominators which permits a lower scaling of the triples equations [25]. We stress that OVOS can be combined with these approaches and increase its efficiency even more. Very topical seem Natural orbitals (NO) which were believed [26] to be an optimal orbital basis for fastest convergence of the CI (Configuration Interaction) expansion. Not only that this statement was never proved, Bytautas at al. [27] showed few counter-examples using their split-localized orbitals, which led to even faster CI correlation energy convergence with respect to number of determinants.

For applications in CC calculations, use of approximate natural orbitals called FNO (Frozen Natural Orbitals) [28, 29, 30] is more convenient than traditional NO's. As it was shown in papers by Adamowicz [13] and Bartlett *at al.* [31] and also according to our experience the efficiency of OVOS and FNO is very similar except when the virtual space is reduced drastically (to less than 30–40% of the full virtual space) or when the core-valence correlation is considered. In such cases OVOS would be preferable. We should mention that implementations for more general WF reference (RHF/ROHF/UHF) and analytic gradients is already available for FNO.

2 Theoretical background

Canonical HF virtual orbitals are ordered according to their orbital energy, which roughly reflects their individual importance in correlated WF expansion. Besides this ordering, they are by no means "optimized" for further use in the correlation energy calculation.

In the OVOS method we transform VOs in such a way, that the correlated WF expansion in these rotated orbitals can be truncated without significantly affecting its overall accuracy and flexibility for description of the studied system. From the "energy" point of view, we rotate VOs so that the correlation energy converges faster with the dimension of truncated OVOS compared to the truncation of original VOS (virtual orbital space). This transformation of VOs comprises mixing and "rearranging" of the HF VOs according to their importance in the total correlated WF. In our simplest implementation (see also [14] is the importance of optimized VOs in the total correlated WF estimated by their importance in the Møller-Plesset first order perturbation WF.

Standard WF methods for calculation of the correlation energy (configuration interaction, coupled clusters, etc.) are invariant to unitary rotations within the space of OOs and VOs separately, i. e. the semicanonical transformation. That means, we have a possibility to modify the virtual HF orbitals purposely to enhance the efficiency of the electron correlation step.

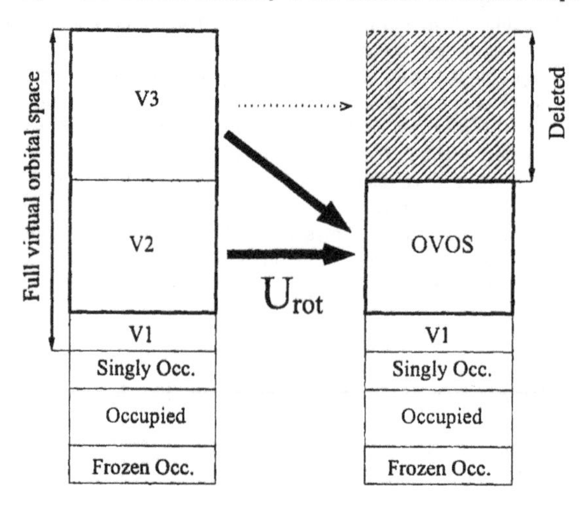

Figure 1: Basic idea of virtual orbitals transformation in the OVOS procedure.

The idea behind the OVOS method is schematically presented in Fig. 1. V1 denotes the space of inactive virtual orbitals, i.e. VOs which are not changed by the rotation. V2 and V3 represent the rest of the original, full VOS. As indicated by thick arrows in Fig. 1, we are trying to rearrange the major portion of the electron correlation contribution related to VOs from the V3 subspace to the V2 subspace, thus forming a new, OVOS space having dimension of V2. In the subsequent correlated calculation we can delete rotated VOs from the formal V3 subspace, because their contribution to the final correlated WF should be negligible.

This goal is technically achieved by searching for the maximum overlap of the first order MBPT(2) wave function in the original VOS and the truncated, optimized VOS :

$$L = \langle \Phi_0 | T_{\mathrm{MBPT(2)}} | T^{\mathrm{OVOS}}_{\mathrm{MBPT(2)}} | \Phi_0 \rangle \tag{1}$$

where excitation operators in a closed-shell case are given by :

$$T_{\mathrm{MBPT(2)}} = \frac{1}{4} \sum_{i,j,a,b} t_{ij}^{ab} a^{\dagger} i b^{\dagger} j, \quad T^{\mathrm{OVOS}}_{\mathrm{MBPT(2)}} = \frac{1}{4} \sum_{i,j,a^*,b^*} t_{ij}^{a^*b^*} a^{*\dagger} i b^{*\dagger} j \tag{2}$$

Indices with asterisk label VOs from the truncated optimized space, while indexes without asterisk span the full, original VOS. In the restricted open-shell case, or in cases when more sophisticated form of the wave function is used in L (for instance a coupled-clusters WF), T contains also the single excitation operator.

Although we formulate our optimization criterion in the WF language (in contrary to the original work of Adamowicz, Sadlej and Bartlett [11] [12], where they rely on minimizing the second order energy Hylleraas functional), we expect that our form of the functional is able to reproduce the electron correlation energy in the truncated space accurately enough.

FNO method [30] can be viewed as a special arrangement of the OVOS approach. If we put all VOs into the V2 subspace, our functional becomes the overlap of the full VOS first order WF:

$$L = \langle \Phi_0 | T_{\mathrm{MBPT(2)}} | T_{\mathrm{MBPT(2)}} | \Phi_0 \rangle \tag{3}$$

The major difference between our approach and FNOs is that after carrying out the differentiation of the functional with respect to the rotation coefficients, see eq. (1), we arrive at an iterative scheme in the OVOS case :

$$\mathbf{U}^{\mathrm{T}}\mathbf{Q}\mathbf{U} = \lambda \tag{4}$$

where $\mathbf{Q}$ depends on $\mathbf{U}$,

$$Q_{a^*b^*} = \sum_{ij}\sum_{cd} t_{ij}^{ca^*} t_{ij}^{db^*} P_{cd} \tag{5}$$

$$P_{ab} = \sum_{p^*} U_{p^*a} U_{p^*b} \tag{6}$$

In the FNO approach, in which p^* runs over all "active" virtuals (thus $\mathbf{P} = \mathbf{1}$), same equations as for the first order density matrix are obtained,

$$Q_{ab} = \sum_{ij}\sum_{c} t_{ij}^{ac} t_{ij}^{cb} \tag{7}$$

Now, matrix $\mathbf{Q}$ does not depend on the rotation matrix $\mathbf{U}$ and can be thus diagonalized in a single step, see eq. (4). In both approaches, Fock operator is transformed into new OVOS/FNO orbitals and diagonalized afterwards to satisfy the semicanonical character of the post Hartree–Fock electron correlation treatment.

Selection of the V2 space is to some extent ambiguous. The simplest way of splitting the VOS would be based on orbital energies. More physically relevant selection can be done, if we first construct FNOs as a starting point for OVOS orbitals, and populate the V2 subspace according to the FNO occupation numbers. Further, to avoid convergence problems, selection of VO's should respect the degeneracy of orbitals within different symmetry group representations as well as the "shell" structure outlined by the FNO occupation numbers. This means, that symmetry equivalent virtual orbitals must all either be deleted or retained. When we will be saying that our VOS is reduced to, say, 50%, due to the above "symmetry" rule this is frequently just approximate. Unfortunately, even if we consider all these "rules" in the selection of VOs the OVOS procedure may converge into a non-optimal local minimum in some cases. This may cause that the CCSD(T) correlation energy deviates roughly by $10^5 - 10^6$ a.u. from the true energy minimum. Error of this magnitude can be tolerated in calculation of the properties which are not too sensitive to the energy accuracy. However, in cases like the finite field calculation of dipole polarizabilities such error could cause a disaster. A cure of this pathological cases would be an implementation of a "globally convergent" algorithm, which is in progress.

Approaches based either on OVOS or FNO reduced, to say typically 50% of the full VOS, are capable of retrieving more than 98% of the MBPT(2)/CCSD/CCSD(T) correlation energy. This is obviously not enough to achieve desired accuracy for properties (particularly those which are very sensitive to the energy accuracy). If we analyze the CCSD(T) energy obtained in the truncated OVOS by its "components", i. e. the second order contribution (MBPT(2)), higher order contributions (X_2, defined as CCSD – MBPT(2)) and perturbative triples (T), we see that the major source of error originates from the MBPT(2) energy. Since the optimization procedure is based on the knowledge of the first order wave function in the full VOS, we can refine our OVOS/FNO energy defining $CCSD(T)_2$ as

$$X_2 = CCSD(T)^{(OVOS)} - MBPT(2)^{(OVOS)} \tag{8}$$

and

$$CCSD(T)_2^{(OVOS)} = MBPT(2)^{(Full\ VOS)} + X_2 \tag{9}$$

Alternative formulation of the same correction can be expressed also via Y_2

$$Y_2 = MBPT(2)^{(Full\ VOS)} - MBPT(2)^{(OVOS)} \tag{10}$$

and

$$CCSD(T)_2^{(OVOS)} = CCSD(T)^{(OVOS)} + Y_2 \tag{11}$$

Formulation using Y_2 has an advantage, that its value can serve as an indicator of the potential failure of the OVOS approach. Small value of Y_2 certifies the quality of the OVOS truncation and optimization. Larger Y_2 value does not automatically mean that the result is incorrect, however attention should be given particularly to the proper structure of the V2 virtual orbital space. More detailed discussion can be found in our previous paper [15]. We note that in chapter 3.1 and in conclusions of the cited paper we erroneously used X_2 instead of Y_2, see *erratum* [32].

Using the full space MBPT(2) "correction" we are able to push the CCSD or CCSD(T) energy to the value almost identical with the energy in the full space with OVOS reduced to about 50–60% particularly with (aug)–cc-pVTZ or better basis sets.

3 Dependence of the CCSD(T) energy on the basis set and OVOS

Due to the unfavorable scaling of all correlated WF methods with number of virtual orbitals practitioners of *ab initio* methods try to reconcile the accuracy of required result with the basis set quality and its size. In this part we wish to demonstrate the magnitude of the electron correlation contribution for the fluorine molecule using a systematic sequence of the correlation consistent polarized basis sets (cc-pVXZ) and their augmented counterparts (aug-cc-pVXZ) as defined by Dunning and his collaborators [33, 34, 35]. Aug-cc-pVXZ basis sets (or doubly, triply etc. augmented bases) are designated for highly accurate calculations. A typical feature of these basis sets is a possibility of the extrapolation of results to the complete basis set (CBS) limit with respect to the basis set cardinal number X [36]. In Fig. 2 we present the CCSD(T) electron correlation energy of the fluorine molecule with aug-cc-pVXZ (X=3,4,5) basis sets using the full virtual space and with OVOS reduced to 40–90 % of the full space. We observe that more CCSD(T) electron correlation energy (i.e. energy closer to the full space limit with the aug-cc-pV5Z basis) is recovered with basis sets having higher X when the same number of virtuals N_v is used with OVOS . Even more visible is the usefulness of OVOS in Fig. 3 where the X_2 correction (eq. (8)) is employed. Note also the stability of the electron correlation with N_v reduced down to at least 50%. Obviously, starting from lower CCSD(T) energy is profitable for calculations of properties and processes in which we rely on the error cancellation when incomplete basis sets are used (which is the case in almost all highly correlated calculations).

Now, let us demonstrate the stability of the portion of the CCSD(T) electron correlation for distances around the equilibrium bond length of F_2 using OVOS reduced to about 50% . Such energies can be used for the numerical fit from which the vibrational harmonic frequencies, ω_e, can be obtained. The performance of OVOS in comparison to the full virtual space is excellent,

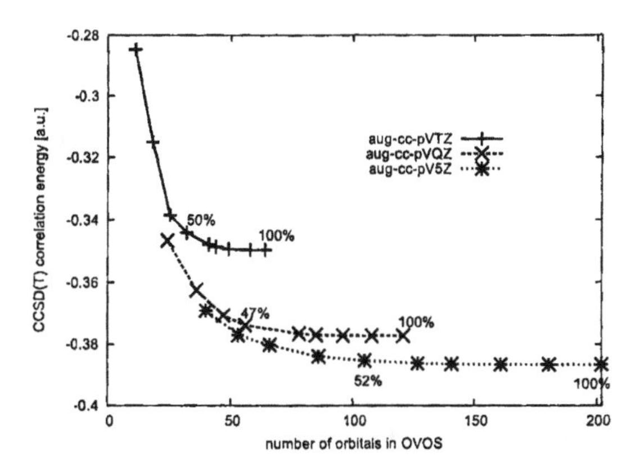

Figure 2: CCSD(T) correlation energy of the F_2 molecule in aug-cc-pVXZ, X=T, Q, 5 basis sets with different number of virtual orbitals using OVOS. 1s orbitals are frozen. Inserted numbers represent percentage of the OVOS size with respect to the full virtual space.

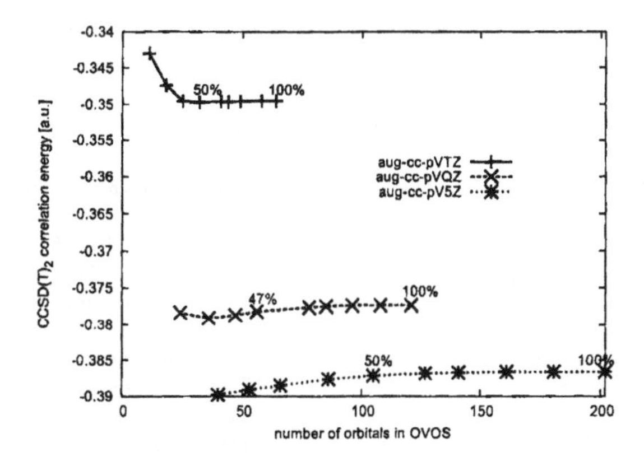

Figure 3: $CCSD(T)_2$ correlation energy of the F_2 molecule in aug-cc-pVXZ, X=T, Q, 5 basis sets with different number of virtual orbitals using OVOS. 1s orbitals are frozen. CCSD(T) energies are corrected using X_2 (see eq. (8), (9))

.

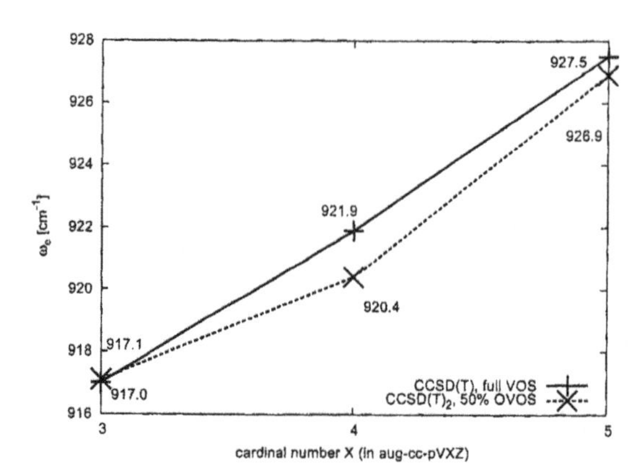

Figure 4: $CCSD(T)_2$ harmonic vibrational frequencies ω_e [cm^{-1}] of the F_2 molecule using OVOS truncated to 50% and the full virtual orbital space. 1s orbital are frozen. Inserted numbers represent ω_e using the full space and the OVOS space, respectively, with aug-cc-pVXZ, X=T, Q and 5 basis sets.

see Fig. 4. The agreement of ω_e with OVOS and the full VOS for aug-cc-pVTZ basis sets is fortuitous. Very important is the finding that the error introduced by cutting the virtual space to one halve is much lower (less than 2 cm^{-1}) than is the basis set effect. The basis set trend is fully reproduced by OVOS. With largest basis set experimental ω_e = 916.6 cm^{-1} is overestimated by about 11 cm^{-1}, in accord with other basis-set convergence studies (see, e.g. Ref. [37]). Higher excitations than those considered in CCSD(T) or considering the multi reference CC methods [38] are needed to reach experimental accuracy of ω_e for F_2.

4 Reaction energies

Explicit cancellation of errors within OVOS and the full virtual space calculations can be demonstrated by calculations of reaction energies [14]. Our first example is the isomerization energy HCN $\leftrightarrow$ HNC. First, we present the basis set dependence of the total ΔE in Fig. 5. Starting from the cc-pVTZ basis set reaction energy changes with increasing X very little. This trend is completely reproduced with OVOS reduced to 50% of the full VOS. The CBS limit for ΔE calculated from cc-pVQZ and cc-pV5Z data is practically the same with the full VOS and OVOS reduced to 50% for which we need less than 10% of the computer time in comparison with the full VOS calculation. Pure electron correlation contribution to the HCN $\leftrightarrow$ HNC isomerization is presented in Fig. 6. Main message provided by this Figure is that the performance of OVOS improves with increasing cardinal number X and that starting with the cc-pVTZ basis set OVOS reproduces basis set effects excellently. We should also note that the final ΔE as presented in Fig. 5, namely apparent insensitivity of ΔE on X results partly from compensation of the basis set effect at the SCF and the CCSD(T) levels.

Similar compensation of the basis set effect is also observed for the dissociation of pentene to propene and ethane, Fig. 7. Final ΔE with the cc-pVDZ and cc=pVTZ basis sets differ by only 2.7 kJ/mol but the full space CCSD(T) correlation contributions differ by 4.9 kJ/mol. OVOS reproduces this difference to within 1 kJ/mol. With the cc-pVTZ basis set the full space CCSD(T) result and the result with OVOS reduced to 50% differ by negligible 0.2 kJ/mol demonstrating, once again, improving accuracy of OVOS with extending the basis set. Our CCSD(T)$_2$ reaction energies presented in Fig. 7 employ the X$_2$ correction, see eq. (8), in which we use the full virtual space for calculating the MBPT(2) energy. It is thus interesting to note that the SCF + MBPT(2) reaction energy itself is in error larger than 10 kJ/mol with both cc-pVDZ and cc-pVTZ basis sets.

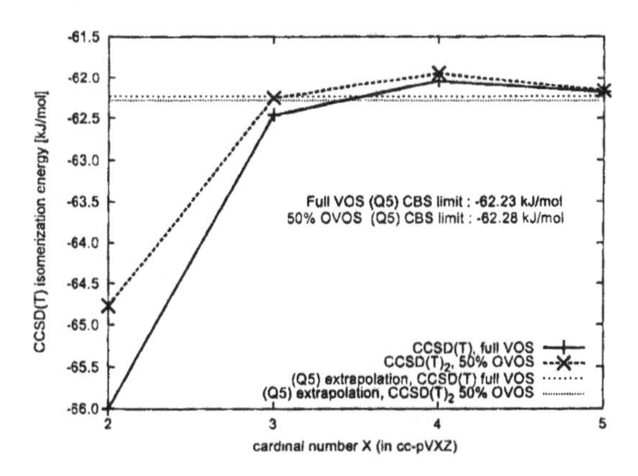

Figure 5: CCSD(T)$_2$ isomerization energy of HCN $\leftrightarrow$ HNC reaction with cc-pVXZ, X=T, Q, 5 basis sets, 1s orbitals of C and N frozen. Full VOS, 50% OVOS and the CBS limit extrapolated from the cc-pVQZ and cc-pV5Z (Q5) reaction energies are shown.

5 Intermolecular interaction energies

In this part we present typical applications of the OVOS technique to calculations of the interaction energies (ΔE) of the hydrogen-bonded system, namely the cyclic dimer of formamidine (FI...FI), and a stacking system, the formaldehyde...formaldehyde (FO...FO) dimer. Both are typical structural units resembling interactions present in the DNA structure. The FI...FI dimer is planar and has two N-H...N hydrogen bonds in its cyclic structure taken from Ref. [39]. The planes of the two formaldehyde molecules are separated by 3.3Å. Such distance is typical for the stacking part of interacting species in DNA [39]. Dipole moments of individual monomers in the FO...FO dimer are antiparallel which leads to an attractive component of the electrostatic interaction. Both dimers were calculated previously by Pittner and Hobza [39] by CCSD(T) and also by iterative CCSDT using smaller basis sets. They determined the CC correction to ΔE by evaluating the difference [MBPT(2) - CC] with a smaller basis set and adding this correction to the MBPT(2) interaction energy with a larger basis set. Such procedure resembles our X$_2$ correction, see eq. (8).

Interaction energies for the FI...FI presented in Table 1 indicate that the SCF part of the interaction energy is rather insensitive to the basis set quality. However, the electron correlation part varies with the basis set significantly. With the cc-pVDZ basis the CCSD(T) contribution represents 13% of the total ΔE while with the aug-cc-pVTZ basis set this part increases to 30%. The difference of ΔE with these two basis sets is 12 kJ/mol. OVOS, when the virtual space is reduced to 50%, reproduces the basis set trends perfectly saving more than one order of magnitude of the computer time. The differences between the full-space and the OVOS CCSD(T) results are

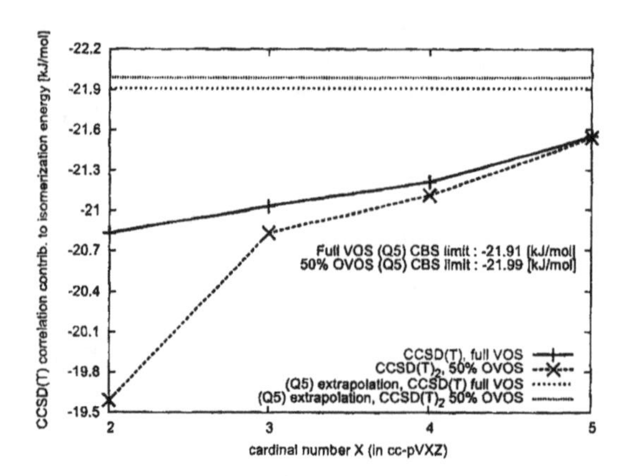

Figure 6: CCSD(T)$_2$ correlation contribution to isomerization energy of HCN ↔ HNC reaction in cc-pVXZ, X=T, Q, 5 basis sets, 1s orbitals of C and N frozen. Full VOS, 50% OVOS the CBS limit extrapolated from the cc-pVQZ and cc-pV5Z (Q5) reaction energies are shown.

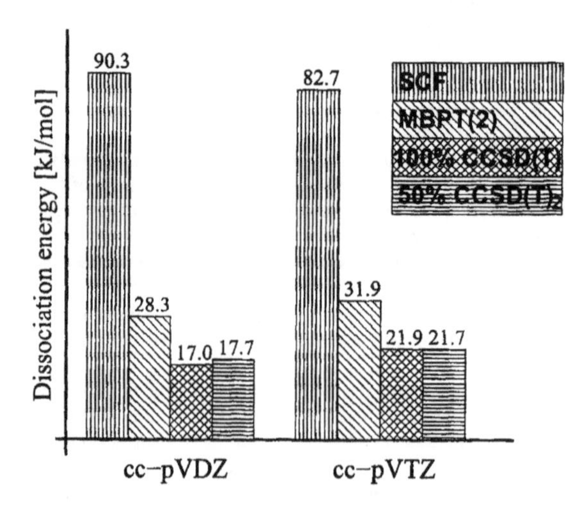

Figure 7: SCF and the correlation contribution to the dissociation energy, ΔE, of pentane to propene and ethane [kJ/mol] with cc-pVDZ and cc-pVTZ basis sets. 1s orbital of C are frozen. ΔE components were calculated in the full VOS and OVOS truncated to 50%.

Table 1: SCF and CCSD(T) interaction energies of the planar double hydrogene bonded cyclic dimer of formamidine with the full VOS and OVOS reduced to 50%. BSSE corrrected [40] interaction energies in kJ/mol.

Basis set	SCF	CCSD(T) 100%	$CCSD(T)_2^{(OVOS)}$ 50%
cc-pVDZ	-44.9	-51.8	-53.5
cc-pVTZ	-44.6	-60.3	-59.9
aug-cc-pVDZ	-43.9	-57.9	-57.1
aug-cc-pVTZ	-44.8	-63.6	-63.0

less than 1.0 kJ/mol (just for the cc-pVDZ basis it is 1.7 kJ/mol). Such error is much smaller than is the basis set effect.

For reasonably accurate description of the electron correlation contribution to the interaction energy of the stacking structure of the FO...FO dimer it is crucial to use augmented basis sets. Non–augmented cc-pVDZ and cc-pVTZ basis sets (Table 2) and basis sets of the type like 6-31G** give even repulsive electron correlation contribution to ΔE [39]. This means that large basis sets combined with OVOS is particularly important in similar applications. Data presented in Table 2 show that errors introduced by OVOS reduced to 50% are much smaller than are the basis set effects. At the same time, using OVOS reduces the computer time considerably.

More detailed analysis of the performance of OVOS for hydrogen–bonded systems and weak intermolecular interactions will be published separately [41].

Table 2: SCF and CCSD(T) interaction energies of the stacking formaldehyde dimer with the full VOS and OVOS reduced to 50%. BSSE corrrected [40] interaction energies in kJ/mol.

Basis set	SCF	CCSD(T) 100%	$CCSD(T)_2^{(OVOS)}$ 50%
cc-pVDZ	-7.09	-5.06	-5.36
cc-pVTZ	-8.43	-7.99	-7.91
aug-cc-pVDZ	-9.14	-9.46	-9.41
aug-cc-pVTZ	-8.99	-10.17	-10.03

6 Dipole moments and dipole polarizabilities

Calculation of electric properties is very demanding test of the performance of OVOS [42]. Since the gradient technique for electric properties is not yet available within OVOS, we calculated dipole moments and dipole polarizabilities using the finite–field method. After a few numerical tests we used the strength of the external field 0.003 a.u. This means that for polarizability accurate to one decimal point energies should be accurate to 10^{-6} a.u., or at least the error compensation must apply at this level of accuracy. Numerical demands can be alleviated by using a series of external fields using numerical recipes [43] which allow to eliminate the higher order contributions to the power expansion of the energy with respect to an external field. Such procedures are unavoidable for calculations of hyperpolarizabilities.

In Figures 8 and 9 we present dipole moments for CO and the thiophene molecule, respectively. The performance of OVOS with the virtual space reduced to 50% is excellent, particularly for the largest basis set, aug-cc-pV5Z. The CBS limit for the dipole moment μ_e using the two largest basis sets, aug-cc-pVQZ and aug-cc-pV5Z with the full and the OVOS space, respectively, agree to within 0.002 a.u. Both values agree with experiment, $\mu_e = 0.048$ a.u. excellently. The correlation contribution to the CO dipole moment is crucial. It is notoriously known that the SCF method leads to incorrect polarity of CO [44, 45].

Concerning the performance of OVOS, essentially the same as for CO can also be said for the dipole moment of thiophene, see Fig. 9. The agreement of the OVOS with the full space result with the aug-cc-pVDZ basis set must be considered as accidental. CBS limits from aug-cc-pVTZ and aug-cc-pVQZ basis sets are essentially the same, agreeing with the experimental value of 0.21 a.u. [46] excellently. We note that the full virtual space calculation with the aug-cc-pVQZ basis set, 566 virtuals, approaches our computer capabilities. With OVOS, of course, we could use even larger basis sets than aug-cc-pVQZ. Detailed analysis of μ_e of thiophene with different VOS reductions is presented in Table 3. This time we also compare plain OVOS CCSD(T) results with results following from OVOS corrected by X_2, see eq. (8), (9). Resulting CCSD$(T)_2$ dipole moments agree excellently with the full-space result even if VOS is reduced down to 40% of the full virtual space. Plain OVOS CCSD(T) dipole moment is slightly worse, even if the error introduced by OVOS remains smaller than is the basis set effect. Apparently μ_e varies very little when extending the basis set from aug-cc-pVTZ to aug-cc-pVQZ. Actually there is very significant basis set dependence as is seen form comparison of our data with those published so far [47]. Only large basis sets can represent the dipole moment of thiophene accurately enough. Essentially the same can be told about the dipole polarizability of thiophene, see Table 4. Due to the lower symmetry when applying the perpendicular external electric field to the symmetry plain of the thiophene molecule we could not obtain the full-space CCSD(T) results for the α_{xx} and α_{yy} polarizability components. Excellent performance of OVOS with the virtual space reduced down to 40% of the full space is clearly documented for the aug-cc-pVTZ basis set.

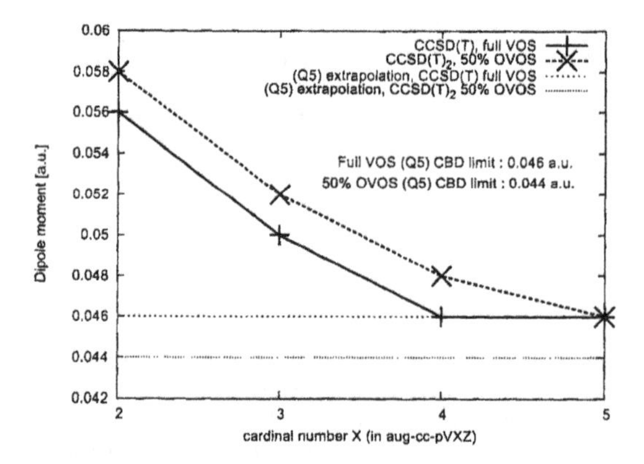

Figure 8: Dipole moment μ_e [a.u.] of the CO molecule (r(CO)=1.128 $\AA$) calculated with the full virtual orbital space and CCSD$(T)_2$ calculations using OVOS truncated to 50%. Aug-cc-pVXZ, X=D, T, Q and 5 basis sets, with 1s orbital of O and C frozen. Horizontal lines represent the CBS limit (Q5) obtained with aug-cc-pVQZ and aug-cc-pV5Z basis sets.

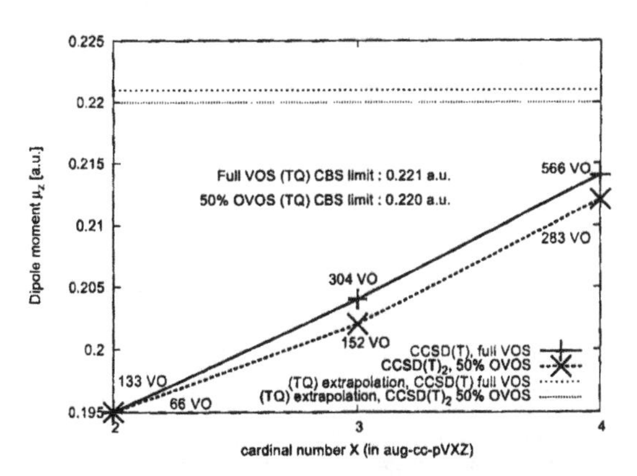

Figure 9: Dipole moment μ_e [a.u.] of the thiophene molecule calculated with the full virtual orbital space and $CCSD(T)_2$ calculations using OVOS truncated to 50%. Aug-cc-pVXZ, X=D, T, and Q basis sets, with 1s orbitals of C and 1s, 2s and 2p orbitals frozen. Horizontal lines represent the CBS limit (TQ) obtained using basis sets with X=T and Q. Inserted are numbers of VOs used in the full space and the OVOS calculations, respectively.

Table 3: CCSD(T) dipole moment μ_z [a.u.] of thiophene[(a)] with aug-cc-pVXZ, X=T, Q basis sets. 1s orbital of C and 1s, 2s and 2p orbitals of S are frozen. Compared are the full VOS results with a series of truncated OVOS results and other theoretical and experimental results. Presented are plain CCSD(T) results and the X_2 corrected $CCSD(T)_2$ results (see eq. (8), (9)).

% of OVOS	No. of VOs	CCSD(T)	$CCSD(T)_2^{(OVOS)}$
aug-cc-pVTZ			
100.0	304	0.204	0.204
59.9	182	0.210	0.204
50.0	152	0.210	0.202
39.8	121	0.214	0.203
aug-cc-pVQZ			
100.0	566	0.214	0.214
59.9	339	0.216	0.212
50.0	283	0.217	0.212
39.9	226	0.218	0.213
Kamada at al.[(b)] [47]		CCSD(T)/6-31G+pd	0.400
		CCSD(T)/6-31G+pdd	0.362
		MP4SDTQ/6-31G(3d)+pd	0.218
Friguelli at al. [46]		Exp.	0.21

[(a)] - MBPT(2)/aug-cc-pVDZ optimized geometry
[(b)] - SCF optimized geometry

Table 4: $CCSD(T)_2$ dipole polarizability components, average value and anisotropy [a.u.] of thiophene[(a)] with aug-cc-pVXZ, X=T, Q basis sets. 1s orbital of C and 1s, 2s and 2p orbitals of S are frozen. Compared are the full VOS results with a series of truncated OVOS results and other theoretical and experimental results.

% of OVOS	No of VOs	α_{zz}	α_{xx}	α_{yy}	α_s	$\Delta\alpha$
aug-cc-pVTZ						
100.0	304	77.75	44.83	71.54	64.71	30.29
59.9	182	77.77	44.88	71.54	64.73	30.26
50.0	152	77.78	44.88	71.54	64.73	30.27
39.8	121	77.67	44.82	71.71	64.73	30.32
aug-cc-pVQZ						
100.0	566	77.76	-	-	-	-
59.9	339	77.74	-	-	-	-
50.0	283	77.78	-	-	-	-
39.9	226	77.87	-	-	-	-
Kamada at al.[(b)] [47]	CCSD(T)/6-31G+pd	72.93	40.38	64.80	59.37	29.34
	CCSD(T)/6-31G+pdd	77.34	45.67	69.21	64.07	28.49
	MP4SDTQ/6-31G(3d)+pd	75.50	44.62	68.32	62.81	27.99
Friguelli at al. [46]	Exp.	-	-	-	65.18	-

[(a)] - MBPT(2)/aug-cc-pVDZ optimized geometry
[(b)] - SCF optimized geometry

7 OVOS in quasi–relativistic CCSD(T) calculations

Quasi-relativistic calculations represent another interesting area of applications of the OVOS technique. Most frequently used Douglas-Kroll-Hess (DKH) method [48, 49, 50] is based on the Foldy-Wouthuysen transformation [51] of the Dirac Hamiltonian. Its simplest form, the second–order one component spin–averaged DKH2 approach represents nowadays one of the most popular quasi-relativistic techniques. Recently, also higher order one– and two–component versions of the DKH method were formulated and successfully applied [52, 53, 54, 55]. One–component DKH method combines description of the most important scalar relativistic effects with the methodological simplicity. Relativistic effects are treated in this method via an additional one electron term in the Hamiltonian. Thus, quasi–relativistic molecular orbitals may differ from their non–relativistic counterpart quite significantly. Consequently, non–relativistic contraction of the basis set is not suitable for DKH calculations. Proper treatment would be either using the uncontracted basis sets, or we must employ specific relativistically modified contracted basis sets. Of course, basis sets including famous Pol_dk or HyPol_dk were contracted [56, 57, 58] having all this in mind. These basis sets were developed particularly for the one–component DKH2 calculations. Oddly enough, specific contraction should be applied to any specific Hamiltonian, that means, for any one– or two–component version of the DKH method, if we stay within this model. Obviously, this is completely impractical. Another possibility, that is using uncontracted basis sets represents a serious limitation, due to their size, as compared with the reasonable size of available contracted DKH quasi-relativistic basis sets. Under these circumstances the idea of using OVOS in DKH calculations appears to be very fruitful. We can run the SCF step using an original uncontracted basis set producing proper quasi-relativistic MO's which serve as a proper reference for subsequent time consuming correlation step. Starting from such properly constructed one–electron reference, we can subsequently reduce the virtual orbital space for correlated calculations to an acceptable

Table 5: Polarizability of Au^- (in a.u.). Spin free DKH2 calculations

% of VO[a]	shells	SCF	Y_2^b	CCSD[c]	$CCSD_2$	(T)[d]	*corr.*[e]	$CCSD(T)_2$
u-HyPol, *no OVOS reduction*								
100.0	15s13p8d8f2g	199.57	—	-93.29	—	-15.29	-108.58	90.99
HyPolX_dk, *no OVOS reduction*								
56.0	7s7p4d4f2g	198.49	—	-92.85	—	-14.89	-107.75	90.74
u-HyPol, *reduced by OVOS*								
70.8	6s7p5d7f2g	199.57	-0.002	-93.29	-93.29	-15.29	-108.58	90.99
59.5	—[f]		-0.065	-93.35	-93.28	-15.29	-108.57	91.00
48.2	5s5p3d4f1g		-0.850	-92.09	-92.94	-15.24	-108.18	91.39
32.7	3s4p2d3f1g		8.933	-100.78	-91.85	-16.35	-108.20	91.37
31.0	3s3p2d3f1g		-1.437	-96.03	-94.59	-16.33	-110.92	88.65

[a] Related to the size of u-HyPol virtual orbital space (168 v.o.)
[b] The Y_2 indicator defined by eq. (10)
[c] Pure (X_2 uncorrected) CCSD correlation contribution to α.
[d] The noniterative triples correlation contribution to α.
[e] Complete $CCSD(T)_2$ correlation contribution to α.
[f] The orbital shell structure could not be recognized.

size employing the OVOS technique. Due to the nature of the OVOS approach, resulting reduced VOS of an uncontracted basis set has usually different structure (i.e. the number of *s*, *p*, *d* ... shells) than would have a chosen contracted basis set even if both may have the same size.

Let us illustrate this idea using electric properties of the gold element containing species [15]. We employ uncontracted u-HyPol basis sets, for which quasi-relativistic HyPol_dk contracted basis sets are known [56, 57, 58] along with the smallest contracted basis sets of this kind, namely Pol_dk. For our illustrations we will use the simplest second-order spin-averaged DKH2 approach. Results for the polarizability of Au^- are summarized in Table 5. The size of all basis sets (including the contracted ones) is related to the size of the starting uncontracted u-HyPol basis set. Standard contraction of this basis set is denoted as HyPolX_dk [57]. We see, that the polarizability α with the contracted HyPolX_dk basis is lower than α with the uncontracted basis by only 0.25 a.u. Thus, contraction modifies α very little. However, this favorable result follows partly from the compensation of the positive SCF contribution (the difference due to the contraction is -1.08 a.u.) with the negative CC correlation contribution (the difference is 0.83). The size of VOS in the contracted HyPolX_dk basis represents 56% of the size of the full VOS of the uncontracted basis. OVOS correlation contributions to α of Au^- are actually more stable than that with the HyPolX_dk basis down to reduction of OVOS of 33% (note however, large Y_2 correcting factor for this last reduction). With OVOS which is 59.5% of the full uncontracted space is α almost identical with that with the uncontracted basis. From this point of view, results obtained with OVOS reduction of the u-HyPol basis set must be viewed as acceptable down to the level of reduction $\approx 1/3$ of the u-HyPol VOS, i.e. much less than is the size of the HyPol_dk (56%).

As a molecular examples we present AuH and AuF. The dipole moment of AuH with OVOS reduced even to 40% of the full VOS of the uncontacted basis set still deviates from the u-HyPol basis set result by only 0.3%, in contrast to larger HyPolX_dk basis with the difference 1.5%. Polarizabilities with OVOS and HyPolX_dk basis are approximately the same, the former ones are slightly better. For the dipole moment of AuF OVOS performs excellently provided that the reduction with respect to the full space of the u-HyPol basis is not more drastic than 50%. The same holds for the α_{xx} polarizability component. The performance of OVOS for the parallel component,

Table 6: Electric properties (in a.u.) of AuH and AuF molecules[a]. Spin free DKH2 CCSD(T)$_2$ calculations

% of VO[b]	μ_z	%(μ_z)[c]	α_{zz}	%(α_{zz})[c]	α_{xx}	%(α_{xx})[c]
AuH						
u-HyPol, *no OVOS reduction*						
100.0	-0.5002	—	39.80	—	35.80	—
HyPolX_dk, *no OVOS reduction*						
59.2	-0.4929	-1.5	39.84	0.1	35.82	0.1
PolX_dk, *no OVOS reduction*						
39.8	-0.5214	4.2	39.14	-1.7	34.66	-3.2
u-HyPol *reduced by OVOS*						
69.9	-0.5004	0.0	39.83	0.1	35.79	0.0
59.2	-0.5002	0.0	39.79	0.0	35.79	0.0
50.0	-0.5010	0.2	39.66	-0.3	35.80	0.0
39.8	-0.4988	-0.3	39.90	0.3	35.76	-0.1
29.6	-0.5001	0.0	40.04	0.6	35.74	-0.2
AuF						
u-HyPol, *no OVOS reduction*						
100.0	1.652	—	36.13	—	29.57	—
HyPolX_dk, *no OVOS reduction*						
58.8	1.645	-0.4	36.18	0.2	29.65	0.3
PolX_dk, *no OVOS reduction*						
39.4	1.683	1.9	35.05	-3.0	28.52	-3.6
u-HyPol *reduced by OVOS*						
73.9	1.652	0.0	36.13	0.0	29.58	0.0
64.6	1.653	0.1	36.11	-0.1	29.58	0.0
58.8	1.655	0.2	36.20	0.2	29.59	0.1
50.0	1.658	0.4	35.83	-0.8	29.60	0.1
39.8	1.666	0.8	35.62	-1.4	29.62	0.2
30.1	1.670	1.1	35.50	-3.0	29.61	0.1

[a] r(AuH)=1.52385Å, r(AuF)=1.917449Å

[b] Related to the size of u-HyPol virtual orbital space (196 v.o. for AuH and 226 v.o. for AuF)

[c] The relative error related to the non-reduced u-HyPol value in %.

α_{zz}, is slightly worse but still remains reasonably good. Altogether data collected in Tab. 6 clearly show that OVOS in connection with uncontracted basis sets is a promising alternative for the treatment of relativistic calculations when there is no reasonably contracted basis set available for a selected relativistic Hamiltonian.

Calculations of the Au^- anion polarizabilities enable us to demonstrate some specific aspects of the OVOS procedure. First, this example illustrates, that OVOS is applicable also for negatively charged system, for which α is very sensitive to the basis set selection. Another specific feature that needs to be kept under control is the necessity of the balanced reduction of VOS, already mentioned in the Introduction. In general, we can not delete desired portion of the virtual space (say 50%) arbitrarily, irrespective of the distribution of reduced virtual orbitals among irreducible representations (irreps). First, we must ensure, that employed distribution of reduced virtual orbitals among irreps for selected extent of the reduction yields the largest possible overlap (when using the overlap functional). This basic condition is necessary, but not sufficient. When reducing specific type of orbital (or group of orbitals) in one irrep, we must simultaneously exclude all their counterparts in other irreps, what is not guaranteed automatically. In our example of Au^-, deleting complete p, d ... shells is required for balanced description. Different $2l+1$ components of these shells belong to different irreps. Thus, incorrectly chosen OVOS space may have broken shell structure when a mixture of different components of shells is deleted (i.e. some of the shells are not deleted/preserved completely). "Unbalanced" reductions may give incorrect results particularly when OVOS is reduced to a high extend. With modest reductions of VOS the optimization procedure may still recover some deficiencies of the selected space (see e.g. 59.5% OVOS reduction of u-HyPol basis set in table 5).

Atomic species are a good systems for demonstrating a proper role of a balanced truncations with the OVOS technique [15]. With the lowering of the symmetry of the system under study such problems rapidly diminish.

8 Summary

The essence of the technique for obtaining the optimized reduced space of virtual orbitals suitable for effective electron correlation calculations is summarized. In the present paper we deal with the first-order overlap wave function in the full virtual space and the reduced OVOS space, respectively, as an optimization functional [14]. An analysis of the electron correlation energy of the fluorine molecule with a series of the aug-cc-pVXZ basis sets (X=T, Q, 5) demonstrates the convergence with respect to a systematic improvement of the basis set with the full and the reduced OVOS. For example, CCSD(T) electron correlation contribution with the VOS of the aug-cc-pV5Z basis set reduced to the size of the full virtual space of the aug-cc-pVQZ basis is significantly larger in the former case with approximately the same computer time needed for both calculations. The performance of OVOS can be further enhanced using the X_2 correction, defined as a difference [CCSD(T)^{OVOS} - MBPT(2)^{OVOS}]. CCSD(T)^{OVOS} and MBPT(2)^{OVOS} energies are calculated in the reduced optimized space of virtual orbitals. Since MBPT(2) is a dominating contribution to the correlation energy, we add the X_2 correction to MBPT(2) calculated using the full virtual orbitals space. Resulting energies on F_2 provide, for example, vibrational harmonic frequencies which deviate from the full VOS by less than 1 cm^{-1} using OVOS reduced to 50%.

We have demonstrated the usefulness of the OVOS technique for a variety of applications. Our VOS is typically reduced to one half of the full virtual space of any selected basis set, which reduces the computer time needed for the CCSD(T) calculation more than ten times. The first group of applications concerns the reaction energy of the HCN $\leftrightarrow$ HNC isomerization and the dissociation of pentene to propene and ethane. We have shown that ΔE calculated using OVOS reduced to 50% deviates from the target full VOS result by less than 0.2 kJ/mol. This is well beyond chemical accuracy. The full VOS interaction energies of the hydrogen–bonded systems (like the cyclic formamidine dimer) are reproduced by using OVOS reduced to 50% to within 1 kJ/mole. The accuracy of OVOS results improves with increased basis set size. Similar accuracy is reached in weak intermolecular interactions of, e.g. the stacking formaldehyde dimer. The performance of OVOS is better when the basis set superposition error in ΔE is eliminated using the common Boys–Bernardi method [40], i.e. when the same basis set is used for the supersystem and its subsystems. Plain interaction energies calculated by OVOS suffer from the unbalanced truncation of the virtual orbitals within OVOS. Calculations of ΔE using OVOS indicate that the expected error cancellation is better satisfied when the SCF energy and the electron correlation contribution is closer to the accurate result. This is better fulfilled when the virtual space of a larger basis set is reduced by OVOS rather than when we stay with the full virtual space of a smaller basis set.

Demanding test of the performance of OVOS represent finite field calculations of dipole moments and dipole polarizabilities. Error cancellation must be well balanced when energies are calculated without and with an external field. This is demanding particularly for polarizabilities particularly when an external field reduces the symmetry. Yet the accuracy of OVOS reduced to 50% in presented cases is about 0.002–0.006 a.u. for dipole moments and about 0.02–0.1 a.u. for the dipole polarizabilities. For other examples see [42].

We stress that the virtual space must be truncated taking into account the symmetry of the system under consideration. In other words, symmetry equivalent orbitals must all be truncated or retained within OVOS.

In all applications the magnitude of the basis set effect in standard full VOS calculation was larger than was the error introduced by OVOS when the virtual space was reduced to, say, 50%. Almost exact results are obtained with OVOS reduced to about 70–80%. The computer time saving in CCSD(T) is, however, only about 30–40 % of the full VOS calculation.

OVOS may also serve as an alternative to the contraction of the primitive basis sets in relativistic calculations. Any approximate relativistic Hamiltonian needs in principle specific contraction scheme and contraction coefficients. We suggest, that the uncontracted basis may be used in the relativistic calculation corresponding to the SCF step. A reference orbital basis obtained within relativistic scheme can than be used in the electron correlation step with reduced VOS by using the OVOS technique. Comparison of results within the DKH2 model shows that CCSD(T) calculations with VOS of the uncontracted basis set reduced to the size of a standard contracted basis sets, like HyPolX_dk or PolX_dk are still sufficiently accurate and efficient. Examples include calculations of the dipole polarizabilities and dipole moments of the Au^- anion and AuH and AuF molecules.

Future development will be oriented towards improving the optimization procedures, improved convergence of the optimization process and better control of the accuracy. This last item concerns a possible failure of the optimization process due to, for example, landing in the local extreme of the optimization functional.

Further extensions will be devoted to implementation of OVOS to open shell systems which will allow calculations of radicals that can be represented by the single–determinant reference using the Restricted open–shell HF reference [59] and the two–determinant reference for open–shell singlets [60]. This will also open straightforward applications to effective obtaining of ionization energies and electron affinities. Using OVOS in accurate or approximate iterative CC procedures including higher than the single and double excitation operators, i.e. iterative CCSDT or CCSDTQ may increase the efficiency of such CC methods much more than is the case with CCSD(T). The reason is obvious, and follows from much worse scaling of the computer time with number of virtual orbitals. Important area is using OVOS in multi–reference CC and other CC methods [61, 38, 62, 63] aimed at calculations of systems which manifest quasidegeneracy.

Acknowledgment

The support of the Slovak Grant Agency VEGA (Contract No. 1/3560/06) and the APVV grant agency (Contract No. APVV-20-018405) is gratefully acknowledged.

References

[1] R.G. Parr and W. Yang, *Density-Functional Theory of Atoms and Molecules*, Oxford Science Publications, Oxford, 1989

[2] R.J. Bartlett, V.F. Lotrich and I.V. Schweigert, J. Chem. Phys. (2005) **123**, article num. 062205.

[3] J. Čížek, J. Chem. Phys. (1966) **45**, 4256.

[4] J. Paldus and X. Li, Adv. Chem. Phys. (1999) **110**, 1.

[5] R.J. Bartlett in *Modern Electronic Structure Theory, Part II.*, ed. by D.R. Yarkony, World Scientific, Singapore, (1995), 1047.

[6] M. Urban, I. Černušák, V. Kellö, J. Noga in *Methods in Computational Chemistry, Vol. 1.*, ed. by S. Wilson, Plenum Press, New York, (1987), 117.

[7] T.J. Lee and G.E. Scuseria, in *Quantum Mechanical Electronic Structure Calculations with Chemical Accuracy*, ed. by S.R. Langhoff, Kluwer Academic Publishers, Dordrecht, (1995), 47.

[8] J. Gauss in *Encyclopedia of Computational Chemistry*, ed. by P. von Ragué Schleyer, John Wiley & Sons, Chichester, (1998), 615.

[9] K. Raghavachari, G.W. Trucks, J.A. Pople and M. Head-Gordon, Chem. Phys. Letters (1989) **157**, 479.

[10] M. Urban, J. Noga, S.J. Cole and R.J. Bartlett, J. Chem. Phys. (1985) **83**, 4041.

[11] L. Adamowicz and R.J. Bartlett, J. Chem. Phys. (1987) **86**, 6314.

[12] L. Adamowicz, R.J. Bartlett and A.J. Sadlej, J. Chem. Phys. (1988) **88**, 5749.

[13] L. Adamowicz, J. Phys. Chem. (1989) **93**, 1780.

[14] P. Neogrády, M. Pitoňák and M. Urban, Mol. Phys. (2005) **103**, 2141.

[15] M. Pitoňák, P. Neogrády, V. Kellö and M. Urban, Mol. Phys. (2006) **104**, 2277.

[16] W. Klopper, J. Noga, H. Koch and T. Helgaker, Theor. Chem. Acc. (1997) **97**, 164.

[17] H.J.Aa. Jensen, P. Jøorgensen, H. Agren and J. Olsen, J. Chem. Phys. (1988) **88**, 3834.

[18] J. Wasilewski, Int. J. Quantum Chem. (1991) **39**, 649.; J. Wasilewski, S. Zelek and M. Wierzbowska, Int. J. Quantum Chem. (1996) **60**, 1027.

[19] W.D. Laidig, G.D. Purvis and R.J. Bartlett, J. Phys. Chem. (1985) **89**, 2161.

[20] P. Pulay and S. Saebø, Theor. Chim. Acta (1986) **69**, 357.; J.W. Boughton and P. Pulay, J. Comput. Chem. (1993) **14**, 736.

[21] M. Schütz, G. Hetzer and H-J. Werner, J. Chem. Phys. (1999) **111**, 5691.

[22] M. Schütz, and H-J. Werner, J. Chem. Phys. (2001) **114**, 661.

[23] T. Korona, K. Pflüger and H-J. Werner, Phys. Chem. Chem. Phys. (2004) **6**, 2059.

[24] H. Koch and A. Sánches de Merás, J. Chem. Phys. (2000) **113**, 508.; H. Koch, A. Sánches de Merás and T.B. Pedersen, J. Chem. Phys. (2003) **118**, 9481.

[25] P. Constans and G.E. Scuseria, Collect. Czech. Chem. Commun. (2003) **68**, 357.; P. Constans, P.Y. Ayala and G.E. Scuseria, J. Chem. Phys. (2000) **113**, 10451.

[26] P.O. Löwdin, Phys. Rev. (1955) **97**, 1474.

[27] L. Bytautas, J. Ivanic and K. Ruedenberg, J. Chem. Phys (2003) **119**, 8217.

[28] C. Edmiston and M. Krauss, J. Chem. Phys. (1966) **45**, 1833.

[29] T.L. Barr and E.R. Davidson, Phys. Rev. A (1970) **1**, 644.

[30] A.G. Taube and R.J. Bartlett, Collect. Czechoslov. Chem. Commun. (2005) **70**, 837.

[31] C. Sosa, J. Geertsen, G.W. Trucks and R.J. Bartlett, Chem. Phys. Lett. (1989) **159**, 148.

[32] M. Pitoňák, P. Neogrády, V. Kellö and M. Urban, *erratum*, Mol. Phys., in press.

[33] T.H. Dunning, Jr. and K.A. Peterson in *Encyclopedia of Computational Chemistry*, ed. by P. von Ragué Schleyer, John Wiley & Sons, Chichester, (1998).

[34] D.E. Woon and T.H. Dunning, Jr., J. Chem. Phys. (1994) **100**, 2975.

[35] K.A. Peterson, R.A. Kendall and T.H. Dunning, Jr., J. Chem. Phys. (1993) **99**, 9790.

[36] W. Klopper and T. Helgaker, Theor. Chem. Acc. (1998) **99**, 265.; A. Halkier, T. Helgaker, P. Jørgensen, W. Klopper, H. Koch, J. Olsen and A.K. Wilson, Chem. Phys. Lett. (1998) **286**, 243.

[37] F. Pawlowski, A. Halkier, P. Jørgensen, K.L. Bak, T. Helgaker and W. Klopper, J. Chem. Phys. (2003) **118**, 2539.

[38] J. Pittner, J. Šmydke, P. Čársky and I. Hubač, THEOCHEM–J. Mol. Struct. (2001) **547**, 239.

[39] J. Pittner and P. Hobza, Chem. Phys. Lett. (2004) **390**, 496.

[40] S. F. Boys and F. Bernardi, Mol. Phys. (1970) **19**, 553.

[41] P. Dedíková, M. Pitoňák, P. Neogrády, I. Černušák and M. Urban, to be published.

[42] M. Pitoňák, P. Neogrády, F. Holka and M. Urban, THEOCHEM–J. Mol. Struct., in press.

[43] G. Maroulis, J. Chem. Phys. (1991) **94**, 1182.

[44] M. Medveď, M. Urban and J. Noga, Theor. Chem. Acc. (1997) **98**, 75.

[45] K.L. Bak, J. Gauss, T. Helgaker, P. Jørgensen and J. Olsen, Chem. Phys. Lett. (2000) **319**, 563.

[46] F. Friguelli, G. Marino and A. Taticchi, Adv. Heterocykl. Chem. (1977) **21**, 119.

[47] K. Kamada, M. Ueda, H. Nagao, K. Tawa, T. Sugino, Y. Shmizu and K. Ohta, J. Phys. Chem. (2000) **104**, 4723.

[48] M. Douglas and N.M. Kroll., Ann. Phys. (N.Y.) (1974) **82**, 89.

[49] B.A. Hess., Phys. Rev. A, (1986) **33**, 3742.

[50] G. Jansen and B.A. Hess., Phys. Rev. A, (1989) **39**, 6016.

[51] L.L. Foldy and S.A. Wouthuysen., Phys. Rev., (1950) **78**, 29.

[52] M. Barysz and A.J. Sadlej., THEOCHEM–J. Mol. Struct. (2001) **573**, 181.

[53] M. Barysz and A.J. Sadlej., J. Chem. Phys. (2002) **116**, 2696.

[54] M. Barysz. in *Theoretical chemistry and physics of heavy and superheavy elements*, ed. by U. Kaldor and S. Wilson, Kluwer Academic, London (2003), 349.

[55] D. Kedziera and M. Barysz., J. Chem. Phys. (2004) **121**, 6719.

[56] V. Kellö and A.J. Sadlej., Theor. Chim. Acta (1994) **94**, 93.

[57] T. Pluta and A.J. Sadlej., Chem. Phys. Lett. (1998) **297**, 391.

[58] Polarized basis sets family: PolX, PolX_dk, HyPolX, HyPolX_dk can be found on the web page http://www.qch.fns.uniba.sk/Baslib.

[59] M. Urban, P. Neogrády and I. Hubač. in *Recent Advances in Computational Chemistry, Vol. 3*, ed. by R.J. Bartlett, World Scientific, Singapore (1997), 275.

[60] P. Neogrády, P. G. Szalay, W. P. Kraemer, and M. Urban, Collect. Czechoslov. Chem. Commun. (2005) **70**, 951.

[61] X. Li and J. Paldus, J. Chem. Phys. (1997) **107**, 6257.

[62] M. Musial, S.A. Kucharski and R.J. Bartlett, J. Chem. Phys. (2002) **116**, 4382.

[63] P. Piecuch, S.A. Kucharski and R.J. Bartlett, J. Chem. Phys., (1999) **110**, 6103.

Brill Academic Publishers
P.O. Box 9000, 2300 PA Leiden
The Netherlands

Lecture Series on Computer and Computational Sciences
Volume 6, 2006, pp. 286-293

General spin orbital density functional study of transition metal clusters and complexes

S. Yamanaka, M. Shoji, K. Koizumi, R. Takeda, Y. Kitagawa, H. Isobe, K. Yamaguchi[1]

Department of Chemistry,
Faculty of Sciences,
Osaka University,
Toyonaka, Osaka, Japan

Received 21 June, 2006; accepted 25 June, 2006

Abstract: Generalized spin orbital version of density functional approach is reviewed. We discuss the fundamental characteristics of GSO-DFT from the viewpoint of group theoretical aspect, in relation to time-dependent DFT, and information entropy. As typical types of broken symmetry solutions, such as noncollinear magnetism of multicenter transition metal clusters, e.g. icosidodecahedron, Fe_{30}, are presented. Cubane-type Mn-O and Fe-S clusters in biological systems are also examined as typical examples.

Keywords: General spin orbital, symmetry-broken approach, Heisenberg model, information entropy, multicenter clusters, biological systems

1. Introduction

The electronic properties of the multicenter transition metal (TM) clusters and complexes has attracted much attention in relation to the particular properties such as reactivities of metalloprotein and the transition of electronic structure between small molecules and bulk [1]. The electronic structures of these species are often investigated by the measurements of magnetic susceptibility, which can be calculated by using Heisenberg spin Hamiltonian,

$$\hat{H} = -2\sum_{ab} J_{ab}\hat{\mathbf{S}}_a \cdot \hat{\mathbf{S}}_b \tag{1}$$

where $\hat{\mathbf{S}}_c$ (c=a,b) and J_{ab} are the spin operator at site c and effective exchange integrals, respectively. In the field of computational chemistry and physics, the Kohn-Sham version of density functional theory (KS-DFT) [2] enables us to treat such a systems based on *ab initio* quantum theory. Although KS-DFT is another type of *exact* electronic structure theory, its mathematical structure is similar to the mean-field approximation, leading to classical magnetic solutions. Such classical solutions of KS-DFT provide the useful information like stable spin state, magnetic moment, and molecular orbitals. Further our mapping schemes [3-5] to spin Hamiltonian enables us to estimate magnetic interactions between transition metal sites For instance, the effective exchange integrals between two sites is given by [3]

$$J_{ab} = \frac{E^{HS}_{KS-DFT} - E^{LS}_{KS-DFT}}{\left\langle \hat{\mathbf{S}}^2 \right\rangle^{LS}_{KS-DFT} - \left\langle \hat{\mathbf{S}}^2 \right\rangle^{HS}_{KS-DFT}}. \tag{2}$$

for a simple system like triangular and tetrahedral TM compounds, and

[1] Corresponding author. Prof. Kizashi Yamaguchi, E-mail: yama@chem.sci.osaka-u.ac.jp

$$J = \frac{E_{KS-DFT}^{HS} - E_{KS-DFT}^{LS}}{2N\left[\left\langle \hat{\mathbf{S}}_i \cdot \hat{\mathbf{S}}_{nn\,of\,i}\right\rangle_{KS-DFT}^{LS} - \left\langle \hat{\mathbf{S}}_i \cdot \hat{\mathbf{S}}_{nn\,of\,i}\right\rangle_{KS-DFT}^{HS}\right]} \tag{3}$$

for any system described by a homogeneous nearest-neighbor interacted Heisenberg model [4] where N is the number of nearest-neighbor spin-pairs. In addition, E_{KS-DFT}^{X} , $\left\langle \hat{\mathbf{S}}^2 \right\rangle_{KS-DFT}^{X}$ and $\left\langle \hat{\mathbf{S}}_i \cdot \hat{\mathbf{S}}_{nn\,of\,i}\right\rangle_{KS-DFT}^{X}$, are energy, spin expectation values, and nearest-neighbor (nn) spin-correlation functional of symmetry-broken KS-DFT solutions, respectively. Further the estimation of the anisotropic parameters like D and E originated from the magnetic dipole interactions and spin-orbit interactions are also possible using KS-DFT solutions [5]. Indeed, Shoji et al. investigated synthesized, modeled, and native transition metal complexes such as, oxo-and-hydroxo diiron compounds, iron-sulfur clusters, Ni_9 complexes [6] using spin unrestricted KS-DFT approach as illustrated in figure 1.

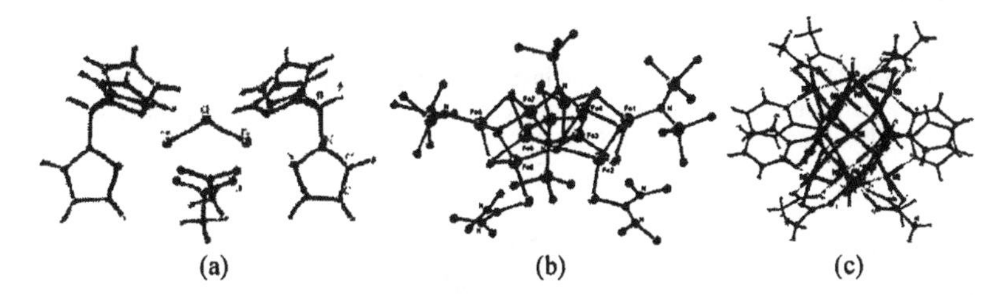

Figure 1 Geometries of (a) $[Fe_2O(O_2CCH_3)(HBpz_3)_2]$, (b) model complex of P cluster (Fe_8S_7), (c) $Ni_9(OH)_2(O_2CMe)_8$

2. General spin orbital density functional theory

From the viewpoint of molecular magnetism, one interesting point is that the triangular and tetrahedral units of metal skeleton are ubiquitous among these native TM complexes and synthesized TM clusters and complexes [7]. Since such units often exhibit spin-frustration, noncollinear magnetism might be inevitable in some cases. Further, the mixed valency of these compounds is another source of the noncollinear magnetism. To investigate such particular magnetic structure, we have to apply is general spin orbital (GSO) version of KS-DFT [8].

In the GSO-DFT, the fundamental variable is 2×2 spin density,

$$\rho_s(\mathbf{r}) = \begin{pmatrix} \rho_{\alpha\alpha}(\mathbf{r}) & \rho_{\alpha\beta}(\mathbf{r}) \\ \rho_{\beta\alpha}(\mathbf{r}) & \rho_{\beta\beta}(\mathbf{r}) \end{pmatrix} = \frac{\rho(\mathbf{r})}{2}\mathbf{E} + \sum_m^{x,y,z} \rho_m(\mathbf{r})\sigma_m . \tag{4}$$

Here E and $\{\sigma_m\}_{m=x,y,z}$ are a unit and Pauli matrices, respectively. Thus, the minimization of KS energy functional is taken over the extended search region,

$$E_0 = \mathrm{Min}_{\rho_s}\left[T_s[\rho_s(\mathbf{r})] + U_{clmb}[\rho_s(\mathbf{r})] + \int d\mathbf{r}\rho_s(\mathbf{r})V_{ext}(\mathbf{r}) + E_{XC}[\rho_s(\mathbf{r})]\right]. \tag{5}$$

In order to exploit the search region of eq. (4), we have to employ general spin orbital (GSO),

$$\phi_i(\mathbf{x}) = \phi_i^{\alpha}(\mathbf{r})\alpha + \phi_i^{\beta}(\mathbf{r})\beta = \begin{pmatrix} \phi_i^{\alpha}(\mathbf{r}) \\ \phi_i^{\beta}(\mathbf{r}) \end{pmatrix}, \tag{6}$$

for KS orbitals, together with localized spin treatment of localized spin density approximation for XC functional,

$$E_{XC}[\rho_+(\mathbf{r}),\rho_-(\mathbf{r})] = \int d\mathbf{r}(\rho_+(\mathbf{r})+\rho_-(\mathbf{r}))\varepsilon_{XC}[\rho_+(\mathbf{r}),\rho_-(\mathbf{r})], \tag{7}$$

proposed by Barth and Hedin [9]. In this form, the essential point is that the localized up- (ρ_+) and down- (ρ_-) spin densities are expressed not by ρ_α and ρ_β, but by

$$\rho_\pm(\mathbf{r}) = \rho(\mathbf{r}) \pm \left|Tr\vec{\sigma}\rho_s(\mathbf{r})\right|/2. \tag{8}$$

Here we should note that $Tr\vec{\sigma}\rho_s(\mathbf{r})\,/\,2 = (\rho_x, \rho_y, \rho_z)$ is a local spin. The extensions to general gradient approximation, self-interacted correction, GW approximation, and Hybrid DFT are straightforward [10]. Then, eq. (7) leads to the GSO version of KS equation given by,

$$\sum\nolimits_{\sigma_2}\left[\delta_{\sigma_1\sigma_2}\left\{-\frac{\Delta_i}{2} + \int d\mathbf{r}'\frac{\rho(\mathbf{r}')}{|\mathbf{r}-\mathbf{r}'|}\right\} + \frac{\delta E_{XC}}{\delta\rho_{\sigma_2\sigma_1}}\right]\phi_i^{\sigma_2}(\mathbf{r}) = \varepsilon_i\phi_i^{\sigma_1}(\mathbf{r}). \tag{9}$$

Indeed, using ab initio GSO-DFT approach, we found that three-dimensional stable solutions for tetrahedral clusters as shown in figure 2, and spin-canting character of electronic structure along the cleavage of oxygen molecule as shown in figure 3.

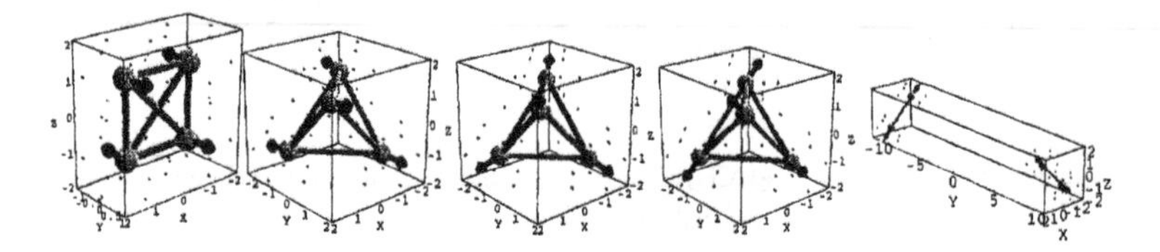

Figure 2. Spin-structural changes of GSO-DFT solutions along D_{2d} reaction path

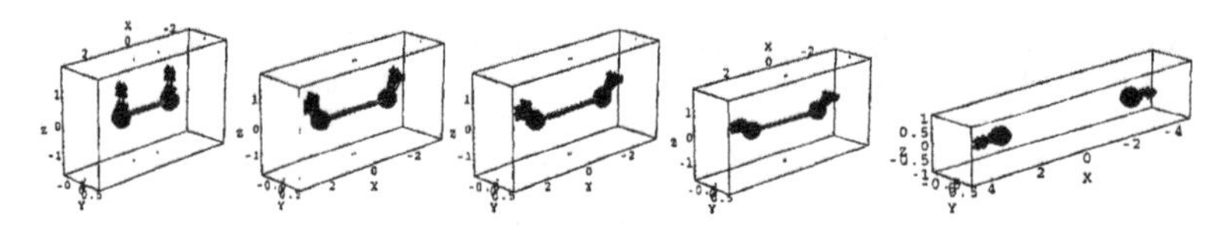

Figure 3. Spin-structural changes of GSO-DFT solutions of O_2

These two types of GSO solutions are fundamental building blocks of noncollinear magnetic structures of TM compounds.

3. Group-theoretical classification of GSO-DFT

Since GSO-DFT approach is symmetry-broken approach as in the case of generalized Hartree-Fock (GHF) theory [11], the group theoretical work of GHF can be transferred to GSO-KS-DFT as follows:if we notice that the fundamental variable of GSO-DFT is given by $\rho_s(\mathbf{r})$, the original symmetry framework is given by, $SO(3) \times P_I$, instead of the direct product of spin rotation and time-reversal groups, i.e., $S \times T$, for operations on spin density. Here $SO(3)$ is a (unimodular) three-dimensional (3D) rotation group and P_I is spatial inversion group. We can determine sets of GSO-DFT solution and subgroup of $SO(3) \times P_I$ by which the corresponding $\rho_s(\mathbf{r})$ is invariant. The symmetry-adapted (spinless) solutions have original symmetry of $SO(3) \times P_I$, literally. The usual spin-unrestricted KS-DFT (UKS-DFT) solutions having one-dimensional (1D) spin density, for instance in homolysis at dissociation limit of hydrogen molecules is called as axial spin density wave (ASDW). This UKS-DFT solution is treated as "symmetry-broken", but there remains a part of $SO(3) \times P_I$, i.e., A(**e**)M(**e'**). Here A(**e**) is the subgroup of SO(3) consisting of all spin rotations around a spin-polarized axis **e** and M(**e'**) is the group of order two generated by the spatial inversion with the π rotation around a axis, **e'** which is perpendicular to **e**. On the other hands, helical spin density wave (HSDW) has canted spins with a two-dimensional (2D) spin-polarized direction. If we take **e'** to be perpendicular the 2D plane, this type of spin densities is invariant for operations of M(**e'**). Finally, three-dimensional spin structures, which is denoted as torsional spin wave (TSW), has no symmetry within $SO(3) \times P_I$ except identity element 1.

. We note here that the inclusion relation,

$$SO(3) \times P_I \supset A(\mathbf{e}) \times M(\mathbf{e}') \supset M(\mathbf{e}') \supset \mathbf{1}\,. \tag{10}$$

holds. These symmetries restriction reflects the relation among those solutions concerning the constrained search region as

$$R_{SA} \subset R_{ASDW} \subset R_{HSDW} \subset R_{TSW}\,. \tag{10}$$

Here R_{SA}, R_{ASDW}, R_{HSDW}, and R_{TSW} are sets constituted of zero-, one-, two-, and three-dimensional spin densities, respectively.

$$R_{SA} \equiv \left\{\rho_s(\mathbf{r}) = \frac{1}{2}\rho(\mathbf{r})\mathbf{E}\right\}, \tag{11}$$

$$R_{ASDW} \equiv \left\{\rho_s(\mathbf{r}) = \frac{1}{2}\rho(\mathbf{r})\mathbf{E} + \rho_z(\mathbf{r})\sigma_z\right\}, \tag{12}$$

$$R_{HSDW} \equiv \left\{\rho_s(\mathbf{r}) = \frac{1}{2}\rho(\mathbf{r})\mathbf{E} + \rho_x(\mathbf{r})\sigma_x + \rho_z(\mathbf{r})\sigma_z\right\}, \tag{13}$$

$$R_{TSW} \equiv \left\{\rho_s(\mathbf{r}) = \frac{1}{2}\rho(\mathbf{r})\mathbf{E} + \rho_x(\mathbf{r})\sigma_x + \rho_y(\mathbf{r})\sigma_y + \rho_z(\mathbf{r})\sigma_z\right\}. \tag{14}$$

The relation between search region GSO-DFT and symmetry-framework is depicted in Figure 4. We see that the search region becomes narrower as symmetry restriction becomes more tight.

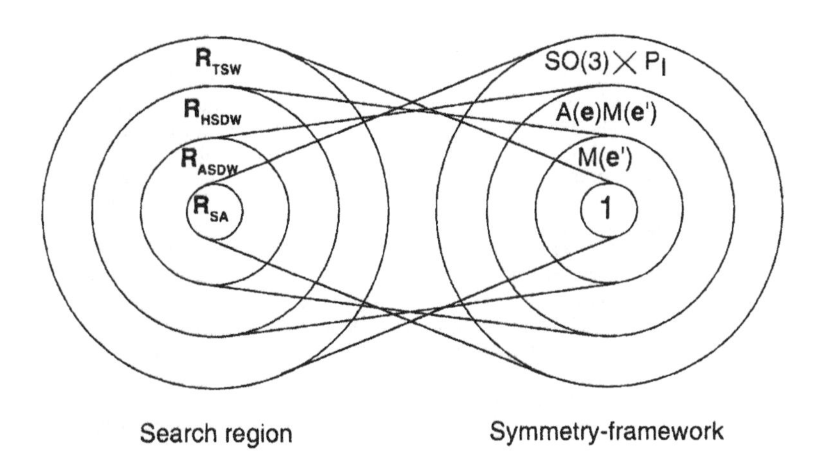

Figure 4. The relationship between GSO solutions and corresponding symmetry framework.

Note that in classification of GHF solutions, there are other four types of solutions. Those types explore off diagonal ($\mathbf{r} \neq \mathbf{r}'$) term of $\rho_s(\mathbf{r};\mathbf{r}')$, so we omitted in classification of GSO-DFT, in which fundamental variable is $\rho_s(\mathbf{r}) \equiv \rho_s(\mathbf{r};\mathbf{r})$.

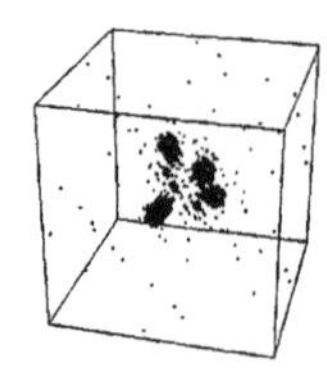
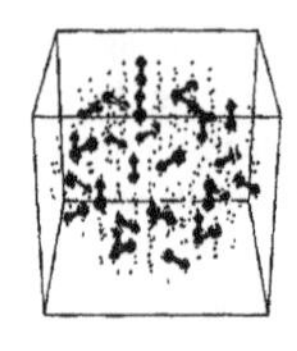

Figure 5. Stable spin structures of cubane Mn_4O_4 and icosidodecahedron cluster

Figure 5 illustrates the 3D spin structure of the cubane-type Mn4O4 cluster by GSO-DFT and icosidodecahedron cluster, Fe_{30}, which is a new type of single molecular magnet. The noncollinear structures also appear in several Fe-S clusters in biological systems.

4. Computational schemes and procedures

The most stable solution can be found by following the stability condition starting from usual restricted solution [12]. This can be understood in the context of random phase approximation (RPA) theory or time-dependent DFT (TD-DFT) [13] as shown in figure 6. Starting from a RHF or RKS-DFT solution, eigenvalues of the so-called response or instability matrix,

$$\begin{bmatrix} \mathbf{A} & \mathbf{B} \\ \mathbf{B}^* & \mathbf{A}^* \end{bmatrix} \begin{bmatrix} \mathbf{X} \\ \mathbf{Y} \end{bmatrix} = \omega \begin{bmatrix} 1 & 0 \\ 0 & -1 \end{bmatrix} \begin{bmatrix} \mathbf{X} \\ \mathbf{Y} \end{bmatrix} \tag{15}$$

provide the criteria of whether there are any stabile solutions or not: any negative ω detects a lower state in the wider search region. Here **A** and **B** are expressed by orbital energies differences, two-electron integral, and exchange (and correlation) by

$$\mathbf{A}_{ia,jb} = \delta_{ij}\delta_{ab}(\varepsilon_a - \varepsilon_i) + (ia|jb) + (ia|f_{XC}|jb)$$

$$\mathbf{B}_{ia,jb} = (ia|bj) + (ia|f_{XC}|bj). \tag{16}$$

So if we obtain a negative ω of TD-RDFT equation, we have to employ UKS-DFT with $\mathbf{R}_{ASDW}$ in figure 4. On the other hand, a positive ω value corresponds to a usual excited spectrum in TD-RDFT. Further, the TD-UDFT equation might disclose the instability of the UKS-DFT solution and the existence of GSO-DFT (noncollinear magnetic) solutions. Further, the instability for the superposition of electron and hole amplitudes of KS orbital corresponds to the existence of Bogoliubov quasiparticle [12,14], so we proceeds to Bogoliubov de Gennes (BdG) equation, responsible for superconductivity. These steps are summarized in figure 6.

Note that all of nontrivial UKS-DFT, GSO-DFT, BdG solutions are symmetry-broken ones. If the system is finite, the state usually favors symmetric solution in the world of the nonrelativistic quantum mechanics. In such cases, SB solutions at any level can be mapped to the corresponding natural orbitals, then multireference treatment [15-17] is possible, which is also shown in figure 6. The resonating approach based on such SB solutions are also possible in relation to the resonating valence bond (RVB) theory [18].

The chemical bonds in several metal clusters in biological systems are found to be labile, and therefore variable with environment effects. This property would be related to functions of these active sites.

The anisotropic exchange terms D and E arising from the spin-orbit (SO) interaction were calculated by the perturbation method based on spin unrestricted DFT. The SO interaction is also included in the GSO HF (DFT) scheme to calculate the Dzyaloshinskii-Moriya (DM) type interaction in spin-frustrated systems.

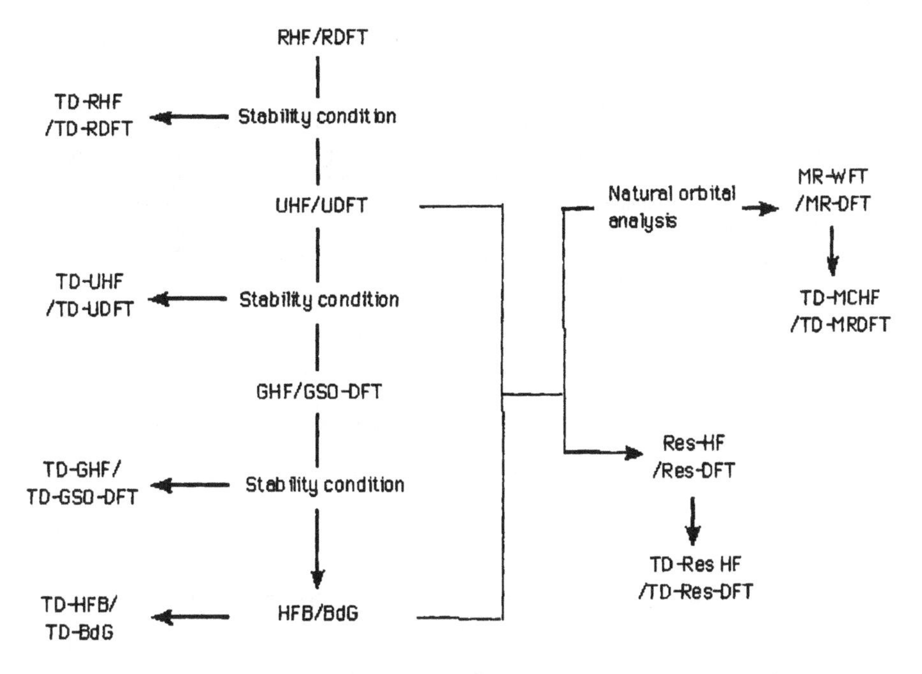

Figure 6. Schematic illustrations of relationship between hierarchy of several broken-symmetry solutions and the corresponding time-dependent theories, together with the connection to mutlireference and resonating approaches.

5. Bond indices and information entropy

As is well-known, the natural orbitals and those occupation numbers of spin unrestricted solutions show indices for instability of chemical bonds [15]. As the covalent bond becomes tighter, the occupation number reduces to 2 (double-occupation in bonding orbital). On the other hand, the occupation number decreases to be 1 as the covalent bond cleavages. Thus, we can define the effective bond order of natural orbitals of projected spin-unrestricted solution using difference between i-th bonding (n_i) and antibonding (n_i^*) as

$$b_i = \frac{n_i - n_i^*}{2} \tag{17}$$

Further the normalized information entropy can be defined as [19],

$$I = \frac{-2N_c \ln 2 + \sum_i n_i \ln n_i}{-2N_c \ln 2} \tag{18}$$

detecting the instability of chemical bond. Here Nc is the number of closed shell. Note that I defined in eq. (18) can be applied for any class of symmetry broken solutions discussed in this article, dictating a measure of the electron correlation involving with such solutions. The information entropy is equally available by the symmetry-adapted (SA) MCSCF and MRCI methods. Then it is a useful bridge between SB and SA methods. Indeed, the natural orbitals by the BS methods are symmetry-adapted, as alternatives of MCSCF orbitals for large systems. The unpaired electron density and spin density are also calculated to elucidate electron localizations via strong electron-electron interactions in metal clusters in fig.1.

6. Magnetic properties at finite temperature

The mapping to Heisenberg model enables us to determine temperature dependent magnetic properties via energy spectra obtained by the exact diagonalization of Heisenberg Hamiltonian in eq. (1). Indeed, Shoji et al. have computed the magnetic dependent magnetic moment of the modeled 8Fe-7S cluster [6]. As shown in figure 7, the temperature dependent magnetic moment using BLYP results shows the good agreement with the experimental results. Then our approach is one of the multi-scale multi-physics approaches in molecular material science.

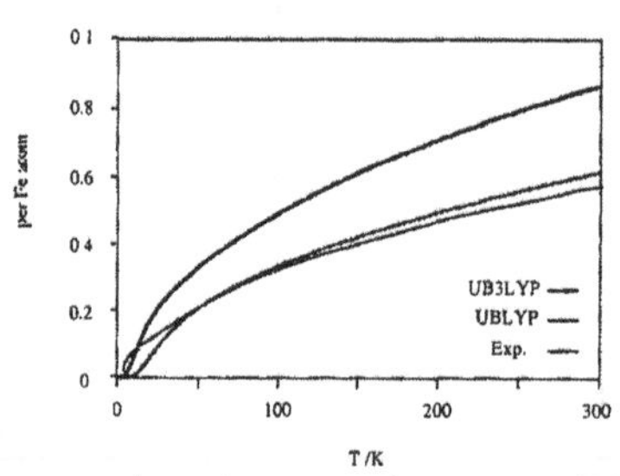

Figure 7. Temperature –dependent magnetic moment of [8Fe-7S] complex.

7. Concluding remarks

In this study, we present collinear and noncollinear magnetic structures of some transition metal (TM) clusters and complexes using spin unrestricted KS-DFT and GSO-DFT. The results are discussed from the viewpoint of the competition of collinear and noncollinear, which are induced by valencies, and structural changes of TM complexes that determine the balance among kinetic exchange, potential exchange, and double exchange interactions.

Acknowledgments

This work has been supported by Grand-in-Aid for Scientific research (16750049, 1835008, 18205023) from Japan-Society for the promotion of Science. This work is partially supported by JST-CREST and NEDO.

References

[1] L. Cur Jr. (Eds) *Metal Clusters in Proteins*, American Chemical Society ,1988;P. Ballone and W. Andrenoi, (Eds.) *Metal clusters*,1999..

[2] W. Kohn and L. J. Sham : *Phys Rev* 1965, 1140, A1133.

[3] K.Yamaguchi, F. Jensen, A. Dorigo, K. N. Houk, *Chem Phys Lett*, 149, 537 (1988); S. Yamanaka, T. Kawakami, H. Nagao, K. Yamaguchi, *Chem Phys Lett*, 231, 25 (1994).

[4] S. Yamanaka, R. Takeda, M. Shoji, Y. Kitagawa, H. Honda, *Int. J. Quantum Chem*. 105, 605 (2005).

[5] R. Takeda, M. Shoji, S. Yamanaka, K. Yamaguchi, *Polyhedron*, 24, 2238 (2005);M. Shoji, et al. *Polyhedron*, 24, 2708 (2005).

[6] M. Shoji et al., *Int. J. Quantum Chem*. 100, 887 (2004); *Int. J. Quantum Chem*. 105, 628 (2005);*Chem. Phys. Lett*., 421, 483 (2006); *Int. J. Quantum Chem*. in press.

[7] K. Yamaguchi et al. *Theoret. Chem. Acc.* 102, 328 (1999).

[8] S. Yamanaka, et al. *Int. J. Quantum Chem.* 80, 664 (2000).

[9] U. von Barth and L. Hedin, *J. Phys. C, Solid State Phys.*, **5**, 1629 (1972).

[10] S. Yamanaka et al, *Int. J. Quantum Chem.* 85, 421 (2001);*Int. J. Quantum Chem.* 84, 369 (2000); *Polyhedron.* 22, 2013 (2003); *J. Mag. Mag. Mater.* 272, E255 (2004); *Bull Chem. Soc. Jpn.* 77, 1269(2004).

[11] H. Fukutome, *Prog. Theor. Phys.*, 1974, **52**, 115;M. Ozaki and H. Fukutome, *Prog. Theor. Phys.*, 1978, **60**, 1322; K. Yamaguchi and H. Fukutome, *Prog. Theor. Phys.*, 1975, **54**, 1599.

[12] J-P. Blaizot and G. Ripka, *Quantum theory of finite systems,* MIT Press, Cambridge, Mass. 1986.

[13] M. E. Casida, in *Recent Advances in Density Functional Methods, Part I*, edited by D. P. Chong World Scientific, Singapore, 1995

[14] K. Yamaguchi, et al. *Int. J. Quantum Chem.* (2006). 106, 1052 and references therein.

[15] K. Yamaguchi, *Int J. Quantum Chem* (1980) S14, 269.

[16] R. Takeda, S. Yamanaka, K. Yamaguchi, *Chem. Phys. Lett.* (2002)366, 321; *Int J. Quantum Chem.* (2004) 96 463.

[17] S. Yamanaka et al., *Chem. Lett.* (2006) 35, 242; *Int. J. Quantum Chem.* in press; K. Nakata, et al., *Int. J. Quantum Chem.* in press.

[18] R. Takeda, S. Yamanaka, K. Yamaguchi, *Int J. Quantum Chem*, to be published.

[19] D. M. Collins, *Z Naturforsch* (1993) 48, 68

Brill Academic Publishers
P.O. Box 9000, 2300 PA Leiden
The Netherlands

Lecture Series on Computer and Computational Sciences
Volume 6, 2006, pp. 294-307

A systematic study of the linear and non-linear optical properties of small molecules and clusters: The correlation, vibrational and relativistic contributions

M. G. Papadopoulos[*,a], H. Reis[a], A. Avramopoulos[a] and A. Alparone[b]

[a] Institute of Organic and Pharmaceutical Chemistry, National Hellenic Research Foundation, 48 Vas. Constantinou Ave., Athens 116 35, Greece.
[b] Dipartimento di Scienze Chimiche, Università di Catania, Viale A. Doria 8, Catania 95125, Italy.

Received 28 July, 2006; accepted 3 August, 2006

Abstract. We review some results of the polarizability, first and second hyperpolarizability of a series of small molecules and clusters, which we recently published. First, the linear and non-linear optical (L&NLO) properties of CH_3CN are discussed. A systematic study of the basis set effect is reported. The vibrational contributions as well as the frequency dependence of those properties are presented. Second, the electric and vibrational contributions to the L&NLO properties of C_4H_4XH, where X=N, P, As, Sb and Bi are presented and commented upon. Third, the effect of lithiation of the amino group of 3-amino-acrolein on the polarizability and hyperpolarizabilities is investigated. Fourth, the polarizability of a set of Zn and Cd clusters are reported. We consider how the change of size of the cluster affects the polarizability as well the effect of correlation and relativistic corrections on this property. The present results of the second hyperpolarizability of Zn_m and Cd_n show that, at the MP2 level, the above property can be approximated by constant increments.

Keywords :(hyper)polarizability, vibrational contributions, electron correlation, relativistic correction, lithiation effect.

PACS: 31.15Ar, 42.65 -k

1. Introduction

The non-linear optical (NLO) properties of materials have attracted the interest of many research groups in the recent years [1]. The reported studies involve the design and synthesis of efficient photon manipulating materials as well as the determination (theoretical and/or experimental) of the NLO properties of a large number of derivatives [2]. The calculation of these properties presents a great challenge for the computational methods (rigorous and more approximate) [3]. In addition the NLO properties are connected with a large number of significant technological applications (e.g. devices to store or transfer data) [4-6] and provide valuable information for many important effects (e.g. second harmonic generation, third harmonic generation four-wave mixing etc.) [7].

Our interest in this area is focused on the development and implementation of methods for the computation of the linear and non-linear (L&NLO) properties of (i) Molecules considered as isolated species [8-14] and (ii) Molecules in condensed phases (solids and liquids) [15-18]. For the reliable computation of these molecular properties, we develop techniques for the computation of the local field effect, which allows to take into account the effect of the environmental interactions on the properties of interest. These local field dependent molecular properties are used to compute the macroscopic L&NLO properties of solids and liquids.

In this article we review some of our recent work on the systematic study of the L&NLO properties of small molecules and clusters [10, 19-22]. In some pilot cases we have reported all the

* Corresponding author: mpapad@eie.gr

important contributions, that is, the effect of correlation, vibrations and relativistic correction. So this article involves: First, a study of the L&NLO properties of acetonitrile, second, the L&NLO properties of the series C_4H_4XH, where X=N, P, As, Sb and Bi. Third, the L&NLO properties $NLi_2.C_2H_2.CHO$, which are compared with those of $NH_2.C_2H_2.CHO$. The lithiated derivatives have been studied because they have large NLO properties. It is known that there is a great need for derivatives with large non-linearities, since these compounds have many applications (e.g. in the photonic industry). Fourth, we consider the polarizability and hyperpolarizability of some Zn and Cd clusters and in particular how the change of size affects their properties. A relevant question is how the size of cluster affects the bonding between its atoms, which in turn affects the electric properties. We have used a series of basis sets, some of which have been developed in the framework of this study and some other have been taken from the literature. In addition we have employed a hierarchy of methods to take into account the effect of the correlation of electrons on the polarizability, as well as appropriate methods in order to calculate the effect of vibrations and the relativistic correction.

2. Computational Methods

When a molecule is set in a uniform electric field **F**, its energy, E, may be expanded in the following way:

$$E= E^0 - \mu_i F_i - (1/2)\alpha_{ij} F_i F_j - (1/6)\beta_{ijk} F_i F_j F_k - (1/24)\gamma_{ijkl} F_i F_j F_k F_l - ..., \quad (1)$$

where E^0 is the field free energy of the molecule, F_i, F_j, F_k are the electric field components and μ_i, α_{ij}, β_{ijk}, γ_{ijkl} are the components of the dipole moment, dipole polarizability, first and second hyperpolarizability, respectively. Summation over repeated indices is implied. The average polarizability is given by:

$$\alpha = (\alpha_{xx} + \alpha_{yy} + \alpha_{zz})/3. \quad (2)$$

If the molecule is oriented so that its dipole moment is along the z axis, then the vector component of the first hyperpolarizability is given by:

$$\beta = \frac{1}{5} \sum_{i=x,y,z} (\beta_{zii} + 2\beta_{izi}) \quad (3)$$

The average second hyperpolarizability is given by:

$$\gamma = \frac{1}{15} \sum_i \sum_j (2\gamma_{iijj} + \gamma_{ijji}) \quad (4)$$

The vibrational contributions to the properties of interest have been computed by employing the Bishop and Kirtman perturbation theory (BKPT) [23]. Within this approach the zero-point vibrational averaging correction of a given property P is given by :

$$P^{zpva} = [P^e]^{(1,0)} + [P^e]^{(0,1)} \quad (5)$$

$$[P^e]^{(0,1)} = -\frac{\hbar}{4} \sum_\alpha \frac{1}{\omega_\alpha^2} \left(\sum_b \frac{F_{\alpha bb}}{\omega_b}\right)\left(\frac{\partial P^e}{\partial Q_\alpha}\right) \quad (6)$$

and

$$[P^e]^{(1,0)} = \frac{\hbar}{4} \sum_\alpha \frac{1}{\omega_\alpha} \left(\frac{\partial^2 P^e}{\partial Q_\alpha^2}\right), \quad (7)$$

where, ω_α is the harmonic frequency, $F_{\alpha bb}$ is the cubic force constant and Q_α is the normal coordinate. The pure vibrational contributions to the dipole polarizability and dipole first hyperpolarizability are given by the following expressions :

$$\alpha^{pv} = [\mu^2]^{(0,0)} + [\mu^2]^{(2,0)} + [\mu^2]^{(1,1)} + [\mu^2]^{(0,2)} \quad (8)$$

$$\beta^{pv} = [\mu\alpha]^{(0,0)} + [\mu\alpha]^{(2,0)} + [\mu\alpha]^{(1,1)} + [\mu\alpha]^{(0,2)} + [\mu^3]^{(1,0)} + [\mu^3]^{(0,1)} \quad (9)$$

$$\gamma^{pv} = [\alpha^2]^{(0,0)} + [\alpha^2]^{(2,0)} + [\alpha^2]^{(1,1)} + [\alpha^2]^{(0,2)} + [\mu\beta]^{(0,0)} + [\mu\beta]^{(2,0)} + [\mu\beta]^{(1,1)} + [\mu\beta]^{(0,2)} + [\mu^2\alpha]^{(1,0)}$$

$$+[\mu^2\alpha]^{(0,1)}+[\mu^4]^{(2,0)}+[\mu^4]^{(1,1)}+[\mu^4]^{(0,2)}. \quad (10)$$

In the notation $[A]^{n,m}$, the superscripts n and m denote the level of the electrical and mechanical anharmonicities, respectively. We shall use the notation *mnopq* to define the order of the energy (m), dipole moment (n), polarzabillity (o), first hyperpolarizability (p) and second hyperpolarizability (q) derivatives, with respect to normal coordinates, which have been employed in the present work.

For the computation pf the properties we have used a large variety of basis sets (e.g. those developed by Dunning et al.[24-28], Sadlej et al.[29-31]), a hierarchy of techniques in order to take into account correlation and some effective core potentials for Zn and Cd [32-34]. The relativistic correction has been computed by employing the Douglas-Kroll approximation [35]. The results presented here have computed by employing the following packages: Gaussian 98 [36], DALTON [37], CADPAC 5.0 [38] and 6.0[39] and SPECTRO [40].

3. Results and discussion

3.1. Acetonitrile

An extensive report of the electric properties of CH_3CN have been presented elsewhere [10]. The computations reported in Table 1 have been performed by employing the experimental geometry [41]. The molecular z-axis is along the C-C-N axis (pointing from N to C) . One of the C-H bonds lies on the yz plane. We have employed the correlation-consistent basis sets reported by Dunning et al. [24-28]. For the dipole moment and polarizability convergence is attained with the DZ basis set. Larger basis sets are required to attain convergence for β and γ. The effect of correlation on μ and α is rather small, but on β (in particular) and γ is remarkable. It is noted that CC2 overshoots α, β and γ. It is observed that d-aug-cc-pVDZ yields converged results for all electric properties.

Table 1 The static electronic dipole moment, polarizability, first and second hyperpolarizability of acetonitrile, computed at various theory levels with different basis sets. All reported values are in a.u.

Property Method	μ_z	α	β	γ
SCF				
aug-cc-pvDZ	1.676	28.48	-3.92	2330
aug-cc-pvTZ	1.674	28.75	1.18	2673
aug-cc-pvQZ	1.674	28.79	3.23	2861
aug-cc-pv5Z	1.673	28.79	3.39	2931
d-aug-cc-pvDZ	1.673	28.78	4.14	2922
d-aug-cc-pvTZ	1.673	28.80	3.71	3017
MP2				
aug-cc-pvDZ	1.539	28.97	15.88	3018
aug-cc-pvTZ	1.545	29.18	20.74	3453
aug-cc-pvQZ	1.548	29.17	22.66	3585
d-aug-cc-pvDZ	1.537	29.28	24.16	3775
d-aug-cc-pvTZ	1.544	29.25	23.52	3814
CC2				
d-aug-cc-pvDZ		30.02	23.12	4350
CCSD				
aug-cc-pvDZ		29.39	22.60	3873
CCSD(T)				
aug-cc-pvDZ	1.525	29.45	20.83	3856

The frequency dependent results are given in Table 2. A rather remarkable difference between the γ results computed at CC2 [42a] and CCSD [42b] is observed. A similar difference has been observed for the static results as well.

Table 2. Average frequency dependent values of electronic (hyper)polarizabilites of acetonitrile, at λ= 514.5 nm[a]. All values are in a.u.

Property	RPA daDZ	RPA daTZ	CC2 daDZ	CCSD daDZ
$\alpha(-\omega;\omega)$	29.58	29.60	30.90	30.23
$\beta(-2\omega;\omega,\omega)$	7.33	6.59	32.71	32.54
$\gamma(-2\omega;\omega,\omega,0)$	4328	4449	6691	5855

[a] daDZ=d-aug-cc-pVDZ and daTZ=d-aug-cc-pvTZ. RPA:Random Phase Approximation.

The pure vibrational contributions to α, β and γ of CH_3CN are presented in Table 3. The aug-cc-pVDZ has been employed, while correlation has been taken into account at the MP2 level. Two methods have been used for the calculation of the property derivatives, required for the computation of the vibrational contributions, that is, they have been calculated analytically and numerically. In the latter approach finite differences of the nuclear displacements or the external electric fields have been considered [42c]. We observe that all employed techniques for the calculation of the derivatives give similar values [10]. The double harmonic approximation at both the SCF and the MP2 levels, for all the considered properties, give satisfactory results.

Table 3 Static pure vibrational corrections to the dipole moment, polarizability, first and second hyperpolarizability of acetonitrile.[a] All reported values are in a.u.

	SCF					MP2	
	analytic			displ	field	displ	field
mnopq[b]	43210	32210	21110[c]	32210	32222	32210	21110[c]
α	0.37	0.40	0.36	0.41	0.41	0.20	0.17
β	34.37	35.04	35.98	35.92	35.92	15.85	15.71
γ	1019	1021	1124	1004	1012	1148	1238

[a]The derivatives were calculated analytically (analytic) and numerically using geometry displacements (displ) or the finite field (field) approach.The aug-cc-pVDZ basis set has been used for the computations.
[b] Order of derivatives (mnopq)
[c] Double harmonic approximation

The ZPVA correction to μ,α, β and γ of CH_3CN are given in Table 4. The remarkable effect of correlation on β and, in particular, γ is observed. Comparison of the theoretical and the experimental results is given in Table 5. The satisfactory agreement of the theoretical and the experimental results for $\alpha(-\omega;\omega)$ and $\beta(-2\omega;\omega,\omega)$ is observed. A larger discrepancy for $\gamma(-2\omega;\omega,\omega,0)$ has been found.

Table 4 Static ZPVA corrections to the dipole moment, polarizability, first and second hyperpolarizability of acetonitrile.[a] All reported values are in a.u.

Property	aug-cc-pVDZ		
	SCF		MP2
	displ	field	field
μ_z	-0.003	-0.003	-0.01
α	0.84	0.84	0.80
β		-3.68	-2.33
γ		187	316

[a] The derivatives were calculated numerically using geometry displacements (displ) or finite field (field) approach [10].

Table 5 Comparison of experimental and calculated (hyper)polariza- bilites of acetonitrile. All reported values are in a.u.

λ/nm	d-aug-cc-pVDZ		
	514.5		
	α	β	γ
Expt	30.43[a]	26.3[b]	4619[b]
CCSD(T)	31.13	26.7	6200

[a] Ref. 43a
[b] Ref. 43b.

3.2. C_4H_4XH

A detailed report for the polarizability and hyperpolarizabilities of C_4H_4XH, where X=N, P, As, Sb and Bi has been presented in ref. 19. For these computations we have employed various methods (HF, MP2, DFT, CCSD(T)). Here we will only consider the HF and the CCSD(T) [44] results. The Pol basis set has been employed for the computations [29-31]. For C_4H_4BiH an ECP has also been employed [45].

Table 6 Electronic dipole (hyper)polarisabilities (a.u.) of C_4H_4XH (X = N, P, As, Sb, Bi)[a].

	α	γ
C_4H_4NH		
HF/POL	53.70	16345
CCSD(T)/Pol	54.87	18838
Exp.	55.80 [b]	
C_4H_4PH		
HF/POL	70.42	18022
CCSD(T)/Pol	71.39	23027
C_4H_4AsH		
HF/POL	75.21	19578
CCSD(T)/Pol	76.12	25696
C_4H_4SbH		
HF/POL	87.61	27584
CCSD(T)/Pol	88.74	35874
C_4H_4BiH		
HF/POL	94.43	32650
CCSD(T)/ECP-Pol[c]	93.99	39306

[a] Calculations are carried out on the B3LYP/Pol equilibrium geometry.
[b] Reference [46a]
[c] Calculations are carried out on the B3LYP/ECP-Pol equilibrium geometry.

Electronic polarizability results. The effect of correlation at the CCSD(T) level is small (Table 6). A satisfactory agreement for α of C_4H_4NH with the corresponding experimental value has been found. An increase of α with the atomic number of X has been found.

Electronic hyperpolarizability results. There are no experimental results for γ of C_4H_4XH. The effect of correlation is larger in comparison to that we have observed for α (Table 6). The corresponding effect is 15.3%, 27.8%, 31.2%, 30.1% and 20.4% for X=N, P, As, Sb and Bi, respectively (C_4H_4XH).

The vibrational contributions, PV and ZPVA to α and γ of C_4H_4XH have also been studied. Here we present the results for the PV contributions. The ZPVA contributions to α are small (the largest ZPVA contribution is 2.16 a.u for C_4H_4SbH; method: HF). The ZPVA contributions to γ of C_4H_4XH are negligible, except of C_4H_4SbH (the reported value is 1146 a.u., at the HF level).

Table 7 Static pure vibrational (PV) contributionsto the (hyper)polarizabilities (a.u) of C_4H_4XH (X = N, P, As, Sb,)[a].

	α	γ
C_4H_4NH		
PV/SCF/43210[b]	6.89	29933
PV/SCF/DH[c]	7.25	1468
C_4H_4PH		
PV/SCF/43210[b]	3.27	5517
PV/SCF/DH[c]	3.22	4334
C_4H_4AsH		
PV/SCF/43210[b]	3.42	6661
PV/SCF/DH[c]	3.36	5426
C_4H_4SbH[d]		
PV/SCF/32222[e]	5.57	12420
PV/SCF/DH[c]	5.08	8067

[a] Calculations are carried out on the equilibrium geometry 6-311G** basis set.
[b] Property derivatives of order mnopq=43210
[c] Double -harmonic approximation.
[d] Calculations are carried out on the B3LYP/Pol equilibrium geometry with the Pol basis set.
[e] Property derivatives mnopq=32222.

Pure vibrational contributions (PV) to α. We have considered derivatives at the order 43210 for X=N, P, As and 32222 for X=Sb. The geometries and the vibratonal properties have been computed by employing the 6-311G** basis set. It is observed that the PV contribution is small, but not negligible (Table 7). The performance of the double harmonic approximation is very satisfactory.

Pure vibrational contributions to γ. The PV contribution is larger than the corresponding electronic one only for C_4H_4NH. For the other considered molecules the PV contribution is smaller than the corresponding electronic one, but not negligible. The double harmonic approximation fails for C_4H_4NH, but for the other molecules it is in satisfactory agreement with the higher approximation we have employed (Table 7).

The CCSD(T) polarizabiity and second hyperpolarizability of C_4H_4BiH has been computed by employing a quasi-relativistic effective core potentials (ECP) with the corresponding valence electron basis set [45].

3.3. NLi_2-C_2H_2-CHO: The effect of lithiation

There is a great demand for derivatives with large NLO properties, because these compounds may be useful in many applications. In addition the significant NLO response is associated with interesting physical phenomena and processes. As a contribution to this topic we have shown that lithiation usually leads to the required significant increase of the NLO properties [46b]. Here we present the L&NLO properties of NLi_2-C_2H_2-CHO, which are compared with those of the reference derivative NH_2-C_2H_2-CHO (Figure 1, Table 8). The -NH_2 group has been selected for the lithiation, because there is an extensive literature on the lithiated derivatives of NH_3 [47].

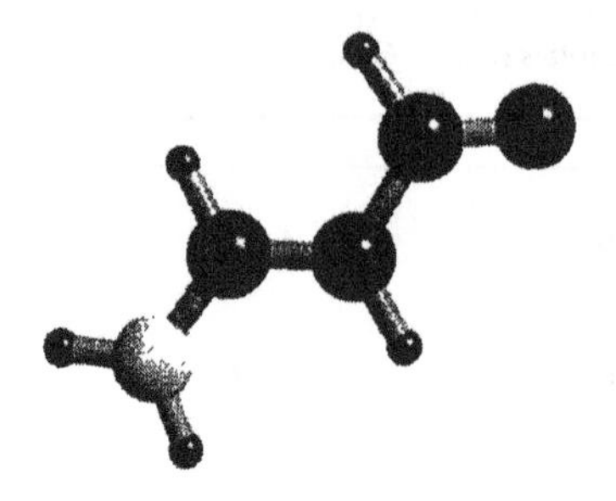

Figure 1. The structure of 3-amino-acrolein.

Würthwein et al. [48] have shown that the planarity of the isolated species NLi_mH_{3-m}, m=1-3, is due to the σ-donating and the π-electron-accepting character of lithium. They suggested that N-Li bonds are partially covalent and discussed some important differences between Li and H: (i) hydrogen is much less electropositive than lithium, (ii) bonds involving H are less ionic than the corresponding ones involving Li and (iii) lithium has low-lying vacant p-orbitals, which can participate in bonding. Lazicki et al. [49] noted that Li_3N is one of the most ionic nitrides and the only known thermodynamically stable alkali metal nitride. Lithium nitride has been extensively studied as a potential candidate for hydrogen storage [50, 51]. It can theoretically absorb about 10.4 wt % hydrogen [52].

The fully lithiated compound of 3-aminoacrolein (donor: NLi_2), has a β value, which is 6.8 larger and with opposite sign, in comparison with the corresponding value of the non-lithiated derivative (Table 8). The remarkable influence of the lithiation on the first hyperpolarizability has also been noted by Chen et al. [53].

Table 8 The electronic dipole moment, polarizability and first hyperpolarizability of H_mLi_nN-C_2H_2-CHO, m,n=0,2, m+n=2. The properties were computed using the Pol basis set at the corresponding optimized geometries[a,b]. All values are in a. u.

	NH_2-C_2H_2-CHO[a]		NHLi-C_2H_2-CHO[b]		NLi_2-C_2H_2-CHO[b]	
	HF	**MP2**	**HF**	**MP2**	**HF**	**MP2**
μ_z[c]	2.673	2.423	3.198	3.133	5.466	5.213
α_{xx}	43.84	47.61	59.98	70.61	62.06	75.56
α_{yy}	33.75	35.74	39.05	42.36	48.41	54.70
α_{zz}	84.09	93.08	88.29	101.36	121.10	149.83
α	53.89	58.81	62.44	71.44	77.19	93.36
β_{zxx}	43.9	43.6	-143.1	-65.7	-407.3	-845.1
β_{zyy}	-14.2	-23.4	-61.1	-113.7	-438.5	-811.6
β_{zzz}	112.7	549.5	168.9	140.4	-1053.3	-2234.9
β	85.4	341.8	-21.2	-23.4	-1139.4	-2334.9

[a] Planar optimized geometry by using the 6-31G**/MP2 method. The properties were computed using the Pol basis set at the corresponding optimized geometries

[b] Fully optimized geometry by using the aug-cc-pVDZ/MP2 method.

[c] The dipole moment is oriented along z-axis.

In order to explain the significant increase of the first hyperpolarizability, as well as the reversal of the sign of the mean value, we will employ the sum-over-states model (SOS) approach, according to which the first hyperpolarizability is given by:

$$\beta(0) \propto \frac{3(\mu_{ee} - \mu_{gg})(\mu_{ge})^2}{E_{ge}^2}, \qquad (11)$$

where μ_{ee} and μ_{gg} are the excited and ground state dipole moments, respectively, μ_{ge} is the transition dipole moment and E_{ge} is the transition energy. In this application we shall employ five excited states. The SOS model allows to discuss some trends in terms of the excited states and to comment on their contribution to first hyperpolarizability. We observe that the SOS value is larger, but in qualitative agreement, with the Hartree-Fock value (Table 9).

Table 9 Excitation energies (ΔE_{ge}), transition dipole moments (μ_{ge}) and excited dipole moments (μ_e) for the five lower energy transitions, A''-> A', for NH_2-C_2H_2-CHO and NLi_2 -C_2H_2-CHO. The parallel first hyperpolarizability was computed by using the SOS model[a]. The property values are presented in a.u.

NH_2- C_2H_2-CHO[b]			**NLi_2 -C_2H_2-CHO**[c]		
μ_e	μ_{ge}	ΔE_{ge}	μ_e	μ_{ge}	ΔE_{ge}
3.099	2.278	0.215	2.088	-1.102	0.152
3.231	-0.527	0.279	2.439	-1.137	0.180
3.861	0.401	0.306	3.725	-1.114	0.193
2.977	-0.209	0.326	2.312	-1.247	0.222
-1.027	0.278	0.337			
β_{zzz}	148.8[d] (112.7)[e]		-1383.8[d] (-1053.3)[e]		

[a] For the calculations the Pol basis set at the MP2/aug-cc-pVDZ optimized geometry has been employed. The property values were evaluated using the CIS method , unless otherwise specified.

[b] μ_g(HF) = 2.673 a.u.

[c] μ_g (HF)= 5.466 a.u.

[d] β_{zzz}: SOS value .

[e] β_{zzz}: HF value.

The difference $\Delta\mu = \mu_e - \mu_g$, between the excited and the ground state dipole moment of NLi_2-C_2H_2-CHO is negative. This result may be used to explain the negative sign of β. In the case of NH_2-C_2H_2-CHO this difference is positive, except of the one which corresponds to the last presented excited state dipole moment (Table 9). Moreover, the remarkable decrease of the excitation energies of the lithiated compound (Table 9), in connection with the larger transition dipole moments, lead to an increase of β_{zzz}.

Finally we may conclude that lithiation has a significant effect on the excitation spectrum of a compound. This leads to a remarkable increase of the first hyperpolarizability. Similar observations have been made by other authors [54-55]. They have shown that the substantial change of the excited state dipole moment and the remarkable decrease of the excitation energies lead to an increase of the first hyperpolarizability.

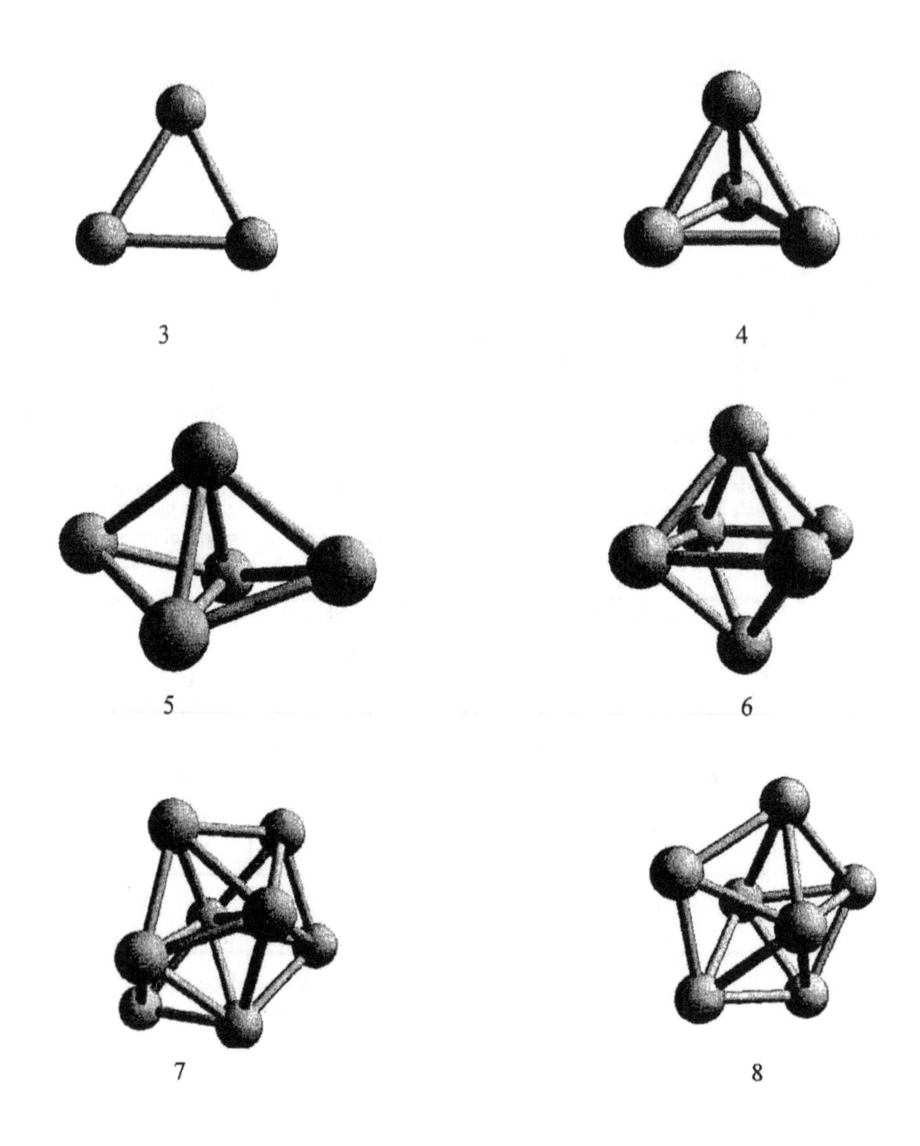

Figure 2. The structure of Zn_m, m=2-8 [21].

3.4. Zn_m, Cd_n and Zn_mCd_n

We have studied a series of clusters of Zn (Figure 2) and Cd [21,22], because we would like to find out: (i) how the change of the cluster size affects its properties and (ii) the effect of *correlation* and the *relativistic correction* on the properties of interest. Several basis sets and methods have been used in order to take into account correlation (e.g. MP2, CCSD(T)). In addition we have used some effective core potentials (ECP). The first goal was to demonstrate the adequacy of the employed basis sets as well as the chosen ECP. It has been found (Table 10) that both MP2/ECP28 and B3LYP/SBK/Z4 satisfactorily describe the change of the polarizability with m (Zn_m) [21].

Table 10. Average polarizability (a.u.) of Zn_m, m=2-8, computed by employing a series of methods.[a]

Method	α	Method	α
Zn_2		MP2/ECP28	203.17
HF/ECP28	81.69	B3LYP/SBK/Z4	217.52
MP2/ECP28	73.09	**Zn_6**	
B3LYP/SBK/Z4	83.57	HF/ECP28	249.71
CCSD(T)/SBK/Z4	82.30	MP2/ECP28	233.43
Zn_3		B3LYP/SBK/Z4	247.97
HF/ECP28	128.17	**Zn_7**	
MP2/ECP28	116.77	HF/ECP28	298.51
B3LYP/SBK/Z4	133.88	MP2/ECP28	282.31
Zn_4		B3LYP/SBK/Z4	295.39
HF/ECP28	170.58	**Zn_8**	
MP2/ECP28	159.23	*HF/SBK/Z4*	420.51
B3LYP/SBK/Z4	172.35	MP2/ECP28	326.46
Zn_5		B3LYP/SBK/Z4	340.63
HF/ECP28	216.97		

[a]The structures have been optimized by Erkoç et al. [56].

It has been reported that in the clusters of Zn there is a change of bonding with the change of the cluster size, that is around Zn_8, there is a transition from van der Waals to covalent bonding and around Zn_{17} there is a transition to metallic bonding [57]. In Figure 3 we present the variation of $R_m=\alpha[Zn_m]/m$ vs m, where m=2-20. This ratio allows to probe the strength of the interaction in the cluster. We have observed two local minima in the variation of R_m vs m. These correspond to m=9 and 18. The rather sharp changes of R_m may be correlated with the change of the bonding from van der Waals to covalent and from covalent to metallic bonding. We have considered the effect of correlation and the relativistic correction on the polarizabilities of Cd_n, n=1-3, at the HF and CC2 levels (Table 11). The relativistic correction has been computed by employing the Douglas-Kroll (DK) [35] approximation. The basis set employed for the computations of Table X have developed by Kellö and Sadlej [56]. It is observed (Table 11) that *the correlation* contribution at the CC2 level reduces α at both the non-relativistic (NR) and the relativistic (R) level. This trend has been confirmed at the CCSD level [22].

It has also been found that *the relativistic* correction reduces α at the HF and CC2 levels. DFT/DK results for α of ZnCd and Zn_2Cd have also been computed. The reported property values for the mixed clusters confirm that the relativistic effect reduces the polarizability.

We have also used two ECPs (Table 12). The first is a small core one and has been developed by Stevens et al. [32, 33], and is denoted by SBK. This ECP is associated with a valence basis set (4s4p3d) and describes the $(n-1)s^2p^6d^{10}ns^2$ electrons of Zn and Cd, where n=4 for Zn and 5 for Cd. The second ECP, is a large core one, has been developed by Schautz, Flad and Dolg and is denoted by SFD [34]. The employed basis set involves 6s6p3d and treats Cd as an atom of two electrons. It is observed (Table 12) that both SFD and SBK give results for Cd_m, m=4-6, at the HF level, which are in satisfactory agreement, but at the MP2 level there is a considerable discrepancy between the results computed by the SFD and SBK ECPs.

In Figure 4 we present the average second hyperpolarizability per atom γ/N for Zn_m and Cd_n. At the MP2 level the functional dependence on N is small, therefore γ of homonuclear clusters can be described at the this level by a constant increament. A multilinear regression analysis yields:

$$\gamma=(42\pm2)\times10^3 N_{Zn} + (84\pm2)\times10^3 N_{Cd}.$$

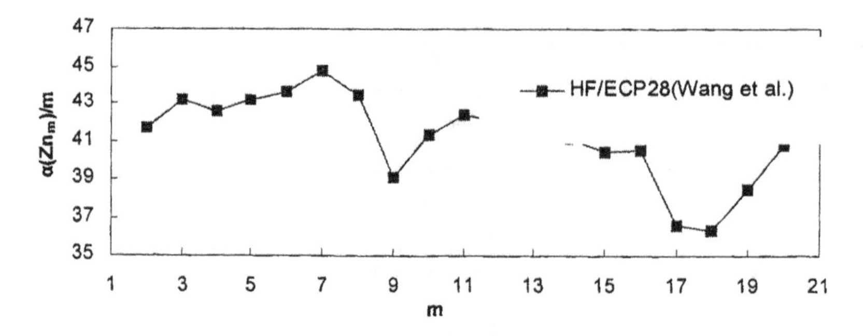

Figure 3 . Variation of $\alpha(Zn_m)/m$ as a function of m.

Table 11. Polarizability components of Cd_m, m=1-4 and Zn_nCd_m, n,m=1,2, computed with relativistic (R) and non-relativistic methods (NR)[a]. All values are in a.u.

Cluster	NR	R	Cluster	NR	R
Method[a]	α		Method[a]	α	
Cd			**ZnCd**		
HF	71.95	62.91	DFT	101.97[c]	91.58[c]
CC2	55.78[b]	46.89[b]			
			$ZnCd_2$		
			DFT		148.55[c]
Cd_2					
HF	149.95	131.40	**Zn_2Cd**		
CC2	116.30[b]	97.52[b]	DFT	151.68[c]	138.82[c]
Cd_3					
HF	230.57	204.75			
CC2	182.50[b]	154.11[b]			

[a] All computations have been performed by employing the basis sets developed by Kellő and Sadlej [58]. The relativistic correction has been computed using the Douglas – Kroll approximation [35].
[b] 12 electrons for each Cd atom were correlated : $(n\text{-}1)d^{10}ns^2$
[c] B3LYP: a hydrid functional .

Table 12. The dipole polarizability of Cd_m, m=1-11. All values are in a.u.

	HF		MP2	
m	SFD[a]	SBK[b]	SFD[a]	SBK[b]
4	277.66	277.57	262.02	226.36
5	351.32	350.06	331.92	283.94
6	403.56	398.53	380.40	323.03

[a] This method corresponds to the large core pseudopotential used.
[b] Ref. 22.

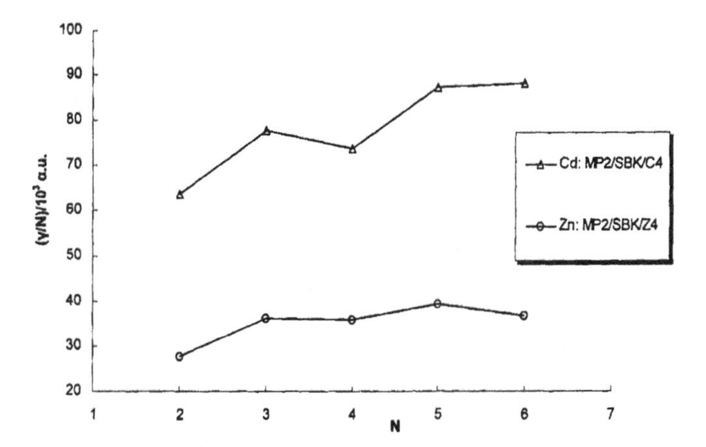

Figure 4. Variation of γ/N as a function of N for Cd_N and Zn_N.

4. Summary

We have presented a review of some of our recent results on the polarizability and hyperpolarizabilities of molecules and clusters. We presented the L&NLO properties of (i) CH_3CN, (ii) C_4H_4XH, where X=N, P, As, Sb and Bi, (iii) NLi_2-C_2H_2-CHO and (iv) Zn_m, Cd_n and Zn_mCd_n. We reported a systematic study of the effect of the basis set and the contributions due to electron correlation, vibrations and the relativistic correction.

References

[1] S. Eisler, A. D. Slepkov, E. Elliott, T. Luu, R. McDonald, F. A. Hegmann and R. R. Tykwinski, *J. Am. Chem Soc.*, **127**, 2666 (2005); S. Keinan, M. A. Ratner and T. J. Marks, *Chem. Phys. Letters*, **417**, 293 (2006); M. A. Belkin and Y. R. Shen, *Int. Rev. Phys. Chem.*, **24**, 257 (2005); E. Annoni, M. Pizzotti, R. Ugo, S. Quici, T. Morotti, M. Bruschi and P.Mussini, *Eur. J. Inorg. Chem.*, **19**, 3857 (2005); B. J. Coe, J. A. Harris, B. S. Brunschwig, I. Asselberghs, K. Clays, J. Garin and J. Orduna, *J. Am. Chem. Soc.*, **127**, 13399 (2005).

[2] D. Kanis, M. A. Ratner, T. J. Marks, *Chem. Rev.*, **94**, 196 (1994).

[3] M. Makowska-Janusik, H. Reis, M. G. Papadopoulos, I. G. Economou and N. Zacharopoulos, *J. Phys. Chem. B*, **108**, 588 (2004); M. Spassova, B. Champagne and B. Kirtman, *Chem. Phys. Letters*, **412**, 217 (2005); L. Frediani, H. Ågren, L. Ferrighi and K. Ruud, *J. Chem. Phys.*, **123**, 144117 (2005); J. Kongsted, T. B. Pedersen, L. Jensen, A. E. Hansen, K. V. Mikkelsen, *J. Am Chem. Soc.*, **128**, 976 (2005).

[4] P. Sarkar and M. Spingborg, *Phys. Rev. B* **68**, 235409 (2003).

[5] F. Zernike, J. Mideinter, *Applied Nonlinear Optics*, Wiley, New York (1973).

[6] L. R. Dalton, W. H. Steier, B. H. Robinson, C. Zhang, A.t Ren, S. Garner, A. Chen, T. Londergan, L. Irwin, B. Carlson, L. Fifield, G. Phelan, C. Kincaid, J. Amend and A.Jen, *J. Mater. Chem.*, **9**, 1905 (1999).

[7] M. Peltz, J. Bartschke, A. Borsutzky, R. Wallenstein, S. Vernay, T. Salva, D. Rytz, *Appl. Phys. B-Lasers and Optics* , **81**, 487 (2005); L. Gao, *Phys. Rev. E*, **71**, 67601 (2005); XG Luo, T. Ishihara, *Opt. Materials*, **27**, 713 (2005).

[8] A. Avramopoulos, H. Reis, J. Li and M. G. Papadopoulos, *J. Am. Chem. Soc.*, **126**, 6179 (2004).

[9] K. Jug, S. Chiodo, P. Calaminici, A. Avramopoulos and M. G. Papadopoulos, *J. Phys.Chem. A*, **107**, 4172 (2003).

[10] H. Reis, M. G. Papadopoulos and A. Avramopoulos *J. Phys. Chem. A*, **107**, 3907 (2003).

[11] A. Avramopoulos and M. G. Papadopoulos, *Mol. Phys.*, **100**, 821 (2002)

[12] H. Reis, S. G. Raptis, *Phys. Chem.Chem. Phys.*, **3**, 3901 (2001).

[13] U. Eckart, V. E. Ingamells, M. G. Papadopoulos and A. J. Sadlej, *J. Chem Phys.*, **114**, 735 (2001).

[14] A. Avramopoulos, V. E. Ingamellls, M. G. Papadopoulos, A. J. Sadlej, *J. Chem. Phys.*, **114**, 198 (2001).

[15] H. Reis, A. Grzybowski and M. G. Papadopoulos, *J. Phys. Chem. A*, **109**, 10106 (2005).

[16] H. Reis, M. Makowska-Janusika and M. G. Papadopoulos, *J. Phys. Chem. A*, **108**, 8931 (2004).

[17] H. Reis, S. G. Raptis and M. G. Papadopoulos, *Chem. Phys.*, **263**, 301 (2001).

[18] H. Reis, M. G. Papadopoulos and D. N. Theodorou, *J. Chem. Phys.*, **114**, 876 (2001).

[19] A. Alparone, H. Reis and M. G. Papadopoulos, *J. Phys. Chem. A*, **110**, 5909 (2006).

[20] A. Avramopoulos, M. Jabłoński, M. G. Papadopoulos and A. J. Sadlej, Chem. Phys., in press.

[21] M. G. Papadopoulos, H. Reis, A. Avramopoulos, Ş. Erkoç and L. Amirouche, *J. Phys. Chem. B*, **109**, 18822 (2005).

[22] M. G. Papadopoulos, H. Reis, A. Avramopoulos, Ş. Erkoç and L. Amirouche, *Mol. Phys.*, **104**, 2027 (2006).

[23] D. M. Bishop, *Adv. Chem. Phys.*, **104**, 1 (1998); D. M. Bishop, J. M. Luis and B. Kirtman, *J. Chem. Phys.*, **108**, 10013 (1998).

[24] D. E. Woon and T. H. Dunning, *J. Chem. Phys.*, **100**, 2975 (2004).

[25] T.H. Dunning, *J. Chem. Phys.*, **90**, 1007 (1989).

[26] R.A. Kendall, T. H. Dunning and R. J. Harris, *J. Chem. Phys.*, **96**, 6796 (1992).

[27] D. E. Woon and T.H. Dunning, *J. Chem. Phys.*, **98**, 1538 (1994).

[28] D. E. Woon and T. T. H. Dunning, *J. Chem. Phys.*, **103**, 4572 (1995).

[29] A. J. Sadlej, *Collect. Czech. Chem. Commun.*, **53**, 1995 (1988); A. J. Sadlej, *Theor. Chim. Acta.*, **79**, 123 (1991); A. J. Sadlej, *Theor. Chim. Acta*, **81**, 45 (1991); A. J. Sadlej and M. Urban, J. Mol. Struct. (THEOCHEM), **234**, 147 (1991).

[30] A. J. Sadlej, *Theor. Chim. Acta*, **83**, 351 (1992).

[31] A. J. Sadlej, *Theor. Chim Acta*, **81**, 339 (1992).

[32] W. Stevens, H. Basch and J. Kraus, *J. Chem Phys.*, **81**, 6026 (1984).

[33] W. Stevens, H. Basch, J. Kraus and P. G. Jasien, *Can. J. Chem.*, **70**, 612 (1992).

[34] F. Schautz, H. -J. Flad and M. Dolg, *Theor. Chim. Acta*, **99**, 231 (1998).

[35] M. Douglas N. M. Kroll, *Ann. Phys.*, **82**, 89 (1974); B. A. Hess, *Phys. Rev. A*, **33**, 3742 (1986); M. Barysz and A. J. Sadlej, *J. Mol. Struct. (THEOCHEM)*, **573**, 181 (2001).

[36] M. J. Frish, et al., Gaussian 98, revision, A9, Gaussian Inc, Pittsburgh, P.A. 1998.

[37] T. Helgaker et al. DALTON release 1.2 (2001) and 2.0 (2005).

[38] R. D. Amos et al., CADPAC, 5.0, Cambridge, UK, 1992.

[39] R. D. Amos et al., CADPAC 6.0, Cambridge, UK, 1995.

[40] A. Willetts, et al., SPECTRO, A theoretical spectroscopy package, Version 3.0, Cambridge (1994).

[41] C. C. Costain, *J. Chem. Phys.*, **29**, 864 (1958).

[42] (a) O. Christiansen, H. Koch, P. Jørgensen, *Chem. Phys. Lett*, **244**, 75 (1995). (b) G. D. Purvis, and R. J. Bartlett, *J. Chem. Phys.*, **76**, 1910 (1982). (c) J. M. Luis, B. Champagne, B. Kirtman, *Int. J. Quantum. Chem.*, **80,** 471 (2000).

[43] (a) G. R. Alms, A. K. Burnham and W. H. Flygare, J. Chem. Phys., 63, 3321 (1975). (b) M. Stahelin, C. R. Moylan, D. M. Burland, A. Willetts, J. E. Rice, D. P. Shelton and E. A. Donley, J. Chem. Phys., 98, 5595 (1993).

[44] K. Ragavachari, G. W. Trucks, J. A. Pople, M. Head-Gordon, Chem. Phys. Lett., 157, 479 (1989).

[45] A. Bergner, M. Dolg, W. Kueche, H. Stoll, H. Preuss, *Mol. Phys.*, **80**, 1431 (1993); W. Kuechle, M. Dolg, H. Stoll and H. Preuss, *Mol. Phys.*, **74**, 1245 (1991).

[46] (a) K. E. Calderbank, R. L. Calvert, P. B. Lukins, G. L. P. Ritcie, *Aust. J. Chem.*, **34**, 1835 (1981). (b) S. G. Raptis, M. G. Papadopoulos and A. J. Sadlej, *Phys. Chem. Chem. Phys.*, **2**, 3393 (2000); M. Theologitis, G. C. Screttas, S. G. Raptis, and M. G. Papadopoulos, *Int. J. Quant Chem.* **72**, 177 (1999); M. G. Papadopoulos, S. G. Raptis, I. N. Demetropoulos and S. M. Nasiou, *Theor. Chem. Acta*, **99**, 124 (1998); M. G. Papadopoulos, S. G. Raptis and IN. Demetropouos, *Mol. Phys.*, **92**, 547 (1997).

[47] J. F. Herbst and L. G. Hector, Jr., *Phys. Rev. B*, **72**, 125120 (2005); Z. Stoeva, R. I. Smith and D. H. Gregory, *Chem. Mater.*, **18**, 313 (2006); K. Miwa, N. Ohba, S. Towata, Y. Nakamori and S. Orimo, *Phys. Rev. B*, **71**, 195109 (2005).

[48] E. –L. Würthwein, K. D. Sen, J. A. Pople, P. von. R. Schleyer, Inorg. Chem., 22, 496 (1983).

[49] A. Lazicki, B. Maddox, W. J. Evans, C. –S. Yoo, A. K. McMahan, W. E. Pickett and R. T. Scalettar, *Phys. Rev. Letters*, **95**, 165503 (2005).
[50] Y. H. Hu, N. Y. Yu and E. Ruckestein, *Ind. Eng. Chem. Res.*, **44**, 4304 (2005); T. Ichikawa, N. Hanada, S. Isobe, H. Y. Leng and H. Fujii, *J. Phys. Chem. B*, **108**, 7887 (2004);
[51] Y. H. Hu and E. Ruckenstein, *Ind. Eng. Chem. Res.*, **Å**, 2464 (2004).
[52] Y. H. Hu and E. Ruckestein, *J. Phys. Chem. A*, **107**, 9737 (2003).
[53] (a) W.Chen, Z. R. Li, D. Wu, R.Y. Li, C.C. Sun, *J. Phys. Chem. B*, **109**, 601 (2005); (b) W.Chen, Z. R. Li, D. Wu, R.Y. Li, C.C. Sun, F. L. Gu, *J. Am. Chem. Soc.*, **127**, 10977 (2005); (c) W. Chen, Z. R. Li, D. Wu, R.Y. Li, C.C. Sun, F. L. Gu, Y. Aoki, *J. Am. Chem. Soc.*, **128**, 1072 (2006).
[54] L. Serrano-Andres, R. Pou-Amerigo, M. P. Fulscher, A.C. Borin, *J. Chem. Phys.*, **117**, 1649 (2002).
[55] M. Utinans, O. Neilands, *Adv. Mat. For Optics and Elec.*, **9**, 19 (1999).
[56] Ş. Erkoç, *Chem. Phys. Lett.*, **369**, 605 (2003); L. Amirouche and Ş. Erkoç, *Phys. Rev. A*, **68**, 043203 (2003); L. Amiirouche and Ş. Erkoç, *Int. J. Mod. Phys. C*, **14**, 905 (2003).
[57] J. Wang, G. Wang and J. Zhao, *Phys. Rev. A*, **68**, 013201 (2003).
[58] V. Kellő and A. J. Sadlej, Theor. Chim. Acta, 91, 353 (1995); V. Kellő and A. J. Sadlej, Theor. Chim. Acta, 94, 93 (1996).

Brill Academic Publishers
P.O. Box 9000, 2300 PA Leiden,
The Netherlands

Lecture Series on Computer and Computational Sciences
Volume 6, 2006, pp. 308-316

Transition Metal Clusters Polarizabilities

Patrizia Calaminici[1]

Departamento de Quimica, CINVESTAV,
Av. Instituto Politecnico Nacional 2508
A.P. 14-740 Mexico D.F. 07000, MEXICO

Received 16 July, 2006; accepted July 22, 2006

Abstract: The static polarizability and polarizability anisotropy of small transition metal systems, such as Cu_n ($n \leq 9$) and Fe_n ($n \leq 4$) clusters calculated in the framework of density functional theory are presented. The calculations were of all-electron type and performed using a finite field approach implemented in the density functional program deMon2k. Newly developed first-order field induced basis sets for copper and iron atoms for density functional calculations were employed. All cluster structures were fully optimized. The stability of the structures was tested by a frequency analysis. For the copper clusters, structure optimization and frequency analysis were performed on the local density approximation level with the exchange correlation functional by Vosko, Wilk and Nusair. Subsequently improved calculations for the stability were based on the generalized gradient approximation, where the exchange correlation functionals of Perdew and Wang was used. For the iron clusters the structure optimization and frequency analysis were performed on the generalized gradient approximation employing the Perdew, Burke and Ernzerhof functional. For the polarizability calculations non-local exchange-correlation functionals were used. The calculated polarizabilities trends of the studied transition metal clusters are compared with the available experimental polarizabilities of alkali metal clusters.

Keywords: Transition metal clusters, Structures, DFT, First-order field induced basis sets, Polarizability.

1 Introduction

In the last years the static polarizabilities of atoms and free clusters have been extensively studied both theoretically and experimentally (see e.g. Refs. [1, 2] and references therein). The static polarizability (α) represents one of the most important observables for the understanding of the electronic properties of clusters, since it is very sensitive to the delocalization of valence electrons, as well as the structure and shape. Despite numerous investigations on metal clusters, static polarizability measurements have been available over long time only for alkali-metal clusters such as sodium, lithium and potassium [3, 4, 5]. Because of their particular configuration, homo-nuclear alkali metal clusters are often considered to be the simplest metal clusters, so that they have become the prototype systems for understanding size effects in metal clusters. The experimental work of Knight *et al.* [3] by electric deflection techniques has shown that the optical properties of alkali metal aggregates, such as sodium and potassium clusters, follow a general trend toward the bulk value. Knight *et al.* have also shown the existence of a pronounced size dependency in the polarizability of small clusters. In fact, for the sodium aggregates, from the atom to the pentamer, the polarizability per atom has an oscillating behavior. After the pentamer it decreases to a minimum

[1]E-mail: pcalamin@cinvestav.mx

for the octamer and it increases then again for the nonamer. Potassium-cluster polarizabilities follow closely the same pattern [3]. More recently, Benichou *et al.* [4] have measured static electric polarizabilities of lithium clusters up to 22 atoms by deflecting a well-collimated beam through a static inhomogeneous transverse electric field. In order to avoid any systematic error in their experiment, Benichou *et al.* carried also out measurements of sodium cluster polarizabilities. The values they have obtained were in close agreement with those previously measured by Knight *et al.* [3]. The work of Benichou *et al.* shows that the trend of the polarizability per atom of small lithium clusters differs from those of small sodium and potassium clusters. In fact a sharp decrease by about a factor of two from the monomer to the trimer is observed. For larger sizes, $n \geq 4$, the polarizability per atom decreases slowly. For $n \geq 15$ a marked oscillation is observed. The sudden transition of the polarizability from atom to trimer and the following slow variation was interpreted by Benichou *et al.* [4] as a sign that the electronic delocalization already appears in Li_4 or Li_5 clusters.

The electronic configuration of the noble metals Cu, Ag, and Au is characterized by a closed *d* shell and a single valence electron and therefore closely related to that of the alkali metals. In view of this prominent characteristic, clusters of noble metals are expected to exhibit certain similarities to alkali metal clusters. Only very recently static electric polarizabilities of Cu_n clusters with $n \geq 9$ have been measured [6]. Therefore the polarizabilities of smaller noble metal clusters can be only studied with theoretical methods at the moment. The comparison of reliable theoretical polarizabilities of noble metal clusters with those obtained experimentally for alkali metal clusters can give insight into the electronic structure of these systems.

However, mostly of the transition metals are characterized by an open *d* shell and therefore the prediction of their electronic properties is even more complicate. Moreover, these calculations also provide a way of testing computational methods and are, therefore, very interesting for theoreticians. From a theoretical point of view the calculation of properties of iron containing systems present a challenging goal for any quantum chemical approach. The investigation of electronic and geometric structure of these systems was subject of several density functional theory (DFT) investigations in the last 10 years or so and it would be here not appropriate to review all of them. Two of the most recent works are those published by Salahub and Chrétien [7] and Gutsev and Bauschlicher [8], respectively. Both works present results of extensive studies performed on neutral, cationic and anionic small iron clusters concluding that the determinations of the ground state structure of small iron clusters depends on the choice of the correlation functional. Despite many works on these systems are available, additional experimental informations are required in order to assign unequivocally the ground state structure of small iron clusters. Since the static polarizability is a property directly related to the electronic properties of clusters, polarizability measurements would be highly appreciate in order to give a definitive answer about the electronic structure of these systems. However, no many studies on the electric properties of these systems are available. Theoretical studies on the static polarizability of small iron clusters are therefore useful in order to gain more insight into their electronic structure and to guide future experimental investigations.

In this work the results of the first theoretical studies of all-electron gradient-corrected density functional investigations of static polarizability (α) and polarizability anisotropy ($|\Delta\alpha|$) for copper clusters up to the nonamer [9] and for iron clusters up to the tetramer [10, 11] are presented. For these calculations newly developed first-order field induced copper and iron basis sets for density functional calculations were employed. The obtained results of the static polarizabilities are compared with available experimental data for sodium and lithium clusters.

The work is organized as the following. Computational details will be given in section II. The results will be presented and discussed in section III and the conclusions will be summarized in section IV.

2 Computational details

All calculations of the polarizability tensor α and of the polarizability anisotropy $|\Delta\alpha|$ have been carried out using the finite field method of Kurtz *et al.* [12] implemented in the DFT program deMon2k [13]. For the copper clusters the generalized gradient approximation (GGA) by Becke [14] and Perdew (B88-P86) [15] has been used while for the iron clusters the generalized gradient approximation (GGA) by Perdew and Wang (PW86) [16, 15] has been employed. The exchange-correlation potential was numerically integrated on an adaptive grid [17]. The Coulomb energy was calculated by the variational fitting procedure proposed by Dunlap, Connolly and Sabin [18, 19]. For the fitting of the density the auxiliary function set A2 [20] was used in all calculations. The SCF energy convergence criterion [21] was set to 10^{-9} a.u.
It is well known that a general characteristic required for basis sets to perform well for polarizability calculations is that they should contain diffuse functions [22]. An economical strategy for constructing these kinds of basis sets is to augment valence basis sets of reasonable good quality with additional polarization functions [23, 24, 25, 26]. For copper and iron atoms we have chosen as valence triple zeta basis sets (TZVP) which were optimized for local DFT calculations [20]. This basis set was then augmented with seven field-induced polarization (FIP) functions, two *p* and five *f*, which were derived following the work of Zeiss *et al.* [24]. The exponents of the added FIP functions have been optimized by maximizing the polarizability of the copper and iron atoms. These exponents are listed in Table I of Refs. [9, 11]. In order to avoid the contamination of the valence basis set with the diffuse *p*- and *f*-type Gaussians of the FIP functions, spherical basis functions are used in all calculations. We have abbreviated the resulting basis set by TZVP-FIP1. The copper structures have been optimized with the local functional proposed by Vosko, Wilk and Nusair [27]. The initial iron structures were taken from the literature [8] considering the structures optimized with the Perdew, Burke and Ernzerhof functional (PBEPBE) [28] and have been re-optimized with the same nonlocal exchange and correlation functional. For the geometry optimizations a TZVP basis set optimized for density functional calculations [20] was used. The optimization was performed with a quasi-Newton update method [29] using analytic energy gradients. The structure optimization convergence was based on the gradient and displacement vectors with a threshold of 10^{-4} and 10^{-3} a.u., respectively.
In order to discriminate between minima and transition states on each studied potential energy surface (PES) a vibrational analysis was performed at the optimized geometries. The second derivatives were calculated by numerical differentiation (two-point finite difference) of the analytic energy gradients using a displacement of 0.001 a.u. from the optimized geometry for all $3N$ coordinates. The harmonic frequencies were obtained by diagonalizing the mass-weighted Cartesian force constant matrix.
A field strength of 0.001 a.u. for the calculations of α and $|\Delta\alpha|$ was used. The mean polarizability was calculated from the polarizability components as :

$$\overline{\alpha} = \frac{1}{3}\left(\alpha_{xx} + \alpha_{yy} + \alpha_{zz}\right) ,$$

and the polarizability anisotropy as :

$$\begin{aligned} |\Delta\alpha|^2 &= \frac{3\mathrm{tr}\,\alpha^2 - (\mathrm{tr}\,\alpha)^2}{2} \quad \text{(general axes)} \\ &= \frac{1}{2}\left[(\alpha_{xx} - \alpha_{yy})^2 + (\alpha_{xx} - \alpha_{zz})^2 + (\alpha_{yy} - \alpha_{zz})^2\right] \quad \text{(principal axes)} . \end{aligned}$$

3 Results and Discussion

3.1 Polarizability of Cu clusters

The static mean polarizabilities and the polarizability anisotropies of copper clusters up to the nonamer were calculated at the optimized geometries [31] obtained by using a local functional [27]. The structures of these minima are illustrated in Fig. 1.

The geometrical parameters and the electronic states of these minima are given in Ref. [31]. As

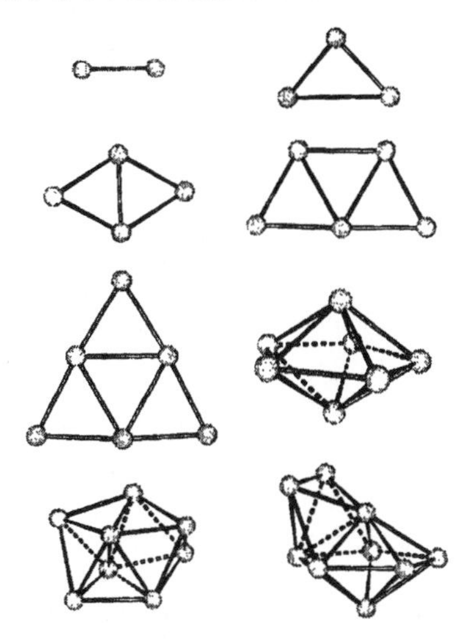

Figure 1: Structures of the ground state structure of Cu_n (n=2-9) clusters.

Fig. 1 shows the ground state structures of the the studied copper clusters are found to be planar up to the hexamer. From the heptamer on, the structures became three-dimensional.
From our experience a good work strategy is to use geometrical structures optimized at a local level of theory and to perform the electronic properties calculations at the gradient corrected level of theory [30, 31]. Therefore, for the calculations of the electronic properties of copper clusters we used a GGA functional. The obtained results of $\overline{\alpha}$ and $|\Delta\alpha|$ are collected in Table II of Ref. [9].
In Fig. 2 the mean polarizability per atom of copper clusters is plotted.

As this figure shows, going from the atom up to the pentamer, the mean polarizability per atom has an oscillating behavior. From the pentamer, the polarizability per atom decreases with a minimum value for the octamer. It increases then again, going from the octamer to the nonamer.
In Figs. 3 and 4 the experimental polarizability per atom of sodium and lithium clusters up to the nonamer are plotted, respectively.

The values of the experimental polarizability per atom of sodium clusters are those reported by Knight *et al.* [3], while the experimental polarizability data of lithium clusters are those reported by Benichou *et al.* [4].
As it was already discussed in the introduction the trend of the experimental mean polarizability of small sodium and lithium clusters is different (see Figs. 3 and 4). The comparison of Figs. 2, 3 and 4 shows that the calculated polarizability per atom of copper clusters, going from the atom to the nonamer, presents the same trend as experimentally observed by Knight *et al.* [3] for

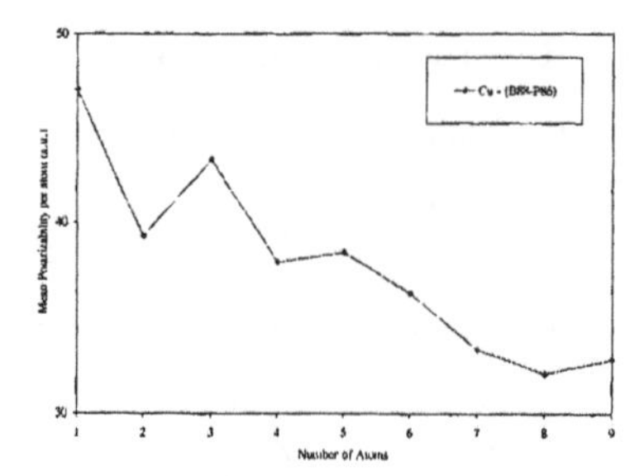

Figure 2: Calculated mean polarizability per atom of Cu_n (n=2-9). The calculations were performed using the B88-P86 functional.

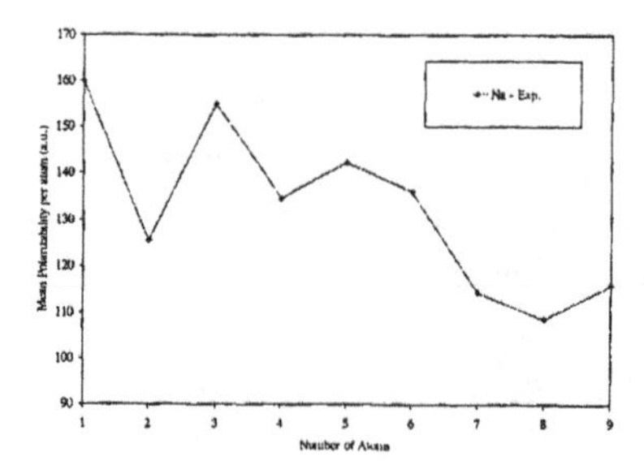

Figure 3: Experimental mean polarizability per atom of Na_n (n=2-9). Values are from Ref. [3].

sodium clusters. However, the mean polarizability per atom of copper clusters is about three times smaller. This result indicates that in copper clusters the electrons are more strongly attracted by the nuclei as in the sodium clusters. Therefore their electronic structure is more compact. We are confident that the reliability of the absolute values for the calculated polarizabilities of the copper clusters are in the same range as for the previous study we have carried on small sodium clusters [32]. Thus deviations of less than 10 % for the absolute values can be expected. The theoretical prediction of the trend of the mean polarizabilities per copper atom is believed to be reliable, too. The calculated polarizability anisotropies for Cu_n (n=2-9) are also given in Table II of Ref. [9]. The polarizability anisotropy increases from the dimer to the hexamer and decreases from the hexamer to the octamer. Similar to the calculated mean polarizability per atom, a minimum value for the polarizability anisotropy was found at the octamer. From these data and the topologies of the optimized copper cluster, it is obvious that the polarizability anisotropy is directly related to the particular cluster structure. In fact, in the planar clusters it increases with an increasing number of copper atoms. For the pentamer and the hexamer very similar values for the polarizability anisotropy are calculated. When the cluster structure becomes three-dimensional the polarizability anisotropy decreases to a minimum value for the octamer, which is the cluster with the most compact structure. It increases again as the cluster structure becomes more open, as in the case of the nonamer.

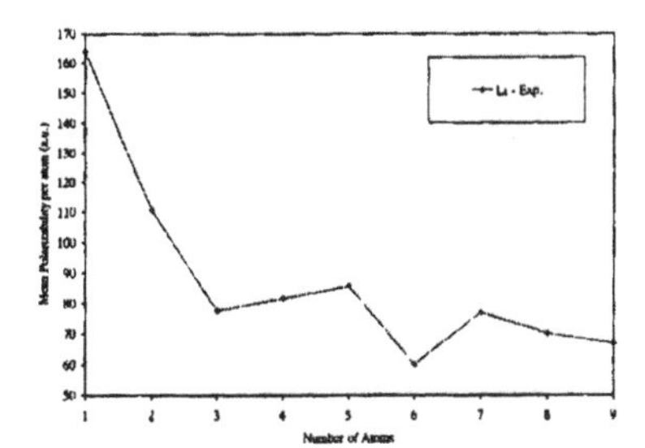

Figure 4: Experimental mean polarizability per atom of Li_n (n=2-9). Values are from Ref. [4].

3.2 Polarizability of Fe clusters

The static mean polarizabilities and the polarizability anisotropies of iron clusters up to the tetramer were calculated at the optimized geometries [13] obtained by using the PBEPBE functional [28]. For the structure optimization we considered as initial structures on the PES the ground state structures of Ref. [8] which have been obtained using the same functional.
The topologies, geometries and electronic states of our optimized structures are reported in Fig. 5.

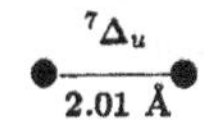

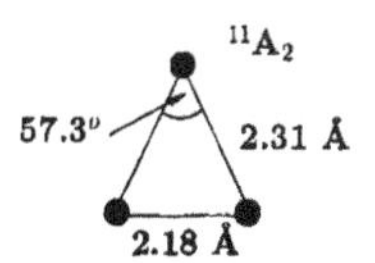

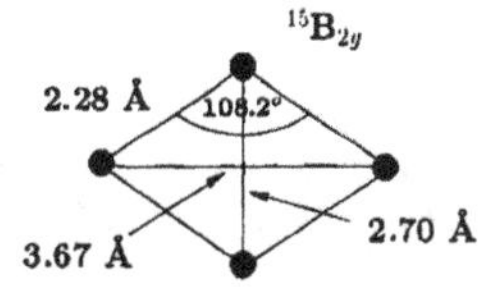

Figure 5: Topologies, geometries and electronic states of the optimized Fe clusters. The optimization was performed with the PBE-PBE functional.

Our results of Fe_2 and Fe_3 are very similar to the ground state structures reported in the literature [7, 8]. For Fe_4, a $^{15}B_{2g}$ state was obtained. This state was reported in Ref. [7] lying at 24.2 kJ/mol above the found ground state structure. In the present work no three-dimensional structures for the tetramer were investigated. The vibrational analysis confirmed that our optimized structures are minima on the PESs since no imaginary frequencies were obtained. The calculated frequencies are: 417 cm^{-1} for the dimer, 58, 239 and 351 cm^{-1} for the trimer, 72, 169, 190, 199, 275 and 286 cm^{-1} for the tetramer. In Fig. 6 the mean polarizability per atom of the studied iron clusters is plotted. These results were obtained with the PW86 functional. Very similar results, with deviations of only about 1-4 a.u. are obtained with PBE-PBE functional considering the same structures. The effect of different GGA functionals in the calculations of polarizabilities of transition metal clusters is a current topic of investigation in our laboratory.

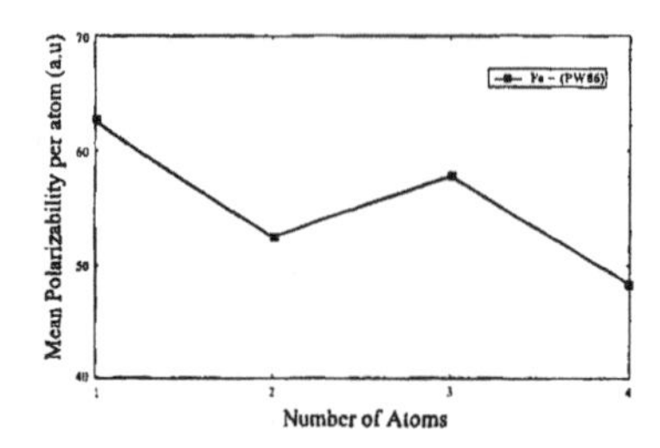

Figure 6: Calculated mean polarizability per atom of Fe_n (n=1-4).

The calculated polarizabilities of iron clusters Fe_n ($n \leq 4$) have been compared with the experimental polarizability per atom of sodium and lithium clusters. It was observed that the trend of the polarizability per atom of iron clusters is similar to that obtained in sodium clusters. However, the calculated static mean polarizabilities of iron clusters result to be around two times smaller of those calculated for the corresponding sodium clusters.
The polarizability anisotropy increases from the dimer to the trimer of about more than two times and increases slightly going from the trimer to the tetramer.
It is well known that in iron clusters many low-lying states lie close in energy. Therefore, a more extensive study considering different isomers should be done in order to gain more insight into the electronic properties of such systems. Also, it would be interesting to investigate how the polarizability trend of iron clusters will change going to larger clusters.

4 Conclusions

In this work we reported the results of the first theoretical study of static polarizabilities and polarizability anisotropies of copper clusters up to the nonamer and of iron clusters up to the tetramer. The calculations have been carried out in the framework of density functional theory at a gradient-corrected level. For the calculations of $\overline{\alpha}$ and $|\Delta\alpha|$ newly developed first-order polarized basis sets for density functional calculations were employed. The exponents of these basis sets have been optimized by maximizing the polarizability of the copper and iron atoms. The calculated polarizabilities of copper clusters Cu_n ($n \leq 9$) and iron clusters Fe_n ($n \leq 4$) have been compared with the experimental polarizability per atom of sodium and lithium clusters. It was observed that the trend of the polarizability per atom of small copper and iron clusters is similar to that obtained in sodium clusters. However, the calculated static mean polarizabilities result to be around three and two times smaller of those calculated for the corresponding sodium clusters, for copper and iron clusters, respectively. It would be important to investigate the trend of static polarizabilities of the here studied homo-nuclear iron systems considering different low-lying isomers as well as larger clusters size and to investigate other transition metal clusters characterized by an open *d* shell. These calculations are currently in progress in our laboratory.

References

[1] U. Kreibig and M. Vollmer, *Optical Properties of Metal Clusters* (Springer, Berlin, 1995)

[2] K.D. Bonin and V.V. Kresin, *Electric-Dipole Polarizabilities of Atoms, Molecules and Clusters* (World Scientific, Singapore, 1997)

[3] W.D. Knight, K. Clemenger, W.A. de Heer and W.A. Saunders, *Phys. Rev. B* **31**, 2539 (1985)

[4] E. Benichou, R. Antoine, D. Rayane, B. Vezin, F.W. Dalby, Ph. Dugourd, M. Broyer, C. Ristori, F. Chandezon, B.A. Huber, J.C. Rocco, S.A. Blundell and C. Guet, *Phys. Rev. A* **59**, R1 (1999)

[5] R. Antoine, D. Rayane, A.R. Allouche, M. Aubert-Frécon, E. Benichou, F.W. Dalby, Ph. Dugourd, M. Broyer and C. Guet, *J. Chem. Phys.* **110**, 5568 (1999)

[6] M.B. Knickelbein, *J. Chem. Phys.* **120**, 10450 (2004)

[7] S. Chrétien and D.R. Salahub, *Phys. Rev. B* **66**, 155425 (2002)

[8] G.L. Gutsev and C.W. Bauschlicher Jr., *J. Phys. Chem. A* **107**, 7013 (2003)

[9] P. Calaminici, A.M. Köster, A. Vela and K. Jug, *J. Chem. Phys.* **113**, 2199 (2000)

[10] P. Calaminici, *Chem. Phys. Lett.* **374**, 650 (2003)

[11] P. Calaminici, *Chem. Phys. Lett.* **387**, 253 (2004)

[12] H.A. Kurtz, J.J.P. Stewart and K.M. Dieter, *J. Comp. Chem.* **11**, 82 (1990)

[13] A.M. Köster, P. Calaminici, R. Flores-Moreno, G. Geudtner, A. Goursot, T. Heine, F. Janetzko, S. Patchkovskii, J.U. Reveles, A. Vela and D.R. Salahub, deMon2k, The deMon developers (2004)

[14] A.D. Becke, *Phys. Rev. A* **38**, 3098 (1988)

[15] J.P. Perdew, *Phys. Rev. B* **33**, 8822 (1986)

[16] J.P. Perdew and Y. Wang, *Phys. Rev. B* **33**, 8800 (1986); erratum: *Phys. Rev. B* **40**, 3399 (1989)

[17] A.M. Köster, R. Flores-Moreno and J.U. Reveles,*J. Chem. Phys.* **121**, 681 (2004)

[18] B.I. Dunlap, J.W.D. Connolly and J.R. Sabin, *J. Chem. Phys.* **71**, 4993 (1979)

[19] J.W. Mintmire and B.I. Dunlap, *Phys. Rev. A* **25**, 88 (1982)

[20] N. Godbout, D.R. Salahub, J. Andzelm and E. Wimmer, *Can. J. Phys.* **70**, 560 (1992)

[21] A.M. Köster, *Habilitation thesis*, Universität Hannover, 1998

[22] H.-J. Werner, W. Meyer, *Mol. Phys.* **31**, 855 (1976)

[23] A.J. Sadlej, *Collection Czechoslovak Chem. Commun.* **53**, 1995 (1988)

[24] G.D. Zeiss, W.R. Scott, N. Suzuki, D.P. Chong and S.R. Langhoff, *Mol. Phys.* **37**, 1543 (1979)

[25] M. Jaszuński and B.O. Roos, *Mol. Phys.* **52**, 1209 (1984)

[26] B.O. Roos and A.J. Sadlej, *Chem. Phys.* **94**, 43 (1985)

[27] S.H. Vosko, L. Wilk and M. Nusair, *Can. J. Phys.* **58**, 1200 (1980)

[28] J.P. Perdew, K. Burke and M. Ernzerhof, *Phys. Rev. Lett.* **77**, 3865 (1996)

[29] Reveles J.U. and Köster A.M., *J. Comput. Chem.* **25**, 1109 (2004)

[30] P. Calaminici, A.M. Köster, N. Russo and D.R. Salahub, *J. Chem. Phys.* **105**, 9546 (1996)

[31] K. Jug, B. Zimmermann, P. Calaminici and A. M. Köster, *J. Chem. Phys.* **116**, 4497 (2002)

[32] P. Calaminici, K. Jug and A.M. Köster, *J. Chem. Phys.* **111**, 4613 (1999)

Brill Academic Publishers
P.O. Box 9000, 2300 PA Leiden
The Netherlands

Lecture Series on Computer and Computational Sciences
Volume 6, 2006, pp. 317-323

Ab Initio Methods for Simulating and Interpreting hyper-Raman Spectra of Molecules

Benoît Champagne[1]

Laboratoire de Chimie Théorique Appliquée, Groupe de Chimie-Physique
Facultés Universitaires Notre-Dame de la Paix,
Rue de Bruxelles, 61, B-5000 Namur, Belgium

Received 29 July, 2006; accepted 2 August, 2006.

Abstract: the hyper-Raman effect is a nonlinear inelastic scattering phenomenon, which can provide complementary information to IR and Raman spectroscopies. Simulating hyper-Raman spectra is however a challenge because it requires the evaluation of high-order response properties. This paper reviews methods for simulating and interpreting the hyper-Raman spectra of molecules and illustrates them with recent applications, carried out in parallel with experimental studies.

Keywords: hyper-Raman effect, time-dependent Hartree-Fock, geometrical derivatives, vibrational spectroscopies, *ab initio* simulations.

PACS: 31.15.-p, 31.25.-v, 33.20.Fb, 33.20.Tp

1. Introduction

The (vibrational) hyper-Raman phenomenon was first observed in 1965 by Terhune and co-workers [1]. It is a three-photon process in which the system is excited by two photons of pulsation ω and emits one photon at the pulsation $2\omega \pm \omega_a$, where ω_a is the pulsation associated with a vibrational transition [2]. This effect is associated with the quadratic term ($\chi^{(2)}$) in the expansion of the polarization (P) of the medium induced by the incident light (E is the electric field),

$$P = P_0 + \chi^{(1)} E + \chi^{(2)} E^2 + \ldots \quad (1)$$

It is therefore a nonlinear inelastic scattering phenomenon. The hyper-Raman scattering is ruled by different selection rules from Raman scattering and infrared (IR) absorption and is therefore complementary to IR and Raman [3]. In particular, it can provide information on low-frequency vibrational transitions, which are difficult to detect with the IR and Raman spectroscopies as well as to detect modes that are silent in both IR and Raman spectroscopies.

Experimental recording of hyper-Raman spectra has often been hampered by the weakness of the signal – experimental spectra are used to exhibit important noise – but the latest improvements in the spectral detection and analysis of scattering processes within the last decade (spectrographs with high luminosity efficiencies including CCD detectors) allow to overcome this intrinsic difficulty as demonstrated by the recent developments of Vincent RODRIGUEZ in Bordeaux [4].

Since the hyper-Raman phenomenon is a quadratic effect [Eq. (1)], within the harmonic approximation its intensities depend on the derivatives with respect to the vibrational normal coordinates of the frequency-dependent first hyperpolarizability, $(\partial\beta(-2\omega;\omega,\omega)/\partial Q_a)_0$, whereas the corresponding IR and Raman intensities are given by the corresponding derivatives of the dipole moment and polarizability, respectively. Simulating the hyper-Raman spectra requires therefore substantial theoretical and

[1] Corresponding author. benoit.champagne@fundp.ac.be

computational efforts. Several approaches are described in Section 2. In addition, theoretical modeling can help in interpreting the spectra of complex molecules. In particular, like for IR and Raman spectroscopies, simple rules of thumb can be deduced in order to help spectral assignments. This is particularly true because many modes are active in hyper-Raman spectroscopy and therefore because the spectra are rich in information. Section 3 briefly illustrates these aspects.

2. Theoretical aspects and computational methods

Within the clamped nucleus and harmonic approximations, the intensity of the Stokes hyper-Raman scattered incoherent (monomolecular) radiation reads [2]

$$\Im_{hR,t}(2\omega-\omega_t) \propto N\Im_0^2(2\omega-\omega_t)^4 \left|\left\langle \Psi_{init}\left|\hat{\beta}(\omega,Q)\right|\Psi_{fin}\right\rangle\right|^2 \tag{2}$$

where ω is the angular frequency of the incident radiation, N is the number of molecules in the initial configuration state, $\Im_0$ is the intensity of the incident radiation, and $\hbar(2\omega-\omega_t)$ is the scattering energy. The bracket term is a transition matrix element of the first hyperpolarizability operator between the initial and final wavefunctions, Ψ_{init} and Ψ_{fin}. In these conditions, only fundamental transitions are considered, *i.e.* the vibrational transitions correspond to changes by one unit of one vibrational quantum number at a time. Then, for liquids and gases, the molecules are free to assume all orientations with respect to the observer with equal probability so that the observed scattered intensity is an average, which depends on the geometry of the experimental setup. Generally, the incident light is vertically(V)- or horizontally(H)-plane-polarized, the detection is made in the direction perpendicular to both the V-polarized electric field and its propagation direction, and the V-polarized scattered light is recorded. In particular, the hyper-Raman VV scattering intensity reads:

$$\Im_{hR,a}^{VV}(2\omega-\omega_a) \propto N\hbar\Im_0^2 \frac{(2\omega-\omega_a)^4}{2\omega_a\left[1-\exp(-\hbar\omega_a/kT)\right]}\left\langle \tilde{\beta}_{\zeta\zeta\zeta}^2(\omega,Q_a)\right\rangle \tag{3}$$

k is the Boltzmann constant, T is the temperature, and the energy of transition amounts to $\hbar\omega_t = \hbar\omega_a$ with ω_a the circular frequency of the ath vibrational normal mode. The square bracket in the denominator accounts for the vibrational state populations while the rotational average reads:

$$\begin{aligned}\left\langle \tilde{\beta}_{\zeta\zeta\zeta}^2\right\rangle = {} & \frac{1}{7}\sum_{\zeta}^{xyz}\beta_{\zeta\zeta\zeta}^{'2} + \frac{4}{35}\sum_{\zeta\neq\eta}^{xyz}\beta_{\zeta\zeta\eta}^{'2} + \frac{2}{35}\sum_{\zeta\neq\eta}^{xyz}\beta'_{\zeta\zeta\zeta}\beta'_{\zeta\eta\eta} + \frac{4}{35}\sum_{\zeta\neq\eta}^{xyz}\beta'_{\eta\zeta\zeta}\beta'_{\zeta\zeta\eta} + \frac{4}{35}\sum_{\zeta\neq\eta}^{xyz}\beta'_{\zeta\zeta\zeta}\beta'_{\eta\eta\zeta} + \frac{1}{35}\sum_{\zeta\neq\eta}^{xyz}\beta_{\eta\zeta\zeta}^{'2} \\ & + \frac{4}{105}\sum_{\zeta\neq\eta\neq\xi}^{xyz}\beta'_{\zeta\zeta\eta}\beta'_{\eta\xi\xi} + \frac{1}{105}\sum_{\zeta\neq\eta\neq\xi}^{xyz}\beta'_{\eta\zeta\zeta}\beta'_{\eta\xi\xi} + \frac{4}{105}\sum_{\zeta\neq\eta\neq\xi}^{xyz}\beta'_{\zeta\zeta\eta}\beta'_{\xi\xi\eta} + \frac{2}{105}\sum_{\zeta\neq\eta\neq\xi}^{xyz}\beta_{\zeta\eta\xi}^{'2} + \frac{4}{105}\sum_{\zeta\neq\eta\neq\xi}^{xyz}\beta'_{\zeta\eta\xi}\beta'_{\eta\zeta\xi}\end{aligned} \tag{4}$$

with

$$\beta'_{\zeta\eta\xi} = \left(\frac{\partial\beta_{\zeta\eta\xi}(-2\omega;\omega,\omega)}{\partial Q_a}\right) \tag{5}$$

Similarly, the hyper-Raman HV (and VH) scattering intensity is given by

$$\Im_{hR,a}^{HV}(2\omega-\omega_a) \propto N\hbar\Im_0^2 \frac{(2\omega-\omega_a)^4}{2\omega_a\left[1-\exp(-\hbar\omega_a/kT)\right]}\left\langle \tilde{\beta}_{\zeta\eta\eta}^2(\omega,Q_a)\right\rangle \tag{6}$$

where

$$\begin{aligned}\left\langle \tilde{\beta}_{\zeta\eta\eta}^2\right\rangle = {} & \frac{1}{35}\sum_{\zeta}^{xyz}\beta_{\zeta\zeta\zeta}^{'2} + \frac{4}{105}\sum_{\zeta\neq\eta}^{xyz}\beta'_{\zeta\zeta\zeta}\beta'_{\zeta\eta\eta} - \frac{2}{35}\sum_{\zeta\neq\eta}^{xyz}\beta'_{\zeta\zeta\zeta}\beta'_{\eta\eta\zeta} + \frac{8}{105}\sum_{\zeta\neq\eta}^{xyz}\beta_{\zeta\zeta\eta}^{'2} + \frac{3}{35}\sum_{\zeta\neq\eta}^{xyz}\beta_{\zeta\eta\eta}^{'2} - \frac{2}{35}\sum_{\zeta\neq\eta}^{xyz}\beta'_{\zeta\zeta\eta}\beta'_{\eta\zeta\zeta} \\ & + \frac{1}{35}\sum_{\zeta\neq\eta\neq\xi}^{xyz}\beta'_{\zeta\eta\eta}\beta'_{\zeta\xi\xi} - \frac{2}{105}\sum_{\zeta\neq\eta\neq\xi}^{xyz}\beta'_{\zeta\zeta\xi}\beta'_{\eta\eta\xi} - \frac{2}{105}\sum_{\zeta\neq\eta\neq\xi}^{xyz}\beta'_{\zeta\zeta\eta}\beta'_{\eta\xi\xi} + \frac{2}{35}\sum_{\zeta\neq\eta\neq\xi}^{xyz}\beta_{\zeta\eta\xi}^{'2} - \frac{2}{105}\sum_{\zeta\neq\eta\neq\xi}^{xyz}\beta'_{\zeta\eta\xi}\beta'_{\eta\zeta\xi}\end{aligned} \tag{7}$$

The depolarization ratio, which gives information on the symmetry of the modes, reads:

$$\rho_{hR,a}(\omega)=\frac{\Im^{HV}_{hR,a}(2\omega-\omega_a)}{\Im^{VV}_{hR,a}(2\omega-\omega_a)}=\frac{\langle\tilde{\beta}^2_{\zeta\eta\eta}\rangle}{\langle\tilde{\beta}^2_{\zeta\zeta\zeta}\rangle} \quad (8)$$

In most simulations of vibrational spectra, the transition frequencies are calculated within the harmonic approximation. In our investigations, they are determined at the coupled-perturbed Hartree-Fock (CPHF) level or by including electron correlation effects, either at DFT levels or using the Møller-Plesset perturbation theory and coupled cluster approaches. In all these cases, for comparing with fundamental transition frequencies, the vibrational frequencies are scaled to account for missing electron correlation and anharmonicity effects [5]. Typical scaling factors are 0.91 and 0.96 at the HF and MP2 levels of approximation, respectively.

Last century, only a few works have been devoted to the evaluation of the hyper-Raman intensities, *i.e.* to the determination of the quantity in Eq. 5 within different approximations. In order to address the orientation of pyridine molecules adsorbed onto silver electrodes, Golab *et al.* [6] calculated the dynamic first hyperpolarizability using the π-electron Pariser-Parr-Pople (PPP) method and differentiated it numerically with respect to empirically determined normal coordinates. Improvements of this approach include the use of the semi-empirical intermediate neglect of differential overlap (INDO) scheme [7] or the *ab initio* Hartree-Fock method [7]. Although in the later case the vibrational characteristics and electrical properties are determined at a similar level of approximation, only the static values were evaluated. Nørby Svendsen and Stroyer-Hansen [8] employed a variational approach within the CNDO/2 approximation and an extended basis set with polarization functions. Moreover, they addressed the effect of taking into account the frequency of the incident light on the hyper-Raman intensities for methane, ethylene, and ethane. Difficulties in calculating these intensities can probably be traced back to the high order of the energy (or density matrix) derivatives and to their frequency dependence, which partly explains why there have been so few theoretical modeling studies.

A few years ago, we have initiated a research project on simulating vibrational hyper-Raman spectra. First, we elaborated analytical procedures based on the time-dependent Hartree-Fock (TDHF) scheme to evaluate the first-order derivatives of the dynamic first hyperpolarizability with respect to atomic Cartesian coordinates, $\partial\beta(-2\omega;\omega,\omega)/\partial X$, and subsequently to simulate the nonresonant hyper-Raman spectra within the double harmonic approximation [9]. In this case, the $\partial\beta(-2\omega;\omega,\omega)/\partial Q_a$ quantities are evaluated by projecting on the vibrational normal modes the $\partial\beta(-2\omega;\omega,\omega)/\partial X$ quantities, which enables to employ different methods for determining the vibrational normal modes and frequencies and for calculating the electrical properties. In order to reduce the computational task, the algorithms have been developed to satisfy the $2n+1$ rule [10] and to take advantage of the interchange relations [11]. The $2n+1$ rule states that, for a variational wave-function, the $2n+1^{st}$ energy derivatives can be determined from the nth-order (and lower orders) derivatives of the LCAO coefficients. The $\partial\beta(-2\omega;\omega,\omega)/\partial X$ quantities being fourth-order derivatives of the energy, first- and second-order derivatives of the LCAO coefficients are sufficient. On the other hand, the interchange relations enable to replace derivatives taken with respect to the $3N$ atomic Cartesian coordinates by derivatives taken with respect to the more convenient 3 components of the electric fields. In summary, the hyper-Raman intensities can be evaluated from calculating i) the first-order derivatives of the LCAO coefficients with respect to the dynamic electric fields and the atomic Cartesian coordinates and ii) the second-order derivatives of the LCAO coefficients with respect to dynamic electric fields. These methods have been implemented in the GAMESS quantum chemistry package [12].

Anharmonicity effects were later considered to deal with the Fermi resonance in the CCl_4 molecule [13]. Terms up to second order in electrical and/or mechanical anharmonicity were included in the perturbation expansion of the hyper-Raman intensity:

$$[\Im]=[\Im]^{0,0}+[\Im]^{1,0}+[\Im]^{01}+[\Im]^{2,0}+[\Im]^{11}+[\Im]^{02} \quad (9)$$

where the first (second) superscript defines the order in electrical (mechanical) anharmonicity and where the leading term corresponds to Eq. 3. To evaluate Eq. 9, the cubic force constants are evaluated by using a numerical distortion procedure. These quantities also enable to estimate the amplitude of the Fermi splitting. In the case of the CCl_4 molecule, our HF/aug-cc-pvdz calculations give a value of 18 cm^{-1} in comparison with the experimental value of 23 cm^{-1} [13]. On the other hand, the second-order derivatives of the dynamic first hyperpolarizability are determined analytically at the TDHF level following the scheme described in Ref. 14 and implemented in GAMESS [12].

Electron correlation plays also an important role on the hyper-Raman intensities, both indirectly by modifying the vibrational normal coordinates and directly by affecting the electrical properties. The first studies where electron correlation effects are included adopt however the finite distorting scheme because analytical schemes are still under development [15-17]. Time-dependent density functional theory (TDDFT) was applied with several exchange-correlation functionals [15] while coupled-cluster response methods were used for benchmarking $\partial\beta/\partial X$ values [16]. Table 1 describes the effects of the exchange correlation functional on the relative hyper-Raman intensities as well as depolarization ratios of the methane molecule. Going from LDA to different GGA functionals has a reduced impact on the relative intensities but a clear effect on the depolarization ratio of the triply-degenerate bending mode. In addition to these studies that account for frequency dispersion, hyper-Raman intensities were also calculated in the static limit at the MP2 level of approximation [17]. As an illustration, electron correlation is of paramount importance to reproduce the hyper-Raman intensity of the CN stretching mode of acetonitrile (but not for its Raman counterpart) [17].

Table 1: Effect of the exchange-correlation functional on the relative hyper-Raman intensities and depolarization ratios (in parentheses) in methane. The calculations have been performed using the ADF package and the ET-pVQZ basis set [15]. λ = 1064 nm. The vibrational normal modes have been labeled according to Herzberg [18] – the vibrational frequencies correspond to fundamental (experimental) transitions.

	LDA	BLYP	PBE	RPBE
ω_1 (A_1, 2914 cm^{-1})	0.30 (2/3)	0.34 (2/3)	0.32 (2/3)	0.32 (2/3)
ω_3 (T_2, 3020 cm^{-1})	1.00 (0.24)	1.00 (0.26)	1.00 (0.24)	1.00 (0.24)
ω_4 (T_2, 1306 cm^{-1})	0.05 (0.39)	0.04 (0.45)	0.05 (0.32)	0.05 (0.27)

3. Applications

Illustrations of the reliability of these simulation methods are given below and concentrate on three molecules, ethylene, acetonitrile, and benzene. The scattering peaks are described by Lorentzians. All calculations have been performed using the aug-cc-pvdz basis set. For ethylene, the reported experimental spectra have been taken from the work of Verdieck *et al.* [19] whereas the experimental spectra of acetonitrile were recorded in Bordeaux by Rodriguez [17].

Ethylene is a good example to demonstrate the complementarity of hyper-Raman spectroscopy with IR and Raman. The top part of Figure 1 represents the experimental and simulated hyper-Raman spectra for a conventional experimental setup (plane-polarized incident light and detection of the total scattered light) while the bottom part to an alternative setup where the incident radiation is circularly polarized and the scattered light is analyzed with polarization parallel to the scattering plane. In the former case, the intensity is given by the sum of the VV and HV intensities (Eqs. 3-4, 6-7) whereas in the latter case, the $\langle\ \rangle$ bracket has to be replaced by:

$$\begin{aligned}\langle\tilde{\beta}^2_{\zeta\eta\xi}\rangle = &\frac{4}{105}\sum_{\zeta}^{xyz}\beta'^2_{\zeta\zeta\zeta} - \frac{4}{105}\sum_{\zeta\neq\eta}^{xyz}\beta'_{\zeta\zeta\zeta}\beta'_{\zeta\eta\eta} - \frac{8}{105}\sum_{\zeta\neq\eta}^{xyz}\beta'_{\zeta\zeta\zeta}\beta'_{\eta\eta\zeta} + \frac{4}{21}\sum_{\zeta\neq\eta}^{xyz}\beta'^2_{\zeta\zeta\eta} + \frac{4}{35}\sum_{\zeta\neq\eta}^{xyz}\beta'^2_{\zeta\eta\eta} - \frac{8}{105}\sum_{\zeta\neq\eta}^{xyz}\beta'_{\zeta\zeta\eta}\beta'_{\eta\zeta\zeta}\\ &-\frac{2}{21}\sum_{\zeta\neq\eta\neq\xi}^{xyz}\beta'_{\zeta\eta\eta}\beta'_{\zeta\xi\xi} - \frac{4}{35}\sum_{\zeta\neq\eta\neq\xi}^{xyz}\beta'_{\zeta\zeta\xi}\beta'_{\eta\eta\xi} + \frac{16}{105}\sum_{\zeta\neq\eta\neq\xi}^{xyz}\beta'_{\zeta\zeta\eta}\beta'_{\eta\xi\xi} + \frac{22}{105}\sum_{\zeta\neq\eta\neq\xi}^{xyz}\beta'^2_{\zeta\eta\xi} - \frac{4}{35}\sum_{\zeta\neq\eta\neq\xi}^{xyz}\beta'_{\zeta\eta\xi}\beta'_{\eta\zeta\xi}\end{aligned} \quad (10)$$

The characteristics of the hyper-Raman spectra are listed in Table 2. In general, the agreement between the experimental and simulated spectra is good. Of course, the noise is not negligible, especially in the case of the non conventional setup. For both setups, our calculations confirm that the broad peak at 3040 wavenumbers results from the overlap between the ω_9 (B_{2u}) and ω_{11} (B_{3u}) C-H stretching bands. Nevertheless, as shown by the shoulder towards high frequencies, the relative hyper-Raman intensities of the *anti* and *cis* C-H stretching modes are inverted in the simulation. The intensity of the ω_{12} (B_{3u})

mode corresponding to a scissor motion is too weak and experimentally not observed. Theory also predicts that this mode has a weak hyper-Raman activity. Then, in the case of the conventional setup, the strong band at 980 wavenumbers is due to the out-of-plane CH_2 wagging motion as well as, to a lower extend, to the ω_4 (A_u) twisting mode. The ω_{10} (B_{2u}) rocking mode predicted at 798 wavenumbers is not detected. On the other hand, for the alternative CV (C stands for circularly-polarized) setup, the strong band around 1000 cm^{-1} turns out to be mostly due to the ω_4 twisting mode. This is particularly interesting because the ω_4 (A_u) twisting mode is silent in both IR and Raman (harmonic) spectroscopies and it cannot be located in the conventional setup experiment due to the proximity with the intense ω_7 (B_{1u}) mode.

Table 2 : Theoretical scaled harmonic frequencies (ω_a, in cm^{-1}) and relative hyper-Raman scattering intensities ($I_{hR,a}$, in %) of ethylene in relation with the experimental peak positions (ω, in cm^{-1}). λ = 694.3 nm.

	ω_a	$I_{hR,a}$ (VV + VH) Eqs. 4 + 7	$I_{hR,a}$ (CH) Eq. 10	ω
ω_{10} (B_{2u})	798	6	11	
ω_7 (B_{1u})	979	100	28	980
ω_4 (A_u)	1017	52	100	
ω_{12} (B_{3u})	1422	1	3	/
ω_{11} (B_{3u})	2972	21	39	3040
ω_9 (B_{2u})	3070	11	16	

In the case of acetonitrile, the TDHF/aug-cc-pvdz spectrum differs substantially from the experimental spectrum (Table 3, columns 2 and 4). Indeed, calculations attribute non-negligible intensities to modes 1, 5, and 7 (labeled according to Herzberg) whereas these modes are not visible on the hyper-Raman spectrum. Furthermore, the relative intensities of several other modes (2, 3, and 8) are underestimated. Part of this discrepancy vanishes when accounting for electron correlation effects in the evaluation of the electric properties and their derivatives with respect to normal coordinates (Table 3, column 5). Indeed, at the MP2 level, the intensity of mode 2 is in much better agreement with experiment. The intensities of modes 5 and 7 decrease and the intensity of mode 3 increases but these corrections are still incomplete and the intensity of mode 1 remains unchanged. Ref. 17 suggests that this major discrepancy for hyper-Raman, while the Raman spectrum is well reproduced at the same levels of approximation, originates from the hard experimental hyper-Raman conditions that may exalt the translational and rotational motions. In order to explain the discrepancy between the theoretical and observed spectra, it was hypothesized in Ref. 17 that the molecule does not really belong to the C_{3v} but rather to a pseudo $C_{\infty v}$ point group. This enables to reproduce an experimental feature: the hyper-Raman active modes involve variations of the longitudinal components of the dipole moment whereas, with the exception of the ω_8 mode, the perpendicular modes are not observed. Indeed, when considering the $C_{\infty v}$ symmetry with the inclusion of electron correlation effects at the MP2 level for the calculations of β derivatives and at the B3LYP level for the evaluation of the normal modes, the agreement between the experimental and theoretical hyper-Raman spectra improves substantially (Table 3, Figure 2). There remain however *smaller* differences associated with modes 1 and 7. They probably find their origin in the missing anharmonicity effects or in the fact that frequency dependence and electron correlation effects on the property derivatives are considered separately rather than simultaneously.

Table 3 : Experimental and theoretical relative hyper-Raman scattering intensities, $\Im^{VV}_{hR,a}$ (in %), of acetonitrile [λ = 1064 nm, T = 350 K]. The calculations have been carried out at different levels of approximation for different molecular symmetries. X//Y stands for "method X" for property determination while "method Y" for geometry optimization and evaluation of the normal mode coordinates and frequencies.

	$\Im^{VV}_{hR,a}$	ω_a	$\Im^{VV}_{hR,a}$, C_{3v}	$\Im^{VV}_{hR,a}$, C_{3v}	$\Im^{VV}_{hR,a}$, $C_{\infty v}$	$\Im^{VV}_{hR,a}$, $C_{\infty v}$
	Experiment	**CPHF**	**TDHF//CPHF**	**MP2//CPHF**	**TDHF//CPHF**	**MP2//B3LYP**
ω_1 (a_1)	-	2914	13.9	10.2	13.4	10.7
ω_2 (a_1)	24	2353	2.0	25.4	2.1	24.4
ω_3 (a_1)	11	1373	1.5	4.4	1.6	5.2
ω_4 (a_1)	-	879	0.9	1.6	1.1	2.9
ω_5 (e)	-	2993	29.1	16.9	0.5	0.4
ω_6 (e)	-	1428	0.7	1.1	0.2	0.1
ω_7 (e)	-	1041	7.9	5.3	13.6	9.7
ω_8 (e)	65	389	44.0	35.2	67.4	46.7

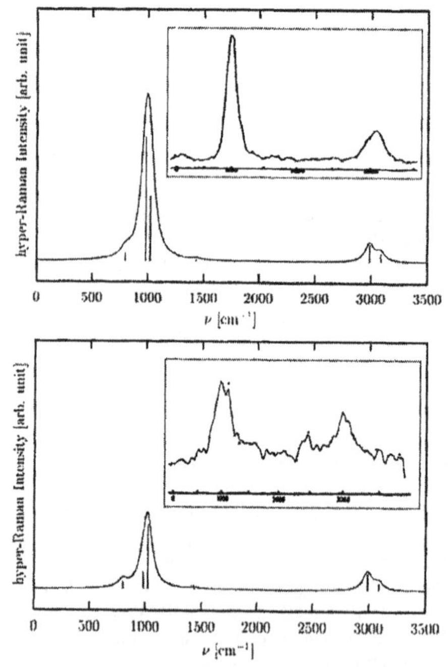

Figure 1: Simulated hyper-Raman scattering spectrum of ethylene based on TDHF results (λ = 694.3 nm, FWHM = 100 cm^{-1}, T = 298.15K) [20] in comparison with the spectrum (inset) of Verdieck *et al.* [19]. TOP: for plane-polarized incident light and detection of the total scattered light; BOTTOM: for circularly-polarized incident light and detection of the scattered light polarized parallel to the scattering plane.

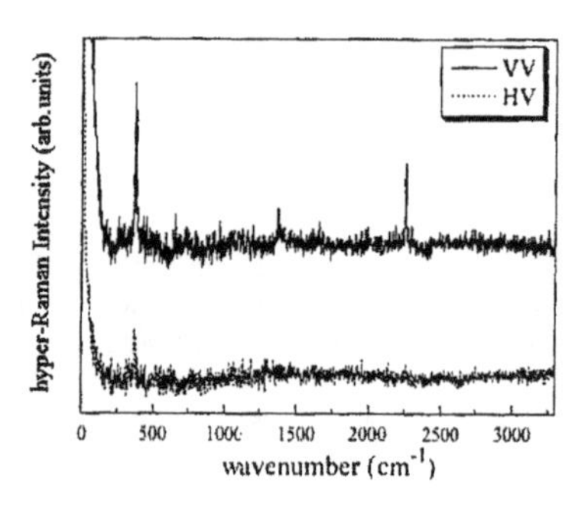

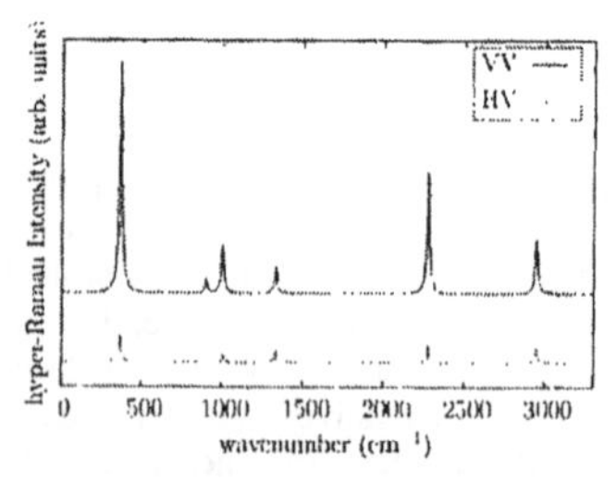

Figure 2: Experimental (TOP) and simulated (BOTTOM) hyper-Raman scattering spectra of CH_3–C≡N according to the light polarization (VV, HV) [17]. The simulation is based on B3LYP vibrational property calculations combined with MP2 evaluations of the static electrical properties, and accounting for $C_{\infty v}$ symmetry. FWHM = 20 cm^{-1}, T = 350 K.

Figure 3 gives the simulated hyper-Raman spectrum of benzene as determined at the TDHF/aug-cc-pvdz level of approximation. The most intense band located at 681 cm^{-1} is associated with the $\omega_4(A_{2u})$ CH bending mode. Experimentally, the hyper-Raman spectrum only displays a single band at 668 cm^{-1} [21]. In fact, the agreement between theory and experiment is rather good since it has been reported in Ref. 21 that the signal/noise ratio is such that one cannot detect transitions having frequencies smaller than 1/5 of the 668 cm^{-1} band. Obviously, further experimental investigation is needed.

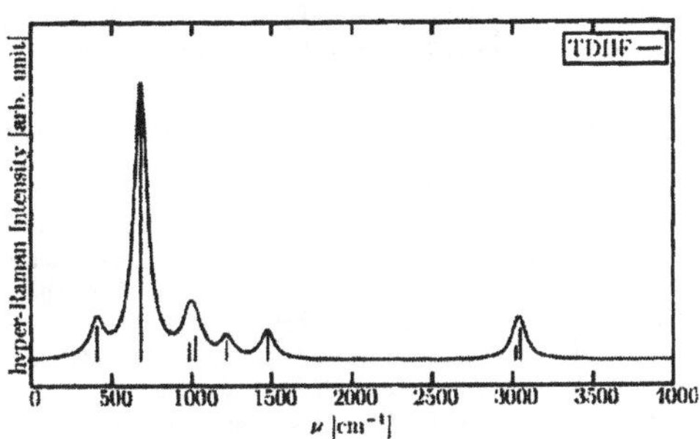

Figure 3: Simulated hyper-Raman scattering spectrum of benzene based on TDHF/aug-cc-pvdz results (λ = 694.3 nm, FWHM = 100 cm^{-1}, T = 298.15K).

Acknowledgments

The author thanks the Belgium National Fund for Scientific Research for his Research Director position. He is also grateful to Prof. Vincent RODRIGUEZ and Dr. Olivier QUINET for fruitful collaboration.

References

[1] R.W. Terhune, P.D. Maker, and C.M. Savage, *Phys. Rev. Lett.* **14**, 681 (1965).
[2] D.A. Long, *The Raman Effect* (Wiley, New York, 2002).
[3] S.J. Cyvin, J.E. Rauch, and J.C. Decius, *J. Chem. Phys.* **43**, 4083 (1965).
[4] V. Rodriguez, *Thθse d'Habilitation θ Diriger des Recherches*, Bordeaux (France), 2003.
[5] P.A. Scott and L. Radom, *J. Phys. Chem.* **100**, 16502 (1996); K.K. Irikura, R.D. Johnson III, and R.N. Kacker, *J. Phys. Chem. A* **109**, 8430 (2005).
[6] J.T. Golab, J.R. Sprague, K.T. Carron, G.C. Schatz, and R.P. Van Duyne, *J. Chem. Phys.* **88**, 7942 (1988).
[7] W.H. Yang and G.C. Schatz, *J. Chem. Phys.* **97**, 3831 (1992); W.H. Yang, J. Hulteen, G.C. Schatz, and R.P. Van Duyne, *J. Chem. Phys.* **104**, 4313 (1996).
[8] E. Nørby Svendsen and T. Stroyer-Hansen, *J. Molec. Struct.* **266**, 423 (1992).
[9] O. Quinet and B. Champagne, *J. Chem. Phys.* **117**, 2481 (2002).
[10] J.L. Silverman and J.L. van Leuven, *Phys. Rev.* **162**, 1175 (1967).
[11] A. Dalgarno and A.L. Stewart, *Proc. R. Soc. London, Ser. A* **242**, 245 (1958). J.O. Hirschfelder, W. Byers Brown, and S.T. Epstein, *Adv. Quantum Chem.* **1**, 255 (1964).
[12] M.W. Schmidt, K.K. Baldridge, J.A. Boatz, S.T. Elbert, M.S. Gordon, J.H. Jansen, S. Koseki, M. Matsunaga, K.A. Nguyen, S.J. Su, T.L. Windus, M. Dupuis, and J.A. Montgomery, *J. Comput. Chem.* **14**, 1347 (1993).
[13] O. Quinet, B. Champagne, and V. Rodriguez, *J. Chem. Phys.* **121**, 4705 (2004).
[14] O. Quinet, B. Champagne, and B. Kirtman, *J. Chem. Phys.* **118**, 505 (2003).
[15] O. Quinet, B. Champagne, and S.J.A. van Gisbergen, *Int. J. Quantum Chem.* **106**, 599 (2006).
[16] M. Pecul, F. Pawlowski, P. Jørgensen, A. Kohn, and C. Hättig, *J. Chem. Phys.* **124**, 114001 (2006).
[17] O. Quinet, B. Champagne, and V. Rodriguez, *J. Chem. Phys.* **124**, 244312 (2006).
[18] G. Herzberg, *Molecular Spectra and Molecular Structure, II. Infrared and Raman Spectra of Polyatomic Molecules* (VNR, New York, 1945).
[19] J.F. Verdieck, S.H. Peterson, C.M. Savage, and P.D. Maker, *Chem. Phys. Lett.* **7**, 219 (1970).
[20] O. Quinet and B. Champagne, *Theor. Chim. Acta*, **111**, 390 (2004).
[21] J.P. Neddersen, S.A. Mounter, J.M. Bostick, and C.K. Johnson, *J. Chem. Phys.* **90**, 4719 (1989).

Brill Academic Publishers
P.O. Box 9000, 2300 PA Leiden
The Netherlands

Lecture Series on Computer and Computational Sciences
Volume 6, 2006, pp. 324-331

Towards a First Observation of Molecular Parity Violation by Laser Spectroscopy

C. Chardonnet[1], C. Daussy, O. Lopez, A. Amy-Klein

Laboratoire de Physique des Lasers UMR CNRS 7538,
Universitι Paris 13, FR-93430 Villetaneuse, France

Received August 21, 2006; accepted August 23, 2006

Abstract: The observation of a parity violation in molecules is still a dream since the first proposition in 1974. The weakness of the effect at the molecular scale represents a very difficult experimental challenge. We propose to observe a frequency difference in the rovibrational spectrum of the enantiomers of a chiral molecule observed by two photon Ramsey fringes in a supersonic beam. This project is made possible by a strong collaboration between quantum chemists, chemists specialists in synthesis and spectroscopists in order to optimize the choice of the molecule and to prepare the two species in large quantity. New organometallic molecules which could present effects larger than 10^{-13} let us hope a first observation of parity violation since a realistic sensitivity of our experimental scheme is of the order of 10^{-15}.

Keywords: Parity Violation, Molecule, Laser Spectroscopy.
PACS: 06.30.Ft, 33;15.-e, 33.20.Ea, 39.30.+w

1. Introduction

The parity non-conservation (PNC) in molecules was first suggested by Rein in 1974 [1]. This PNC originates in the weak interaction with a typical range at the nuclear scale. In 1975, Letokhov [2] proposed to observe a frequency difference in the infrared spectrum of enantiomers of a simple chiral molecule, CHFClBr. Thirty years later, no experiment could yet reveal parity violation in molecules. The main reason is the smallness of the searched effect: For instance, the frequency difference in the spectrum of the C-F stretching band of CHFClBr at 930 cm^{-1} has been calculated to be 1.7 mHz [3], i.e. the required relative sensitivity has to be better than 5.10^{-17} which up to now is much beyond the experimental state-of-the-art.

Though the lack of experimental observation, molecular PNC gave rise to strong activity in the scientific community. It has been suggested that the very small energy difference induced by the weak interaction between right- and left-handed molecules could be at the origin of the left-right symmetry breaking observed in biomolecules and living systems [4]. Models of molecular dynamics were developed, which describe the possible amplification of the enantiomeric excess of one species over the other [5], and giving credit to the thesis that nature would have privileged one orientation by a deterministic way. This is a very controversial hypothesis against non deterministic approaches: the origin of this symmetry breaking is certainly one of the most fundamental open questions. In parallel, numerous PNC quantum chemistry calculations were performed on various biomolecules. Specific computing programs were developed and significantly improved over the last 25 years and now a fully relativistic version is available which permits to calculate the PNC on molecules with heavy atoms for which the effect is larger.

On the experimental side, very few experiments were carried out. A pioneer experiment was performed in 1977 on camphor with a sensitivity of 10^{-8} [6] but a recent calculation estimated that the effect is not larger than 10^{-19} [7]! In 1999, because of the availability of the enantiomers of CHFClBr, our group could attempt to observe PNC in this molecule for the first time. However, the improved sensitivity of 10^{-13} was still insufficient, by 3 orders of magnitude for this molecule. These experiments are based on laser spectroscopy in a vibrational band of separated chiral molecules. There are very few alternative proposals: we can mention for example NMR spectroscopy [9]. Quack proposed also to use chiral molecules in the ground electronic state but which have an achiral configuration in an excited state. This can be used to prepare a coherent superposition of right and left-handed states and study the

[1] Corresponding author.. E-mail: chardonnet@lpl.univ-paris13.fr

quantum beats between these states [10]. Stodolsky proposed to study the light polarization propagating in a chiral medium, the tunnel barrier of which is comparable to the PNC energy difference [11]. To our knowledge, no experiment has been tried so far based on one of these schemes. Finally, let us mention the announced observation in 2000 of a PNC by Mφssbauer spectroscopy in solid phase, but in contradiction with theoretical calculations, which makes the result highly questionable [12].
The project which is presented here should allow for the first time the observation of a parity non-conservation effect in the spectrum of molecules. We keep the general method of measuring a frequency difference in the vibrational spectrum of enantiomers of a chiral molecule by laser spectroscopy. The key points which can make the project successful is the choice of a molecule for which the effect is strongly enhanced, and the choice of a much more powerful laser spectroscopy method based on a two-photon Ramsey fringes experiment on a supersonic molecular beam. We will also point towards the difficulties that we will face, mainly due to the necessity to prepare a supersonic beam of molecules which are in the solid phase at room temperature.

2. Overview of the project

The project is made possible through a strong collaboration between different research groups of various competences which has been built from our previous experiments on CHFClBr [8]. The group of A. Collet and J. Crassous synthesized the two enantiomeric species of CHFClBr [13]. This work stimulated numerous theoretical works to calculate the PNC on this molecule. This is especially the case of the research group of P. Schwerdtfeger [14] who adapted the DIRAC program of relativistic quantum chemistry developed mainly by T. Saue and L. Visscher [15].
The principle of the PNC test performed on CHFClBr was the following: with a CO_2 laser, we record quasi-simultaneously the saturation spectrum of the same hyperfine component of a rovibrational line of the two enantiomeric species of CHFClBr filling at low pressure (a few tenths of Pa) two identical Fabry-Perot cavities. The sensitivity of the experiment relies on the precision with which the line center of each spectrum is measured and the enantiomeric excess of the samples (56% and 70%). The best sensitivity was a few 10^{-13}, 3-4 orders of magnitude larger than the expected PNC value. The conclusions were:

- the residual pressure shift limits the sensitivity of such experiments at that level of a few 10^{-13};
- to drastically reduce this systematic effect, an experiment to be successful must be performed on a molecular beam;
- the choice of the molecule must be optimized in order to have the strongest PNC effect and to be able to synthesize a large quantity of molecules with the highest enantiomeric excess.

After proposing the sister molecules, CHFClI and CHFBrI, for which the calculated effect is 30 and 50 times larger than for CHFClBr, which is still not sufficient; P. Schwerdtfeger and R. Bast proposed Rhenium and Osmium complexes which present a strong vibrational band in the 30 THz region of the CO_2 laser and a large PNC effect between 1 and 5 Hz [16]. This consists of a real breakthrough for our project. In addition, for synthetical chemists, it is conceivable to produce such molecules in large quantity and with an enantiomeric excess of 100% which, of course, optimizes the sensitivity but also the precision in the comparison with theoretical calculations. The choice of the exact molecule for the new project is not finalized yet since it still requires interaction between experimentalists and theoreticians. Once the molecule is chosen, it is necessary to perform moderately high resolution spectroscopy to identify the vibrational band and the best candidate line to be studied. For that purpose, we have to face the most difficult task: these molecules are solid at room temperature. Recent developments however permit to prepare solid state molecules in gas phase: the simplest method is to heat the substance. For instance, the sublimation temperature of these Rhenium complexes is around 140°C. The second option is laser ablation: a laser pulse of typically 1MW per shot evaporates a matrix in which the desired molecule is trapped. In both cases, the gas is the carried by a rare gas to form a seeded supersonic beam. The interest is that the rotational temperature is strongly reduced down to 0.1-1 K during the expansion of the beam. This is a key point for performing high-resolution spectroscopy of molecules of large degrees of freedom, which present an unfavorable partition function. This intermediate step is probably the most critical point to be solved and must orientate the choice of the molecule: smallest number of atoms, smallest momentum of inertia. Once a spectroscopic signal in the infrared region is observed, it will be necessary to detect a Doppler-free two photon transition adapted for the ultra-precise measurement of the PNC frequency difference. Doppler-free two-photon spectroscopy is with saturation spectroscopy the major method to suppress the Doppler broadening. For that, a molecule interacts with a standing wave and absorbs a first photon propagating in one direction and a second photon in the opposite direction. It is easy to verify that the resonance condition is independent of the velocity of the molecule. However, this two-photon transition is efficient if there is an intermediate level in the vicinity of the middle of the lower and upper level of the transition. There is a practical and crucial question however: how difficult is it to find the position of such a transition? In our case, a two-photon transition will join a state in the fundamental vibrational level, v=0 and the v=2 level of a certain vibrational band. The intermediate state belongs to the v=1 level. These states must obey the usual selection rules of an electric dipole transition. Practically, for a molecule like a Rhenium complex, it is hopeless in the short term to gain enough knowledge of the rotational and vibrational spectroscopy to be able to predict the position of these two-photon transitions. More reasonably, we can roughly calculate the theoretical spectrum. A comparison with the experimental spectra can fix at least the band center and identify the rotational structure of the vibrational bands. First simulations confirm that this should be easily accessible by the present instrumentation. A Doppler free two photon transition is an extremely narrow line compared to the resolution of vibrational spectra which will be obtained first by Fourier Transform spectroscopy; this will make it quite difficult to find. We know two-photon transitions of only very few molecules: mainly SF_6 and, very recently, OsO_4. Our experience let us hope that, statistically, we should be able to find such transitions by a systematic scan of the CO_2 laser frequency in the vicinity of the rotational spectrum of the v=1 band if the anharmonicity of the band is not too strong. The quantum numbers of the line that we will select for the PNC experiment will not be fully known. Only the vibrational quantum number and maybe J, the total rotational quantum number, will be identified. In particular, the hyperfine structure that we will partly resolve cannot be interpreted without considerable efforts. This will be sufficient for a comparison with theory since the major contribution, by far, of the PNC effect that we plan to observe comes from the vibration. So, we expect a global and essential constant shift of the whole vibrational band due to PNC, any modulation coming from a rotational contribution will probably not be observable. After the selection of the best two-photon transition of the molecule, we then can start the search of our PNC effect. This is presented in the next section.

3. Search for a PNC frequency shift by Doppler-free two-photon Ramsey fringes spectroscopy

The principle of the proposed experiment is simple: we want to compare the position of the center of the same line for the two enantiomers of the chosen molecule. Any frequency difference is interpreted as a parity violation effect which can only be due to the weak interaction. In ultra-precise experiments as here, statistical and systematic errors are the limiting factors to be reduced as much as possible. The first type of error is reduced usually as the square root of the accumulating time of the experiment, for a white noise: its reduction is just a question of patience. The second type is usually more delicate to control and evaluate. In the present case, we perform a difference experiment. By a well thought-through design of the experiment, we can hope to have a strong fraction of identical systematic errors in the measurement of the two lines which are thus cancelled. In our former experiment on CHFClBr, the line width was about 100 kHz due to an unresolved hyperfine structure and the pressure. A systematic shift due to imperfection of the alignment of 2 Fabry-Perot cavities of 50-100 Hz could be measured and compensated. However, a residual differential pressure shift maintained the uncertainty at the level of about 10 Hz, which is only 10^{-4} of the line width. In a frequency measurement experiment, a thumb rule says that the systematic errors on the measurement of a line center are mostly proportional to the line width (or the resolution). Clearly, observing a PNC effect of a few Hz requires an experiment with a much higher resolution. A molecular beam experiment is required for reducing strongly the collisional effects, and this is fortunate as this is the only way to perform the spectroscopy of molecules like rhenium complexes. In such a case the resolution of a sub-Doppler spectroscopy is imposed by the inverse of the transit time of the molecule through the excitation laser beam. This transit broadening is typically of the order of 100 kHz. To overcome this limit, we need to use the method of separated fields introduced by Ramsey [17]. In that case, the line shape (in its central part) is a sine wave with a periodicity equal to the inverse of the transit time between the laser beams, which can be much smaller than the transit broadening in one beam only. The sine wave apart the central region is progressively damped because of the velocity distribution of the molecules in the beam which usually limits the number of periods to a few units. Associated with the saturation spectroscopy, the required geometry is in fact three perfectly parallel and equidistant standing waves or two pairs of equidistant traveling waves. In practice, specific optics must be developed and the gain in resolution is not so high. The scheme for an association with a Doppler-free two-photon spectroscopy is much more straightforward since the geometry of two standing waves, where parallelism is not required, is just needed to observe a Ramsey signal. This is the choice we made to reach the highest resolution. Finally, we propose to use one molecular supersonic beam and to feed it alternatively with right-handed and left-handed chiral molecules which we will have selected and prepared before. We will record the Doppler-free two-photon Ramsey fringes signal of the same line and measure the frequency difference. The estimation of the ultimate precision of such a measurement is, of course, a key question. We must distinguish between systematic and statistical errors. A realistic estimation of the systematic errors is difficult to do but at least, at this stage, we can compare with the previous experiment on CHFClBr: we identify several significant advantages. The background pressure in a molecular beam experiment is 4 to 5 orders of magnitude lower than in the cell experiment performed on CHFClBr (0.1 Pa) which limited the ultimate precision to a few Hz; the optical set-up is unique and perfectly identical for recording the two kind of spectra: the geometry of the supersonic beam of the two molecules is defined essentially by a nozzle and a skimmer, and the properties of the gas carrier will be exactly the same because it will be the same machine. Thus, the interaction of the molecules with the laser zones will be, by construction, identical. So, we can expect a high degree of rejection of the systematic errors with this scheme. In our previous experiment which used two cells, we observed systematic effects due to that, which could be measured and compensated, but the compensation cannot be guaranteed to better than 1 Hz. Finally, as we will see below, the resolution will be of the order of 100 Hz instead of 100 kHz, which reduces a lot of systematic errors in the same proportion. In conclusion, reducing the systematic errors below 0.01 Hz (3.10^{-16}) is certainly realistic with this scheme and should not be a problem.

As a result, the statistical errors will be the main problem. First, the signal-to-noise ratio (S/N) is the most direct point. Suppose that S/N of a certain experiment leads, for the line center, to a precision of 100 Hz on 1 sec. By accumulating the signal, the precision will reach 1Hz after 10^4 sec, which is still a poor result as this is the order of magnitude for the searched effect. We need to accumulate 100 times more to reach a precision on the PNC effect of 10%, i.e. more than 12 days of continuous runs of the experiment. For such a long period, we have to face the general stability of the various parameters of a heavy experiment, which is very difficult to maintain. The conclusion is that we will need a frequency precision better than 10 Hz on 1s to hope to measure the PNC effect with a precision better than 10%. This precision does not depend only on the S/N of the fringes. It depends also on the performance of the frequency standard which is used to calibrate the spectra to be measured. In the previous experiment with CHFClBr, the recordings of the spectra whose difference had to be measured, were completely entangled: for spectra of 300 data points recorded over about 2 min, the time delay between points of the 2 spectra was 200 ms, alternatively before and after the other one. Thus, even a drift of the reference frequency of 1 Hz over 1 min had a negligible impact much below 0.01 Hz. In the present scheme, this entanglement will probably not be possible because we will have to switch as fast as possible from a beam of left handed molecules to a beam of right-handed ones and vice-versa. Various valves will have to be switched on and off which takes time. Most likely, we will have an incompressible delay of a few minutes between the spectra recordings. Thus the stability of the frequency reference over 100 sec and more will affect directly the precision of the experiment, although one can imagine to measure and partly compensate a drift of this reference. In fact, we developed recently an experiment of Doppler-free two photon Ramsey fringes on a supersonic beam of SF_6. In parallel, we renewed completely the control of the frequency reference. We present now these results that are very promising for the PNC project.

4. Measurement of the absolute frequency of the $P(4)E^0$ line of the $2\nu_3$ band of SF_6 by Doppler-free two-photon Ramsey fringes spectroscopy

Figure 1 shows the experimental set up. The two-photon Ramsey fringes experiment employing a supersonic beam of SF_6 has been developed over number of years and is already described in some detail [18]. In brief, the molecular beam interacts successively with two standing waves, which are tuned to a two-photon resonance. The excitation probabilities for the two zones of interaction interfere so that the population of the upper level oscillates with a periodicity P which depends on the transit time between the two zones: $P = u/2D$, where u is the mean velocity of the beam and D the distance between zones. In order to obtain the central fringe in exact coincidence with the two-photon resonance the relative phase of the two standing waves must cancel. This is ensured by using a single folded Fabry-Perot cavity, comprising four mirrors in a U configuration. This was mounted in a cylindrical invar structure and surrounded by a μ-metal shield to eliminate the Zeeman effect in the earth's magnetic field. The two-photon transition is the $P(4)E^0$ line of the $2\nu_3$ band of SF_6, excited by the P(16) CO_2 laser line. Fringes in the two-photon absorption are detected by stimulated emission in a separate detection cavity, using a frequency comb technique as described earlier [19], and which is crucial in obtaining a good S/N.

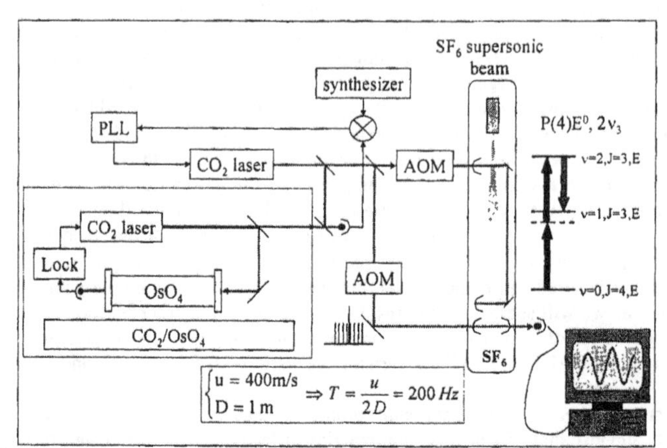

Figure 1: Experimental set up of the Doppler-free two-photon Ramsey fringes experiment on a supersonic beam of SF_6. (PLL: phase lock loop, AOM: acousto-optic modulator)

The current experiment incorporates a folded cavity with a separation of 1 m between absorption zones. The first is locked to the P(55) line in OsO_4 while the second is phase locked to the first with a radio frequency offset. The spectral purity for the whole experiment is thus determined by the first laser, while the second confers tunability under computer control. An electro-optic and three acousto-optic modulators are incorporated to shift the laser frequencies to the exact points required and to introduce various modulations. The clean phase modulation introduced at the electro-optic is notably significant for the quality of the OsO_4 lock [20].

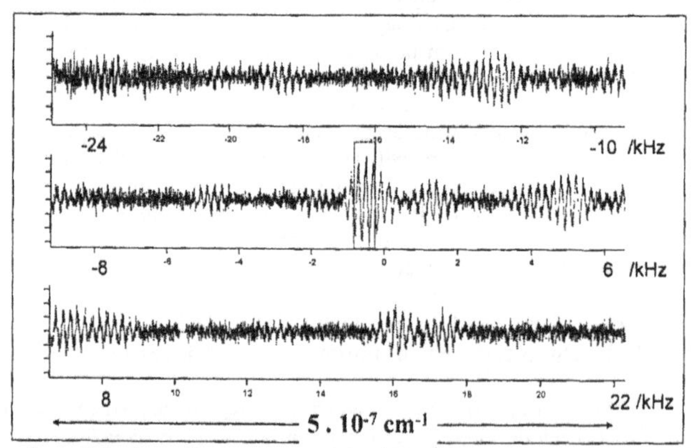

Figure 2: Overview of the hyperfine structure of the P(4)E^0 line of the $2\nu_3$ band of SF_6. The central part (rectangle) is used for ultra-precise measurements.

Figure 2 shows an overview of the rich hyperfine structure which extends only on 50 kHz. It is noticeable that the resolution is not affected by the possible overlapping of hyperfine components, which is an obvious advantage compared to other kinds of spectroscopy. Their relative positions can only affect the amplitude of the apparent Ramsey signal, depending on the exact value of the periodicity of the fringes, which can be optimized.

Figure 3 shows an example of a recording of the central fringes as obtained using a beam of pure SF_6 [21], generated from a reservoir pressure of $5X10^5$ Pa, 10 scans of 200 data points, lock-in time constant 100 ms. The fringe periodicity is 200 Hz and the S/N on 1 sec is 20. The recording is thus obtained in less than 4 minutes, using 10 seconds of averaging per point. The notably high S/N reflects a combination of patient attention to detail and a particularly stable laboratory environment. However, it fulfills the requirements of S/N mentioned in the previous section for a PNC observation.

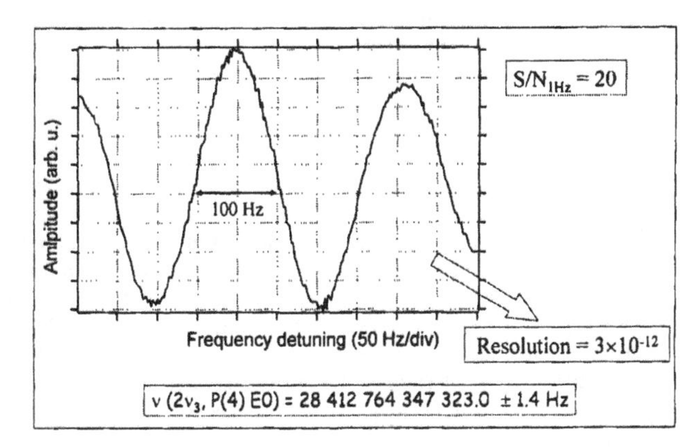

Figure 3: Central fringes obtained with a pure SF_6 supersonic beam (10 sec/point). Its absolute frequency was measured against the SYRTE Cs fountain located in Paris Center by mean of an optical link of 43 km long.

In obtaining the highest resolution the most critical component is the first laser locked to OsO_4. To obtain fringes of periodicity 200 Hz at good contrast a stability of better than 20 Hz is required, which is difficult to obtain starting from a OsO_4 reference of FWHM 20 kHz. The most significant requirement is spectral purity over times of order 1s, which can certainly be obtained using OsO_4 as the discriminator [20]. Stability, however, is also required over longer times, 100 sec and more for data collection and averaging. The OsO_4 lock reaches its limit and an additional active long-term stabilization is clearly necessary. This has now been implemented by using a femtosecond comb and a connection to a Cs fountain located in Paris which currently provides the best frequency calibration.
The principle of frequency measurements using a femtosecond laser comb has been widely described [22]. A mode-locked Ti:Sa laser produces a train of fs pulses. Equivalently in the frequency domain these comprise a comb of modes perfectly equidistant and spaced by frequency f_{rep}, which is also the repetition rate of the pulses and depends on the length of the fs cavity. This comb operates like a ultra-precise frequency ruler. The CO_2 laser frequency is smaller than the span of the modes of the fs laser, and this can be exploited to control the separation between two of these modes by the CO_2 laser frequency.
The basic technique used now [23] consists in mixing the fs laser and the CO_2 laser frequencies in a non linear crystal of $AgGaS_2$ which generates the sum or the difference and to detect the beat note with the fs laser again on a photodetector. A radiofrequency signal equal to the difference between the CO_2 laser and the difference between two modes of the fs laser can be detected and used to phase lock the repetition rate of the fs laser by acting on the length of the cavity. Thus, the CO_2 laser frequency is exactly a multiple (approx. 28400) of the repetition rate of the order of 1 GHz.
The repetition rate, f_{rep}, has to be measured precisely to obtain the CO_2 laser frequency and, finally, the SF_6 line position. It is counted against a highly stable oscillator which is phase locked on the long-term to a RF standard (at 100 MHz or 1 GHz) at SYRTE. This standard is generated from a hydrogen maser or a cryogenic oscillator, itself compared to a Cs fountain [24]. The signal arrives at LPL via 43 km of optical fiber as amplitude modulation on a 1.55 μm carrier generated by a laser diode. The accuracy of the Cs fountain reaches now a few parts in 10^{-16}. This will not limit our experiment. The quality of the frequency chain to measure the SF_6 signal can be a limitation. After some improvement of the measurement of the repetition rate, we demonstrated a resolution of the chain better than 1 Hz (3.10^{-14}) on 1 sec time scale (including optical fiber, measurement of the repetition rate, CO_2/OsO_4 laser connection) [25]. Since each step of the frequency chain involves a phase lock loop, the resolution improves like the inverse of measurement time and can reach 0.01Hz after only 100 sec. This description does not take into account the transfer of the frequency of the reference signal along the 43 km optical fiber from SYRTE to LPL. In fact, phase fluctuations are superimposed to the reference frequency, due to noise of human activity in Paris and its suburbs and, on longer periods, by thermal

expansion due to day-to-night or weather changes. At the level of 10^{-14}-10^{-16} precision, a careful check of the influence of this effect is absolutely necessary. In fact the free running fiber gave already very good results with typical short term stability better than 5.10^{-14}. However, on longer term, the stability was not better than 10^{-15}, a result which is not sufficient for the PNC experiment, for example. Thus, in collaboration with SYRTE, we developed several stabilization systems of the phase of the reference signal [26, 27]. By using the two identical fibers which link the two labs, we were able to check the efficiency of the stabilization. We just present the results on Figure 4 and 5 which correspond in fact to the performances obtained on twice the length between the two labs. Figure 4 shows a relative stability of 10^{-17} obtained after one day, 2 orders of magnitude better than for the free-running fiber (open loop). The stability on 100 sec is already better than 10^{-15}. However, these results were worse than expected by the gain of the servo-loop. We attributed this to the residual birefringence of the fiber which is an old standard telecom fiber. We could improve again our results by a fast scrambling of the light polarization at the entrance of the fiber. This is displayed on the right part of Figure 4 which shows stability of 10^{-18} over one day and of $1.5.10^{-16}$ (0.005Hz @ 30 THz) over 100 sec. This impressive result demonstrates that we are able to transport the best frequency standard from SYRTE to LPL without degrading its performance. In conclusion, the calibration of the spectra of the Ramsey signals can benefit of the performance of the SYRTE frequency standard which provides a relative stability better than 10^{-15} over 100 sec.

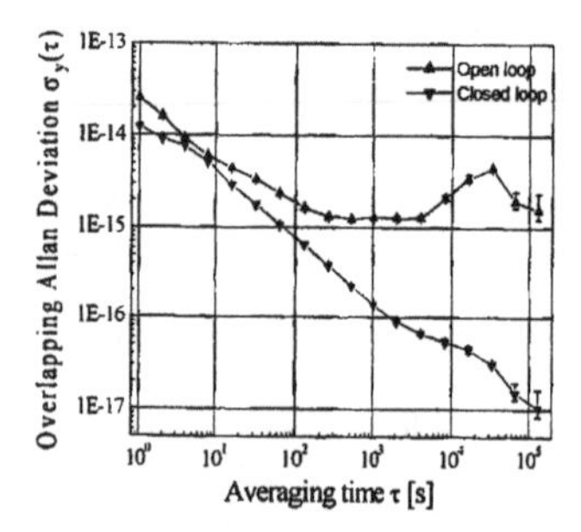

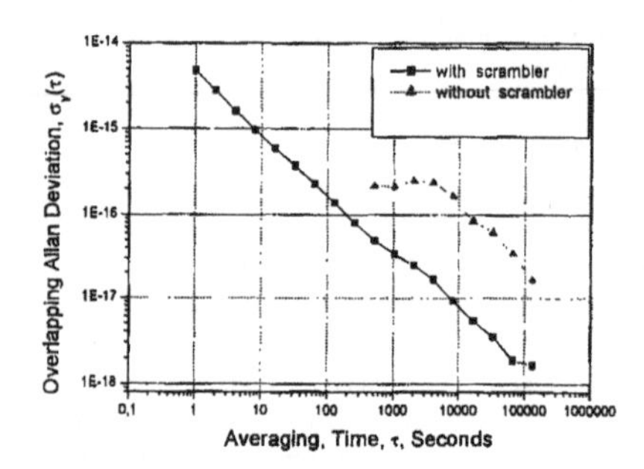

Figure 4: Allan Deviations in terms of frequencies due to 86 km optical fiber linking SYRTE and LPL. Left: comparison between the optical fiber without control and that with a stabilized optical length. Right: comparison with the addition of a scrambler of the light polarization (stabilized fiber).

5. Conclusion

In this paper, we presented the new scheme considered to observe for the first time PNC in molecules. Severe (but solvable) difficulties have still to be overcome before attempting this observation. One of the greatest difficulties is the preparation of the supersonic beam of chiral molecules and, then, the detection of a spectroscopic signal. However, if these difficulties are solved, we showed that the sensitivity of this experiment should be able to permit the observation of PNC with a high precision. Beyond the performance of such an observation, this is important for the comparison with the calculations of relativistic quantum chemistry. It would be of great interest to reach a precision of about 1%, which is conceivable, to perform a fruitful evaluation of the weak interaction itself, which will be compared to the results obtained in atomic physics. In that case, we will require the same precision for the calculations of quantum chemistry.

Acknowledgments

The authors wish to thank the collaborators this project: T. Saue and P. Schwerdtfeger for the theoretical calculations, J. Crassous for the synthesis of chiral molecules, T. Huet and P. Asselin for the preparation of the molecules in a supersonic beam and the rotational and vibrational spectroscopy. This project is supported by Agence Nationale de la Recherche (ANR) and by CNRS.

References

[1] D. W. Rein, *J. Mol. Evol.* 4, 15 (1974).
[2] V. S. Letokhov, *Phys. Lett.* 53A, 275 (1975).
[3] P. Schwerdtfeger, J. K. Laerdahl, Ch. Chardonnet, *Phys. Rev. A* 65, 042508 (2002).
[4] P. Frank, W. A. Bonner, R. N. Zare, in *Chemistry for the 21st Century*, E. Keinan and I. Schechter (ed.), Wiley-VCH, Germany, (2001), pgs.173-208.
[5] D. K. Kondepudi, G.W. Nelson, *Nature* 314, 438 (1985).
[6] E. Arimondo, P. Glorieux and T. Oka, *Opt. Commun.* 23, 369 (1977).
[7] P. Schwerdtfeger, A. Kóhn, R. Bast, J. K. Laerdahl, F. Faglioni, P. Lazzeretti, *Chem. Phys. Lett.* 383, 496 (2004).
[8] a) C. Daussy, T. Marrel, A. Amy-Klein, C.T. Nguyen, C.J. Bordı, Ch. Chardonnet, *Phys. Rev. Lett.* 83, 1554 (1999). b) M. Ziskind, T. Marrel, C. Daussy, C. Chardonnet, *Eur. Phys. J. D* 20, 219 (2002).
[9] L. Barra, J. B. Robert, *Mol. Phys.* 88, 875 (1996).
[10] M. Quack, *Angew. Chem., Int. Ed. Engl.* 28, 571 (1989).
[11] R. A. Harris and L. Stodolsky, *Phys. Lett. B* 78, 313, (1978).
[12] A. Lahamer, S.M. Mahurin, R.N. Compton, D. House, J.K. Laerdahl, M. Lein, P. Schwerdtfeger, *Phys. Rev. Lett.* 85, 4470 (2000).
[13] J. Crassous, F. Monier, J.-P. Dutasta, M. Ziskind, C. Daussy, C. Grain, C. Chardonnet, *ChemPhysChem* 4, 541 (2003).
[14] P. Schwerdtfeger, T. Saue, J. N. P. van Stralen, L. Visscher, *Phys. Rev. A* 71, 012103 (2005).
[15] *Dirac, a relativistic ab initio electronic structure program, Release 3.2* (2000), written by T. Saue, V. Bakken, T. Enevoldsen, T. Helgaker, H. J. Aa. Jensen, J. K. Laerdahl, K. Ruud, J. Thyssen, L. Visscher (http://dirac.chem.sdu.dk)
[16] P. Schwerdtfeger, R. Bast, *Large Parity Violation Effects in the Vibrational Spectrum of Organometallic Compounds*, *J. Am. Chem. Soc.* 126, 1652 (2004).
[17] N. F. Ramsey, Phys. Rev., **78**, 695-699 (1950).
[18] A. Shelkovnikov, C. Grain, C. T. Nguyen, R. J. Butcher, A. Amy-Klein, C. Chardonnet, *Appl. Phys. B* 73, 93 (2001) and references therein.
[19] C. Grain, A. Shelkovnikov, A. Amy-Klein, R. J. Butcher, C. Chardonnet, *IEEE J. of Quant. Electron.* 38, 1406 (2002).
[20] V. Bernard, C. Daussy, G. Nogues, L. Constantin, P. E. Durand, A. Amy-Klein, A. van Lerberghe, C. Chardonnet, *IEEE J. of Quant. Electron.* QE-33, 1282 (1997).
[21] A. Shelkovnikov, C. Grain, R.J. Butcher, A. Amy-Klein, A. Goncharov, C. Chardonnet, IEEE Quant. Electron. 40, 1023 (2004).
[22] J. Reichert, M. Niering, R. Holzwarth, M. Weitz, Th. Udem, and T. W. Hδnsch, *Phys. Rev. Lett.* 84, 3232 (2000).
[23] A. Amy-Klein, A. Goncharov, M. Guinet, C. Daussy, O. Lopez, A. Shelkovnikov and C. Chardonnet, *Optics Letters* 30, 3320 (2005)
[24] C. Vian, P. Rosenbusch, H. Marion, S. Bize, L. Cacciapuoti, S. Zhang, M. Abgrall, D. Chambon, I. Maksimovic, P. Laurent, G. Santarelli, A. Clairon, A. Luiten, M. Tobar, and C. Salomon, *IEEE Trans. Instrum. Meas.* 54, 833 (2005).
[25] A. Amy-Klein, A. Goncharov, C. Daussy, C. Grain, O. Lopez, G. Santarelli, C. Chardonnet, *Appl. Phys. B* 78, 25 (2004).
[26] C. Daussy, O. Lopez, A. Amy-Klein, A. Goncharov, M. Guinet, C. Chardonnet, F. Narbonneau, M. Lours, D.Chambon, S. Bize, A. Clairon, G. Santarelli, M.E Tobar, A.N. Luiten, *Phys. Rev. Lett.* 94, 203904 (2005).
[27] F. Narbonneau, M. Lours, O. Lopez, A. Amy-Klein, C. Daussy, C. Chardonnet, S. Bize, A. Clairon, G. Santarelli, *Rev. of Scient. Instrum.* 77, 064701 (2006).

Brill Academic Publishers
P.O. Box 9000, 2300 PA Leiden,
The Netherlands

Lecture Series on Computer and Computational Sciences
Volume 6, 2006, pp. 332-349

Global Optimization of 1- and 2-Dimensional Nanoscale Structures

Cristian V. Ciobanu [1]

Division of Engineering,
Colorado School of Mines
Golden, Colorado 80401, U.S.A.

Received 10 July, 2006; accepted 15 July, 2006

Abstract: In the cluster structure community, global optimization methods are common tools for arriving at the atomic structure of molecular and atomic clusters. The large number of local minima of the potential energy surface (PES) of these clusters, and the fact that these local minima proliferate exponentially with the number of atoms in the cluster simply demands the use of fast stochastic methods to find the optimum atomic configuration. Therefore, most of the development work has come from (and mostly stayed within) the cluster structure community. Partly due to wide availability and landmark successes of scanning tunneling microscopy (STM) and other high resolution microscopy techniques, finding the structure of periodically reconstructed semiconductor surfaces was not posed as a problem of stochastic optimization until recently, when we have shown that high-index semiconductor surfaces can posses a rather large number of local minima with such low surface energies that the identification of the global minimum becomes problematic. We have therefore set out to develop global optimization methods for systems other than clusters, focusing on periodic systems in one- and two- dimensions as such systems currently occupy a central place in the field of nanoscience. In this article, we review some of our recent work on global optimization methods (the parallel-tempering Monte Carlo method and the genetic algorithm) and show examples/results from two main problem categories: (1) the two-dimensional problem of determining the atomic configuration of clean semiconductor surfaces, and (2) finding the structure of freestanding nanowires. While focused on mainly on atomic structure, our account will show examples of how these development efforts contributed to elucidating several physical problems and we will attempt to make a case for widespread use of these methods for structural problems in one and two dimensions.

Keywords: parallel tempering Monte Carlo, genetic algorithm, surface reconstructions, nanowires, silicon, germanium

PACS: 68.35.Bs, 68.35.Md, 68.47.Fg, 68.65.La, 05.10.Ln

[1] Contact information: Phone: 303-384-2119, Fax: 303-273-3602, E-mail: cciobanu@mines.edu

1 Introduction

The discovery of carbon nanotubes (CNTs) [1] sparked arduous interest in nanotube science and applications. The fascination with nanotubes, now still at an all-time high, is beginning to slow down as researchers have found another class of materials (nanowires) with superior potential to impact science at the nanometer scale, as well as our everyday lives. While CNTs are inert, nanowires (NWs) have "chemistry" which opens up unprecedented avenues for controlling their structure as well as their electronic, optical, mechanical and magnetic properties [2]. The scientific curiosity about nanostructures in general and about NWs in particular comes from the realization that at such small scales the structure, properties, and phenomena cannot be straightforwardly inferred from our knowledge of the bulk forms. The appeal of the NWs is also driven by the continuous miniaturization of electronics and optoelectronics industry, which has achieved the limit in which the interconnection of devices in a reliable and controllable way is particularly challenging. Fervent strides are underway in the preparation of NWs for molecular and nano-electronics applications [2, 3]: such wires (possibly doped or functionalized) can operate both as nanoscale devices and as interconnects [4]. Silicon nanowires (SiNWs) offer, in addition to their appeal as building blocks for nanoscale electronics, the benefit of simple fabrication techniques compatible with the currently well-developed silicon technology.

While remarkable progress has been achieved in the synthesis and device applications of SiNWs, atomic-level knowledge of the structure of ultrathin nanowires remains largely speculative. Stating the obvious, when the diameters are as small as 1 nm, the atomic structure of the NW is the single most important factor that determines its electronic, optical, and mechanical properties, as well as the ensuing phenomena and technological applications. The importance of atomic structure has been emphasized in a sequence of recent high-profile publications [5, 6, 7, 8, 9, 10], which bring strongly plausible arguments and simulation evidence for various configurations in certain diameter regimes. The current proposals for SiNW configurations fall into several main categories: fused-fullerenes [5], fused-clathrates [9, 11], diamond structure single-crystals with reconstructed nanofacets [8, 12], polycrystals [7] and high-density phases [10], with each of the categories representing a novelty with respect to previous work. However, when considered together, the works [5, 6, 7, 8, 9, 10, 11, 12] appear to collectively suggest that the procedures currently used to investigate the structure of thin nanowires are not reliable. One may be determined to draw this conclusion because seemly simple questions regarding the SiNW diameter ("what is thinnest stable Si nanowire? " [7]) and its core structure are still under investigation.

The reason for the current situation of the SiNW problem is that this problem is exponentially complex (NP-complete, in the language of complexity theory), as we could recognize from a quick comparison with the problem of structure of atomic clusters: for a given number and type of atoms, the only mathematical difference between the two problems (wire vs. cluster) is the periodic boundary condition necessary to simulate the quasi 1-dimensional wires. Yet, in most theoretical approaches to date the structures proposed are not derived from the kind of robust search procedures that have been employed for atomic clusters. The problem is exacerbated by the fact that in the few-nanometer diameter regime experimental characterization of NWs with atomic resolution is extremely difficult, and thus theoretical proposals for NW structures cannot be easily confirmed or refuted. Since the problem of structure determination is NP complete and the current approaches are not dealing with this aspect, it is likely that a set of structures based on physical intuition (albeit refined using electronic structure relaxations) may not include many low-energy minima which could end up being the global minimum. Even if we assume (purely for the sake of the argument) that *all* the thermodynamically relevant structures of *silicon* NWs have already been reported in previous theoretical studies over the last 7 years, such a solution for the case of silicon will not readily transfer into methodologies or knowledge about NWs made of *other materials*, and we would again have to resort to trying numerous intuitive structures over many years.

Another problem, analog in principle with the above problem of finding the structure of nanowires, is searching for the structure of semiconductor surfaces. Under conditions of ultra-high vacuum, these surfaces reorganize their atomic configuration to minimize the surface energy, and in the process create periodic reconstructions which can repeat almost flawlessly over thousands of Angstroms in the nominal plane of the surface. The determination of atomic structure of crystalline surfaces is a long-standing problem in surface science. Despite major progress brought by experimental techniques such as scanning tunneling microscopy (STM) and advanced theoretical methods for treating the electronic and ionic motion, the commonly used procedures for finding the atomic structure of surfaces still rely to a large extent on one's intuition in interpreting STM images. While these procedures have proven successful for determining the atomic configuration of many low-index surfaces [e.g., Si(001) and Si(111)], in the case of high-index surfaces their usefulness is limited because the number of good structural models for high-index surfaces is rather large, and may not be exhausted heuristically. An illustrative example is Si(5 5 12), whose structure has been the subject of intense dispute [13, 14, 15, 16] since the publication of the first atomic model proposed for this surface [17]. There are also other stable surfaces of silicon such as (113) [18, 19] and (105)[20, 21, 22, 23, 24, 25], which required a long time for their correct structures to be revealed.

The high-index surfaces attract a great deal of scientific and technological interest since they can serve as natural and inexpensive templates for the fabrication of low-dimensional nanoscale structures. Knowledge about the template surface can lead to new ways of engineering the morphological and physical properties of these nanostructures. The main technique for investigating atomic-scale features of surfaces is STM, although, as pointed out in a recent review, STM alone is only able to provide "a range of speculative structural models which are increasingly regarded as solved surface structures" [26]. The role of theoretical methods for structural optimization of high-index surfaces has been largely reduced to the relaxation of these speculative models. However, the publication of numerous studies that report different structures for a given stable high-index silicon surface (see, e.g., [13, 14, 15, 16, 17]) indicates a need to develop methodologies capable of actually searching for the atomic structure in a way that does not predominantly rely on the heuristic reasoning associated with interpreting STM data. We are thus facing the same problem as described above for the case of nanowires, which is why we develop global optimization methods as useful tools to complement experimental data (when exists or is readily obtainable) or to make more robust predictions of structure in case experiments are not available.

A truly general and robust way of predicting the atomic structure of 1-D and 2-D systems surfaces takes sustained effort over many years. It is not entirely clear that such robust atomic-scale predictions about semiconductor surfaces can even be ventured, since theoretical efforts have been somewhat tempered by the lack of empirical or semiempirical potentials that are ***both sufficiently fast and sufficiently transferable*** for surface or nanowire calculations. However, the long process that lead to the discovery of the reconstruction of the (105) surface [20, 27, 22, 23, 24, 25] indicates a clear need for a search methodology that does not rely on human intuition. In Section 2 of this paper we will present two such strategies (the parallel-tempering Monte Carlo and the genetic algorithm) for finding the lowest-energy reconstructions for an elemental crystal surface. Our initial efforts will be focused on the silicon because of its utmost fundamental and technological importance; nonetheless, the same strategies could be applied for any other material surfaces provided suitable models for atomic interactions are available. In Section 3 we will show how the genetic algorithm can be modified to address the structure of nanowires, and describe one recent result on the magic configurations of the H-passivated SiNWs oriented parallel to the [110] direction. Section 4 summarizes our results on various surfaces and nanowires obtained so far, and identifies a number of future directions that can tremendously benefit from the application of global search methods.

2 Reconstruction of silicon surfaces as a problem of global optimization

In choosing a methodology that can help predict the surface reconstructions, we have taken into account the following considerations. First, the number of atoms in the simulation slab is large because it includes several subsurface layers in addition to the surface ones. Moreover, the number of local minima of the potential energy surface is also large, as it scales roughly exponentially[28, 29] with the number of atoms involved in the reconstruction; by itself, such scaling requires the use of fast stochastic search methods. Secondly, methods that are based on the modification of the potential energy surface (PES) (such as the basin-hoping[30] algorithm), although very powerful in predicting global minima, have been avoided as our future studies are aimed at predicting not only the correct lowest-energy reconstructions, but also the thermodynamics of the surface. Lastly, the calculation of interatomic forces is expensive, so the method should be based on Monte Carlo algorithms rather than molecular dynamics. We mention, however, that recent advances in molecular dynamics algorithms, especially the parallel replica[31] and temperature accelerated dynamics[32] developed by Voter and coworkers, may constitute viable alternatives to Monte Carlo parallel tempering for the sampling of low-temperature systems.

These considerations, coupled with a desire for simplicity and robustness of implementation, prompted us to choose the parallel-tempering Monte Carlo (PTMC) algorithm [33, 34] for finding the reconstruction of semiconductor surfaces. While we describe the salient features of the PTMC for crystal surfaces in Section 2.1, the reader is refer to the original work [35] for full implementation details.

The PTMC simulations, however, have a broader scope that the global minimum search, as they are used to perform a thorough thermodynamic sampling of the surface systems under study. Given their scope, such calculations [35] are very demanding, usually requiring several tens of processors that run canonical simulations at different temperatures and exchange configurations in order to drive the low-temperature replicas into the ground state. If we focus only on finding the reconstructions at zero Kelvin (which can be representative for crystal surfaces in the low-temperature regimes achieved in laboratory conditions), it is then justified to explore alternative methods for finding the structure of high-index surfaces. In Section 2.2, we will address the problem of surface structure determination at zero Kelvin, and report a genetically-based strategy for finding the reconstructions of elemental semiconductor surfaces. Our choice for developing this genetic algorithm (GA) was motivated by its successful application for the structural optimization of atomic clusters [36, 37]. We have designed and tested the algorithm for Si(105)[38], but we tested it on other surfaces as well [39, 40, 41]. Both Sec. 2.1. and 2.2. deal with the Si(114)-(2×1) surface, as an illustrative example of how the two methodologies fare in the quest for finding low energy structures.

2.1 The Parallel-tempering Monte Carlo

The reconstructions of semiconductor surfaces are determined not only by the efficient bonding of the surface atoms, but also by the stress created in the process [17]. Therefore, we retain a large number of subsurface atoms when performing a global search for the lowest energy configuration: this way the surface stress is intrinsically considered when reconstructions are sorted out. The number of local minima of the potential energy is also large, as it scales roughly exponentially [28, 29] with the number of atoms involved in the reconstruction; by itself, such scaling requires the use of fast stochastic search methods. One such method is the parallel-tempering Monte Carlo (PTMC) algorithm [33, 34], which was shown to successfully find the reconstructions of a vicinal Si surface when coupled with an exponential cooling [35]. Before outlining the procedure, we discuss briefly the computational cell and the empirical potential used.

The simulation cell [of dimensions $3a \times a\sqrt{2}$ for Si(114)] in the plane of the surface) has a single-face slab geometry with periodic boundary conditions applied in the plane of the surface, and no periodicity in the direction normal to it. The "hot" atoms from the top part of the slab (10–15 Å thick) are allowed to move, while the bottom layers of atoms are kept fixed to simulate the underlying bulk crystal. The area of the simulation cell and the number of atoms in the cell are kept fixed during each simulation. Under these conditions, the problem of finding the most stable reconstruction reduces to the global minimization of the total potential energy $V(\mathbf{x})$ of the atoms in the simulation cell (here $\mathbf{x}$ denotes the set of atomic positions). In terms of atomic interactions, we are constrained to use empirical potentials because the highly accurate ab-initio or tight-binding methods are prohibitive as far as the search itself is concerned. Since this work is aimed at finding the *lowest* energy reconstructions for arbitrary surfaces, the choice of the empirical potential is important. After numerical experimentation with several empirical models, we chose to use the highly optimized empirical potential (HOEP) recently developed by Lenosky *et al.* [42]. HOEP is fitted to a large database of ab-initio calculations using the force-matching method, and provides a good description of the energetics of all atomic coordinations up to $Z = 12$.

The parallel tempering Monte Carlo method (also known as the replica-exchange Monte-Carlo method) consists in running parallel canonical simulations of many statistically independent replicas of the system, each at a different temperature $T_1 < T_2 < \ldots < T_N$. The set of N temperatures $\{T_i,\ i = 1, 2, ..., N\}$ is called a temperature schedule (or schedule for short). The probability distributions of the individual replicas are sampled with the Metropolis algorithm [43], although any other ergodic strategy can be employed [44]. Irrespective of what sampling strategy is being used for each replica, the key feature of the parallel tempering method is that swaps between replicas of neighboring temperatures T_i and T_j ($j = i \pm 1$) are proposed and allowed with the conditional probability [33, 34] given by

$$\min\left\{1, e^{(1/T_j - 1/T_i)[V(\mathbf{x}_j) - V(\mathbf{x}_i)]/k_B}\right\}, \tag{1}$$

where $V(\mathbf{x}_i)$ represents the energy of the replica i and k_B is the Boltzmann constant. The conditional probability (1) ensures that the detailed balance condition is satisfied and that the equilibrium distributions are the Boltzmann ones for each temperature.

In the limit of low temperatures, the PTMC procedure allows for a geometric temperature schedule [46, 47]. To show this, we note that when the temperature drops to zero, the system is well approximated by a multidimensional harmonic oscillator, so the acceptance probability for swaps attempted between two replicas with temperatures $T < T'$ is given by the incomplete beta function law [47]

$$Ac(T, T') \simeq \frac{2}{B(d/2, d/2)} \int_0^{1/(1+R)} \theta^{d/2-1}(1-\theta)^{d/2-1} d\theta\ , \tag{2}$$

where d denotes the number of degrees of freedom of the system, B is the Euler beta function, and $R \equiv T'/T$. Since it depends only on the temperature ratio R, the acceptance probability (2) has the same value for any arbitrary replica running at a temperature T_i, provided that its neighboring upper temperature T_{i+1} is given by $T_{i+1} = RT_i$. The value of R is determined such that the acceptance probability given by Eq. (2) attains a prescribed value p. Thus, the (optimal) schedule that ensures a constant probability p for swaps between neighboring temperatures is a geometric progression:

$$T_i = R^{i-1} T_{min}, \quad 1 \leq i \leq N, \tag{3}$$

where $T_{min} = T_1$ is the minimum temperature of the schedule.

The typical Monte Carlo simulation done in this work consists of two main parts that are equal in terms of computational effort. In the first stage of the computation, we perform a parallel tempering run for a range of temperatures $[T_{min}, T_{max}]$. The configurations of minimum energy are

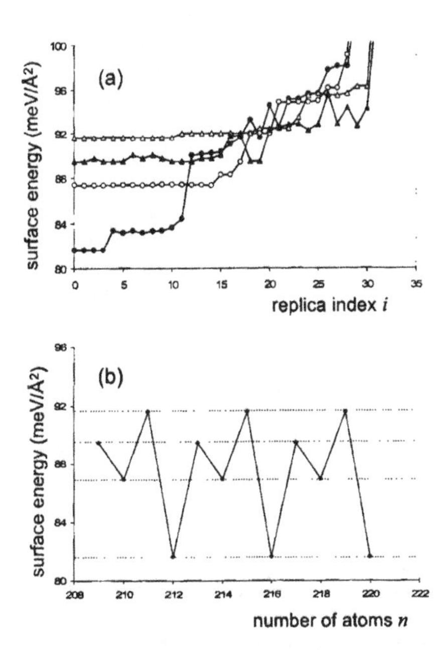

Figure 1: (a) Surface energies of the relaxed parallel tempering replicas i, $(0 \leq i \leq 31)$ with total number of atoms $n = 216$ (solid circles), 215 (open triangles), 214 (open circles) and 213 (solid triangles). For clarity, the range of plotted surface energies was limited from above at 100 meV/Å^2. (b) Surface energy of the global minimum structure showing a periodic behavior as a function of n, with a period of $\Delta n = 4$; this finding helps narrowing down the set of values for n that need to be considered for determining the Si(114) reconstructions that have a $3a \times a\sqrt{2}$ periodic cell. (Reproduced from Ref. [39], permission from Elsevier requested and pending).

retained for each replica, and used as starting configurations for the second part of the simulation, in which replicas are cooled down exponentially until the largest temperature drops below a prescribed value. As a key feature of the procedure, the parallel tempering swaps are not turned off during the cooling steps. Thus, in the second part of the simulation we are in fact using a combination of parallel tempering and simulated annealing, rather than a simple cooling. At the k-th cooling step, each temperature from the initial temperature schedule $\{T_i, i = 1, 2, ..., N\}$ is decreased by a factor which is independent of the index i of the replica, $T_i^{(k)} = \alpha_k T_i^{(k-1)}$. Because the parallel tempering swaps are not turned off, we require that at any cooling step k all N temperatures must be modified by the same factor α_k in order to preserve the original swap acceptance probabilities. We have used a cooling schedule of the form [35]

$$T_i^{(k)} = \alpha T_i^{(k-1)} = \alpha^{k-1} T_i \quad (k \geq 1), \tag{4}$$

where $T_i \equiv T_i^{(1)}$ and $\alpha = 0.85$.

The third and final part of the minimization procedure is a conjugate-gradient optimization of the last configurations attained by each replica. The relaxation is necessary because we aim to classify the reconstructions in a way that does not depend on temperature, so we compute the surface energy at zero Kelvin for the relaxed slabs i, $i = 1, 2, ..., N$. The surface energy γ is defined as the excess energy (with respect to the ideal bulk configuration) introduced by the presence of the surface:

$$\gamma = (E_m - n_m e_b)/A \quad (5)$$

where E_m is the potential energy of the n_m atoms that are allowed to move, $e_b = -4.6124$eV is the bulk cohesion energy given by HOEP, and A is the surface area of the slab.

At the end of the simulation, we analyze the energies of the relaxed replicas. Typical plots showing the surface energies of the structures retrieved by the PTMC replicas are shown in Fig. 1(a), for different numbers of particles in the computational cell. To exhaust all the possibilities for the numbers of particles corresponding to the supercell dimensions of $3a \times a\sqrt{2}$, we repeat the PTMC simulation for different values of n ranging from 208 to 220, and look for a periodic behavior of the lowest surface energy as a function of n. For the case of Si(114), this periodicity occurs at intervals of $\Delta n = 4$, as shown in Fig. 1(b). Therefore, the (correct) number of atoms n at which the lowest surface energy is attained is $n = 216$, up to an integer multiple of Δn. As we shall show in Section 2.2 below, the repetition of the simulation for different values of n in the simulation cell can be avoided within a genetic algorithm approach.

2.2 Genetic Algorithm

Like the previous method, the genetic algorithm also circumvents the intuitive process when proposing candidate models for a given high-index surface. An advantage of this algorithm over most of the previous methodologies used for structural optimization is that the number of atoms involved in the reconstruction, as well as their most favorable bonding topology, can be found within the same genetic search [38].

This search procedure is based on the idea of evolutionary approach in which the members of a generation (pool of models for the surface) mate with the goal of producing the best specimens, i.e. lowest energy reconstructions [38]. "Generation zero" is a pool of p different structures obtained by randomizing the positions of the topmost atoms (thickness d), and by subsequently relaxing the simulation slabs through a conjugate-gradient procedure. The evolution from a generation to the next one takes place by mating, which is achieved by subjecting two randomly picked structures from the pool to a certain operation (mating) $\mathcal{O}$:(A,B)$\longrightarrow$C. The mating operation $\mathcal{O}$ produces a child structure C from two parent configurations A and B, as follows. The topmost parts of the parent models A and B (thickness d) are separated from the underlying bulk and sectioned by an arbitrary plane perpendicular to the surface. The (upper part of the) child structure C is created by combining the part of A that lies to the left of the cutting plane and the part of slab B lying to the right of that plane: the assembly is placed on a thicker slab, and the resulting structure C is subsequently relaxed.

A mechanism for the survival of the fittest is implemented as a defining feature of the genetic evolution. In each generation, a number of m mating operations are performed. The resulting m children are relaxed and considered for the possible inclusion in the pool based on their surface energy. If there exists at least one candidate in the pool that has a higher surface energy than that of the child considered, then the child structure is included in the pool. Upon inclusion of the child, the structure with the highest surface energy is discarded in order to preserve the total population p. As described, the algorithm favors the crowding of the ecology with identical metastable configurations, which slows down the evolution towards the global minimum. To avoid the duplication of members, we retain a new structure only if its surface energy differs by more than δ when compared to the surface energy of any of the current members p of the pool. Relevant values for the parameters of the algorithm are given in [38]: $10 \leq p \leq 40$, $m = 10$, $d = 5$Å, and $\delta = 10^{-5}$meV/Å^2.

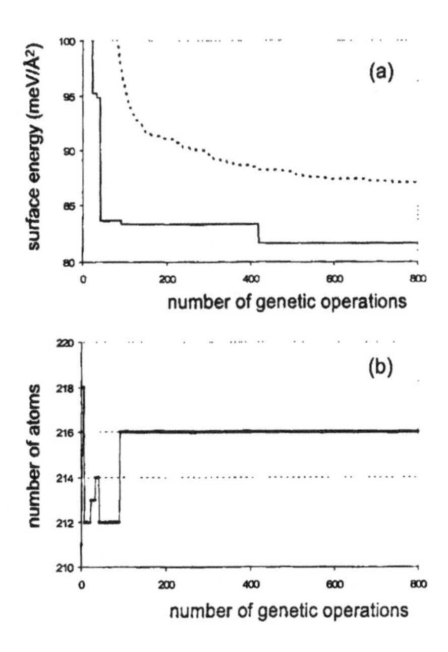

Figure 2: (a) Evolution of the lowest surface energy (solid line) and the average energy (dash line) for a pool of $p = 30$ structures during a genetic algorithm (GA) run with variable n ($210 \leq n \leq 222$). (b) Evolution of the number of atoms n that corresponds to the model with the lowest energy from the pool, during the same GA run. Note that the lowest energy structure of the pool spends most of its evolution in states with numbers of atoms that are compatible with the global minimum, i.e. $n = 212$ and $n = 216$.(Reproduced from Ref. [39], permission from Elsevier requested and pending).

We have developed two versions of the algorithm. In the first version, the number of atoms n is kept the same for every member of the pool by automatically rejecting child structures that have different numbers of atoms from their parents (mutants). In the second version of the algorithm, this restriction is not enforced, i.e. mutants are allowed to be part of the pool: in this case, the procedure is able to select the correct number of atoms for the ground state reconstruction without any increase over the computational effort required for one single constant-n run. The results of a variable-n run are shown in Fig. 2(a) which shows how the lowest energy and the average energy from a pool of $p = 30$ structures decreases as the genetic algorithm run proceeds. The plot in Fig. 2(a) displays typical features of the evolutionary approach: the most unfavorable structures are eliminated from the pool rather fast (initial steep transient region of the graphs) and a longer time is taken for the algorithm to retrieve the most stable configuration. The lowest energy structure is retrieved in less than 500 mating operations. The correct number of atoms [refer to Fig. 2(b)] is retrieved much faster, within approximately 100 operations. It is worth noting that even during the transient period, the lowest-energy member of the pool spends most of its evolution in a state with a number of atoms ($n = 212$) that is compatible with the global minimum structure.

Table 1: Surface energies of selected Si(114) reconstructions, sorted by the number of atoms n in the $3a \times a\sqrt{2}$ periodic cell. The second column shows the number of dangling bonds (counted for structures relaxed with HOEP) per unit area. The last two columns list the surface energies given by the HOEP interaction model [42] and by density functional calculations (DFT) [45] with the parameters described in text. (Adapted from Ref.[39], permission from Elsevier requested and pending)

n	Bond counting ($db/3a^2\sqrt{2}$)	HOEP (meV/Å^2)	DFT (meV/Å^2)
216	8	81.66	89.48
	8	83.16	90.34
	8	83.31	91.29
	8	83.39	88.77
	8	83.64	94.68
	8	84.42	92.16
215	8	91.61	97.53
	8	91.82	95.30
	8	92.00	94.20
214	6	86.95	95.17
	10	87.32	99.58
	10	87.39	98.47
	10	87.49	93.88
213	4	89.46	90.43
	6	89.76	94.01

The two independent algorithms (PTMC and GA) presented here are able to retrieve a set of possible candidates for the lowest energy surface structure. We use both of the algorithms in this work in order to assess how robust their structure predictions are. As it turns out, the two methods not only find the same lowest energy structures for each value of the total number of atoms n, but also most of the other low-energy reconstructions – a finding that builds confidence in the quality of the configuration sampling performed here. Since the atomic interactions are modelled by an empirical potential [42], it is desirable to check the relative stability of different model structures using higher-level calculations based on density functional theory (DFT. The details of these density functional calculations can be found, e.g., in Ref. [39].

2.3 Selected results on Si(114)

At the end of the global search procedures (PTMC and GA), we obtain a set of model structures which we sort by the number of atoms in the simulation cell and by their surface energy. Since the empirical potentials may not be fully transferable to different surface environments, we study not only the global minima given by the model for different values of n, but also most of the local minima that are within 15 meV/Å^2 from the lowest energy configurations. After the global optimizations, the structures obtained are also relaxed by density functional theory (DFT) methods [39, 40]. The results are summarized in Table 1 below.

Table 1 lists the density of dangling bonds (db per area), as well as the surface energies of several different models calculated using the HOEP potential and DFT. The configurations have been listed in increasing order of the surface energies computed with HOEP, as this is the actual outcome of the global optimum searches. The data shows clearly that the density of dangling bonds at the Si(114) surface is, in fact uncorrelated with the surface energy. The lowest number of dbs per area reported here is 4, and it corresponds to $n = 213$ and $\gamma = 90.43$ meV/Å^2 at the DFT level. The optimum structure, however, has twice as many dangling bonds but its surface energy is smaller, 88.77 meV/Å^2. Furthermore, for the same number of atoms in the supercell ($n = 216$) and the same dangling bond density ($8db/3a^2\sqrt{2}$), the different reconstructions obtained via global searches span an energy interval of at least 5 meV/Å^2. These findings constitute a clear example that the number of dangling bonds can not be used as a criterion for selecting model reconstructions for Si(114). We expect this conclusion to hold for many other high-index semiconductor surfaces as well.

The HOEP surface energy and the DFT surface energy also show very little correlation, indicating that the transferability of the interaction model [42] for Si(114) is not as good as, for instance, in the case of Si(001) and Si(105) [35]. The most that can be asked from this model potential [42] is that the observed reconstruction [50] is amongst the lower lying energetic configurations -which, in this case it is. We have also tested the transferability of HOEP for the case of Si(113), and found that, although the ad-atom interstitial models [18] are not the most stable structures, they are still retrieved by HOEP as local minima of the surface energy. We found that the low-index (but much more complex) Si(111)-(7×7) reconstruction is also a local minimum of the HOEP interaction model, albeit with a very high surface energy. Other tests indicated that, while the transferability of HOEP to the Si(114) orientation is marginal in terms of sorting structural models by their surface energy, this potential [42] performs much better than the more popular interaction models [48, 49], which sometimes do not retrieve the correct reconstructions even as local minima. Therefore, HOEP is very useful as a way to find different local minimum configurations for further optimization at the level of electronic structure calculations.

A practical issue that arises when carrying out the global searches for surface reconstructions is the two-dimensional periodicity of the computational slab. In general, if a periodic surface pattern has been observed, then the lengths and directions of the surface unit vectors may be determined accurately through experimental means (e.g., STM or LEED analysis): in those cases, the periodic vectors of the simulation slab should simply be chosen the same as the ones found in experiments. When the surface does not have two-dimensional periodicity, or when experimental data is difficult to analyze, then one should systematically test computational cells with periodic vectors that are integer multiples of the unit vectors of the bulk truncated surface, which are easily computed from knowledge of crystal structure and surface orientation. There is no preset criterion as to when the incremental testing of the size of the surface cell should be stopped -other than the limitation imposed by finite computational resources; nevertheless, this approach gives a systematic way of ranking the surface energies of slabs of different areas, and eventually finding the global minimum surface structure.

In this section we have reviewed the PTMC and GA methods of global optimization, and exemplified their application using the case case of Si(114), a stable high-index orientation of silicon. The PTMC and GA procedures coupled with the use of a highly optimized interatomic potential for silicon have lead to finding a set of possible models for Si(114), whose energies have been recalculated via ab initio density functional methods. The most stable structure obtained here without experimental input coincides with the structure determined from scanning tunneling microscopy experiments and density functional calculations in Ref. [50]. Motivated by these results for 2-dimensional systems, we have set out to adapt the less computationally intensive of the two methods (GA) for the study of quasi-1-dimensional nanowire systems. The blue print of this genetic algorithm for 1-d systems, along with an example application for the case of hydrogenated SiNWs is presented in the next section.

3 The Structure of Freestanding Nanowires

Interestingly, the application of GA for the study of nanowires has been around for the last few years, and as we have found one research group and their close collaborators [51, 52, 53] who used GA for finding the structure of metallic nanowires. To the best of our determination from Refs. [51, 52, 53], the method can be traced to the original article of Deaven and Ho on molecular clusters [36], since very little has been reported in terms of actual GA procedure for nanowires. While the spirit of the algorithm is general in that it is based on the natural evolution of living ecosystems, the adaptation of GA for artificial systems of atoms requires a certain amount of inspiration, design, as well as intense testing of the versatility of the genetic operations (cross-overs, mutations) for specific boundary conditions. Once GA is fully set up for 1-dimensional boundary conditions (possibly including variable unit-cell period and variable numbers of atoms), then it can be used for NWs made of any material provided that suitable (i.e. fast and sufficiently accurate) atomic interaction models are available. In what follows we describe the blueprint of such algorithm (Sec. 3.1) and show its application for hydrogenated SiNWs (Sec. 3.2).

3.1 A genetic algorithm for 1-D nanowire systems

As in the 2-dimensional case, the GA uses concept of a genetic pool to search for low energy structures using principles inspired by the evolution of biological systems. In biological evolution, offspring generations inherit traits from the older generations that may or may not help them survive. Similarly, during GA simulations, new NW configurations (children) inherit diverse structural motifs (genes) from their "parent" structures that may lower their formation energies, case in which the NW children live. It may also happen that new NW structures have too high energies to "survive", case in which they would not be accepted in the genetic pool. We have shown above an example of how GA can be used to find the structure of 2-dimensional reconstructions. The key to extending the use of GA to NWs is to realize that these algorithms must be system-specific. In other words, the type of genetic moves (e.g. mating operations) that define the evolution cannot be identical for e.g., clusters [36] and surfaces [38]. On the other hand, the moves and their sequence should not be so radically different that each new physical system would require a prohibitive amount of programming and testing.

In Sec. 2.2. we described a simple GA scheme in 2-D where the main assumptions were that the periodic lengths are a priori known, and that the number of Si atoms in the supercell was either fixed or left to vary. If we consider the vast number of proposals for SiNWs [5, 6, 7, 8, 9, 10], it becomes apparent that an algorithm with constant periodic length cannot sort through such multitude of structures with different periods along the wire. We therefore describe here a GA with variable periodic length and variable number of atoms (e.g., with fewer than 100 atoms per unit cell), so that highly unfavorable numbers of atoms can be eliminated quickly and naturally during the genetic evolution. It is our hope that such implementation of GA will prove very versatile as it will be able to simultaneously find the number, the period and the atomic structure of thin NW structures.

We now describe the salient features of the proposed GA. We again start with a pool of p structures, in which the atoms are placed at random (but connected) positions. The key modification is that we allow the boundaries of the periodic cells to relax along with the atomic positions. The "Generation Zero" itself (see Sec. 2.2) will have members of different numbers of atoms and different periodic lengths. The evolution proceeds as follows (refer to Fig. 3). Two members are randomly chosen from the pool (parent structures A and B), scaled to have the same length (e.g., the geometric mean of their periodic lengths) and translated so that they lie between the same spatial bounds in the periodic direction (z) with their centers of mass projecting in the same location in a plane perpendicular to z. A child structure C is produced from A and B through a cross-over

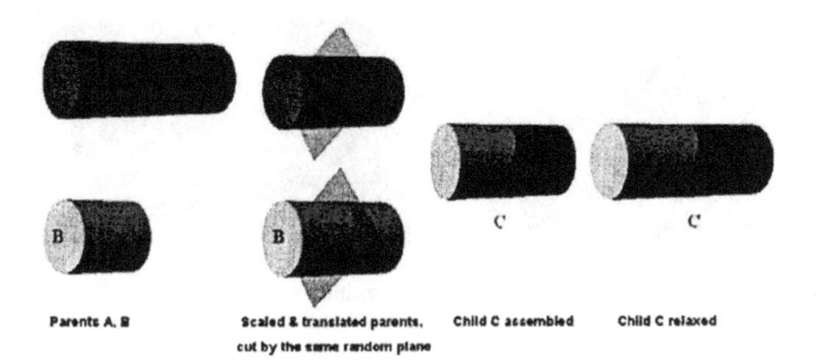

Figure 3: Schematics of the cross-over operation in a genetic algorithm design to tackle variations of the periodic cell along its axis.

(mating) operation in which the parents are sectioned by the same random plane (Fig. 3) then portions of parents that lie on different sides of the cutting plane are assembled to form C. There are two possibilities to form the child structure, and one of them will be discarded at random without an energy calculation. The child structure C is relaxed (both atomic positions and periodic boundaries), and placed in the pool if its total energy per atom is smaller than that of the most unfavorable pool member. Upon placing the structure in the pool, the highest energy structure is removed in order to preserve the total population p. The procedure is repeated for thousands of generations and stopped according to a set of desired criteria. Since we are interested in diverse types of low energy structures, we would not want the pool crowded with the same structure. For this reason, we allow a child to enter the pool not only when its formation energy is favorable, but also if its structure is different from all the other structures in the pool.

The boundary conditions. The periodicity can still be fixed from the start when it is known from experiments [55] or when we attempt to study wires with crystalline core [54]. In the latter case, one subtlety arises as we need to consider integer multiples of the period of the core in order to allow for various reconstructions to form along the wire [8]. In the case of moving boundary conditions, we will have to ensure from the beginning that our system is connected, as otherwise connectivity may be hard to achieve during the GA evolution.

The cross-over operations. In its most simple form, the cross-over operation is the one shown in Fig. 3 We can readily imagine other ways to perform cross-over operations such as using two random planes to cut and assemble parts of two or more parents. Based on extensive testing for the high-index surfaces [38, 39, 40], we found that cross-over operations with 2 parents and either 1 or 2 cutting planes are sufficiently robust and efficient for a wide range of aspect ratios of the structures, and there is little need to go beyond these values for the thickness regimes that we are interested in. For the crystalline-core nanowires the planes need not be random, but rather parallel to the axis of the nanowire in order for the supercell boundaries not to act like cutting planes themselves. This compatibility between periodic boundary conditions and the mating operations only appears necessary for wires with known core crystal structures: in all other cases it may negatively affect the course of the evolution as the parents would be limited in terms of the kind of offspring they could produce using cutting planes parallel to the axis.

The mutations. As described, there are no explicit mutations in the algorithm. One can envision the simplest standard mutation as selecting one atoms and moving it arbitrarily by a small distance. Many of these standard mutations will have to be forcefully accepted to diversify

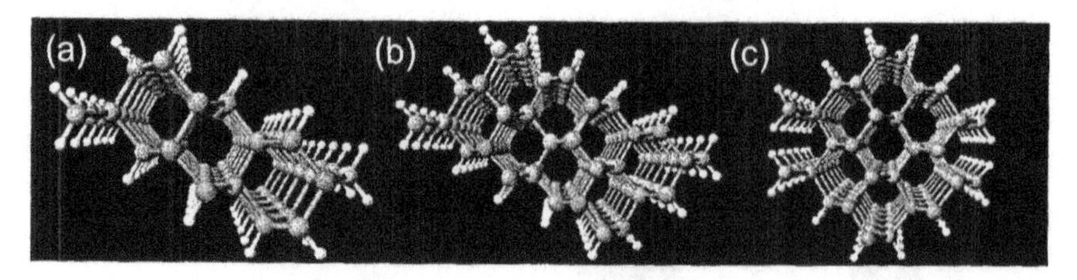

Figure 4: Magic nanowires (perspective view) found as minima of the formation energy per atom, Eq. 6: (a) chain, (b) double-chain, and (c) hexagonal.

in the pool, since random mutations almost always increase the energy per atom. More efficient moves that are aimed at improving the diversity of the parents participating in mating operations are zero-penalty, global "mutations" that amount to rotating a child around its axis, shuffling its atomic coordinates through boundary conditions, or taking the mirror image of the child with respect to a plane perpendicular to its axis.

3.2 The magic structures of H-passivated Si-[110] nanowires

Our choice to first study this particular NW system (i.e. hydrogenated SiNW oriented along the [110] direction) was motivated by recent experiments [55] that succeeded in characterizing, with atomic-scale resolution, H-SiNWs with diameters between 2 and 7 nm. The authors of Ref. [55] have showed that the H-SiNWs are single-crystals with axis orientations along [110], [112] or [111], and they have also reported STM imaging of the NW facets. Comparison with these experiments leaves therefore little wiggle room for theory, and thus constitutes a solid testing bed for our approach. It should be mentioned that in experiments what is being controlled is the number of H atoms on the surface and the wire diameter through timed exposure to the HF environment. What we control in our simulations is the chemical potential of H atoms, and determine the most stable thermodynamic state for a given chemical potential [56].

During a GA optimization run, a pool of at $p = 60$ structures (initially just random collections of atoms with periodic boundary conditions with a fixed period of 3.84Å , corresponding to the [110] crystal axis) is evolved by performing genetic (mating) operations. For this particular SiNW system, the mating operations consist in selecting two random parent structures from the pool, cutting them with the *same plane parallel to the wire axis*, then combining parts of the parent structures that lie on the opposite sides on the cutting plane to create a new structure (child). The child structure is then passivated by satisfying all its dangling bonds with H atoms, then relaxed with the Hansel-Vogl (HV) empirical model [57]. We include the child structure in the genetic pool based on its formation energy f defined as

$$f = (E - \mu_H n_H)/n - \mu, \qquad (6)$$

where E is the total energy of the computational cell with n Si atoms and n_H hydrogen atoms, μ is the bulk cohesive energy of Si in its diamond structure, and μ_H is the chemical potential of hydrogen. The H chemical potential is set such that certain hydrogenation reactions at surfaces are thermodynamically allowed [56]. The pool is divided into two equal subsets, one for each values of H. The mating operations are performed both with parents in the same subset and with parents in different subsets, in order to ensure a superior sampling of the potential energy landscape. The mating operation is carried out 15 times during a generation, and a typical GA run has 50,000

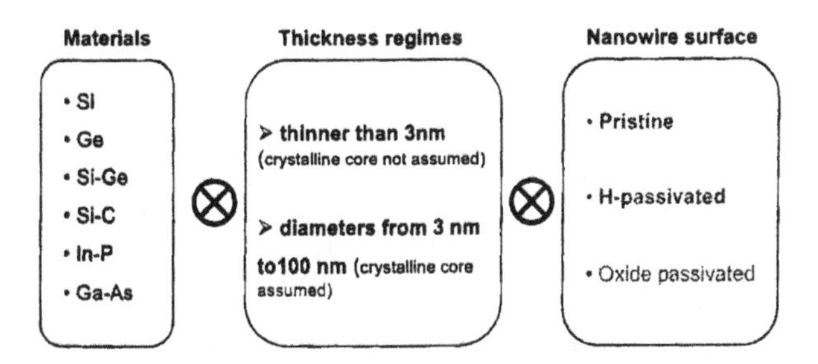

Figure 5: An overview of the nanowire semiconductor systems that might be tackled by global optimization methods. With the exception of the oxide-passivated wires (included here only for completness), the systems presented should be fairly amenable to study via genetic algorithms for diameters smaller than 10 nm.

generations. At the end of each run, all structures are relaxed at DFT level using the VASP package [58]. The chemical potential H used to compute the DFT formation energies is determined so that it maximizes the correlation with the HV formation energies for a few hundred configurations [56]. While some energetic reordering does occur after the DFT calculations, most of the low-energy structures found with the HV model remain relevant at the DFT level. GA runs with numbers of atoms in the range $9 < n < 31$ revealed three classes of spatially closed structures with relatively low formation energies, which we described as 6-atom-ring chains, double-chains (fused pairs of 6-ring chains) and hexagons [refer to Fig. 4]. As the number of atoms per unit of wire length increases, we found that the most stable cross section of the nanowire evolves from chains of six-atom rings to double-chains to hexagons bounded by 001 and 111 facets. Our calculations predict that hexagonal wires become stable starting at about 1.2 nm diameter, which is consistent with recent experiments on NWs with diameters of about 3 nm [55]. Pursuing further the comparison with Ref. [55], we computed STM images for the facets of hexagons with diameters in the range of 2-3 nm. Our calculation is in agreement with the STM experiments, which also showed the exclusive presence of dihydride species on the 001 facets of [110] H-SiNWs.

4 Future directions

We have so far described, in certain detail, two methodologies (PTMC and GA) for the finding the structure of semiconductor surfaces. One of the two methods (GA) has also been designed for the study of nanowire structures. At this stage, the problem of pristine SiNWs structure that has triggered the development of GA for 1-dimensional system is still not solved, as we have chosen to study first a system (H-passivated Si [110] nanowire for which experimental observations [55] are available for comparison. The algorithm is to be developed over time, as one can add capabilities to it in order to study an increasing number of 1-dimensional material systems. An overview of the 1-D systems that can be tackled in the future is summarized in Fig. 5, which shows the possible combinations between the materials targeted in this study, wire diameter regimes, and experimentally relevant surface terminations.

In terms of the semiconductor materials, silicon is perhaps the most important, and it is the material that drew our attention to the nanowire structure problem in the first place. Since, at its core, our methodology remains the same irrespective of the material chosen, we will consider not only Si but other materials as well. The binary materials (e.g., Si-Ge, In-P) in the chart above have been less scrutinized than silicon, and thus carry a tremendous potential for scientific novelty. The presence of a second atomic species in a sizeable proportion is bound to change the NW structure and will give insight into how the optical, electronic and mechanical properties can be tailored by changing the wire material and/or its composition.

The two regimes of wire diameters are delineated according to the current capabilities of the genetic algorithm, which turns out to be very efficient when dealing with less than 100 atoms per periodic cell along the wire axis. Defining the NW diameter as that of the smallest cylinder that includes all atoms, this number of atoms roughly corresponds to 3 nm diameter in the case of Si. For larger numbers of atoms per unit cell, the global optimization methods become much slower, as the difficulty of the structure determination problem increases exponentially; the methods should be further developed and refined to address, for example, a tripling of the number of atoms. Fortunately, in the regimes of diameters thicker than 3 nm, current experiments have been able to precisely identify a crystalline core with specific axis orientations [55, 54]. Therefore, for thick wires we will most likely not have to apply the GA, but instead study the structure and energetics of NWs starting from facet and facet-edge energies.

Most of the recent experiments show that the surface of the nanowires can fall into three main categories: clean [5], H-terminated [55] or oxide-terminated [59, 60]. Of these categories, the oxide surface is the least tractable with the currently available atomic interaction models. We cannot reasonably apply DFT methods either, because the structure of the oxide is mostly amorphous, thus hard to predict or assume. On the other hand, the clean and H-passivated NWs are very important for our fundamental understanding of the NWs structure and its effect on NW properties and applications.

Since in the nanometer-thin regime the structure of the NWs is crucial in determining novel phenomena, properties and applications, we believe that the impact of using global optimization methods for finding the structure of low-dimensional nanostructures should be significant, especially in the very foreseeable future when the experimental community will achieve the stage at which it can routinely fabricate devices based on ultra-thin wires (smaller than 3 nm diameter). At that point, research in (e.g.) device conduction, optical phenomena, chemical sensing at the nanoscale, and nano-electromechanical properties will require immediate and detailed knowledge about the atomic structure as a starting point for any robust understanding of the operation of such devices. The methodologies that we have reviewed here have strong predictive capabilities, and therefore we hope that they will robustly complement the experimental techniques and provide the needed structure input for investigations of novel properties and functionalities of ultra-thin NWs.

Acknowledgments

The author has benefited from collaborations with Prof. Kai-Ming Ho, Prof. Vivek B. Shenoy and Prof. Cai-Zhuang Wang, as well as with Dr. Cristian Predescu, Prof. Mihaela Predescu, and Dr. Dhananjay T. Tambe. Joint work with Dr. Tzu-Liang Chan, Prof. Feng-Chuan Chuang, Mr. Ning Lu and Dr. Mihail M. Popa has been very fruitful, and a substantial part of it is reviewed in this lecture paper for the 2006 International Conference on Computational Methods in Science and Engineering. The author thanks Ryan M. Briggs, Teresa Davies, Aaron S. Kofford and Damon Lytle, students at the Colorado School of Mines, for their important contributions to this research. Computational support from the National Center for Supercomputer Applications Urbana-Champaign (through grant no. DMR-050031N) and from the Center for Scientific Computation and Visualization at Brown University (through generous permission of Prof. Jim Doll) is gratefully acknowledged.

References

[1] S. Iijima, Nature 354, 56 (1991)

[2] D. Appell, Nature 419, 553 (2002)

[3] H. Kind, H.Q. Yan, B. Messer, M. Law, P.D. Yang, Adv. Mater. 14, 158 (2002)

[4] Y. Cui and C.M. Lieber, Science 291, 851 (2001)

[5] B. Marsen and K. Sattler, Phys. Rev. B. 60, 11593 (1999)

[6] M. Menon and E. Richter, Phys. Rev. Lett. 83, 2973 (1999)

[7] Y.F. Zhao and B.I. Yakobson, Phys. Rev. Lett. 91, 035501 (2003)

[8] R. Rurali and N. Lorente, Phys. Rev. Lett. 94, 026805 (2005)

[9] I. Ponomareva, M. Menon, D. Srivastava, A.N. Andriotis, Phys. Rev. Lett. 95, 265502 (2005)

[10] R. Kagimura, R.W. Nunes, and H. Chacham, Phys. Rev. Lett. 95, 115502 (2005)

[11] M. Durandurdu, Phys. Stat. Solidi B, 243, R7 (2006)

[12] A.K. Singh, V. Kumar, R. Note, and Y. Kawazoe, Nano Lett. 5, 2302 (2005)

[13] W. Ranke, Y.R. Xing, Surf. Sci. **381**, 1 (1997).

[14] T. Suzuki, H. Minoda, Y. Tanishiro, K. Yagi, Surf. Sci. **358**, 522 (1996); *ibid.*, Surf. Sci. **348**, 335 (1996).

[15] J. Liu, M. Takeguchi, H. Yasuda, K. Furuya, J. Cryst. Growth **237-239**, 188 (2002).

[16] S. Joeng, H. Jeong, S. Cho, J.M. Seo, Surf. Sci. **557**, 183 (2004).

[17] A.A. Baski, S.C. Erwin, L.J. Whitman, Science **269**, 1556 (1995).

[18] J. Dabrowski, H. J. Müssig, G. Wolff, Phys. Rev. Lett. **73**, 1660 (1994).

[19] G.D. Lee, E. Yoon, Phys. Rev. B **68**, 113304 (2003).

[20] Y.W. Mo, D.E. Savage, B.S. Swartzentruber, M.G. Lagally, Phys. Rev. Lett. **65**, 1020 (1990).

[21] R.G. Zhao, Z. Gai, W. Li, J. Jiang, Y. Fujikawa, T. Sakurai, W.S. Yang, Surf. Sci. **517**, 98 (2002).

[22] Y. Fujikawa, K. Akiyama, T. Nagao, T. Sakurai, M.G. Lagally, T. Hashimoto, Y. Morikawa, K. Terakura, Phys. Rev. Lett. **88**,176101 (2002).

[23] P. Raiteri, D.B. Migas, L. Miglio, A. Rastelli, H. von Känel, Phys. Rev. Lett. **88**, 256103 (2002).

[24] V.B. Shenoy, C.V. Ciobanu, L.B. Freund, Appl. Phys. Lett. **81**, 364 (2002).

[25] C.V. Ciobanu, V.B. Shenoy, C.Z. Wang, K.M. Ho, Surf. Sci. **544**, L715 (2003).

[26] D.P. Woodruff, Surf. Sci. **500**, 147 (2002).

[27] K.E. Khor, S. Das Sarma, J. Vac. Sci. Technol. B **15**, 1051 (1997) 1051

[28] F.H. Stillinger and T.A. Weber, Phys. Rev. A **28**, 2408 (1983).

[29] F.H. Stillinger, Phys. Rev. E **59**, 48 (1999).

[30] D.J. Wales and J.P.K. Doye, J. Phys. Chem. A **101**, 5111 (1997).

[31] A.F. Voter, Phys. Rev. B **57**, R13985 (1998).

[32] M.R. Sørensen and A.F. Voter, J. Chem. Phys. **112**, 9599 (2000).

[33] C.J. Geyer and E.A. Thompson, J. Am. Stat. Assoc. **90**, 909 (1995).

[34] K. Hukushima and K. Nemoto, J. Phys. Soc. Jpn. **65**, 1604 (1996).

[35] C.V. Ciobanu and C. Predescu, Phys. Rev. B **70**, 085321 (2004).

[36] D.M. Deaven and K.M. Ho, Phys. Rev. Lett. **75**, 288 (1995).

[37] K.M. Ho, A.A. Shvartsburg, B.C. Pan, Z.Y. Lu, C.Z. Wang, J. Wacker, J.L. Fye, and M.F. Jarrold, Nature **392**, 582 (1998).

[38] F. C. Chuang, C. V. Ciobanu, V. B. Shenoy, C. Z. Wang, K. M. Ho, Surf. Sci. **573**, L375 (2004).

[39] F.C. Chuang, C.V. Ciobanu, C.Z. Wang, and K.M. Ho, J. Appl. Phys. **98**, 073507 (2005).

[40] F.C. Chuang, C.V. Ciobanu, C. Predescu, C.Z. Wang, and K.M. Ho, Surf. Sci. **578**, 183 (2005).

[41] C.V. Ciobanu, F.C. Chuang, and D. Lytle, Reconstructions of Si(103), in preparation (2006).

[42] T.J. Lenosky, B. Sadigh, E. Alonso, V.V. Bulatov, T. Diaz de la Rubia, J. Kim, A.F. Voter, and J.D. Kress, Modelling Simul. Mater. Sci. Eng. **8**, 825 (2000).

[43] N. Metropolis, A.W. Rosenbluth, M.N. Rosenbluth, A.M. Teller, E. Teller, J. Chem. Phys. **21**, 1087 (1953).

[44] U.H.E. Hansmann, Chem. Phys. Lett **281**, 140 (1997).

[45] S. Baroni, A. Dal Corso, S. de Gironcoli, and P. Giannozzi, http://www.pwscf.org.

[46] Y. Sugita, A. Kitao, and Y. Okamoto, J. Chem. Phys. **113**, 6042 (2000).

[47] C. Predescu, M. Predescu, and C.V. Ciobanu, J. Chem Phys. **120**, 4119 (2004).

[48] F. H. Stillinger and T. A. Weber, Phys. Rev. B **31**, 5262 (1985).

[49] J. Tersoff, Phys. Rev. B **38**, 9902 (1988);*ibid* Phys. Rev. B **37**, 6991 (1988).

[50] S. C. Erwin, A. A. Baski, L. J. Whitman, Phys. Rev. Lett. **77**, 687 (1996).

[51] L. Hui, B.L. Wang, J.L. Wang, and G.H. Wang, J. Chem. Phys. 121, 8990 (2004).

[52] B.L. Wang, G.H. Wang, J.J. Zhao, Phys. Rev. B 65, 235406 (2002)

[53] B.L. Wang, S.Y. Yin, G.H. Wang, A. Buldum, J.J. Zhao, Phys. Rev. Lett. 86, 2046(2001).

[54] Y. Wu, Y. Cui, L. Huynh, C.J. Barrelet, D.C. Bell, C.M. Lieber, Nano Lett. 4, 433 (2004).

[55] D.D.D. Ma, C.S. Lee, F.C.K. Au, S.Y. Tong, S.T. Lee, Science 299, 1874 (2003).

[56] T.L. Chan, C.V. Ciobanu, F.C. Chuang, N. Lu, C.Z. Wang, and K.M. Ho, Nano Lett. **6**, 277 (2006).

[57] U. Hansen and P. Vogl, Phys. ReV. B **57**, 13295 (1998).

[58] VIENNA ab initio simulation package, Technische Universitat Wien, 1999; G. Kresse, J. Hafner, Phys. Rev. B 47, R558 (1993); G. Kresse, J. Furthmuller, Phys. Rev. B **54**, 11169 (1996).

[59] Y. Cui, L.J. Gudiksen, M.S. Wang, and C.M. Lieber, Appl. Phys. Lett. 78, 2214 (2001)

[60] N. Wang, Y.H. Tang, Y.F. Zhang, C.S. Lee, I. Bello, S. T. Lee, Chem. Phys. Lett. 299, 237 (1999).

Brill Academic Publishers
P.O. Box 9000, 2300 PA Leiden,
The Netherlands

Lecture Series on Computer and Computational Sciences
Volume 6, 2006, pp. 350-356

Periodic Orbits in Biological Molecules: Phase Space Structures and Selectivity

S. C. Farantos[a,b,1]

[a] **Institute of Electronic Structure and Laser, Foundation for Research and Technology-Hellas, Iraklion 711 10, Crete, Greece.**
[b]**Department of Chemistry, University of Crete, Iraklion 710 03, Crete, Greece.**

Received 1 August, 2006; accepted 5 August, 2006

Abstract: Small and large molecules may localize their energy in specific bonds or generally in vibrational modes for extended periods of time, an effect which may have dramatic consequences in reaction dynamics. We can trace such localization regions in phase space hierarchically by following families of periodic orbits which emanate from minima and transition states of the potential energy surface. Furthermore, we can study the stability of nearby trajectories and we can detect bifurcations which determine new characteristic motions of the molecule as energy or other parameters vary. In this article we demonstrate that our techniques to locate periodic orbits that were developed for small molecules can be extended to biological molecules such as alanine dipeptide.

Keywords: Periodic orbits, phase space dynamics, peptides.

Mathematics Subject Classification: 0.2.60-x

PACS: 33.15.Hp, 33.20.Tp

1 Introduction

Many body complex systems are studied by two different approaches. Either by using statistical mechanics methods or by the systematic methods of non-linear mechanics [1]. In the latter case, models of complex dynamical systems are explored by locating hierarchically classical mechanical stationary objects, such as equilibrium points (minima, maxima and saddles of the potential function), periodic orbits and their bifurcations, tori, reduced dimension tori as well as stable and unstable manifolds [2]. These multidimensional stationary objects reveal the structure of phase space and they assist us to understand and predict non-linear effects such as resonances and chaos. The progress of non-linear mechanics in the last decades is immense and the mathematical theories and numerical techniques which have been developed are now powerful tools for the computer exploration of realistic systems.

Molecules are complex many-particle systems and they are usually studied by quantum, classical and semiclassical mechanical theories. Chemical reactions involve the break and the formation of chemical bonds after the excitation of the molecule above potential barriers. Non-linear phenomena, like resonances, are expected and they have been observed spectroscopically [3]. Selectivity and specificity are well established concepts in elementary chemical reactions when the role of mode

[1]E-mail: farantos@iesl.forth.gr

excitation in the reactant molecules and the energy disposal in the products are investigated. Triatomic molecules have been used as prototypes to develop theories as well as to build sophisticated experimental apparatus to study elementary chemical reactions at the molecular level and at the femtosecond time scales [4]. The small number of degrees of freedom in these systems has allowed a detailed analysis of the correspondence between quantum and classical theories [5].

Moving to more complex systems with many more degrees of freedom such as biological molecules, the application of the systematic methods becomes a challenge, since not only more computer power is needed but also the development of concepts and techniques to extract the physics from the calculations. It is not surprising that up to now statistical mechanics methods have mainly been used, implemented either by averaging over phase space points or transition paths [6]. The latter method is promising for studying rare events in large dynamical systems. On the other hand, the systematic approach to explore large systems is usually exhausted by the location of equilibrium points (minima and saddles), to be followed with the calculation of phase space averages.

The hierarchical detailed exploration of the molecular phase space structure requires the location of periodic orbits (POs), the tori around stable POs, stable and unstable manifolds for the unstable POs, and even transition state objects such as normally hyperbolic invariant manifolds [7]. Such a program has been applied up to now to two and three degrees of freedom models for triatomic molecules [8, 9, 10]. This work has revealed the importance of periodic orbits in elucidating non-linear effects in spectroscopy and the good correspondence between classical and quantum mechanics. They have also motivated the development of semiclassical theories.

Efforts to find localized motions in infinite periodic or random anharmonic lattices have led to the concept of discrete breathers [11, 12]. The initial observations of localized motions in the work of Sievers and Takeno [13] triggered the discovery of significant mathematical theorems. However, most of the potential functions employed in the numerical studies were rather simple to describe realistic systems. In this lecture we show how to extend the methods of locating POs to biological molecules, such as peptides described with the widely used empirical potential functions. It is shown, that we can systematically trace regions in phase space where the trajectories stay localized in specific vibrational modes of a minimum or a transition state. In this way, the road is opened for elucidating localization phenomena and selectivity in biological systems [12, 14].

2 Computational Methods

To locate periodic orbits in a dynamical system is equivalent of finding the roots of the non-linear equations which describe the return of the trajectory to its initial point in phase space after the time period T. If $q_1, q_2, ..., q_N$ are the generalized coordinates of a dynamical system of N degrees of freedom and $p_1, p_2, ..., p_N$ their conjugate momenta, we define the column vector

$$\vec{x} = (\vec{q}, \vec{p})^+, \tag{1}$$

where + denotes the transpose matrix. Using $\vec{x}$ we can write Hamilton's equations of motion in the form

$$\frac{d\vec{x}(t)}{dt} = J\nabla H[\vec{x}(t)] \quad (0 \leq t \leq T), \tag{2}$$

where H is the Hamiltonian function, and J is the matrix

$$J = \begin{pmatrix} 0_N & I_N \\ -I_N & 0_N \end{pmatrix}. \tag{3}$$

0_N and I_N are the zero and unit $N \times N$ matrices respectively. $J\nabla H(\vec{x})$ is a vector field, and a matrix M which has the symplectic property satisfies the relation,

$$JMJ^+ = M. \tag{4}$$

If $\vec{x}(0)$ denotes the initial conditions of a trajectory at time $t_1 = 0$, then this trajectory is periodic if it returns to its initial point in phase space after the time $t_2 = T$

$$\vec{x}(T) - \vec{x}(0) = 0. \tag{5}$$

Thus, to find periodic solutions it is necessary to solve Eqs (2) subject to the 2-point boundary conditions, Eqs (5). The roots are found by iterative methods which require the linearization of the equations of motions around properly selected initial conditions. Powerful existence theorems for POs [15, 16] have converted the art of solving non-linear equations to science. We use multiple shooting techniques and the algorithms and computer codes have been described in previous publications [17]. However, the challenge to extend these methods to many degrees of freedom systems such as alanine dipeptide ($H_3CCONHCH(CH_3)CONHCH_3$) and larger molecules requires the adoption of new practices. We use cartesian coordinate systems and empirical force fields to describe the forces among the atoms. We have adopted the Molecular Mechanics suit of programs, TINKER [18], to our computer code for locating periodic orbits, POMULT [19], in order to calculate the potential and its first and second derivatives analytically. For alanine dipeptide we use the parameters charmm27 adopted in TINKER.

3 Results

In this article we look for periodic orbits emerging from the two lowest minima, that we have found in a search for stationary points, denoted Min1 and Min2, and the saddle point between them representing the transition state (TS) of the isomerization reaction Min2 → Min1. In Table I we tabulate the energies of the three equilibrium points, the distance of nitrogen atoms in the two aminoacids and the harmonic frequencies of two characteristic normal modes to be discussed below. We believe that this low barrier conformational change is typical for the class of the molecules we want to examine.

Figure 1 shows the structures of the molecule at the stationary points. Figure 2 depicts the minimum potential energy pathway from Min2 to Min1. It was calculated with the method of Czerminski and Elber [20] implemented in TINKER. On the same figure we show the monotonic increase of the nitrogen-nitrogen distance.

Table 1: Energies in kcal/mol, the distance of the two nitrogen atoms in Å and the harmonic frequencies in cm^{-1} for the 23rd and 24th normal modes of three stationary points of the potential energy surface of alanine dipeptide examined in this work.

	Energy	N-N Distance	H.f. 23	H.f. 24
Min1	-16.49	3.068	662.02	738.26
Min2	-15.58	3.640	668.78	704.59
TS	-15.00	3.566	655.12	700.66

A significant difference of 0.57 Å is found for the nitrogen-nitrogen distance between the two minimum structures. Contrary to that, the harmonic normal mode frequencies show small variations for the three equilibrium points. The 23rd and 24th normal modes shown in Table I, are two out of plane motions that mainly involve those atoms of the dipeptide that are enclosed in the squares of Fig.1. The NH and CO bonds oscillate in phase executing the largest displacements. It is expected that such motions on the transition state will affect the isomerization of the molecule and this is the reason that we have chosen them in our investigation.

Figure 1: The two minimum conformations and the transition state for the isomerization reaction of alanine dipeptide. The two squares drawn on the transition state enclose the atoms which execute the largest motions in the h23 (left) and h24 (right) periodic orbits (see text). From left to right the balls correspond to the atoms of the chemical structure $H_3CCONHCH(CH_3)CONHCH_3$.

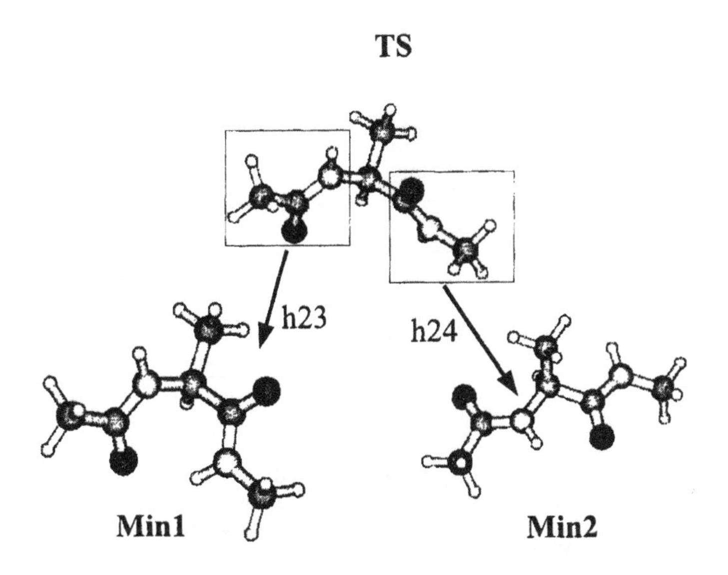

Figure 2: The minimum energy path connecting two minima of alanine dipeptide and the variation of the nitrogen-nitrogen distance along the reaction coordinate.

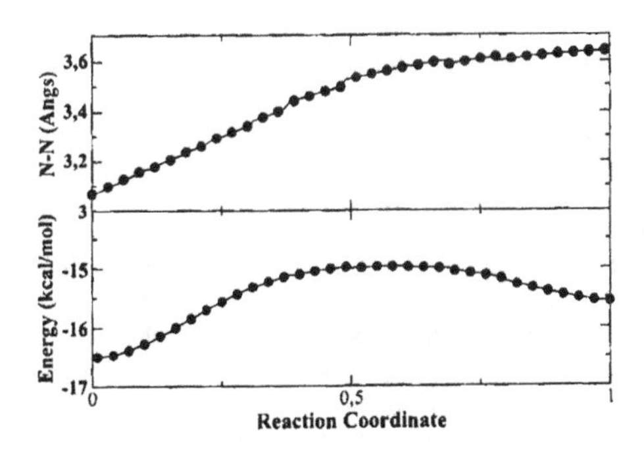

The harmonic normal mode analysis at each equilibrium point helps us to find good initial conditions for the periodic orbits. At least, an equal number of POs to the number of normal modes exists for each equilibrium point according to Weinstein and Moser theorems [15, 16].

Figure 3: Continuation diagram of the periodic orbits originated from the minimum Min2.

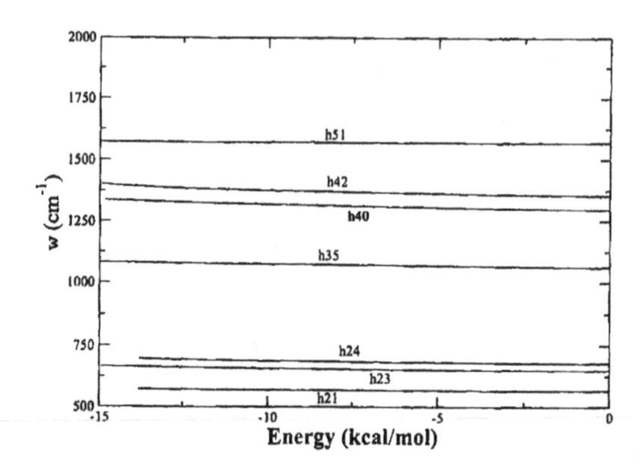

In Fig. 3 the continuation diagram of representative periodic orbits for Min2 is shown. Similar plots have been obtained for Min1 and the transition state. The numbers used to assign the families are the enumeration of the normal modes with increasing frequency. We can see that all families show small anharmonicities. After closing the orbit we carry out a linear stability analysis to find the eigenfrequencies and the eigenvectors of the PO from which we can determine its stability. Those POs which originate from the minima remain stable in most of the directions perpendicular to the periodic orbit in the examined energy range, except to the appearance of minor single or complex instabilities. For the transition state we have always one pair of real eigenvalues, thus single instability. The calculated instabilities are small and this justifies the localization of the nearby trajectories as we discuss below. High frequency families not shown in Fig. 3, correspond mainly to stretches with very small anharmonic behavior.

The location of periodic orbits and their continuation in energy allows one to select trajectories from regions of phase space that are associated with the normal modes of the molecule. For example, the behavior of the nearby trajectories to periodic orbits can be followed by calculating autocorrelation functions. By sampling a Gaussian distribution of 1000 trajectories centered on the periodic orbit we have produced Fig. 4, which shows the autocorrelation functions and their power spectra for two POs, h23 and h24, emanated from the transition state and at total energies -7.14 kcal/mol and -6.99 kcal/mol, respectively.

The regularities in the plots reveal the localization and the regularity of the selected trajectories. The frequency of the highest peak in the power spectra is that of the periodic orbit whereas the side peaks is the result of the non-linear coupling among the normal modes.

Although for a few degrees of freedom systems we can visualize the POs by projections on coordinate planes this is not practical with many degrees of freedom systems. Instead, the motions of the atoms along the periodic orbit are best visualized by using the graphics available for molecular mechanics. We have visually examined the motions of the atoms for all families of POs at several

Figure 4: Correlation functions and spectra averaged over one thousand trajectories selected from a Gaussian distribution. The center of the distributions are the periodic orbits of type h23 and h24 at energies -7.137 kcal/mol and -6.987 kcal/mol, respectively. 1 time unit is equal to 0.01018 ps.

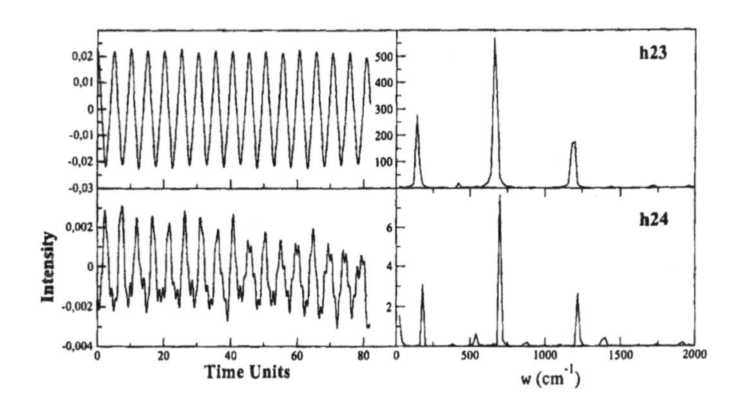

energies. We confirmed that the h23 and h24 modes are mainly local type motions involving the atoms enclosed in the squares of Fig. 1 even at high excitation energies. Furthermore, by minimizing the energy starting from phase space points along the periodic orbits, we found that every point in the region of h23 leads to Min1, whereas by quenching from the region of h24 we converge to Min2 and in a few times to the saddle point.

4 Conclusions

Families of periodic orbits associated with equilibrium points, the fundamental ones, of an empirical force field potential function of alanine dipeptide have been calculated with continuation techniques and using analytical first and second derivatives. Linear analysis of the POs demonstrate their stability for high excitation energies, and thus the long time localization of the nearby trajectories even for vibrational modes of the transition state. *We have demonstrated that with the periodic solutions of the classical equations of motion we can climb to high energy regions of phase space and select systematically trajectories which can lead the molecule to specific conformations.* Currently, larger peptides are studied to examine the robustness of our methods for locating periodic orbits.

Acknowledgment

The author is grateful to Dr. Reinhard Schinke and Dr. Sergy Grebenshchikov for their comments and stimulating discussions. Financial support from the Ministry of Education and European Union in the frame of the program Pythagoras II (EPEAEK) is kindly acknowledged.

References

[1] S. Wiggins, *Introduction to Applied Nonlinear Dynamical Systems and Chaos*, (Second Edition, Springer, New York, 2003).

[2] S. Wiggins, *Normally Hyperbolic Invariant Manifolds in Dynamical Systems* (Springer, 1994).

[3] H. Ishikawa, R. W. Field, S. C. Farantos, M. Joyeux, J. Koput, C. Beck, and R. Schinke, *HCP - CPH Isomerization: Caught in the Act*, **50**, 443-484, (Annual Review of Physical Chemistry, 1999).

[4] A. H. Zewail, *J. Phys. Chem. A* **104**, 5660 (2000).

[5] M. Joyeux, S. Yu. Grebenshchikov, J. Bredenbeck, R. Schinke, and S. C. Farantos, *Intramolecular Dynamics Along Isomerization and Dissociation Pathways*, in *"Geometrical Structures of Phase Space in Multi-Dimensional Chaos"*, **130** Adv. Chem. Phys., 267-303 (2005).

[6] Ch. Dellago, P. G. Bolhuis and P. L. Geissler, *Transition Path Sampling*, **123** Adv. Chem. Phys., 1-78 (2002).

[7] T. Uzer, C. Jaffé, J. Palacián, P. Yanguas, and S. Wiggins, *Non-linearity* **15**, 957 (2002).

[8] J. Main, C. Jung, and H. S. Taylor, *J. Chem. Phys.* **107**, 6577 (1997).

[9] J. Svitak, Z. Li, J. Rose, and M. E. Kellman, *J. Chem. Phys.* **102**, 4340 (1995).

[10] M. Joyeux, S. C. Farantos, and R. Schinke, *J. Phys. Chem.* **106**,5407 (2002).

[11] S. Aubry, *Physica D* **103**, 201 (1997).

[12] D. K. Campbell, S. Flach, and Y. S. Kivshar, *Physics Today* (January issue), 43 (2004).

[13] A. J. Sievers and S. Takeno, *Phys. Rev. Letters* **61**, 970 (1988).

[14] A. Xie, L. van der Meer, W. Hoff, and R. H. Austin, *Phys. Rev. Letters* **84**, 5435 (2000).

[15] A. Weinstein, *Inv. Math.* **20**, 47 (1973).

[16] J. Moser, *Commun. Pure Appl. Math.* **29**, 727 (1976).

[17] S. C. Farantos, *THEOCHEM J. Mol. Struct.* **341**, 91 (1995).

[18] J. W. Ponder, http://dasher.wustl.edu/tinker/ (2004).

[19] S. C. Farantos, *Comp. Phys. Comm.* **108**, 240 (1998).

[20] R. Czerminski, and R. Elber, *J. Chem. Phys.* **92**, 5580 (1990).

Brill Academic Publishers
P.O. Box 9000, 2300 PA Leiden,
The Netherlands

Lecture Series on Computer and Computational Sciences
Volume 6, 2006, pp. 357-368

The Beauty of Spinors

I. P. Grant[1]

University of Oxford,
Mathematical Institute,
24/29 St. Giles',
Oxford OX1 3LB, UK

Received 17 July, 2006; accepted 22 July, 2006

Abstract: Relativistic atomic and molecular structure theory can appear formidable: many of the procedures adopted in existing four component codes are complicated and it is easy to lose sight of the underlying physics. A strategy which treats four-component spinors as computational units rather than as unstructured four-component arrays leads to compact algorithms which are easy to interpret in physical terms. This tactic, first used at the time when electronic digital computers were just becoming accessible, was exploited in the first relativistic atomic self-consistent field calculations of the 1960s, and made possible widely used codes such as GRASP and DARC in the 1980s. Spinor properties have been exploited even more effectively in the recent molecular structure code BERTHA. This paper explains the ideas underlying these developments.

Keywords: Atoms, molecules, relativistic electronic structure, Dirac spinors, angular momentum algebra, electron interaction integrals, G-spinors

PACS: 03.65.Pm, 12.20.-m, 31.15.-p

1 Relativistic atomic and molecular structure

We take for granted the need for approximation in the theory of atomic and molecular structure. The traditional approach [1, 2] uses nonrelativistic quantum mechanics. Schrödinger's equation allows us to describe the dynamics of a single particle moving in some external field, very often without mathematical approximation, though not always without numerical approximation. The majority of many-electron calculations use a wavefunction that is an antisymmetrized product of single-particle functions, a (linear combination of) Slater determinant(s). The interaction between the particles ensures that such a wavefunction can never satisfy Schrödinger's equation for the many-particle problem exactly. Variational methods using them as trial wavefunctions and perturbation theories are commonly used to extract expectation values and transition rates which can be compared with experiment.

Symmetry plays a major role in atomic and molecular calculations. The indistinguishability of electrons in accordance with Pauli's exclusion principle is incorporated by antisymmetrization with repect to particle coordinates. The particle spins are not coupled to the dynamics in strictly nonrelativistic models, a fact which can be exploited when constructing totally antisymmetric wavefunctions using group theoretical methods as in [2, Chapter 4]. An isolated atom has no preferred orientation, so that its wavefunctions can be classified by angular momentum theory using irreps of the rotation group SO(3). The symmetry of the ambient electrostatic field acting

[1] E-mail: ipg@maths.ox.ac.uk

on a transition metal ion in an ionic crystal will split the degenerate electronic states of the isolated atom in a manner which can be studied by group theoretical procedures [3]. Similarly the nuclear skeleton of a symmetric molecule will be reflected in orbital symmetry and in the structure of electronic energy levels. The elaboration of these powerful techniques often exploits elegant mathematics, supported by elegant computational algorithms, leading to very satisfying interpretations of the physics of electronic structures and processes.

A good deal of this scheme's attractiveness is often lost when we try to incorporate relativistic effects. The Breit-Pauli effective Hamiltonian, defined for example in [4, §39], [5, §21.1,§31.3.2], gives a set of perturbation operators of order $O(\alpha^2)$, where $\alpha \approx 1/137$ is the fine structure constant. The effects can be tricky to calculate, and some of the less important terms are often ignored in quantum chemical calculations [6]. One notable consequence is that the spin and orbital motion of the electrons can no longer be treated as independent, so that total electron spin is no longer a constant of the motion. Whilst it is reasonably straightforward to account for most of the Breit-Pauli terms in atomic calculations [7, Chapter 7], the treatment of relativistic corrections in molecular calculations usually involves approximate semi-quantitative and complex computational schemes [8, Chapters 11-14] whose reliability is often difficult to assess impartially.

The $O(\alpha^2)$ Breit-Pauli (BP) Hamiltonian postulates a nonrelativistic starting point. Its construction leads to terms of order $O(\alpha^4)$ and higher containing operators which have infinite expectation values and so are not defined on the domain of the nonrelativistic zero order wavefunctions. The BP Hamiltonian can therefore only be used consistently as a first order perturbation of a nonrelativistic model. Methods such as the generalized Douglas-Kroll-Hess (DKH) scheme [8, Chapter 11] avoid introducing inadmissible operators, and in spin-free form can be readily incorporated into nonrelativistic molecular structure packages. However spin dependent interactions are not easy to handle and the need to justify approximations by way of numerical benchmark calculations is not very satisfying.

The Breit-Pauli corrections are of order $O(\alpha^2 Z^2)$ relative to nonrelativistic energies, so that their relative importance increases rapidly with the atomic number Z. A non-perturbative treatment of relativistic effects is aesthetically more appealing, although those familiar with the traditional nonrelativistic viewpoint may have difficulty with it at the outset. Relativistic effects originate in regions near the nuclei where electrons in penetrating orbitals can move at relativistic speeds, so that relativistic dynamical effects dominate in inner shell atomic physics of the heavier elements. The electron-electron interaction propagates the relativistically induced changes in the electron distribution across the atom or molecule, so that predictions derived from a truly relativistic model of electronic structure are more likely to be trustworthy. This paper therefore focuses on the way in which the structure of Dirac spinors is used in relativistic electronic structure theory and suggests how spinor properties are essential to turn this into an effective tool for physicists and chemists alike.

2 The atomic central field model

For simplicity, we shall use Hartree atomic units [5, §1.2]. The simplest atomic model [9, Chapter 6] assumes that electrons move independently in a spherically symmetric electrostatic potential field, so that each orbital satisfies the Schrödinger equation

$$i\frac{\partial \psi}{\partial t} = H_S\, \psi(t, \mathbf{r}), \quad H_S = \frac{\mathbf{p}^2}{2m} + V(r) \tag{2.1}$$

where H_S is the Schrödinger Hamiltonian operator, $\mathbf{r}$ is the electron position vector, $r = |\mathbf{r}|$, $V(r)$ is some spherically symmetric potential, $\mathbf{p} = -i\nabla$ is the momentum operator and E is the energy. The operator H_S is unchanged by rotation of the coordinate axes, so that it commutes with the infinitesimal generators (*alias* orbital angular momentum operators, $\mathbf{l}$) of the rotation group. H_S does not act on electron spin. It follows that stationary state solutions of (2.1) can be written as a product of functions of time, t, spherical polar coordinates (r, θ, φ) and spin

$$\psi_{E,l,m_l,m_s}(t, \mathbf{r}) = \text{Const.} \times e^{-iEt} \frac{P_{El}(r)}{r} Y_l^{m_l}(\theta, \varphi) . \phi_{m_s}, \tag{2.2}$$

where $\mathbf{l}^2 Y_l^{m_l} = l(l+1) Y_l^{m_l}$, $l = 0, 1, 2, \ldots$ and $l_z Y_l^{m_l} = m_l Y_l^{m_l}$, $m_l = -l, -l+1, \ldots, +l$. The phase relations required for the spherical harmonics $Y_l^{m_l}$ to be angular momentum eigenfunctions are given by Condon and Shortley [9, §4[3], page 52]. The wavefunction (2.2) has *parity* $(-1)^l$, the change of sign after replacing $\mathbf{r}$ by $-\mathbf{r}$. Similarly the spin-1/2 functions satisfy $\mathbf{s}^2 \phi_{m_s} = 3/4 . \phi_{m_s}$ and $s_z \phi_{m_s} = m_s \phi_{m_s}, \; , m_s = \pm 1/2$.

Relativistic quantum mechanics involves functions of coordinates x in Minkowski space-time, $x = \{x^\mu\}$, $\mu = 0, 1, 2, 3$ or $x = \{x^0 = ct, \mathbf{x}\}$. Equation (2.1) is replaced by Dirac's equation [4, §10], [9, §5[5]], [10, Chapter XI], etc., so that

$$i \frac{\partial \psi}{\partial t} = H_D \psi, \quad H_D = c \alpha \cdot \mathbf{p} + \beta m c^2 + V(r) \tag{2.3}$$

where c is the speed of light ($c \approx 137$ in atomic units), $\alpha = (\alpha_x, \alpha_y, \alpha_z)$ and β are 4×4 matrices and m is the electron rest mass ($m = 1$ in atomic units). The operator's matrix character requires $\psi(x)$ to be a four component *spinor*. As in the nonrelativistic case, H_D is a scalar with respect to coordinate rotations and commutes with the infinitesimal generators of the rotation group, which in this case are the components of the *total* angular momentum, $\mathbf{j} = \mathbf{l} + \mathbf{s}$, rather than the orbital angular momentum $\mathbf{l}$. The structure of central field Dirac spinors is therefore more complicated,

$$\psi_{E,\kappa,m}(x) = \text{Const.} \times e^{-iEt} \frac{1}{r} \begin{pmatrix} P_{E,\kappa}(r) \chi_{\kappa m}(\theta, \varphi) \\ i Q_{E,\kappa}(r) \chi_{-\kappa m}(\theta, \varphi) \end{pmatrix} \tag{2.4}$$

where the angular quantum number $\kappa = (j + 1/2)\eta$, $\eta = \pm 1$ will be explained below. Electron states with eigenvalues in the range $0 < E < c^2$, belong to the discrete spectrum, and those with $E > c^2$ are unbound. The upper radial component $P_{E,\kappa}(r)$ approximates the Schrödinger radial function $P_{El}(r)$ in the nonrelativistic limit $c \to \infty$, and to a first approximation, the lower component is related to it by the Pauli approximation

$$Q_{E,\kappa}(r) = \frac{1}{c} \left(\frac{dP_{E,\kappa}}{dr} + \kappa \frac{P_{E,\kappa}}{r} \right) + O\left(\frac{1}{c^2} \right). \tag{2.5}$$

For this reason, it is usual to refer to $P_{El}(r)$ as the "large" and $Q_{El}(r)$ as the "small" component. For positron states (negative eigenvalues $E < -c^2$) the roles are reversed. The functional relation of the components is non-trivial and must be respected to get sensible results.

The angular factors in (2.2) span the reducible SO(3) product representation $\mathcal{D}^{(l)} \times \mathcal{D}^{(1/2)}$. The spin space $\mathcal{D}^{(1/2)}$ is two-dimensional, and we can choose a basis

$$\phi_{1/2} = \begin{pmatrix} 1 \\ 0 \end{pmatrix}, \quad \phi_{-1/2} = \begin{pmatrix} 0 \\ 1 \end{pmatrix}. \tag{2.6}$$

The representation space $\mathcal{D}^{(l)}$ spanned by the spherical harmonics $Y_l^{m_l}$ is $(2l+1)$-dimensional. The product representation is reducible, with the Clebsch-Gordan decomposition

$$\mathcal{D}^{(l)} \times \mathcal{D}^{(1/2)} = \mathcal{D}^{(l+1/2)} \oplus \mathcal{D}^{(l-1/2)} \tag{2.7}$$

into two even order irreps $\mathcal{D}^{(j)}$, $j = l \pm 1/2$. The two possible irreps $\mathcal{D}^{(j)}$ are spanned by the functions $\chi_{\kappa,m}$ of (2.4):

$$\chi_{\kappa m}(\theta,\varphi) = \sum_{m_l, m_s} (l, m_l, 1/2, m_s \,|\, l, 1/2, j, m)\; Y_l^{m_l}(\theta,\varphi)\phi_{m_s} \tag{2.8}$$

where $l = j + \frac{1}{2}\eta$, $\eta = \mathrm{sgn}\ \kappa$. Using the representation (2.6) gives the more explicit formula[2]

$$\chi_{\kappa m}(\theta,\varphi) = \begin{pmatrix} (l, m-1/2, 1/2, 1/2 \,|\, l, 1/2, j, m)\; Y_l^{m-1/2}(\theta,\varphi) \\ (l, m+1/2, 1/2, -1/2 \,|\, l, 1/2, j, m)\; Y_l^{m+1/2}(\theta,\varphi) \end{pmatrix}. \tag{2.9}$$

Thus for both signs of κ, the 4-spinors $\psi_{E,\kappa,m}(x)$, $m = -j, \ldots, +j$ span the irrep $\mathcal{D}^{(j)}$. However, the upper and lower 2-spinors $\chi_{\kappa m}(\theta,\varphi)$ have opposite parity $(-1)^l$; also if

$$K = -1 - 2\mathbf{s}.\mathbf{l} = -1 + \mathbf{l}^2 + \mathbf{s}^2 - \mathbf{j}^2, \quad \text{then} \quad K\,\chi_{\kappa m}(\theta,\varphi) = \kappa\,\chi_{\kappa m}(\theta,\varphi).$$

The two sets of 2-spinors are related by the important involution

$$\sigma_r\,\chi_{\kappa m}(\theta,\varphi) = -\chi_{-\kappa m}(\theta,\varphi), \quad \sigma_r = \sigma.\mathbf{e}_r = 2\mathbf{s}.\mathbf{e}_r \tag{2.10}$$

where $\mathbf{e}_r = \mathbf{r}/r$. This enables us to write

$$\sigma.\mathbf{p} = -i\sigma_r\left(\frac{\partial}{\partial r} + \frac{K+1}{r}\right), \tag{2.11}$$

with the frequently exploited result that

$$\sigma.\mathbf{p}\,\frac{f(r)}{r}\,\chi_{\kappa m}(\theta,\varphi) = \frac{i}{r}\left(\frac{df}{dr} + \frac{\kappa f}{r}\right)\chi_{-\kappa m}(\theta,\varphi).$$

Dirac's original derivation of (2.3), as presented in [10, §67], was motivated by the search for a linear wave equation involving the time-like and space-like components of momentum in Minkowski space-time on the same footing. For many purposes, it is better to think of the Dirac equation in the context of a unitary representation of the inhomogeneous Lorentz (or Poincaré) group; the determination of all such unitary representations is equivalent to the determination of all possible relativistic wave equations [11, p. 17]; for details see [12, 13, 14]. The 10 infinitesimal generators of the Poincaré algebra for a free particle include the components, P^μ of the 4-momentum operator, and the irreps can be classified in terms of three invariants (Casimir operators). One of these is $C_1 = P^\mu P_\mu = m^2c^2$, which is positive if P^μ is a *time-like* vector, and we can then interpret m as a particle rest-mass. A second invariant is $C_2 = m^2c^2\mathbf{J}^2$, where $\mathbf{J}$ is the angular momentum operator in the particle's rest frame (i.e. its *spin*). Thus the Hamiltonian $H = cP^0$ is the infinitesimal generator of displacements in time, the space-like components $P^i, i = 1, 2, 3$ are the infinitesimal generators of displacements in space, and the remaining 6 operators $\mathbf{J}$ and $\mathbf{K}$ are respectively infinitesimal generators of spatial rotations and boosts to another inertial frame. The central potential $V(r)$ in (2.3) fixes the frame of reference, and it is the remaining rotational symmetries that we exploit in atomic and molecular structure theory.

A similar viewpoint could be used to set up nonrelativistic equations, although this is seldom done.

---

[2] Compare [5, §2.12]; Louck uses $\mathcal{Y}^{(j\pm 1/2, 1/2)jm}$ for our $\chi_{\pm|\kappa|m}$

3 Relativistic and nonrelativistic atomic structure

Most atomic structure calculations are still based on Slater's approach [15] using determinantal wavefunctions built from one-electron central field wavefunctions. The essentials can easily be understood by considering a simple closed shell model in which the wavefunction is approximated by a single Slater determinant. The total energy can be written [16, Chapter 6]

$$E = \sum_{a} \left\{ \langle a \,|\, h \,|\, a \rangle + \frac{1}{2} \sum_{b} [\langle a, b \,|\, g \,|\, a, b \rangle - \langle a, b \,|\, g \,|\, b, a \rangle] \right\} \tag{3.1}$$

where $a, b, \ldots$ denote one-electron central field orbitals of the form (2.2), h is a bare-nucleus Hamiltonian, and g is the electron-electron interaction. In nonrelativistic models, the one-electron part can be written immediately in terms of radial amplitudes only:

$$\langle a \,|\, h \,|\, a \rangle = \int_0^\infty P_a^*(r) \left\{ \frac{1}{2} \frac{d^2}{dr^2} - \frac{l_a(l_a+1)}{2r^2} - \frac{Z}{r} \right\} P_a(r)\, dr, \tag{3.2}$$

whilst, when $g = 1/|\mathbf{r}_1 - \mathbf{r}_2|$, the two-electron integrals are

$$\langle a, b \,|\, g \,|\, c, d \rangle \equiv (ac \,|\, bd) = \iint \frac{D_{ac}(\mathbf{r}_1).D_{bd}(\mathbf{r}_2)}{|\mathbf{r}_1 - \mathbf{r}_2|} d\mathbf{r}_1\, d\mathbf{r}_2 \tag{3.3}$$

in which

$$D_{ac}(\mathbf{r}) = \psi_a^*(\mathbf{r}) \psi_c(\mathbf{r}) \tag{3.4}$$

is an overlap charge density of two orbital wavefunctions. Using the orbital expressions (2.2) along with the expansion

$$1/|\mathbf{r}_1 - \mathbf{r}_2| = \sum_k U_k(r_<, r_>) \sum_q C_q^{k*}(\theta_1, \varphi_1).C_q^k(\theta_2, \varphi_2), \quad U_k(r_<, r_>) = r_<^k / r_>^{k+1} \tag{3.5}$$

where $r_< / r_> = \min/\max(r_1, r_2)$ and $C_q^k(\theta, \varphi) = \sqrt{4\pi/2k+1}\, Y_q^k(\theta, \varphi)$, permits separation of angular and radial integrals leading to the general expression [9, §7[6]],

$$\langle a, b \,|\, g \,|\, c, d \rangle = \delta(q, m_l a - m_{l_c}) \delta(-q, m_{l_b} - m_{l_d}) \sum_k c^k(l_a, l_c) c^k(l_b, l_d).R^k(a, b, c, d) \tag{3.6}$$

with the radial *Slater integral*

$$R^k(a, b, c, d) = \int_0^\infty \int_0^\infty \rho_{ac}(r_1)\, U_k(r_<, r_>)\, \rho_{bd}(r_2)\, dr_1\, dr_2, \quad \rho_{ac}(r) = P_a^*(r) P_c(r).$$

The delta-functions restrict the sum over q to a single term provided $m_{l_a} + m_{l_b} = m_{l_c} + m_{l_d}$. The angular coefficients $c^k(lm, l'm')$ were constructed by Gaunt [17] in terms of integrals over products of three associated Legendre polynomials, expressed explicitly as hypergeometric algebraic sums. Today, we would make use of the seminal papers of Racah [18] to write them in terms of Wigner's covariant $3j$-symbols and reduced matrix elements of the $\mathbf{C}^k$ tensor operator as

$$c^k(lm, l'm') = \begin{pmatrix} m & k & l' \\ l & q & m' \end{pmatrix} (l \,\|\, \mathbf{C}^k \,\|\, l') \tag{3.7}$$

with

$$(l \,\|\, \mathbf{C}^k \,\|\, l') = (-1)^l [(2l+1)(2l'+1)]^{1/2} \begin{pmatrix} l & k & l' \\ 0 & 0 & 0 \end{pmatrix}.$$

The usefulness of exploiting group theoretical properties of the nonrelativistic wavefunctions is obvious.

Much of the theory of angular momentum had not been developed when Swirles [19] wrote the first paper on relativistic self-consistent fields in 1935, and her formulae for the relativistic version of (3.6) involve epressions taken from Gaunt's paper [17]. In order to use the spinor orbitals (2.4) in place of (2.2) in evaluating (3.6), she was forced to treat each spinor component as if it were a single nonrelativistic wavefunction, so that her result was a lengthy expression in which each of the orbital angular momentum arguments appearing in the 2-spinors $\chi_{\pm\kappa,m}$, (2.9) were involved. Angular momentum theory suggests that as Dirac central field spinors are eigenfunctions of $\mathbf{j}^2$, j_z and parity, one might expect only j, m and parity labels to appear in matrix elements. Using this notion, I was able to derive [20] a formula very like (3.6), with angular coefficients

$$d^k(lm, l'm') = \begin{pmatrix} m & k & j' \\ j & q & m' \end{pmatrix} (j \parallel \mathbf{C}^k \parallel j').\Pi^e(\kappa\kappa'k), \tag{3.8}$$

where

$$(j \parallel \mathbf{C}^k \parallel j') = (-1)^{j+1/2}[(2j+1)(2j'+1)]^{1/2} \begin{pmatrix} j & k & j' \\ 1/2 & 0 & -1/2 \end{pmatrix},$$

and the parity factor

$$\Pi^e(\kappa\kappa'k) = [1 - \eta\eta'(-1)^{j+j'+k}]/2$$

takes care of the two possible spin-orbit coupling arrangements of each 4-spinor (2.9) through $\eta =$ sign κ. A similar parity factor is implicit in the nonrelativistic formula (3.7). The simplicity of this result depends on the important fact that the *angular density distribution*, $\chi^\dagger_{\kappa,m}(\theta,\varphi).\chi_{\kappa,m}(\theta,\varphi)$ is independent of the sign of κ, so that we can combine large and small component contributions in a Dirac overlap *radial density distribution*

$$\rho_{ac}(r) = P_a^*(r)P_c(r) + Q_a^*(r)Q_c(r). \tag{3.9}$$

for substitution in the Slater integral of (3.6).

In the Dirac-Coulomb model, in which only the *electrostatic* interaction is taken into account, the theory of atomic structure thus can be cast in very similar form to the nonrelativistic theory. The main differences concern the approximation of radial amplitudes, which has some features which are not present in nonrelativistic models, the fact that QED introduces magnetic as well as electric forces, and the need to use jj-coupling in treating open shell systems, all of which make the relativistic calculations more complex, although not necessarily prohibitively expensive, in relation to nonrelativistic equivalents. Relativistic QED couples the electron field, with a 4-vector *current density*, $j^\mu(x)$, to electromagnetic fields with a covariant *4-potential*, $a_\mu(x)$, by means of the interaction Hamiltonian

$$H_{int} = \frac{1}{c}\int j^\mu(x) a_\mu(x)\, d^3x \tag{3.10}$$

where, in terms of Dirac γ^μ matrices, the electron current density vector is

$$j^\mu(x) = -ec\,\overline{\psi}(x)\,\gamma^\mu\,\psi(x), \quad \overline{\psi}(x) = \psi^\dagger(x)\gamma^0. \tag{3.11}$$

where the dagger denots Hermitian conjugation. When the electron field is expanded in terms of 4-spinor orbitals, we get overlap charge densities (*ordinary space-time functions*)

$$\rho_{ac}(x) = [j^0(x)]_{ac}/c = \psi_a^\dagger(x)\,\psi_c(x) \tag{3.12}$$

and overlap current densities

$$\mathbf{j}_{ac}(x) = c\,\psi_a^\dagger(x)\,\alpha\,\psi_c(x) \tag{3.13}$$

where α was defined in (2.3). These in turn can be regarded as the source terms of Maxwell's equations which generate, along with electrostatic and vector potentials, the internal electric and magneticd fields, $\mathbf{E}_{ac}$ and $\mathbf{B}_{ac}$, to which we shall return later.

4 Radial amplitudes

Once the angular parts of atomic calculations have been dealt with using group theoretical methods, the task of determining the radial amplitudes remains. There are three main approaches to solving the radial equations in mean field potential or self-consistent field (SCF) atomic models:

- Finite difference approximation -- the oldest approach.
- Finite element approximation – B-splines are the most popular.
- Roothaan-style basis sets – predominantly STO and GTO.

The book of Fischer *et al.* [7] covers the use of finite difference methods in nonrelativistic atomic calculations. A discretized version of the radial Schrödinger equation together with boundary conditions at $r = 0$ and as $r \to \infty$ produces approximations to $P_a(r)$ on a radial grid. Starting values at the origin can be obtained by taking one or two terms of the power series solution at the first few grid points. Finite element schemes, for example [21], expand the radial functions in terms of splines, defined piecewise on a radial grid, imposing continuity conditions at the *knots*. Variational methods are usually used to define relations for the expansion coefficients, although collocation schemes have also been used. Hall [22] and Roothaan [23] pioneered the now widely used matrix Hartree-Fock method, in which the radial amplitudes are expanded in sets of analytic functions, typically of STO- or GTO-type. The last generate the most practical and widely used method for studying molecules.

The same approximation schemes are all applicable in relativistic calculations. Finite difference (and to a lesser extent) finite element schemes automatically respect the functional interdependence of the 4-spinor radial components. The first finite difference Dirac-Hartree calculation (for Cu^+) was performed *by hand* by Williams [24] in 1940. The first published Dirac-Hartree calculation on a computer dates back to 1957 [25] and this was followed by a succession of calculations based on the new formulation in [20] over the next 15 or so years. Kim [26] published the first successful Roothaan-style calculation on Be in 1967 using STO basis functions. These had the right point-nucleus Dirac cusp conditions, but used the same basis functions for both large and small components without investigating the validity of this prescription. Progress with this approach was slow; eventually, attempts some 25 years ago (for example by Schwartz and Wallmeier [27] and many others) to extend nonrelativistic Roothan-style calculations to molecules revealed unsuspected problems. Although many papers were written subsequently alleging that such calculations were doomed to fail, we now know that this was due to the fact that no account was taken of the 4-spinor internal structure on which this paper focuses. This has been remedied by defining *spinor* basis functions [8, Chapter 3], [5, Ch22]: L-spinors are analogues of nonrelativistic Coulomb Sturmian functions as defined by Rotenberg [28], whilst S-spinors have more in common with STO functions. Both are useful for point nucleus atomic models. G-spinors, like GTO functions, are easier to use in molecular calculations.

The key to the relativistic Roothaan method is the recognition that the components $M[\beta, \mu, \mathbf{x}]$ of a basis spinor μ must be *matched* in the sense that

$$M[-1, \mu, \mathbf{x}] = \text{const.}\ \sigma \cdot \mathbf{p}\, M[+1, \mu, \mathbf{x}] \tag{4.1}$$

in the nonrelativistic limit, $c \to \infty$, for which (2.11) is useful. The (time-independent) orbital spinors can then be written

$$\psi_a(\mathbf{x}) = \begin{pmatrix} \sum_{\mu=1}^{N} c_{\mu a}^{+1}\, M[+1, \mu, \mathbf{x}] \\ i \sum_{\mu=1}^{N} c_{\mu a}^{-1}\, M[-1, \mu, \mathbf{x}] \end{pmatrix}. \tag{4.2}$$

The $2N \times 2N$ Dirac Hamiltonian matrix $\mathbf{H}$ can be partitioned into $N \times N$ blocks $\mathbf{H}^{\beta\beta'}$ given by

$$\mathbf{H} = \begin{bmatrix} c^2\mathbf{S}^{++} + \mathbf{V}^{++} & c\mathbf{\Pi}^{+-} \\ c\mathbf{\Pi}^{-+} & -c^2\mathbf{S}^{--} + \mathbf{V}^{--} \end{bmatrix} \tag{4.3}$$

where the different matrix blocks are constructed from the 2-spinor components. In particular, the kinetic matrices $\mathbf{\Pi}^{-+} = \mathbf{\Pi}^{+-\dagger}$ are defined as

$$\mathbf{\Pi}^{\beta,-\beta}_{\mu,\nu} = \int M[\beta,\mu,\mathbf{x}]^\dagger\, \sigma\cdot\mathbf{p}\, M[-\beta,\nu,\mathbf{x}]\, d\mathbf{x}, \quad \beta = \pm, \tag{4.4}$$

which, by virtue of (4.1) is proportional to the overlap matrix element $\mathbf{S}^{--}_{\bar{\mu},\bar{\nu}}$. This has the great advantage that $\mathbf{\Pi}^{+-}(\mathbf{S}^{--})^{-1}\mathbf{\Pi}^{-+} \equiv [\mathbf{p}^2]^{++}$, so that the correct nonrelativistic kinetic energy matrix in the large component basis is retrieved exactly in the Pauli approximation. The 1-1 matching ensures that there are no spurious eigenfunctions of the matrix Hamiltonian. If $\mathbf{S}^{++}$ is well-conditioned, then so is $\mathbf{S}^{--}$. Moreover, the prescription (4.1) automatically ensures that the angular parts of the 4-spinors are properly constructed as in (2.4) and (2.10), so that the orbitals belong to the correct representation of the rotation group. Time reversal symmetry, which relates Kramers paired orbitals, is built in to the formalism, and there is no need to impose it later as in other formulations [29, 30]. These spinor basis sets are compatible with the TSYM software [31] for generating symmetry orbitals of relativistic double point groups.

5 Relativistic and nonrelativistic molecular structure

G-spinors, analogues of Gaussian basis sets, are the most convenient for molecular calculations. If $\eta_\mu = \mathrm{sgn}\, \kappa_\mu = \pm 1$, $l_\mu = j_\mu + \frac{1}{2}\eta_\mu$, and $j_\mu = |\kappa_\mu| - \frac{1}{2}$, the upper components of basis function μ can be written

$$M[+1,\mu,\mathbf{r}_{A_\mu}] = N^+_\mu \begin{bmatrix} -\eta_\mu C^{-\eta_\mu}_{l_\mu m_\mu}\, S[a_\mu, \mathbf{r}_{A_\mu}\,;\, 0, l_\mu, m_\mu - 1/2] \\ C^{+\eta_\mu}_{l_\mu m_\mu}\, S[a_\mu, \mathbf{r}_{A_\mu}\,;\, 0, l_\mu, m_\mu + 1/2] \end{bmatrix}, \tag{5.1}$$

where

$$S(a, \mathbf{r}_A; n, l, m) = r_A^{2n}\mathcal{Y}_{lm}(\mathbf{r}_A)\exp(-ar_A^2), \tag{5.2}$$

where N^+_μ is a normalizing constant,

$$\mathcal{Y}_{lm}(\mathbf{r}_A) = s_{lm} r^l P_l^{|m|}(\cos\theta_A)\, e^{im\varphi_A}, \quad s_{lm} = (-1)^{(m+|m|)/2}\left[\frac{2l+1}{4\pi}\frac{(l-|m|)!}{(l+|m|)!}\right]^{1/2} \tag{5.3}$$

is a *normalized solid harmonic* with standard normalization and phase [5, §2.1], and

$$C^\eta_{lm} = s_{lm}\left(\frac{l + 1/2 + \eta m}{2l+1}\right)^{1/2}. \tag{5.4}$$

combines this with the CG coefficient of the Dirac spinor component. The set of Gaussian exponents, a_μ, can often be taken, with minor modifications, from nonrelativistic calculations. When we apply (4.1), which is valid in this case for finite values of c provided we use a finite size nuclear model, then

$$M[-1,\mu,\mathbf{r}_{A_\mu}] = N^-_\mu \begin{bmatrix} \eta_\mu C^{\eta_\mu}_{\bar{l}_\mu, m_\mu}\, \{t_\mu\, S[a_\mu, \mathbf{r}_{A_\mu}\,;\, 0, \bar{l}_\mu, m_\mu - 1/2] \\ \quad -2a_\mu\, S[a_\mu, \mathbf{r}_{A_\mu}\,;\, 1, \bar{l}_\mu, m_\mu - 1/2]\,\} \\ C^{-\eta_\mu}_{\bar{l}_\mu, m_\mu}\, \{t_\mu\, S[a_\mu, \mathbf{r}_{A_\mu}\,;\, 0, \bar{l}_\mu, m_\mu + 1/2] \\ \quad -2a_\mu\, S[a_\mu, \mathbf{r}_{A_\mu}\,;\, 1, \bar{l}_\mu, m_\mu + 1/2]\,\} \end{bmatrix}, \tag{5.5}$$

where $\bar{l}_\mu = l_\mu - \eta_\mu$, $t_\mu = \kappa_\mu + l_\mu + 1$ and $\mathbf{r}_{\mathbf{A}_\mu} = \mathbf{r} - \mathbf{A}_\mu$. Notice that $t_\mu = 0$ when $\kappa_\mu < 0$ so that only one term survives in each row of (5.5); the fact that two SGTF are needed when $\kappa_\mu > 0$ is one of the reasons why early attempts to devise a relativistic Roothaan scheme failed. All interaction integrals are built from charge-density components of (3.11–3.13) so that

$$\begin{aligned} \varrho_{\mu\nu}(\mathbf{x}) &= \varrho_{\mu\nu}^{++}(\mathbf{x}) + \varrho_{\mu\nu}^{--}(\mathbf{x}), \text{ where } \varrho_{\mu\nu}^{\beta\beta}(\mathbf{x}) = -eM^\dagger(\beta,\mu,\mathbf{x}).\, M(\beta,\nu,\mathbf{x}), \\ \mathbf{j}_{\mu\nu}^{+-}(\mathbf{x}) &= \left(\mathbf{j}_{\nu\mu}^{-+}\right)^* = -iecM^\dagger(+1,\mu,\mathbf{x})\,\sigma M(-1,\nu,\mathbf{x}). \end{aligned} \tag{5.6}$$

These components are therefore linear combinations of products of SGTF on different centres, which can be expressed in terms of Hermitian Gaussian (HGTF) intermediates using the well-known Gaussian product lemma [32, See article by V.R.Saunders, p.1]. The product of SGTF on sites $\mathbf{r}_A, \mathbf{r}_B$ with Gaussian exponents a, b respectively is [33]

$$\begin{aligned} &S(a,\mathbf{r}_A;n,l,m)\,S(b,\mathbf{r}_B;n',l',m') \\ &\quad= \sum_{\mathbf{k}\in\mathcal{T}_\Lambda} E[n,l,m\,;n',l',m'\,;\mathbf{k}]\,.\,H(p,\mathbf{r}_P;\mathbf{k}) \end{aligned} \tag{5.7}$$

where $E[n,l,m\,;n',l',m'\,;\mathbf{k}]$ is a numerical coefficient, $p = a+b$, $\mathbf{P} = (a\mathbf{A}+b\mathbf{B})/p$, and the HGTF intermediates are here denoted

$$H(p,\mathbf{r}_P;\mathbf{k}) = \left(\frac{\partial}{\partial P_x}\right)^\rho \left(\frac{\partial}{\partial P_y}\right)^\sigma \left(\frac{\partial}{\partial P_z}\right)^\tau \exp(-p\mathbf{r}_P^2), \quad \text{where } \mathbf{k} = (\rho,\sigma,\tau), \tag{5.8}$$

where $\mathbf{k}$ runs over all integer-valued triples in the set

$$\mathcal{T}_\Lambda = \{(\rho,\sigma,\tau)\,|\,0 \le \rho+\sigma+\tau \le \Lambda = 2n+2n'+l+l'\} \tag{5.9}$$

The method of evaluating interaction integrals over products of HGTF is well-known to quantum chemists. The relativistic generalization of this scheme [34] expresses the relativistic charge density in the analogous form

$$\varrho_{\mu\nu}^{\beta\beta}(\mathbf{r}) = -e \sum_{\mathbf{k}\in\mathcal{T}_\Lambda} E_0^{\beta\beta}(\mu,\nu;\mathbf{k})\,.\,H(p,\mathbf{r}_P;\mathbf{k}) \tag{5.10}$$

and the (spherical) components, $q = 0, \pm 1$, of the current density vector as

$$\left(\mathbf{j}_{\mu\nu}^{\beta,-\beta}(\mathbf{r})\right)_q = -iec \sum_{\mathbf{k}\in\mathcal{T}_\Lambda} E_q^{\beta,-\beta}(\mu,\nu;\mathbf{k})\,.\,H(p,\mathbf{r}_P;\mathbf{k}). \tag{5.11}$$

All relativistic interaction integrals (ERI) can therefore be expressed as sums of products of relativistic E-coefficients with standard interaction integrals over HGTF. An efficient recurrence scheme for generating the relativistic E-coefficients implemented in the BERTHA code [8, Chapter 3], [35, pp. 199–215] is currently being written up. Several versions of this code exist, of which a relativistic DFT version [36] is probably the most advanced, with a growing number of applications. The recent ERI algorithms proposed by the Tokyo group [37] use the same central field basis functions but make little use of the structures exploited in this paper.

Finally we note a promising development [38] and [35, pp. 199–215], which allows us to replace the four-index ERI algorithms with equivalent integrals over the internal electromagnetic fields generated by the two-index charge-current densities (5.10) and (5.11). Thus the general electrostatic ERI, notation of (3.3), can be rewritten

$$(\mu\beta,\nu\beta|\sigma\beta',\tau\beta') = \epsilon_0 \sum_{\beta\beta'} \int d\mathbf{x}\; \mathbf{E}_{\mu\nu}^{\beta\beta}(\mathbf{x}) \cdot \mathbf{E}_{\sigma\tau}^{\beta'\beta'}(\mathbf{x}). \tag{5.12}$$

The Cartesian component of the field along the unit vector $\mathbf{e}_q$ is

$$\left[\mathbf{E}^{\beta\beta'}_{\mu\nu}(\mathbf{x})\right]_q = \frac{e}{4\pi\epsilon_0}\delta_{\beta\beta'}\sum_{\mathbf{k}} E_0^{\beta\beta}(\mu,\nu;\mathbf{k})\,(a_{\mu\nu},\mathbf{A}_{\mu\nu};\mathbf{k}+\mathbf{e}_q|\mathbf{x}_{A_{\mu\nu}}) \tag{5.13}$$

where, in terms of the intermediate HGTF,

$$(a_{\mu\nu},\mathbf{A}_{\mu\nu};\mathbf{k}|\mathbf{x}_{A_{\mu\nu}}) = \int H(a_{\mu\nu},\mathbf{y}_{\mathbf{A}_{\mu\nu}};\mathbf{k})/|\mathbf{x}-\mathbf{y}|\,dy. \tag{5.14}$$

The total electrostatic field at the point $\mathbf{x}$ is then

$$\mathbf{E}(\mathbf{x}) = \sum_{\sigma\tau}\sum_{\beta}\mathbf{E}^{\beta\beta}_{\sigma\tau}(\mathbf{x})\,D^{\beta\beta}_{\sigma\tau} \tag{5.15}$$

where $D^{\beta\beta}_{\sigma\tau}$ is an element of the $\mathbf{D}^{\beta\beta}$ density matrix. The Breit interaction gives analogous expressions in terms of magnetic fields $\mathbf{B}^{\beta,-\beta}_{\sigma\tau}$. For a closed shell system in the Dirac-Coulomb approximation, the Fock equations become

$$\mathbf{F}\,\mathbf{c} = \varepsilon\,\mathbf{S}\,\mathbf{c}, \quad \mathbf{F} := \mathbf{h}+\mathbf{G} \tag{5.16}$$

where $\mathbf{h}$ has the structure of (4.3) and the electron-electron interaction energy is

$$G^{\beta\beta'}_{\mu\nu} = \epsilon_0\int d\mathrm{x}\left\{\delta_{\beta\beta'}\,\mathbf{E}(\mathbf{x})\cdot\mathbf{E}^{\beta\beta}_{\mu\nu}(\mathbf{x}) - \sum_{\sigma\tau}\mathbf{E}^{\beta\beta}_{\mu\tau}(\mathbf{x})\cdot\mathbf{E}^{\beta'\beta'}_{\sigma\nu}(\mathbf{x})D^{\beta'\beta}_{\sigma\tau}\right\}. \tag{5.17}$$

Notice that we need only the *total* field $\mathbf{E}(\mathbf{x})$ to construct the direct interaction, whereas the second, exchange term cannot be simplified in this way. The amount of work is therefore comparable with DFT, making for a very fast algorithm. There are $O(N^4)$ ERI, whereas this method requires $O(3N^2)$ components to be evaluated at M space integration points. We have adapted Becke's cell integration scheme [39] from DFT for the space integration. One advantage is that the fields are well approximated by asymptotic expansions over much of the region occupied by a molecule, and are therefore very cheap to construct. Also the self-interaction terms in (5.17) cancel exactly, unlike in DFT, where the incomplete cancellation of direct and exchange self-interaction when using approximate exchange functionals is a major source of error.

6 Conclusions

The internal structure of Dirac 4-component spinors is a source of strength in relativistic electronic structure calculations for atoms and molecules. The theory is shaped by the physics so that its meaning is transparent and the internal spinor structure can be exploited to devize efficient algorithms. The formalism is effective for both atoms and molecules, and may also be useful in the future for calculations on solid matter.

References

[1] L. Pauling and E. B. Wilson, *Introduction to Quantum Mechanics.* McGraw-Hill, New York, 1935.

[2] R. McWeeny, *Methods of Molecular Quantum Mechanics*, 2nd edition. Academic Press, London, 1989.

[3] B. R. Judd, *Operator Techniques in Atomic Spectroscopy* McGraw-Hill, New York, 1963.

[4] H. A. Bethe and E. E. Salpeter, *Quantum Mechanics of One- and Two-electron Atoms* (1st edn.) Springer-Verlag, Berlin, 1957.

[5] G. W. F. Drake (ed.) *Springer Handbook of Atomic, Molecular, and Optical Physics* Springer Science+ Business Media, Inc., New York, 2006.

[6] G. Tarczay, A. G. Császár, W. Klopper and H. M. Quiney, Anatomy of relativistic energy corrections in light molecular systems, *Mol. Phys.* **99**, 1769–1794 (2001).

[7] C. Froese Fischer, T Brage and P. Jönsson, *Computational Atomic Structure. An MCHF Approach* Insitute of Physics Publishing Ltd., Bristol, 1997.

[8] P. Schwerdtfeger, *Relativistic Electronic Structure Theory. Part I. Fundamentals* Elevier, Amsterdam, 2002.

[9] E. U. Condon and G. H. Shortley, *Theory of Atomic Spectra.* Cambridge University Press, Cambridge, 1935.

[10] P. A. M. Dirac, *The Principles of Quantum Mechanics* (4th edn.) Clarendon Press, Oxford, 1958.

[11] S.. S. Schweber, *An Introduction to Relativistic Quantum Field Theory* Harper and Row, New York,1961.

[12] E. P. Wigner, *Ann. Math.* **40**, 149, (1939).

[13] L. L. Foldy,*Phys. Rev.* **102**, 368, (1956).

[14] Yu. M. Shirokov, *Sov. Phys. JETP*, **6**, 664, (1958).

[15] J. C. Slater, *Phys. Rev.* **34**, 1293, (1929).

[16] E. U. Condon and G. H. Shortley, *The Theory of Atomic Spectra* Cambridge University Press, 1935.

[17] J. A. Gaunt, *Phil. Trans. Roy, Soc. A* **228**, 151 (1929).

[18] G. Racah, *Phys. Rev.* **62**, 438 (1942).

[19] B. Swirles, *Proc. Roy, Soc. A* **152**, 625 (1935).

[20] I. P. Grant, *Proc. Roy, Soc. A* **262**, 555 (1961).

[21] C. Froese Fischer and M. Idrees, *J. Phys. B: At. Mol. Opt. Phys.* **23**, 679 (1990).

[22] G. G. Hall and J. E. Lennard-Jones, *Proc. Roy, Soc. A* **202**, 155 (1950); see also *ibid.* **205**, 541 (1951) and **208**, 328 (1951).

[23] C. C. J. Roothan, *Rev. Mod. Phys.* **23**, 69 (1951); *ibid.* **32**, 179 (1960).

[24] A. O. Williams, *Phys. Rev.* **58**, 723 (1940).

[25] D. F. Mayers, *Proc. Roy, Soc. A* **241**, 93 (1957).

[26] Y.-K. Kim, *Phys. Rev.* **159**, 190 (1967).

[27] W. H. E. Schwarz and H. Wallmeier, *Mol. Phys.* **46**, 1045 (1982).

[28] M. Rotenberg, *Ann. Phys. (N.Y.)* **19**, 262 (1962); *Adv. At. Mol. Phys. 6*, 233 (1970).

[29] H. J. Aa. Jensen, K. G. Dyall, T. Saue and K. Fægri, *J. Chem. Phys.* **104**, 4083 (1996).

[30] T. Saue, H. J. Aa. Jensen, *J. Chem. Phys.* **111**, 6211 (1999).

[31] J. Meyer, W.-D. Sepp, B. Fricke and A. Rosen, *Comput. Phys. Commun.* **96**, 263 (1996).

[32] G. H. F. Diercksen and S. Wilson (editors), *Methods of Computational Molecular Physics*, Vol. 1. Dordrecht, Reidel, 1983.

[33] L. E. McMurchie and E. R. Davidson, *J. Comput. Phys.* **26**, 218 (1978).

[34] H. M. Quiney, H. Skaane and I. P. Grant, *J. Phys. B: At. Mol. Opt. Phys.* **30**, L829 (1997).

[35] J.-P. Julien, J. Maruani, D. Mayou, S. Wilson and G. Delgado-Barrio (editors), *Recent Advances in the Theory of Physical and Chemical Systems.* New York, Springer (2006).

[36] H. M. Quiney and P. Belanzoni, *J. Chem. Phys.* **117**, 5550 (2002; also L. Belpassi, L. Storchi, F. Tarantelli, A. Sgamellotti and H. M. Quiney, *Future Generation Computer Systems* **20**, 739 (2003).

[37] T. Yanai, T. Nkajima, Y. Ishikawa and K. Hirao, *J. Chem. Phys.* **114**, 6526 (2001); ibid. *J. Chem. Phys.* **115**, 8267 (2001); ibid. *J. Chem. Phys.* **116**,10122 (2002).

[38] H. M. Quiney and I. P. Grant, *Int. J. Quant. Chem.* **99**, 198 (2004).

[39] A. D. Becke, *J. Chem. Phys.* **88**, 2547 (1988).

Brill Academic Publishers
P.O. Box 9000, 2300 PA Leiden,
The Netherlands

Lecture Series on Computer and Computational Sciences
Volume 6, 2006, pp. 369-380

Problems in the experimental determination of higher-order dipole-polarizabilities

Uwe Hohm[1]

Institute of Physical and Theoretical Chemistry,
Technical University of Braunschweig
Hans-Sommer-Str. 10
D-38106 Braunschweig
Germany

Received June 20, 2006; accepted June 25, 2006

Abstract: In this paper experimental methods for obtaining the higher-order linear polarizabilities are discussed in view of their accuracy and compatibility to quantum-chemical ab initio calculations. It is shown that despite of the improvement of experimental and theoretical techniques still some points remain which have not been fully taken into account yet.

Keywords: Polarizabilities, Intermolecular interaction, Light scattering

PACS: 32.10.Dk, 33.15.Kr, 33.20.Fb, 34.20.Gj

1 Introduction

The induction energy U_{ind} of a molecule in an applied electric field can be defined via a series expansion in terms of the field strength as [1]

$$\begin{aligned} U_{ind} &= -\frac{1}{2}\hat{\alpha}:\vec{F}\vec{F} - \frac{1}{6}\hat{\beta}\vdots\vec{F}\vec{F}\vec{F} - \frac{1}{24}\hat{\gamma}\,\vdots\,\vec{F}\vec{F}\vec{F}\vec{F} - \ldots \\ &- \frac{1}{3}\hat{A}\vdots\nabla\vec{F}\vec{F} - \frac{1}{15}\hat{E}\,\vdots\,\nabla\nabla\vec{F}\vec{F} - -\frac{1}{6}\hat{C}\,\vdots\,\nabla\vec{F}\nabla\vec{F} - \ldots, \end{aligned} \tag{1}$$

where $\vec{F}$ and $\nabla\vec{F}$ etc. are the field and field gradient at the molecular origin. $\hat{\alpha}$ is the dipole-polarizability, $\hat{\beta}$ and $\hat{\gamma}$ are the first and second dipole hyperpolarizabilities. $\hat{A}$ and $\hat{E}$ are the dipole-quadrupole and dipole-octopole polarizabilities, whereas $\hat{C}$ is the quadrupole-quadrupole polarizability. The polarizabilities are responsible for a large number of phenomena, such as intermolecular interactions, absorbtion and scattering of light, and also some aspects of chemical kinetics. After a long period where experimentalists and theoreticians have concentrated on the linear dipole-dipole polarizability $\hat{\alpha}$ at present most studies focus on the non-linear optical properties ($\hat{\beta}$ and $\hat{\gamma}$) of molecules. Only some are devoted to the higher-order polarizabilities $\hat{A}$ and $\hat{E}$, although they might be important for a rational explanation of e.g. surface-enhanced Raman scattering [2], electric-field induced birefringence [3, 4], the electric-field induced differential scattering effect [5], collision-induced absorption [6] and scattering of light [7], or bonding in van der Waals molecules and ions [1, 8].

[1] E-mail: u.hohm@tu-bs.de

In this contribution I will address some problems in the experimental determination of the higher order linear polarizabilities $\hat{A}$ and $\hat{E}$. The following section starts, however, with some simple and straightforward considerations on the dipole-polarizability $\hat{\alpha}$ of the rare-gas atoms.

2 The dipole-polarizability $\hat{\alpha}$

2.1 Rare-gas atoms

The rare-gases from helium to xenon are inert and for symmetry reasons their polarizability tensor $\hat{\alpha}$ reduces to a single pure scalar quantity α. These gases are available with a very high purity and accordingly their polarizability as a function of the frequency should be measurable with high accuracy via the Lorentz-Lorenz equation.

$$\frac{n(\omega,p,T)^2-1}{n(\omega,p,T)^2+2}=\frac{4\pi}{3}N_1(p,T)\alpha(\omega)+B_R(\omega,T)N_1^2(p,T)+\ldots \tag{2}$$

Here $n(\omega,p,T)$ is the refractive index at pressure p, temperature T and frequency ω, $N_1(p,T)$ is the number density and $B_R(\omega,T)$ is the second optical virial coefficient describing the influence of pair interactions on the polarizability. By using interferometric techniques $n(\omega,p,T)-1$ in the gas phase can be obtained with a relative precision of less than 0.1% . But as was clearly stated by Bulanin and Kislyakov [9, 10] the polarizability value derived by Eq.(2) depends also on a proper determination of the number density $N_1(p,T)$. This consideration, although simple and straightforward, has not always received proper attention. The consequence is that polarizabilities of the rare-gas atoms determined by different authors in many cases do not agree within their error limits (see e.g. Bulanin and Kistyakov [9, 10] for an extensive discussion).

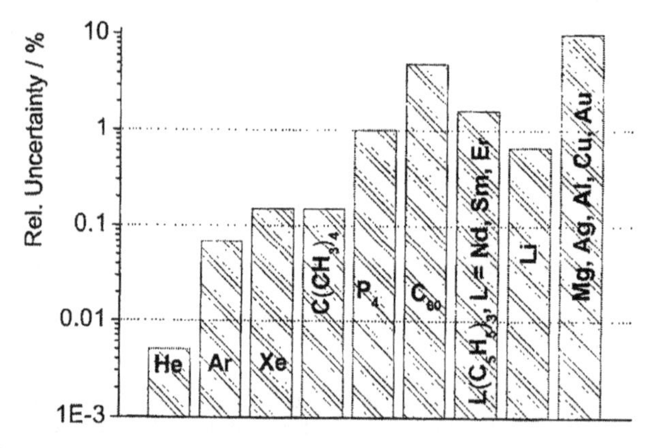

Figure 1: Uncertainties in the determination of the mean linear dipole-dipole polarizability α. The experimental data are taken from Refs. [11, 12, 13, 14, 15, 16, 17, 18, 19, 20].

2.2 Other atoms and molecules

In the case of molecules the situation gets much more complicated. The number of independent tensor components of $\hat{\alpha}$ increases [1, 21] so that different experimental techniques must be combined to get information on the individual tensor components. But even if interest is only in the rotational invariants of $\hat{\alpha}$, the mean dipole-polarizability $\alpha = \mathrm{Tr}(\hat{\alpha})/3$ and the dipole-polarizability anisotropy $\kappa = \{(3\hat{\alpha} : \hat{\alpha} - 9\alpha^2)/2\}^{1/2}$, the situation is more complicated than expected. I only mention the proper determination of the dispersion in α and κ. It was shown that in the case of e.g. CO_2 not only the electronic part $\alpha_{xy}^{(el)}$ but also the vibrational part $\alpha_{xy}^{(vib)}$ of the polarizability tensor elements $\alpha_{xy} = \alpha_{xy}^{(vib)} + \alpha_{xy}^{(el)}$ has to be taken into account if experimental results determined at optical frequencies are to be compared to calculations [22]. Moreover, especially for low volatile species the question about the precise determination of the number density becomes dominant in many experiments [17, 20]. Application of new experimental techniques like the light-force technique [16] or atom-beam interferometry may overcome these difficulties [19]. However, in the latter case only information on the static polarizability might become available.

A general survey of the accuracies of experimental determinations of α is shown in Fig.1. For comparison the generally lower experimental accuracies of the non-linear dipole polarizabilities β and γ are shown in Fig. 2.

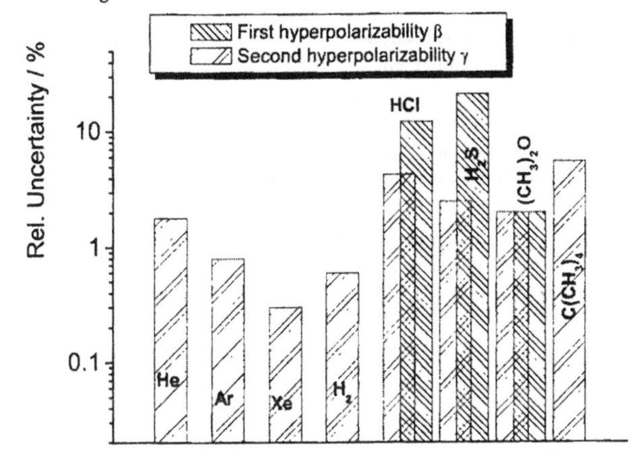

Figure 2: Uncertainties in the determination of the first and second dipole-hyperpolarizabilities β and γ. Except for helium the results for γ are based on theoretical calculations for helium. The data are taken from Shelton and Rice [23].

3 Higher-order (field-gradient) polarizabilities: dipole-quadrupole and dipole-octopole-polarizability $\hat{A}$ and $\hat{E}$

Eq.(2) provides a fairly direct means for measuring the mean linear dipole-polarizability $\alpha(\omega)$. In contrast to this there is no straightforward and direct way for measuring $\hat{A}$ and $\hat{E}$, or at least of some of their individual tensor components and their rotational invariants. At present the most promising technique for obtaining $\hat{A}$ and $\hat{E}$ is collision-induced light scattering, CILS. In this experimental technique a Raman (frequency shifted) signal is observed which is symmetry forbidden for a single atom or molecule. The observed scattering of light is a consequence of the interaction induced change $\Delta\hat{\alpha}(R) = \hat{\alpha}_1 + \hat{\alpha}_2 - \hat{\alpha}_{12}(R)$ in the polarizability $\hat{\alpha}_{12}(R)$ of the interacting pair compared to $\hat{\alpha}_1$ and $\hat{\alpha}_2$ of the individual entities [24]. $\Delta\hat{\alpha}(R)$ is a function of the internuclear separation. Therefore moving particles give raise to an interaction induced Raman signal. In the case of the rare-gas atoms this signal solely comes from a translational contribution, called dipole-induced-dipole (DID) contribution. Interacting molecules also show this effect. But in addition they exhibit a further contribution to the scattered signal which is referred to as collision-induced rotational Raman scattering [7]. The latter effect comes from electric-field gradient induced fluctuating dipole-moments. The fluctuation is caused by the rotation of the interacting molecules. Its strength is determined by the higher-order dipole-polarizabilities $\hat{A}$, $\hat{E}$ etc. of the molecules. In interacting molecules there is also a collision-induced contribution inside a vibrational band. This contribution is determined by the geometrical derivatives $\partial\hat{A}/\partial Q$ and $\partial\hat{E}/\partial Q$. An excellent and comprehensive treatment of how to obtain multipolar polarizabilities from collision-induced light scattering is given by Bancewicz et al. [25]. I do not want to repeat the theory or even details of it in this paper. The following sections are intended to point out some problems in the determination of the higher-order polarizabilities, which from my own experience sometimes are not properly taken into account. It is by no means my purpose to criticize all of the fine and outstanding experimental and theoretical work which has been done in the past.

An overview of the experimental accuracy obtained so far is given in Fig. 3. Inspection of this figure makes it clear that due to the comparatively low accuracy much work has to be done in view of a more precise determination of these molecular quantities.

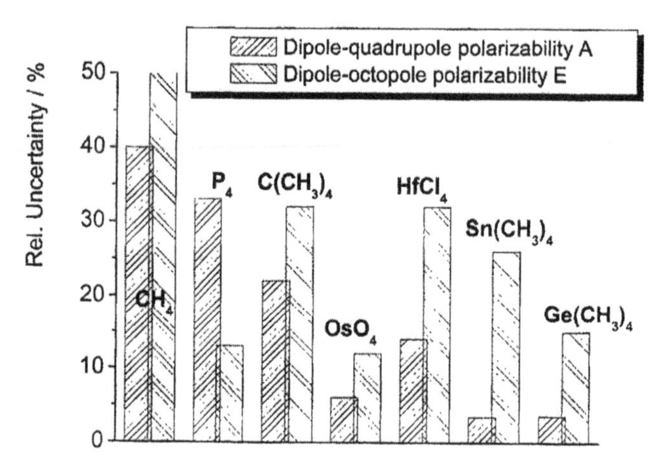

Figure 3: Uncertainties in the experimental determination of the dipole-quadrupole and dipole-octopole polarizabilities A and E [26, 27, 28, 29, 30, 31, 32].

1. Comparison between experiment and theory

   (a) Effect of surroundings
   Theoretical calculations are performed with a definite number of particles and well-defined surroundings. In real experiments the molecules under study are more or less influenced by the surroundings which can consist of matter [33] or black-body radiation photons [34]. Although two-body interaction induced spectra can be, and of course are, extracted via a virial expansion of the scattered intensity [35] the effect of black-body radiation photons has never been taken into account. Except for P_4 where the measuring temperature is about 1000 K [27] its influence on the CILS spectra should be undetectable. Nevertheless, the black-body radiation effect has been observed in polarizability measurements [36, 37] and is continuously discussed in radiation-induced

fragmentation of ionic species [38]. Needless to say that it must be taken into account in ultra-precise determinations of the hyperfine structure of atomic levels which are used as the standard of the clock transition frequency in cesium [39].

(b) Frequency dependence
The frequency-dependence of the higher polarizability tensors $\hat{A}$ and $\hat{E}$ has not been studied experimentally. It is to be expected that the dispersion at the usual frequencies employed in the experiments (514.5 nm and 458 nm) is much smaller than the experimental errors. Any serious comparison between experiment and theory must account for the dispersion of these properties, however. Sum-over-states formulae describing the influence of the measuring frequency on $\hat{A}$ are given by Buckingham [1]. Recently Kedziora and Schatz [2] and Quinet et al. [40] have made a quantum-chemical investigation on the frequency dependence of $\hat{A}$. Its relative change is in the same order as for the linear dipole polarizability α.

(c) Intramolecular motions
Experimental investigations with real molecules at finite temperatures necessarily incorporate the influence of nuclear motions on the measured properties. Therefore, measurements of $\hat{A}$ and $\hat{E}$ at different temperatures should yield different values. However, no systematic investigations of a possible temperature-dependence of $\hat{A}$ and $\hat{E}$ are known, although CILS spectra of methane at different temperatures are reported in the literature [41]. Also from a theoretical point of view the influence of nuclear motions on the field gradient polarizabilities are, with a few exceptions given in e.g. [2, 40, 42, 43], not very well known. Since calculations are mostly carried out with a frozen geometry a comparison between experiment and theory is not straightforward.

2. Performance and evaluation of collision-induced light scattering (CILS) experiments

(a) Intermolecular interaction potential
Beside the influence of the spectroscopic and electric properties of the molecules under study, CILS spectra depend on the intermolecular interaction potential of the molecules. This means that a proper evaluation of $\hat{A}$ and $\hat{E}$ is not possible unless the intermolecular interaction potential $U(R)$ is known. This is examplified in Fig. 4, where line shape calculations of the DID term of the CILS spectrum of CCl_4 with different functions of $U(R)$ are presented. The difference between the spectra is striking. Only $U_1(R)$ is able to reproduce experimentally recorded CILS spectra of CCl_4 [44]. This dependence on the potential energy function was also discussed in several papers by Le Duff et al. [26, 45, 46] and El-Kader [47, 48] who concentrated on the evaluation of the higher-order polarizabilities of CH_4, CF_4 and SF_6.

(b) Mutual dependence of parameters
The higher polarizabilities $\hat{A}$ and $\hat{E}$ as well as their derivatives $\partial\hat{A}/\partial Q$ and $\partial\hat{E}/\partial Q$ are obtained by fitting simulated spectra to the experimental recordings. Different procedures are in use. Common to all of them is that these parameters cannot be determined independently. This is demonstrated most clearly by Le Duff and co-workers [52, 46], who used a set inversion technique for the determination of the molecular properties. One result is presented in Fig.5 for SF_6 where the area of possible solutions is given that reproduces the experimental results.

(c) Pair-polarizability model
Beside the intermolecular interaction potential $U(R)$ the incremental pair polarizability tensor $\Delta\hat{\alpha}_{12}(R)$ determines the intensity and shape of the CILS spectra. It is common practice to devide $\Delta\hat{\alpha}_{12}(R)$ in two parts, a short-range and a long-range contribution.

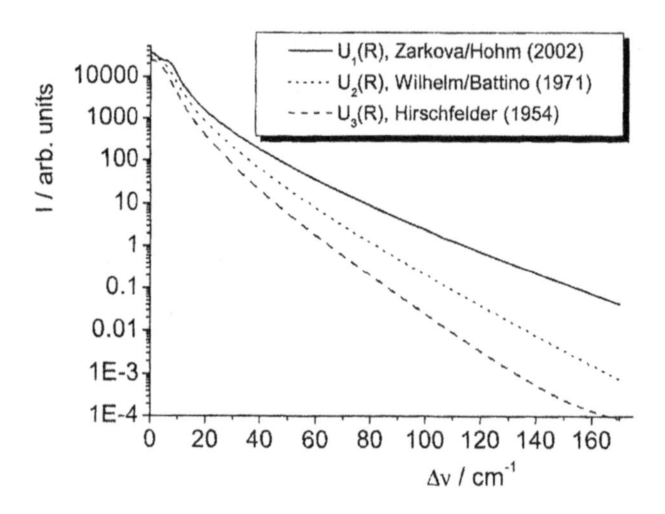

Figure 4: Influence of the potential energy function $U(R)$ of CCl_4 on the calculated collision-induced lineshapes of the DID contribution (Zarkova/Hohm (2002): Ref. [49], Wilhelm/Battino (1971): Ref. [50], Hirschfelder (1954): Ref. [51]).

i. Long-range contributions
It seems that at least in the case of atoms the long range contributions to the incremental pair-polarizability tensor are more or less fully understood [25, 53, 54, 55]. However the effect of retardation seems not yet be included in theories of $\Delta\hat{\alpha}(R)$.

ii. Short-range contributions
In order to evaluate CILS experiments the functional form of the incremental pair-polarizability anisotropy $\Delta\hat{\alpha}(R)$ as a function of the geometry of the interacting pair is necessary. However, especially the short range part dominated by electron repulsion is generally not known with sufficient accuracy. It contributes mostly to the wings of the CILS spectra.
In recent studies on the collision-induced *absorption* of methane the effect of frame distortion in molecular collisions was discussed [56, 57]. Up to now this effect was not explicitly taken into account when studying CILS spectra of molecules. However, this might explain the increasing discrepancy between experiment and theory when the number of intramolecular modes increases, like in $C_{10}H_{16}$, $Ge(CH_3)_4$ and $Sn(CH_3)_4$ [31, 32, 58, 59].

(d) Scattered intensity
Most experimental studies on the CILS spectra do invoke a thorough determination of the scattered intensity in absolute units. This is done by calibration with external (e.g. rotational Raman lines of nitrogen or hydrogen [60, 61]) and internal (ν_1 vibrational mode of the molecule under study [44]) standards. However, there are some investigations where such an internal standard due to a lack of experimental and theoretical data does not exist and where the application of an external standard is experimentally

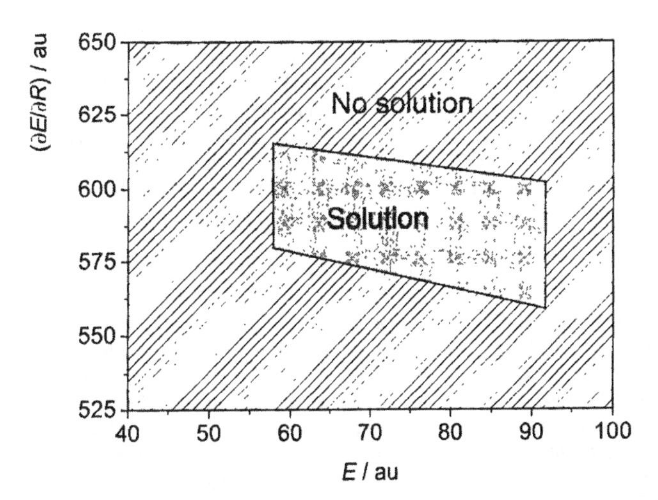

Figure 5: Area of possible solutions for E and $\partial E/\partial R$ describing the collision-induced contribution to the ν_1-band of SF_6 [46].

not possible [31]. Unfortunately, in such cases only a fit to the experimentally recorded line-shape is possible. This leads to a higher uncertainty in the experimentally determined properties like the higher order polarizabilities and the intermolecular interaction potential.

(e) Symmetry and number of independent tensor components
CILS experiments are mostly carried out with molecules of T_d, O_h or other suitable symmetries (see e.g. [30, 45, 46, 62, 63, 64, 65, 66, 67]) where the polarizability tensors $\hat{A}$ and $\hat{E}$ reduce to a single scalar quantity [68]. Due to the lower symmetry in some cases only combinations of various tensor-elements become available in experiments [69, 70]. It is questionable if in the case of molecules with lower symmetry all independent elements could be determined. Moreover, the strict symmetry requirements in the theoretical treatment of the molecules are not always be present in real experiments. The problem of the influence of free rotation or libration of e.g. methyl groups in molecules like $C(CH_3)_4$ or $Sn(CH_3)_4$ on their CILS spectra is still not examined and might explain the observed discrepancy between experiment and theory in that cases [31, 32, 59].

(f) Purity
The purity of the sample in CILS experiments is of special importance. Usually collision-induced effects are small and can easily be masked by allowed Raman signals produced by impurities. Sometimes like in CF_4 the purity of the sample can be enhanced to values exceeding 99.995% [61]. In other cases like OsO_4 [29] such purities cannot be produced with standard methods and care has to be taken in evaluating the measured CILS spectra.

3. Intermolecular interactions

(a) Polarizabilities from long-range intermolecular interactions
By probing intermolecular interactions via spectroscopy [71, 72], thermophysical properties [73], or scattering experiments [74] information on the attractive dispersion-

interaction energy $U_{\text{Disp}}(R)$ becomes available. This has been done for methane, CH_4. By analyzing $U_{\text{Disp}}(R)$ the R^{-7} part of the interaction energy can be shown to be proportional to dipole-quadrupole polarizability A [1]. Therefore suitable fitting techniques might extract A from these experiments. In my opinion the difficulty is that the coefficients must be extracted from a truncated power series in $1/R$. This truncation might influence the numerical values obtained for the expansion coefficients [75], possibly leading to small and unpredictable errors in the dipole-quadrupole polarizability A.

4 Conclusion

Although much theoretical and experimental work has been done improvements in the determination of the higher-order polarizabilities are still possible. Care has to be taken when comparing experimental and theoretical data. Especially in the case of molecules it is not always clear if one is comparing ' apples to oranges '. It may be so that in the near future the theoretical calculations of these delicate quantities surpass the experimental capabilities, at least as long as one is interested in properties determined for fixed intramolecular geometries. However, experiments still reflect the real behaviour of molecules in a more direct way, although sometimes not all parts of the experiments are fully understood.

References

[1] A. D. Buckingham, Permanent and Induced Molecular Moments and Long-Range Intermolecular Forces, *Adv. Chem. Phys.* **12** 107-142 (1967).

[2] G. S. Kedziora and G. C. Schatz, Calculating dipole and quadrupole polarizabilities relevant to surface enhanced Raman spectroscopy, *Spectrochimica Acta A* **55** 625-638 (1999).

[3] A. D. Buckingham, Direct Method for Measuring Molecular Quadrupole Moments, *J. Chem. Phys.* **30** 1580-1585 (1959).

[4] O. L. De Lange and R. E. Raab, On the Theory of the Buckingham Effect, *Mol. Phys.* **104** 607-611 (2006).

[5] A. D. Buckingham and R. E. Raab, Electric-field-induced differential scattering of right and left circularly polarized light, *Proc. Roy. Soc. A* **345** 365-377 (1975).

[6] L. Frommhold, *Collision-Induced Absorption in Gases*, Cambridge University Press, Cambridge 1993.

[7] A. D. Buckingham and G. C. Tabisz, Collision-induced rotational Raman scattering by tetrahedral and octahedral molecules, *Mol. Phys.* **36** 583-596 (1978).

[8] D. Bellert and W. H. Breckenridge, Bonding in Ground-State and Excited-State $A^+ \cdot$ Rg van der Waals Ions (A= Atom, Rg=Rare-Gas Atom): A Model-Potential Analysis, *Chem. Rev.* **102** 1595-1622 (2002).

[9] M. O. Bulanin and I. M. Kislyakov, Dynamic Polarizabilities of Rare-Gas Atoms. Krypton and Xenon, *Opt. Spectrosc.* **85** 819-825 (1998).

[10] M. O. Bulanin and I. M. Kislyakov, Dynamic Polarizabilities of Rare-Gas Atoms. Helium, Neon, and Argon, *Opt. Spectrosc.* **86** 632-639 (1999).

[11] D. Gugan and G. W. Michel, Measurements of the polarizability and of the second and third virial coefficients of ^{4}He in the range 4.2-27.1K, *Mol. Phys.* **39** 783-785 (1980).

[12] D. Gugan and G. W. Michel, Dielectric constant gas thermometry from 4.2 to 27.1 K, *Metrologia* **16** 149-167 (1980).

[13] U. Hohm and K. Kerl, Interferometric measurements of the dipole polarizability α of molecules between 300 K and 1100 K. I. Monochromatic measurements at λ=632.99 nm for the noble gases and H_2, N_2, O_2, and CH_4, *Mol. Phys.* **69** 803-817 (1990).

[14] U. Hohm, Experimental determination of the dispersion in the mean linear dipole-polarizability $\alpha(\omega)$ of small hydrocarbons and evaluation of Cauchy moments between 325 and 633 nm, *Mol. Phys.* **78** 929-941 (1993).

[15] U. Hohm, A. Loose, G. Maroulis, and D. Xenides, Combined experimental and theoretical treatment of the dipole polarizability of P_4 clusters, *Phys. Rev. A* **61** 053202-1 - 053202-6 (2000).

[16] A. Ballard, K. Bonin, and J. Louderback, Absolute measurement of the optical polarizability of C_{60}, *J. Chem. Phys.* **113** 5732-5735 (2000).

[17] U. Hohm and A. Loose, Mean dipole-polarizability of the tris(cyclopentadienides) of neodymium, samarium and erbium, $Nd(C_5H_5)_3$, $Sm(C_5H_5)_3$, and $Er(C_5H_5)_3$, *Chem. Phys. Lett.* **348** 375-380 (2001).

[18] A. Miffre, M. Jacquey, M. Büchner, G. Trénec, and J. Vigué, Atom interferometry measurement of the electric polarizability of lithium, *Eur. Phys. J. D* **38** 353-365 (2006).

[19] A. Miffre, M. Jacquey, M. Büchner, G. Trénec, and J. Vigué, Measurement of the electric polarizability of lithium by atom interferometry, *Phys. Rev. A* **73** 011603-1 - 011603-4 (2006).

[20] G. S. Sarkinov, I. L. Beigman, V. P. Shevelko, and K. W. Struve, Interferometric measurements of dynamic polarizabilities for metal atoms using electrically exploding wires in vacuum, *Phys. Rev. A* **73** 042501-1 - 042501-8 (2006).

[21] S. Kielich, General Molecular Theory and Electric Field Effects in Isotropic Dielectrics, *Dielectric and Rel. Mol. Processes* **1** 192-277 (1972).

[22] A. Chrissanthopoulos, U. Hohm, and U. Wachsmuth, Frequency-dependence of the polarizability anisotropy of CO_2 revisited, *J. Mol. Struct.* **526** 323-328 (2000).

[23] D. P. Shelton and J. E. Rice, Measurements and Calculations of the Hyperpolarizabilities of Atoms and Small Molecules in the Gas Phase, *Chem. Rev.* **94** 3-29 (1994).

[24] L. Frommhold, Collision-Induced Scattering of Light and the Diatom Polarizabilities, *Adv. Chem. Phys.* **46** 1-72 (1981).

[25] T. Bancewicz, Y. Le Duff, and J.-L. Godet, Multipolar polarizabilities from interaction-induced Raman scattering, *Adv. Chem. Phys.* **119** 267-307 (2001).

[26] T. Bancewicz, K. Nowicka, J.-L. Godet, and Y. Le Duff, Multipolar polarizations of methane from isotropic and anisotropic collision-induced light scattering, *Phys. Rev. A* **69** 062704-1 - 062704-8 (2004).

[27] U. Hohm, Higher-order dipole-polarizabilities and intermolecular interaction potential of P_4 clusters obtained from collision-induced light scattering measurements, *Chem. Phys. Lett.* **311** 117-122 (1999).

[28] H. A. Posch, Collision induced light scattering from fluids composed of tetrahedral molecules. III. Neopentane vapour, *Mol. Phys.* **46** 1213-1230 (1982).

[29] U. Hohm and G. Maroulis, Dipole-quadrupole and dipole-octopole polarizability of OsO_4 from depolarized collision-induced light scattering experiments, ab initio and density functional theory calculations *J. Chem. Phys.* **121** 10411-10418 (2004).

[30] U. Hohm and G. Maroulis, Experimental and theoretical determination of the dipole-quadrupole and dipole-octopole polarizabilities of the group IV tetrachlorides $TiCl_4$, $ZrCl_4$, and $HfCl_4$, *J. Chem. Phys.* **124** 124312-1 - 124312-6 (2006).

[31] U. Hohm, Dipole-quadrupole - and dipole-octopole polarizability of $Sn(CH_3)_4$: Experiment and bond polarizability model, *Chem. Phys. Lett.* **425** (2006) 242-245.

[32] G. Maroulis and U. Hohm, in preparation (2006).

[33] D. M. Bishop, Effect of the surroundings on atomic and molecular properties, *Int. Rev. Phys. Chem.* **13** 21-39 (1994).

[34] U. Hohm, The Influence of Matter and Black-Body Radiation Photons on the Dipole Polarizabilities α and γ of atoms, *Z. Naturforsch.* **51a** 805-808 (1996).

[35] M. Moraldi, M. Celli, and F. Barocchi, Theory of virial expansion of correlation functions and spectra: Application to interaction-induced spectroscopy, Phys. Rev A **40** 1116-1126 (1989).

[36] U. Hohm and U. Trümper, Temperature dependence of the dipole polarizability of xenon ($^1S_0$ due to dynamic non-resonant Stark effect caused by black body radiation, *Chem. Phys.* **189** 443-449 (1994).

[37] U. Hohm and U. Trümper, Frequency Dependence of the Polarizability Derivative $[\partial\alpha(\omega)/\partial r]_0$ of Chlorine from Temperature-Dependent Gas-Phase Refractive Index Measurements, *J. Raman Spectrosc.* **26** 1059-1065 (1995).

[38] B. S. Fox, M. K. Beyer, and V. E. Bondybey, Black Body Fragmentation of Cationic Ammonia Clusters, *J. Phys. Chem. A* **105** 6386-6392 (2001).

[39] S. Ulzega, A. Hofer, P. Moroshkin, and A. Weis, Stark effect of the cesium ground state: electric tensor polarizability and shift of the clock transition frequency Los Alamos National Laboratory, Preprint Archive, Physics (2006), 1-4, arXiv:physics/0604233.

[40] O. Quienet, V. Liégeois, and B. Champagne, TDHF Evaluation of the Dipole-Quadrupole Polarizability and Its Geometrical Derivatives, *J. Chem. Theor. Comp.* **1** 444-452 (2005).

[41] F. Barocchi, A. Guasti, M. Zoppi, S. M. El-Sheikh, G. C. Tabisz, and N. Meinander, Temperature dependence of the collision-induced light scattering by CH_4 gas, *Phys. Rev. A* **39** 4537-4544 (1989).

[42] G. Maroulis, A systematic study of basis set, electron correlation, and geometry effects on the electric multipole moments, polarizability, and hyperpolarizability of HCl, *J. Chem. Phys.* **108** 5432-5448 (1998).

[43] D. M. Bishop, Molecular vibrational and rotational motion in static and dynamic electric fields, *Rev. Mod. Phys.* **62** 343-374 (1990).

[44] U. Hohm, Depolarized collision-induced light scattering of gaseous CCl_4, *Chem. Phys. Lett.* **379** 380-385 (2003).

[45] J.-L. Godet, A. Elliasmine, Y. Le Duff, and T. Bancewicz, Isotropic collision-induced scattering by CF_4 in a Raman vibrational band, *J. Chem. Phys.* **110** 11303-11308 (1999).

[46] Y. Le Duff, J.-L. Godet, T. Bancewicz, and K. Nowicka, Isotropic and anisotropic collision-induced light scattering by gaseous sulfur hexafluoride at the frequency region of the ν_1 vibrational Raman line, *J. Chem. Phys.* **118** 11009-11016 (2003).

[47] M. S. A. Kader, Empirical intermolecular potential from depolarized interaction-induced light scattering spectra for tetrafluoromethane, *Chem. Phys.* **281** 49-60 (2002).

[48] M. S. A. El-Kader, S. M. El-Sheikh, and M. Omran, Contributions of Multipolar Polarizabilities to the Isotropic and Anisotropic Light Scattering Induced by Molecular Interactions in Gaseous Sulfur Hexafluoride, *Z. Phys. Chem.* **218** 1197-1212 (2004).

[49] L. Zarkova and U. Hohm, pVT-Second Virial Coefficients $B(T)$, Viscosity $\eta(T)$, and Self-Diffusion $\rho D(T)$ of the Gases: BF_3, CF_4, SiF_4, CCl_4, $SiCl_4$, SF_6, MoF_6, WF_6, UF_6, $C(CH_3)_4$, and $Si(CH_3)_4$ Determined by Means of an Isotropic Temperature-Dependent Potential, *J. Phys. Chem. Ref. Data* **31** 183-216 (2002).

[50] E. Wilhelm and R. Battino, Estimation of (6,12) Lennard-Jones Potential Parameters From Gas Solubility Data, *J. Chem. Phys.* **55** 4012-4017 (1971).

[51] J. O. Hirschfelder, C. F. Curtiss, and R. B. Bird, *Molecular Theory of Gases and Liquids*, John Wiley & Sons, New York 1954.

[52] L. Jaulin, J.-L. Godet, E. Walter, A. Elliasmine, and Y. Le Duff, Light scattering data analysis via set inversion, *J. Phys. A: Math. Gen.* **30** 7733-7738 (1997).

[53] M. Chrysos, S. Dixneuf, and F. Rachet, On a singularity-free pair polarizability anisotropy-model of atomic gases, *J. Chem. Phys.* **124** 234303-1 - 234303-4 (2006).

[54] S. M. El-Sheikh, G. C. Tabisz, and A. D. Buckingham, Collision-induced light scattering by isotropic molecules: the role of the quadrupole polarizability, *Chem. Phys.* **247** 407-412 (1999).

[55] M. Chrysos and S. Dixneuf, Accurate pair-anisotropy models out of a set of input experimental moments via a novel non-linear inversion method, *J. Raman Spectr.* **36** 158-164 (2005).

[56] M. Buser and L. Frommhold, Infrared absorption by collisional $CH_4 + X$ pairs, with X=He, H_2, or N_2, *J. Chem. Phys.* **122** 024301-1 - 024301-7 (2005).

[57] M. Buser and L. Frommhold, Collision-induced rototranslational absorption in compressed methane gas, *Phys. Rev. A* **72** 042715-1 - 042715-7 (2005).

[58] G. Maroulis, D. Xenides, U. Hohm, and A. Loose, Dipole, dipolequadrupole, and dipoleoctopole polarizability of adamantane, $C_{10}H_{16}$, from refractive index measurements, depolarized collision-induced light scattering, conventional ab initio and density functional theory calculations, *J. Chem. Phys.* **115** 7957-7967 (2001)

[59] G. Maroulis, private communication (2006).

[60] T. Bancewicz, V. Teboul, and Y. Le Duff, High-frequency interaction-induced rototranslational wings of isotropic nitrogen spectra, *Mol. Phys.* **81** 1353-1372 (1994).

[61] A. Elliasmine, J.-L. Godet, Y. Le Duff, and T. Bancewicz, Collision-induced depolarized light scattering by CF_4, *Mol. Phys.* **90** 147-157 (1997).

[62] A. Elliasmine, J.-L. Godet, Y. Le Duff, and T. Bancewicz, Isotropic collision-induced light scattering by gaseous CF_4, *Phys. Rev. A* **55** 4230-4237 (1997).

[63] S. M. El-Sheikh and G. C. Tabisz, Perturber dependence of the collision-induced light scattering by SF_6 and CF_4, *Mol. Phys.* **68** 1225-1238 (1989).

[64] D. P. Shelton and G. C. Tabisz, Binary collision-induced light scattering by isotropic molecular gases. II. Molecular spectra and induced rotational Raman scattering, *Mol. Phys.* **40** 299-308 (1980).

[65] D. P. Shelton and G. C. Tabisz, A comparison of the collision-induced light scattering by argon and by isotropic molecular gases, *Can. J. Phys.* **59** 1430-1433 (1981).

[66] N. Meinander, An isotropic intermolecular potential for sulfur hexafluoride based on the collision-induced light scattering spectrum, viscosity, and virial coefficient data, *J. Chem. Phys.* **99** 8654-8667 (1993).

[67] J.-L. Godet, F. Rachet, Y. Le Duff, K. Nowicka, and T. Bancewicz, Isotropic collision induced light scattering spectra from gaseous SF_6, *J. Chem. Phys.* **116** 5337-5340 (2002).

[68] P. W. Fowler, H. M. Kelly, and N. Vaideh, A model for multipole polarizabilities and dispersion coefficients, *Mol. Phys.* **82** 211-225 (1994).

[69] T. Bancewicz, V. Teboul, and Y. Le Duff, High-frequency interaction-induced rototranslational scattering from gaseous nitrogen, *Phys. Rev. A* **46** 1349-1356 (1992).

[70] A. De Lorenzi, A. De Santis, R. Frattini, and M. Sampoli, Induced contributions in the Rayleigh band of gaseous H_2S, *Phys. Rev. A* **33** 3900-3912 (1986).

[71] S. Rajan, K. Lalita, and S. V. Babu, Nuclear spin-lattice relaxation in CH_4-insert gas mixtures, *J. Magn. Res.* **16** 115-129 (1974).

[72] H. Hoshina, T. Wakabayashi, T. Momose, and T. Shida, Infrared spectroscopic study of rovibrational states of perdeuterated methane (CD_4) trapped in parahydrogen crystal, *J. Chem. Phys.* **110** 5728-5733 (1999).

[73] P. Isnard, D. Robert, and L. Galatry, On the determination of the intermolecular potential between a tetrahedral molecule and an atom or a linear or a tetrahedral molecule-application to CH_4 molecule, *Mol. Phys.* **31** 1789-1811 (1976).

[74] U. Buck, J. Schleusener, D. J. Malik, and D. Secrest, On the argon-methane interaction from scattering data, *J. Chem. Phys.* **74** 1707-1717 (1981).

[75] U. Hohm, On Polynomial Gaussian Least-Squares Fits and Interpretation of the Resulting Fit-Parameters, *Z. Naturforsch.* **48a** 878-882 (1993).

Brill Academic Publishers
P.O. Box 9000, 2300 PA Leiden,
The Netherlands

Lecture Series on Computer and Computational Sciences
Volume 6, 2006, pp. 381-384

Lagrange-like error formula in exponential fitting

L. Gr. Ixaru[1]

"Horia Hulubei" National Institute of Physics and Nuclear Engineering,
Department of Theoretical Physics,
P.O. Box MG-6, Platforma Magurele,
Bucharest, Romania

Received July 13, 2006; accepted July 20, 2006

Abstract: Here we report on a recent result of Coleman and Ixaru [1] that the error for linear approximations constructed by exponential fitting for oscillatory problems can be expressed as a sum of two Lagrange-like terms, which in some cases can be reduced to only one term. The new expression enables a correct prediction for the asymptotic behaviour of the error when the frequency is increased, in contrast to the prediction produced by the leading term in the formal series of the error, as largely used in the existing literature. Two quadrature rules are investigted in the new frame, that is the extended Newton-Cotes and the Gaussian rule, respectively, to show that the latter is clearly more advantageous at big frequencies. In contrast the analysis based on the leading term does not distinguish between these rules.

Keywords: Error formula, exponential fitting

Mathematics Subject Classification: AMS: 65D30, 65D32, 65D20, 65L70

1 Formulation of the problem and the main theoretical result

When the value of a function f at $X+h$ is approximated by a truncated Taylor expansion about X, that is by $f_K(X+h) = \sum_{k=0}^{K} h^k f^{(k)}(X)/k!$, the resulting error may be expressed in the Lagrange form

$$err := f(X+h) - f_K(X+h) = \frac{h^{K+1}}{(K+1)!} f^{(K+1)}(\eta)\,, \tag{1}$$

for some $\eta \in (X,\, X+h)$, if $f^{(K+1)}(x)$ is continuous on $[X,\, X+h]$. That error may also be written, less usefully, as the formal expansion

$$err = \sum_{k=K+1}^{\infty} \frac{h^k}{k!} f^{(k)}(X)\,. \tag{2}$$

Expressions of Lagrange type are also available for the truncation errors in some polynomial-based approximations. For example, the error of the simplest approximation for the first derivative,

$$f'(X) \approx \frac{1}{2h}[f(X+h) - f(X-h)]\,, \tag{3}$$

[1] E-mail: ixaru@theor1.theory.nipne.ro

has the Lagrange-like expression

$$err = -\frac{1}{6}h^2 f^{(3)}(\eta)\,, \tag{4}$$

for $\eta \in (X-h, X+h)$. A formal expansion as in eq.(2) can also be written and, in both cases considered above, the expression of the leading term is the same as that in the Lagrange form except that η should be replaced by X.
The exponential fitting (ef for short) is a powerful technique for the construction of linear approximation formulae when f has some special behaviour, in particular when it is an oscillatory function. If

$$f(x) = f_1(x)\sin(\omega x) + f_2(x)\cos(\omega x) \tag{5}$$

with smooth f_1 and f_2, then the slightly modified formula

$$f'(X) \approx \frac{1}{2h}\frac{\theta}{\sin(\theta)}[f(X+h) - f(X-h)],\ \text{where}\ \theta = \omega h\,, \tag{6}$$

becomes appropriate; it tends to the former when $\omega \to 0$. As a matter of fact, this is perhaps the simplest example of a formula to be derived by exponential fitting. The leading term of the error is now

$$lte = h^2 T_0^*[f^{(3)}(X) + \omega^2 f'(X)]\,,\ T_0^* = \frac{\sin(\theta) - \theta}{\theta^2 \sin(\theta)}\,, \tag{7}$$

see [3], but is it any basis to expect that the true err is of the same form with X replaced by η ? The problem is important because the characterisation of the error in terms of the leading term may often fail: if X is such that the factor in square brackets happens to vanish, then eq.(7) will suggest that the computed $f'(X)$ is exact but this may not be true. In spite of this only the expressions of the leading terms are provided for most of the ef-based approximation formulae developed in the literature for a large variety of operations including numerical differentiation, quadrature, solving differential equations, interpolation, see [4].
The problem of whether a Lagrange-like formula for the error can be derived for these approximations has been considered recently by Coleman and Ixaru [1]. To fix the ideas, let us focus on a linear approximation formula over some interval $[a, b]$ which uses data at N points situated in this interval. We use $X = (b+a)/2$, $h = (b-a)/2$ and identify the positions of the mesh points by $x_n = X + x_n^* h$. We assume that the formula has been constructed upon condition that this is exact if f is any solution of the differential equation $Lf = 0$ where L is the linear operator

$$L = \sum_{k=0}^{m} a_k(x) D^{m-k}\,,\ x \in [a, b]\,,\ \text{where}\ \ D^p = \frac{d^p}{dx^p}\,,$$

with $a_0(x) = 1$. By the very construction the leading term of the error reads

$$lte = h^{m+l} T_0^* Lf(X)\,,$$

where T_0^* depends on coefficients a_k, weights x_n^* (hereinafter collected in vectors $\mathbf{a}$ and $\mathbf{x}^*$, respectively) and h, and l depends on the formula type: $l = -1, -2$ for first or second order derivative, respectively, 0 for interpolation, 1 for quadrature etc. Just for illustration, in formula (6) the interval is $[a = X-h, b = X+h]$. It consists in two equidistant steps, and we also have $m = 3$, $l = -1$, $a_0 = 1$, $a_1 = a_3 = 0$, $a_2 = \omega^2$. The same holds for the classical approximation (3) except for that now $a_2 = 0$.
The result of Coleman and Ixaru is:
The error whose leading term is of the mentioned form can be always written as a sum of two Lagrange-like terms

$$err = h^{m+l}[T_+^* Lf(\eta_+) + T_-^* Lf(\eta_-)], \tag{8}$$

for some $\eta_+, \eta_- \in (a, b)$, which depend on $\mathbf{a}$, $\mathbf{x}^*$, h *and on the function* f. *The functions* $T_+^*(\mathbf{a}, \mathbf{x}^*, h)$ *and* $T_-^*(\mathbf{a}, \mathbf{x}^*, h)$, *associated to* $T_0^*(\mathbf{a}, \mathbf{x}^*, h)$, *are uniquely determined by some procedure explained in that paper. Two important properties of them are*

$$T_+^*(\mathbf{a}, \mathbf{x}^*, h) \geq 0\,, \; T_-^*(\mathbf{a}, \mathbf{x}^*, h) \leq 0 \tag{9}$$

and

$$T_+^*(\mathbf{a}, \mathbf{x}^*, h) + T_-^*(\mathbf{a}, \mathbf{x}^*, h) = T_0^*(\mathbf{a}, \mathbf{x}^*, h)\,. \tag{10}$$

When one of these functions is strictly zero the error gets the standard one-term Lagrange-like form.

This result, whose derivation uses for start the work of Ghizzetti and Ossicini [2] concerning a class of quadrature formulae, is more general than needed in this contribution. It assumes that the coefficients a_k depend on x, while in the exponential fitting these are simply constants.

2 Applications

We consider two ef-based quadrature rules.

- Extended Newton-Cotes rule, [5], [4]:

$$\int_a^b f(x)dx = \int_{X-h}^{X+h} f(x)dx \approx h \sum_{n=1}^{N} [a_n^{(0)} f(X + x_n^* h) + h a_n^{(1)} f'(X + x_n^* h)]\,, \tag{11}$$

on evenly-spaced abscissas $x_n^* = 2(n-1)/(N-1) - 1$ $(n = 1, 2, \ldots, N)$. The rule is called extended because it uses the values of f and its derivative.

- Gauss rule, [6], [4]:

$$\int_a^b f(x)dx = \int_{X-h}^{X+h} f(x)dx \approx h \sum_{n=1}^{N} a_n^{(0)} f(X + x_n^* h)\,. \tag{12}$$

The $2N$ coefficients, that is $a_n^{(0)}$, $a_n^{(1)}$ for the first rule, and $a_n^{(0)}$, x_n^* for the second $(n = 1, \ldots, N)$ are determined from the condition that the rule is exact if f satisfies $Lf = 0$ for

$$L = (D^2 + \omega^2)^N = h^{-2N}(D^{*2} + \theta^2)^N.$$

In the last member we have used the dimensionless $x^* = x/h$ and $D^{*p} = d^p/dx^{*p} = h^p D^p$. Both rules are exact if the integrand f is of form (5) where f_1, f_2 are polynomials of degree $N-1$ or less.

The coefficients of each rule will depend on θ only and the same holds for function T_0^*. Its expression is the same in both rules,

$$T_0^*(\theta) = \frac{2 - \sum_{n=1}^{N} a_n^{(0)}(\theta)}{\theta^{2N}}\,,$$

and therefore

$$lte = hT_0^*(\theta)(D^{*2} + \theta^2)^N f(X)\,.$$

Let us keep h fixed and examine the behaviour of lte when ω (or θ) tends to infinity. As shown in [4] the last factor increases as θ^N, while $T_0^*(\theta)$ decreases as θ^{-2N} (because $a_n^{(0)}(\theta) \to 0$) such that the prediction is that the error should decrease as θ^{-N}.

However, the two-term form of the error, eq.(8), leads to a different picture because the asymptotic behaviours of $T_0^*(\theta)$, on one hand, and those of its components $T_\pm^*(\theta)$, on the other, are not necessarily similar. Suppose that for large values of θ the functions $T^*_\pm(\theta)$ are well described by the approximation

$$T^*_\pm(\theta) \approx \pm c(\theta)\theta^{-(2N-\bar{N})} + c_\pm(\theta)\theta^{-2N}, \tag{13}$$

where $0 < \bar{N} < 2N$ and the functions $c(\theta)$ and $c_\pm(\theta)$, with $c_+(\theta) \neq -c_-(\theta)$, are oscillating between constant limits; think, for example, of the case where $c(\theta) = c_+(\theta) = 1 + \cos\theta$ and $c_-(\theta) = -1 + \cos\theta$. Then $T^*_\pm(\theta)$ will damp out as $\theta^{-(2N-\bar{N})}$, that is more slowly than their sum $T^*_0(\theta)$ which decays as θ^{-2N}. The determination of this $\bar{N}$ therefore represents a key issue if we want to characterize the asymptotic behaviour of the error by means of equation (8).
The investigations of Coleman and Ixaru have shown that $\bar{N} = N - 2$ is the right value for the extended Newton-Cotes rule but $\bar{N} = \lfloor (N-1)/2 \rfloor$ for the Gauss rule, where $\lfloor u \rfloor$ is the biggest integer less than or equal to u. For $N \leq 6$ these values are $\bar{N} = 0$ for $N = 2$, $\bar{N} = 1$ for $N = 3$, 4, and $\bar{N} = 2$ for $N = 5$, 6. Thus for $N \geq 4$ they are smaller than the corresponding values for the Newton-Cotes rules.

In short the prediction resulted from the new error formula is that the error should decrease as $\theta^{-(N-\bar{N})}$ which is slower than the rule θ^{-N} suggested by the behaviour of the leading term. It also indicates that the decrease is as θ^{-2} irrespective of N for the extended Newton-Cotes rule but it is faster and faster when N is increased for the Gauss rule, viz.: θ^{-2} for $N = 2$, 3, θ^{-3} for $N = 4$, 5 etc. [Numerical evidence (not provided in this short text but to be included in an extended version) clearly shows that these predictions are correct.] This suggests that exponential-fitting quadrature methods, and particularly those of Gaussian type, are worthy of further investigation as practical methods for the numerical integration of oscillatory integrands.

References

[1] J. P. Coleman and L. Gr. Ixaru, Truncation errors in exponential fitting for oscillatory problems, *SIAM J. of Numerical Analysis* (in press)

[2] A. Ghizzetti and A. Ossicini, *Quadrature Formulae*, Birkhaüser, Basel, 1970.

[3] L. Gr. Ixaru, Operations on oscillatory functions, *Comput. Phys. Commun.* **105** 1–19(1997).

[4] L. Gr. Ixaru and G. Vanden Berghe, *Exponential Fitting*, Kluwer, Dordrecht, 2004.

[5] J. K. Kim, R. Cools and L. Gr. Ixaru, Quadrature rules using first derivatives for oscillatory integrands, *J. Comput. Appl. Math.* **140** 479–497(2002).

[6] L. Gr. Ixaru and B. Paternoster, A Gauss quadrature rule for oscillatory integrands, *Comput. Phys. Commun.* **133** 177–188(2001).

Brill Academic Publishers
P.O. Box 9000, 2300 PA Leiden
The Netherlands

Lecture Series on Computer and Computational Sciences
Volume 6, 2006, pp. 385-394

Computational Approaches to Supramolecular Functions

[a]S. J. Lee and [b]J. Y. Lee[1]

[a] Center for Computational Biology and Bioinformatics,
Korea Institute of Science and Technology Information,
Daejeon 305-806, Korea
and
[b] Department of Chemistry,
Sungkyunkwan University,
Suwon 440-746, Korea

Received 31 May 2006; accepted 2 June 2006.

Abstract: Supramolecular chemistry now has become a central of the research activities. Basically, it mostly concerns the self aggregation by non-covalent intermolecular interactions, such as hydrogen bonding and π-π stacking, and van der Waals interactions. Until very recently, computational applications were very limited because of the large systems. Several examples of the computational approaches to understand supramolecular functions will be discussed. On-Off molecular switching behavior based on the excimer emission can be understood by the frontier molecular orbitals of the host, and its guest-complexes. The fullerene-porphyrin-fullerene triad was predicted to show interesting conformational changes and different electron transport behavior upon the reduction (electron addition). Finally, the intermolecular interactions can be understood by replicating the monomer unit and manipulating the translation and rotation to predict the self-assembled structures.

Keywords: supramolecular chemistry, computational, intermolecular interactions, non-covalent, functions

Mathematics Subject Classification: 31.15.Ar

PACS: 32.50.+d, 34.20.Gj

1. Introduction

Supramolecular chemistry, molecular self-assembly, molecular recognition, crystal engineering, and other similar research areas have been extensively growing in the recent past. The packing of molecules into arrays of varying and beautiful structures is the result of non-covalent interactions such as H-bonding, □-□ stacking, and other weak interactions.[1,2] For example, the H-bonding has been used with considerable success as a stereochemical force or vector in designing molecular arrays of various shapes.[3] Many chemists have used the H-bonding as a design tool in placing the H-bond donors and acceptors in the right places on the molecular frame to achieve desirable results.[3,4] Crystal structures obtained as such determine key solid-state properties such as the bioavailability of pharmaceutical solids and the performance of electric materials. Therefore, predicting and understanding the phenomenon of crystallization or self-assembly is of great importance. However, unfortunately, the ability to routinely predict three-dimentional crystal structures of even the simplest molecules in common space groups remains a distant goal despite an intensive effort.[5] Recently, a reductionist approach has been applied to understand the crystallization which decreases the dimensionality of the problem.[6] Another active research area that needs some computational help is to understand the fluorescence and photo-induced charge transfer phenomena. Chemosensors are molecules capable of binding a given analyte selectively such that the binding event induces a measurable signal, most often

[1] Corresponding author. E-mail: jinylee@skku.edu

a change in the spectral properties of a chromophore or fluorophore.[7] In one particular design, a reporter molecule occupies the binding site for the target analyte within the receptor and signalling occurs upon competitive displacement of this reporter molecule from the binding site by the analyte. Following the pioneering work of Aviram and Ratner, there has been a recent resurgence of interest in molecular electronic devices because of their utility as electronically active elements in nano-devices.[8] Additionally, advances in manipulation of a single molecule have led to the fabrication of molecular devices[9,10] and the measurement of electronic transport properties in individual or a small number of organic molecules connected to metal electrodes. The current-voltage (I-V) characteristics of the benzene-based molecular systems were also investigated in the presence of an externally manipulated gate voltage. The observed I-V characteristics give rise to the possibility that the benzene-based molecular systems can be used as models of molecular transistors.[11] A number of chemosensors have been synthesized and interesting molecular electronic properties have been reported.

Supramolecular chemistry and self-assembly processes in particular have been applied to the development of new materials. Large structures can be readily accessed using bottom-up synthesis as they are composed of small molecules requiring fewer steps to synthesize. Thus most of the bottom-up approaches to nanotechnology. are based on supramolecular chemistry. Supramolecular chemistry is often pursued to develop new functions that cannot appear from a single molecule. These functions include magnetic properties, light responsiveness, catalytic activity, self-healing polymers, chemical sensors, etc. Supramolecular research has been applied to develop high-tech sensors, processes to treat radioactive waste, compact information storage devices for computers, high-performance catalysts for industrial processes, and contrast agents. Supramolecular chemistry is also important to the development of new pharmaceutical therapies by understanding the interactions at a drug binding site. In addition, supramolecular systems have been designed to disrupt protein-protein interactions that are important to cellular function. Research in supramolecular chemistry also has application in green chemistry where reactions have been developed which proceed in the solid state directed by non-covalent bonding. Such procedures are highly desirable since they reduce the need for solvents during the production of chemicals.

Recent advances in synthetic methods and experimental techniques produced a number of interesting chemical phenomena. However, many of those phenomena are waiting for the detailed understanding for their function at the molecular level. To this end, both theoreticians and experimentalists need to be cooperative. I have applied computational approaches to understand molecular functions observed experimentally in several subjects such as "on-off" molecular switch,[12] electrochemical machinery,[13] molecular self-assembly.[14] In this presentation, such experimental phenomena will be introduced and explained based on our computational results as representatives.

2. Computational methods

In all of our calculations and structure optimization we have used the DFT method using the non local exchange correlation functional of Beck's three parameters employing the Lee-Yang-Parr functional (B3LYP) using the Gaussian 98[15a] or 03[15b] programs. The interaction energy is simply obtained by the energy of the complex subtracted by the sum of energies of individual constituents. Generally, basis set superposition error (BSSE) correction should be included to obtain an accurate interaction energies.[16] In the case of our on-off molecular switching we have used the same exchange functional (B3LYP) and 3-21G basis set have been used except for the case of Pb^{2+} where the LANL2DZdp effective core potential (ECP) was used.

3. Recent studies in supramolecular chemistry

The field of "supramolecular chemistry" is advancing very rapidly. The knowledge gained thus far has allowed some remarkably sophisticated and effect system to be developed, however, there are many underlying principles and design rules have not been discovered. Dendrimers has been used as building blocks for self-assembly processes using both hydrophilic and hydrophobic moieties.[17] The self-organization of hyperbranched polymers in liquid crystalline phases has been explored by several research groups.[18] Stoddart and others have used the strong interaction between p-donor and a p-acceptor in the design of a variety of self-assembling systems such as rotaxanes and catenanes. Of course, hydrogen bonding interactions are ubiquitous in biological self-assembly processes and also widely employed in studies of abiotic self-assembly. Recently, Ercolani presented a model treating the competition between self-assembly and nonlinear random polymerization under thermodynamic control in solution based on the effective molarity (EM) of the assembly and the equilibrium constant for the intermolecular model reaction between monofunctional reactants (K_{inter}).[19]

There is considerable interest in the development of supramolecular systems that have the ability to bind, identify, and signal the presence of negatively charged ions as well as positively charged ions. Many theoretical studies have been devoted to understand the intermolecular interactions for those systems. It is still not easy to treat theoretically the entire experimental systems though the computational power has been exponentially grown during the last decades. Now it is the time for theoretical and computational chemist to challenge such systems by utilizing many experimental and computational results along with the help of enhanced computational capacity.

4. On-Off Molecular Switch[12]

A new fluorescent chemosensor with two different types of cation binding sites on the lower rims of a 1,3-alternate calix[4]arene (**1**) is synthesized. Two pyrene moieties linked to a cation recognition unit composed of two amide groups form a strong excimer in solution. For **1**, the excimer fluorescence is quenched by Pb^{2+}, but revived by addition of K^+ to the Pb^{2+}-ligand complex. Thus metal ion exchange produces an on-off switchable fluorescence and the chemosensor functions like a "molecular bulb." Ab initio calculations give a reasonable explanation for the function of molecular bulb observed in the host complexation experiments.

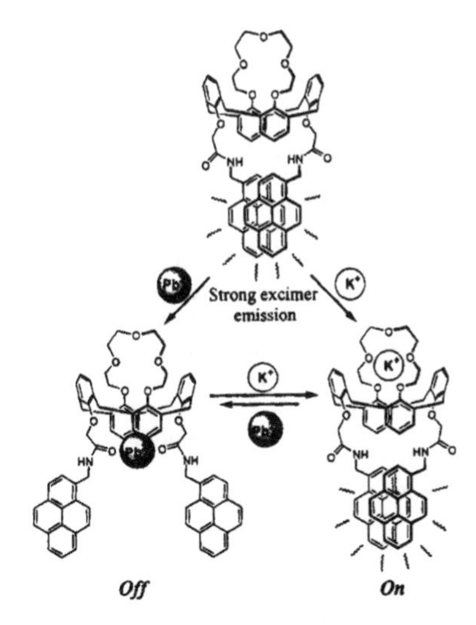

Figure 1. Schematic representation of "On-off" switching mechanism based on excimer emission of **1**.

From the various fluorescence studies of pyrene moieties, it is well-established that the monomer bands appear at 370-430 nm with wellresolved vibronic features and a broad featureless emission band peak occurs at approximately 480 nm.[20] Also, pyrene excimer fluorescence has been established as a useful spectroscopic tool to probe intra- and intermolecular interactions of exchangeable apoliproteins upon binding to lipids. Host molecules with more than one pyrenyl group exhibit intramolecular excimer emission by two different mechanisms. One results from □-□ stacking of the pyrene rings in the free state, which results in a characteristic decrease of the excimer emission intensity and a concomitant increase of monomer emission intensity. The other mechanism is due to interaction of an excited pyrene (Py*) with a ground-state pyrene (Py). Some pyrene-containing hosts exhibit excimer emission due to the former mechanism[21,22] and some due to the latter mechanism.[23]

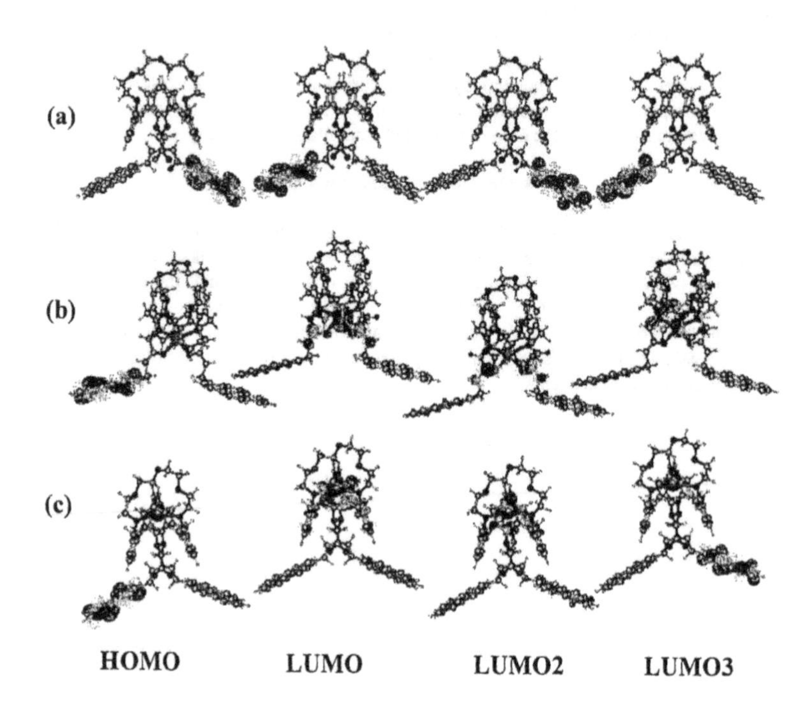

Figure 2. B3LYP/6-31G* predicted structures and molecular orbitals of (a) **1**, (b) **1•Pb^{2+}**, and (c) **1•K^{+}**

As explained in the previous subject, the mechanism of "on-off" switchable fluorescent sensor can be understood by the molecular orbitals and ground state structures. In all cases, the ground state structures do not show the stacked pyrene dimer which can easily forms an excimer upon the excitation. Thus the excimer emission mechanism does not originate from the ground state stabilization of pyrene dimer. The fluorescence intensity can be considered to be the excimer emission. Therefore, chemical systems whose excited states are stabilized can give strong emission. In this regard, we first assume that the excited states of the host molecule and its K^+ complex are stabilized through the pyrene-pyrene interactions, while its Pb^{2+} complex is not upon the excitation. This assumption can be justified from the frontier molecular orbitals because the general electronic transitions under UV irradiation (with an excitation wavelength of 344 nm) are HOMO → LUMO, HOMO → LUMO2, HOMO → LUMO3, or some transitions near them. (LUMO2 and LUMO3 represent the second and third LUMO, respectively.) In our previous work, it was found that these molecular orbitals could explain the electronic transitions quite reasonably.[24] In **1** and its complexes, the HOMO shows that most electron density resides in one of the pyrene moieties. However, the LUMOs for the Pb^{2+} complex reveal a remarkable difference from those for **1** and its K^+ complex. It is seen in Figure 5 that the excited states of **1** under UV irradiation can be stabilized by □-□ interactions, which can be evidenced by the HOMO → LUMO and HOMO → LUMO3 transitions. Similarly, the excited states for the K^+ complex can be stabilized as implicated in the HOMO → LUMO3 transitions. From this investigation of the molecular orbitals, it is apparent that the excited state (LUMO) of one of the two pyrene units shows a strong interaction with the ground state (HOMO) of the other pyrene unit through the π-π stacking in **1** and the **1•K$^+$** complex. A steady-state fluorescence spectroscopy experiment showed that the intensities of both monomer and excimer emissions increase upon addition of K^+, which excludes intramolecular excimer emission from π-π stacking of the pyrene rings in the free state and supports intramolecular excimer emission by interaction of an excited pyrene (Py*) with a ground-state pyrene (Py).[25,26] However, the Pb^{2+} complex does not show such π-π interactions up to LUMO7.

5. Axially coordinated fullerene-porphyrin-fullerene triad[13]

In our previous study, we reported the X-ray structures and spectroscopic characterizations of axially coordinated (C_{60}CHCOO)$_2$-Sn(IV) porphyrin.[27] The X-ray crystal structure revealed that the two fullerenes have *trans* conformation. There are two possible conformations in each fullerenoacetato group, *anti* and *syn*, considering the OC-CX where oxygen is axially linked to Sn(IV) and X is fullerene. (See the Scheme below) Thus, there are four possibilities in (C_{60}CHCOO)$_2$-Sn(IV) porphyrin, *anti-anti*, *anti-syn*, *syn-anti*, and *syn-syn*. Among them, *anti-syn* and *syn-anti* conformations are equivalent by symmetry, and give the *cis* conformer. Similarly, both *anti-anti* and *syn-syn* conformations are equivalent to give the *trans* one. In our previous experiment, the *trans* conformer was identified to have *syn-syn* conformation. However, the relative stability of the *cis* and *trans* conformers is not apparent. Thus, it is interesting to study on the conformational stability and electronic structures for them. Here, we try to compare the stability of the *syn-syn* and *syn-anti* conformations of (C_{60}CHCOO)$_2$-Sn(IV) porphyrin as representatives, and study their electrochemical properties by *ab initio* calculations. In the present study, we do not include the *anti-anti* conformation, which seems to be less important because the *syn-syn* was obtained as a *trans* conformer in our previous experiment, and it takes a lot of computational time due to the large molecular size.

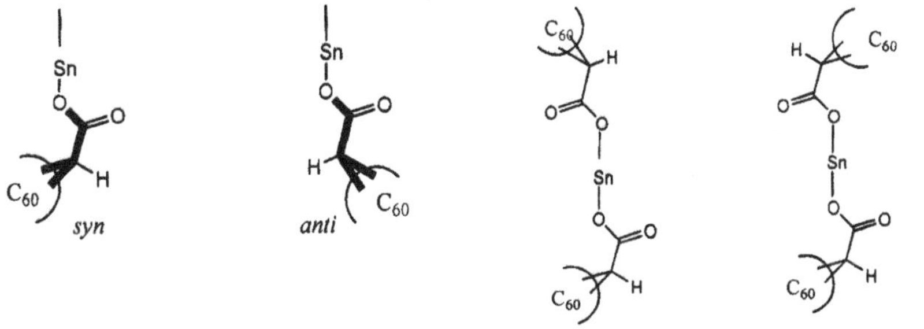

syn *anti* *syn-anti*(*cis*) *syn-syn* (*trans*)

It is very interesting to find that the *cis* conformer is found to be more stable than the *trans* one by 3.79 kcal/mol at the HF/3-21G optimization though we observed the *trans* conformer in the experiment. The HF calculations often mislead to more or less higher energy conformers due to the inherent inaccuracy in involving weak interactions such as π-π and CH...π interactions.[28] Thus, the B3LYP/3-21G* single point energy calculations have been performed at the HF/3-21G optimized geometries, and the *cis* conformer is still found to be more stable than the *trans* one by 1.38 kcal/mol. Considering that we did not carry out the full optimization at the B3LYP level of theory due to the computational limit, the relative energy of the *cis* and *trans* conformers is expected to be quite similar. In our previous experiment, the *trans* conformation was obtained in the crystallization process. It is not unusual to observe in experiment a conformer which is energetically more unstable in the calculations. In an earlier study, for example, Reinhoudt et al. reported structural studies of a different group of di(thio)urea-substituted calixarenes and found that the *syn–syn* configuration was stabilized in solution and in the gas phase.[29] However, in the case of solid state, it was found that the similar molecular family adopted the *syn–anti* configuration as a result of the steric requirements of the hydrogen bonding.[30]

We investigated the structural features and electronic structures of the *cis* and *trans* conformers of axially coordinated fullerene-porphyrin-fullerene triad (C_{60}CHCOO)$_2$-Sn(IV) porphyrin. The *cis* conformer was found to be slightly more stable than the *trans* one by 1.38 kcal/mol at the B3LYP/3-21G* single point calculation at the HF/3-21G optimized geometries. Upon the addition of an electron to the compound, the relative stability of the *cis* conformer was found to be higher (3.29 kcal/mol) than in the neutral one. Considering the relative stability is significantly changed at the B3LYP energy calculation from the HF results in the neutral compound, the relative stability of the *cis* conformer may further decrease if the full B3LYP optimization were performed, or even the *trans* conformer may become more stable. As a matter of fact, in our previous crystallization experiment, the *trans* conformer was obtained. However, the relative stability in the anionic compound does not vary much at both HF and B3LYP calculations, signifying that the relative stability of the *cis* conformer in the anionic compound may not change much. By setting up a proper experimental equipment to adopt the *trans* conformer in the neutral and to adopt the *cis* one in the anion, our title compound could be a potential candidate for a electrochemical machinery. From the investigation of frontier molecular orbitals, for the *cis* conformer, it was found that the electrons are localized at porphyrin in HOMO, while the electrons are localized at the *syn*-fullerene in LUMO. This implies that the electron can transfer from the porphyrin to the *syn*-fullerene upon the photoexcitation rather to the *anti*-fullerene. For the *trans* conformer, it was found that the electrons are localized at porphyrin in HOMO, while the electrons are localized at one of the two fullerene moieties in LUMO, the electrons are localized at the other fullerene moiety in LUMO2. But, the LUMO and LUMO2 have the same orbital energy and they are degenerate, thus the PET may take place from the porphyrin to either way through the both fullerene moieties.

6. Self-assembled structures by intermolecular interactions[14]

The stereochemical series of amino acid groups not only act as effective unit to the helical morphological control of the self-assembly, but also give a deriving force to form the self-assembled organogel by the intermolecular hydrogen-bonding interaction. With those objects in mind, we have newly synthesized a stereochemical series of *p*-nitro-azobenzene-coupled bis-alanines **1**, **2**, and **5** as shown in SI Schemes 1 and 2(Supporting Information).[14] Compounds **3** and **4** in the absence of alanine units or long alkyl chain groups were synthesized as references to confirm the critical roles for the intermolecular hydrogen-bonding interaction of the numbers of amide groups and the hydrophobic interaction between alkyl chain groups in gel formation, respectively. We herewith report on morphological control and the self-assembled behaviors of the organogels **1-3**, respectively.

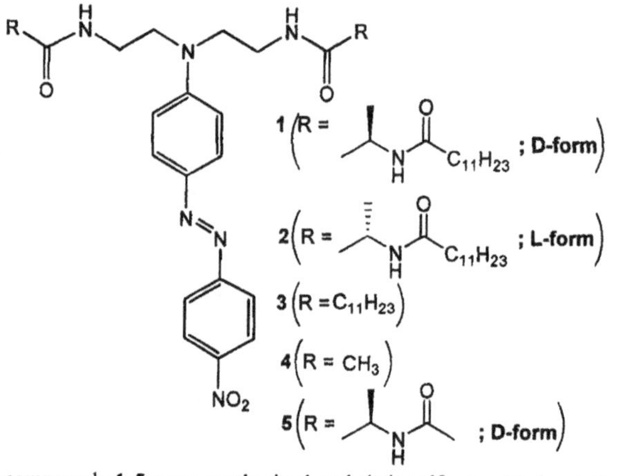

R N H N N H R O O N N NO2

1 (R = N H O $C_{11}H_{23}$; D-form)

2 (R = N H O $C_{11}H_{23}$; L-form)

3 (R = $C_{11}H_{23}$)

4 (R = CH_3)

5 (R = N H O ; D-form)

Five azobenzene compounds **1-5** were synthesized and their self-assembled gelation abilities were characterized in organic solvents. Organogelators **1-3** could gelate common organic solvents. The CD spectra of self-assembled **1** and **2** exhibit a positive and a negative sign in the first Cotton effect, respectively. Furthermore, the self-assembled organogels **1** and **2** are found to have right- and left-handed helical structures, respectively, suggesting that the macroscopic helicities of the alanine-appended gels **1** and **2** are reflected from corresponding microscopic helicities. In particular, formations of intermolecular hydrogen-bonding in the alanine unit of organogelators **1** and **2** play an intrinsic role in their fiber self-assemblies along with an efficient gelation. On the other hand, the organogels **3** shows the linear fiber structure with 50-200 nm diameter and with micrometer in length. According to FT-IR, powder-XRD, and ^{1}H-NMR experiments, the intermolecular hydrogen-bonding interaction between amide groups, and π-π stacking between two azobenzene moieties act as an effective deriving force to form gel in organic solvents.

To understand the H-bonding and the π-π stacking interactions operating in the self-assembled structures, we carried out ab initio calculations and molecular modeling. The molecules **1** and **2** are our main interest, but the molecules are quite large. Furthermore, we believe that the long aliphatic chains at the ends of **1** and **2** are involved in the hydrophobic interactions, which will be discussed later. So, we carried out ab initio calculations for the molecule **5** to obtained energy minimized geometry using a suite of Gaussian 98 program.[1] The density functional theory calculations with B3LYP exchange correlation functional were employed with 6-31G* basis sets.

First, we obtained the energy-minimum geometry which has an intramolecular hydrogen bonding between carbonyl oxygen at the terminal amide and the amide hydrogen at the inner amide. Thus, this geometry can not grow to a supramolecule by intermolecular hydrogen bonding. However, we already know that, in some cases, the intermolecular H-bonding networks can be realized in the condensed phase though the intramolecular H-bonding networks are favored in the gas phase.[31] Therefore, we have tried to model an assembled structure with eight monomers by translation using a proper translation operations based on our experimental observations such as π-π stacking interactions as previously treated.[32] We have done many trials by rotation of possible dihedral angles, that is, we have tried the modeling for many rotamers. Among them, we finally found a fancy assembled structure, which is in agreement with our experimental observations.

It can be supposed, from the assembled structure in Figure 3, that the terminal aliphatic chains may not change the roles of H-bonding and π-π stacking interactions and they may play an important role in forming the assembled structure by hydrophobic interactions. Thus, our proposition that the modeled assembled structure using **5** would not change the H-bonding and π-π stacking in **1** and **2** would be valid. The interplanar distance between two azobenzene moieties is about 4.6 Å and the interatomic distance in the H-bonding between the oxygen and nitrogen atom should be less than 3.4 Å, and these values are close to the typical distances in π-π stacking and H-bonding.[33] Of course, it is apparent that the H-bonding distance and π-π stacking distance should be modified if the whole molecular system is fully optimized, which is almost unpractical due to the size limitation.

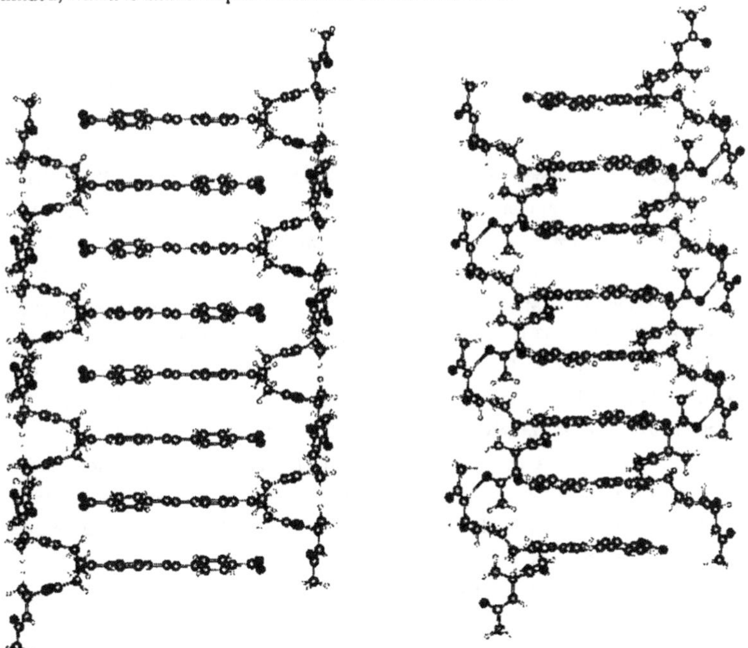

Figure 3. Assembled structures of eight molecules of **5** at different angle showing the networks with H-bonding and π-π stacking interactions.

7. Concluding Remarks

The computational approaches are very helpful to understand the molecular functions originated from the optical, structural, and electronic natures of molecules. Especially, owing to the advancement of electronics, theories, and computational methodologies, the state-of-the-art ab initio calculations can handle large molecular systems up to 1000 atoms and about 3000 number of basis sets. This enables the theoreticians to keep face with the experimental advancement. I have presented only a few examples, "on-off" molecular switch, electron transport through the fullerene-porphyrin-fullerene triad, and molecular self-assembly. A number of other examples have been successfully applied. Furthermore, much more interesting phenomena are still waiting for reasonable explanations, and they can be supported by computational approaches. Finally, I would like to close my talk by addressing again the impact of communications and collaborations by theoreticians and experimentalists.

Acknowledgement

This research could be successful with the financial support by Sungkyunkwan University. We also acknowledge the Supercomputing Center at Korea Institute of Science and Technology Information.

References

(1) (a) Kim, K. S.; Lee, J. Y.; Lee, S. J.; Ha, T.-K.; Kim, D. H. *J. Am. Chem. Soc.* **1994**, *116*, 7399. (b) Lee, J. Y.; Lee, S. J.; Choi, H. S.; Cho, S. J.; Kim, K. S.; Ha, T.-K. *Chem. Phys. Lett.* **1995**, *232*, 67. (c) Tarakeshwar, P.; Lee, S. J.; Lee, J. Y.; Kim, K. S. *J. Chem. Phys.* **1998**, *108*, 7217. (d) Tarakeshwar, P.; Kim, K. S.; Brutschy, B. *J. Chem. Phys.* **1999**, *110*, 8501. (e) Kim, K. S.; Lee, J. Y.; Choi, H. S.; Kim, J.; Jang, J. H. *Chem. Phys. Lett.* **1997**, *265*, 497. (f) Tarakeshwar, P.; Lee, J. Y.; Kim, K. S. *J. Phys. Chem. A* **1998**, *102*, 2253. (g) Tarakeshwar, P.; Lee, S. J.; Lee, J. Y.; Kim, K. S. *J. Phys. Chem. B* **1999**, *103*, 184. (h) Kim, K.S.; Park, I.; Lee, S.; Cho, K.; Lee, J. Y.; Kim, J.; Joannopoulos, J. D. *Phys. Rev. Lett.* **1996**, *76*, 956.

(2) For leading references, see the special issue dedicated to Margaret Etter, "Structure and Chemistry of the Organic Solid State": *Chem. Mater.* **1994**, *6*, 1087. See also the following reviews: (a) MacDonald, J. C.; Whitesides, G. M. *Chem. ReV.* **1994**, *94*, 2383. (b) Webb, T. H.; Wilcox, C. S. *Chem. Soc. ReV.* **1993**, 383. (c) Zaworotko, M. J. *Chem. Soc. ReV.* **1994**, 283. (d) Subramanian, S.; Zaworotko, M. J. *Coord. Chem. ReV.* **1994**, *137*, 357. (e) Aakero¨y, C. B.; Seddon, K. R. *Chem. Soc. ReV.* **1993**, 397. (f) Jurema, M. W.; Shields, G. C. *J. Comput. Chem.* **1993**, *14*, 89. (g) Dannenberg, J. J.; Evleth, E. M. *Int. J. Quantum Chem.* **1992**, *44*, 869. (h) Scheiner, S. In *ReViews in Computational Chemistry II*; Lipkowitz, K. B., Boyd, D. B., Ed.; VCH: New York, 1991; pp 165. (i) Dykstra, C. E. *Acc. Chem. Res.* **1988**, *10*, 355.

(3) (a) Subramanian, S.; Zaworotko, M. J. *Can. J. Chem.* **1995**, *73*, 414. (b) Hanessian, S.; Simard, M.; Roelens, S. *J. Am. Chem. Soc.* **1995**, *117*, 7630. (c) Kane, J. J.; Liao, R.-F.; Lauher, J. W.; Fowler, F. W. *J. Am. Chem. Soc.* **1995**, *117*, 12003. (d) Russell, V. A.; Etter, M. C.; Ward, M. D. *J. Am. Chem. Soc.* **1994**, *116*, 1941. (e) Zerkowski, J. A.; MacDonald, J. C.; Seto, C. T.; Wierda, D. A.; Whitesides, G. M. *J. Am. Chem. Soc.* **1994**, *116*, 2382. (f) Chang, Y.-L.; West, M.-A.; Fowler, F. W.; Lauher, J. W. *J. Am. Chem. Soc.* **1993**, *115*, 5991. (g) Yank, J.; Fan, E.; Geib, S. J.; Hamilton, A. D. *J. Am. Chem. Soc.* **1993**, *115*, 5214. (h) Etter, M. C.; Reutzel, S. M.; Choo, C. G. *J. Am. Chem. Soc.* **1993**, *115*, 4411. (i) Pascal, R. A., Jr.; Ho, D. M. *J. Am. Chem. Soc.* **1993**, *115*, 8507. (j) Wyler, R.; Mendoza, J. D.; Rebek, J. *Angew. Chem., Int. Ed. Engl.* **1993**, *32*, 1699. (k) Simard, M.; Su, D.; Wuest, J. D. J. *J. Am. Chem. Soc.* **1991**, *113*, 4696.

(4) (a) Etter, M. C. *J. Phys. Chem.* **1991**, *95*, 4601. (b) Etter, M. C. *Acc. Chem. Res.* **1990**, *23*, 120. (c) Bernstein, J.; Davis, R. E.; Shimoni, L.; Chang, N.-L. *Angew. Chem., Int. Ed. Engl.* **1995**, *34*, 1555.

(5) (a) Motherwell, W. D. S.; Ammon, H. L.; Dunitz, J. D.; Dzyabchenko, A.; Erk, P.; Gavezzotti, A.; Hofmann, D. W. M.; Leusen, F. J. J.; Lommerse, J. P. M.; Mooij, W. T. M.; Price, S. L.; Scheraga, H.; Schweizer, B.; Schmidt, M. U.; van Eijck, B. P.; Verwer, P.; Williams, D. E. *Acta Crystallogr., Sect. B: Struct. Sci.* **2002**, *58*, 647. (b) Lommerse, J. P. M.; Motherwell, W. D. S.; Ammon, H. L.; Dunitz, J. D.; Gavezzotti, A.; Hofmann, D. W. M.; Leusen, F. J. J.; Mooij, W. T. M.; Price, S. L.; Schweizer, B.; Schmidt, M. U.; van Eijck, B. P.; Verwer, P.; Williams, D. E. *Acta Crystallogr., Sect. B: Struct. Sci.* **2000**, *56*, 697. (c) Desiraju, G. R. *Nat. Mater.* **2002**, *1*, 77. (d) Dunitz, J. D. *Chem. Commun.* **2003**, 545.

(6) (a) De Feyter, S.; De Schryver, F. C. *Chem. Soc. ReV.* **2003**, *32*, 139. (b) Giancarlo, L. C.; Flynn, G. W. *Acc. Chem. Res.* **2000**, *33*, 491. (c) De Feyter, S.; Gesquie`re, A.; Abdel-Mottaleb, M. M.; Grim, P. C. M.; De Schryver, F. C.; Meiners, C.; Sieffert, M.; Valiyaveettil, S.; Mu¨llen, K. *Acc. Chem. Res.* **2000**, *33*, 520. (d)

Giancarlo, L. C.; Flynn, G. W. *Annu. ReV. Phys. Chem.* **1998**, *49*, 297.

(7) de Silva, A. P.; Gunaratne, H. Q. N.; Gunnlaugsson, T.; Huxley, A. J. M.; McCoy, C. P.; Rademacher, J. T.; Rice, T. E. *Chem. Rev.* **1997**, *97*, 1515.

(8) (a) Aviram, A.; Ratner, M. A. *Chem. Phys. Lett.* **1974**, *29*, 277. (b) Aviram, A. *J. Am. Chem. Soc.* **1988**, *110*, 5687.

(9) (a) Andres, R. P. Et al. *Science* **1996**, *272*, 1323. (b) Metzger, R. M. Et al. *J. Am. Chem. Soc.* **1997**, *119*, 10455. (c) Troisi, A.; Ratner, M. A. *J. Am. Chem. Soc.* **2002**, *124*, 14528. (d) Davis, W. B.; Svec, W. A.; Ratner, M. A.; Wasielewski, W. A. *Nature* **1998**, *396*, 60.

(10) (a) Yao, Z.; Postma, H. W. C.; Balents, L.; Dekker, C. *Nature*, 1999, *402*, 273. (b) Bachtold, A.; Hadley, P.; Nakanishi, T.; Dekker, C. *Science* **2001**, *294*, 1317. (c) Collier, C. P. et al. *Science* **2000**, *289*, 1172.

(11) Di Ventra, M. ; Pantelides, S. T. ; Lang, N. D. *Phys. Rev. Lett.* **2000**, *84*, 979.

(12) Kim, S. K.; Lee, S. H.; Lee, J. Y.; Lee, J. Y.; Bartsch, R. A.; Kim, J. S. *J. Am. Chem. Soc.* **2004**, 126, 16499.

(13) Lee, J. Y.; Lee, S. J.; Kim, H. J.; Kim, H.-J. *J. Phys. Chem. B* **2006**, *110*, 5337.

(14) Lee, S. J.; Lee, S. S.; Kim, J. S.; Lee, J. Y. *Chem. Mater.* **2005**, *17*, 6517.

(15) (a) Frisch, M. J. et al. *Gaussian 98*, Revision A6; Gaussian Inc.: Pittsburgh, PA, 1999. (b) Frisch, M. J. et al. *Gaussian 03*, Rev. C.02, Gaussian, Inc., Wallingford CT, 2004.

(16) (a) Kim, K. S.; Cui, C.; Cho, S. J. *J. Phys. Chem. B* **1998**, *102*, 461. (b) Yamaguchi, S.; Akiyama, S.; Tamao, K. *J. Am. Chem. Soc.* **2001**, *123*, 11372. (c) Beer, P. D.; Drew, M. G. M.; Hesek, D.; Nam, K. C. *Chem. Commun.* **1997**, 107. (d) Nam, K. C.; Kang, S. O.; Jeong, H. S.; Jeon, S. *Tetrahedron Lett.* **1999**, *40*, 7343. (e) Beer, P. D.; Hesek, D.; Nam, K. C.; Drew, M. G. M. *Organometallics* **1999**, *18*, 3933. (f) Kang, S. O.; Jeon, S.; Nam, K. C. *Supramol. Chem.* **2002**, *14*, 405. (g) Jeong, H. S.; Choi, E.M.; Kang, S. O.; Nam, K. C.; Jeon, S. *J. Electroanal. Chem.* **2000**, *485*, 154.

(17) (a) Escamilla, G. H.; Newkome, G. R. *Angew. Chem., Int. Ed. Engl.* **1994**, *33*, 1937. (b) Newkome, G. R.; Baker, G. R.; Saunders, M. J.; Russo, P. S.; Gupta, V. K.; Yao, Z.-Q.; Miller, J. E.; Bouillion, K. *J. Chem. Soc., Chem. Commun.* **1986**, 752. (c) Newkome, G. R.; Yao, Z.-Q.; Baker, G. R.; Gupta, V. K.; Russo, P. S.; Saunders, M. J. *J. Am. Chem. Soc.* **1986**, *108*, 849. (d) Newkome, G. R.; Baker, G. R.; Arai, S.; Saunders, M. J.; Russo, P. S.; Theriot, K. J.; Moorefield, C. N.; Rogers, L. E.; Miller, J. E.; Lieux, T. R.; Murray, M. E.; Phillips, B.; Pascal, L. *J. Am. Chem. Soc.* **1990**, *112*, 8458. (e) Newkome, G. R.; Moorefield, C. N.; Baker, G. R.; Behera, R. K.; Escamillia, G. H.; Saunders, M. J. *Angew. Chem., Int. Ed. Engl.* **1992**, *31*, 917. (f) Newkome, G. R.; Lin, X.; Chen, Y.; Escamilla, G. H. *J. Org. Chem.* **1993**, *58*, 3123.

(18) (a) Kim, Y. H. *J. Am. Chem. Soc.* **1992**, *114*, 4947. (b) Percec, V.; Kawasumi, M. *Polym. Prepr.* **1992**, *33*, 221. (c) Percec, V.; Chu, P. W.; Kawasumi, M. *Macromolecules* **1994**, *27*, 4441. (d) Percec, V. *Pure Appl. Chem.* **1995**, *67*, 2031.

(19) Ercolani, G. *J. Phys. Chem. B* **2003**, *107*, 5052.

(20) Winnik, F. M. *Chem. ReV.* **1993**, *93*, 587.

(21) Matsui, J.; Mitsuishi, M.; Miyashita, T. *J. Phys. Chem. B* **2002**, *106*, 2468.

(22) Nakahara, Y.; Kida, T.; Nakatsuji, Y.; Akashi, M. *J. Org. Chem.* **2004**, *69*, 4403.

(23) (a) Jones, G., II; Vullev, V. I. *J. Phys. Chem. A* **2001**, *105*, 6402. (b) Araujo, E.; Rharbi, Y.; Huang, X.; Ainnik, M. A. *Langmuir* **2000**, *16*, 8664.

(24) (a) Cho, E. J.; Moon, J. W.; Ko, S. W.; Lee, J. Y.; Kim, S. K.; Yoon, J.; Nam, K. C. *J. Am. Chem. Soc.* **2003**, *125*, 12376. (b) Lee, J. Y.; Cho, E. J.; Mukamel, S.; Nam, K. C. *J. Org. Chem.* **2004**, *69*, 943.

(25) Kleerekoper, M. *Endocrinol. Metab. Clin. North Am.* **1998**, *27*, 441.

(26) (a) Mukamel, S.; Wang, H. X. *Phys. Rev. Lett.* **1992**, *69*, 65. (b) Mukamel, S.; Takahashi, A.; Wang, H. X.; Chen, G. *Science* **1994**, *266*, 250. (c) Takahashi, A.; Mukamel, S. *J. Chem. Phys.* **1994**, *100*, 2366.

(27) Kim, H. J.; Park, K.-M.; Ahn, T. K.; Kim, S. K.; Kim, K. S.; Kim, D.; Kim, H.-J. *Chem. Commun.* **2004**, 2594.

(28) Kim, K. S.; Tarakeshwar, P.; Lee, J. Y. *Chem. Rev.* **2000**, *100*, 4145.

(29) Scheerder, J.; Vreekamp, R. H.; Engbersen, J. F. L.; Verboom, W.; van Duynhoven, J. P. M.; Reinhoudt, D. N. *J. Org. Chem.* **1996**, *61*, 3476.

(30) Kim, S. J.; Jo, M.-G.; Lee, J. Y.; Kim, B. H. *Org. Lett.* **2004**, *6*, 1963.

(31) Kim, S. J.; Jo, M.-G.; Lee, J. Y.; Kim, B. H. *Org. Lett.* **2004**, *6*, 1963.

(32) van Esch, J.; Schoonbeek, F.; de Loos, M.; Kooijman, H.; Spek, A. L.; Kellogg, R. M.; Feringa, B. L. *Chem. Eur. J.* **1999**, *5*, 937.

(33) Kim, K. S.; Mhin, B. J.; Choi, U-S.; Lee, K. *J. Chem. Phys.* **1992**, *97*, 6649.

Brill Academic Publishers
P.O. Box 9000, 2300 PA Leiden
The Netherlands

Lecture Series on Computer and Computational Sciences
Volume 6, 2006, pp. 395-404

Electric properties for HCCH, H_2CC, H_2CSi and H_2CGe

G.Maroulis*, D.Xenides[1] and P.Karamanis[2]
Department of Chemistry, University of Patras, GR-26500 Patras, Greece

Recived July 20, 2006; accepted 1 August, 2006

Abstract: We have calculated electric dipole moments, static polarizabilities and hyperpolarizabilities for HCCH, H_2CC, H_2CSi and H_2CGe. All properties have been obtained from finite-field Møller-Plesset perturbation theory and coupled cluster calculations with large, purpose-oriented basis sets. We have examined the effect on the electric properties of the acetylene-vinylidene isomerization and the evolution of the (hyper)polarizability in the H_2CX sequence, where X=C, Si, Ge. The mean second dipole hyperpolarizability of vinylidene is $\bar{\gamma} / e^4a_0^4E_h^{-3} = 5878$, slightly higher than the 5794 found for acetylene. $\bar{\gamma}$ (H_2CGe) is significantly higher than $\bar{\gamma}$ (H_2CCe) but comparable to $\bar{\gamma}$ (H_2CSi)

Keywords: Acetylene, vinylidene, silylidene, germylidene, dipole moment, polarizability, hyperpolarizability.

1. Introduction

The elegant theory of electric polarizability [1,2] has been the starting point of significant progress in molecular science. It is now routinely associated with molecular characteristics as hardness [3,4], softness [1] and hypersoftness [5], stiffness [6] and compressibility [7]. This basic theory is now the cornerstone of the rational approach to the interpretation of phenomena in fields ranging from intermolecular interactions [9,10] to nonlinear optics [11]. Computational quantum chemistry can now provide reliable estimates of polarizabilities for wide classes of systems of all sizes, from atoms, molecules and clusters to nanoparticles [11, 12]. Particular research fronts that benefit from this expansion include the simulation of fluids [13,14], the analysis of spectroscopic observations [15,16] and systematic studies of molecular architectures with potential use as nonlinear optical materials [17,18]. Electric (hyper)polarizability is also used as a molecular descriptor in pharmacological QSAR studies [19,20].

In this work we report electric properties for acetylene (H-C≡C-H), vinylidene ($H_2C=C$), silylidene ($H_2C=Si$) and germylidene ($H_2C=Ge$). Our goal is to investigate, primo, the effect of the acetylene-vinylidene isomerization on the electric (hyper)polarizability and secundo, the effect of substitution for the sequence H_2CX, X= C, Si and Ge.

2. Theory

Our calculation of the electric properties relies on the finite-field method [21]. The energy and the electric multipole moments of an uncharged molecule perturbed by a weak, static electric field can be expanded in terms of the permanent properties (in bold) of the free molecule and the components of the field, as [1,2,22]

* Corresponding author. E-mail: maroulis@upatras.gr

[1] Present address: Department of Computer Science and Technology, University of the Peloponnese, GR-221 00 Tripolis, Greece.

[2] Present address: Laboratoire de Chimie Structurale, UMR 5624, Université de Pau et des Pays de l'Adour, F-64000 Pau, France.

$$E^p \equiv E^p(F_\alpha, F_{\alpha\beta}, F_{\alpha\beta\gamma}, F_{\alpha\beta\gamma\delta}, ...)$$
$$= E^0 - \mu_\alpha F_\alpha - (1/3)\Theta_{\alpha\beta}F_{\alpha\beta} - (1/15)\Omega_{\alpha\beta\gamma}F_{\alpha\beta\gamma} - (1/105)\Phi_{\alpha\beta\gamma\delta}F_{\alpha\beta\gamma\delta} + ...$$
$$- (1/2)\alpha_{\alpha\beta}F_\alpha F_\beta - (1/3)A_{\alpha,\beta\gamma}F_\alpha F_{\beta\gamma} - (1/6)C_{\alpha\beta,\gamma\delta}F_{\alpha\beta}F_{\gamma\delta}$$
$$- (1/15)E_{\alpha,\beta\gamma\delta}F_\alpha F_{\beta\gamma\delta} + ...$$
$$- (1/6)\beta_{\alpha\beta\gamma}F_\alpha F_\beta F_\gamma - (1/6)B_{\alpha\beta,\gamma\delta}F_\alpha F_\beta F_{\gamma\delta} + ...$$
$$- (1/24)\gamma_{\alpha\beta\gamma\delta}F_\alpha F_\beta F_\gamma F_\delta + ... \quad (1)$$

$$\mu_\alpha{}^p = \mu_\alpha + \alpha_{\alpha\beta} F_\beta + (1/3)A_{\alpha,\beta\gamma} F_{\beta\gamma} + (1/2)\beta_{\alpha\beta\gamma} F_\beta F_\gamma + (1/3)B_{\alpha\beta,\gamma\delta} F_\beta F_{\gamma\delta}$$
$$+ (1/6)\gamma_{\alpha\beta\gamma\delta} F_\beta F_\gamma F_\delta + ... \quad (2)$$

$$\Theta_{\alpha\beta}{}^p = \Theta_{\alpha\beta} + A_{\gamma,\alpha\beta}E_\gamma + C_{\alpha\beta,\gamma\delta} F_{\gamma\delta} + (1/2)B_{\gamma\delta,\alpha\beta} F_\gamma F_\delta + ... \quad (3)$$

$$\Omega_{\alpha\beta\gamma}{}^p = \Omega_{\alpha\beta\gamma} + E_{\delta,\alpha\beta\gamma} F_\delta + ... \quad (4)$$

where F_α, $F_{\alpha\beta}$, etc. are the field, field gradient, etc. at the origin. E^0, $\mu_\alpha{}^0$, $\Theta_{\alpha\beta}{}^0$, $\Omega_{\alpha\beta\gamma}{}^0$ and $\Phi_{\alpha\beta\gamma\delta}{}^0$ are the energy and the dipole, quadrupole, octopole and hexadecapole moment of the free molecule. The second, third and fourth-order properties are the dipole and quadrupole polarizabilities and hyperpolarizabilities $\alpha_{\alpha\beta}$, $\beta_{\alpha\beta\gamma}$, $\gamma_{\alpha\beta\gamma\delta}$, $A_{\alpha,\beta\gamma}$, $C_{\alpha\beta,\gamma\delta}$, $E_{\alpha,\beta\gamma\delta}$ and $B_{\alpha\beta,\gamma\delta}$. The subscripts denote Cartesian components. A repeated subscript implies summation over x, y and z. The number of independent components needed to specify the non-vanishing tensors is regulated by symmetry [1].

For sufficiently weak fields the expansions of Eqs 1-4 converge rapidly. Thus, the finite-field method offers the possibility of a direct approach to the calculation of electric moments, polarizabilities and hyperpolarizabilities. In the case of a homogeneous electric field, Eq 1 reduces to a much simpler one,

$$E = E^0 - \mu_\alpha{}^0F_\alpha - (1/2)\alpha_{\alpha\beta}F_\alpha F_\beta - (1/6)\beta_{\alpha\beta\gamma}F_\alpha F_\beta F_\gamma - (1/24)\gamma_{\alpha\beta\gamma\delta}F_\alpha F_\beta F_\gamma F_\delta + ... \quad (5)$$

The properties of interest in this work are the dipole ones $\mu_\alpha{}^0$, $\alpha_{\alpha\beta}$, $\beta_{\alpha\beta\gamma}$ and $\gamma_{\alpha\beta\gamma\delta}$. The independent components of the respective tensors have been specified in previous work [23,24]. In addition to the Cartesian components, other properties of interest in this work are the mean ($\bar{\alpha}$) and the anisotropy ($\Delta\alpha$) of the dipole polarizability, the mean of the first ($\beta_{\alpha\beta\gamma}$) and the second ($\gamma_{\alpha\beta\gamma\delta}$) hyperpolarizability ($\bar{\gamma}$). All the above are directly associated to measurable quantities. For H_2CC, H_2CSi and H_2CGe they are defined (with z as the C_2 axis) as [1],

$$\bar{\alpha} = (\alpha_{xx} + \alpha_{yy} + \alpha_{zz})/3$$
$$\Delta\alpha = (1/2)^{1/2}[(\alpha_{xx}-\alpha_{yy})^2 + (\alpha_{yy}-\alpha_{zz})^2 + (\alpha_{zz}-\alpha_{xx})^2]^{1/2}$$
$$\bar{\beta} = (3/5)(\beta_{zxx} + \beta_{zyy} + \beta_{zzz})$$
$$\bar{\gamma} = (1/5)(\gamma_{xxxx} + \gamma_{yyyy} + \gamma_{zzzz} + 2\gamma_{xxyy} + 2\gamma_{yyzz} + 2\gamma_{zzxx}) \quad (6)$$

In the case of a linear molecule, as HCCH, these definitions reduce to

$$\bar{\alpha} = (2\alpha_{xx} + \alpha_{zz})/3$$
$$\Delta\alpha = \alpha_{zz}-\alpha_{xx}$$
$$\bar{\beta} = (3/5)(2\beta_{zxx} + \beta_{zzz})$$

$$\bar{\gamma} = (1/15)(3\gamma_{zzzz} + 8\gamma_{xxxx} + 12\gamma_{zzxx}) \quad (7)$$

Our approach to the calculation of electric properties from Eqs 1-4 has been presented in rich detail in previous work [24-30]. The theoretical methods used in this paper are self-consistent field (SCF), Møller-Plesset perturbation theory (MP) and coupled cluster theory (CC). Essential presentations of these methods are available in standard textbooks [31-34].

Electron correlation effects on the molecular properties are accounted for via MPn (n=2,3,4, second, third and fourth order many-body perturbation theory), CCSD (single and double excitation coupled cluster theory) and CCSD(T) (single and double excitation coupled cluster theory including an estimate of connected triple excitations by a perturbational treatment). We restrict our presentation to the definition of the various orders of MP

$$\begin{aligned} MP2 &= SCF + D2 \\ MP3 &= MP2 + D3 \\ DQ\text{-}MP4 &= MP3 + D4 + QR4 = MP3 + DQ4 \\ SDQ\text{-}MP4 &= DQ\text{-}MP4 + S4 \\ MP4 &= SDQ\text{-}MP4 + T4 \\ &= SCF + D2 + D3 + S4 + D4 + T4 + QR4 \end{aligned} \quad (8)$$

(where the fourth-order terms S4, D4, T4 and Q4 are contributions from single, double, triple and quadruple substitutions from the reference, zeroth-order wavefunction and R4 is the renormalization term). For the CC levels of theory,

$$\begin{aligned} CCSD &= SCF + \Delta CCSD \\ CCSD(T) &= CCSD + T \end{aligned} \quad (9)$$

By virtue of Eqs 6 and 7, analogous decompositions are adopted for the mean polarizability or hyperpolarizability, as $\bar{\alpha}$, $\bar{\beta}$ and $\bar{\gamma}$ are linear in the Cartesian components. The same holds true for the anisotropy [illegible] in the case of acetylene. For the other three systems the anisotropy [illegible] at a given level of theory is computed by inserting in Eq 6 the respective quantities for the Cartesian components of [illegible].

3. Computational Details

Large, carefully optimized Gaussian basis sets were used for all molecules. The details of their construction and composition are given elsewhere [35]. The basis sets are [9s6p5d2f/6s4p2d] (188 contracted GTF) for acetylene, [9s6p4d1f/6s4p1d] (156 CGTF) for vinylidene, [8s6p3d1f/6s4p3d1f/4s3p1d] (124 CGTF) for silylidene and [8s7p6d2f/6s4p4d2f/4s4p2d] (177 CGTF) for germylidene. 5d and 7f GTF were used in all calculations.

All calculations were performed at the molecular geometries selected from the available literature. Our source is Martin et al [36] for acetylene, Chang et al [37] for vinylidene, Hilliard and Grev [38] for silylidene and Stogner and Grev [39] for germylidene. The respective parameters are shown in Figure 1. The molecular axis for acetylene is z. For vinylidene, silylidene and germylidene the molecule is on the xz plane, with z as the C_2 axis and the hydrogen nuclei on the negative part of the z axis.

The 2 innermost MO were kept froze in all post-Hartree-Fock calculations for acetylene. Similarly, 2 MO were kept frozen for vinylidene, 6 for silylidene and 10 for germylidene.

All calculations were performed with GAUSSIAN 94 [40] and GAUSSIAN 98 [41].

Atomic units are used throughout this paper. Conversion factors to SI units are, Energy, 1 E_h = 4.3597482 x 10^{-18} J, length, 1 a_0 = 0.529177249 x 10^{-10} m, μ, 1 ea_0 = 8.478358 x 10^{-30} Cm, α, 1 $e^2a_0^2E_h^{-1}$ = 1.648778 x 10^{-41} $C^2m^2J^{-1}$, β, 1 $e^3a_0^3E_h^{-2}$ = 3.206361 x 10^{-53} $C^3m^3J^{-2}$, γ, 1 $e^4a_0^4E_h^{-3}$ = 6.235378 x 10^{-65} $C^4m^4J^{-3}$. Property values are in most cases given as pure numbers, i.e. [illegible]$/ea_0$, $\alpha_{\alpha\beta}/e^2a_0^2E_h^{-1}$, $\beta_{\alpha\beta\gamma}/e^3a_0^3E_h^{-2}$ and $\gamma_{\alpha\beta\gamma\delta}/e^4a_0^4E_h^{-3}$

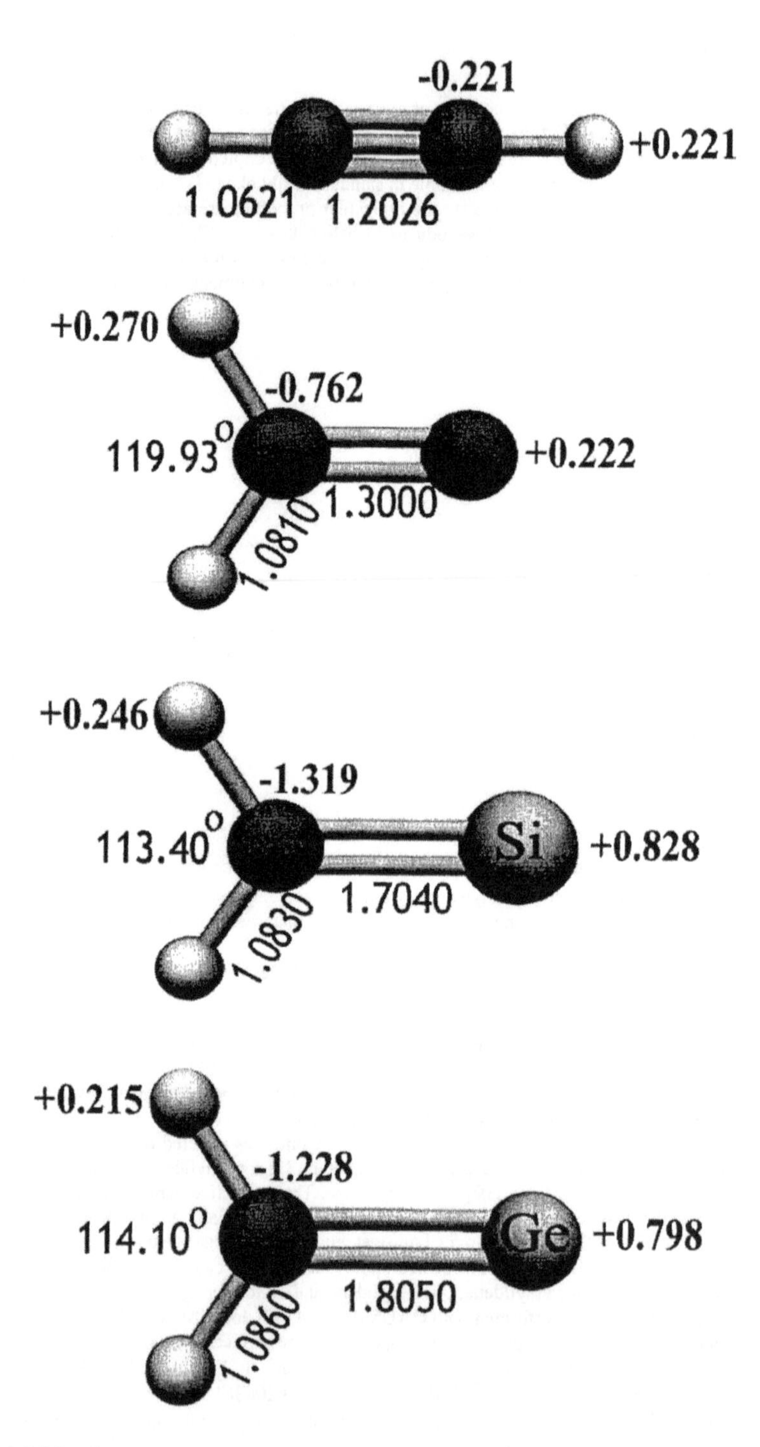

Fig 1. Molecular geometry specification and atomic charges from a NBO population analysis.

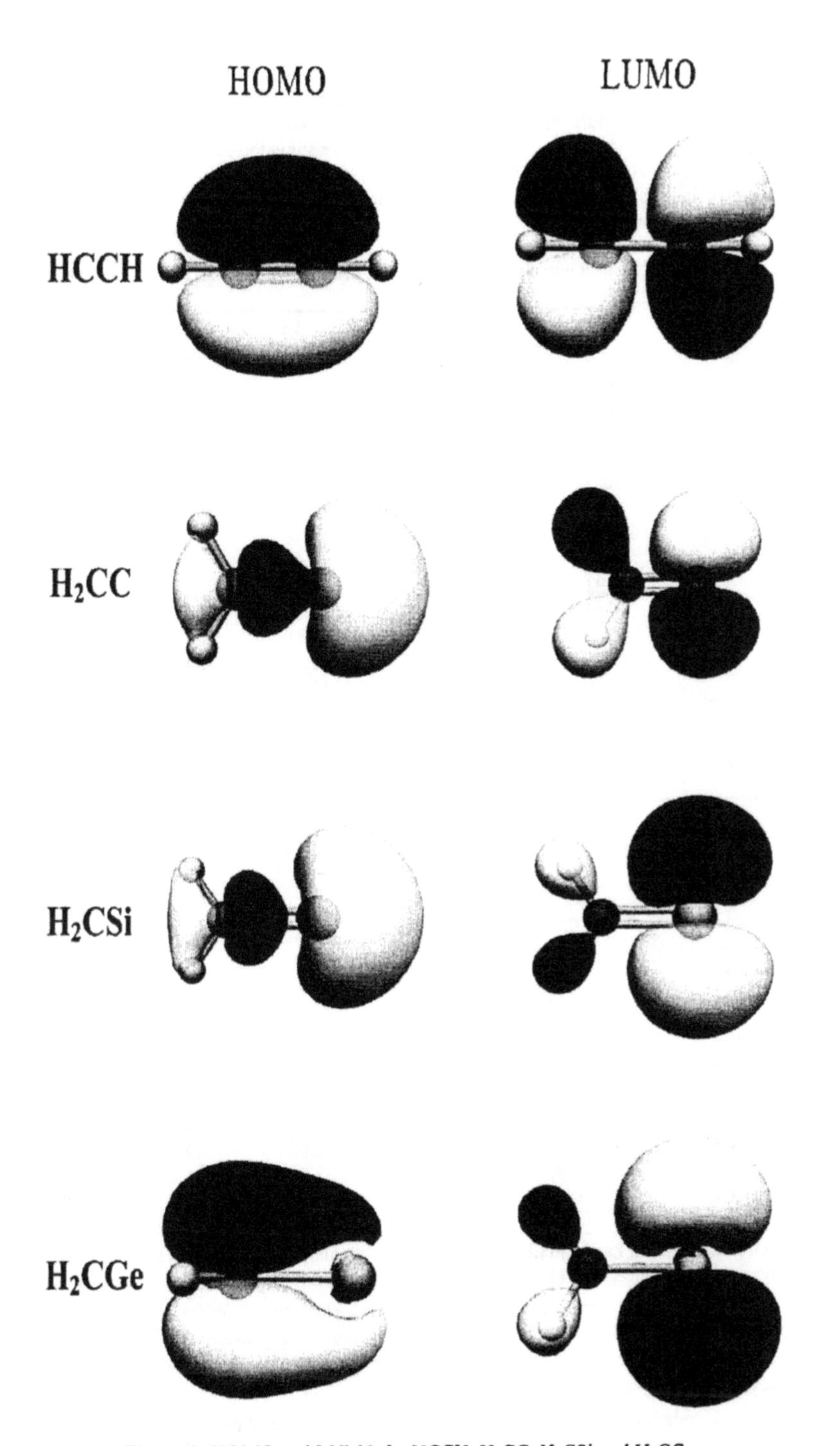

Figure 2. HOMO and LUMO for HCCH, H_2CC, H_2CSi and H_2CGe .

4. Results and Discussion

In Figure 1, in addition to the geometrical parameters of the molecules, we show also atomic charges obtained from a NBO analysis at the MP2/cc-pvdz level of theory for acetylene, vinylidene and silylidene and MP2/[8s7p4d/6s4p2d/4s2p] for germylidene. We observe that there is a negative charge on the C atom both for acetylene and H_2CX. In Figure 2 we show the evolution of the HOMO-LUMO in the sequence HCCH→H_2CC→H_2CSi→H_2CGe.

We give in Tables 1 and 2 a detailed analysis of electron correlation effects on the electric properties of acetylene and vinylidene.

Table 1. Electron correlation effects on the dipole (hyper)polarizability of acetylene.

Method	$\bar{\alpha}$	$\Delta\alpha$	$\bar{\gamma}$
SCF	23.41	12.04	5299
D2	-0.58	-0.66	833
D3	-0.03	0.21	-489
S4	0.01	0.12	-51
D4	-0.12	-0.05	9
T4	0.35	0.06	670
QR	-0.12	0.08	-311
ΔCCSD	-0.86	-0.44	113
T	0.24	0.06	382
MP2	22.83	11.38	6133
SDQ-MP4	22.80	11.59	5644
MP4	22.92	11.80	5960
CCSD	22.55	11.59	5413
CCSD(T)	22.78	11.65	5794
ECC	-0.62	-0.39	495

Table 2. Electron correlation effects on the dipole moment and (hyper)polarizability of vinylidene.

Method	μ	$\bar{\alpha}$	$\Delta\alpha$	$\bar{\beta}$	$\bar{\gamma}$
SCF	0.9318	23.86	9.15	38.7	4481
D2	0.0232	0.25		3.8	1224
D3	-0.0311	-0.32		-1.4	-485
S4	0.0134	0.15		1.7	137
D4	-0.0069	-0.05		-0.8	56
T4	0.0081	0.27		2.4	470
QR	0.0002	-0.06		-0.3	-209
ΔCCSD	-0.0100	0.09		4.0	933
T	0.0041	0.27	0.19	2.7	465
MP2	0.9550	24.12	8.07	42.5	5705
SDQ-MP4	0.9239	23.79	8.20	41.2	5220
MP4	0.9386	24.10	8.41	44.1	5674
CCSD	0.9218	23.95	8.92	42.7	5414
CCSD(T)	0.9259	24.23	8.95	45.4	5878
ECC	-0.0059	0.36	-0.20	6.7	1397

Electron correlation has a small effect on the dipole polarizability of acetylene. The magnitude of both invariants is reduced by electron correlation at all levels of theory. The electron correlation correction (ECC), defined as ECC = CCSD(T) – SCF is -0.62 and -0.39 for $\bar{\alpha}$ and $\Delta\alpha$, respectively. This corresponds to a reduction of the SCF value by 2.7 and 3.2%, respectively. The post-Hartree-Fock values of the mean hyperpolarizability vary significantly. The presumably most accurate CCSD(T) methods yields $\bar{\gamma} = 5794$, or 9.3% above the SCF value of 5299.

Electron correlation has a small effect for the dipole moment and polarizability of vinylidene. The dipole moment is predicted to be $\mu = 0.9318$ at the SCF level but the CCSD(T) is lower at 0.9259. The mean polarizability $\bar{\alpha}$ increases from 23.86 (SCF) to 24.23 (CCSD(T)), a change of only 1.5%. A reduction of the magnitude of the anisotropy is predicted by all post-Hartree-Fock methods but the overall effect is rather small. The CCSD(T) value of the mean first hyperpolarizability is $\bar{\beta}$= 45.43. This represents an strong increase of 17.3% over the SCF result. The effect is quite strong for the second dipole hyperpolarizability, as the CCSD(T) value is $\bar{\gamma}$= 5878, or 31.2% above the respective SCF.

Overall, vinylidene is characterized by a mean dipole polarizability slightly higher than that of acetylene. This is reversed for the anisotropy. At the SCF level of theory $\bar{\gamma}$(HCCH) > $\bar{\gamma}$(H_2CC). Electron correlation inverses this inequality. Thus, taking into account the effect on $\bar{\alpha}$ and $\bar{\gamma}$, vinylidene is more (hyper)polarizable than acetylene.

The calculated electric properties for silylidene and germylidene are shown in Tables 3 an 4, respectively.

Table 3. Electric properties for silylidene.

Method	μ	$\bar{\alpha}$	Δα	$\bar{\beta}$	$\bar{\gamma}$
SCF	0.1076	48.19	18.70	-268.4	24124
MP2	0.1217	47.66	16.71	-130.1	26945
SDQ-MP4	0.1125	47.09	18.08	-167.6	24793
MP4	0.0918	47.51	17.94	-159.7	26591
CCSD	0.1143	47.07	18.32	-173.5	25048
CCSD(T)	0.0907	47.39	18.28	-171.6	26639
ECC	**-0.0169**	**-0.80**	**-0.42**	**96.8**	**2515**

The SCF dipole moment of silylidene is significantly smaller than that of vinylidene. Electron correlation reduces further this value by 15.7%. In absolute terms this change is rather small. It is worth noticing that MP2, SDQ-MP4 and CCSD predict a positive correlation effect, in disagreement with MP4 and CCSD(T). At the CCSD(T) level we obtain $\bar{\alpha}$ =47.39 and Δα = 18.28, lower by 1.7 and 2.2% lower than the respective SCF values. Electron correlation effects are very important for the mean first hyperpolarizability but considerably lower for the second one. We observe an overestimation of the correlation effect by MP2 for both properties.

Table 4. Electric properties for germylidene.

Method	μ	$\bar{\alpha}$	Δα	$\bar{\beta}$	$\bar{\gamma}$
SCF	0.1853	52.58	24.56	-290.2	28949
MP2	0.1510	50.25	21.02	-48.2	28597
SDQ-MP4	0.1095	50.05	22.97	-97.5	25055
MP4	0.0796	50.32	22.66	-81.0	27189
CCSD	0.1271	50.23	22.83	-124.5	26824
CCSD(T)	0.0981	50.35	22.65	-120.3	28186
ECC	**-0.0872**	**-2.23**	**-1.91**	**169.9**	**-763**

The dipole moment of germylidene is higher than that of silylidene. Electron correlation more than halves this property. We observe a rather larger electron correlation effect on the dipole polarizability: at the CCSD(T) level of theory $\bar{\alpha}$ = 50.35 and Δα = 22.65, lower by 4.2 and 8.8% than the respective SCF values. As in the case of silylidene the electron correlation effect on the first hyperpolarizability is very pronounced. The MP methods are in disagreement with both CC ones. Nevertheless, in the case of the second hyperpolarizability all post-Hartree-Fock methods do not change significantly the mean $\bar{\gamma}$. The CCSD(T) value is just 2.6% lower than the SCF one.

The mean dipole polarizability increases, as expected, for the three systems: $\bar{\alpha}$ = 24.23 (H_2CC), 47.39 (H_2CSi), 50.35 (H_2CGe). We observe that although the value of this property nearly doubles from vinylidene to silylidene, the increase is not as large from slilylidene to germylidene.

In Figures 3 and 4 we have traced the evolution of the mean first and second hyperpolarizability for the vinylidene, silylidene and germylidene.

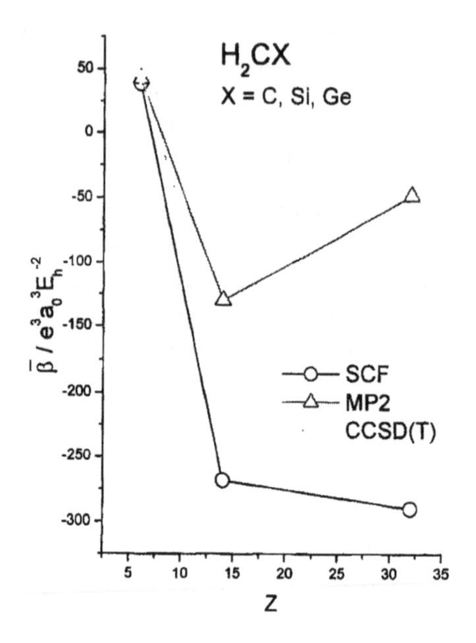

Figure 3. Evolution of the mean first dipole hyperpolarizability in the sequence H_2CX, X=C,Si,Ge with the nuclear charge of X.

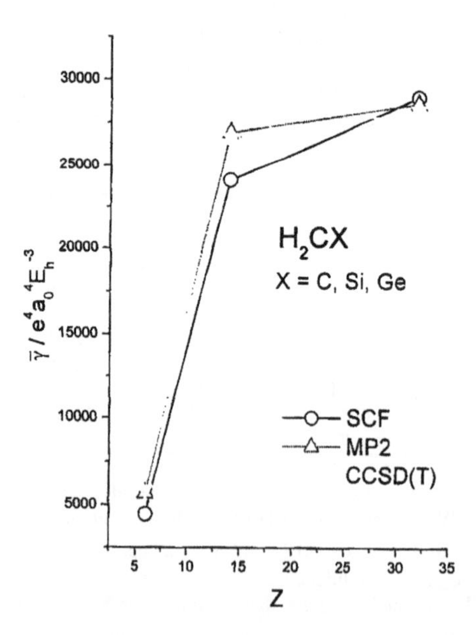

Figure 4. Evolution of the mean second dipole hyperpolarizability in the sequence H_2CX, X=C,Si,Ge with the nuclear charge of X.

At the SCF level of theory the magnitude of the mean first hyperpolarizability increases monotonically. Due to the very strong electron correlation effects, the monotonicity is destroyed at the post-Hartree-Fock level. This is not the case of the mean second hyperpolarizability, where $\bar{\gamma}(H_2CC) < \bar{\gamma}(H_2CSi) < \bar{\gamma}(H_2CGe)$. As in the case of the mean dipole polarizability, the second hyperpolarizability of silylidene is much larger than that of vinylidene. The change from silylidene to germylidene is less important.

We are not aware of previous electric property values for vinylidene, silylidene and germylidene. We have compared our findings for acetylene to those of other workers [42-44] in another paper [35].

5. Conclusions

We have calculated electric (hyper)polarizabilities for acetylene, vinylidene, silylidene and germylidene. We have found that vinylidene is more (hyper)polarizable than acetylene. The electric (hyper)polarizability increases in the sequence $H_2CC \rightarrow H_2CSi \rightarrow H_2CGe$. The mean polarizability and second hyperpolarizability of silylidene is considerably larger than that of vinylidene. The same properties of germylidene are not significantly larger than those of silylidene.

References

[1] A.D.Buckingham, Adv. Chem. Phys. 12, 107 (1967).

[2] A.D.McLean and M.Yoshimine, J.Chem. Phys. 47, 1927 (1967).

[3] M.Berkowitz and R.G.Parr, J.Chem.Phys. 88, 2554 (1988).

[4] T.K.Ghanty and S.K.Ghosh, J.Phys.Chem. 100, 12295 (1996).

[5] P.H.Liu and K.L.C.Hunt, J.Chem.Phys. 103, 10597 (1995).

[6] M.Torrent-Succarat, F. De Proft and P.Geerlings, J.Phys.Chem. A 109, 6071 (2005).

[7] K.J.Donald, J.Phys.Chem. A 110, 2283 (2006).

[8] A.J.Thakkar, in *Encyclopedia of Chemical Physics and Physical Chemistry*, J.Moore and N.Spencer (Eds) (IOP Publishing, Bristol, 2001), Vol I, pp. 181-186

[9] G.Birnbaum (Ed.), *Phenomena induced by intermolecular interactions* (Plenum, New York, 1985).

[10] D.C.Hanna, M.A.Yuratich and D.Cotter, *Nonlinear optics of free atoms and molecules* (Springer, Berlin, 1979).

[11] G.Maroulis (Guest-Editor), Special issue, *Computational Aspects of Electric Polarizability Calculations. Atoms, Molecules and Clusters*, J.Comput.Meth.Sci.Eng. 4, 235-764 (2004).

[12] G.Maroulis (Ed.), *Atoms, Molecules and Clusters in Electric Fields. Theoretical Approaches to the Calculation of Electric Polarizability* (Imperial College Press, London, 2006).

[13] C.G.Gray and K.E.Gubbins, *Theory of molecular fluids* (Clarendon Press, Oxford, 1984) Vol. 1.

[14] G.Ruocco and M.Sampoli, Mol. Phys. **82**, 875 (1994).

[15] B.S.Galabov and T.Dudev, *Vibrational Intensities* (Elsevier, Amsterdam, 1996).

[16] J.L.Godet, A.Elliasmine, Y.Le Duff and T.Bancewicz, J.Chem.Phys. 110, 11303 (1999).

[17] S.P.Karna and A.T.Yeates (Eds.), *Nonlinear Optical Materials: Theory and Modeling*, ACS Symposium Series 628, (ACS, Washington DC, 1996).

[18] H.S.Nalwa and S.Miyata (Eds.), *Nonlinear Optics of Organic Molecules and Polymers* (CRC Press, Boca Raton, 1997).

[19] M.Karelson, V.S.Lobanov and A.E.Katritzky, Chem.Rev. 96, 1027 (1996).

[20] C.Hansch, W.E.Steinmetz, A.J.Leo, S.B.Mekapati, A.Kurup and D.Hoekman, J.Chem.Inf.Comput.Sci. 43, 120 (2003).

[21] H.D.Cohen and C.C.J.Roothaan, J.Chem.Phys. 43S, 34 (1965).

[22] P.Isnard, D.Robert and L.Galatry, Mol.Phys. 31, 1789 (1976).

[23] G.Maroulis, J.Chem.Phys. 94, 1182 (1991).

[24] G.Maroulis, Chem.Phys.Lett. 289, 403 (1998).

[25] G.Maroulis, J.Chem.Phys. **108**, 5432 (1998).

[26] G.Maroulis, J.Chem.Phys. **111**, 6846 (1999).

[27] G.Maroulis, J.Chem.Phys. **118**, 2673 (2003).

[28] G.Maroulis and D.Xenides, J.Phys.Chem. A **107**, 712 (2003).

[29] G.Maroulis, J.Phys.Chem A 107, 6495 (2003).

[30] G.Maroulis, J.Chem.Phys. 121, 10519 (2004).

[31] R.J.Bartlett, Annu.Rev.Phys.Chem. 32, 359 (1981).

[32] A.Szabo and N.S.Ostlund, *Modern Quantum Chemistry* (MacMillan, New York, 1982)

[33] S.Wilson, *Electron Correlation in Molecules* (Clarendon, Oxford, 1984).

[34] T.Helgaker, P.Jørgensen and J.Olsen, *Molecular Electronic-Structure Theory* (Wiley, Chichester, 2000).

[35] G.Maroulis, D.Xenides and P.Karamanis, to be submitted.

[36] J.M.L.Martin, T.J.Lee and P.R.Taylor, J.Chem.Phys. 108, 676 (1998).

[37] N.Y.Chang, M.Y.Shen and C.H.Yu, J.Chem.Phys. 106, 3237 (1997).

[38] R.K.Hilliard and R.S.Grev, J.Chem.Phys. 107, 8823 (1997).

[39] S.M.Stogner and R.S.Grev, J.Chem.Phys. 108, 5458 (1998).

[40] M.J.Frisch, G.W.Trucks, H.B.Schlegel et al., GAUSSIAN 94, Revision E.1 (Gaussian, Inc, Pittsburgh PA, 1995).

[41] M.J.Frisch, G.W.Trucks, H.B.Schlegel et al., *GAUSSIAN 98*, Revision A.7 (Gaussian, Inc., Pittsburgh PA, 1998).

[42] M.Jaszunski, P.Jørgensen, H.Koch, H.Ågren and T.Helgaker, J.Chem.Phys. 98, 7229 (1993).

[43] A.Rizzo and N.Rahman, Laser Physics 9, 1 (1999).

[44] O.Quinet and B.Champagne, Int.J.Quant.Chem. 80, 871 (2000).

Brill Academic Publishers
P.O. Box 9000, 2300 PA Leiden
The Netherlands

Lecture Series on Computer and Computational Sciences
Volume 6, 2006, pp. 405-409

On Zagreb Matrices and Derived Descriptors*

Dušanka Janežič,[a] Ante Miličević[b] and Sonja Nikolić[c1]

[a]National Institute of Chemistry, Hajdrihova 19, SI-1000 Ljubljana, Slovenia,
[b]The Institute of Medical Research and Occupational Health, P. O .Box 291, HR-10002 Zagreb, Croatia and [c]The Rugjer Bošković Institute, P. O. Box 180, HR-10002 Zagreb, Croatia

Received 5 June, 2006; accepted 10 June, 2006

Abstract: The Zagreb matrices can be considered as the vertex- and edge-weighted matrices. They can be formulated in terms of the vertex- or edge-degrees. The degree of a vertex in a (molecular) graph is equal to the number of adjacent vertices and the degree of an edge is equal to the number of adjacent edges.. The Zagreb matrices can serve as sources for deriving the Zagreb indices.

Keywords: Zagreb matrices, diagonal matrices, sparse matrices, dense matrices

Mathematics Subject Classification: 05C10, 05C50, 68R10, 65F50

1. Zagreb matrices in terms of the vertex-degrees

The Zagreb matrices in terms of the vertex-degrees are the $V \times V$ matrices of six kinds: (i) two are *diagonal* matrices - vertex-Zagreb matrix and modified vertex-Zagreb matrix, (ii) two are *sparse* matrices – edge-Zagreb matrix and modified edge-Zagreb matrix and (ii) two are *dense* matrices – path-Zagreb matrix and modified path-Zagreb matrix.

The *vertex-Zagreb matrix*, denoted by $^{v}\mathbf{ZM}$, is a diagonal $V \times V$ matrix defined by:

$$\left[{}^{v}\mathbf{ZM}\right]_{ij} = \begin{cases} [d(i)]^2 & \text{if } i = j \\ 0 & \text{otherwise} \end{cases} \tag{1}$$

where $d(i)$ is the degree of a vertex i [1].

The $^{v}\mathbf{ZM}$ matrix for a tree T representing the carbon skeleton of 2-methylpentane is as follows.

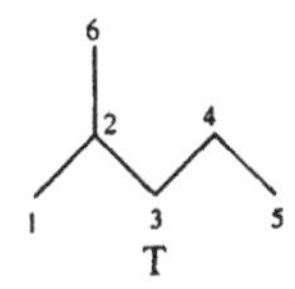

*Dedicated to Professor Nenad Trinajstić on the occasion of his 70th birthday.

[1] Corresponding author. E-mail: sonja@irb.hr

$$
{}^{v}\mathbf{ZM}(\mathrm{T}) = \begin{bmatrix} 1 & 0 & 0 & 0 & 0 & 0 \\ 0 & 9 & 0 & 0 & 0 & 0 \\ 0 & 0 & 4 & 0 & 0 & 0 \\ 0 & 0 & 0 & 4 & 0 & 0 \\ 0 & 0 & 0 & 0 & 1 & 0 \\ 0 & 0 & 0 & 0 & 0 & 1 \end{bmatrix}
$$

The summation of the diagonal elements gives the first Zagreb index $ZG(1)$ [2-4].

$$
ZG(1) = \sum_i \left[{}^{v}\mathbf{ZM} \right]_{ii} \tag{2}
$$

The *modified* vertex-Zagreb matrix, denoted by ${}^{mv}\mathbf{ZM}$, is defined as:

$$
\left[{}^{mv}\mathbf{ZM} \right]_{ij} = \begin{cases} 1/[d(i)]^2 & \text{if } i = j \\ 0 & \text{otherwise} \end{cases} \tag{3}
$$

The ${}^{mv}\mathbf{ZM}$ matrix of T is as follows.

$$
{}^{mv}\mathbf{ZM}(\mathrm{T}) = \begin{bmatrix} 1 & 0 & 0 & 0 & 0 & 0 \\ 0 & 1/9 & 0 & 0 & 0 & 0 \\ 0 & 0 & 1/4 & 0 & 0 & 0 \\ 0 & 0 & 0 & 1/4 & 0 & 0 \\ 0 & 0 & 0 & 0 & 1 & 0 \\ 0 & 0 & 0 & 0 & 0 & 1 \end{bmatrix}
$$

The summation of the diagonal elements gives the modified first Zagreb index ${}^{m}ZG(1)$ [5,6].

$$
{}^{m}ZG(1) = \sum_i \left[{}^{v}\mathbf{ZM} \right]_{ii} \tag{4}
$$

The *edge-Zagreb matrix*, denoted ${}^{e}\mathbf{ZM}$, is defined by:

$$
\left[{}^{e}\mathbf{ZM} \right]_{ij} = \begin{cases} d(i)d(j) & \text{if vertices } i \text{ and } j \text{ are adjacent} \\ 0 & \text{otherwise} \end{cases} \tag{5}
$$

The edge-Zagreb matrix ${}^{e}\mathbf{ZM}$ of T is presented below.

$$
{}^{e}\mathbf{ZM}(\mathrm{T}) = \begin{bmatrix} 0 & 3 & 0 & 0 & 0 & 0 \\ 3 & 0 & 6 & 0 & 0 & 3 \\ 0 & 6 & 0 & 4 & 0 & 0 \\ 0 & 0 & 4 & 0 & 2 & 0 \\ 0 & 0 & 0 & 2 & 0 & 0 \\ 0 & 3 & 0 & 0 & 0 & 0 \end{bmatrix}
$$

The summation of the off-diagonal elements in the upper (or lower) matrix-triangle produces the second Zagreb index $ZG(2)$ [2,3,5-7].

$$ZG(2) = (½) \sum_{i,j} \left[{}^{v}\mathbf{ZM} \right]_{ij} \tag{6}$$

Finally, the *modified* edge-Zagreb matrix, denoted by ${}^{me}\mathbf{ZM}$, is defined as:

$$\left[{}^{me}\mathbf{ZM} \right]_{ij} = \begin{cases} 1/[d(i)d(j)] & \text{if vertices } i \text{ and } j \text{ are adjacent} \\ 0 & \text{otherwise} \end{cases} \tag{7}$$

As an example, the modified edge-Zagreb matrix ${}^{me}\mathbf{ZM}$ of T is presented below.

$${}^{me}\mathbf{ZM}(\mathrm{T}) = \begin{bmatrix} 0 & (3)^{-1} & 0 & 0 & 0 & 0 \\ (3)^{-1} & 0 & (6)^{-1} & 0 & 0 & (3)^{-1} \\ 0 & (6)^{-1} & 0 & (4)^{-1} & 0 & 0 \\ 0 & 0 & (4)^{-1} & 0 & (2)^{-1} & 0 \\ 0 & 0 & 0 & (2)^{-1} & 0 & 0 \\ 0 & (3)^{-1} & 0 & 0 & 0 & 0 \end{bmatrix}$$

The summation of the off-diagonal elements in the upper (or lower) matrix-triangle produces the modified second Zagreb index ${}^{m}ZG(2)$ [5,6].

$${}^{me}ZG(2) = (½) \sum_{i,j} \left[{}^{me}\mathbf{ZM} \right]_{ij} \tag{8}$$

The ${}^{m}ZG(2)$ index correlates with the Randić vertex-connectivity index [8]. The corresponding graph-theoretical matrix, called the vertex-connectivity matrix, denoted by ${}^{v}\mathbf{X}$, is defined as [9]:

$$\left[{}^{v}\mathbf{X} \right]_{ij} = \begin{cases} [d(i)d(j)]^{-½} & \text{if vertices } i \text{ and } j \text{ are adjacent} \\ 0 & \text{otherwise} \end{cases} \tag{9}$$

The vertex-connectivity matrix of T is given as follows.

$${}^{v}\mathbf{X}(\mathrm{T}) = \begin{bmatrix} 0 & (3)^{-½} & 0 & 0 & 0 & 0 \\ (3)^{-½} & 0 & (6)^{-½} & 0 & 0 & (3)^{-½} \\ 0 & (6)^{-½} & 0 & (4)^{-½} & 0 & 0 \\ 0 & 0 & (4)^{-½} & 0 & (2)^{-½} & 0 \\ 0 & 0 & 0 & (2)^{-½} & 0 & 0 \\ 0 & (3)^{-½} & 0 & 0 & 0 & 0 \end{bmatrix}$$

2. Zagreb matrices in terms of the edge-degrees

The Zagreb matrices in terms of the edge-degrees are the $E \times E$ matrices. There are also six kinds of these matrices. However, it should be noted that the Zagreb matrices of a (molecular) graph G in terms of the edge-degrees are the Zagreb matrices of the corresponding line graph of G, L(G), in terms of the vertex-degrees. . This must be so because the edges in G are replaced by vertices in L(G) [10]. Since below is given the line graph of T, L(T), the above can easily confirmed.

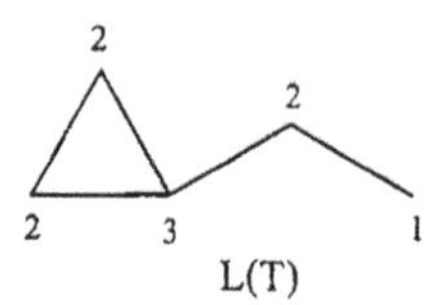

L(T)

The edge-connectivity index correlates with the Estrada edge-connectivity index [11]. The corresponding matrix, called ther edge-connectivity matrix [12], denoted by $^e\mathbf{X}$, of a graph T is the vertex-connectivity matrix of the corresponding line graph L(T). As an example, we give $^e\mathbf{X}$ of L(T), given above together with the vertex-degrees:

$$^e\mathbf{X}(\mathrm{T}) = \begin{bmatrix} 0 & (4)^{-1/2} & (6)^{-1/2} & 0 & 0 \\ (4)^{-1/2} & 0 & (6)^{-1/2} & 0 & 0 \\ (6)^{-1/2} & (6)^{-1/2} & 0 & (6)^{-1/2} & 0 \\ 0 & 0 & (6)^{-1/2} & 0 & (2)^{-1/2} \\ 0 & 0 & 0 & (2)^{-1/2} & 0 \end{bmatrix}$$

It is interesting to note that the line graph concept, though not in the explicit mathematical formalism, may be traced back to the beginnings of structural chemistry. Thus, van't Hoff represented simple organic molecules in terms of points and lines, that is, in terms of the line graphs of the modern structural formulas [*e.g.*, 13].

It should be also noted that the Zagreb indices found moderate, but persistent use in the structure-property modeling [*e.g.*, 14].]. In this respect, a recent contribution by Peng *et al.* [15] is important, because these authors showed how to improve the use of the Zagreb indices in modeling molecular properties.

Acknowledgments

This research was supported by Grant No. 0098034 from the Ministry of Science, Education and Sports of Croatia and by a grant from the Croatian-Slovenian joint research collaboration in computational and mathematical chemistry.

References

[1] N. Trinajstić, *Chemical Graph Theory*, 2nd edition, CRC, Boca Raton, FL, 1992.

[2] I. Gutman and N. Trinajstić, Graph theory and molecular orbitals. Total π-electron energy of alternant hydrocarbons, *Chem. Phys. Lett.* **17** (1972) 535-538.

[3] I. Gutman, B. Ruščić, N. Trinajstić and C.F. Wilcox, Jr., Graph theory and molecular orbitals. XII. Acyclic polyenes, *J. Chem. Phys.* **62** (1975) 3399-3405.

[4] I. Gutman and K. C. Das, The first Zagreb index 30 years after, *MATCH Commun. Math. Comput. Chem.* **50** (2004) 83-92.

[5] S. Nikolić, G. Kovačević, A. Miličević and N. Trinajstić, The Zagreb indices 30 years after, *Croat. Chem. Acta* **76** (2003) 113-124.

[6] A. Miličević, S. Nikolić and N. Trinajstić, On reformulated Zagreb indices, *Mol. Diversity* **8** (2004) 393-399.

[7] K. C. Das and I. Gutman, Some properties of the second Zagreb index, *MATCH Commun. Math. Comput. Chem.* **52** (2004) 103-112.

[8] M. Randić, On characterization of molecualr branching, *J. Am. Chem. Soc.* **97** (1975) 6609-6615.

[9] M. Randić, Similarity based on extended basis descriptors, *J. Chem. Inf. Comput. Sci.* **32** (1992) 686-692.

[10] F. Harary, *Graph Theory*, 2nd printing, Addison-Wesley, Reading, MA, 1971.

[11] E. Estrada, Edge adjacency relationships and a novel topological index related to molecular volume, *J. Chem. Inf. Comput. Sci.* **35** (1995) 31-33.

[12] D. Janežič, A. Miličević, S. Nikolić and N. Trinajstić, *Graph-Theoretical Matrices in Chemistry*, Mathematical Chemistry Series, Kragujevac, 2006.

[13] I. Gutman, Lj. Petrović, E. Estrada and S.H. Bertz, The line graph model. Predicting physico-chemical properties of alkanes, *ACH-Models in Chemistry* **135** (1998) 147-155.

[14] R. Todeschini and V. Consonni, *Handbook of Molecular Descriptors*, Wiley-VCH, Weinheim, 2000.

[15] X.-L. Peng, K.-T. Fang, Q.-N. Hu and Y.-Z. Liang, Impersonality of the connectivity index and recomposition of topological indices according to different properties, *Molecules* **9** (2004) 1089-1099.

Brill Academic Publishers
P.O. Box 9000, 2300 PA Leiden
The Netherlands

Lecture Series on Computer and Computational Sciences
Volume 6, 2006, pp. 410-415

Anharmonic calculation of vibrational spectra for P_4O_6 and P_4O_{10} systems.

P. Carbonnière[1] and C. Pouchan

Université de Pau et des Pays de l'Adour
Laboratoire de Chimie Théorique et Physico-Chimie Moléculaire
UMR 5624 – FR CNRS IPREM 2606
IFR Rue Jules Ferry – BP 27540
64075 Pau - France

Received 17 July, 2006; accepted 20 July 2006

Abstract: This paper presents the first anharmonic vibrational results obtained for the tetraphosphorus hexoxide and the tetraphosphorus decaoxide from B3LYP and PBE0, DFT calculations using 6-311+G(2d) and 6-31+G(d) basis sets. These theoretical results are faced with the existing Raman and IR data for a new vibrational analysis.
Keywords: DFT model, phosphorus oxide, anharmonic vibrational calculations.

1. Introduction

The reactions of oxygen atoms with phosphorus compounds are among the most intense chemiluminescent reactions and form the ground of possible laser systems [1-2]. This is the case of P_4O_{10} largely used in the development of high-energy laser glasses [3]. Belonging at the group V oxides, several of these coumponds have generated considerable interest due to their cage inorganic structures used as molecular models for network solids [4]. In particular, the adamantoïd cages present interesting chemical characteristic because their formally empty 3d orbitals is known to participate in bonding. This is particularly relevant in case of phosphorus coumpounds as P_4O_6 and P_4O_{10}. The first one P_4O_6 named tetraphosphorus hexoxide formed with white phosphorus heated in a shortage of oxygen, has largely examined and experimental informations including electron diffraction studies [5] on vapor phase and infrared and Raman spectroscopic properties [6] at room temperature are available. The second P_4O_{10}, named tetraphosphorus decaoxide formed in the same condition in presence of oxygen excess, has recently focused attention due to its use in the development of high-energy laser glasses [3] and its application as a host material for the vitrification of nuclear wastes [7]. Relevant informations concerning diffraction [8], X ray [9] and IR and Raman [10-13] studies have provided useful complementary structured information to investigate changes in structure as a function of the phase.

Theoretical studies support the experimental works in particular in the vibrational analysis [14-15] but these studies are limited to simple harmonic approach known to give, following the wave-function used, an approximate description needing the use of scaling factor to obtain the vibrational frequencies. The aim of this paper is to present from an anharmonic approach using a DFT quartic force field reliable values of the vibrational spectra for the two systems P_4O_6 and P_4O_{10} considered.

2. Method and computational details

In the basis of curvilinear coordinates S_k and their conjugate moment $P(S_k)$ the quantum mechanical pure vibrational Hamiltonian is written as

[1] Corresponding author : philippe.carbonniere@univ-pau.fr

$$H_v = \frac{1}{2}\sum_{i,j} g_{ij}(s) P_{S_i} P_{S_j} + V(s)$$

where g_{ij} is the element of the **G** matrix described by Wilson and $V(s)$ the potential function expressed from a complete set of quadratic, cubic and quartic force constants by

$$V(S) = \frac{1}{2}\sum_{ij} f_{ij} S_i S_j + \frac{1}{6}\sum_{ij} f_{ijk} S_i S_j S_k + \frac{1}{24}\sum_{ij} f_{ijkl} S_i S_j S_k S_l$$

For systems in which anharmonicity is weak, it is possible to write the potential function $V(s)$ as a Taylor expansion series in terms of curvilinear displacement coordinates. Quadratic (f_{ij}), cubic (f_{ijk}), and quartic (f_{ijkl}) force constants are generally obtained either by fitting the electronic energy calculated by ab initio methods for various nuclear configurations close to the optimized geometry or by a finite difference procedure including first and second derivatives of the electronic energy with respect to the nuclear coordinates. In both cases, the required number of ab initio or DFT data points grows strongly with increasing size of the molecule, so that it becomes difficult to determine accurately a complete quartic force field for current organic and inorganic systems. To reduce the number and the time of calculations, we have used in this work a parallel procedure implemented in our previously described code REGRESS EGH [15]. Briefly, this strategy includes in the same process of linear regression of values the energy and, if accessible, the corresponding first analytical derivatives obtained at each point of a well-suited grid. Thus, this approach allows the determination of the analytical form of the potential with a significantly reduced number of points to be calculated without affecting the accuracy of the results. Moreover the tetramodes interactions ($i \neq j \neq k \neq l$), particularly weak, have been discarded.

The kinetic part of the Hamiltonian also can be written as a Taylor expansion in terms of curvilinear coordinates:

$$T = \frac{1}{2}\sum_{ij} g_{ij}(0) + \frac{1}{2}\sum_{ij} g_{ijk} P_{S_i} S_{S_j} P_{S_k} + \ldots$$

in which $g_{ijk} = \left(\partial g_{ij} / \partial S_k\right)_0$ and where $g_{ij}(0)$ are the terms calculated in the harmonic approximation. This approach can be retained for the kinetic operator when large amplitude displacements are involved. In our calculations, both the kinetic and potential parts of the vibrational Hamiltonian are expressed in the basis of dimensionless normal coordinates q_i and conjugate momentum operators.

Once the potential energy function is obtained, the second step in the vibrational energy-level calculations consists of solving the vibrational Schrödinger equation. This can be treated by using the perturbational [16], variational [17] or mixed perturbation-variation approaches [18].

The perturbational process is the simplest and fastest way to obtain the vibrational energy levels for large molecules, however this method often encounters problems for near-degenerate vibrational states as Fermi and Darling-Dennison resonances. An improvement can be achieved by treating the strongest interactions by diagonalizing the corresponding part of the Hamiltonian matrix [19]. The variational method has often been restricted to small systems due to the size of the matrix to be diagonalized. When a configuration interaction (CI) is guessed by a preliminary perturbative treatment [18], the term mixed perturbation-variation approach can be used. Indeed, the vibrational Hamiltonian representation is "in fine" diagonalized in a subspace built iteratively by means of a second-order Rayleigh-Schrödinger perturbation theory where only those configurations with weights greater than a given threshold are included in the primary subspace for the following iterations. The multireference vibrational function is then corrected to first order by the remaining states which interact weakly.

The molecular energies, gradients and Hessians of P_4O_6 and P_4O_{10} around their optimized geometry were carried out by the Gaussian 03 package [20] using the B3LYP and PBE0 models with the 6-311+G(2d) and the 6-31+G(d) basis sets. Second order perturbative approach has then been retained for the vibrational treatment since we are aimed on a reliable description of fundamental transitions exclusively. Harmonic IR and Raman intensities have also been computed.

3. Results and discussion

The structures of P_4O_6 and P_4O_{10} in tetrahedral (Td) symmetry are calculated with the DFT method using two exchange-correlation function (B3LYP and PBE0) and two polarized basis sets augmented by diffuse functions [6-31+G(d) and 6-311+G(2d)]. The two structures are displayed in fig. 1 and the optimized parameters are reported in Table 1 in comparison with the more recent experimental data. As can be seen from the results all calculated angles are very close to experiment whatever the functional and the basis set chosen since the theoretical results do not deviate more than 0.5° from the electron diffraction studies. For the bond lengths, the extension of the basis set coupled with the use of PBE0 functional in place of B3LYP decrease the interatomics P-O distances. In the two cases, the best results relatively to the experimental data are given by the PBE0/6-311+G(2d) calculations.

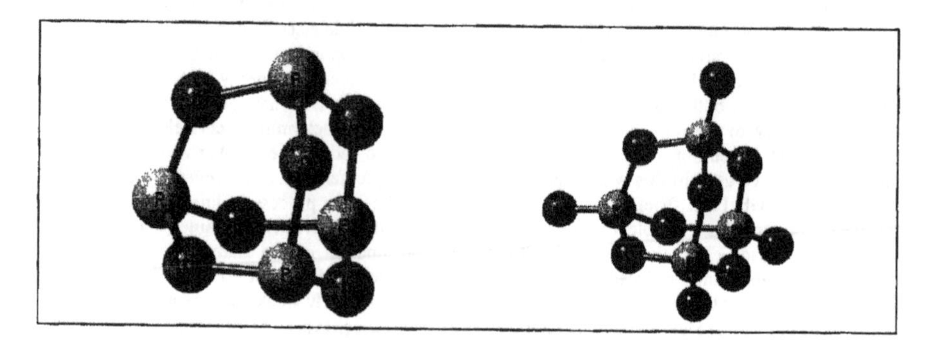

Figure 1 : Geometrical structure of P_4O_6 and P_4O_{10}.

Table 1 : Calculated and Measured Bond length (in E) and Angles (in °) for P_4O_6 and P_4O_{10}

P_4O_6	B3LYP/ 6-31+G(d)	B3LYP/ 6-311+G(2d)	PBE0/ 6-31+G(d)	PBE0/ 6-311+G(2d)	experiment
R(PO)	1,680	1,665	1,668	1,664	1,638 ± 0,003
a(OPO)	99,7	99,9	99,7	100	99,8 ± 0,8
a(POP)	126,7	126,3	126,6	126,1	126,7 ± 0,7
P_4O_{10}					
R(PO)	1,636	1,620	1,625	1,610	1,604 ± 0,003
R(PO)[a]	1,455	1,440	1,450	1,436	1,429 ± 0,003
a(OPO)	101,52	101,8	101,7	102	101,6 ± 0,8
a(OPO)[a]	116,6	116,3	116,5	116,15	116,5 ± 0,7
a(POP)	123,7	123,2	123,4	122,8	123,5 ± 0,7

[a] Terminal oxygen atom.

The calculated harmonic frequencies for the two molecules are presented in Table 2. One can observe that the minimal frequencies are given by the B3LYP/6-31+G(d) calculations and the maximal ones by the PBE0/6-311+G(2d) computations. The mean deviation between the two approaches is only 3.3 % for each system i.e about 20 cm^{-1}. The PBE0/6-311+G(2d) calculations are expected to give, as for the geometrical parameters, the best harmonic frequencies. For this reason, IR and Raman intensities are calculated only at this level of calculation

Table 2 : Calculated Harmonic Vibrational Frequencies (cm^{-1}), IR (Km/mol, in parenthesis) and Raman Intensities (Å4/amu, in square bracket) for P_4O_6 and P_4O_{10} and Experimental Frequencies (in cm^{-1}) observed in Infrared and Raman Spectra.

P_4O_6

Symmetry	Label	B3LYP/6-31+G(d)	B3LYP/6-311+G(2d)	PBE0/6-31+G(d)	PBE0/6-311+G(2d)		
A1	ν_1	709	712	715	714	(0)	[0,5]
	ν_2	588	587	610	609	(0)	[40,4]
E	ν_3	635	640	663	668	(0)	[1,7]
	ν_4	283	294	286	297	(0)	[2,2]
T1	ν_5	570	565	621	618	(0)	[0]
	ν_6	268	267	272	269	(0)	[0]
T2	ν_7	915	908	950	943	(700,4)	[3,3]
	ν_8	619	621	645	646	(62,2)	[6,8]
	ν_9	527	534	534	538	(0,0)	[0,0]
	ν_{10}	386	388	391	392	(18,9)	[2,3]

P_4O_{10}

Symmetry	Label	B3LYP/6-31+G(d)	B3LYP/6-311+G(2d)	PBE0/6-31+G(d)	PBE0/6-311+G(2d)		
A1	ν_1	1409	1424	1437	1449	(0)	[55,0]
	ν_2	684	696	695	706	(0)	[15,6]
	ν_3	523	530	534	541	(0)	[30,1]
E	ν_4	798	813	823	838	(0)	[1,0]
	ν_5	319	331	323	333	(0)	[333,0]
	ν_6	237	241	242	244	(0)	[4,1]
T1	ν_7	794	802	831	842	(0)	[0]
	ν_8	387	397	395	404	(0)	[0]
	ν_9	251	253	253	254	(0)	[0]
T2	ν_{10}	1378	1393	1406	1418	(400,6)	[11,35]
	ν_{11}	979	985	1012	1018	(661,2)	[0,2]
	ν_{12}	733	746	756	769	(135,2)	[0,4]
	ν_{13}	548	553	553	556	(11,2)	[0,2]
	ν_{14}	392	398	396	401	(25,9)	[3,9]
	ν_{15}	257	261	261	264	(18,2)	[1,3]

Table 3 contains the calculated anharmonic vibrational frequencies in comparison with the observed IR and Raman spectra for each system. It should be noted that the computational results are in fair agreement with the most reliable observed data since the mean deviation is about 13 cm^{-1} and 11 cm^{-1} for P_4O_6 and P_4O_{10} respectively. In all cases the anharmonicity correction is found to be weak. These contributions reach 26 cm^{-1}, 20 cm^{-1} and 17 cm^{-1} for ν_5, ν_3 and (ν_7,ν_8) respectively in the case of P_4O_6 and are found smaller than 10 cm^{-1} for all other vibrations. For P_4O_{10}, except for ν_1 and ν_{11}, the anharmonic effect often is negligible.

Table 3 : Calculated Anharmonic Vibrational Frequencies (cm-1) for P_4O_6 and P_4O_{10} and Experimental Frequencies (in cm-1) Observed in Infrared and Raman Spectra.

P_4O_6		theory	experiment		
Symmetry	Label	PBE0/6-31+G(d)	Beattie et al.	Chapman	Mowrey et al.
A_1	ν_1	708		718	718
	ν_2	602	620	613	613
E	ν_3	643		643	691
	ν_4	281	305	302	302
T_1	ν_5	595		569	702
	ν_6	268		302	285
T_2	ν_7	926	959	919	919
	ν_8	629	642	643	643
	ν_9	527	562	549	549
	ν_{10}	386	408	407	407

P_4O_{10}		theory	experiment		
Symmetry	Label	PBE0/6-311+G(2d)	Beattie et al.	Konings et al,	Chapman
A_1	ν_1	1432	1440		1413
	ν_2	704	717		721
	ν_3	537	553		556
E	ν_4	829			
	ν_5	329			
	ν_6	241	254		258
F_1	ν_7	835			
	ν_8	401			
	ν_9	251			
F_2	ν_{10}	1412	1406	1404	1390
	ν_{11}	(977 : F)		1012	1010
	ν_{12}	758		764	760
	ν_{13}	551		575	573
	ν_{14}	395	411	409	424
	ν_{15}	261	264	270	278

Acknowledgments

The author wishes to thank the anonymous referees for their careful reading of the manuscript and their fruitful comments and suggestions.

References

[1] M.E Fraser, d. h. Stedman? J. Chem. Soc. Faraday Trans I 79 (1983) 527.

[2] D. G. Harais, M. S. Chou, T. A. Cool, J. Chem. Phys. 82 (19985) 3502 and reference therein.

[3] M. Karabulut, G. K. Marainghe, C. A. Click, E. Metwalli, R. K. Brow, J. J. Booth, C. H. Bucher, D. K. Shuh, T. I. Suratwala, J. H. Campbell, J. Am. Ceram. Soc. 85 (2002) 1093.

[4] C. L. Bowes, W. U. Huynh, S. J. Kirby, A. Malek, G. A. Ozin, S. Petrov, M. Twardowski, D. Young, R. L. Bedard, R. Broach. Chem. Mater. 8 (1996) 2147.

[5] B. Beagley, D. W. J. Cruickshank, T. G. Hewitt, K. H. Jost, Trans. Faraday Soc. 65 (1969) 1219.

[6] A. C. Chapman, Spectrochim. Acta A24 (1968) 1687.

[7] G. K. Marasinghe, M. K. Karabulut, C. S. Ray, D. E. Day, D. K. Shuh, P.G. Allen, M. L. Saboungi, M. Grimsditch, D. J. Haeffner, J. Non-Cryst. Solids, 263 (2000) 146.

[8] B. Beagley, D. W. J. Cruickshank, T. G. Hewitt, Trans. Faraday Soc. 63 (1967) 836.

[9] H. C. J. Dedecker, C. H. MacGillavry, Recl. Trav. Chim. 60 (1941) 153.

[10] I. R. Beattie, K. M. S. Livingston, G. A. Ozin, D. J. Reynolds, J. Chem. Soc. A3 (1970) 449.

[11] Z. Mielke, L. J. Andrews, J. Phys. Chem. 93 (1989) 2971.

[12] R. J. M. Konings, E. H. P. Cordfluke, A. S. Booij, J. Mol. Spectrosc. 152 (1992) 29.

[13] S. J. Gilliam, S. J. Gilliam, S. J. Kirkby, C. N. Merrow, D. Zeroka, A. Banerjee, J. O. Jensen, J. Phys. Chem. B, 107 (2003) 2892.

[14] J. O. Jensen, A. Banerjee, C. N. Merrow, D. Zeroka, J. M. Lochner, J. Mol. Struct. (Theochem) 531 (2000) 323.

[15] J. O. Jensen, A. Banerjee, D. Zeroka, C. N. Merrow, S. J. Gilliam, S. J. Kirkby, Spectrochim. Acta, A60 (2004) 1947.

[16] D. Papousek and R. Aliev, *Molecular Vibrational Rotational Spectra.* (Elsevier: Amsterdam 1982).

[17] P. Cassam-Chenaï and J. Lievin, Int. J. of Quantum Chem. **93** 245 (2003).

[18] C. Pouchan and K. Zaki, J. Chem. Phys. **107** 342 (1997).

[19] P. Carbonnière, T. Lucca, C. Pouchan, N. Rega and V. Barone, J. Comput. Chem. **26** 384 (2005).

Brill Academic Publishers
P.O. Box 9000, 2300 PA Leiden
The Netherlands

Lecture Series on Computer and Computational Sciences
Volume 6, 2006, pp. 416-425

How accurately can structural, spectroscopic and thermochemical properties be predicted by ab initio computations?

Cristina Puzzarini[1]

Dipartimento di Chimica "G. Ciamician,
Università di Bologna,
Via F. Selmi 2
I-40126, Bologna Italy

Received July 1, 2006; accepted 10 July, 2006

Abstract: Nowadays, ab initio calculations are able to provide very accurate predictions of molecular properties and energetics not only for compounds containing light atoms (such as first-row elements) but also for heavy atom (from second-row elements to transition metals) containing systems. The predictive capabilities of ab initio computations are consolidated in various fields:
- On one hand, theoretical computations themselves are able to provide very accurate equilibrium geometries by performing highly correlated calculations in conjunction with extrapolation to the complete basis set limit and accurate treatments of core-valence correlation. On the other hand, the combination of experimental ground-state rotational constants and calculated vibration-rotation interaction constants leads to one of the most successful approaches to the determination of molecular equilibrium structures: the most accurate way for obtaining equilibrium geometry for large molecules.
- Theoretical predictions recover a fundamental role in the high resolution spectroscopy field. For example, it is even possible to accurately predict the fine- and/or hyperfine-structure of rotational spectra due to electric and/or magnetic interactions; that is to say that spin-rotation and nuclear quadrupole coupling constants can be accurately evaluated by ab initio calculations.
- High-level ab initio calculations can be competitive with experiment in the precise determination of thermochemical properties, such as ionization potential, electron affinity, enthalpy of formation, ... By accounting for model and basis set truncations as well as core-valence, spin-orbit and, in case, relativistic effects the chemical accuracy (<1 kcal/mol) can be reliably obtained.

Keywords: molecular structure, spectroscopic properties, thermochemical properties,

1. Methodology

1a. Molecular structure determination and energy evaluation

In order to obtain highly accurate results, first of all it is important to reduce as much as possible errors due to the wave function model truncation. From the literature, it is well known that the coupled-cluster level of theory with single and double excitations, and a quasiperturbative account for triples substitutions [CCSD(T)] is able fulfill this requirement. Therefore, the CCSD(T) method has been used for the closed-shell species, whereas for the open-shell molecules the variants denoted R/RCCSD(T) or R/UCCSD(T), which are based on restricted open-shell Hartree-Fock (ROHF) orbitals but spin restricted or unrestricted, respectively, in the solution of the CCSD equations, have been employed.
To account for basis set truncation errors, hierarchical series of correlation consistent basis sets have been employed in all the investigations here reported. More precisely, the standard cc-pV*n*Z sets [1] have been employed for the first-row atoms, the *d*-augmented valence cc-pV(*n*+*d*)Z bases [2] for the second row atoms, and small-core relativistic pseudopotential cc-pV*n*Z-PP sets [3-5] for heavier atoms and transition metals. The frozen core approximation has been adopted in the computations involving these basis sets.

[1] Corresponding author. E-mail: cristina.puzzarini@unibo.it

Geometry optimizations have been carried out using the MOLPRO suite of programs [6].
Since hierarchical sequences of bases have been considered, the systematic trend of geometrical parameters can be exploited to estimate the infinite (or complete) valence basis set (CBS) limit. Making use of the assumption that the convergence of the structural parameters has the same functional form as the correlation energy contribution, the consolidated $1/n^3$ extrapolation form has been used to describe this convergence behavior [7]:

$$\Delta r^{corr}(n) = \Delta r_{\infty}^{corr} + An^{-3} , \quad (1)$$

and it has been applied to the two largest values of n considered. To obtain the extrapolated structure (denoted r(CBS,*valence*)), the CBS limit value of the correlation contribution has then been added to the HF-SCF CBS limit, which is assumed to be reached at the HF-SCF/cc-pV6Z level.
Complete basis set limits of the total energies have been obtained following two different extrapolation schemes. The first approach is the same followed for the extrapolation of the geometrical parameters, i.e., the CBS limit has been obtained by extrapolating the correlation contributions to the complete basis limit by $1/n^3$ [7]:

$$\Delta E^{corr}(n) = \Delta E_{\infty}^{corr} + A'n^{-3} , \quad (2)$$

and adding the Hartree-Fock complete basis limit, evaluated by the formula [8]

$$\Delta E^{SCF}(n) = \Delta E_{\infty}^{SCF} + B'\exp(-C'n) . \quad (3)$$

The second extrapolation of energies to the complete basis set limit has been performed by using the mixed exponential/Gaussian extrapolation formula [9]

$$E(n) = E_{CBS} + B''e^{-(n-1)} + C''e^{-(n-1)^2} . \quad (4)$$

Making use of the additivity approximation, corrections for taking into account core-valence (CV) correlation effects have then been added to the CBS limit of energies and geometries. This involved carrying out both energy evaluations and geometry optimizations using the core-valence correlation consistent cc-pwCVnZ basis set [10] (cc-pwCV(n+d)Z for second-row atoms, cc-pwCVnZ-PP for heavier atoms and transition metals). In regards to the geometrical parameters, the core correlation corrections have been added to the extrapolated structures r(CBS,*valence*) as:

$$r_e \cong r(CBS, valence) + r(wCVnZ, all) - r(wCVnZ, valence) , \quad (5)$$

where r(wCVnZ,*all*) and r(wCVnZ,*valence*) are the geometries optimized at the CCSD(T)/cc-pwCVnZ level correlating all and only valence electrons, respectively.
The core-valence corrections to the total energies have been calculated as

$$\Delta E_{CV} = E_{core+val} - E_{val} , \quad (6)$$

where $E_{core-val}$ is the CCSD(T) total energy obtained by correlating all electrons and E_{val} is the CCSD(T) total energy obtained in the frozen core approximation, both in the same cc-pwCVnZ basis set.
In order to obtain accurate results for molecular structures and energies, it should be clear that when heavy elements are involved scalar relativistic (SR) as well as spin-orbit (SO) effects should also be taken into account. As concerns the first ones, the use of pseudopotentials for heavy atoms usually allows to completely recover such corrections; for compounds containing less heavy elements the Douglas-Kroll-Hess approach can be employed [11,12].
Quantum chemical calculations not only allow the pure theoretical determination of equilibrium structures, but also are able to accurately evaluate vibrational corrections to rotational constants. The combination of experimental ground-state rotational constants with calculated vibrational corrections has turned out to be a very powerful approach to accurately determine "experimental" structures for polyatomic molecules (see for example Ref. [13]). Due to the computational contribution, these structures are usually referred to as "empirical'", "mixed experimental/theoretical'", or "semiexperimental" The accuracy of such a joint experimental and theoretical procedure for the

determination of equilibrium structures is well known to be typically better than 0.001 Å [13], as long as an electron-correlated approach is used in the calculation of the vibrational corrections.
Empirical equilibrium structures have been obtained by least-squares fits of the molecular structural parameters to the given equilibrium rotational constants. These fits require at least as many independent rotational constants as there are structural degrees of freedom, but the use of more constants is preferred, as in this way the consistency of the available experimental data can be checked. "Experimental" equilibrium rotational constants B_e^i have been derived from the experimentally derived ground-state constants B_0^i by correcting them for vibrational effects:

$$B_e^i = B_0^i + \frac{1}{2}\sum_r \alpha_r^i \tag{7}$$

using the computed vibration-rotation interaction constants α_r^i, where r runs over the vibrational modes and i over the inertial axes. Therefore, for each isotopic species involved in the fit, the cubic force field should be evaluated to compute the vibration-rotation interaction constants using expressions from the usual second-order perturbation treatment of the rovibrational problem.
The anharmonic (cubic) force fields have been evaluated at the MP2/cc-pVTZ and CCSD(T)/cc-pVTZ levels using the Mainz-Austin-Budapest version of the ACES2 program package [14], with MP2 denoting second-order Möller-Plesset perturbation theory. The harmonic force fields have been obtained using analytic second derivatives of the energy [15], and the corresponding cubic force fields have been determined in a normal-coordinate representation via numerical differentiation of the analytically evaluated force constants as described in Ref. [16]. The cubic force field, initially obtained for the main isotopic, has then been transformed to the normal-coordinate representations of all of the isotopologues for which experimental ground-state rotational constants are available.

1b. Molecular and spectroscopic properties

The dipole moment components have been evaluated as a first derivative of the total energy with respect to a homogeneous electric field at zero field strength. More precisely, the results have been obtained from computations in which finite perturbations with electric field strengths of ±0.0001 a.u. have been applied and thus the dipole moments have been calculated from central-differences numerical differentiation of the energy. Since inclusion of diffuse functions in the basis set is particularly important for an accurate evaluation of dipole moments, the calculations have been carried out with the aug-cc-pV*n*Z basis sets for first-row atoms, the aug-cc-pV(*n*+*d*)Z series of bases for second-row atoms and the pseudopotential aug-cc-pV*n*Z-PP sets for heavier atoms and transition metals (frozen core), and with the aug-cc-pwCV*n*Z (aug-cc-pwCV(*n*+*d*)Z for second-row atoms and aug-cc-pwCV*n*Z-PP for heavy elements) basis sets (where *n*=T or Q) for accounting for core-valence effects (frozen core and all electrons). The MOLPRO package [6] has been used in such computations.
Since hierarchical sequences of bases have been used, the dipole moment components have also been extrapolated to the CBS limit using the $1/n^3$ extrapolation form [17], analogously to what has been done for structural parameters. Best estimates of the dipole moment have then been obtained by adding to the CBS limit the core-valence corrections determined using the aug-cc-pwCV*n*Z basis set as analogously made for geometrical parameters.
Electric and magnetic properties that have a particular relevance as spectroscopic properties have also been evaluated at the CCSD(T) level of theory in conjunction with valence and core-valence correlation consistent basis sets of triple, quadruple and eventually quintuple zeta quality. To this purpose, the Mainz-Austin-Budapest version of the ACES2 program package [14] has been employed. More precisely, the nuclear quadrupole coupling and spin-rotation tensors, that recover a fundamental role in the high resolution spectroscopy field in order to accurately predict the fine- and/or hyperfine-structure of spectra, have been computed.
A nuclear magnetic dipole interacting with the effective magnetic field of a rotating molecule leads to a splitting of the lines in the rotational spectrum. The effective Hamiltonian responsible for this splitting can be expressed in terms of the nuclear spin I_K of the nucleus K and the total rotational angular momentum J

$$\boldsymbol{H}_{NSR} = \sum_{K=1}^{N} \boldsymbol{I}_K \cdot \boldsymbol{C}_K \cdot \boldsymbol{J} \tag{8}$$

with $\boldsymbol{C}_K$ as the nuclear spin-rotation (NSR) tensor of nucleus K. $\boldsymbol{C}_K$ is non-vanishing only for nuclei with $\boldsymbol{I}_K \geq 1/2$. In that case the number of non-vanishing tensor elements depends on the so-called site symmetry, i.e., the actual symmetry seen by the given nucleus in its position.
In evaluating $\boldsymbol{C}_K$ the use of perturbation-dependent basis functions (often referred to as rotational London Atomic Orbitals [18]) is recommended to improve basis-set convergence.
For nuclei with a quadrupole moment ($\boldsymbol{I}_K \geq 1$), a contribution to the hyperfine structure of rotational spectra arises also from the interaction of the nuclear quadrupole moment ($-eQ_K$) and the electric field gradient at that nucleus, $\boldsymbol{V}^K$

$$H_{NQC} = \frac{1}{2}\sum_K \frac{-eQ_K}{I_K(2I_K-1)} \boldsymbol{I}_K \cdot \boldsymbol{V}^K \cdot \boldsymbol{I}_K \quad . \tag{9}$$

The elements of the nuclear quadrupole coupling (NQC) tensor for nucleus K are defined as $\chi_{ij} = -eQ_K V^K_{ij}/\hbar$ where i,j refer to the inertial axes.
To monitor basis set convergence and to investigate the effect of additional diffuse functions as well as of core correlation, a wide selection of basis sets and basis-set sequences has been employed. Best estimates of these properties have been obtained from the largest valence basis set computations by adding diffuse functions as well as core-valence corrections. Vibrational corrections have also been considered: they have been obtained using the perturbational approach described in Ref. [19] for NMR shielding tensors. The vibrational corrections, defined as the difference between the vibrationally averaged and the equilibrium values obtained in the same basis, have been added to the best estimated values.

1c. Thermodynamics

Adiabatic ionization potentials

$$IP^e_{ad} = E_e(cation) - E_e(neutral) \tag{10}$$

have been obtained at the coupled-cluster level by using equilibrium total energies E_e of the neutral and cationic species. Analogously, adiabatic electron affinities have been derived by the following expression:

$$EA^e_{ad} = E_e(neutral) - E_e(anion) \quad . \tag{11}$$

From the energies computations, equilibrium dissociation energies, D_e, have also been determined as the difference between the minimum energy of the reactant and the sum of the energy of the separate fragments at their equilibrium geometry:

$$RH \rightarrow R\cdot + H\cdot \tag{12a}$$

$$RX \rightarrow R\cdot + X\cdot \quad . \tag{12b}$$

Finally, from harmonic frequencies calculations for all the species involved in Equations (10)-(12) the zero-point vibrational (ZPV) energy corrections have been obtained at the harmonic approximation:

$$E^{harm}_{ZPV} = \frac{1}{2}\sum_i d_i\omega_i \quad , \tag{13}$$

where d_i and ω_i are the degeneracy and the harmonic frequency of the i-th vibrational mode, respectively. The computations have usually been performed employing the CCSD(T) method in conjunction with quadruple-zeta basis set. Consequently, IP_0, EA_0 and D_0 have been derived from the corresponding equilibrium values, IP_e, EA_e and D_e, by adding the appropriate zero-point energy differences.

2. Results and discussion

In the present section some significant results are reported and commented out. The examples chosen collect results for compounds containing only first-row atoms, compounds containing also second-row atoms and/or heavier elements and/or transitions metals.

2a. Molecular structure

Table 1: Best estimate equilibrium structures compared to available experimental data. Distances in Å and angles in degrees.

XBS	r(X-B)	r(B-S)	ϑ(XBS)
X=H			
Best estimate [a]	1.1695	1.5978	180.0
Experiment [b]	1.1698(4)	1.5978(1)	180.0
X=F			
Best estimate [a]	1.2766	1.6084	180.0
Experiment [b]	1.2762(2)	1.6091(2)	180.0
X=Cl			
Best estimate [a]	1.6806	1.6046	180.0
Experiment [b]	1.6806(1)	1.6049(1)	180.0
XCP	r(X-C)	r(C-P)	ϑ(XCP)
X=H			
Best estimate [a]	1.0701	1.5388	180.0
Experiment [b]	1.0702(10)	1.5399(2)	180.0
X=F			
Best estimate [a]	1.2755	1.5429	180.0
Experiment [b]	--	--	180.0
X=Cl			
Best estimate [a]	1.6383	1.5472	180.0
Experiment [b]	1.6341(62)	1.5526(61)	180.0
HCS	r(H-C)	r(C-S)	ϑ(HCS)
Best estimate [c]	1.0854	1.5550	132.41
Experiment [d]	1.079(3)	1.56228(3)	132.8(3)
HCS⁺			
Best estimate [c]	1.0805	1.4741	180.0
Experiment [d]	1.080686(13)	1.475869(3)	180.0
HSC	r(H-S)	r(C-S)	ϑ(HSC)
Best estimate [c]	1.3642	1.6367	103.01
Experiment [d]	1.379(3)	1.6343(5)	104.2(2)
HSC⁺			
Best estimate [c]	1.3960	1.6072	73.15(20)
Experiment [d]	--	--	--
XPO	r(X-P)	r(P-O)	ϑ(XPO)
X=H			
Best estimate [e]	1.4520	1.4789	104.33
Experiment [f]	1.455(7)	1.4800	104.57
X=F			
Best estimate [e]	1.5718	1.4516	110.17
Experiment [f]	1.5727(1)	1.4528(2)	110.16(2)
X=Cl			
Best estimate [e]	2.0527	1.4581	109.85
Experiment [f]	2.0571(1)	1.4609(5)	109.98(3)
X=Br			
Best estimate [e]	2.2261	1.4592	110.07
Experiment [f]	--	--	--

[a] Evaluated from Eq. (5) (n=Q). Ref. [20].
[b] Experimental equilibrium structure r_e. For details see Ref. [20].
[c] Evaluated from Eq. (5) (n=5), scalar relativistic effects included. Ref. [21].
[d] Experimental r_0 or equilibrium structure r_e. For details see Ref. [21].
[e] Evaluated from Eq. (5) (n=T,Q), scalar relativistic effects included. Ref. [22].
[f] Experimental r_s or equilibrium structure r_e. For details see Ref. [22].

For compounds containing second-row atoms and/or transition metals pure theoretical equilibrium structures are given in Tables 1 and 2, respectively, and compared to the available experimental data. From these Tables a very good agreement between theory and experiment, when experimental r_e is available, is apparent. Therefore, the results reported in these two Tables provide a typical example of the accuracy that nowadays can be reached by ab initio calculations even for compounds containing heavy elements. It has to be noted that extrapolation to the valence CBS limit and inclusion of CV corrections play a fundamental role in such an accuracy.

Table 2: Best estimate equilibrium structures compared to available experimental data. Distances in Å and angles in degrees.

XAuC [a]	r(X-Au)	r(Au-C)	ϑ(XAuC)
X=F			
Best estimate	1.9060	1.7741	180.0
X=Cl			
Best estimate	2.2360	1.8022	180.0
X=Br			
Best estimate	2.3606	1.8077	180.0
X=I			
Best estimate	2.5310	1.8183	180.0
YC_2	r(Y-C)	r(C-C)	ϑ(CYC)
Best estimate [b]	2.1998	1.2690	33.53
Experiment [c]	2.1946	1.2697	33.63
Cu_2	r(Cu-Cu)		
Best estimate [d]	2.214		
Experiment [e]	2.2193		
Ag_2	r(Ag-Ag)		
Best estimate [d]	2.522		
Experiment [e]	2.5303		
Au_2	r(Au-Au)		
Best estimate [d]	2.472		
Experiment [e]	2.4715		
Zn_2	r(Zn-Zn)		
Best estimate [d]	3.847		
Experiment [e]	4.19		
Cd_2	r(Cd-Cd)		
Best estimate [d]	3.894		
Experiment [e]	4.07		
Hg_2	r(Hg-Hg)		
Best estimate [d]	3.665		
Experiment [e]	3.63(4)		

[a] Evaluated from Eq. (5) (n=T), scalar relativistic effects included using pseudopotentials. Ref. [23].
[b] Evaluated from the CBS+CV+SR potential energy surface. Ref. [24].
[c] Experimental r_0 structure. For details see Ref. [24].
[d] Evaluated from the CBS+CV+SR+SO potential energy function. Ref. [5].
[e] Experimental structure. For details see Ref. [5].

In Table 3 the results for semiempirical equilibrium structures are collected for some substituted ethylenes. From the uncertainties reported in this Table the accuracy characterizing this kind of structure is clear: as detailed shown for instance in Refs. [13,25], the semiempirical equilibrium structure is the most accurate experimental equilibrium geometry that can be derived for polyatomic molecules.
In Refs. [26,27] it is shown how accurate results for pure theoretical as well as semiempirical equilibrium structures may guide the observation and assignment of rotational spectra of species for which experimental data were missing. In particular, Ref. [27] shows that accurate theoretical predictions play a crucial role in elucidating the experimental misassignments.

Table 3: Empirical equilibrium structure of substituted ethylenes[a]. Distances in Å and angles in degrees.

***t*-CHX=CHX**	r(C-H)	r(C-X)	r(C-C)			ϑ(CCH)	ϑ(CCX)		
X=F	1.078(1)	1.339(1)	1.324(1)			125.1(1)	119.8(1)		
X=Cl	1.0793(6)	1.7131(10)	1.3367(15)			122.86(27)	121.32(5)		
***c*-CHX=CHX**	r(C-H)	r(C-X)	r(C-C)			ϑ(CCH)	ϑ(CCX)		
X=F	1.075(1)	1.334(1)	1.323(1)			122.5(1)	122.3(1)		
X=Cl	1.0796(5)	1.7103(5)	1.3301(15)			120.85(5)	124.15(2)		
***t*-CHF=CHCl**	r(C_1-H)	r(C_1-F)	r(C-C)	r(C_2-H)	r(C_2-Cl)	$\vartheta(C_2C_1H)$	$\vartheta(C_2C_1F)$	$\vartheta(C_1C_2H)$	$\vartheta(C_1C_2Cl)$
Empirical	1.0784(1)	1.3383(12)	1.324	1.0772(1)	1.7187(9)	125.80(6)	120.15(2)	123.09(5)	120.54(1)
Pure theor. [b]	1.0785	1.3376	1.3240	1.0775	1.7163	125.82	120.14	122.95	120.63
***c*-CHF=CHCl**	r(C_1-H)	r(C_1-F)	r(C-C)	r(C_2-H)	r(C_2-Cl)	$\vartheta(C_2C_1H)$	$\vartheta(C_2C_1F)$	$\vartheta(C_1C_2H)$	$\vartheta(C_1C_2Cl)$
Empirical	1.0802(6)	1.3317(3)	1.3240(14)	1.0776(4)	1.7128(6)	123.50(2)	122.61(6)	120.74(9)	123.07(1)
Pure theor. [b]	1.0787	1.3310	1.3249	1.0764	1.7107	123.43	122.53	120.43	123.10

[a] For more details see Ref. [25].
[b] Evaluated from Eq. (5) (n=Q). For details see Ref. [25].

2b. Molecular and spectroscopic properties

In Table 4 results for molecular dipole moment components are reported and compared to experiment. The main conclusion that can be derived from this Table is that vibrational effects should be taken into account when comparison to experiment is performed. It should also be noted that when permitted by the extent of the molecule it is important to account for basis set truncation as well as core-valence corrections. To get reliable results, it is also fundamental to use basis sets including diffuse functions for such computations.

Table 4: Theoretical equilibrium and vibrational averaged dipole moment components compared to experiment (absolute value).

CH_3-CD_3 [a]			μ_z (D)
Equilibrium			0.0
Vibrational averaged			-0.0111
Experiment			0.0108617(5)
SiH_3-SiD_3 [a]			μ_z (D)
Equilibrium			0.0
Vibrational averaged			0.0044
Experiment			--
F_2CNH [b]	μ_x (D)	μ_y (D)	
Equilibrium	0.427	-1.381	
Vibrational averaged	0.515	-1.281	
Experiment	0.415(1)	1.330(1)	
CF_3-CFH_2 [c]	μ_x (D)	μ_y (D)	
Equilibrium	0.361	2.033	
Vibrational averaged	0.378	2.034	
Experiment	0.4049(14)	2.0254(39)	

[a] z is the molecular symmetry axis. Vibrational averaging performed at T=100 K. For more details see Ref. [28].
[b] The molecule lies on the xy plane. For more details see Ref. [29].
[c] xy is the symmetry plane. For more details see Ref. [30].

Table 5: Theoretical equilibrium and vibrational averaged spin-rotation constants of X=F or ^{35}Cl compared to experiment. Values in kHz.

CX_2 [a]	C_{xx}	C_{yy}	C_{zz}	C_{xy}	C_{yx}
X=F					
Equilibrium	380.86	33.14	13.83	±122.87	±16.76
Vibrational averaged	380.17	33.12	13.44	±119.12	±16.50
Experiment	379(11)	31(4)	11(4)		
X=^{35}Cl					
Equilibrium	56.56	2.69	1.23	±20.05	±0.71
Vibrational averaged	56.83	2.69	1.25	±20.46	±0.73
Experiment	58.2(9)	2.8(2)	1.4(3)		
HXCl [b]					
X=C					
Equilibrium	734.00	20.24	2.42	-222.79	-7.18
Vibrational averaged	743.70	21.04	2.53	-232.91	-8.06
Experiment		23.07(47)[c]			
X=Si					
Equilibrium	115.13	5.14	-0.13	-41.62	-1.06
Vibrational averaged	114.67	5.25	-0.09	-42.20	-1.06
Experiment		5.2(7)[c]			
SiF_2 [b]					
X=F					
Equilibrium	64.09	11.62	9.88	±34.05	±10.75
Vibrational averaged	64.18	11.67	9.92	±34.70	±10.76
Experiment	65.13(36)	11.85(22)	10.12(19)		

[a] For more details see Ref. [31].
[b] For more details see Ref. [32].
[c] Experimental value is $C_{yy}+C_{zz}$: see Ref. [32].

In Table 5 theoretical spin-rotation constants are compared to the available experimental data. From this Table one can deduce the great accuracy that can be obtained when the CCSD(T) method is

employed for evaluating this property. Additionally, when large valence basis sets are employed and CV as well as vibrational corrections are accounted for, then almost quantitative results are derived. It is interesting to notice that the results for CF_2 even allowed to put on evidence that wrong experimental results were previously published (for details see Ref. [31]).
Theoretical spin-rotation constants for some rare isotopologues of HCN are reported in Table 6, where they are also compared to experiment. It is relevant to point out that the ab initio evaluation of these hyperfine parameters strongly supported the experimental determination: the theoretical values of the spin-rotation constants provided an essential support to the experimental observation in providing data for the fit as well as in establishing independent reference values for comparison (see Refs. [32,33]).

Table 6: Theoretical spin-rotation constants for some rare isotopic species of HCN compared to experiment. Values in kHz.

	C(H)	C(D)	$C(^{13}C)$	$C(^{14}N)$	$C(^{15}N)$
$H^{13}CN$					
Theory [a]	-4.56		17.68	9.72	
Experiment [b]	--		17.50(27)	10.002(50)	
$HC^{15}N$ [a]					
Theory	-4.55				-13.58
Experiment	--				-13.74(19)
$DC^{15}N$ [a]					
Theory		-0.57			-11.09
Experiment		-0.79(15)			-10.84(18)

[a] For more details see Ref. [32].
[b] For more details see Ref. [33].

As an example, the theoretical ^{35}Cl nuclear quadrupole coupling constants for a few compounds are reported and compared to experiment in Table 7. From this Table it is apparent the very good accuracy that can be obtained even for this property. The principal requirements for such an accuracy are the CCSD(T) method and the account for CV effects. In particular, the results for trans-CHF=CHCl played a fundamental role in the experimental observation and assignment of the rotational spectra [26,27].

Table 7: Theoretical equilibrium nuclear quadrupole coupling constants[a] of ^{35}Cl compared to experiment. Values in MHz.

HCCl [b]	χ_{xx}	χ_{yy}	χ_{xy}
Equilibrium	-44.48	52.91	-4.02
Experiment	-45.7907(20)	54.1(2.1)	
HSiCl [b]			
Equilibrium	-26.77	28.80	-0.79
Experiment	-27.303(3)		
$SiCl_2$ [b]			
Equilibrium	-7.75	5.37	±23.15
Experiment	-8.061(80)	5.52(11)	
***trans*-CHF=CHCl** [c]			
Equilibrium	-63.13	27.52	-33.34
Experiment	-63.586(58)	27.53(27)	

[a] The nuclear quadrupole coupling tensor is traceless.
[b] For more details see Ref. [34].
[c] For more details see Ref. [27].

2c. Thermochemical properites

Table 8: Ionization potentials, electron affinities and dissociation energies of HCNN and HNCN

HCNN [a]	IP_{ad} (eV)	Ea_{ad} (eV)	D(C-H) (kcal/mol)
Equilibrium	9.653	1.698	83.76
Zero-point corrected	9.601	1.683	77.20
Experiment	--	1.685(6)	78(4)
HNCN [a]	IP_{ad} (eV)	Ea_{ad} (eV)	D(N-H) (kcal/mol)
Equilibrium	10.808	2.667	89.27
Zero-point corrected	10.799	2.627	82.32
Experiment	--	2.624(9)	81.6(1.9)

[a] For details see Ref. [35].

In Table 8 some thermochemical properties, such as ionization potentials, electron affinities and dissociation energies, of HCNN and HNCN are collected and compared to the available experimental data [35]. From this Table it can be concluded that ab initio calculations can be really competitive with experiment in the precise determination of thermochemical properties; in fact, by accounting for model and basis set truncations as well as core-valence, spin-orbit and, in case, relativistic effects the chemical accuracy (<1 kcal/mol) can be reliably obtained.
In Table 9 dissociation energies of XPO (X=H,F,Cl,Br) compounds (X-P bond) [22] and of transition metal group 11 (Cu,Ag,Au) and 12 (Zn,Cd,Hg) diatomic species [5] are collected and compared to experiment. It should be noted that the great accuracy observed for molecules containing only first-row atoms can also be obtained for species containing second-row elements or even transition metals. This excellent result has been made feasible thanks to suitable basis sets recently developed (see for instance Ref. [5] and references therein).

Table 9: Theoretical equilibrium and zero-point corrected dissociation energies compared to experiment. Values in kcal/mol (values in cm^{-1} for Zn_2, Cd_2 and Hg_2).

X—PO [a]	X=H	X=F	X=Cl	X=Br		
Equilibrium	70.07	133.48	90.61	76.72		
Zero-point corr.	65.22	130.76	88.19	72.08		
Experiment (D_0)	--	--	--	72.9(2)		
X_2 [b]	X=Cu	X=Ag	X=Au	X=Zn	X=Cd	X=Hg
Equilibrium	46.6	39.4	53.2	226	325	398
Experiment (D_e)	47.93(57)	38.0(8)	53.5(1)	279.1	330.5	380(15)

[a] For details see Ref. [22].
[b] For details see Ref. [5].

3. Conclusions and remarks

In conclusion, it has been shown how ab initio calculations are able to provide very accurate predictions of molecular, spectroscopic and thermochemical properties not only for compounds containing light atoms (i.e., first-row elements) but also for heavy atoms (from second-row elements to transition metals) containing systems. It has also been demonstrated the importance of the predictive capabilities of ab initio computations in various fields: from high-resolution spectroscopy to thermochemistry.

Acknowledgments

This work has been supported by 'PRIN 2005' funds (project "Trasferimenti di energia e di carica a livello molecolare"), and by University of Bologna (funds for selected research topics and ex-60% funds).
The author wishes to thank Prof. G. Cazzoli (University of Bologna), Prof. J. Gauss (University of Mainz), Prof. K. A. Peterson (Washington State University) and Prof. P. R. Taylor (University of Warwick) for the fruitful collaboration.

References

[1] T. H. Dunning, Jr., *J. Chem. Phys.* **90** 1007 (1989).

[2] T. H. Dunning, Jr., K. A. Peterson, and A. K. Wilson, *J. Chem. Phys.* **114** 9244 (2001).

[3] K. A. Peterson, D. Figgen, E. Goll, H. Stoll, M. Dolg, *J. Chem. Phys.* **119** 11113 (2003).

[4] K. A. Peterson, A. K. Wilson and K. A. Peterson KA (Eds.): *Recent Advances in Electron Correlation Methodology*, ACS, 2005.

[5] K. A. Peterson and C. Puzzarini, *Theor. Chem. Acc.* **114** 283 (2005).

[6] MOLPRO is a package of *ab initio* programs written by H. -J. Werner and P. J. Knowles, with contributions of R. D. Amos et al.

[7] T. Helgaker, W. Klopper, H. Koch, and J. Noga, *J. Chem. Phys.* **106** 9639 (1997).

[8] D. Feller, *J. Chem. Phys.* **96** 6104 (1992).

[9] K. A. Peterson, D. E. Woon, and T. H. Dunning, Jr., *J. Chem. Phys.* **100** 7410 (1994).

[10] K. Peterson and T. H. Dunning, Jr., *J. Chem. Phys.* **117** 10548 (2002).

[11] M. Douglas and N. M. Kroll, *Ann. Phys.* (Leipzig) **82** 89 (1974).

[12] B. A. Hess, *Phys. Rev.* **32A** 756 (1985).

[13] F. Pawłowski, P. Jørgensen, J. Olsen, F. Hegelund, T. Helgaker, J. Gauss, K. L. Bak, and J. F. Stanton, *J. Chem. Phys.* **116** 6482 (2002).

[14] ACES2 (Mainz-Austin-Budapest version), a quantum-chemical program package for high-level calculations of energies and properties by J.F. Stanton et al., see http://www.aces2.de.

[15] J. Gauss and J. F. Stanton, *Chem. Phys. Lett.* **276** 70 (1997).

[16] J. F. Stanton and J. Gauss, *Int. Rev. Phys. Chem.* **19** 61 (2000).

[17] A. Halkier, W. Klopper, T. Helgaker, and P. Jørgensen, *J. Chem. Phys.* **111** 4424 (1999).

[18] J. Gauss, K. Ruud and T. Helgaker, *J. Chem. Phys.* **105** 2804 (1996).

[19] A. A. Auer, J. Gauss, and J. F. Stanton, *J. Chem. Phys.* **118** 10407 (2003).

[20] C. Puzzarini, *Phys. Chem. Chem. Phys.* **6** 344 (2004).

[21] C. Puzzarini, *J. Chem. Phys.* **123** 024313 (2005).

[22] C. Puzzarini, *J. Mol. Struct.* **780-781** 238 (2006).

[23] C. Puzzarini and K. A. Peterson, *Chem. Phys.* **311** 177 (2005).

[24] C. Puzzarini and K. A. Peterson, *J. Chem. Phys.* **122** 084323 (2005).

[25] C. Puzzarini, G. Cazzoli, A. Gambi, and J. Gauss, *J. Chem. Phys.* in press (2006).

[26] G. Cazzoli, C. Puzzarini, and A. Gambi, *J. Chem. Phys.* **120** 6495 (2004).

[27] G. Cazzoli, C. Puzzarini, A. Gambi, and J. Gauss, *J. Chem. Phys.* in press (2006).

[28] C. Puzzarini and P. R. Taylor, *J. Chem. Phys.* **122** 054315 (2005).

[29] C. Puzzarini and A. Gambi, *J. Phys. Chem.* **108A** 4138 (2004).

[30] C. Puzzarini, L. Dore, L. Cludi and G. Cazzoli, *Phys. Chem. Chem. Phys.* **5** 1519 (2003).

[31] C. Puzzarini, S. Coriani, A. Rizzo, and J. Gauss, *Chem. Phys. Lett.* **409** 118 (2005).

[32] G. Cazzoli, C. Puzzarini, and J. Gauss, *Astrophys. J. Suppl.* **159** 181 (2005).

[33] G. Cazzoli and C. Puzzarini, *J. Mol. Spectrosc.* **233** 280 (2005).

[34] A. Rizzo, C. Puzzarini, S. Coriani, and J. Gauss, *J. Chem. Phys.* **124** 064302 (2006).

[35] C. Puzzarini and A. Gambi, *J. Chem. Phys.* **122** 064316 (2005).

Brill Academic Publishers
P.O. Box 9000, 2300 PA Leiden,
The Netherlands

Lecture Series on Computer and Computational Sciences
Volume 6, 2006, pp. 426-440

Recent advances in the computation of linear and nonlinear optical susceptibilities of polymers, liquids, solutions and crystals using discrete local field theory

H. Reis[1], M. G. Papadopoulos[2]

Institute of Organic and Pharmaceutical Chemistry,
National Hellenic Research Foundation,
Vasileos Constantinou 48, GR–11635 Athens, Greece.

Received August 25, 2006; accepted in revised form August 25, 2006

Abstract: The application of a discrete local field theory to the computation of linear and nonlinear optical properties of molecular macroscopic materials is presented. The flexibility of the approach is demonstrated by its application to different classes of molecular materials: crystals, liquids, solutions and glasses.

Keywords: Nonlinear optical properties, molecular materials, discrete local field theory.

PACS: 42.65.Am, 78.20.-e, 71.15.Pd, 77.22.-d

1 Introduction

Quantitative microscopic theories of linear and nonlinear optic suseptibilities of molecular materials are important for interpreting and controlling of material properties which may be of practical significance for prospective applications in future technologies. Relevant properties include linear and nonlinear refraction, second- and third-harmonic generation and others. A successful approach for the computation of these properties, which will be reviewed here, is discrete local field theory (DLFT). This method combines accurately calculated molecular response with an electrostatic computation of intermolecular interactions.

The role of the local electric field induced by an external applied electric field has long been recognized as essential to successful treatments of electric susceptibilities in condensed materials. Additionally, it has also been recognized that molecules respond to the induced local field through *effective* (hyper)polarizabilities appropriate to the molecular environment. These effective properties generally differ from those of the free molecules, and one important factor for this difference is what may be called the *permanent* local field, which is the field due to the permanent charges distributions of neighboring molecules in the material. Both of these local fields are taken into account in DLFT.

The DLFT approach started from the treatment of molecular crystals, where it has been employed for a wide range of other effects apart from the prediction of optical suseceptibilities. Examples include the prediction of intensities of lattice vibrational spectra, electro-absorption spectra, charge-carrier generation, transport and trapping (see Ref. [37] for a review). The first

[1]Corresponding author. E-mail: hreis@eie.gr
[2]E-mail: mpapad@eie.gr

applications of DLFT to molecular crystals were concerned with the extraction of effective (hyper)polarizabilities from experimental susceptibilities. Later the approach was applied for the *prediction* of macroscopic susceptibilities starting from free-molecule properties. By combining the discrete description of the local field with molecular dynamics or Monte Carlo simulations, the approach could also be used to predict the susceptibililities of liquids, solutions and polymers.

2 Local field and susceptibilities

This section summarizes the basic theory for the treatment of the local field induced by an externally applied field and the susceptibilities of a molecular material. We begin with a crystal in a one-point treatment. Extensions to disordered materials and distributed approaches are discussed briefly afterwards.

Assuming the unit cell, with volume v, contains Z molecules labeled k, with effective polarizabilities α_k and first and second hyperpolarizabilities β_k and γ_k. All responses are frequency dependent, but this will not be written explicitly. The local electric field $\underline{F}_k$ at molecule k is in linear response related to the macroscopic electric field $\underline{E}$ by

$$\underline{F}_k = \underline{\underline{d}}_k \cdot \underline{E} \tag{1}$$

where $\underline{\underline{d}}_k$ is the local field tensor, given by

$$\underline{\underline{d}}_k = \sum_{k'} (\underline{\underline{I}} - (\epsilon v)^{-1} \underline{\underline{L}} \cdot \underline{\underline{\alpha}}_k)^{-1}_{kk'} \equiv \sum_{k'} \underline{\underline{D}}_{kk'}, \tag{2}$$

where $\underline{\underline{I}}$ is the $3Z$ unit matrix, $\underline{\underline{L}}$ is the $3Z$ matrix of the Lorentz factor tensors $\underline{\underline{L}}_{kk'}$ which can be calculated from the crystal structure, and $\underline{\underline{\alpha}}$ is a block-diagonal $3Z$ matrix of effective polarizabilities $\underline{\underline{\alpha}}_k$. The linear electric susceptibility is then:

$$\underline{\underline{\chi}}^{(1)} = (\epsilon v)^{-1} \sum_k \underline{\underline{\alpha}}_k \cdot \underline{\underline{d}}_k. \tag{3}$$

Note that the molecules are considered to be fixed in the crystal, even in a static or low-frequent electric field. In a liquid or gas, a reorientational term needs to be added to this term for polar molecules.

For the macroscopic non-linear response, it is necessary to consider also non-linear local field contributions. Nevertheless, it is still possible to express the susceptibilities in terms of the local-field tensors $\underline{\underline{d}}_k$. The first and the direct contribution to the second non-linear susceptibility are:

$$\underline{\underline{\underline{\chi}}}^{(2)} = (2\epsilon v)^{-1} \sum_k \underline{\underline{d}}_k^T \cdot \underline{\underline{\underline{\beta}}}_k : \underline{\underline{d}}_k \underline{\underline{d}}_k \tag{4}$$

$$\underline{\underline{\underline{\underline{\chi}}}}^{(3),d} = (6\epsilon v)^{-1} \sum_k \underline{\underline{d}}_k^T \cdot \underline{\underline{\underline{\underline{\gamma}}}}_k : \underline{\underline{d}}_k \underline{\underline{d}}_k \underline{\underline{d}}_k. \tag{5}$$

The factors 1/2 and 1/6 ocurring here are a consequence of the fact that the susceptibilities are the expansion terms of the *perturbation* series for the macroscopic polarization $\underline{P}$, while the (hyper)polarizabilities are the terms of the *Taylor* expansion for the induced dipole moment $\underline{p}_k$.

Apart from the direct term a *cascading* term contributes to $\chi^{(3)}$, as a consequence of coupling of first and second order electric fields. This term is [26, 35, 36]:

$$\underline{\underline{\underline{\underline{\chi}}}}^{(3),c} = (2\epsilon v)^{-1} \sum_\omega K(\omega) \sum_{kk'k''} \underline{\underline{d}}_k^T \underline{\underline{d}}_k^T : \underline{\underline{\underline{\beta}}}_k : \underline{\underline{D}}_{kk'} \cdot \underline{\underline{L}}_{k'k''} \cdot \beta_{k''} \underline{\underline{d}}_k \underline{\underline{d}}_{k''} \underline{\underline{d}}_{k''}, \tag{6}$$

where the summation over ω indicates that all frequency combinations leading to the same third order frequency have to be considered [35], and the factor $K(\omega)$ depends on the degeneracy factors of the different processes involved [36].

For disordered materials, the expressions can still be applied directly, if the structure of the material is obtained from molecular simulations. Then the unit cell is replaced by the simulation box, and averaging over trajectories or Monte Carlo snapshots is performed to obtain ensemble averages.

As is well known, the multipole expansion becomes invalid for points close to large molecules which deviate strongly from the spherical form. The best solution in such cases is to use multi-centred multipolar expansions, i.e. to distribute the multipoles and response functions over the molecules. Accurate methods to do so are in principle available [34, 33], but are generally computationally quite expensive and/or do no allow computation of hyperpolarizabilities. A simple approach, which can easily be accomodated into the one-point formalism described above, is the submolecule treatment, introduced originally by Luty [32]. In this approach, the molecular electric properties are distributed equally over point ***submolecules***. These submolecules were originally supposed to be chemically equivalent [32], but later this restriction was abandoned as being too restrictive, thus allowing the treatment of any molecules within this approach, but also introducing a certain arbitrariness in the choice of submolecules. This approach has been shown for example to suppress unphysical negative polarizabilities required to fit one point expansions to experimental refractive indices [42].

In order to connect to the more familiar continuum approaches for the local fields, we mention here that the local field tensors $\underline{\underline{d}}_k$ in the expressions have a one-to-one correspondence with the usual local field factors obtained for example in the Onsager-Lorentz local field treatment, although then, of course, the explicit dependence on the individual molecular environment, expressed by the index k in DLFT, is lost.

3 Molecular Crystals: Permanent local field and Molecular Response

Molecular crystals constitute the class of materials to which the DLFT approach is most easily applied, as long as the crystal structure is known from X-ray or neutron scattering experiments. On the other hand, we note that, computations of crystal structures are still very difficult, which means that, if the structure is not known experimentally, it is practically not possible to apply DLFT to compute optical susceptibilitities.

In order to connect macroscopic susceptibilities with properties of the free molecule, not only the induced response due to the externally applied field needs to be computed but also the effect of the environment on the molecules in field-free macroscopic sample has to be known. In other words, the effective multipole moments and (hyper)polarizabilities ocurring in the expressions of the previous section have to be related with corresponding free molecule properties. The most important environmental effect in molecular materials is probably the permanent local or reaction field effect, that is, the effect due to the electric field $\underline{E}_k^{(0)}$ caused by the permanent charge distributions of the surrounding molecules. If the molecules in the crystal have effective dipole moments $\underline{\mu}_k$ this field is, in dipolar approximation, given by:

$$\underline{E}_k^{(0)} = \sum_{k'} \underline{\underline{L}}_{kk'} \cdot \underline{\mu}_{k'}/(\epsilon_0 v). \tag{7}$$

The effective dipole moment, in turn, is given by

$$\underline{\mu}_k = \underline{\mu}_k^{(0)} + \alpha_k^{(0)} \cdot \underline{E}_k^{(0)} + \beta_k^{(0)} : \underline{E}_k^{(0)} \underline{E}_k^{(0)} + \ldots \tag{8}$$

Thus it becomes clear that the field has to be computed in an iterative fashion [27]. After computing a first approximation with the polarizability computed a zero field, the quantum-mechanical calculations of the electric properties are repeated including now the first estimate of the field, etc. Fortunately, in most of the cases considered up to now, it turned out that the polarizabilities α depend little on the field, so that one iteration was generally enough to reach convergence. This approach has been applied to the urea crystal, where it had been found that the permanent local field had little effect on the polarizability and the linear susceptibility, with good agreement with the experimental refractive indices. The effect on the first hyperpolarizablity was large, with some components doubling in magnitude and changing in sign. Due to a fortuitous cancellation, however, a much smaller effect on the quadratic susceptibility was found. Using the molecular properties at the SCF level, good agreement with experimental second-harmonic generation results was found, but the refractive indices were too small. With electronic correlation taken into account at the MP2 level, the quadratic susceptibility was too large by a factor of nearly 2, while the refractive indices were nearly quantitatively reproduced.

The DLFT approach as applied here, takes into account only electrostatic effects describable by multipole moments and (hyper)polarizabilities. Other effects not based on these properties are ignored. To test for effects of the extended hydrogen bonding network and intermolecular charge-transfer effects in urea crystal on the linear and nonlinear suscpetibilities, the electric properties of the linear urea dimer as ocurring in the crystal, were computed quantum-mechanically at the correlated MP2 level. The crystal susceptibilities were then recalculated using the dimer as the "buiding block", and were found to differ very little from those obtained at the single molecule level.

3.1 Applications to other molecular crystals

The approach described above for urea has been applied to compute the linear and nonlinear susceptibilities of several other molecular crystals, among them the ice polymorphs iceIh, iceII, ice IX and iceVIII [40]. In this work, vibrational contributions to the electric properties were taken into account, too. These contributions have a two-fold influence on the susceptibilities: first, the static, (i.e. at zero frequency) properties determine together with the static electronic multipoles and (hyper)polarizabilities the permanent local field $\underline{F}^{(0)}$, while the dynamic vibrational contributions add to the corresponding dynamic electronic (hyper)polarizabilities determining the response to the externally induced local field and thus the macroscopic susceptibilities. Ideally, the vibrational contributions in the latter case should be computed in the presence of $\underline{F}^{(0)}$, but this is usually not done due to the generally much larger computational burden compared with the electronic contributions. Additionally, if only macroscopic fields at optical frequencies are considered, the second effect mentioned above is usually small (mainly due to the so-called zero-point vibrational average (ZPVA) levels) and the main effect is then via the permanent local field. In the case of water, the vibrational contributions haave been found to be quite small compared to the electronic contributions, and thus the influence on the suscpetibiilties was also small.

Another aspect, the representation of the charge distribution of larger molecules was explored in work on naphthalene, anthracene and m–nitroaniline crystals [41]. As mentioned above, for larger molecules the question of shape and size becomes important and may require distributed schemes for the electric properties instead of the one-point treatment. In order to investigate this question, the submolecule treatment was applied, which had been shown previously to lead to physically reasonable results and is easily tractable [42].

In the case of the urea crystal it was found that the distribution of the molecular response over submolecules centred at the heavy atoms did not have an appreciable effect on the computed susceptibilities [27]. The same was found for the benzene crystal [44]. For the hydrocarbons naphthalene and anthracene, on the other hand, refractive indices in satisfactory agreement with experiment could only be obtained by distributing the response over all heavy atoms in the molecules. The predicted values were also compared with resultss obtained with the anistropic Lorentz-factor approximation (ALFFA), which assumes that the result of the Lorentz local field treament obtained for cubic crystals (or completely disordered systems) is approximately valid also for each crystal axis in other crystal symmetries [45]. For naphthalene, ALFFA provides values of similar quality as the 10 submolecule distribution, but for anthracene the values predicted by ALFFA were considerably worse than those of the most distributed DLFT model. The third-order susceptibilities were also computed using the DLFT approach, and very strong effects of the distribution model employed were observed. For example, the component $\chi^{(3)}_{bbbb}$ of anthracene in the one-point treatment was about fifty times larger than in the 14 submolecule treatment. No experimental data are available, however, to assess the reliability of the computed values.

The two hydrocarbon crystals were treated in dipolar approximation, where the permanent local field for these systems is zero. This is not so for the *m*-nitroaniline crystal. The molecular properties of the molecule were computed using two approaches, first, as for anthracene and naphthalene, with DFT, using a basis set specially designed for the computation of (hyper)polarizabilities, called TZVP-FIP [25], and secondly with the standard 6-31++G** basis set at the MP2 level. Both frequency dispersion and vibrational contributions were neglected. As shown in Table 1, only consideration of the permanent local field *and* a large distribution leads to reasonable agreement of computed linear and first nonlinear (second harmonic generation SHG) susceptibilities with experiment. In variance of the original paper [41], the experimental SHG have been multiplied here with 0.6 in order to be in accord with the generally accepted latest value of the standard quarz-crystal [24]. With this change it is seen that the 6-31++G**/MP2 values agree reasonably well with experiment, while the DFT values overshoot them considerably, especially considering that inclusion of frequency dispersion would most probably enhance the computed values even more.

Table 1: Comparison of computed first, $\chi^{(1)}_{ii}$ and second, $\chi^{(2)}_{iij}/10^{-12}$ [m/V] macroscopic susceptibility components of mNA crystal, obtained using different partition models in the respective permanent fields $|F^{(0)}|/\mathrm{GVm}^{-1}$, with experimental values obtained from refractive indices and SHG measurements.

		MP2/6–31++G**			DFT/TZVP-FIP			
# submols	$\lvert F^{(0)}\rvert$	$\chi^{(1)}_{aa}$	$\chi^{(1)}_{bb}$	$\chi^{(1)}_{cc}$	$\lvert F_0\rvert$	$\chi^{(1)}_{aa}$	$\chi^{(1)}_{bb}$	$\chi^{(1)}_{cc}$
1	10.2	1.78	1.91	2.32		No convergence		
10	3.8	1.77	1.68	1.46	4.4	2.15	1.86	1.79
		Experiment						
λ [nm]		$\chi^{(1)}_{aa}$	$\chi^{(1)}_{bb}$	$\chi^{(1)}_{cc}$				
1540		1.89	1.78	1.61				
		MP2/6-31++G**			DFT/TZVP-FIP			
	$\lvert F^{(0)}\rvert$	$\chi^{(2)}_{caa}$	$\chi^{(2)}_{cbb}$	$\chi^{(2)}_{ccc}$	$\lvert F_0\rvert$	$\chi^{(2)}_{caa}$	$\chi^{(2)}_{cbb}$	$\chi^{(2)}_{ccc}$
1	10.2	29	15	62		No convergence		
10	3.8	10	3	8	4.4	22	4	21
		Experiment						
λ [nm]		$\chi^{(2)}_{caa}$	$\chi^{(2)}_{cbb}$	$\chi^{(2)}_{ccc}$	Ref.			
SHG	1319	16±4	-	15±3	[38]			
SHG	1064	23±2	2±1	25±2	[38]			

4 Liquids

4.1 Pure liquids: benzene and water

The next level of complexity to which the DLFT approximation may be applied are pure molecular liquids. To treat these, it is necessary to obtain first the necessary structure information, which may be obtained by molecular simulation techniques. Using trajectories obtained by molecular dynamics or liquid configurations computed by Monte Carlo methods, each snapshot can be analysed in the same way as crystal structures discussed previously. The results may then be just averaged with a unit weight function [3] or, more accurately, averaged using a Boltzmann distribution as weight function [2]. Comparisons of the two methods suggest that they lead to very similar results [3].

One of the objectives of the study in Ref. [3] was a comparison of the performance of continuum and discrete models in the computation of the refractive index $n(\omega)$ and the third harmonic generation $\chi^{(3)}(-3\omega;\omega,\omega,\omega)$ susceptibility of liquid benzene and water, for which experimental values are available. As a first step, the (hyper)polarizabilities of the free molecules were computed using high-level *ab-initio* methods and were found to be in good agreement with corresponding experimental gas-phase values. Using DLFT, the refractive indices and susceptibilities presented in Table 2 were then computed. The refractive indices needed to compute the local field factor in the Lorentz and Onsager continumm models [39] were taken from experiment [6, 13, 4], except for the value $n(3\omega)$ of benzene, for which the computed DLFT value was used. In general all the employed models yield satisfactory refractive indices compared with experiment. On the other hand, we observe a considerable difference for the third-order susceptibilities computed by the various models. In the original work, the Onsager model was applied as advocated by Wortmann and Bishop [39]. It was later shown by Munn et al. [5] that in their formulas a cavity field factor $f^C(3\omega)$ was missing, this has been corrected in the Table. Additionally, the experimental values have been rescaled to comply with the latest calibration value for fused silica, as determined by Bosshard et al. [12] and based on the currently accepted SHG value for quartz [24]. With these changes it can be seen that all models are able to reproduce the experimental, rescaled $\chi^{(3)}(-3\omega;\omega,\omega,\omega)$ value of benzene, with the DLFT value showing the largest deviation (7%), but the value for water is not well reproduced. The best approximation is obtained by the Onsager model with the cavity radius choice O2, but this model yields the worst performance for the refractive indices. Generally, the choice of the cavity radius in the Onsager model is a delicate issue, with a very strong influence on the obtained value. As a consequence of the general failure of the models for $\chi^{(3)}$ of water, the ratio $\chi^{(3)}(C_6H_6)/\chi^{(3)}(H_2O)$ is also not well predicted. The reason for the failure is likely to be associated with non-electrostaic interactions on the second hyperpolarizability not taken into account by the models.

Finally we note that the linear response of C_6H_6 and H_2O as well as the THG susceptibility of C_6H_6 were successfully computed using both continuum models and the DLFT approach, but a larger discrepancy between theory and experiment still persists for $\chi^{(3)}(H_2O)$.

More results of other one-component liquid systems will be mentioned in the next subsection.

4.2 Solutions: p-Nitroaniline in Cyclohexane, 1,4-Dioxane, Tetrahydrofuran and Acetonitrile

At the next level of complexity, dilute solutions were considered with DLFT. Specifically, the local fields on and nonlinear response of the widely investigated molecule *p*-nitroaniline in solvents of different multipolar character were computed. One goal of that work was a clarification of the effect of the so-called "dioxane-anomaly" on the nonlinear macroscopic response. The anomaly refers to the fact that 1,4-dioxane generally behaves like a solvent of medium polarity, although the pure liquid has the same dielectric constant as a nonpolar liquid like cyclohexane. This anomaly

Table 2: Calculated and experimental refractive indices $n(\omega)$ and THG susceptibilities $\chi^{(3)}(-3\omega;\omega,\omega,\omega)/10^{-24}\text{m}^2\text{V}^{-2}$ of liquid benzene and water, and the ratio $\chi^{(3)}(C_6H_6)/\chi^{(3)}(H_2O)$. (Abbreviations: Lo=Lorentz model, O=Onsager model with a=3.28 Å, O1=Onsager with a=1.93 Å, O2=Onsager with a=2.37 Å, MS=Molecular simulation (DLFT)).

	C_6H_6				
	Exp.	Lo	O		MS
$n(\omega)$	1.483[c]	1.490	1.495		1.488
$\chi^{(3)}$	1675[ad],698[e]	693	705		748
	H_2O				
	Exp.	Lo	O1	O2	MS
$n(\omega)$	1.326[a]	1.328	1.327	1.317	1.334
$\chi^{(3)}$	390[ac],163[e]	195	218	171	272
	$\chi^{(3)}(C_6H_6)/\chi^{(3)}(H_2O)$				
	Exp.	Lo	O1	O2	MS
	4.3±0.4[a]	3.6	3.2	4.1	2.8

stems probably from large quadrupole moments of 1,4-dioxane. Thus the dipolar approach to the permanent local field of Eqs 7–8 had to be extended to include higher multipoles as well as higher order polarizabilities. For consistency reasons, the approach was also extended to include local field gradients. As solvents the predominantly nonpolar cyclohexane (CH), quadrupolar 1,4-dioxane (DI) and dipolar tetrahydrofuran (THF) were chosen [29]. Additionally, in this work we report results of a solution of pNA in acetonitrile (ACN). As for the pure liquids, classical molecular dynamics simulation were performed to obtain trajectories which could be analysed *a posteriori* by DLFT. All molecular simulations were performed with the non-polarizable OPLS-AA force field developed by Jorgensen and coworkers [7]. The original parameters for the partial charges and some of the torsional parameters for pNA had to be adjusted.

In order to validate the quality of the obtained in-liquid properties, the single molecule properties of pNA were computed, too. Comparing the optimized geometry of the free molecule with that of the solvated molecule computed with different continuum field models it was found that the solvent has a non-negligible effect on the geometry, with ensuing considerable differences in the electric properties. Although in all cases the average properties of the solvated species were larger then those of the free molecule, the effects depended considerably on basis sets, quantum-mechanical level (RHF, DFT or MP2) and continuum solvation method.

The experimental EFISH datum $<\gamma>_{\text{av}}$ is defined by

$$<\gamma>_{\text{av}}=\frac{\mu\beta_{||}}{3kT}+\gamma_{\text{av}} \tag{9}$$

where μ is the dipole moment and $\beta_{||}=3/5\sum_i\beta_{zii}$, with z the direction of the dipole moment. The experimental value of $<\gamma>_{\text{av}}$ of pNA, measured in the gas phase [31], could nearly be reproduced quantitatively by the computations, when the frequency dependence of β was computed at a correlated level (CC2), and the effect of vibrational averaging due to torsional motions of the nitro and the amino group as well as inversional motion of the amino group were taken into account, albeit only in a classical manner.

In order to compute the influence of the local field and field gradients on the linear and non-linear optical properties of pNA in solution, trajectories computed by molecular dynamics were analysed with the DLFT approach. Additional checks were performed by simulating and analysing the pure solvents employing DLFT. Comparison of simulated molecular radial distribution functions (MRDF) with corresponding experimental results showed that the positions of the maxima and minima were good reproduced, but the corresponding absolute values of the MRDF at these points were too large, in other words, the simulated liquids are ordered a bit too strongly. Applying the DLFT approach to the simulated trajectories, the static relative permittivities ϵ_r, refractive indices $n(\omega)$ and second nonlinear susceptibilities corresponding to third harmonic generation $\chi^{(3)}(-3\omega;\omega,\omega,\omega)$ were predicted (Table 3), which were all found to be in reasonable or good agreement with experimental data, with the exception of the relative permittivity of THF, which was found to be too low (5.64 versus 7.58). This discrepancy is probably due to shortcomings of the nonpolarizable force field used in the molecular simulation step.

Table 3: Computed relative permittivities ϵ_r, first $\chi^{(1)} \equiv \chi^{(1)}(3\omega)$, and third order (THG) $\chi^{(3)} \equiv \chi^{(3)}(-3\omega;\omega,\omega,\omega)/(10^{-24}\ \mathrm{m^2/V^2})$ susceptibilities at $\omega = 0.023896$ au of liquid CH, DI, THF and ACN at $T = 298$ K.

	CH		DI		THF		ACN	
	Calc	Exp	Calc	Exp	Calc	Exp	Calc	Exp
ϵ_r	2.040	2.025[a]	2.333	2.209[b]	5.64	7.58[c]	33.2[d]	37.5
$\chi^{(1)}$	1.047	1.028[b,e]	1.047	1.0123[e]	0.9951	0.9760[f]	0.806[g]	0.803[c]
$\chi^{(3)}_{iiii}$	327±1	325 [h,i]	360±2	277[e,i],320[j,i]	286±1	292[g,i]	(194)[k]	148[c]

[a]Ref. [8], 293 K
[b]Ref. [9]
[c]Ref. [10]
[d]Solution pNA in ACN
[e]$\omega = 0.02579$ au
[f]Ref. [11]
[g]ω=0.0284
[h]Ref. [13]
[i]Experimental $\chi^{(3)}$ values were rescaled with the calibration value of Ref. [12]
[j]Ref. [15]
[k]Computed with $\gamma(-2\omega;\omega,\omega,0)$ at $\omega = 0.04283$ au

Having shown the overall reliability of the combined MD/DLFT approach with the pure solvent checks, the method was applied to the solutions of pNA. The molecular electric properties of pNA with the two most different geometries obtained with different continuum solvation models, basis sets and levels of theory were employed in the computations in order to assess the importance of the differences. The first geometry was obtained with the polarized continuum model in the integral equation formalism (IEFPCM) with the aug-cc-pVDZ basis set at the MP2 level, the second was obtained with the dipolar spherical Onsager model with DFT and the 6-311G** basis set (OSCRF).

Although most of the experimental results of pNA in solution available concern EFISH experiments, either directly or indirectly as reference values as in some Hyper-Rayleigh scattering reports, it was not considered viable in Ref. [29] to predict EFISH results directly. In EFISH a static external field is applied which destroys the isotropy of the liquid, and is the cause for the first term in the expression for $< \gamma >_{\mathrm{av}}$ (Eq. 9). The effect of this field should be included in the molecular simulation step. However, with only one molecule of pNA in the simulation box, it is doubtful that a reasonable signal-noise ratio could be obtained in attainable simulation times and orienting fields of realistic magnitudes. Additionally, the use of non-polarizable force fields makes it unlikely that a a good simulation of the static field effect can be obtained for this conjugated push-pull substituted aromatic system. Therefore, only a field–free solution was simulated, for which then the average local fields and gradients effective on pNA were computed. Reasonable results were obtained employing a single point multipole expansion on the solvent molecules, but a distribution of all moments and polarizbilities over ten submolecules on pNA. The results for the permanent local fields on pNA for the different solutions are shown in Table 4. In all cases it was found that the dipole and quadrupole terms contribute in non-negligible amounts to the fields, but the effect of higher order permanent multipoles (octopoles and hexadecapoles) is negligible. As expected, in 1,4-dioxane the quadrupole terms dominate the local fields, leading to total average fields which are larger than in the dipolar solvent THF, although the latter are probably underestimated, as indicated by the underestimation of the relative permittivity for the pure THF solvent. Although the values of the permanent local field on pNA in ACN are the largest among all the solvents, the difference to DI and THF is surprisingly small.

Table 4: Average local fields $F_x/10^9$ V/m, in quadrupolar (Q) and hexadecapolar (H) approximation for pNA in CH, DI, THF, ACN obtained with the submolecule distribution m/n; $n = 10$ for pNA, $m = 1$ for the solvent molecules.

	CH		DI		THF		ACN	
Geom	Q	H	Q	H	Q	H	Q	H
OSCRF	1.08	1.00	2.91	2.91	2.78	2.83	3.78	3.54
IEFPCM	0.96	0.90	2.76	2.77	2.66	2.73	-	-

Computing the electric properties of pNA quantum-mechanically, adding the obtained average permanent local fields and field gradients to the Hamiltonian, the so-called "solute" [39] properties are obtained. It turns out that the local field gradients have a practically negligible effect on the properties, so that only the local fields need to be considered. Assuming that the usual Osager-Lorentz treatment of induced local fields is reasonably accurate, the solute quantities can be compared with experimental values. The results are shown in Table 5. The computed values for the low to medium polar solvents CH, DI, THF are in good to reasonable agreement with the experimental values, considering the quite severe approximations made in the computational model. The slightly larger discrepancies found for THF may again be caused by the same reasons which lead to the underestimated ϵ_r for the free solvent.

On the other hand, for ACN solution, rather strong disagreement with experimental data is observed, which cannot be explained by incorrect simulation of the pure solvent, as this property is simulated reasonably well (see Table 3). As the obtained value of the permanent local field shows, it seems that the electrostatic description of the pNA/ACN system is too inaccurate to yield correct results. The reason for this may lie in the molecular dynamics step or in DLFT step. It should be mentioned, however, that the experimental results for this system show a rather large spread, too ($\beta_{\mathrm{pNA/ACN}}/\beta_{\mathrm{pNA/DI}}$=1.38 [21] – 1.59 [19]), although this spread does not include the computed value.

Finally, the results of DLFT on pNA in CH, DI and THF solutions were compared with predictions of the continuum approach using Onsager-Lorentz local field factors, employing an ellipsoidal cavity to accomodate the pNA molecule [14]. Both the permanent local fields (in DI and THF) and the local field factors were found to be in good to very good agreement in DLFT and the continuum Onsager-Lorentz approach.

In related work, Jensen and van Duijnen [28] simulated pNA in 1,4-dioxane, using the discrete

Table 5: Comparison of computed and experimental microscopic EFISH signal $<\gamma(2\omega)>_E/(10^{-60}\ C^4\ m^4\ J^{-3})$ and relative (for CH, THF and ACN) or absolute HRS (for DIi, in au) signal in VV geometry $<\beta_{ZZZ}(2\omega)>_{rel/abs}$ of pNA in CH, DI, THF and ACN solution.

Solvent	Geom.	ω/au	$<\gamma>_E$ [a]	Exp.	$<\beta_{ZZZ}>_{rel/abs}$	Exp.
CH	IEFPCM	0.0428	112.7		0.71[b], 0.76[c]	0.85[d],0.76[e]
DI	IEFPCM	0.0239	104.4	86.1[f]		
DI	OSCRF		121.4			
DI	IEFPCM	0.0428	148.4	152.2[f],165.2[g]	1610	1755[i]
DI	OSCRF		175.4	150.0 [h]	1747	
THF	IEFPCM	0.0428	147.9	219.5[g]	0.93[b], 1.0[c]	1.12[d],
THF	OSCRF		173.4		0.99[b],1.06[c]	1.17±0.07[j]
ACN	OSCRF	0.0428	192.8	262.3[g]	1.10[b],1.30[c]	1.38[j],1.48[d]

[a] $10^{-60}\ C^4\ m^4\ J^{-3}=1.6036\ 10^5$ au
[b] With OSCRF-optimized geom. of pNA in DI
[c] With IEFPCM-optimized geom. of pNA in DI
[d] 10% 1-nitropropane in cyclohexane, Ref. [20] .
[e] Extrapolated from data in Ref. [20], see text.
[f] Ref. [18], assuming μ=2.44 au (Refs. [14, 15])
[g] Ref. [19]
[h] Ref. [14]
[i] Obtained from 2470x1.88x$\sqrt{1/7}$, see Refs. [16, 17, 22, 23]
[j] Ref. [21]

reaction field (DRF) method, where the linear and nonlinear response of the solute is directly computed by density functional theory (DFT) during the simulation. The electrostatic solvent effectsare modeled by polarizable point charges. The total molecular response is computed in an extended Lorentz local field model, combined with a statistical average. In this approach, the effects of the surrounding solvent molecules as well as those caused by a macroscopic field due to an externally applied field can be considered. Comparing the nonlinear response of pNA with and without external field, it was found that the macroscopic field strongly *decreases* the solute response, in contrast to the predictions of the usual Lorentz-Onsager local field approach and in contrast to the DLFT results. The reason for this discrepancy is still unclear and warrants further investigation.

5 Poled Polymers: p-N,N-Dimethylamino-nitrobenzene, N,N'-di-n-propyl-1,5- diamino-2,4-dinitrobenzene, p-dimethylamino-p'-nitrostilbene in poly(methyl)methacryalate

At the next higher level of system complexity it was attempted to model realistic guest–host polymer systems by DLFT. In the first part of a two-part study the poling process of a polymethylmethacrylate (PMMA) matrix doped with one of three different chromophors (*N,N*–dimethyl-*p*–nitroaniline (DPNA), 4-(dimethylamino)-4'-nitrostilbene (DMANS) and *N,N*–di-*n*-propyl-1,5-diamino- 2,4–dinitrobenzene (DPDADNB)) was simulated by molecular dynamics methods [46]. In the second part of the study [47], on which we will concentrate here, the simulated poled (i.e. glassy) doped systems were analysed with the DLFT approach.

The first two chromophors belong to what are commonly called "one-dimensional" NLO-chromophors, that is their first hyperpolarizability tensor is strongly dominated by one diagonal component, while the third one is a "two-dimensional" chromophor, with one dominant non-diagonal component, assuming Kleinmann symmetry. The permanent local field $\underline{F}^{(0)}$ on the chromophors obtained from Eq. 7 by DLFT was generally directed parallel to the molecular dipole moment, although perpendicular components were also found to be of unusually large magnitude, as compared to what was generally found in liquids and crystals. Overall, the field values were rather small, in comaprison to values computed for crystals (e.g. in urea [27]). On the other hand, they are comparable to local field values found in liquid nitrobenzene and benzene [2, 1]. The local field effect on the polarizabilities was small in all cases, while the β_z values of DPNA and DPDADNB increased by 10% (scaled MP2 level). The corresponding increase for DMANS was ~25%, comp. Table 6.

Table 6: Average local fields $\underline{F}/\mathrm{GVm}^{-1}$ on the chromophores in the systems DPNA/PMMA, DMANS/PMMA and DPDADNB/PMMA and electrical properties α_{ii}, β_{ijk} (in au), of the chromophores in this field computed at the MP2 and TDDFT/GRAC levels. The changes to the field–free properties are given in parentheses.

	DPNA/PMMA		DMANS/PMMA	DPDADNB/PMMA	
F_x, F_y, F_z	0.13,0.12,0.73		−0.07,−0.39,0.65	0.14,0.61,1.30	
	MP2/Pol	GRAC	MP2/6-31G**	MP2/6-31G**	GRAC
	$\lambda = 1064$ nm		$\lambda = 1907$ nm	$\lambda = 1064$ nm	
$\alpha_{av}(2\omega)$	140.14(0.4%)	157.85(2.5%)	225.43(1.4%)	196.71(0.7%)	260.45(1.1%)
β_{yyz}	−179	−279	−66	1791	2789
β_{zyy}	−201	−418	−71	1728	2078
β_{zzz}	3601	6244	12873	−39	−33
β_z	2011(15%)	3478(12%)	7663(26%)	1007(8.8%)	1391(21%)

As found generally for most systems investigated with DLFT, the computed refractive indices are in reasonable agreement with available experimental values. For the second-order susceptibilities $\chi^{(2)}(-2\omega;\omega,\omega)$ (see Table 7) reasonable values are obtained for DPNA/PMMA already in the one-point dipole description ($(n/m) = (1/1)$, where n is the number of submolecules per chromophor and m the number of submolecules per *monomer* of the PMMA matrix), but for DPDADNB/PMMA the minimum submolecule distribution is $(n/m) = (5/1)$, while for DMANS/PMMA even the monomer of the matrix need to be further partitioned: only $(n/m) = (20/5)$ was found to yield reasonable $\chi^{(2)}$ values.

The DLFT values were also compared with predictions from the Lorentz local field approximation (LFFA), which were computed using the simulated averaged order parameter $< \cos\theta >$ and $< \cos^3\theta >$. It was found that the two schemes agreed inside the, rather large, error margins of the order parameters. In the case of DPDADNB the ordering of non-=diagonal adnd diagonal values were reversed in the two schemes, although the computed values still agree within the error margins. Considering the surprising similarity between the predicted results of LLFA adn DLFT, as well as the ease of application of the former, a remarkable advantage of LLFA over DLFT was deduced. However, it was anticipated that with more polarizable and/or more dipolar polymer hosts, discrepancies between DLFT and LLFA may become much larger. Such hosts will probably lead due to much stronger changes of the in-phase electrical properties of the chromophors as compared to the free molecule properties. These changes cannot be described by the the LLFA. In such cases one may apply the Onsager continuum approxiamtion, which is still much easier than DLFT.

All the foregoing conclusions pertain to low chromophore concentrations, as investigated in

Table 7: Refractive indices $n_\perp(\omega)$, $n_3(\omega)$ and SHG susceptibilities $\chi^{(2)}_{333}(-2\omega;\omega,\omega)$, $\chi^{(2)}_{\perp\perp 3}(-2\omega;\omega,\omega)$, $\chi^{(2)}_{3\perp\perp}(-2\omega;\omega,\omega)$ (in pm/V) of guest-host systems poled in 3-direction, computed using input electrical properties at different computational levels and applying different partitioning schemes n/m.

Level	n/m	$n_\perp$ [a]	n_3	$\chi^{(2)}_{\perp\perp 3}$ [b]	$\chi^{(2)}_{3\perp\perp}$ [b]	$\chi^{(2)}_{333}$
			DPNA, $\lambda = 1064$ nm			
GRAC	1/1	1.478±0.004	1.506±0.010	0.43±0.12	0.36±0.13	1.78±0.31
GRAC	3/1	1.477±0.005	1.508±0.010	0.38±0.11	0.32±0.11	2.02±0.26
			DMANS, $\lambda = 1907$ nm			
MP2/6-31G**	1/1	1.6±0.1	1.48±0.03	300±300	900±600	2000±2000
MP2/6-31G**	20/5	1.473±0.005	1.505±0.005	0.72±0.20	0.72±0.20	4.90±0.93
			DPDADNB, $\lambda = 1064$ nm			
GRAC	1/1	1.540±0.015	1.570±0.039	9.9±2.4	4.3±0.7	2.2±4.4
GRAC	5/1	1.499±0.007	1.524±0.008	2.09±0.14	1.42±0.14	0.62±0.15

[a] $n_\perp = (n_1 + n_2)/2$
[b] $\chi^{(2)}_{\perp\perp 3} \equiv (\chi^{(2)}_{113} + \chi^{(2)}_{223})/2$

Refs. [46, 47]. Work by other groups [51] has shown that approximations as LLFA or Onsager, which predict a linear relationship between $\chi^{(2)}$ and $\mu\beta$ at low external field strengths, may break down at high chromophore concentrations.

Overall the results show that the Lorentz approximation may be adequate as a routine screening tool for checking the suitability of guest-host polymeric systems, for low chromophore concentrations and as long as the polymer will not create large permanent local fields.

6 Summary and Future Work

As shown by the various examples in this work, DLFT is a versatile method to compute the linear and nonlinear optical properties of condensed molecular systems. Although it performs quite succesfull in several occasions, there is still room for improvement. Extensions planned for the future include:

- The inclusion of the effect of static fields on phonons in molecular crystals. This is important in order to predict susceptibilities which include the effect of quasi-static fields, as the relative permittivity, Pockels-effect, etc. Work in progress on the molecular crystal of 3-methyl-4-nitro-pyridin-N-oxide (POM) [48] suggests that by including the field effect on *intramolecular* vibrations explains a substantial part of the relative permittivity and the Pockels effect of this crystal, but there are still larger discrepancies with experimental values than for pure optical susceptibilities (refractive indices, SHG susceptibility). These differences may be due to the field effect on phonons. However, the description of these effects is not trivial, because neither Hartree-Fock nor standard DFT approaches, which are the two methods commonly employed in computations of solid-state properties, are able to describe intermolecular interactions in molecular materials well.

- Use of more accurate distribution schemes than the simple submolecule scheme, in order to be able to describe the distribution of the electric properties over the molecule better. This may be especially important for phases containing conjugated charge-transfer molecules. In the case of disordered phases, both the simulation step and the DLFT analysis would then need to be extended. For the simulations, force fields would need not only to be polarizable, but also be able to describe intermolecular charge-transfer effects in conjunction with intramolecular structure changes accurately. In the DLFT part, it would be necessary to describe effects of structure changes on multipole moments and (hyper)polarizabilities. Both extensions may be possible by combining accurate multipole distribution approaches as Stone's DMA or Bader's AIM approach with dipole or monopole-dipole interaction models as developed by Jensen et al. [49] or Applequist [50].

Acknowledgment

We gratefully acknowledge financial support from the European Commission from the HPC Europa Transnational Access Programme (Contract Nr. RII3-CT-2003-506079). H. R. acknowledges support from the European Center for Parallelism of Barcelona (CEPBA).

References

[1] R.H.C. Janssen, D.N. Theodorou, S. Raptis, M.G. Papadopoulos *J. Chem. Phys.* **111**, 9711 (1999).

[2] R.H.C. Janssen, J.-M. Bomont, D.N. Theodorou, S. Raptis, M.G. Papadopoulos *J. Chem. Phys.* **110**, 6463 (1999).

[3] H. Reis, M.G. Papadopoulos, D.N. Theodorou *J. Chem. Phys.* **114**, 876 (2001).

[4] W. M. Irvine, J. B. Pollack, *Icarus* **8**, 324 (1968).

[5] R. W. Munn, Yi Luo, P. Macák, H. Ågren, *J. Chem. Phys.* **114**, 3105 (2001).

[6] C.J.F. Boettcher and P. Bordewijk, *Theory of Electronic Polarization*, Vol. 1, Elsevier, New York, 1978.

[7] W.L. Jorgensen, *OPLS All-Atom Force Field, parameter files oplsaa.par and oplsaa.sb*, distributed with the BOSS program, Yale University, March 2004.

[8] L. Hartshorn, J.V.L. Parry, L. Essen, *Proc. Phys. Soc. B* **68**, 422 (1955).

[9] *Handbook of Chemistry and Physics*, 62nd edition, CRC: Boca Raton, 1981.

[10] M. Stähelin, C.R. Moylan, D.M. Burland, A. Willetts, J.E. Rice, D.P. Shelton, E.A. Donley, *J. Chem. Phys.* **98**, 5595 (1993).

[11] G.R. Meredith, B. Buchalter, C. Hanzlik *J. Chem. Phys.* **78**, 1543 (1983).

[12] C. Bosshard, U. Gubler, P. Kaatz, W. Mazerant, U. Meier, *Phys. Rev. A* **61**, 10688 (2000).

[13] F. Kajzar, J. Messier, *Phys. Rev. A* **32**, 2352 (1985).

[14] R. Wortmann, P. Krämer, C. Glania, S. Lebus, N. Detzer, *Chem. Phys.* **99**, 173 (1993).

[15] L.-T. Cheng, W. Tam, S.H. Stevenson, G.R. Meredith, G. Rikken, S.R. Marder, *J. Phys. Chem.* **95**, 10631 (1991).

[16] R. Bersohn, Y.-H. Pao, H.L. Frisch, *J. Chem. Phys.* **45**, 3184 (1966).

[17] H. Reis, *J. Chem. Phys.* **125**, 014506 (2006).

[18] C.C. Teng, A.F. Garito, *Phys. Rev. B* **28** , 6766 (1983).

[19] M. Stähelin, D.M. Burland, J.E. Rice, *Chem. Phys. Lett.* **191**, 245 (1992).

[20] F.L. Huyskens, R.L. Huyskens, A. Persoons, *J. Chem. Phys.* **108**, 8161 (1998).

[21] T. Kodaira, A. Watanabe, O. Ito, M. Matsuda, K. Clays, A. Persoons, *J. Chem. Soc., Faraday Trans.* **93**, 3039 (1997).

[22] P. Kaatz, D.P. Shelton, *J. Chem. Phys.* **105**, 3918 (1996).

[23] R.D. Pyatt, D.P. Shelton, *J. Chem. Phys.* **114**, 9938 (2001).

[24] D.A. Roberts, *IEEE J. Quant. Electronics* **28**, 2057 (1992).

[25] P. Calaminici, K. Jug, A. M. Köster, *J. Chem. Phys.* **109**, 7756 (1998).

[26] M. Hurst, R.W. Munn, 1986, *J. Mol. Electron.* **2**, 35, 43 (1986).

[27] H. Reis, M.G. Papadopoulos, R.W. Munn, *J. Chem. Phys.* **109**, 6828 (1998).

[28] L. Jensen, P.Th. van Duijnen, *J. Chem. Phys.* **123**, 074307 (2005).

[29] H. Reis, M.G. Papadopoulos, A. Grzybowski, *J. Phys. Chem. submitted.*

[30] H. Reis, A. Grzybowski, M.G. Papadopoulos, *J. Phys. Chem. A* **109**, 10106 (2005).

[31] P. Kaatz, E.A. Donley, D.P. Shelton, *J. Chem. Phys.* **108**, 849 (1998).

[32] T. Luty, *Chem. Phys. Letters* **44**, 335 (1976).

[33] A.J. Stone, *The Theory of Intermolecular Forces*, Clarendon, Oxford (1996).

[34] R.F.W. Bader, *Atoms in Molecules-A Quantum Theory*, Clarendon, Oxford (1994).

[35] M. Malagoli, R.W. Munn, *J. Chem. Phys.* **107**, 7926 (1997).

[36] H. Reis, M.G. Papadopoulos, in *Nonlinear Optical Responses of Molecules, Solids and Liquids: Methods and Applications*, (Editor: M. G. Papadopoulos), Trivandrum, Research Signpost, 195 (2003).

[37] R.W. Munn, *Mol. Phys.* **64** 1 (1988).

[38] A. Carenco, J. Jerphagnon, A. Perigaud, *J. Chem. Phys.* **66** 3806 (1977).

[39] R. Wortmann, D.M. Bishop, *J. Chem. Phys.* **108** 1001 (1998).

[40] H. Reis, S. Raptis, M.G. Papadopoulos, *Chem. Phys.* **263** 301 (2001).

[41] H. Reis, M.G. Papadopoulos, P. Calaminici, K. Jug, A.M. Köster, *Chem. Phys.* **261** 359 (2000).

[42] P.J. Bounds, R.W. Munn, *Chem. Phys.* **59** 47 (1981).

[43] H. Reis, M.G. Papadopoulos, R.W. Munn, *J. Chem. Phys.* **114** 876 (2001).

[44] H. Reis, S. Raptis, M..G. Papadopoulos, R.H.C. Janssen, D.N. Theodorou, R.W. Munn, *Theor. Chem. Acc.* **99** 384 (1998).

[45] C.J.F. Böttcher, *Theory of Electric Polarization* 2nd ed., Elsevier, Amsterdam, 1973.

[46] M. Makowska-Janusik, H. Reis, M.G. Papadopoulos, I. Economou, N. Zacharopoulos *J. Phys. Chem. B* **108**, 588 (2004).

[47] H. Reis, M. Makowska-Janusik, M.G. Papadopoloulos *J. Phys. Chem. B* **108**, 8931 (2004).

[48] H. Reis, J.M. Luis, *in preparation.*

[49] L. Jensen, K.O. Sylvester-Hvid, K.V. Mikkelsen, P.-O. Åstrand, *J. Phys. Chem. A* **107**, 2270 (2003).

[50] J. Applequist, *J. Chem. Phys.* **83**, 809 (1985).

[51] L.R. Dalton, W.H. Steier, B.H. Robinson, C. Zhang, A. Ren, S. Garner, A. Chen, T. Londergan, L. Irwin, B. Carlson, L. Fifield, G. Phelan, C. Kincaid, J. Amend, A. Jen *J. Mater. Chem.* **1999**, 1905 (1999); L.R. Dalton *J. Phys.: Condens. Matter* **15** , R897 (2003); Y.V. Pereverzev, O.V. Prezdho, L.R. Dalton, *Chem. Phys. Letters* **373**, 207 2003); B.H. Robinson, L.R. Dalton, A.W. Harper, A. Ren, F. Wang, C. Zhang, G. Todorova, M. Lee, R. Ansizfeld, S. Garner, A. Chen, W.H. Steier, S. Houbracht, A. Persoons, I. Ledoux, J. Zyss, A.K.Y. Jen, *Chem. Phys.* **245**, 35 (1999).

Brill Academic Publishers
P.O. Box 9000, 2300 PA Leiden,
The Netherlands

Lecture Series on Computer and Computational Sciences
Volume 6, 2006, pp. 441-452

Ab initio Quantum Mechanical Charge Field (QMCF) Simulations: New Horizons in Solution Chemistry

Bernd M. Rode[1], Thomas S. Hofer, Bernhard R. Randolf, Andreas B. Pribil and Viwat Vchirawongkwin

Theoretical Chemistry Division
Institute of General, Inorganic and Theoretical Chemistry
University of Innsbruck, Innrain 52a, A-6020 Innsbruck, Austria

Abstract: Results of a novel quantum mechanical/molecular mechanical (QM/MM) molecular dynamics framework – the QMCF MD method – are presented and discussed with respect to methodical as well as computational details. Simulations of four different chemical systems with increasing complexity are presented demonstrating the unique capabilites and methodical flexibility of the approach which, combined with the accuracy of a quantum mechanical treatment, will give access to simulations of a wide vaiety of solute systems in the near future.

1 Introduction

Solution chemistry is one of the most important domains of chemistry, as the majority of chemical reactions take place in the liquid phase and hence, accurate theories for the treatment of solutions are highly desirable to understand the nature of liquids and their interaction with all kinds of solutes. Examples for solute systems range from ionic hydrates, inorganic and metal organic complexes to nucleic acids, proteins and composite membrane layers. Besides the use of various experimental techniques, which measure selected properties of the investigated systems, computer simulations offer a challenging alternative to obtain a detailed and universal description of such systems at molecular level. The importance of the accurate description of all interactions within the system for any kind of simulation is obvious. During the past decades a considerable amount of research has been devoted to the improvement of simulation techniques with respect to accuracy as well as computational cost, which are strongly correlated.

While the size of even small biomolecules such as peptides demands, due to the computational cost associated with the number of particles, a treatment based on molecular mechanics (i.e. parametrized potential functions), the more accurate quantum mechanical description is at present already feasible for smaller systems like hydrated ions or complexes in solution.

It has been shown that a quantum mechanical treatment is not only desirable but mandatory for the description of charged systems like hydrated ions [1, 2, 3, 4, 5, 6], as the molecular mechanics treatment based on pair- and three-body potential functions does not reach the accuracy needed for a correct description of n-body effects and thus can produce strongly erroneous data, e.g. for coordination numbers and exchange dynamics of ligands.

Systems accessible to quantum mechanical simulations are still relatively small. Car-Parinello molecular dynamics [7] (CPMD) try to achieve the necessary compromise between computational

[1] Corresponding author, E-mail:bernd.m.rode@uibk.ac.at, Tel.: +43-512-507-5160,d Fax: +43-512-507-2714

effort and accuracy by the use of simple density functionals (for example BLYP [8] or PBE[9, 10]) and the reduction of the investigated system to a moderate size of 100–200 atoms. Hybrid quantum mechanical/molecular mechanical (QM/MM) methods [11, 12, 13, 14] reduce the computational effort to an affordable extent by treating only a sub-region of the system by *ab initio* or DFT quantum mechanics and the remaining system by molecular mechanics potential functions. In this method the accuracy of quantum mechanics employed can be substantially better than in CPMD, and the elementary box can be much larger. Difficulties in combining both subsystems arise if particles are exchanged or chemical bonds are cut by the border between the subregions. In such cases a proper treatment has to be sought for the border region.

The QM/MM formalism requires, besides the quantum mechanical calculation of all interactions inside the QM region and the potential-based force field calculations within the MM region, the evaluation of interactions between species inside and outside the QM region. This is usually achieved on the basis of force fields, in the case of solutions usually consisting of *ab initio* constructed pair and three-body potential functions. The construction of such potential functions is a time-consuming and tedious task: Several thousand single points of the energy hypersurfaces of all interacting species pairs and triples have to be evaluated and fitted to suitable analytical functions representing these surfaces.

Numerous QM/MM MD simulations of hydrated ions have been carried out in the past decade, showing that Hartree-Fock level calculations with double zeta plus polarisation basis sets seem to be the lower limit of accuracy and that the inclusion of the full first and sometimes even the second solvation shell of the solute in the QM region is mandatory [15, 6] for a reliable description of hydrated species. However, it was observed that the potential functions of the solutes rapidly decay and become negligible beyond a certain diameter of the QM region. Hence these potential functions and their elaborate construction can be avoided, except for the solvent–solvent interaction. The only significant long–range solute–solvent contributions result from the Coulombic interactions which are easily implemented in the simulation protocol.

Quantum mechanical methods take into account, besides the automatic inclusion of many–body terms, charge transfer and polarisation effects, and they enable the evaluation of the dynamically changing electron density by means of partial atomic charges. The evaluation of these partial charges in every simulation step is physically meaningful and more realistic than the fixed partial charges of the typical (nonpolarisable) potential functions, as all changes of the chemical surrounding induce charge fluctuations. Thus one can describe the continously changing charges as a dynamical charge field. Similar, the influence of the MM partial charges on the QM region can be incorporated into the quantum mechanical treatment as a perturbation part added to the core hamiltonian. Although these charges are fixed according to the molecular mechanics potentials, the dynamical motion of the MM particles results in a fluctuating charge field as well. These methodical improvements have been employed in a novel QM/MM protocol, the Quantum Mechanical Charge Field (QMCF) MD [16] approach which is the main subject of this paper.

2 Methodology of the QMCF MD approach

The partition of the system into a QM and a MM region as in the common QM/MM scheme is maintained [11, 12, 13, 14], but the QM region is now split into two subregions, namely the 'core region' and the 'solvation layer'. (cf. figure 1). The main difference of the method lies in the treatment of forces between the 'inner' QM subregion ('core') and the MM region. Inside the QM region, all interactions are evaluated by means of quantum mechanics, usually *ab initio* Hartree-Fock which shows despite the neglect of electron correlation good performance with respect to experimental results if at least DZP basis sets are employed. Other levels of theory could be applied as well, e.g. MP/2 [17], but at the price of a much higher computational demand. Suitable

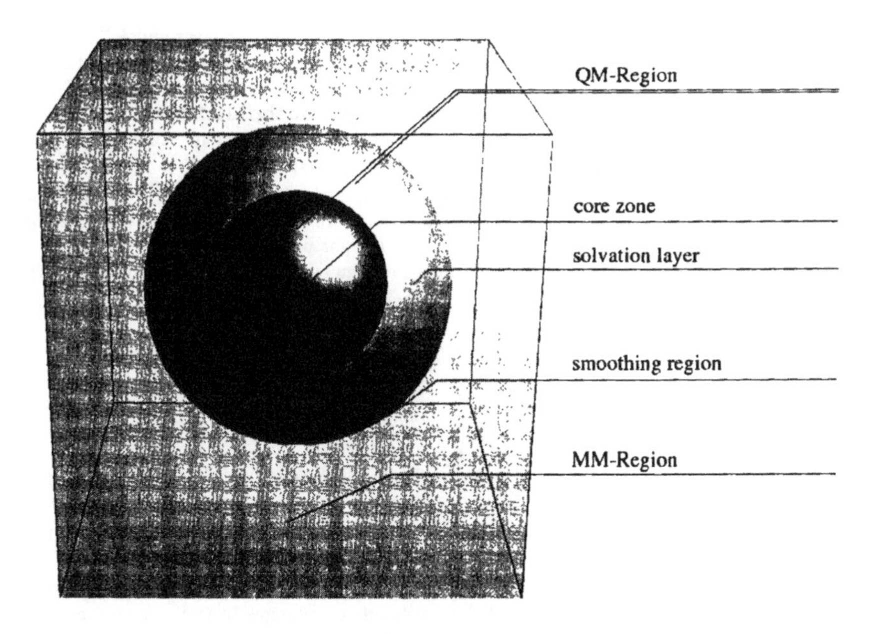

Figure 1: Definition of the QM and MM regions in the QMCF approach.

hybrid density functional methods e.g. B3LYP [18] would require similar computing times as HF and have proven inferiour to HF in several cases [19, 20, 21].

The solute is contained in the core region and can be any type of molecule or composite species, e.g. a metal ion with identical or different ligands, and the solvation layer consists of one complete layer of solvent molecules surrounding the solute. The size of the 'core' region is set to include the complete solute including the first layer of solvent ligands and the 'layer' region is sized to include a further complete solvation shell. The simplest case are hydrated ions: The ion and two hydration spheres would be included in the QM region. PO_4^{3-} is an example for a more complex species which forms itself the core region and, due to the size of the solute and the weakness of interactions, the first hydration shell can be defined as the solvation layer. Outside the QM region, the remaining solvent is treated by molecular mechanics, i.e. by suitable potential functions.

The definition of forces in the different subregion is given by:

$$\vec{F}_J^{\text{core}} = \vec{F}_J^{\text{QM}} \tag{1}$$

$$\vec{F}_J^{\text{layer}} = \vec{F}_J^{\text{QM}} + \sum_{I=1}^{M} \vec{F}_{IJ}^{\text{BJHnC}} \tag{2}$$

$$\vec{F}_J^{\text{MM}} = \sum_{\substack{I=1\\I\neq J}}^{M} \vec{F}_{IJ}^{\text{BJH}} + \sum_{I=1}^{N_1+N_2} \frac{q_I^{\text{QM}} \cdot q_J^{\text{MM}}}{\vec{r}_{IJ}^{\,2}} + \sum_{I=1}^{N_2} \vec{F}_{IJ}^{\text{BJHnC}} \tag{3}$$

F_J^{core} (eq. 1) corresponds to the quantum mechanical force acting on a particle J in the core zone, F_J^{layer} to the forces acting on a particle J located in the solvation layer and F_J^{MM} to the forces acting on a particle J in the MM region. All forces in the core region are obtained by the quantum mechanical treatment. The MM partial charges of all particles M located in the MM region are incorporated as perturbational term (cf. equation 4 and 5) into the quantum mechanical treatment to account for the influence of the MM particles on the entire QM region ('core' plus 'layer'). Due to the fact that the layer region is in close vicinity to the MM region, it is mandatory to apply potential functions for the treatment of the non–Coulombic interactions $\vec{F}_{\text{IJ}}^{\text{BJHnC}}$ of MM particles M (in this case the BJH-CF2 water model [25, 26]), as their contributions are not negligible at these small distances. This implies that only MM solvent particles (for which suitable potential functions have to be employed for the MM treatment anyway) are allowed in this layer region.

$$\hat{H}^c_{QMCF} = \hat{H}^c_{HF} + V' \quad (4)$$

$$V' = \sum_{J=1}^{M} \frac{q_J}{r_{iJ}} \quad (5)$$

The force acting on a particle in the MM region is composed of the forces $\vec{F}_{\text{IJ}}^{\text{BJH}}$ resulting from the interactions with all other MM particles M and the non-Coulombic interactions $\vec{F}_{\text{IJ}}^{\text{BJHnC}}$ with all atoms located in the layer region N_2, according in our case to the BJH water model [25, 26], plus the charge field interaction of all QM particles (core N_1 plus layer N_2).

The latter contribution is calculated employing partial charges assigned to QM atoms which are re–evaluated in every subsequent step to account for changes in the chemical environment. Different partial charges schemes are available: The Mulliken charges showed [22] despite their known deficiencies good compatibility with the MM partial charges of the BJH-CF2 water model. One has to keep in mind that Mulliken charges are considerably basis set dependent and great care has to be taken, therefore, when chosing the basis sets to ensure a proper balancing.

Whenever solvent molecules move between the solvation layer and the MM region, a smoothing factor (based on the molecule's center of mass) is applied to all atoms of the respective molecule:

$$F_j^{\text{smooth}} = S(r) \cdot \left(F_j^{\text{layer}} - F_j^{\text{MM}}\right) + F_j^{\text{MM}} \quad (6)$$

where the smoothing factor $S(r)$ is:

$$\begin{aligned} S(r) &= 1, \text{ for } \mathrm{r} \leq \mathrm{r}_1 \\ S(r) &= \frac{(\mathrm{r}_0^2 - \mathrm{r}^2)^2(\mathrm{r}_0^2 + 2\mathrm{r}^2 - 3\mathrm{r}_1^2)}{(\mathrm{r}_0^2 - \mathrm{r}_1^2)^3}, \text{ for } \mathrm{r}_1 < \mathrm{r} \leq \mathrm{r}_0 \\ S(r) &= 0, \text{ for } \mathrm{r} > \mathrm{r}_0 \end{aligned} \quad (7)$$

r is the distance of a given solvent molecule's centre of mass from the centre of the QM region, r_0 the radius of the QM region and r_1 the inner border of the smoothing region. A thickness of 0.2 Å is usually employed as this value was found to be the minimum distance to ensure smooth transitions of solvent molecules.

By this approach the treatment of solutes of very different nature becomes a much more simple and straightforward process, as only potential functions for solvent-solvent interactions are necessary, whereas all other interactions are dealt with at quantum mechanical level, which in addition ensures a much higher accuracy than molecular mechanics potentials could achieve.

All other conditions mandatory for MD simulations [23] like periodic boundary condition, minimal image convention, long-range forces corrections (here by the reaction field method) and a sufficiently large number of solvent molecules ensure that the system corresponds to the condensed, liquid state.

The inclusion of a relatively large number of species in the QM regions pushes the computational effort to a much higher demand than typical QM/MM molecular dynamics describing only ion plus first hydration shell by quantum mechanics. Due to the constant and rapid progress of processor speed, this increased effort will become a less limiting factor soon. The simulation time of a QMCF MD simulation of an ion is comparable to a conventional QM/MM simulation including two full hydration shells, which on the other hand corresponds to that of a simulation with only one hydration shell a couple of years ago. Simulations of the *ab initio* QMCF MD type for ions are already feasible and the method will thus soon allow the treatment of more complex systems of solution chemistry and biochemistry.

The molecular dynamics simulation protocol employed in the following examples was similar to numerous previous conventional QM/MM MD simulations [15, 6]: an elementary box containing the solute and a sufficient number of water molecules (499 to 1000), periodic boundary condition, NVT ensemble with temperature control by the Berendsen algorithm[24], time step of 0.2 fs, predictor-corrector algorithm to treat the Newtonian equations of motion, cut-off for non-Coulombic interactions in the MM region of 5 and 3 Å for O-O and H-H, respectively, and reaction field correction for long-range Coulombic interactions [23]. For the solvent water, the BJH-CF2 [25, 26] potential was employed, as its intramolecular term ensures the full flexibility of water molecules required for a correct transition from the QM region with its full molecular flexibility into the MM region and vice versa. All simulations reported here were performed on Linux clusters with 4–8 AMD OPTERON 206 processors (according to optimal parallelisability). The CPU time for one timestep of the simulation is, depending on the number of particles, between 1.5 and 3 minutes. As some ten thousand steps are required for a sufficiently long trajectory the net CPU time of one simulation amounts to 6 - 8 months. In the following a few examples of application of the QMCF MD method will be presented, demonstrating its flexibility and accuracy.

3 Results of QMCF MD Simulations

The first example are QMCF MD simulations of hydrated Al(III), performed with two different population analysis charge schemes, the first are according to Mulliken [22], the other one to Löwdin [27]. This hydrate is a very sensitive test case because of the strong polarisation induced by the small and highly charged ion in its surrounding, and noticeable differences could be expected when changing from the QM/MM to the QMCF framework. The basis sets assigned to aluminium, oxygen and hydrogen in the QMCF simulations were the same as employed in a recent conventional QM/MM MD study [28]. The radii of 'core' and 'layer' region were set to 2.5 and 5.0 Å, respectively.

Figure 2 displays the Al(III)–O radial distribution function (RDF) of the QMCF simulations and the QM/MM simulation carried out previously, the structural parameters are listed in table 1. The intensity of the first shell peak decreased noticeably accompanied by an increased tailing of the peak towards larger distances when moving from the QM/MM to the QMCF framework, whereas the first shell distance and the coordination number remain unchanged. The composition of the second hydration shell shows significant differences between the QMCF and QM/MM method.

The second shell peak obtained from the QM/MM simulation shows a more or less exponential decay towards the third shell, instead of the expected Gauss type distribution. Employing the QMCF methodology a shift of the second shell peak by 0.1 Å results, accompanied by a slight decrease in the peak's intensity and a significant change in the peak's shape. The second shell coordination number obtained with Mulliken charges was obtained as 12.0 which is similar to the

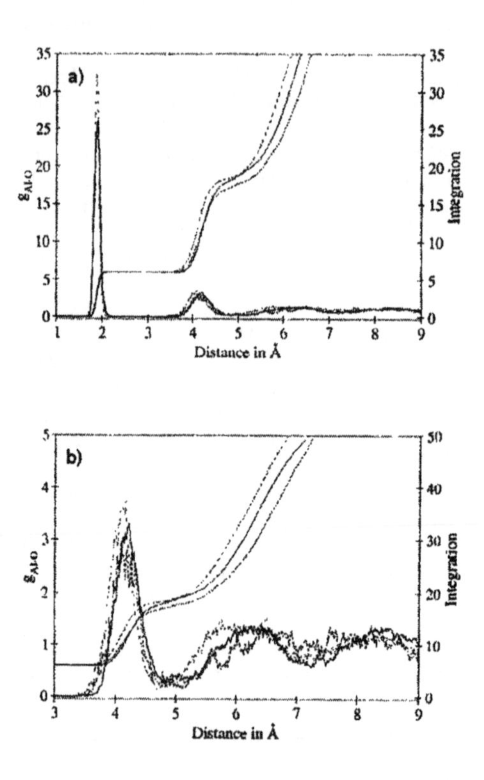

Figure 2: a) Al–O radial distribution functions and their running integration numbers for QMCF MD simulations employing Mulliken (solid line) and Löwdin (dashed line) partial charges compared to a two shell QM/MM MD simulation (dashed–dotted line) of Al(III) in aqueous solution. b) Enlarged illustration of the second and third shell.

	$g_{Al(III)-O}$									
	r_{M1}	r_{m1}	r_{M2}	r_{m2}	r_{M3}	r_{m3}	$CN_{av,1}$	$CN_{av,2}$	$CN_{av,3}$	Ref.
2 shell QM/MM simulation	1.8	2.1	4.1	4.7	~6.1	7.2	6.0	12.2	37.3	[28]
QMCF simulation - Mulliken charges	1.9	2.2	4.2	4.8	6.3	7.3	6.0	12.0	33.5	this work
QMCF simulation - Löwdin charges	1.9	2.2	4.1	4.8	6.5	7.1	6.0	11.0	31.5	this work

Table 1: Maxima r_M and minima r_m of the Al(III)–O radial distribution functions in Å and average coordination numbers of the respective shells.

value of the QM/MM method given as 12.2, whereas it decreased to 11.0 in the case of Löwdin charges. The location of the minimum between second and third shell is very similar in all three cases, but major differences are visible for the third hydration shell. The peak obtained from the QM/MM MD simulation is very broad and forms some kind of plateau rather than a Gaussian distribution. This behaviour can be explained by a too strong contribution of the Al(III)–water coulombic interactions resulting from the fixed charge of +3 assigned to aluminium. A considerable 'pressure' of the MM paticles on the QM region occurs, which is also responsible for the deformation of the second shell peak.

As the partial charge of Al(III) employed in the QMCF MD approach fluctuates around 2.4 to 2.5 in the case of Mulliken charges and around 2.0 to 2.1 for Löwdin charges, this artificial pressure disappears resulting in well structured second and third shell peaks. The charge of Al(III) obtained from the Löwdin scheme is rather low and as a consequence a less defined third shell peak is visible and its separation from the bulk hardly discernable.

As the experimental investigation of second shell structures already poses difficulties, even less data about the third shell is available and hence it is difficult to judge which partial charge scheme is best with respect to structural and dynamical properties. A QMCF MD simulation including also the third hydration shell in the QM region would yield appropriate data, but as the third shell contains more molecules than the first and second shell together, the computational demand would exceed present capabilities. Taking into account the partial charge on Al(III) and the shape of the third shell peak it appears, that Mulliken charges are – in contrast to popular criticism – a satisfactory choice to evaluate the partial charges in the QMCF MD approach. Furthermore, one can expect that the differences between the charge schemes will be less pronounced for di– and monovalent ions.

Overall one can conclude that the QMCF MD method is superior to the conventional QM/MM MD approach, yielding improved results with the great advantage of not having to construct potential functions involving the solute.

The potential functions employed in the case of Al(III) are relatively simple – only a radial dependence occurs for pair and three–body potentials. The advantage to omit these solvent–solute potentials becomes the more relevant the more complicated these interaction potentials are. One prominent example for asymmetry is the Cu(II) ion, with its dynamical Jahn–Teller distortions in aqueous solution. Nevertheless, this system is still treatable by radial potential functions if quantum mechanics are applied to treat the first and second hydration shell [21, 29]. Pd(II) is an even more complicated case: experimental investigations report a planar structure with coordination number four [30, 31, 32, 33], whereas coordination number six was recently predicted by a classical molecular dynamics simulation employing a polarisable shell model [34].

Figure 3 displays the Pd(II)–water radial distribution functions obtained from a QMCF MD simulation employing the Mulliken partial charge scheme, the main structural data are summarized in table 2. The basis sets employed have been the same as in a previous one shell QM/MM MD simulation for Pd(II) in water yielding a five–coordinated pyramidal first shell structure [35]. The radius of the layer region was set to 5.2 Å, that of the core region to 2.3 Å.

It can be immediatly deduced from the Pd(II)–O RDF, that the first shell consists of two structural features, one peak centered at 2.0 Å containing 4 water molecules and a side peak located at 2.7 Å, consisting in average of 2 ligands. These structural features correspond to a strongly distorted octahedron, which is illustrated by a snapshot from the simulation (cf. figure 3). The second shell situated at 4.4 Å is in average composed of eight ligands, which is an unusually small value. Apparently, the water molecules in the axial positions are too labile to form stable hydrogen bonds towards outer solvent molecules. The classical MD study leading to a similar result employed a shell model, treating water molecules by different functions according to their relative position to the Pd(II) ion [34]. This is a good example for the difficulties encountered when treating systems with inherent asymmetry by potential functions – the asymmetry has to

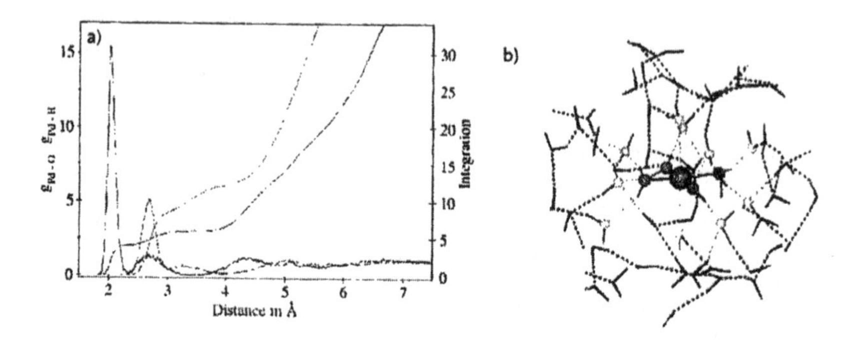

Figure 3: a) Pd–O (solid line) and Pd–H (dashed line) radial distribution functions and their running integration numbers obtained from a QMCF MD simulations employing Mulliken partial charges of Pd(II) in aqueous solution. b) Snapshot from the simulation.

	r_{M1}	r_{m1}	$r_{M,ext}$	$r_{m,ext}$	$CN_{av,1}$	r_{M2}	r_{m2}	$CN_{av,2}$
Pd(II)-O	2.02	2.33	2.7	3.45	4.0+1.94	4.4	4.96	8.0
Pd(II)-H	2.69	3.07	3.3	4.0	12.1	5.0	6.05	31.5

Table 2: Maxima r_M and minima r_m of the Pd(II)–O and Pd(II)–H radial distribution function in Å and average coordination numbers of the respective shells.

be introduced by suitable functions beforehand. Very recent EXAFS and LAXS measurements of aqueous Pd(II) solutions have fully confirmed the results of our QMCF MD simulation [36]. System complexity increases if the solute consists of more than a single atom as it is the case for the Hg(I) species, appearing as Hg_2^{2+} in aqueous solution. The potential functions would require, besides a radial dependene, the consideration of the relative angle between the Hg(I)–Hg(I) and the Hg(I)–water vector, which would require the scanning of many more points on the Hg_2^{2+}–water hypersurface and the construction of a suitable analytical function representing this more complex surface. The accuracy of this pair function would be rather low though and still not take into account even three–body effects. The QMCF MD method offers, due to the renounceability of these potential functions a universal tool for the study of such composite solutes.

Figure 4 depicts the Hg(I)–O and Hg(I)–H radial distribution functions and a snapshot obtained from a QMCF MD simulation employing Mulliken partial charges (relativistic corrected basis set for Hg by Stevens,Krauss and Basch [37], Dunning DZP basis sets for oxygen and hydrogen [38]). The radii of 'core' and 'layer' region were set to 2.3 and 5.2 Å, respectively. The first shell consisting in average of 3.7 ligand peaks at 2.4 Å. EXAFS and LAXS studies of a 0.5M $Hg_2(ClO_4)_2$ yield first shell distances of 2.3 and 2.2 Å, respectively. An excess of perchloric acid (30%) added to prevent possible hydrolysis might explain the minor deviations. The snapshot of the simulation shows the arrangement of the ligands surrounding the Hg atoms. The water molecules are located at both sides of the Hg(I)–Hg(I) dimer, whereas the region perpendicular to the bond is rarely populated. The average Hg(I)–Hg(I) bond length was determined as 2.6 Å, which nearly coincides with the experimentally estimated value of 2.5 Å [39].

A second example for a more complex species is the phosphate ion which, due to its tetrahedral

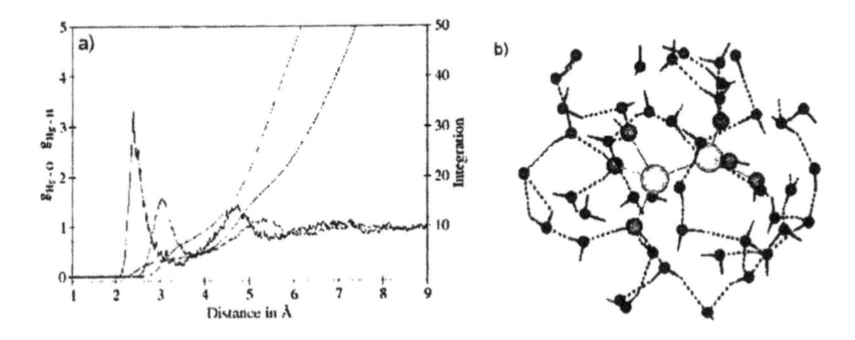

Figure 4: a) Hg–O (solid line) and Hg–H (dashed line) radial distribution functions and their running integration numbers obtained from a QMCF MD simulation. b) Snapshot from the simulation

structure would require very complex potential functions. Figures 5 a and b show the phosphorus – water and O_P – water radial distribution functions and a snapshot from the QMCF MD simulation. Dunning DZP basis sets [38] have been applied to all atoms in the system, with additional diffuse functions assigned to the phosphate oxygens. The radii of 'core' and 'layer' region were set to 3.0 and 5.7 Å, respectively.

Although phosphate is a considerably large solute and occupies the entire core region no significant transition artifact between QM and MM region is observed (cf. figure 5 a) which confirms that the QMCF MD approach is also suitable to properly describe larger solutes as well. The average P-O distance was obtained as 1.58 Å in good agreement with isotope substituted neutron diffraction (NDIS) experiments of a 0.89 M K_3PO_4 solution yielding 1.53 Å [40]. The average P-O distance to first shell oxygens is 3.77 Å in excellent agreement with experimental estimations of 3.6 to 3.9 Å [40, 41]. Each of the phosphate's oxygens is in average linked to solvent molecules by 3.2 hydrogen atoms at an average bond length of 1.7 Å, which is also in good agreement with the value of 1.8 Å obtained from NDIS experiments [40]. Figure 5 c depicts a snapshot from the simulation showing the regular arrangement of the water molecules around each of the phosphate's oxygen atoms.

4 Conclusion and Outlook

The four different examples for QMCF MD simulations presented here clearly indicate not only the methodical superiority of the QMCF MD method over conventional QM/MM MD simulation but also the enormous capabilities of this method to be applied to even larger and more complex systems important for biochemistry and solution chemistry by increasing the computational means to limits feasible in the near future. The inclusion of different solvent species will further contribute to the universality of the methodology, extending its applicability to mixed electrolyte solutions.

Acknowledgments

Financial support of this work by the Austrian Science Foundation (FWF Project No 18429) is gratefully acknowledged.

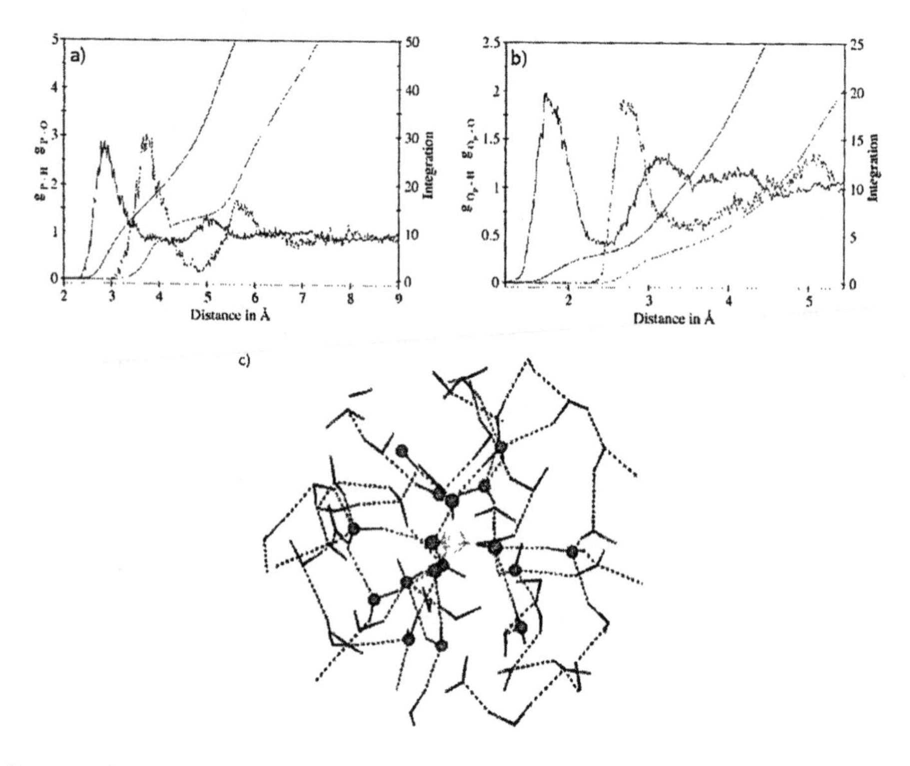

Figure 5: a) P–H (solid line) and P–O (dashed line) and b) O_P–H (solid line) and O_P–O (dashed line) radial distribution functions and their running integration numbers obtained from a QMCF MD simulation of phosphate in aqueous solution. c) Snapshot from the simulation

References

[1] Clementi, E.; Corongui, G. *Int. J. Quant. Chem., Symp.* **1983**, *10*, 31.

[2] Detrich, J. H.; Clementi, E.; Corongui, G. *Chem. Phys. Lett.* **1984**, *112*, 426.

[3] Curtiss, L. A.; Halley, J. W.; Hautman, J.; Rahman, A. *J. Chem. Phys.* **1987**, *86*(4), 2319–2327.

[4] Elrod, M. J.; Saykelly, R. J. *Chem. Rev.* **1994**, *94*, 1974.

[5] Rode, B. M.; Schwenk, C. F.; Tongraar, A. *J. Mol. Liquids* **2003**, *110*, 105.

[6] Rode, B. M.; Schwenk, C. F.; Hofer, T. S.; Randolf, B. R. *Coord. Chem. Rev.* **2005**, *249*, 2993.

[7] Car, R.; Parinello, M. *Phys. Rev. Lett.* **1985**, *55*(22), 2471–2474.

[8] Becke, A. D. *Phys. Rev.* **1988**, *38*, A3098.

[9] Perdew, J. P.; Burke, K.; Ernzerhof, M. *Phys. Rev. Lett.* **1996**, *77*, 3865.

[10] Perdew, J. P.; Burke, K.; Ernzerhof, M. *Phys. Rev. Lett.* **1997**, *78*, 1396.

[11] Warshel, A.; Levitt, M. *J. Mol. Biol.* **1976**, *103*, 227.

[12] Field, M. J.; Bash, P. A.; Karplus, M. *J. Comput. Chem.* **1990**, *11*(6), 700–733.

[13] Gao, J. *J. Am. Chem. Soc.* **1993**, *115*, 2930–2935.

[14] Bakowies, D.; Thiel, W. *J. Phys. Chem.* **1996**, *100*(25), 10580–10594.

[15] Schwenk, C. F.; Tongraar, A.; Rode, B. M. *J. Mol. Liquids* **2004**, *110*, 105.

[16] Rode, B. M.; Hofer, T. S.; Randolf, B. R.; Schwenk, C. F.; Xenides, D.; Vchirawongkwin, V. *Theor. Chem. Acc.* **2006**, *115*, 77.

[17] Møller, C.; Plesset, M. S. *Phys. Rev.* **1934**, *46*, 618–622.

[18] Becke, A. D. *J. Chem. Phys.* **1993**, *98*(7), 5648–5652.

[19] Schwenk, C. F.; Hofer, T. S.; Rode, B. M. *J. Phys. Chem. A* **2004**, *108*, 1509.

[20] Schwenk, C. F.; Löffler, H. H.; Rode, B. M. *J. Chem. Phys.* **2001**, *115*, 10808.

[21] Schwenk, C. F.; Rode, B. M. *J. Chem. Phys.* **2003**, *119*, 9523.

[22] Mulliken, R. S. *J. Chem. Phys.* **1962**, *36*, 3428.

[23] Allen, M. P.; Tildesley, D. J. *Computer Simulation of Liquids;* Oxford Science Publications: Oxford, 1990.

[24] Berendsen, H. J. C.; Postma, J. P. M.; van Gunsteren, W. F.; DiNola, A.; Haak, J. R. *J. Phys. Chem.* **1984**, *81*, 3684–3690.

[25] Stillinger, F. H.; Rahman, A. *J. Chem. Phys.* **1978**, *68*(2), 666.

[26] Bopp, P.; Janscó, G.; Heinzinger, K. *Chem. Phys. Lett.* **1983**, *98*(2), 129.

[27] Löwdin, O. P. *J. Chem. Phys.* **1950**, *18*, 365.

[28] Hofer, T. S.; Randolf, B. R.; Rode, B. M. *PhysChemChemPhys* **2005**, *7*, 1382.

[29] Schwenk, C. F.; Rode, B. M. *J. Chem. Phys.* **2003**, *119*(18), 9523–9531.

[30] Elding, L. *Inorg. Chim. Acta* **1972**, *6*, 647.

[31] Elding, L. *Inorg. Chim. Acta* **1972**, *6*, 683.

[32] Elding, L. *Inorg. Chim. Acta* **1972**, *20*, 65.

[33] Hellquist, B.; Bengtsson, L. A.; Holmberg, B.; Hedman, B.; Persson, I.; Elding, L. I. *Acta Chem. Scand.* **1991**, *45*, 449.

[34] Martinez, J. M.; Torrico, F.; Pappalardo, R. R.; Marcos, E. S. *J. Phys. Chem. A* **2004**, *108*, 15851.

[35] Shah, S. A. A.; Hofer, T. S.; Fatmi, M. Q.; Randolf, B. R.; Rode, B. M. *Chem. Phys. Lett.* **2006**, *in press*, DOI:10.1016/j.cplett.2006.05.132.

[36] Persson, I. *personal correspondence.*

[37] Stevens, W. J.; Krauss, M.; Basch, H.; Jasien, P. G. *Can. J. Phys.* **1992**, *70*, 612.

[38] Dunning, Jr., T. H. *J. Chem. Phys.* **1970**, *53*(7), 2823–2833.

[39] Rosdahl, J.; Persson, I.; Kloo, L.; Ståhl, K. *Acta Chem. Scand.* **1991**, *45*, 449.

[40] Mason, P. E.; Cruickshank, J. M.; Neilson, G. W.; Buchanan, P. *PhysChemChemPhys* **2003**, *5*, 4686.

[41] Ohtaki, H.; Radnai, T. *Chem. Rev.* **1993**, *93*(3), 1157–1204.

Brill Academic Publishers
P.O. Box 9000, 2300 PA Leiden
The Netherlands

Lecture Series on Computer and Computational Sciences
Volume 6, 2006, pp. 453-460

Relativistic Quantum Chemistry – A Historical Overview.

Peter Schwerdtfeger and Christian Thierfelder

Centre of Theoretical Chemistry and Physics,
Institute of Fundamental Sciences, Massey University (Albany Campus),
Private Bag 102904, North Shore MSC, Auckland, New Zealand

Received 19 June, 2006; accepted 22 June, 2006

Abstract: A short historical overview on the development of relativistic quantum chemistry is given as an introduction to the contributions which follow for the ICCMSE symposium on relativistic electronic structure theory.

Keywords: Relativistic Electronic Structure Theory, Dirac Equation, QED, Electroweak Interactions

Nonrelativistic quantum mechanics has been developed in the mid-twenties of the last century and is widely used since by quantum chemists. There are, however, several reasons that the description of microscopic phenomena through nonrelativistic quantum mechanics is incomplete. First, nonrelativistic quantum mechanics cannot describe the fine structure in atoms or molecules (A.Sommerfeld) and the Schrödinger equation is not Lorentz invariant. An ad-hoc introduction of the spin-orbit coupling operator to the nonrelativistic Hamiltonian (W.Pauli) does not lead to the required accuracy of the measured fine-structure in atoms. Second, the Coulomb interaction between charged particles is described by the exchange of photons demanding a Lorentz invariant field theoretical treatment (P.A.M.Dirac [1], W.Heisenberg). And third, the classical limit of the Schrödinger equation (classical energy-momentum relation) is not valid for very fast moving particles as this is the case for inner electrons in heavy atoms (A.Einstein, P.A.M.Dirac). The foundations of relativistic quantum field theory have been laid in the early thirties of the last century.

It was not evident until the early seventies that relativistic effects are important in chemical properties. For heavy elements relativistic effects are most important and significantly change molecular properties and chemical behaviour [2-5]. The chemistry of gold, for example, cannot be understood without the inclusion of relativistic effects [5-8]. Only in the last two decades has the relativistic treatment of atoms and molecules become mainstream, and can now be found in modern chemistry textbooks.

The time-dependent Schrödinger equation is not Lorentz-invariant as we have different orders in space and time differentials. After nonrelativistic quantum mechanics was established in 1925 the aim was therefore to construct a relativistic Hamiltonian and to quantize to a fully Lorentz-covariant form. This automatically leads to the famous Dirac-equation for a charged particle in an external field,

$$i\hbar\frac{\partial}{\partial t}\psi = c\vec{\alpha}\left(-i\hbar\vec{\nabla} + e\vec{A}\right)\psi + \left(\beta mc^2 - eV_{ext}\right)\psi \tag{1}$$

or in the covariant form

$$\left[\gamma^{\mu}\left(i\hbar\frac{\partial}{\partial x^{\mu}} - eA_{\mu}\right) - mc^2\right]\psi = 0 \tag{2}$$

The electron spin turned out to be a necessary consequence of the Dirac equation included in the four-component wavefunction ψ, resulting for example in the correct value for the gyromagnetic ratio. The anomalous Zeeman effect could be explained for the first time. The probability density ρ now was positive definite as first derivatives in ρ are avoided. P.A.M.Dirac therefore noted: *The general theory of quantum mechanics is now almost complete, the imperfections that still remain being in connection with the exact fitting in of the theory with relativity ideas ... The underlying physical laws necessary for the mathematical theory of a large part of physics and the whole of chemistry are thus completely known, and the difficulty is only that the exact application of these laws leads to equations much too complicated to be solvable. It therefore becomes desirable that approximative practical methods of applying quantum mechanics should be developed, which can lead to an explanation of the main features of complex atomic systems without too much computation.*

In the same year, in 1928, Dirac's equation gave results on the fine-structure of hydrogen and ionized helium in good agreement with results from measurements by optical spectroscopy [9,10]. A many-electron formulation was given by Dirac in 1928, and a relativistic Hartree-Fock scheme was proposed by B.Swirles only seven years later, in 1935 [11]. B.Swirles constructed tables for the Slater coefficients and in 1936 the calculations carried out by B.Swirles [12] for a two-electron system including the Breit correction [13] led to a qualitative agreement for the 2^3P fine structure observed for helium.

It became apparent, however, that the unbound differential operator (1) represents a major difficulty with variational techniques like the Ritz procedure. Without implying certain boundary conditions, (1) leads to a variational collapse of the electron into the negative energy states. To avoid the 'downfall' of an electron to negative continuum states (antiparticle states), P.A.M.Dirac had to propose that all negative energy states are filled with electrons (Dirac-sea), which forms the physical vacuum state Dirac (1930): *Let us assume there are so many electrons in the world that all the most stable states are occupied, or more accurately, that all the states of negative energy are occupied except perhaps a few of small velocities. Any electron with positive energy will now have very little chance of jumping into negative-energy states and will therefore behave like electrons are observed to behave in the laboratory. We shall have an infinite number of negative-energy states, and indeed an infinite number per unit volume all over the world, but if the distribution is exactly uniform we should expect them to be completely unobservable. Only the small departures from exact uniformity, brought about by some of the negative-energy states being unoccupied, can we hope to observe.* Dirac calls these vacant states holes. An electron is therefore not able to occupy one of the negative energy states because of the Pauli exclusion principle that holds for all fermions. A removal of an electron from a negative energy state would produce a positively charged hole that can be identified with a particle state of the same mass compared with the electron, but opposite electric charge. Such a particle is known as a positron. Dirac in 1931: *A hole, if there were one, would be a kind of particle, unknown to experimental physics, having the same mass and opposite charge to an electron.* Dirac's theory predicted pair-creation and the existence of antiparticles. This led to the discovery of the positron by C.D.Anderson in 1932 in cloud chamber pictures of cosmic rays, and to a great success of Dirac's theory (Anderson 1932): *On the assumption that the particles producing the tracks are traveling downward through the chamber rather than upward, particles of positive charge appear as well as electrons.* The name positron has been introduced by Anderson in 1933: *Out of the group of 1300 photographs of cosmic-ray tracks in a vertical Wilson chamber 15 tracks were of positive particles which could not have a mass as great as that of the proton... These particles will be called positrons.*

Equation (1) as it stands does not 'know' anything about the filled Dirac-sea and an exact interpretation is only given quantum electrodynamics (QED). With QED the vacuum becomes a physically reasonable state with no particles in evidence (J.Schwinger 1973): *The picture of an infinite sea of negative energy electrons is now best regarded as an historical curiosity, and forgotten. Unfortunately, this episode, and discussions of vacuum fluctuations, seem to have left people with the impression that the vacuum, the physical state of nothingness, is actually the scene of wild activity.* Although still found in many physics textbooks the filled Dirac see is now seen as a curiosum and due to an incomplete description of the interaction between charged particles. Nevertheless, the variational collapse has to be avoided in four-component calculations, that is the variational wavefunction is chosen to be orthogonal on the Hilbert subspace of the negative energy continuum. In other words, the Rayleigh quotient in the variational procedure should be bounded from below and the correct nonrelativistic limit should be obtained. Here elimination of the small component part of the wavefunction provides the basis for the so-called kinetic balance condition (R.E.Moss and H.P.Trivedi [14], Y.S.Lee and A.D.McLean [15], W.Kutzelnigg [16], R.E.Stanton and S.Havriliak [17,18]),

$$\psi_S = c\left(2mc^2 + E - V\right)^{-1} \vec{\sigma}\vec{p}\psi_L \tag{3}$$

A Taylor expansion ($E \ll 2mc^2$) leads to the well-known approximation $\psi_S \approx (2mc)^{-1} \vec{\sigma}\vec{p}\psi_L$. The simplest way to achieve this is by choosing the small component basis such that each large component has a corresponding small component function in the basis set, $\{\chi_S^p\} \supseteq \{\vec{\sigma}\vec{p}\chi_L^p\}$. For more details see ref.19.

The Dirac equation predicts no splitting between the $2^2S_{1/2}$ and $2^2P_{1/2}$ levels in the hydrogen atom, but a small splitting was noted in 1930 by R.Oppenheimer [20] and I.Waller [21] and later in 1938 by S.Pasternack [22]. In 1947 and 1951 W.E.Lamb and R.C.Retherford measured a splitting of only 0.033 cm^{-1} [23,24]. Lamb's discovery of the level splitting in the hydrogen spectrum shift was already predicted by V.F.Weisskopf and B.French, but published much later in 1949 [25]. Also the magnetic moment of an electron was slightly larger (g_s= 2.00244) than predicted by relativistic quantum theory (H.M.Foley and P.Kusch [26]). These findings were major factors to the development of a special discipline in quantum field theory, QED, in which charged particles and the electromagnetic field are treated on the same footing dynamically. Such effects become important in inner shell ionization energies of heavy elements. For the valence shell such effects can safely be neglected as these are below the "chemical accuracy" limit (~ 1 kcal/mol) even for the heaviest elements in the periodic table (Pyykkö *et al.* [27,28]). At most the Lamb shift in the valence s-levels is comparable with the Breit contribution. However, the so-called Breit interaction can become important in accurate calculations of the fine-structure in atoms and molecules. In the Feynman gauge (C.Møller 1931 [29]) the interaction between two electrons *i* and *j* becomes,

$$V_F(r_{ij},\omega_{ij}) = r_{ij}^{-1}\left(1-\vec{\alpha}_i\vec{\alpha}_j\right)e^{ic^{-1}|\omega_{ij}|r_{ij}} \tag{4}$$

Since the frequency of the exchange photon ω_{ij} is small compared to *c*, the frequency dependent exponential in (4) is often neglected (low frequency limit). The addition to the traditional gauge term is called the Gaunt interaction. A gauge transformation from the Feynman to the Coulomb gauge introduces an additional gauge term (G.E.Brown 1952 [30], J. Sucher 1957 [31], H.A.Bethe and E.E.Salpeter 1957 [32]) and the Breit operator becomes

$$V_C(r_{ij},\omega_{ij}) = V_F(r_{ij},\omega_{ij}) - \left(\vec{\alpha}_i\vec{\nabla}_i\right)\left(\vec{\alpha}_j\vec{\nabla}_j\right)f_C(r_{ij},\omega_{ij}) \tag{5}$$

with

$$f_C(r_{ij},\omega_{ij}) = c^{-2}\omega_{ij}^{-2}r_{ij}^{-1}\left(e^{ic^{-1}|\omega_{ij}|r_{ij}}-1\right) \tag{6}$$

The Dirac-Coulomb-Breit Hamiltonian is the exact Hamiltonian, in the Coulomb gauge, to order α^2 of the fine structure constant. The next terms, of order α^2, were identified for the helium atom in 1957-58 by H.Araki and J.Sucher [33,34]. To illustrate the importance of such effects the K-ionization of Hg may be taken (from Dirac-Fock calculations using the program MCDF/BENA by Grant et al.): nonrelativistic Hartree-Fock 75.537 keV, Dirac-Hartree-Fock (DHF) 83.560 keV, DHF + ω-dependent Breit 83.232 keV, DHF + ω-dependent Breit +self-energy + vacuum polarization 83.109 keV, and the experimental value is 83.102 keV (CRC). The remaining difference can be attributed to correlation effects and smaller QED corrections. The Breit interaction between the electron and the proton can be neglected because of the mass of the proton. It should be noted that as early as in 1933 L.Pincherle pointed out the importance of relativistic effects in the calculation of X-ray spectra [35,36]. Nonrelativistic calculations of such spectra did not lead to satisfactory results. In 1937 Ramberg and Richtmyer noted, that *Perhaps the major defect, especially influential in the case of the radiation width of the deepest energy levels, lies in the fact that the treatment is nonrelativistic* [37]. Only if one considers inner shell electrons in heavy elements relativistic corrections may become important: *While this correction may be expected to be negligible for most atoms, it becomes appreciable as the atomic number increases because for very heavy atoms the electron velocities in the vicinity of the nucleus become high. For such atoms the electron density close to the nucleus may be appreciably changed by the relativity correction* (M.S.Vallarta, N.Rosen 1932 [38]). The Breit interaction should be included in accurate calculations of fine structure splittings or in the calculations of core properties in heavy elements such as electric field gradients [39]. For example, for a two-electron atom with Z=60 radiative corrections contribute to ~11% to the 2^3S_1-2^3P_0 splitting (P.J.Mohr 1983 [40]). Radiative corrections contribute only to 1.1% to the total amount in the ^2P splitting in Tl, and Breit interactions are even more important contributing to 3.6% to the ^2P splitting, which is on the same magnitude as electron correlation. For a recent review on QED corrections in heavy atoms see ref.41. We note that Grelland investigated the bond shortening of Au_2 due to relativistic effects originating from the classical Coulomb interaction between charged nuclei, $\Delta H_R = -Z_a^2 Z_b^2 / 2c^2 \mu_{ab} R_{ab}^2$ [42] between nucleus a and b (μ is the reduced mass), which is commonly neglected in all applications. This leads to a bond contraction of only 0.00015 Å for Au_2. We note however that the Hamiltonian ΔH_R used in this analysis is more than questionable for spin 1/2 particles, as it is not derived from the Dirac equation, and has therefore to be taken with care. Nuclear relativistic effects are generally much smaller compared to the electronic counterpart [43].

In 1973, Rajagopal and Callaway extended the Hohenberg–Kohn Theorem in density functional theory (DFT) to the relativistic domain [44]. Here the total energy becomes a functional of the four-current j_μ. Perdew and Cole implemented the Breit term within a local density approximation pointing out that accurate ionization potentials can only be achieved by including the self-interaction term of DFT for the 1s electrons [45].

There were signs, however, that the restriction of relativistic effects to typical core properties is not sufficient. A.O.Williams noted in his Hartree-calculations on the closed shell atom Cu^+ as early as 1940 [46]: *The charge density of each single electron turns out to resemble that for the nonrelativistic case, but with the maxima "pulled in" and raised. ... The size of the relativistic corrections appears to be just too small to produce important corrections in atomic form factors or other secondary characteristics of the whole atom. ... However, it must be noticed that copper is a relatively light ion, and the corrections for such an ion as mercury would be enormously greater.* S.Cohen in 1955 [47] and D.F.Mayers in 1957 [48] pointed out that relativistic effects are very important even in the valence region of heavy elements such as mercury: *The values of the energy parameter ε ... show a substantial increase over the nonrelativistic values, the relation depending more on the l than on the n value of the wave function. The increase is particularly marked for the inner shells, as was expected. For the outer groups, indirect effects have become appreciable; the 5d group, which is known to be very sensitive to slight changes in the potential, has been so affected by the contraction of the inner groups, that its energy parameter has in fact decreased* [48]. Later in 1967 V.M.Burke and I.P.Grant stated that, *Little attention appears to have been paid to the effect of relativity on atomic wave functions ...* [49]. In 1974 A.Rosen and D.E.Ellis pointed out that to attain accurate electron binding energies in molecules relativistic effects have to be taken into account [50]. At this time it was also recognized that for the accurate prediction of chemical properties of superheavy elements (trans-actinides) relativistic codes are required [51]. It is however astonishing that even in 1988 S.L.Glashow stated [52]: *Modern elementary-particle physics is founded upon the two pillars of quantum mechanics and relativity. I have made little mention of relativity so far because, while the atom is very much a quantum system, it is not very relativistic at all. Relativity becomes important only when velocities become comparable to the speed of light. Electrons in atoms move rather slowly, at a mere one percent of light speed. Thus it is that a satisfactory description of the atom can be obtained without Einstein's revolutionary theory.*

In 1971 to 1973, the first Dirac-Fock calculation for a series of elements by J.P.Desclaux [53,54] and by Mann and Waber [55] appeared. MCDF calculations on atoms were first reported by J.P.Desclaux in 1975 [54] and by I.P.Grant in 1976 [56]. It was however Pekka Pyykkö and Jean-Paul Desclaux in the mid seventies who showed by using the Dirac-Hartree-Fock one-centre expansion method that by propagating into the valence region relativistic effects become more important in chemical bonding than originally presumed: *The orbital energies for non-relativistic AgH and AuH are quite similar while the relativistic ones are not. The non-relativistic bond lengths also agree within 3.0% while the experimental R_e differ by 5.8%. This suggests that the chemical difference between silver and gold may mainly be a relativistic effect* [4,5,57]. It is now well accepted that relativistic effects play an important role in heavy element [58] and superheavy element chemistry [59-64]. For example for the relativistic 6s-orbital contraction of Au we get at the Hartree-Fock level: $\Delta_R\langle r\rangle = 1.21$ Å), 0.73 Å for Hg^+, 0.56 Å for Tl^{2+}, and 0.47 Å for Pb^{3+}. In fact, neutral gold shows a significantly large relativistic 6s contraction and stabilization compared to its neighbouring atoms in the periodic table. This is well known as the 'Au-Maximum' of relativistic effects, or more precisely as other Group 11 members show similar large effects, the 'Group 11 Maximum' of relativistic effects (P.Pyykkö and J.P.Desclaux 1979 [5]). To name a few "unusual" properties connected to relativistic effects: The yellow colour of gold [65], the fact that mercury is the only liquid metal at room temperature, the very small dissociation energy of Tl_2 (due to spin-orbit effects), the unusual chain-like structures of gold halides [66], and many more.

Perhaps the last "train from modern physics into chemistry" is the field of electroweak quantum theory. In 1956, T.D.Lee and C.N.Yang predicted that parity is violated by the weak field [67]: *If parity is not strictly conserved, all atomic and nuclear states become mixtures consisting mainly of the state they are usually assigned, together with small percentages of states possessing the opposite parity.* The parity nonconservation (PNC), or P-violation (PV), has been confirmed experimentally in the β-decay of the nucleus. The weak coupling constant (Fermi coupling constant) is only of the order $\sim 10^{-13}$ a.u. and very small compared with the coupling constant of the electromagnetic field. The Hamiltonian used in these calculations is:

$$H^{PNC} = \frac{G_F}{\sqrt{2}} \sum_{i,a} \left\{ \frac{1}{2} Q_{W,a} \gamma^5 \rho_a(\vec{r}_i) - \lambda_a \left(1 - 4\sin^2\theta_W\right) \vec{\alpha} \cdot \vec{I} \rho_a(\vec{r}_i) \right\} \tag{7}$$

Here the indices i and α run over all electrons and nuclei respectively, $G_F = 2.22254\times10^{-14}\, E_h a_0^3$ is the Fermi coupling constant, $Q_{W,\alpha} = Z_\alpha(1 - 4\sin^2 \theta_W) - N_\alpha$ is the weak nuclear charge with Z_α and N_α representing the number of protons and neutrons in nucleus α, respectively, and θ_W is the Weinberg angle with $sin^2\theta_W = 0.2397$. $\gamma^5 = i\gamma^1\gamma^2\gamma^3\gamma^0$ is one of the Dirac matrices (the 4×4 pseudoscalar chirality operator). ρ_α is the normalized nuclear charge density, and λ_a a nucleus-dependent parameter close to unity. Finally, I_a is the spin of nucleus a. In 1974 I.B.Khriplovich suggested [68] to measure the intrinsic optical rotation of the atomic vapor of an element in vicinity of an atomic transition. In the same year C.Bouchiat and M.A.Bouchiat discovered the scaling with the atomic charge of parity violation effects in atoms and proposed the later successful experiments on optical rotation and highly forbidden transitions in atomic spectra of heavy elements [69,70]. First observation came for ^{209}Bi by Barkov and Zolotorev in 1978 [71]. Other observations on Tl, Pb and Cs atoms followed. The most accurate experiments come from 7s-6p amplitude measurements of ^{133}Cs by the Colorado group, who also discovered the anapole moment [72]. High-precision measurements on heavy atoms in traps offer unique tests of the Standard Model [73-75] and even possibilities to search for new physics.

The parity violation in atoms soon led to similar attempts for molecules [76-78]. In 1974 D.W.Rein [79] calculated the difference between enantiomers in electronic transitions with energy E to be $\Delta E_{PV}/E < 10^{-10}$. This was far beyond any experimental accuracy. B.Y.Zel'dovich noted in 1977 that the difference ΔE_{PV} should be proportional Z^5 [80]. This was confirmed in nonrelativistic [81,82] and relativistic [83] calculations. Hence molecules with heavy nuclei are preferred, and due to the single-center theorem more than one heavy atom in close neighborhood is of advantage [81]. Recent theoretical results suggest similar high scaling laws for parity violation differences in chemical shifts in NMR spectroscopy [84]. In 1977 the spectra of d and l camphor were compared and found to be identical within $\Delta\nu_{dl}/\nu = 10^{-8}$ [85]. A recent relativistic theoretical study shows that parity violation effects in camphor is below 10^{-5} Hz [86]. Molecular beam experiments by Chardonnet and co-workers on chiral molecules such as CHFClBr or CHFClI using a high precision tuneable CO_2 laser offers currently the best resolution for future measurements [77,87]. Using saturation spectroscopy in the 9.3 μm spectral range the Paris group reported $\Delta\nu_{PV} = -4.2\pm0.6\pm1.6$ Hz (including statistical and systematic uncertainties) for the C-F stretching mode in CHFClBr, which is due to collision effects under the experimental conditions [77]. All attempts to measure the influence of PV effects on frequency shifts between enantiomers have failed so far, despite a few claims in the past. Current theoretical estimates set such effects in the range of a few mHz for these molecules, although other molecules with larger parity violation effects have been suggested [88,89]. For a recent review see ref.90.

Relativistic electronic structure theory has become a very active field in quantum chemistry in recent years, and the articles following this historical outline will nicely demonstrate this and will therefore not be discussed here. We only mention that efficient algorithms to treat the Dirac equation at the correlated level are currently being developed [91], and two-component and scalar relativistic methods such as the Douglas-Kroll-Hess (DKH) [92-95] or regular approximations like ZORA [96-98] are now in use in standard program packages. Another strategy is to treat relativistic effects perturbatively as done in direct perturbation theory (DPT) developed by W.Kutzelnigg and co-workers [99]. Relativistic density functional theory at the four-component level is also in development [100-102]. And last but not least relativistic pseudopotential theory at the two-component or scalar relativistic level is the most widely used and economic method of sufficient accuracy to treat relativistic effects [103-107]. The reference list below is far from being complete, however, the relativistic literature has been surveyed intensively by P.Pyykkö [108]. Recently, a number of books have been edited as well on relativistic electronic structure theory [109-112]. And a new and excellent textbook written by K.G.Dyall and K.Faegri in 2006 on relativistic quantum chemistry is highly commended [19]. In summary, the field of relativistic electronic structure theory is now well developed with increasing activity over the last decades in the field of heavy element chemistry and physics. This ICCMSE symposium on relativistic electronic structure theory will certainly highlight the enormous advances being made in this exciting field.

Acknowledgments

This work was supported by the Marsden Fund administered by the Royal Society of New Zealand. We thank Pekka Pyykkö (Helsinki) , Robert N. Compton (Tennessee) and J. Gierlich (Marburg) for many useful discussions and criticism.

References

[1] P. A. M. Dirac, Proc. Roy. Soc. Lond. A 117, 610 (1928).
[2] P.Pyykkö, Chem. Rev. 88, 563 (1988).
[3] K.S.Pitzer, Acc. Chem. Res. 12, 271 (1979).
[4] P.Pyykkö, Adv. Quantum Chem. 11, 353 (1978).
[5] P.Pyykkö, J.P.Desclaux, Acc. Chem. Res. 12, 276 (1979).
[6] P. Schwerdtfeger, Heteroatom Chem. 13, 578 (2002).
[7] P. Pyykkö, *Angew. Chem. Int. Ed.* 42, 4412 (2004).
[8] P. Pyykkö, Inorg. Chim. Acta 358, 4113 (2005).
[9] P. M. Dirac, Proc. Roy. Soc. Lond. A 118, 351 (1928); ibid. A123, 714 (1929); A126, 360 (1930).
[10] P. A. M. Dirac, *The Principles of Quantum Mechanic,* Oxford, Clarendon Press, (1958).
[11] B. Swirles, Proc. Roy. Soc. Lond. A 152, 625 (1935).
[12] B. Swirles, Proc. Roy. Soc. Lond. A 157, 680 (1936).
[13] G. Breit, Phys. Rev. 34, 553 (1929); *ibid.* 36, 383 (1930); Phys. Rev. 39 616 (1932).
[14] R.E.Moss, H.P.Trivedi, Mol. Phys. 38, 1611 (1979).
[15] Y. S. Lee, A. D. McLean, J. Chem. Phys. 76, 735 (1982).
[16] W. Kutzelnigg, Int. J. Quantum Chem. 25, 107 (1984).
[17] R. E. Stanton, S. Havriliak, J. Chem. Phys. 81, 1910 (1984).
[18] K. G. Dyall, I. P. Grant, S. Wilson, J. Phys. B 17, 493 (1984).
[19] K. G. Dyall, K. Faegri Jr., *Introduction to relativistic quantum chemistry*, to be published.
[20] J. R. Oppenheimer, Phys. Rev. 35, 461 (1930); *ibid.* 35, 562 (1930); *ibid.* 35, 939 (1930).
[21] I. Waller, Z. Phys. 62, 673 (1930).
[22] S. Pasternack, Phys. Rev. 54, 1113 (1938).
[23] W. E. Lamb, R. C. Retherford, Phys. Rev. 72, 241 (1947).
[24] N. M. Kroll, W. E. Lamb, Phys. Rev. 75, 388 (1949).
[25] J. B. French, V. F. Weisskopf, Phys. Rev. 75, 1240 (1949).
[26] H. M. Foley, P. Kusch, Phys. Rev. 73, 412 (1948).
[27] P. Pyykkö, M. Tokman, L. Labzowski, Phys. Rev. A 57, R689 (1998).
[28] L. Labzowski, I. Goidenko, M. Tokman, P.Pyykkö, Phys. Rev. A 59, 2707 (1999).
[29] C. Møller, Z. Phys. 70, 786 (1931).
[30] G. E. Brown, Phil Mag. 43, 467 (1952).
[31] J. Sucher, Phys. Rev. 107, 1448 (1957).
[32] H. A. Bethe, E. E. Salpeter, *Quantum Mechanics of One- and Two-Electron Atoms*, Academic Press, New York, 1957.
[33] H. Araki, Prog. Theor. Phys. Japan 17, 619 (1957).
[34] J. Sucher, Phys. Rev. 109, 1010 (1958).
[35] L. Pincherle, Nuovo Cimento 10, 344 (1933); *ibid.* 12, 81 (1935); *ibid.* 12, 162 (1935).
[36] L. Pincherle, Physica 2, 597 (1935).
[37] E. G. Ramberg, F. K. Richtmyer, Phys. Rev. 51, 925 (1937).
[38] M. S. Vallarta, N. Rosen, Phys. Rev. 41, 708 (1932).
[39] M. Pernpointner
[40] P. J.Mohr, in *Relativistic Effects in Atoms, Molecules and Solids*, ed. G.L.Malli, Plenum, New York (1983)
[41] P. J. Mohr, G. Plunien, G. Soff, Phys. Rep. 293, 227 (1998).
[42] H. H. Grelland, J. Phys. B 13, L389 (1980). Note that in equations (5) and (6) in this paper a square is missing. Moreover, the 0.007 Å bond contraction for Au_2 given in this paper is wrong.
[43] P. Pyykkö, *QED: The last train from physics to chemistry?*, Lecture I-24, XII ICQC Satellite

Symposium in Tokyo *on Chemical Accuracy and Beyond*, May 17-19 (2006). See also: A. Veita, K. Pachucki, Phys. Rev. A 69, 042501 (2004); E. Borie, Phys. Rev. 28, 555 (1983).
[44] A.K. Rajagopal, J. Callaway, Phys. Rev. B 7, 1912 (1973).
[45] J. P. Perdew, L. A. Cole, J. Phys. C 15, L905 (1982).
[46] A. O. Williams, Phys. Rev. 58, 723 (1940).
[47] S. Cohen, *Ph.D. Thesis*, Cornell (1955).
[48] D. F.Mayers, Proc. Roy. Soc. London A241, 93 (1957).
[49] V. M. Burke, I. P. Grant, Proc. Phys. Soc. 90, 297 (1967).
[50] A. Rosen, D. E. Ellis, J. Chem. Phys. 62, 3039 (1975).
[51] O. L. Keller Jr., C. W. Nestor Jr., Thomas A. Carlson, B. Fricke, J. Phys. Chem. 77 1806 (1973).
[52] S. L. Glashow, *Interactions*, Warner Books, New York (1988).
[53] J.P.Desclaux, At. Data Nucl. Data Tables 12, 311 (1973).
[54] J. P. Desclaux, Comp. Phys. Commun. 9, 31 (1975).
[55] J. B. Mann, J. T. Waber, J. Chem. Phys. 53, 2397 (1968).
[56] I.P.Grant, B.J.McKenzie, P.H.Norrington, D.F.Mayers, N.C.Pyper, Comp. Phys. Commun. 21, 207 (1980).
[57] J. P. Desclaux, P. Pyykkö, Chem. Phys. Lett. 39, 300 (1976).
[58] P. Pyykkö, Chem. Rev. 88, 563 (1988).
[59] B. Fricke, W. Greiner, *Theoret. Chim. Acta* 21, 235 (1971).
[60] B. Fricke, *Struct. and Bond.* 21, 89 (1975).
[61] P. Schwerdtfeger, M. Seth, in *Encyclopedia of Computational Chemistry*, eds. P. von R. Schleyer, N. L. Allinger, T. Clark, J. Gasteiger, P. A. Kollman, H. F. Schaefer III, P. R. Schreiner, Wiley, New York (1997).
[62] V.Pershina, in *Progress in Theoretical Chemistry and Physics, Theoretical chemistry and physics of heavy and superheavy elements*, U. Kaldor, S. Wilson (eds.), Kluwer Academic (2003).
[63] P.Schwerdtfeger, M.Seth, in *Encyclopedia of Computational Chemistry*, P. von R.Schleyer, P. R. Schreiner, N. L. Allinger, T. Clark, J. Gasteiger, P. A. Kollman, H. F. Schaefer III (Eds.), Vol.4, Wiley, New York (1998); p.2480.
[64] M. Schädel, *The Chemistry of Superheavy Elements*, Kluwer, Dordrecht (2003).
[65] N. E. Christensen, B. O. Seraphin, Phys. Rev. B 4, 3321 (1971).
[66] T. Söhnel, H. L. Hermann, P. Schwerdtfeger, Angew. Chem. Int. Ed. 40, 4381 (2001).
[67] T. D. Lee, C. N. Yang, Phys. Rev. Lett. 104, 254 (1956).
[68] I. B. Khriplovich, Zh. Eksp. Teor. Fiz. 20, 686 (1974).
[69] M.-A. Bouchiat, C. Bouchiat, Phys. Lett. B 48, 111 (1974).
[70] M.-A. Bouchiat, C. Bouchiat, J. Phys. France 35, 899 (1974); *ibid.* 36, 493 (1975).
[71] L. M. Barkov, M. S. Solotorev, Zh. Eksp. Teor. Fiz. 79, 713 (1980).
[72] C. S. Wood, S. C. Bennett, D. Cho, B. P. Masterson, J. L. Roberts, C. E. Tanner, C. E. Wieman, Science 275, 1759 (1997).
[73] V. A. Dzuba, V. V. Flambaum, O. P. Sushkov, Phys. Lett. A 141, 147 (1989).
[74] S. A. Blundell, W. R. Johnson, J. Sapirstein, Phys. Rev. Lett. 65, 1411 (1990); Phys. Rev. D 45, 1602 (1992).
[75] A. Derevianko, Phys. Rev. Lett. 85, 1618 (2000).
[76] M. Quack, Angew. Chem. Int. Ed. Engl. 28, 571 (1989).
[77] C. Daussy, T. Marrel, A. Amy-Klein, C. T. Nguyen, C. J. Bordé, C. Chardonnet, Phys. Rev. Lett. 83, 1554 (1999).
[78] A. Lahamer, S.M. Mahurin, R.N. Compton, D. House, J.K. Laerdahl, M. Lein, P. Schwerdtfeger, Phys. Rev. Lett. 85, 4470 (2000).
[79] D. W. Rein, J. Mol. Evol. 4, 15 (1974).
[80] B. Y. Zel'dovich, D. B. Saakyan, I. I. Sobel'man, Pis'ma Zh. Eksp. Teor. Fiz 25, 106 (1977).
[81] R. A. Hegstrom, D. W. Rein, P. G. H. Sandars, J. Chem. Phys. 73, 2329 (1980).
[82] D. W. Rein, R. A. Hegstrom, P. G. H. Sanders, Phys. Lett. A 71, 499 (1979).
[83] J. K. Laerdahl, P. Schwerdtfeger, Phys. Rev. A 60, 4439 (1999).
[84] R. Bast, P. Schwerdtfeger, T. Saue, J. Chem. Phys., *in press*.
[85] E. Arimondo, P. Glorieux,T. Oka, Opt. Commun. 23, 369 (1977).
[86] P. Schwerdtfeger, A. Kühn, R. Bast, J. K. Laerdahl, F. Faglioni, P. Lazzeretti, Chem. Phys. Lett. 83, 496 (2004).
[87] C. Chardonnet, C. Daussy, T. Marrel, A. Amy-Klein, C. T. Nguyen, C. J. Bordé, in *Parity violation in atomic physics and electron scattering*, B. Frois, M.A. Bouchiat (eds.), World Scientific, New-York (1999); p. 325.

[88] P. Schwerdtfeger, J. Gierlich, T. Bollwein, Angew. Chem. Int. Ed. 42, 1293 (2003).
[89] R. Bast, P. Schwerdtfeger, Phys. Rev. Lett. 91, 023001 (2003).
[90] J. Crassous, Ch. Chardonnet, T. Saue, P. Schwerdtfeger, Org. Biomol. Chem. 3, 2218 (2005).
[91] T. Saue, T. Enveldson, T. Helgaker, H.J.A. Jensen, J.K. Laerdahl, K. Ruud, J. Thyssen, L. Visscher, Dirac - A relativistic *ab initio* electronic structure program, Release 3.2.1 (2000).
[92] B. A. Hess, Phys. Rev. A 33, 3742 (1986).
[93] A. Wolf, M. Reiher, B. A. Hess, J. Chem. Phys. 117, 9215 (2002).
[94] M. Reiher, A. Wolf, J. Chem. Phys. 121, 2037 (2004); *ibid.* 121, 10945 (2004).
[95] M. Reiher, Theor. Chem. Acc., in press (2006); and references therein.
[96] C. Chang, M. Pélissier, P. Durand, Phys. Scr. 34, 395 (1986).
[97] E. van Lenthe, E.J. Baerends, J.G. Snijders, J. Chem. Phys. 99, 4597 (1993).
[98] E. van Lenthe, J.G. Snijders, E.J. Baerends, J. Chem. Phys. 105, 6505 (1996); and references therein.
[99] W. Kutzelnigg, Z. Phys. D 11, 15 (1989); ibid. 15, 27 (1990).
[100] E. Engel, R.M. Dreizler, S. Varga, B. Fricke, B.A. Hess, in *Relativistic Effects in Heavy-Element Chemistry and Physics*, Wiley, New York (2001).
[101] J. Anton, B. Fricke, E. Engel, Phys. Rev. A 69, 012505 (2004).
[102] J. Anton, T. Jacob, B. Fricke, E. Engel, Phys. Rev. Lett. 89, 213001 (2002).
[103] P. Hafner, W. H. E. Schwarz, J. Phys. B 11, 217 (1978).
[104] L. Kleinman, Phys. Rev. B 21, 2630 (1980).
[105] G. B. Bachelet, M. Schlüter, Phys. Rev. B 25, 2103 (1982).
[106] R. M. Pitzer, N. W. Winter, Int. J. Quantum Chem. 40, 773 (1991).
[107] P. Schwerdtfeger, in *Theoretical Chemistry and Physics of Heavy and Superheavy Elements*, U. Kaldor, S. Wilson (ed.), in Progress in Theoretical Chemistry and Physics, Vol. 11, Kluwer, Dordrecht (2003); p.399; in references therein.
[108] P. Pyykkö, *Relativistic Theory of Atoms and Molecules*, Vols. I-III, Springer, Berlin, 1986, 1993 and 2000. See also http://www.csc.fi/rtam/.
[109] P. Schwerdtfeger (ed.), *Relativistic Electronic Structure Theory. Part 1: Fundamentals*, Elsevier, Amsterdam (2002).
[110] P. Schwerdtfeger (ed.), *Relativistic Electronic Structure Theory. Part 2: Applications*, Elsevier, Amsterdam (2004).
[111] B. A. Hess (ed.), *Relativistic Effects in Heavy-Element Chemistry and Physics*, Wiley, Chichester (2003).
[112] K. Hirao, Y. Ishikawa (eds.), *Recent Advances in Relativistic Molecular Theory*, in *Recent Advances in Computational Chemistry*, Vol.5, World Scientific, London (2004).

Brill Academic Publishers
P.O. Box 9000, 2300 PA Leiden
The Netherlands

Lecture Series on Computer and Computational Sciences
Volume 1, 2005, pp. 461-471

Nonlinear Optical Spectroscopy of Molecular Chirality

Y. R. Shen

Physics Department, University of California
Berkeley, California, U.S.A. 94720

Received 25 July, 2006; accepted 28 July, 2006

1. Introduction

Chiral molecules have intrinsically three-dimensional structures that cannot be superimposed with their mirror images by rotation and translation. [1] They are the basis of all life forms, and therefore play a very important role in modern chemistry, biology and medicine [2].For example, most biological molecules in our body consist of only left-handed amino-acids and right-handed sugars. Chiral receptor sites in the body interact differently with molecules of different chirality, and can result in marked differences in the pharmacological activities of enantiomers [3].

The conventional spectroscopic technique used to probe molecular chirality is circular dichroism (CD) that can provide chirality-specific spectroscopic information from which both the identity and the absolute structure of chiral molecules can be deduced.[4]. It measures the difference in absorption coefficients of a chiral sample for left and right circularly polarized light. However, its sensitivity is rather limited because the process is not electric-dipole allowed.[5] The reason is simple. Chirality is a 3-D property, the description of which requires rank-3 tensor elements, but under the electric-dipole approximation, linear optical responses are characterized by dielectric tensor $\ddot{\varepsilon}$, which is only rank-2. As chiral probes, they are effective only if higher-order magnetic-dipole and electric-quadrupole contributions are taken into account so that $\ddot{\varepsilon}$ can be extended into a rank-3 tensor. This makes the chiral elements of $\ddot{\varepsilon}$ about three orders of magnitude smaller than the achiral ones. Moreover, CD detection of chiraity also suffers from the ubiquitous presence of the achiral background in the signal. Consequently, the sensitivity of CD is usually not sufficient to detect chirality of monolayers and thin films. On the other hand, the advent of new technologies, such as combinatorial chemistry [6] or the "lab-on-a-chip" [7] requires rapid screening and testing of chemicals of often limited quantities, sometimes down to the monolayer level. Many biological processes involve molecules that either function only when imbedded in a membrane, such as membrane proteins, or accumulate and interact mainly at interfaces.[8] A more sensitive probe that allows *in-situ* study of molecular chirality of such systems would open up new research opportunities and provide new understanding of molecular chirality.

Optical sum frequency generation (SFG) is known to be electric-dipole-allowed in media without inversion symmetry including chiral media, and as a second-order effect, naturally characterized by a rank-3 tensor. One then expects that chiral and achiral elements could be of the same order of magnitude. In fact, the former could even be much larger than the latter if the achiral elements would vanish in a chiral medium by symmetry. It is also known that SFG has sub-monolayer sensitivity. [9] Thus SFG could be a more sensitive chiral probe than CD and suitable for probing chirality of monolayers and thin films. However, although chiral SFG was proposed 40 years ago,[10] its observation was realized only recently through resonance enhancement.[11, 12, 13] The resonance enhancement not only improves the signal to the monolayer detection level but also generates chiral electronic [14] and vibrational [15] spectra for a chiral medium. Like CD, the chiral SFG spectra could be used to deduce chiral structure of molecules in a chiral medium, but being different microscopic processes, CD and chiral SFG provide complementary information.

We review here recent development of chiral SF spectroscopy on chiral liquids, films and monolayers.[16] Emphasis is on the underlying theory for the spectroscopic technique. We are interested in how the chiral SF spectra are related to the chiral molecular structure. We note in passing that third-order nonlinear optical processes such as two-photon excitations and four-wave mixing, could also be used to probe molecular chirality, but they are electric-dipole forbidden by symmetry like CD.[17] For example, chiral Raman scattering as a two-photon process is much weaker than its achiral counterpart.

2. Sum-Frequency Generation in Chiral Media

Sum-frequency generation, schematically shown in Fig. 1a, results from coherent radiation of nonlinear polarization $\vec{P}(\omega = \omega_1 + \omega_2)$ induced in a medium by incoming waves $\vec{E}(\omega_1)$ and $\vec{E}(\omega_2)$.[9]

$$\vec{P}(\omega = \omega_1 + \omega_2) = \vec{\chi}^{(2)} : \vec{E}(\omega_1)\vec{E}(\omega_2) \tag{1}$$

where $\vec{\chi}^{(2)}$ is a rank-3 second-order susceptibility tensor characterizing the medium.

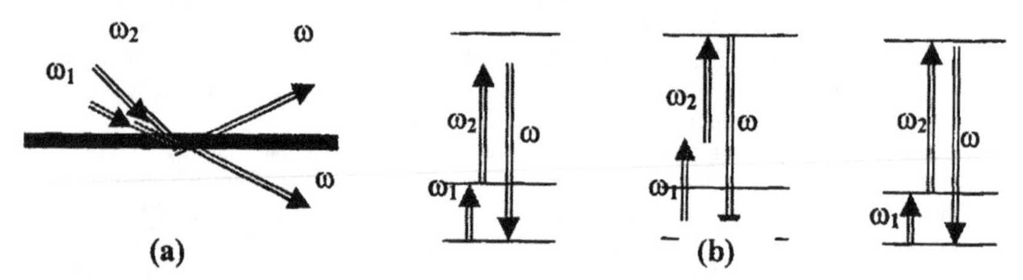

Figure 1. (a) Scematic describing SFG in transmission and reflection from a medium. (b) SFG processes showing single and double resonances with respect to an energy level diagram.

Measurement of the SF output with different input/output beam polarizations allows determination of $\vec{\chi}^{(2)}$. With $\omega_1\,(\omega_2)$ or ω or both approaching resonances (Fig. 1b), $\vec{\chi}^{(2)}$ is resonantly enhanced, and accordingly yields the SF spectrum. The $\vec{\chi}^{(2)}$ tensor has 27 elements, but depending on the structural symmetry of the medium, many elements may vanish or depend on others. We limit our discussion here to isotropic chiral media. Under the electric-dipole approximation, all $\vec{\chi}^{(2)}$ elements should vanish in a medium with inversion symmetry except those that are chiral:

$$\chi^{(2)}_{xyz} = \chi^{(2)}_{yzx} = \chi^{(2)}_{zxy} = -\chi^{(2)}_{yxz} = -\chi^{(2)}_{zyx} = -\chi^{(2)}_{xzy} = \chi^{(2)}_{chiral} \tag{2}$$

The microscopic expression of $\chi^{(2)}_{chiral}$, derived from perturbation calculation with non-resonant terms neglected, is [13, 18]

$$\chi^{(2)}_{chiral} \approx \frac{NL}{6\hbar^2}\sum_{n,n'}\frac{<g|\vec{\mu}|n>\cdot(<n|\vec{\mu}|n'>\times<n'|\vec{\mu}|g>)}{\omega-\omega_{ng}+i\Gamma_{ng}}\left[\frac{1}{\omega_1-\omega_{n'g}+i\Gamma_{n'g}}-\frac{1}{\omega_2-\omega_{n'g}+i\Gamma_{n'g}}\right] \tag{3}$$

where $\vec{\mu}$ is the electric-dipole moment and L the local-field correction factor. From Eq.(3), it is seen that $\chi^{(2)}_{chiral}$ vanishes if $\omega_1 = \omega_2$. Thus second-harmonic generation is not allowed in isotropic chiral media. If both input and output frequencies are far away resonances, such that the frequency denominators in Eq. (3) can be pulled out from the summation as an approximation, then we have

$$\chi^{(2)}_{chiral} \propto \frac{(\omega_1-\omega_2)}{(\omega-\bar{\omega}_{ng})(\omega_1-\bar{\omega}_{n'g})(\omega_2-\bar{\omega}_{n'g})} <g|\vec{\mu}\cdot(\vec{\mu}\times\vec{\mu})|g>=0 \tag{4}$$

This shows that off resonance, $\chi^{(2)}_{chiral}$ must be weak, and explains why earlier experiments on SFG in chiral liquids had failed. Detection of chiral SFG often requires resonant enhancement, but it is also the resonant enhancement that provides us with the chiral spectral information.

The chiral structural information is contained in the matrix elements of $\chi^{(2)}_{chiral}$ associated with the resonances. For comparison, the difference of complex refractive indices for left and right circularly polarized light, responsible for CD and the related optical rotatory dispersion (ORD), has the microscopic expression [4]

$$\Delta n \propto N \sum_{n} \frac{\omega^2 \,\mathrm{Im}[\langle g|\vec{\mu}|n\rangle \cdot \langle n|\vec{m}|g\rangle]}{\omega^2 - \omega_{ng}^2 + i2\omega\Gamma_{ng}} \tag{5}$$

with $\vec{m}$ being the magnetic-dipole moment. Here, it is also the rotatory strength, $R_{ng} \equiv \mathrm{Im}[\langle g|\vec{\mu}|n\rangle \cdot \langle n|\vec{m}|g\rangle]$, that characterizes the CD of the $|g\rangle \rightarrow |n\rangle$ transition. As shown, R_{ng} results from interference between electric-dipole and magnetic-dipole contributions in the transition. Clearly, CD and chiral SFG provide different chiral structural information associated with the same resonance as the matrix elements involved are different. One of the advantages of chiral SFG is that it allows doubly resonant or 2D spectroscopy

In terms of their chromophore structure, chiral molecules or chiral portions of molecules can be divided into two categories: one has intrinsically dissymmetric chromophores such as twisted dimers or polymers with binaphthol (BN) being an example, and the other has a planar or centrosymmetric chromophore situated in a chiral environment. We discuss chiral SF spectroscopy on electronic and virational transitions of the two types of chiral molecules separately in the following.

3. Chiral Sum-Frequency Spectroscopy of Electronic Transition

We take 1-1'-binaphthol (BN, $C_{20}H_{14}O_2$) as a representative of chiral molecules with dissymmetric chromophores. The molecular structure of BN is shown in Fig. 2a. It is dimer-like with C2 symmetry, made of two naphthol monomers with a twist angle of about 100° between them.[19] The twist gives rise to chirality. Coupling between the two monomers lifts the degeneracy of every excited electronic state e_i of the monomers into two exciton-split states e_{i+} and e_{i-}(Fig. 2b) symmetric and anti-symmetric, respectively, with respect to a 180° rotation about the C2 axis.[5] Electric-dipole transitions from the ground state $|0\rangle$ to the first pair of excited states $|1_+\rangle$ and $|1_-\rangle$ produce the first two peaks in the linear absorption spectrum of BN (Fig. 2c).

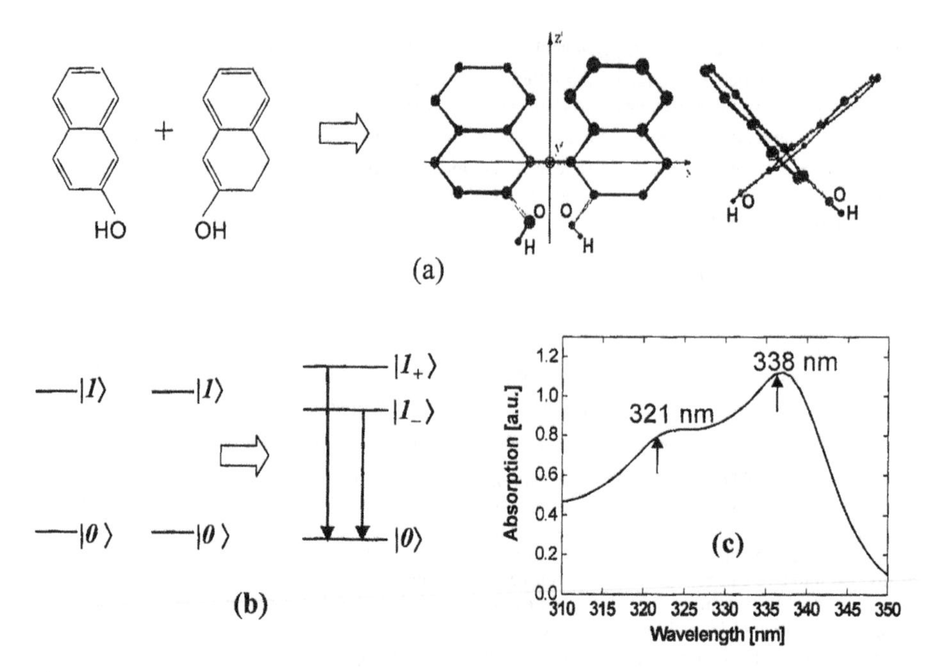

Figure 2. (a) Structure of the 1,1'-bi-2-naphthol molecule as a dimer composed of two linked monomers. (b)Transitions between the lowest exciton-split states and the ground state and (c) the corresponding absorption spectrum of BN in tetrahydrofuran.

Displayed in Fig. 3a are SF spectra of $|\chi^{(2)}_{chiral}|$ for solutions of 0.7M R-BN, S-BN, and racemic mixture in tetrahydorfuran. As expected, those of R-BN and S-BN are identical, exhibiting the two exciton-split peaks, and that of the racemic mixture not detectable. To fit the observed spectra, we can insert exciton-split excited states in Eq. (3) and express it in the form (following the so-called coupled oscillator model) [20]

$$\chi^{(2)}_{chiral} = \frac{A_+}{\omega - \omega_{1_+0} + i\Gamma_{1_+0}} + \frac{A_-}{\omega - \omega_{1_-0} + i\Gamma_{1_-0}} \tag{6}$$

with $A_+ = -A_-$ .If we approximate $(\omega_1 - \omega_{n0})(\omega_2 - \omega_{n0})$ with n≥2 by $(\omega_1 - \omega_{eff})(\omega_2 - \omega_{eff})$ for ω_1 and ω_2 far off resonance, then $A_\pm$ takes the form

$$\begin{aligned} A_\pm = \pm \frac{NL}{6\hbar^2}(\omega_1 - \omega_2)(\vec{\mu}_{11} - \vec{\mu}_{00}) \cdot (\vec{\mu}_{10} \times \vec{\mu}'_{10}) \\ \times [\frac{1}{(\omega_1 - \omega_{10})(\omega_2 - \omega_{10})} - \frac{1}{(\omega_1 - \omega_{eff})(\omega_2 - \omega_{eff})}] \end{aligned} \tag{7}$$

Here $\vec{\mu}_{ij} \equiv \langle i | \vec{\mu} | j \rangle$ refers to the electric-dipole matrix element between the monomer states <i| and |j>. For numerical estimates, values of all quantities in Eq. (7) can be found in the literature except the direction of $(\vec{\mu}_{11} - \vec{\mu}_{00})$ and ω_{eff} . Assuming the former lies perpendicular to the short monomer axis and $\hbar\omega_{eff} = 6\ eV$, we find that the theoretical curve agrees fairly well with the experimental spectra, [20] as shown in Fig. 3b.

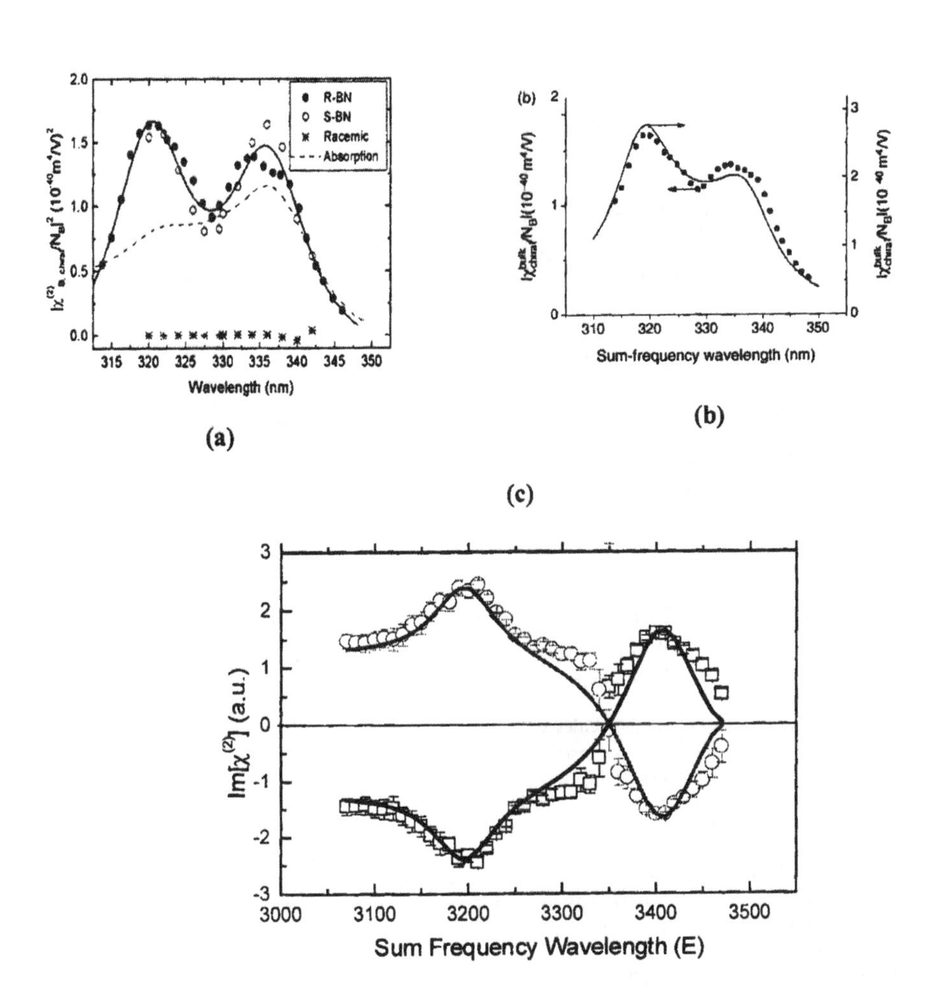

Figure 3. (a) Spectra of $|\chi^{(2)}_{chiral}|^2$ of R-, S-, and racemic mixtures of BN around the first pair of exciton-split transitions, and (b) their agreement with the theoretical spectrum. (c) With the help of an interference scheme, the spectra of Im $\chi^{(2)}_{chiral}$ can be obtained that distihuish R- and S-BN with opposite signs.

We note that from Eq. (5), Δn can also be put into the form [20]

$$\Delta n = \frac{B_+}{\omega - \omega_{1_+0} + i\Gamma_{1_+0}} + \frac{B_-}{\omega - \omega_{1_-0} + i\Gamma_{1_-0}} \tag{8}$$

with $B_\pm \propto \frac{\omega\omega_{10}}{\omega + \omega_{10}} \vec{R}\cdot(\vec{\mu}_{10}\times\vec{\mu}'_{10})$, where $\vec{R}$ is the vector connecting the centers of the two monomers. It is seen that in both Δn and $\chi^{(2)}_{chiral}$, chirality results from the twist between the two monomers, as manifested by $(\vec{\mu}_{10}\times\vec{\mu}'_{10})$, as well as coupling between the two monomers, as manifested by $\omega_{1_+0} \neq \omega_{1_-0}$. In other words, chirality requires the subgroups of a chiral molecule be coupled and oriented in a chiral configurationtion. The difference between Δn and $\chi^{(2)}_{chiral}$ is in the form of their pseudoscalars, $\vec{R}\cdot(\vec{\mu}_{10}\times\vec{\mu}'_{10})$ versus $(\vec{\mu}_{11}-\vec{\mu}_{00})\cdot(\vec{\mu}_{10}\times\vec{\mu}'_{10})$. While Δn depends non-locally on the distance between the two monomers, $\chi^{(2)}_{chiral}$ does not.

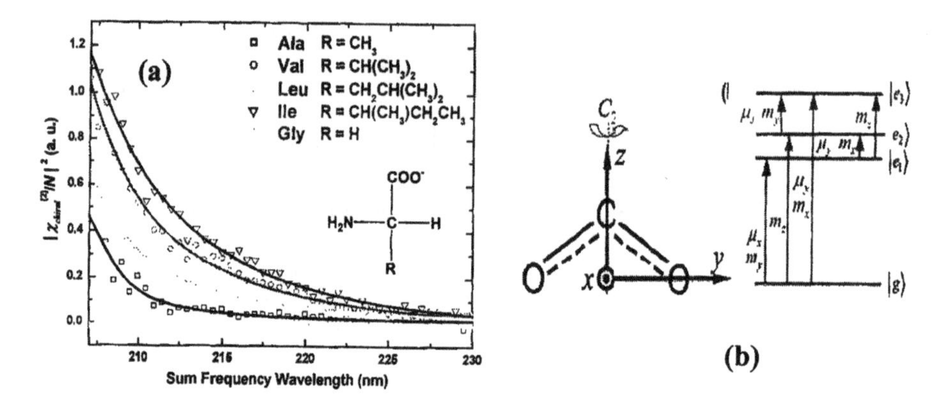

Figure 4. (a) Spectra of $|\chi^{(2)}_{chiral}|^2$ for various amino acids whose molecular structures are shown in the inset. (b) The structure and the four lowest electronic energy levels of a COO group, with allows indicating the allowed electric-dipole and magnetic-dipole transitions.

As representatives of chiral molecules with centrosymmetric chromophore situated in a chiral environment, we take α-amino acids (see inset of Fig. 4a). [21,22] We consider chiral SFG as the SF approaches the electronic resonances of the carboxylate chromophore. In basic solutions, the carboxyl acid group is deprotonated into a carboxylate anion (–COO) with C_{2v} symmetry. The energy-level diagram of –COO are displayed in Fig. 4b, with the allowed electric and magnetic dipole transition moments between them labeled and denoted by arrows. [23] The transitions from ground state g to excited states e1 and e2 are in the 180- to 210-nm region, and that of g to e3 around 160 nm. Figure 4a presents the chiral SF spectra of different amino acids in solution, showing the resonant enhancement toward the electronic resonances as in the CD case. Since –COO is achiral, the observed chiral response toward the –COO resonance must come from chirality induced in –COO through its interactions with the chiral environment.

We can again use Eq.(3) to calculate $\chi^{(2)}_{chiral}$ (or Δn) for –COO. If the –COO group were isolated in space, we would have $\vec{\mu}_{gn}\cdot(\vec{\mu}_{nn'}\times\vec{\mu}_{ng})=0$. Perturbation of the chiral environment modifies the wave functions of the energy states of –COO such that $\vec{\mu}_{gn}\cdot(\vec{\mu}_{nn'}\times\vec{\mu}_{ng})$ no longer all vanish. Various models can be used to carry out the perturbation calculation. For example, in the so-called dynamic coupling model, first proposed by Hohn and Weigang [24] for linear optical activity, the chiral environment is represented by a collection of nonpolar bonds as perturbers at various positions and their effects on the states of -COO are calculated. The interaction potential V between the chromophore and a perturber is Coulombic in nature and can be expanded into multipole-multipole interactions. For a non-polar perturber, the quadrupole-dipole term in V is the lowest-order non-vanishing term. For –COO in a chiral environment, if we consider only the four states of –COO in Fig. 4b, the perturbation calculation yields an expression of the form [22]

$$\chi^{(2)}_{chiral}=f(\omega)\hat{x}\cdot(\hat{y}\times\vec{\mu}_{e_2g})+g(\omega)\hat{x}\cdot(\hat{y}\times\vec{\mu}_{e_2e_1}) \tag{9}$$

where $f(\omega)$ and $g(\omega)$ are functions of input and output frequencies, $\hat{x}$ and $\hat{y}$ denote the molecular axes in the –COO plane, and the electric-dipole matrix elements $\vec{\mu}_{e_2g}$ and $\vec{\mu}_{e_2e_1}$ are only nonvanishing, along $\vec{z}$, through interactions of –COO with the chiral environment. If all bonds are treated as isotropic, the expressions of $\vec{\mu}_{e_2g}$ and $\vec{\mu}_{e_2e_1}$ reduce Eq. (9) to the form

$$\chi^{(2)}_{chiral} = F(\omega)\sum_{\sigma} R_{\sigma}^{-7}\alpha_{\sigma}X_{\sigma}Y_{\sigma}Z_{\sigma} \tag{10}$$

Here, $\vec{R}_{\sigma}$ describes the position of the σth perturber with respect to the center of the chromophore, X_{σ}, Y_{σ} and Z_{σ} are the components of $\vec{R}_{\sigma}$ along the molecular coordinates of the chromophore, and α_{σ} is the polarizability of the perturber. Equation (10) closely resembles the well-known octant rule in CD. [25]

The following general, but physically apparent, results for induced chirality in a plane chromophore like –COO are obtained from the dynamic coupling model.[22] To induce chirality effectively, the perturbative bond must be tilted away from the chromophore plane as well as other symmetric planes of the chromophore. It should be close to the chromophore, and has a large polarizability with properly oriented anisotropy. The theory outlined here can be extended to describe chirality change induced in a molecule by the surrounding molecules in either chiral or achiral arrangement.

4. Chiral Sum-Frequency Spectroscopy of Vibrational Transitions

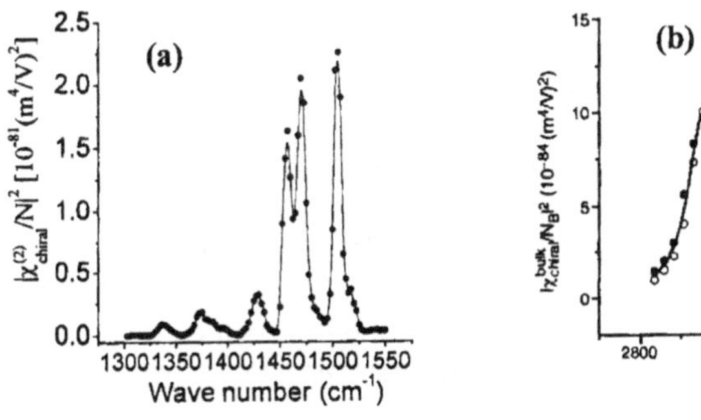

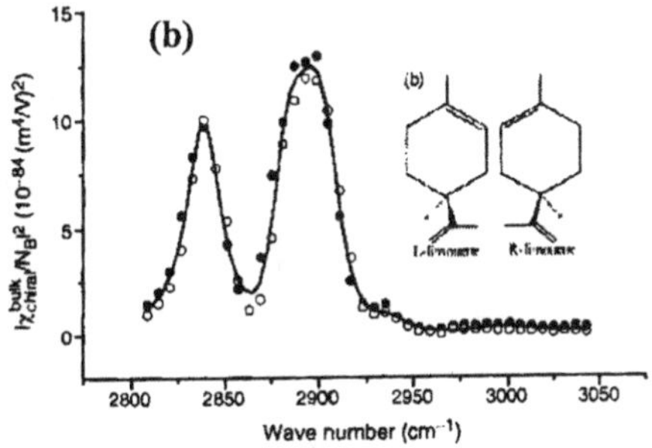

Figure 5. Spectra of $|\chi^{(2)}_{chiral}|^2$ for (a) BN and (b) limone

Chiral SF vibrational spectra for both types of chiral molecules have been observed. We show in Fig. 5 the spectra of BN [15] and lemone [12] as examples. The coupled oscillator model (for molecules like BN) and the perturbative theory (for induced chirality in chromophores like limone) described in the previous section can be extended, in principle, to describe chiral vibrational spectra, although they have not yet been reported. Detailed calculations are expected to be more complicated because the states involved are combined electronic-vibrational states of molecules instead of pure electronic states.

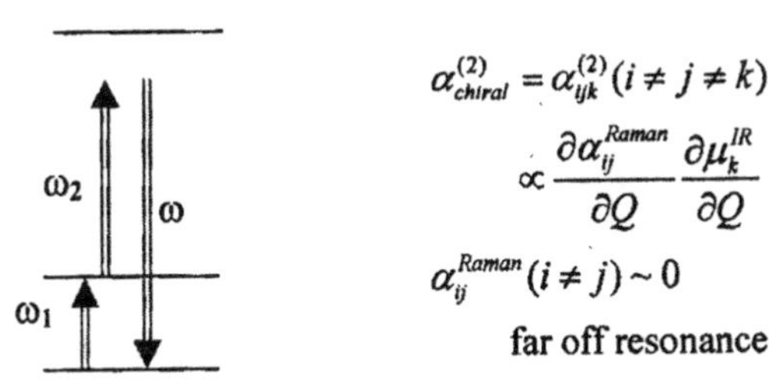

Figure 6. SFG with vibrational resonance is seen as an infrared excitation followed by an antiStokes process, which is weak off resonance if it is antisymmetric.

The experimentally observed $|\chi^{(2)}_{chiral}|$ at vibrational resonances for both BN and limone, however, are about three orders of magnitude smaller than expected from an electric-dipole-allowed one. This surprising result can be understood from the following consideration. As seen in Fig. 6, the SFG from a vibrational resonance can be described as a combined process of infrared excitation followed by anti-Stokes Raman transition. The hyperpolarizability for the SFG can be expressed in the molecular coordinates *(i,j,k)* as [9]

$$\alpha^{(2)}_{ijk} \propto M_{ij}\mu_k \tag{11}$$

where μ_k is the infrared excitation amplitude along k and M_{ij} the Raman tensor element in the (i,j) plane. For chiral SFG in isotropic media, we must have $i \neq j \neq k$, but it is known that far away from electronic resonance, the antisymmetric Raman matrix element nearly vanishes, i.e., $M_{ij} \approx 0$ if $i \neq j$. Therefore, we expect $\alpha^{(2)}_{chiral}/\alpha^{(2)}_{achiral} = M_{ij}(i \neq j)/M_{ii} << 1$. More explicitly, we can show that the perturbation calculation using the Born-Oppenheimer adiabatic approximation plus the first order non-adiabatic correction to take into account the first-order effect of electron-vibration coupling gives the microscopic expression [18, 26]

$$\begin{aligned} M_{ij}(i \neq j) = -\frac{1}{2\hbar^2}\sum_{n\neq g}\sum_{s<n,\neq g} V_{sn}Q_{01}(\mu_{i,ng}\mu_{j,sg} - \mu_{j,ng}\mu_{i,sg}) \\ \times\{\frac{1}{(\omega-\omega_{(n,1)(g,0)}+i\Gamma_{(n,1)(g,0)})(\omega-\omega_{(s,0)(g,0)}+i\Gamma_{(s,0)(g,0)})} \\ -\frac{1}{(\omega-\omega_{(n,0)(g,0)}+i\Gamma_{(n,0)(g,0)})(\omega-\omega_{(s,1)(g,0)}+i\Gamma_{(s,1)(g,0)})}\} \end{aligned} \tag{12}$$

where subindices g, s, and n refer to the electronic states, and 0 and 1 to the vibrational quantum numbers. V_{sn} and Q_{01} denote matrix elements of electron-vibration coupling between s and n and vibrational amplitude between 0 and 1, respectively. From Eq. (12), it is seen that if ω is off-resonance such that the vibrational frequencies in the denominators become negligible, then M_{ij} tends to vanish. With ω approaching electronic resonances, M_{ij} appears increasingly more significant, and can reach a value of the same order of magnitude as M_{ii} on resonance. Crudely, the difference between $M_{ij}(i \neq j)$ and M_{ii} is in the opposite signs appearing between the two terms in the two brackets of Eq. (12). Off resonance, we have [18]

$$\left|\frac{M_{ij}(i \neq j)}{M_{ii}}\right| \sim \left|\frac{\Omega\omega_{(n,0)(s,0)}}{2(\omega-\omega_{(n,0)(g,0)})(\omega-\omega_{(s,0)(g,0)})}\right|_{s>n}$$

which is ~0.01 with Ω being the vibrational frequency. This value is expected to reduce further if $M_{ij}(i \neq j)$ is nonvanishing only because the vibrating moiety interacts with chiral environment in the molecular structure. This then explains the observed $\alpha^{(2)}_{chiral}/\alpha^{(2)}_{achiral} \sim 10^{-3}$ in chiral SF vibrational spectroscopy.

5. Doubly Resonant Chiral Sum-Frequency Spectroscopy

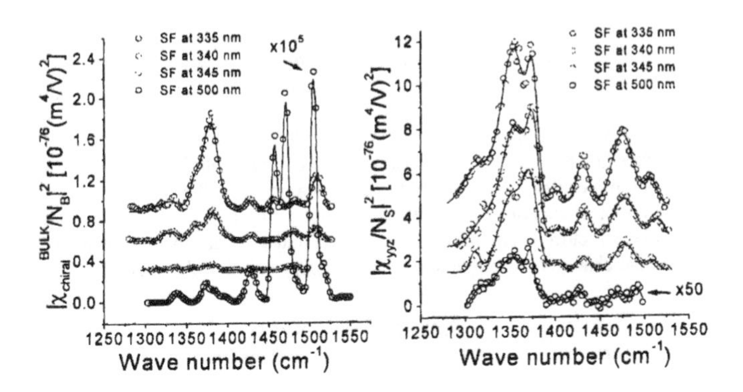

Figure 7. (a) Chiral spectra of $|\chi^{(2)}_{chiral}/N|^2$ of R-BN in acetone and (b)achiral spectra of $|\chi^{(2)}_{achiral}/N|^2$ of an R-BN monolayer on water with sum-frequency at 335 nm (open circles), 340 nm (open squares), 345 nm (open down triangles)and 500 nm (solid diamonds).The 500 nm spectra in (a) and (b) were enhanced by 10^5 and 50,respectively. Exciton resonance is at 335 nm.

Chiral SF spectroscopy probing simultaneous electronic and vibrational resonances is particularly interesting. As discussed in the preceding section, the chiral SF response should experience extraordinarily strong resonant enhancement when the sum frequency ω approaches an electronic resonance, not only because of the usual resonant enhancement but also because the antisymmetric part of the Raman tensor becomes increasingly more significant. This is demonstrated for BN solution in Fig. 7. [15] It is seen that selected vibrational modes in the SF spectrum taken with ω at the first exciton resonance show an enhancement of more than 10^5 with respect to the spectrum taken with ω at 1,25 eV away from resonance. For comparison, we also present in Fig. 7 the achiral SF vibrational spectra of BN taken at the same set of ω values. The resonant enhancement is only about 200. The extraordinarily strong enhancement of the doubly resonant chiral SF vibrational spectroscopy makes the detection of chiral SF vibrational spectra of molecular monolayers possible, as has indeed been demonstrated. [15]

Note that the observed SF vibrational spectra in Fig. 7 exhibit very different resonant enhancements for different vibrational modes. This is especially clear in the chiral SF spectra, and is a manifestation of the very different vibronic couplings of different vibrational modes with the exciton. Selective resonant enhancement of vibrational modes that allows better assignment of the vibrational modes and possible deduction of vibronic couplings for selective modes from the doubly resonant spectrum are known to be the important advantages of doubly resonant spectroscopy like resonant Raman scattering. Here, they could certainly help in attempts to understand chiral structure of molecules.

Concluding Remarks

The potential of optical sum-frequency spectroscopy for *in situ* probing of chirality of surfaces, monolayers and thin films has been demonstrated. The technique is unique in its ability to selectively detect chiral molecules with very high contrast against the achiral background. There are a number of exciting possibilities for applications of this technique. Chiral microscopy for imaging biological systems is an obvious one, and preliminary success on it has already been reported. [27] Others such as studies of chiral separation of molecules, chiral conformational changes, and chiral dynamics are yet to be carried out.

Our exploration of SF spectroscopy as a novel chiral probe is still at the beginning stage. We need to characterize the SF spectral responses of more representative chiral molecules, especially common biological molecules such as DNA and proteins, in order to have a better idea on the sensitivity and range of applications. Theoretical understanding on how to relate a chiral SF spectrum to a chiral molecular structure is crucial for the advance, and so is the spectral response of molecules to chiral conformational change or arrangement of surrounding chiral molecules.

Acknowledgements

This work was supported by the Director,Office of Energy Research,Office of Basic Energy Sciences, Materials Science Division of the US Department of Energy under Contract No.DE-AC03-76SF00098.

References

1. W.Kelvin, *Baltimore Lectures on Molecular Dynamics and the Wave Theory of Light*, (C.J.Clay,London,1904).
2. P.Ball, *Designing the Molecular World*, (Princeton University Press, Princeton, NJ, 1996).
3. G.Blaschke, H.P.Kraft, K.Fickentscher, and F.Kohler, Arzneim.-Forsch./Drug Res. 29,1640 (1979).
4. E. Charney, *The molecular basis of optical activity*. (J. Wiley & Sons; New York,1979); Berova N, Nakanishi K, Woody RW., *Circular Dichroism: Principles and Applications*, (J.Wiley & Sons, New York; 2002).
5. L.D.Barron, *Molecular Light Scattering and Optical Activity*, (Cambridge University Press, Cambridge,1982).
6. L.A.Thompson and J.A.Ellman, Chem Rev 96, 555 (1996).
7. D.Figeys and D.Pinto, Anal Chem 72, 330A (2000).
8. B.Alberts, A.Johnson, J.Lewis, M.Raff, K.Roberts, and P.Walter, *Molecular Biology of the Cell*, (Garland Science, New York, 4th ed, 2002).
9. Y.R.Shen,in "Frontiers in Laser Spectroscopy" ,Proc.Int.School of Physics 'Enrico Fermi ',Course CXX ,edited by T.W.Hansch and M.Inguscio (North Holland, Amsterdam,1994), pp.139 –165; also Y.R.Shen, Proc. Nat. Acad. Sci., 93,12104 (1996).
10. J.A.Giordmaine, Phys.Rev.138,A1599 (1965).
11. Hicks and coworkers first demonstrated the possibility of using second harmonic generation to probe molecular chirality and record the chiral electronic spectrum of a monolayer, but SHG is forbidden in isotropic chiral media. T.Petralli-Mallow,T.M.Wong,J.D.Byers,H.I.Yee,and J.M.Hicks, J.Phys.Chem.97,1383 (1993).
12. M.A.Belkin,T.A.Kulakov,K.-H.Ernst,L.Yan,and Y.R.Shen,Phys.Rev.Lett.85,4474 (2000).
13. M.A.Belkin,S.H.Han,X.Wei,and Y.R.Shen,Phys.Rev.Lett.87,113001 (2001).
14. S.H.Han,N.Ji,M.A.Belkin,and Y.R.Shen,Phys.Rev.B 66,165415 (2002).
15. M.A.Belkin and Y.R.Shen,Phys.Rev.Lett.91,213907 (2003).
16. More extensive review on chiral SF spectroscopy can be found in M.A.Belkin and Y.R.Shen, Int. Rev. Phys. Chem. 24, 257 (2005) and N. Ji and Y.R.Shen, Chirality 18, 146 (2006).
17. G.Wagniere,J.Chem.Phys.77,2786 (1982).
18. M.A.Belkin, Y.R.Shen, and R.A.Harris, J Chem Phys.120, 10118 (2004).
19. I.Hanazaki and H.Akimoto, J Am Chem Soc 94, 4102 (1972).

20. M.A.Belkin,Y.R.Shen,and C.Flytzanis,Chem.Phys.Lett.363,479 (2002).
21. N.Ji and Y.R.Shen, J. Am. Chem. Soc. 126, 15008 (2004).
22. N.Ji and Y.R.Shen, J. Am. Chem. Soc. 127, 12933 (2005).
23. H.J.Maria, D.Larson, M.E.McCarvil, and S.P.McGlynn, Acc. Chem. Res. 3, 368 (1970).
24. E.G.Hohn and O.E.Weigang, J. Chem. Phys. 48, 1127 (1968).
25. W.Moffitt, R.B.Woodward, A.Moscowitz, W.Klyne, and C.Djerassi, J. Am. Chem. Soc. 83, 4013 (1961).
26. F.Liu,J.Phys.Chem.95,7180 (1991); F.Liu and A.D.Buckingham, Chem. Phys. Lett. 207,325 (1993).

Brill Academic Publishers
P.O. Box 9000, 2300 PA Leiden
The Netherlands

Lecture Series on Computer and Computational Sciences
Volume 6, 2006, pp. 472-481

Electronic Response Analysis on Supramolecular Functions. Electronic-Structure Modulation and Molecular Recognition

Manabu Sugimoto[1]

Graduate School of Science and Technology, Kumamoto University
2-39-1 Kurokami, Kumamoto 860-8555, Japan

Received 15 July, 2006; accepted 20 July, 2006

Abstract: Computational analysis on supramolecules is discussed with a particular focus on electronic response to perturbation. It is suggested that definition of supramolecule based on molecular functions is more general and convenient than that based on noncovalent interactions. This idea leads us to suggest that quantitative characterization of supramolecules would be possible by evaluating response properties to an appropriate perturbation. In order to show examples of investigating the perturbation-response relationship, ab initio calculation on a mixed valence compound, $[(NC)_5Os\text{-}CN\text{-}Ru(NH_3)_5]^-$, is reported. The calculation shows that the well known electronic transition called metal-to-metal charge transfer (MMCT) transition is observed because of electronic-structure modulation by solvent polarization. The important perturbation-response relationship is concerned with the resultant electric field and the MMCT excitation energy. The second example of a supramolecular function to be analyzed is molecular shape which can be related to molecular recognition, at least, for qualitative interpretation. For quantitative characterization of the shape of molecule, we propose to use ion-molecule interaction energies. Considering importance of hydrogen-bond-like interactions, it is suggested to adopt the H^+ and F^- ion probes for shape mapping. In this approach, the ion is a perturbation, and the shape is defined as an energy response to the perturbation. These two computational studies clearly show the importance of systematic investigation of the perturbation-response relationship in supramolecular functions.

Keywords: supramolecular chemistry, computational chemistry, electronic structure, ab initio calculations, mixed valence complex, excited state, solvent effect, molecular recognition, molecular shape, perturbation-property relationship

1. Introduction

There is no doubt that the integrate circuit has made a strong impact on human life all over the world. It is an assembly of many electronic devices such as diodes, transistors, etc [1]. Each electronic device consists of solids made of various atoms. It is evident that physics and chemistry of atoms are appreciably different from the solids, electronic devices, and their assembly. It is also true in biological systems. Photosynthesis is achieved in a surprisingly well organized chemical system. The photo energy is efficiently converted to chemical energy in the assembled system, not in a single component [2]. These examples clearly show that assembly of functional units provides various opportunities of yielding new benefits, i.e. useful functional devices. In the assembled system, expected are synergetic effects among components. For functions of the system, an interesting and important issue would be to know how "the system" is different from "individuals". Such characterization is expected provide useful guiding principle for the system design.

Following Lehn [3], a chemical system consisting of molecules may be called supramolecule. The reaction center for photosynthesis is a typical supramolecular system. Another example of a supramolecule is the dye-sensitized solar cell called the Grätzel cell [4]. Photo energy absorption is carried out by TiO_2 particles on which $Ru(dcbpy)_2(NCS)_2$, called N3 dye, is fabricated for

[1] Corresponding author. E-mail: sugimoto@kumamoto-u.ac.jp

photosensitization. By light absorption, electron is transferred from the N3 dye to the conduction band of TiO_2, which is an origin of the photo current. Thus supramolecular chemistry is versatile from fundamental chemistry to the state-of-the-art technology.

A reasonable way of building supramolecular systems would be (i) to choose building blocks including those tabulated in various chemical catalogues, and (ii) to combine them by synthetic reactions into one piece with appropriate linkers. Then (iii) molecular properties are measured, followed by (iv) analyses of structure-property relationships. A set of the processes is repeated until an optimal system is obtained.

Use of computational methods is expected helpful in these processes because many trials are essentially included. To design supramolecules efficiently, what information is necessary ? A plausible answer would be "structure-property relationships". This can be obtained by investigating important characteristics of supramolecules. Thus, one of the important issues for designing supramolecules seems to develop a useful method of analysis which well characterizes the system. For this purpose we need to develop a method of characterization.

In this paper, we point out that important characteristics of supramolecules should be understood on the basis of response properties. The concept of the perturbation-response relationship is of essential importance to define, characterize, understand, and design supramolecules. In Section 2, we propose that a response to an appropriate perturbation should be used as a quantitative measure of a supramolecule. Section 3 shows ab initio calculations on a mixed valence compound, showing that its electronic structure, and hence supramolecular properties are dependent on surrounding medium (solvent). In Section 4, we develop computational algorithm for analyzing molecular recognition, which is one of the important supramolecular functions. Our strategy is to define a shape of molecule as a response to perturbation. This is done by evaluating intermolecular interaction energies with ab initio calculations. The conclusion of the present paper is given in Section 5 with some comments on future perspectives in computational supramolecular chemistry.

2. Definition of Supramolecule and Its Classification

A molecule is an assembly of atoms with a unique structure and functions. This situation might be expressed as "*molecularized*". Lehn suggests that, when molecules are assembled via non-covalent interaction, the system is called "supramolecule". The non-covalent interaction includes electrostatic, hydrogen bond, and van der Waals interactions. Because of the hybrid nature of a supramolecule, its fundamental functions are *molecular recognition*, *transformation*, and *translocation* of chemical species and/or charge carriers. There properties can not be observed in a simple molecular component.

From a viewpoint of three molecular functions, a generalization of the definition seems necessary. To have an idea for plausible modification, let us consider bleomycin. This is a unique molecule that works as antitumor antibiotics. This molecule consists of three functional units: the DNA binding domain, the metal binding moiety, and the carbohydrate domain. The metal binding domain is responsible to activate a substrate, and the activated substrate can react with DNA because of its close contact to the DNA binding domain. Each domain in the molecule separately play a role, but its essential feature as antibiotics is due to the hybrid nature of these components.

It is noted here that bleomycine carries out molecular recognition by the DNA binding domain, transformation at the metal binding domain, and translocation of a reaction product to interact with DNA. Thus the function of bleomycin consists of three phenomena expected for Lehn's supramolecule although the functional units are covalently bonded. From the viewpoint of molecular functions, it seems reasonable to define a supramolecule as molecules, molecular fragments (functional groups) or molecular assembly with unique *functions* realized by hybridization. A molecule might be called "*supramolecularized*" when it forms a "chemical

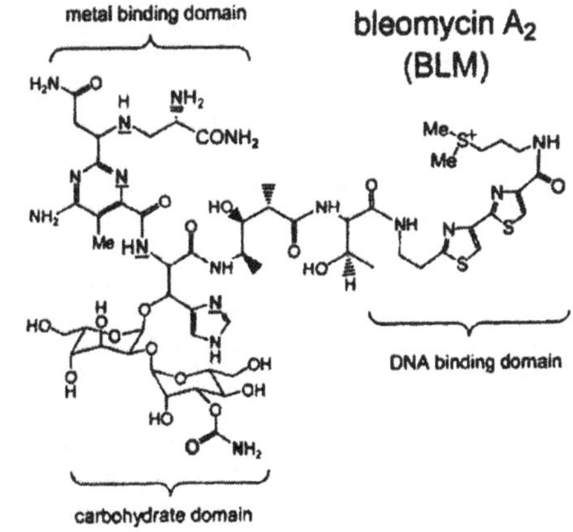

Figure 1. Structure of Blyomicine (ref. [5])

system" with characteristic functions, which are not expected from each component.

A function of molecule is related to a molecular property. It can be further related to response in energy to an applied perturbation. Therefore we suggest that the perturbation-response relationship for characterization should be used for characterization.

Suppose that we have two molecules, A and B, and they are supramolecularized into C. A and B have unique functions (i.e. responses) represented as F_A and F_B, respectively, to an applied perturbation λ. If $F_A = F_C$, it means that the property of A is maintained after supramolecularization. If $F_A > F_C$ ($F_A < F_C$), the function of A is interpreted as being enhanced (suppressed) by supramolecularization. If $F_C = F_A + F_B$, the A and B units are incoherent, and the property is their simple sum. If $F_C \neq 0$ with $F_A = 0$ and $F_B = 0$, C has a unique function as a result of supramolecularization. Thus a set of values of (λ, F_A, F_B, F_C) can be a quantitative measure to characterize a supramolecule. If $F_A = F_B = 0$ or it is practically difficult to define the fragments (such as A and B), the coordinate (λ, F_C) may be used to the quantitative specification. Although, to our knowledge, such a classification of supramolecules has not been suggested so far, a similar one has been adopted for dielectric materials: electric polarization, which is a response property, is used, in the Landau theory, as an order parameter specifying dielectric phases of a solid material [6].

There are, at least, two ways to define energy response (F) for characterization. The simplest one would be to use energy change by perturbation: a supramolecule is characterized by a set of quantity (λ, ΔE_A, ΔE_B, ΔE_C) where ΔE_X (X = A, B, C) is defined as

$$\Delta E_X(\lambda) = E_X(\lambda) - E_X(0). \tag{1}$$

When it is convenient to use a quantity independent of the value of λ, energy derivatives appearing in the Taylor expansion

$$E(\lambda_1, \lambda_2, \cdots, \lambda_N) = E(0) + \sum_{i=1}^{N} \left[\frac{\partial E}{\partial \lambda_i}\right]_{\lambda=0} \lambda_i + \frac{1}{2!}\sum_{i=1}^{N}\sum_{j=1}^{N}\left[\frac{\partial^2 E}{\partial \lambda_i \partial \lambda_j}\right]_{\lambda=0} \lambda_i \lambda_j + \cdots\cdots \tag{2}$$

should be used. When the perturbation is uniform electric field, the first, second, and third derivatives with respect to λ correspond to a dipole moment, electric susceptibility (polarizability), hyperpolarizability, respectively [7].

As a perturbation, we may consider mechanical force (distortion of molecular structure), electric field, magnetic field, electromagnetic wave (light), nuclear spin, electron spin, and so on. The other important perturbation is the electronic one due to chemical species coexisting in the system: when a molecule R reacts with a molecule S, S can be regarded as a perturbation to R. Thus our important task for characterizing supramolecules is to discover appropriate perturbations (their type, numbers, and strengths) and to evaluate resultant responses. Such quantification would be helpful in constructing physical models of supramolecular phenomena.

3. Supramolecular Phenomena. Electronic-Structure Modulation by Environment

Electronic Structure of a Mixed Valence Complex, $[(NC)_5Os\text{-}CN\text{-}Ru(NH_3)_5]^-$:

An asymmetric dinuclear complex, $[(NC)_5M\text{-}CN\text{-}M'(NH_3)_5]^-$ (M, M' = Fe, Ru, Os), has been a target of various spectroscopic and electrochemical studies on electron transfer between two metal centers [8]. From a structural viewpoint, this molecule can be regarded as a supramolecule. A reasonable decomposition into fragments would be to consider $(NC)_5M$ and $M'(NH_3)_5$ moieties. Experimentally it is known that its ground state in a polar solvent is represented as M^{II}-CN-M'^{III}. Here we show that ab initio calculations predict the Os^{III}-CN-Ru^{II} ground state for $[(NC)_5Os\text{-}CN\text{-}Ru(NH_3)_5]^-$ in vacuum, while it changes to Os^{II}-CN-Ru^{III} when an external electric field is included to simulate a polar solvent. This suggests that the electronic structure experimentally observed so far[8] is a result of strong electrostatic interaction in condensed phase, not an intrinsic feature of the isolated molecule.

Ab initio calculations were carried out for $[(NC)_5Os\text{-}CN\text{-}Ru(NH_3)_5]^-$ as follows. Firstly the geometry was optimized by using the Hartree-Fock method for the closed-shell species of $[(NC)_5Os\text{-}CN\text{-}Ru(NH_3)_5]^{2-}$. The calculations described herein were implemented with Gaussian 94 [9a]. The LANL2DZ basis set was used for the optimization. In order to reduce the computational time, bond angles were adjusted so as to lead to the Cs symmetry (C_{4v} when neglecting the hydrogen atoms). The intermetallic distance was calculated to be 5.11 Å, which is 0.1 Å shorter than the experimental estimate reported by Hupp et al. [8g] To calculate ground and excited states of the mixed valence compound with an open shell, we used the symmetry-adapted-cluster (SAC) and symmetry-adapted-cluster configuration-interaction (SAC-CI) method [9,10]. The SAC wave function $|\Phi\rangle$ was based on the closed-shell molecule $[(NC)_5Os\text{-}CN\text{-}Ru(NH_3)_5]^{2-}$, whose Hartree-Fock wave function was used as a reference. The SAC-CI wave function is represented as $|\Psi\rangle = \Sigma\ C_I\ R_I\ |\Phi\rangle$, where C_I is a CI coefficient. In the present calculation, R_I corresponds to an ionization operator, and hence $|\Psi\rangle$ corresponds to the mixed valence state (an ionized state of $[(NC)_5Os\text{-}CN\text{-}Ru(NH_3)_5]^{2-}$). The LANL1DZ and CEP-31G sets were used for metal atoms and first row elements, respectively, in the SAC/SAC-CI calculations, and all the molecular orbitals (MOs) were included in the active space. The perturbation selection method was carried out with the thresholds of 5×10^{-5} hartree for both ground and excited states. Since we are primarily interested in metal-to-metal charge transfer (MMCT) excitations, we did not calculate metal-to-ligand charge transfer (MLCT) excitations which are described as shake-up states of $[(NC)_5Os\text{-}CN\text{-}Ru(NH_3)_5]^{2-}$.

The calculated absorption spectrum of $[(NC)_5Os\text{-}CN\text{-}Ru(NH_3)_5]^-$ in vacuum is shown in Figure 1(a) where a Gaussian band with 0.5 eV width is arbitrarily assumed for each excitation with the peak height equal to the oscillator strength. The highest occupied MO (HOMO) from which an electron is removed is characterized as the Os $d(t_{2g})$ orbital (see Figure 3a), and the ground state is assigned to Os^{III}-CN-Ru^{II} in the mixed valence complex. The absorption spectrum (Figure 2(a)) consists of three bands centered at 2.9, 4.3, and 5.7 eV; they are assigned to ligand-to-metal charge transfer (LMCT) and to metal-to-metal CT (MMCT), respectively. Although the SAC/SAC-CI method has been successful in calculating excited states of a variety of mononuclear complexes [12], the calculated electronic structure and the absorption spectrum in vacuum differs from the experiment: the ground state of $[(NC)_5M\text{-}CN\text{-}M'(NH_3)_5]^-$ has been characterized as M^{II}-CN-M'^{III}, not M^{III}-CN-M'^{II}[8]. The Stark spectrum observed for $[(NC)_5Os\text{-}CN\text{-}Ru(NH_3)_5]^-$ in an ethylene glycol/water matrix showed that the diabatic ET distance associated with the excitation to the lowest excited state is 3.7 Å, indicating that

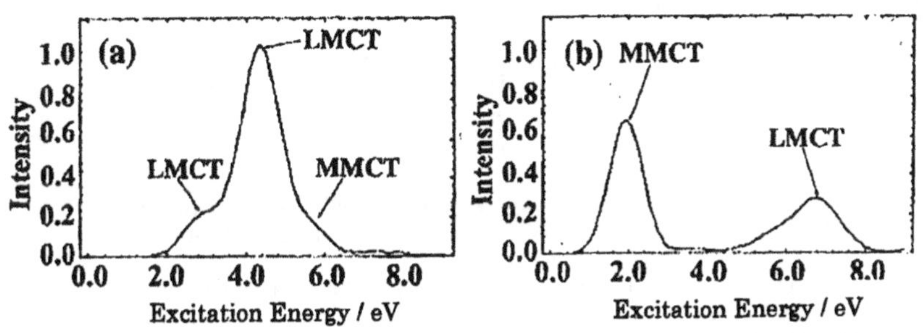

Figure 2: Calculated excitation energy spectra of $[(NC)_5Os\text{-}CN\text{-}Ru(NH_3)_5]^-$: (a) without an electric field; (b) with an electric field (0.04 au).

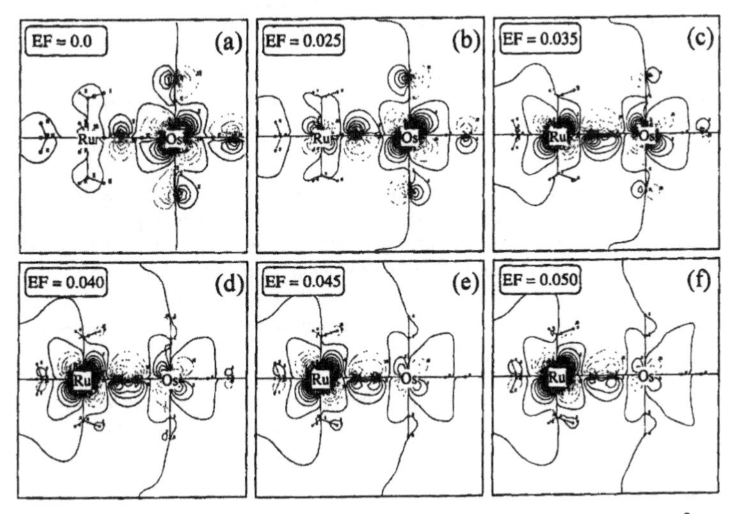

Figure 3: Electric field dependence of the HOMO of $[(NC)_5Os\text{-}CN\text{-}Ru(NH_3)_5]^{2-}$. A value in each figure represents electric field strength in au.

this excited state should be assigned to MMCT [8f].

In the present vacuum calculation, the Ru d(t_{2g}) orbitals are about 7.3 eV lower that the HOMO (Os d(t_{2g})). This result itself is not surprising from the electronic structure viewpoint. The Os orbitals are expected to be destabilized because of the large negative charge of the $Os(CN)_5$ moiety. The Mulliken population analysis for the natural orbitals of the ground state shows that the gross charges on $Os(CN)_5$ and $Ru(NH_3)_5$ moieties are -2.41 and 1.47, respectively. Thus it seems reasonable to think that the discrepancy between theory and experiment implies that some physical effects are essentially required to reproduce the experiment.

A plausible candidate of the missing effect would be an electrostatic effect in the condensed phase. Since the mixed valence compound is highly polarizable, it can be strongly influenced by interaction with the surrounding polar solvent and/or the counter cation. Experimentally it has been know that metal complexes with the CN ligand show marked solvent dependence in metal-to-ligand charge transfer (MLCT) bands [13]. Even for ammine complexes, both theoretical [14] and experimental data [15] suggest appreciable dependences of MLCT energies on the solvent. Interestingly, Zerner et al. [14b] showed that the solvent effect for $[pyRu(NH_3)_5]^{2+}$ was reasonably simulated in their electronic structure calculations by applying an electrostatic field due to point charges which neutralize the total charge of the system. In order to simulate such an electrostatic effect, we applied a uniform electric field to the system, and carried out excited state calculations. The applied field strengths are 0.025, 0.035, 0.040, 0.045, and 0.050 au (1 au is equal to 5.14×10^{11} V/m) in the direction (Os $\rightarrow$ Ru) as shown in Figure 4.

Figure 2b shows the SAC/SAC-CI spectrum of $[(NC)_5Os\text{-}CN\text{-}Ru(NH_3)_5]^-$ in a uniform electric field of 0.04 au. This field strength is close to the optimal value (0.038 au) obtained by linearly interpolating the calculated MMCT energies in comparison with the experimental value. The calculation gives two absorption bands: the MMCT band at 1.92 eV and the LMCT at 6.74 eV. The MMCT excitation energy is in reasonable agreement with the experimental value (1.64 eV) [8]. The calculated adiabatic electron transfer distance is 3.44 Å by using the generalized Mulliken-Hush method [16], which is close to the experimental value of 3.5 Å [8f].

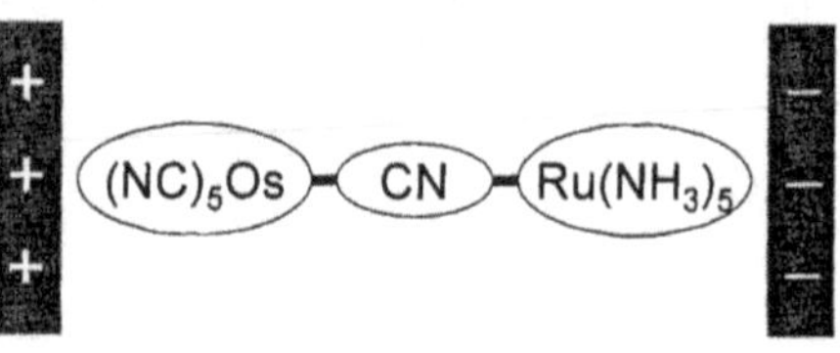

Figure 4: Configuration of two electrodes to apply the external electric field

The change in the excitation spectrum by the electric field results from the large distortion and level shifts of the metal centered molecular orbitals. Figure 3 depicts changes of the HOMO of $[(NC)_5Os\text{-}CN\text{-}Ru(NH_3)_5]^{2-}$, which mainly consists of the d orbitals of the metal atoms. Without the external field, the Os d(t_{2g}) orbital dominates the HOMO. As the field strength increases, the HOMO charge density shifts from Os to Ru. For a field strength of 0.035 au, both the Ru and Os orbitals evenly contribute (a valence-delocalized state). When the field strength is 0.04 au, the Ru orbital dominates, so that the ground state is characterized as Os^{II}-CN-Ru^{III} for the mixed valence (ionized) complex. In this case, the gross charges on the $Os(CN)_5$ and $Ru(NH_3)_5$ moieties are -3.61 and 2.55, respectively. This indicates that external electric field enhances charge polarization which is reflected in destabilization of the $Ru(NH_3)_5$ moiety.

As a physical origin of the electric field, either the solvent or the counter cation may contribute. We cannot rule out the second possibility, since crystallographic data by Hupp et al. [8b] showed that the counter cation (Na^+) for $[(NC)_5Fe\text{-}CN\text{-}Ru(NH_3)_4(py)]^-$ is only 2.33 Å away from the nitrogen atom of the CN ligand on the molecular axis. However, since the Coulombic field ($1/\varepsilon r^2$ where r(Na-Fe) = 5.29 Å [8i]) potential exerted by the cation is roughly estimated as 1.28×10^{-4} au (6.6×10^7 V/m) on Fe in water (ε= 78), it is reasonable to consider that the solvent effect would be more important as the strong electrostatic field inducing the electronic polarization.

In summary, we have shown that medium polarization is the essentially important perturbation. Its modulation can be related to the electronic structure of the supramolecule (the mixed valence complex). A set of the field strength and the energy shift of the MMCT band may be considered of as a measure of characterizing the supramolecule.

Origin of Solvent Dependence Programmed by Fragments:

$[(NC)_5Os\text{-}CN\text{-}Ru(NH_3)_5]^-$ consists of two moieties of $Os(CN)_5$ and $Ru(NH_3)_5$. Considering chemical features of the CN and NH_3 ligands from the viewpoint of the Lewis acid/base, it is anticipated that solvent effect would be somewhat different. This is supported by the experimental observation: the metal-to-ligand charge transfer (MLCT) bands of pentaammine complexes depend on a donor number

Table 1. MLCT excitation energies in eV and oscillator strengths (given in parenthesis) of $[Ru(NH_3)_5L]^{2+}$ and $[Ru(CN)_5L]^{3-}$

L	$[Ru(NH_3)_5L]^{2+}$			$[Ru(CN)_5L]^{3-}$		
	vacuum	water	exptl.[a]	vacuum	water	exptl.[b]
py	3.71 (0.11)	3.21 (0.16)	3.05 (0.16)	2.32 (0.26)	3.15 (0.17)	3.91
pz	3.45 (0.14)	3.00 (0.18)	2.63 (0.20)	2.46 (0.27)	2.81 (0.22)	3.35

a. Ref. [15]; b. Ref. [13]

of the solvent [15], while, in pentacyano complexes, the energy shifts are proportional to an acceptor number of the solvent [13]. Absorption bands in related compounds show unique solvent sensitivity depending on the ligand: $Fe(bpy)_2(CN)_2$ shows a marked solvent dependence, while the electronic spectrum of $Fe(bpy)_3^{2+}$ shows negligible solvent dependence. These facts clearly show that the solvation effect is specific to ligands [13e-g]. Thus it seems interesting to see how the solvent dependence of electronic structure is different between the $Os(CN)_5$ and $Ru(NH_3)_5$ moieties. In order to investigate the solvent dependence, we focus on $[LRu(CN)_5]^{3-}$ and $[LRu(NH_3)_5]^{2+}$ (L = py, pz). In these complexes the MLCT band commonly appears in absorption spectra, and can be a good probe for the solvent effect.

Geometries of $[LRu(NH_3)_5]^{2+}$ and $[LRu(CN)_5]^{3-}$ were fully optimized using the density functional theory (DFT) method. For the analysis of the solvatochromism, we adopted a common geometry optimized for an isolated molecule (i.e. without the solvent effect). We used the three-parameterized Becke-Lee-Yang-Parr (B3LYP) hybrid exchange-correlation functional. For excited state calculations, we used the time dependent density functional theory (TDDFT) method where the solvent effect was taken into account with the polarizable continuum model (PCM). The LANL2DZ basis set was used for Ru and its core electrons were replaced with the accompanied effective core potentials (ECP). For C, N, and H, the 6-31G basis set was adopted. The reason of using DFT rather than the HF and SAC/SACCI methods is to see whether the above mentioned result would depend on the approximate computational methods or not. The computations herein were implemented with Gaussian 98 [9b].

The calculated MLCT excitation energies for $[LRu(NH_3)_5]^{2+}$ in vacuum are tabulated in Table 1. The in-vacuo calculations overestimate the experimental MLCT energy as has been reported previously [14]. The solvent effect represented by the PCM contributes to decrease the MLCT energies, resulting in better agreement with the experiment. The solvent effect on the MLCT excitation for $[LRu(CN)_5]^{3-}$ is in sharp contrast to that for $[LRu(NH_3)_5]^{3-}$. For the in-vacuo calculations, the calculated MLCT excitation energies appreciably underestimate the experimental values. The PCM calculations increase the MLCT excitation energies, and the corresponding values approach to the experimental values. It is interesting that the direction of the solvatochromic shifts is different, depending on the type of ligand.

The solvatochromic shifts of MLCT excitations in $[LRu(NH_3)_5]^{2+}$ and $[LRu(CN)_5]^{3-}$ can be explained by using the orbital energy diagram of the Ru pyridine complexes in Figure 5. The HOMOs are assigned to the Ru 4d orbital, while the LUMOs are assigned to py π^* orbital. In $[pyRu(NH_3)_5]^{2+}$, both

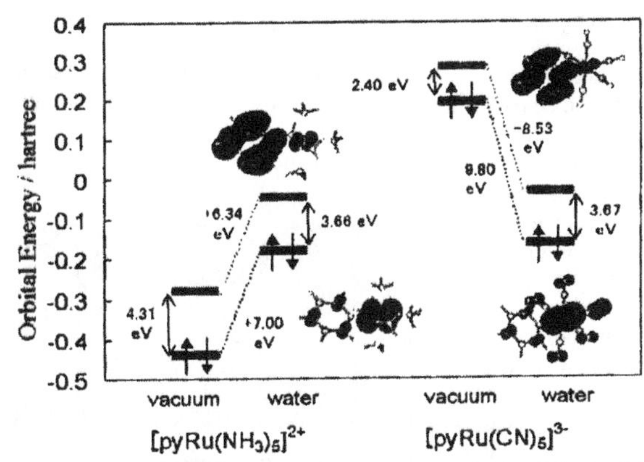

Figure 5: Solvent dependence of HOMO and LUMO of $[pyRu(NH_3)_5]^{2+}$ and $[pyRu(CN)_5]^{3-}$

HOMO and LUMO are destabilized, and, since the destabilization of the HOMO is more appreciable, the energy gap decreases, causing the low-energy shift. In contrast, the HOMO and LUMO are stabilized in $[pyRu(CN)_5]^{3-}$. The high energy shift of the MLCT band is ascribed to the enhanced stabilization of the HOMO, and, hence, the wider orbital gap. Figure 5 clearly shows that the HOMO levels of these two moieties become nearly degenerate in aqueous solution, although they are widely separated in vacuum. This result is consistent with the orbital distortion shown in Figure 5.

4. Computational Analysis of Molecular Recognition: Molecular Surface Probed by Ions

Molecular recognition is a process of forming a specific pair of molecules. This process is a key step in e.g. biological ligand-receptor binding [2]. The driving force to yield the complex is due to intermolecular interactions. Although they arise from quantum mechanical nature of electrons in molecules, empirical potential energy functions are frequently employed for large systems where positions of atoms are their variables [17].

When a docking geometry is computationally optimized or is experimentally elucidated, the easiest way of quantitative interpretation would be to introduce the concept of molecular shape. Usually the shape is defined by van der Waals radii of atoms [17]. It is further used to define molecular volume, which can be a good measure investigating chemical nature of pores in porous materials.

Conceptual importance of molecular shape is well known in biochemistry: the lock-and-key model [2] is a useful one for understanding ligand-receptor interactions. A simple but fundamentally important question about molecular shape is "how the quantum mechanical intermolecular interactions can be modeled by molecular shape". This issue seems important because the concept of molecular shape can be regarded as one of the coarse-grained pictures of complicated quantum mechanical interactions. Here we develop a computational scheme of defining molecular shape with the use of ab initio electronic structure calculations.

Molecular shape can be regarded as one of the molecular properties. It is well known that an electronic property such as electric dipole moment, electric susceptibility, etc. is defined as a response to an applied external field as shown in Eq. (2). Therefore a consistent definition of molecular shape would be to consider energy response to an appropriate perturbation.

Here we propose to define molecular shape by computing quantum mechanical energy for interaction between a target molecule and a probe atom (see Figure 6). From the interest in biological systems and chemistries in aqueous solution, we focus on interaction related to the hydrogen bond. This is done by choosing H^+ and O^{2-} as probe atoms. For a target molecule fixed in the 3D space, electronic-structure calculations are repeatedly carried out for a probe atom on each grid in the same space. By displacing it in the predefined 3D box, energy mapping can be achieved. By extracting an isoenergy contour for an appropriate threshold, we can obtain a surface defining molecular shape. The present method can be called "the computational surface-probing method".

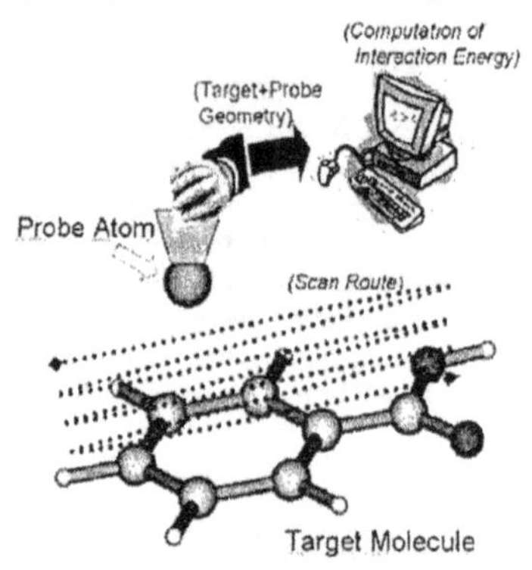

Figure 6: Illustration of the computational surface-probing method.

In the present study, interaction energies were calculated by using the MP2 method with the cc-pVDZ basis set. Gaussian 03 [9c] was used for the calculations. Geometries of target molecules (H_2O, CO_2 and C_2H_2 in this paper) were firstly optimized, and then calculations were repeated for the "target + probe" system where a probe atom was displaced on the uniform 3D grids. The 2 dimensional contours shown below were obtained for the 50×50 grids on the molecular plane of each target. As a probe, we considered H^+ and F^-. The latter ion was used as an alternative to O^{2-}. This is because convergence difficulty was encountered in SCF calculations in case of the O^{2-} probing. The energy contour map was drawn by calculating binding energies (stabilization energies) where the separated limit

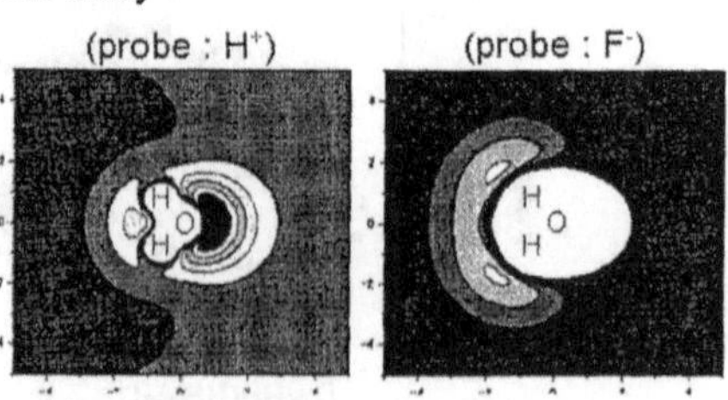

Figure 7: Isoenergy contour of H_2O probed by the H^+ (left) and F^- (right) ions.

was referred to.

Figure 7 shows isoenergy contour map of H_2O. The result indicates that the space can be divided into two parts: one is a region of positive energy representing repulsive interaction. It roughly corresponds to a while region in Figure 7. This region may be called "repulsive shape". It may be approximately represented as superposition of circles centered on atoms when H^+ is used as a probe: three circles on O and two H atoms can be distinguished. For the F^- probe, the repulsive shape is an ellipsoid rather than a circle. The area of the repulsive region is larger for the F^- probe than for the H^+ probe. This may be interpreted as reflecting a difference between ionic radii of H^+ and F^-.

Another part of the energy contour is a region of negative energy (the colored region). This may be called "attractive shape" of the target molecule. An interesting feature of the attractive shape of H_2O is existence of an interaction site near the oxygen atom. From its topological feature, it appears assignable to the oxygen lone pair orbital. This supports that, when interaction sites are determined for deriving empirical potential energy functions, they should be centered on, at least, three atoms (O and two H) and a site corresponding to the lone pair. In the attractive region probed by F^-, we see two energy minima near the hydrogen atoms. This is reasonable to represent an interaction in hydrogen bonding between the electropositive (H) and electronegative atoms (F).

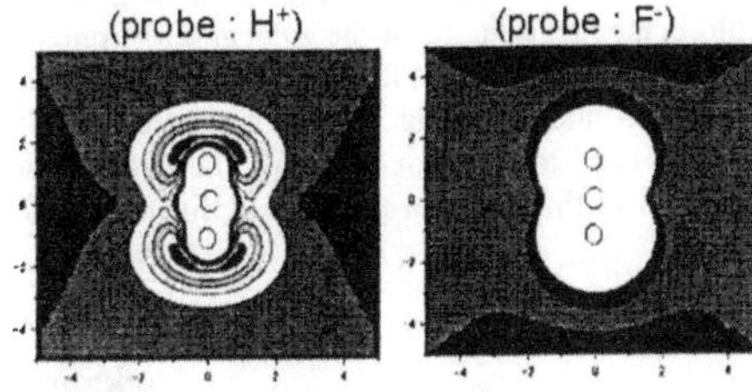

Figure 8: Isoenergy contour of CO_2 probed by the H^+ (left) and F^- (right) ions.

Figure 8 shows contour maps of CO_2. Similar to H_2O, we see interaction sites around two oxygen atoms for the H^+ probe. Reflecting differences of chemical interaction between the C-H^+ and O-H^+ pairs, the radii of the carbon and oxygen atoms are different. The energy contour probed by F^- indicates that CO_2 has no closed region giving energy minima. Thus its shape consists of only repulsive region.

Figure 9 depicts energy contours for acetylene. When it is probed by H^+, energy minima appear on both sides of molecular axis. This seems to be related to the π orbital of acetylene. This feature implies that topological shape of a chemically important orbital can be extracted by the present computational procedure. For the F-probed contour, we see two interaction sites near two hydrogen atoms although the energy stabilization is appreciably small in comparison with that obtained for the H-probed contour.

As seen in the above examples, it is evident that molecular shapes probed by ions show interesting topological features of molecules and look chemically important. The contour is nothing but a potential energy surface, which has clear physical meaning and quantitatively reflects chemical interactions: location and depth of energy minima directly reflect the nature of the ion-molecule interaction.

To summarize this subsection, we have suggested the molecular shape describing molecular recognition should be characterized as an energy response to chemical perturbation (a probe atom). For further applications, numerical analyses of molecular shapes need to be developed.

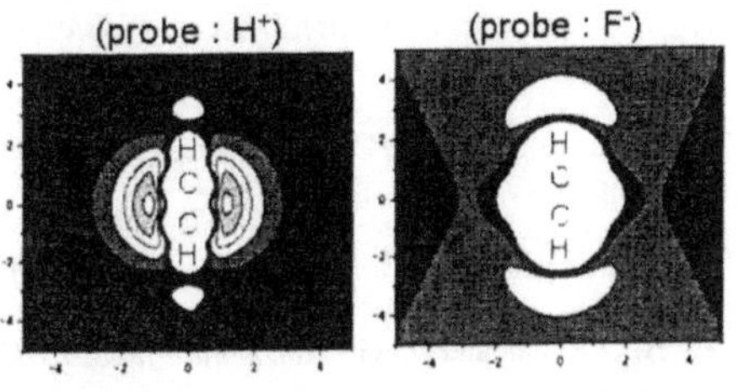

Figure 9: Isoenergy contour of C_2H_2 probed by the H^+ (left) and F^- (right) ions.

6. Concluding Remarks

We have discussed that an interesting approach for understanding supramolecular chemistry, which is of appreciable importance in a variety of science and technology, is to investigate the perturbation-property relationship. This seems useful to define and characterize supramolecules quantitatively. As examples of computational electronic-structure studies on supramolecules, we have shown our computational results on an asymmetric mixed valence compound and molecular shape analysis. In the former, experimentally observed is transport phenomenon which is a characteristic one in supramolecular chemistry. The latter can be related to molecular recognition which is one of the important supramolecular functions as seen in ligand-protein receptor docking.

Through ab initio electronic structure calculations, we have found that the experimentally known electronic structure of a mixed valence compound, $[(NC)_5Os\text{-}CN\text{-}Ru(NH_3)_5]^-$ is reproduced when an appropriate external electric field is applied. This means that the applied electric field is a critical perturbation to probe the electronic structure of the supramolecule.

For the molecular shape analysis, we have proposed the use of ion-molecule interaction energies. In this work, the hydrogen-bond-like interactions were considered by adopting the H^+ and F^- probes. The calculated contour maps have some interesting implications such that the probed surface reflects topological feature of chemical interactions.

Finally we would like to point out that supramolecular chemistry provides many challenging issues for electronic-structure simulations. Since the validity of empirical and reductionists' approaches is not assured in supramolecules, it is necessary to simulate a whole system as it is. Thus a supramolecular target can be big in size for computation, so that fast computation is strongly demanded. Because of the multi component nature of the system, electronic structure can be complicated with quasi-degenerate nature. This implies that accurate computational methods are demanded to obtain reasonable results. For better understanding of the computed results on supramolecules, meaningful and insightful analyses need to be carried out. We hope that investigation of the perturbation-property relationship discussed in this paper would be helpful and useful for understanding and designing supramolecular systems.

Acknowledgements

I acknowledge insightful comments on Sec. 3 by Dr. Marshall D. Newton and Dr. Bruce S. Brunschwig. The kind supply of the SACCI96 program by Prof. Hiroshi Nakatsuji and Prof. Masahiko Hada was essentially important for the present work. Also acknowledged are Ms. Namie Honda and Tomoko Taguchi for their assistance in computing the mixed valence compound. The present work was partially supported by a Grant-in-Aid in Priority Areas "Chemistry of Coordination Space" (No. 16074213) of the Ministry of Education, Culture, Sports, Science, and Technology, Japan.

References

[1] S. M. Sze, *Semiconductor Devices: Physics and Technology*, John Wiley & Sons, New York, 1985.

[2] W. H. Elliott and D. C. Elliott, *Biochemistry and Molecular Biology*, Oxford University Press, London, 1997.

[3] J. –M. Lehn, *Supramolecular Chemistry*, VCH Verlagsgesellschaft mbH, 1995.

[4] B. O'Regan and M. Grätzel, *Nature*, **353**, 737(1991).

[5] (a) U. Galm, M. H. Hager, S. G. Van Lanen, J. Ju, J. S. Thorson, and B. Shen, *Chem. Rev.* **105**, 739(2005). (b) A. Decker, M. S. Chow, J. N. Kemsley, N. Lehnert, E. I. Solomon, *J. Am. Chem. Soc.* **128**, 4719(2006).

[6] F. Jensen, *Introduction to Computational Chemistry*, John Wiley & Sons, Chichester, 1999.

[7] C. Kittel, *Introduction to Solid State Physics*, John Wiley & Sons, New York, 1976.

[8] (a) A. Vogler and J. Kisslinger, *J. Am. Chem. Soc.* **104**, 2311(1982). (b) A. Vogler, A.H. Osman, and H. Kunkely, *Inorg. Chem.* **26**, 2337(1987); (c) A. Vogler and H. Kunkely, *Comments Inorg. Chem.* **19**, 283(1997); (d) A. Burewicz and A. Haim, *Inorg. Chem.* **27**, 1611(1988); (e) A. Haim, In *Electron Transfer Reactions, Inorganic, Organometallic, and Biological Applications*; S. S. Isied, Ed.; Advance in Chemistry Series No. 253; American Chemical Society; Washington DC, 1997. p. 239; (f) G.U. Bublitz, W.M. Laidlaw, R.G. Denning, S.G. Boxer, *J. Am. Chem. Soc.* **120**, 6068(1998); (g) L. Karki, J.T. Hupp, *J. Am. Chem. Soc.* **119**, 4070(1997). (h) F.W. Vance, L. Karki, J.K. Reigle, J.T. Hupp, M.A. Ratner,

J. Phys. Chem. A., **102**, 8320(1998). (i) F.W.Vance, R.V. Slone, C.L. Stern, J.T. Hupp, *Chem. Phys.* **253**, 313(2000).

[9] (a) M.J. Frisch et al., *Gaussian 94*, Rev. E.2, Gaussian, Inc., Pittsburgh PA, 1995; (b) M.J. Frisch et al., *Gaussian 98*, Rev. A.7, Gaussian, Inc., Pittsburgh PA, 1998; (c) M.J. Frisch et al. Gaussian 03, Rev. B.04, Gaussian, Inc., Pittsburgh PA, 2003. See also references in the manual for further details of the computational methods.

[10] (a) H. Nakatsuji, and K. Hirao, *J. Chem. Phys.* **68**, 2053(1979); (b) H. Nakatsuji, *Chem. Phys. Lett.* **59**, 362(1978); (c) H. Nakatsuji, *Chem. Phys. Lett.* **67**, 329(1979); (d) H. Nakatsuji, *Chem. Phys. Lett.* **67**, 334(1979).

[11] H. Nakatsuji et al. SAC/SAC-CI program system (SAC-CI96) for calculating ground, excited, ionized, and electron-attached states having singlet to septet spin multiplicities, 1996. We used a local version of SACCI96 linked with the GAMESS program (M.W. Schmidt et al. *J. Comp. Chem.* **14**, 1347(1993)).

[12] (a) H. Nakatsuji and S. Saito, *J. Chem. Phys.* 93, 1865(1990); (b) H. Nakatsuji and S. Saito, *Intern. J. Quantum Chem.* **39**, 93(1991); (c) H. Nakai, Y. Ohmori, and H. Nakatsuji, *J. Chem. Phys.* **95**, 8287(1991); (d) H. Nakatsuji, M. Ehara, M. H. Palmer, and M. F. Guest, *J. Chem. Phys.* **97**, 2561(1992); (e) H. Nakatsuji and M. Ehara, *J. Chem. Phys.* **101**, 7658(1994); (g) K. Endo, M. Saikawa, M. Sugimoto, M. Hada, and H. Nakatsuji, *Bull. Chem. Soc. Jpn.* **68**, 1601(1995); (f) K. Yasuda, N. Kishimoto, and H. Nakatsuji, *J. Phys. Chem.* **99**, 12501(1995).

[13](a) H.E. Toma and M.S. Takasugi, *J. Solution Chem.* **12**, 547(1983); (b) H.E. Toma and M.S. Takasugi, *J. Solution Chem.* **18**, 575(1989); (c) C.J. Timpson, C.A. Bignozzi, E.M. Kober, and T.J. Meyer, *J. Phys. Chem.* **100**, 2915(1996); (d) L.D. Slep and J.A. Olabe, *J. Am. Chem. Soc.* **123**, 7186(2001); (e) J. Burgess, *Spectrochim. Acta* **26A**, 1369(1970); (f) J. Burgess, *Spectrochim. Acta* **26A**, 1957(1970).

[14] (a) K.K. Stavrev, M.C. Zerner, and T.J. Meyer, *J. Am. Chem. Soc.* **117**, 8684(1995). (b) G.M. Pearl and M.C. Zerner, *J. Am. Chem. Soc.* **121**, 399(1999). (c) Y.-G. Shin, B.S. Brunschwig, C. Creutz, M.D. Newton, and N. Sutin, *J. Phys. Chem.* **100**, 1104(1996); (d) I. Cacelli, A. Ferretti, *J. Chem. Phys.* **109**, 8583(1998). (d) I. Cacelli and A. Ferretti, *J. Phys. Chem. A*, **103**, 4438(1999); (e) N.S. Hush and J.R. Reimers, *Chem. Rev.* **100**, 775(2000).

[15] (a) P. Chen and T.J. Meyer, *Chem. Rev.* **98**, 1439(1998). (b) C.J. Timpson, C.A. Bignozzi, B.P. Sullivan, E.M. Kober, and T.J. Meyer, *J. Phys. Chem.* **100**, 2915(1996); (d) M.E. Gress, C. Creutz, and C.O. Quicksall, *Inorg. Chem.* **20**, 1522(1981). (e) C. Creutz and M.H. Chou, *Inorg. Chem.* **26**, 2995(1987); (f) Y.-K. Shin, B.S. Brunschwig, C. Creutz, and N. Sutin, *J. Phys. Chem.* **100**, 8157(1996).

[16] R.J. Cave and M.D. Newton, *J. Chem. Phys.* **106**, 9213(1997).

[17] J.-P. Doucet and J. Weber, *Computer-Aided Molecular Design*, Academic Press, London, 1996.

Brill Academic Publishers
P.O. Box 9000, 2300 PA Leiden,
The Netherlands

Lecture Series on Computer and Computational Sciences
Volume 6, 2006, pp. 482-491

Spectra of water dimer from *ab initio* calculations

Krzysztof Szalewicz, Garold Murdachaew, Robert Bukowski, Omololu Akin-Ojo
Department of Physics and Astronomy, University of Delaware, Newark, DE 19716, USA

and

Claude Leforestier
Laboratoire de Structure et Dynamique des Systèmes Moléculaires et Solides (LSDSMS)
Unite Mixte de Recherche (UMR-CNRS 5636)-CC 014
Université Montpellier II, 34095 Montpellier, Cedex 05, France

Received 10 July, 2006; accepted in revised form 10 August, 2006

Abstract: Recent theoretical work on water dimer spectra is presented. *Ab initio* calculations for the water dimer have been performed using symmetry-adapted perturbation theory at nearly a quarter million grid points with flexible monomers. A flexible-monomer, 12-dimensional potential has been fitted to the *ab initio* data. Fully quantum mechanical calculations on this surface produced low-energy vibration-rotation-tunneling (VRT) spectra in very good agreement with experiment, with root mean square deviation of 7 cm^{-1}. However, the monomer flexibility effects are smaller than the inaccuracies of the corresponding rigid-monomer potential. For the first time, anharmonic red shifts have been computed from an *ab initio* potential and agreed with experiment very well.

Keywords: water dimer, symmetry-adapted perturbation theory, vibration-rotation-tunneling spectra

PACS: 34.20.-b, 34.30.+h, 36.40.-c

1 Introduction

Intermolecular forces, also known as van der Waals interactions, despite being much weaker than the chemical forces that bind molecules together, are responsible for most properties of ordinary matter, including all living matter. A few exceptions are covalently-bound systems (such as SiO_2 or diamond) or metallic systems. The properties of gases and of condensed phases are related to the intermolecular forces via statistical mechanics [1, 2]. Unfortunately, the potential energy surfaces (PESs) which give rise to the intermolecular forces cannot be measured directly. One can extract some information about PESs from experiments in various indirect ways, always involving significant amounts of theoretical analysis. In fact, the empirical PESs resulting from such procedures are always parametrizations of theoretical models, with the parameters determining the PES. The experimental technique leading to the most accurate PESs is the spectroscopy of atomic and molecular clusters (see, e.g., Ref. [3]). In particular, the experiments performed on dimers allow one to develop true pair potentials, in contrast to empirical pair potentials fitted to bulk properties which also effectively account for many-body effects. The latter potentials have several other limitations, for example can be applied only in a fairly limited range of thermophysical parameters, close to those used in the fitting process.

Another method of obtaining PESs are *ab initio* quantum mechanical calculations for a number of monomers' relative configurations within a dimer, followed by fitting an analytic potential function to reproduce the computed data. One has to use, of course, some approximate electronic-structure approaches, such as many-body perturbation theory (MBPT), coupled-cluster (CC) methods within the supermolecular treatment [4], or symmetry-adapted perturbation theory (SAPT) [5, 6, 7]. The obtained potentials can then be used to predict properties of condensed phases by either performing molecular simulations or by computing some low-dimensionality statistical mechanics integrals. One can also predict the spectra of clusters. For atom-diatom and diatom-diatom interactions, *ab initio* predicted spectra can now be accurate to a fraction of cm^{-1}, for an example see Ref. [8]. Such potentials are usually more accurate than the empirical potentials which can be fitted to the spectra. The *ab initio* potentials can be then tuned, i.e., empirically readjusted by optimizing a small number of PES parameters, in order to still better reproduce the spectra.

The difficulty of *ab initio* calculation, particularly at the highly correlated levels needed for intermolecular interactions, increases rapidly with system size. If a PES is to be obtained, the difficulty also depends on its dimensionality. Therefore, systems beyond dimers consisting of diatoms demand significantly more resources and may require a decrease in the level of accuracy. Although the water dimer is not much bigger than H_2–CO investigated in Ref. [8], the potential energy surface is 12 rather than 6-dimensional. Since the number of configurations of the system which have to be considered grows exponentially with the dimension of the system, the water dimer problem is orders of magnitude more difficult. The difference is smaller if monomers are assumed to be rigid since then the potentials are 6 and 4-dimensional, respectively.

Due to the obvious importance of water, dozens of water pair potentials have been published (see Refs. [9, 10, 11] for comparisons of certain classes of potentials). Most of these potentials are rigid-monomer ones. It appears that a rigid-monomer approximation works rather well for most purposes [12, 13, 14, 15, 16, 17, 18, 19, 20]. However, there are certain quantities that a rigid-monomer potential cannot predict, such as the infrared shifts of the intramonomer transitions, vibrational predissociation, and intramolecular vibrational redistribution, among others. Also, the inclusion of monomer-flexibility effects is necessary at some point to improve the accuracy of theoretical predictions. This paper describes the development of a 12-dimensional PES for the water dimer.

2 Development of the potential energy surface

The potential presented in this work follows a series of rigid-monomer water dimer potentials computed using SAPT [12, 13, 14, 19]. The first such potential [12] was based on 1003 grid points calculated with SAPT in a medium-sized *spdf* basis set, including a set of $3s2p1d$ functions on the midbond. The monomer-centered "plus" basis set ($MC^{+}BS$) format [21] was used, resulting in basis set size of only 83 functions. The full dimer-centered "plus" basis set ($DC^{+}BS$) equivalent size was 112 functions.

The form of the potential applied in Ref. [12] was somewhat too involved to be used in simulations of liquid water. Therefore, a site-site potential labeled SAPT-5s [14] was later developed. The terms in this potential were physically motivated to properly model the long-range behavior of the electrostatic contribution through site charges and of the induction and dispersion contributions through van der Waals asymptotic coefficients. These terms were damped at short-range. The short-range behavior of these components and of the exchange energy were modeled by exponential terms. The potential function used five symmetry-distinct charges per monomer. The fit was performed to asymptotic information and to interaction energies computed using SAPT at 2510 grid points. This set included the grid points from the earlier work of Ref. [12] and additional grid points calculated using the same level of theory and basis set.

The SAPT-5s potential was used to explore the tunneling paths between equivalent minima on the dimer potential surface (geometries and barriers) and to predict the second virial coefficients [14]. The latter were in better agreement [14] with experiment than those predicted with the majority of earlier water potential (only the virials predicted by the ASP-W4 potential [9] were slightly closer to experiment in some regions). The computed spectra [15, 13] were also of high quality and in fact better than those given by any other potential except ones directly fitted to reproduce the spectra, such as the VRT(ASP-W) potential [22]. However, when an extended set of spectral lines was measured [23], the SAPT-5s predictions were found to be better than those of the empirical potentials.

Due to the considerable success of the SAPT-5s potential, we have decided to extend this potential to a fully flexible-monomer, 12-dimensional potential. We have used the same level of theory and the same basis set as in Ref. [14]. SAPT-5s was developed using a modest basis set for present-day standards, therefore a single calculation of the interaction energy takes insignificant computer resources and we were able to compute about a quarter million *ab initio* points.

Since the present potential is an extension of the rigid-monomer SAPT-5s potential, we have used the same set of 2510 dimer grid points as in Ref. [14]. For each dimer grid point, one has to select some number of flexible-monomer geometries associated with it. Since our goal is to obtain a better description of the ground vibrational state of water, we selected, in addition to the original $\langle r \rangle_0$ monomer geometry ($\langle r_{\mathrm{OH}} \rangle_0 = 0.9716257$ Å, $\langle \theta_{\mathrm{HOH}} \rangle_0 = 104.69°$) [12], the monomer geometries which correspond to the classical turning points of the monomer intramolecular potential for an isolated water molecule in its ground vibrational state. We used the symmetry coordinates of the water monomer [24]: the symmetric stretch, the bend, and the asymmetric stretch. For monomer A, these coordinates are defined as

$$s_1 = \frac{\Delta r_1 + \Delta r_2}{\sqrt{2}}, \quad s_2 = \sqrt{r_1 r_2}\Delta\theta_1, \quad s_3 = \frac{\Delta r_1 - \Delta r_2}{\sqrt{2}}, \tag{1}$$

respectively, where

$$\Delta r_1 = r_1 - \langle r_{\mathrm{OH}} \rangle_0, \quad \Delta r_2 = r_2 - \langle r_{\mathrm{OH}} \rangle_0, \quad \Delta\theta_1 = \theta_1 - \langle \theta_{\mathrm{HOH}} \rangle_0, \tag{2}$$

with analogous definitions for s_4, s_5, etc. applying to monomer B. Notice that the standard definition of the s_i uses the equilibrium coordinates as the reference geometry. To identify the turning points in the ground state vibration, we needed a monomer potential for the water molecule. We have chosen the accurate empirical PJT2 [25] water potential. The classical turning points of the ground vibrational state were determined by identifying the energy contour on the PJT2 potential surface with the value of the zero-point energy (ZPE). Thus, additional 14 distorted or flexible-monomer geometries were selected; these geometries all have energies within 1.3 cm^{-1} of ZPE and geometries chosen to correspond to the turning points of the symmetric stretch, the bend, and the asymmetric stretch, as well as combination distortions of symmetric stretches with bends and asymmetric stretches with bends. We decided not to include combination distortions of the type asymmetric stretch with symmetric stretch nor distortions in all symmetry coordinates of a monomer simultaneously. These choices give $15 \times 15 \times 2510 = 564,750$ grid points. Due to computational demands, we had to reduce this number by about half. This was achieved by sorting the grid points into bins according to intermolecular center-of-mass separation R. Since the original set of 2510 rigid-monomer points was unevenly distributed over R, we could reduce the number of points in some bins, aiming at a more uniform distribution.

We have first refitted the rigid-monomer potential using the following site-site functional form which will be denoted as SAPT-5s′:

$$
\begin{aligned}
V &= \sum_{a\in A,\ b\in B}\left[g^{ab}(R_{ab})e^{\alpha_{ab}-\beta_{ab}R_{ab}} + f_1(\delta_1^{ab},R_{ab})\frac{q_a q_b}{R_{ab}} + \sum_{n=6,8,10} f_n(\delta_n^{ab},R_{ab})\frac{C_n^{ab}}{R_{ab}^n}\right] \\
&\quad -\frac{1}{2}\left[\bar{\alpha}^B|\boldsymbol{F}^A|^2 + \bar{\alpha}^A|\boldsymbol{F}^B|^2\right] f_6(\delta_6^d, R),
\end{aligned}
\tag{3}
$$

where

$$g^{ab}(R_{ab}) = \sum_{n=0}^{3} a_n^{ab}R_{ab}^n \quad (a_0^{ab} = 1 \text{ kcal/mol}),$$

$$f_n(\delta_n^{ab},R_{ab}) = 1 - \sum_{i=0}^{n}\frac{(\delta_n^{ab}R_{ab})^i}{i!},$$

and $\bar{\alpha}^X$, X = A or B, is the average polarizability of water molecule. The dipolar field $\boldsymbol{F}^A$ of monomer A on monomer B is

$$\boldsymbol{F}^A = \frac{3\hat{\boldsymbol{R}}(\hat{\boldsymbol{R}}\cdot\boldsymbol{\mu}^A) - \boldsymbol{\mu}^A}{R^3},$$

where $\hat{\boldsymbol{R}} = \boldsymbol{R}/|\boldsymbol{R}|$, with $\boldsymbol{R}$ connecting centers of mass of monomers, and $\boldsymbol{\mu}^A$ is the dipole moment of water, with a similar formula for $\boldsymbol{F}^B$. The functional form of Eq. (3) differs from the SAPT-5s form by the presence of the last term, describing the polarization effects. It reproduces well the major portion of the asymptotic induction energy. We have decided to add this term since it significantly simplifies the inclusion of many-body polarization forces in the molecular simulations. The other terms in Eq. (3) are the same as in the SAPT-5s potential. The first term models the short-range exponentially decaying portion of the interaction energy representing exchange and overlap effects for all components; the second term represents the long-range electrostatic energy; and the third term represents the long-range dispersion energy plus this part of the asymptotic induction energy which is not reproduced by the polarization term. All long-range terms are smoothly damped with the Tang-Toennies [26] functions f_n at shorter separations. Equation (3) consists of a double sum over sites a in monomer A and b in monomer B; each depending on the site-site separation R_{ab} between them. It contains various parameters assigned to a site type (e.g., charges q_a) or a pair of sites—the asymptotic dispersion (and partly induction) coefficients C_n^{ab}, damping parameters δ_n^{ab}, as well as short-range exponential (α_{ab} and β_{ab}) and linear parameters (a_n^{ab})—all obtained by fitting. The q_a and C_n^{ab} parameters were obtained by fitting only to the asymptotic information.

The new fit is only marginally different from and not better than the original SAPT-5s fit. The root mean square deviations (rmsd) in different energetic regions are identical to those given in Ref. [14]. The use of the new form of the induction energy did improve the long-range fit (the rmsd is 65% of the original rmsd), but otherwise there were no effects of the slightly increased flexibility of the fitting function. We will use the SAPT-5s′ potential as the parent rigid-monomer potential for the flexible-monomer fit which will be denoted by SAPT-5s′f.

The flexible-monomer fitting form was obtained by expanding all the parameters in Eq. (3) as functions of the s_i coordinates. For example, for the charges the resulting equations are

$$q_a(s_1,s_2,s_3) = q_a^{(0)} + q_a^{(1)}s_1 + q_a^{(2)}s_2 + q_a^{(3)}s_3 + q_a^{(4)}s_1s_2 + q_a^{(5)}s_2s_3 + q_a^{(6)}s_1^2 + q_a^{(7)}s_2^2 + q_a^{(8)}s_3^2. \tag{7}$$

The coefficients $q_a^{(0)}$ were fixed at the values from the SAPT-5s fit [14] and the remaining ones were fitted to flexible-monomer data. For the coefficients depending on two centers, the resulting expansions are, of course, appropriately more complicated and involve all six s_i coordinates. The fitting functions were properly adapted to the permutational symmetry of the dimer. The final 568-parameter fit to the full set of 239,928 flexible-monomer SAPT points was performed using the singular value decomposition method with an effective condition number flag of 10^{-12}. Some linear combinations of the basis functions, 30 in all, were eliminated automatically by use of this setting. The rmsd of the fit is about twice as large as that for SAPT-5s. This seems to be reasonable taking into account that the ratio of the number of grid points to the number of fit parameters is an order of magnitude larger for the flexible-monomer fit.

3 Spectra of water dimer from flexible-monomer PES

The SAPT-5s′f flexible-monomer PES described above, as well as a variant of it which will be defined below, have been used in 6+6 adiabatic decoupling quantum mechanical calculations. The computational approach was presented in Ref. [27]. The harmonic infra-red (IR) shifts computed from the SAPT-5s′f PES were found to be significantly different from those computed directly from SAPT by numerical differentiation. This shows that the fit is not accurate enough near the minimum. To overcome this problem, we have refitted the *ab initio* data adding 2896 additional points calculated for the purpose of the numerical differentiations in the minimum region (with weights 10^3) and including, as additional fited quantities with weights 10, the 171 second derivatives of the SAPT interaction energies with respect to atomic Cartesian coordinates (the Hessian) computed at the minimum configuration. The resulting PES was dubbed SAPT-5s′fIR. Its rmsd for points with negative energies amounts to 0.3 kcal/mol, i.e. is about 0.1 kcal/mol larger than that of SAPT-5s′f.

Table 1 presents a comparison with the available experimental transitions for the $(H_2O)_2$ dimer. The last two columns contain the results for the two rigid potentials, SAPT-5s′ and SAPT-5s. These two sets of transitions are very close to each other, with the largest difference of only 1 cm^{-1} and the rmsd difference of only 0.2 cm^{-1}. All the differences are negligible compared to the discrepancies of either set with respect to experiment. Since, as discussed earlier, SAPT-5s′ and SAPT-5s differ only in the form of the fitting function, the closeness of the predicted spectra shows that indeed the two PES are nearly equivalent. The small discrepancies between the SAPT-5s transitions given in Table 1 and those published in Ref. [15] result from a different numerical method of solving the rovibrational equations.

The agreement of the SAPT-5s′fIR with experiment is basically at the same level as achieved by the SAPT-5s′ potential. The inclusion of the flexibility effects leads to a relatively minor overall improvement of this agreement. The rmsd was improved by 0.5 cm^{-1}. About 60% of SAPT-5s′fIR transitions agree slightly better than those computed from SAPT-5s′, although in a few cases like the B1+[100] → A1+[200] and A2-[100] → B2-[200] transitions the agreement with experiment became amazingly good. The sum $a_0 + a_1$—which was the largest discrepancy with experiment observed for the SAPT-5s potential—practically does not change (becomes slightly better).

The relatively small improvements observed in Table 1 indicate that the neglect of monomer flexibility effects is not the most important uncertainty in the SAPT-5s′ potential and the main future improvements of the agreement with experiment can be achieved already at the rigid-monomer potential level. Indeed, the SDFT-5s potential [19] (computed using a density-functional based version of SAPT [28]) resulted in several observables improved compared to SAPT-5s, although in the case of the spectra, there was no overall improvement: the small transitions agreed less well with experiment and large transitions agreed better. However, the very recently developed CC-pol potential [20] leads to a significant overall improvement for the majority of transitions, whereas the very few transitions not in this category become only slightly less accurate (these are transitions which were predicted near exactly by SAPT-5s). We expect that when the monomer flexibility effects are included at the level of CC-pol, this will lead to an overall improvement of the agreement with experiment.

Table 2 compares the IR shifts of monomer vibrational frequencies upon dimerization. This is

Table 1: VRT transitions in cm^{-1}. For the labeling of energy levels and references to experimental papers see Ref. [27].

Transition			Exp.	SAPT-5s'f	SAPT-5s'fIR	SAPT-5s'	SAPT-5s
$a_0 + a_1$			13.920	21.373	20.567	20.834	20.404
E+[100]	→	E-[110]	0.411	0.393	0.393	0.396	0.396
E-[100]	→	E+[110]	0.405	0.393	0.394	0.396	0.397
A2-[100]	→	A2+[110]	1.057	1.061	0.969	1.053	1.082
B2+[110]	→	B2-[100]	0.245	0.273	0.180	0.260	0.289
A1+[100]	→	A1-[110]	1.163	1.180	1.034	1.136	1.165
B2+[111]	→	A2+[111]	0.541	0.544	0.459	0.528	0.548
A1-[110]	→	B1-[111]	13.666	16.449	16.333	16.325	16.145
A1+[100]	→	A1-[111]	15.535	18.384	17.979	18.176	18.058
E-[110]	→	E-[211]	14.304	17.012	16.686	16.921	16.781
E+[100]	→	E-[211]	14.715	17.405	17.079	17.317	17.177
A1+[100]	→	B1+[100]	0.752	0.788	0.641	0.741	0.770
A2-[100]	→	B2-[100]	0.652	0.667	0.575	0.657	0.686
B2-[100]	→	B2+[111]	0.290	-4.321	-3.649	-3.956	-3.700
A1-[110]	→	B1-[211]	86.620	80.521	81.830	80.584	80.807
B1-[110]	→	A1-[211]	88.480	82.404	83.230	82.300	82.502
A1+[100]	→	A1-[211]	88.890	82.797	83.623	82.696	82.898
B1+[100]	→	B1-[211]	87.030	80.913	82.224	80.979	81.203
A1+[100]	→	B1+[200]	109.780	112.602	113.298	113.288	112.326
B1+[100]	→	A1+[200]	106.080	101.355	106.577	103.819	103.301
E+[100]	→	E+[300]	107.840	102.688	105.833	105.172	104.747
A2-[100]	→	B2-[200]	98.040	97.175	98.070	99.789	99.213
B2-[100]	→	A2-[200]	97.380	95.526	96.882	98.056	97.503
E-[100]	→	E-[200]	97.540	95.667	97.658	97.639	97.334
A2-[100]	→	A2+[200]	114.030	101.516	103.724	102.423	102.817
B2-[100]	→	B2+[200]	103.970	86.920	92.666	89.465	89.605
E-[100]	→	E+[400]	107.820	98.039	101.705	98.933	98.750
E-[100]	→	E-[300]	103.490	101.530	106.974	102.097	102.084
A2-[100]	→	B2-[300]	143.390	128.133	128.174	128.601	128.365
B2-[100]	→	A2-[300]	141.150	125.877	126.442	126.486	126.152
E-[100]	→	E-[500]	142.450	127.139	127.647	127.459	127.188
A2-[100]	→	A2+[100]	54.550	36.887	39.158	38.573	39.515
B2-[100]	→	B2+[100]	52.010	33.980	36.939	35.901	36.761
E-[100]	→	E+[200]	53.250	35.319	37.614	37.220	38.128
rmsd				8.620	7.399	7.910	7.735

the most critical test of the monomer-flexibility effects since the shifts cannot be predicted at all by a rigid potential. One should mention here that there are significant disagreements between the experimental intramonomer vibrational frequencies (see Table V in Ref. [27]). The experimental shifts listed in Table 2 are based on more recent data.

Whereas the harmonic intramonomer vibrational frequencies are fairly different from the measured ones, on the order of 100 cm^{-1}, due to the anharmonicity effects, these errors partly cancel in calculations of harmonic shifts and the theoretical predictions agree reasonably well with experiment. In particular the results of Tschumper *et al.* [29] are from high-level calculations and their rmsd with respect to experiment is only 12 cm^{-1}. Most likely, the majority of this error is due to the harmonic approximation and further increases of the level of electronic structure theory and of the basis set will lead to very small changes. The older *ab initio* calculations by Xantheas and Dunning [30] have more than twice as large error (note that the authors of Ref. [30] defined some shifts in a different way than we do). The SAPT results computed by a direct differentiation of the SAPT *ab initio* energies (denoted by "SAPT" in Table 2) are in very good agreement with Tschumper *et al.* values and are, probably fortuitously, even closer to experiment: rmsd is only 6 cm^{-1} and all but one shift are better than those of Ref. [29].

Table 2 clearly shows the problem with the SAPT-5s′f fit. With an ideal fit, the harmonic frequencies computed from this potential should be identical to those in the column "SAPT". However, the frequencies are significantly different and the rmsd increases to 16 cm^{-1}. Clearly, such errors lead also to anharmonic shifts which poorly agree with experiment. The refitted SAPT-5s′fIR potential agrees with the numerical SAPT frequencies much better and in particular has the same rmsd of 6 cm^{-1}. Most individual frequencies also agree well between the two cases, with the largest discrepancy equal to 18 cm^{-1} for the bonded OH stretch. Thus, the fit is still not ideal in the region of the minimum. The main reason is that the fitting of the SAPT-5s′IR was performed still within the constraint ("hard-coded" in our functional form) that for undeformed monomers this potential reduces to SAPT-5s′. Thus, to further improve the harmonic frequencies computed from the fit, one would have to change the form of the fitting function or at least use a rigid-monomer potential which has also been fitted to reproduce very well the bottom of the well.

The anharmonic frequencies computed from the SAPT-5s′fIR potential agree with experiment very well, although slightly worse than the harmonic ones given by the same potential: rmsd for the anharmonic case is 2 cm^{-1} larger than for harmonic. The agreement is particularly good, to within 4 cm^{-1}, with the bonded OH shift taken from the new experiment by Buck and Huisken [3]. Thus, our work indicates that this result should be favored over the older value of -127 cm^{-1} by Huang and Miller [31].

The change of IR shifts due to anharmonicity effect ranges from -11 cm^{-1} to +9 cm^{-1} and therefore is quite appreciable. As one might have expected, the acceptor shifts change very little, by only 1 cm^{-1}. The most changed shift is the donor bend. Indeed, this relatively large-amplitude motion should be sensitive to details of the surface farther from the minimum.

The overall agreement with experiment reached by the anharmonic calculations on SAPT-5s′IR potential is rather satisfactory, in particular in view of the potentially large uncertainties of the experimental results. It will be interesting to see whether a flexible-monomer potential based on the rigid CC-pol potential will lead to an improved agreement.

Table 2: IR frequency shifts in cm^{-1}. "5s′f" and "5s′fIR" are calculations with SAPT-5s′f and SAPT-5s′fIR potential, respectively. "Tsch." denotes results of Ref. [29] and "Xan." those of Ref. [30].

	anharmonic			harmonic				
	Exp.	5s′f	5s′fIR	Tsch.	Xan.	SAPT	5s′f	5s′fIR
donor bend	20	8	5	16	34	20	19	16
bonded OH	-56	-33	-52	-83	-107	-65	-30	-47
donor free OH	-21	-9	-15	-29	-48	-27	-4	-24
acceptor bend	6	-1	4	2	3	0	-1	3
acceptor sym. str.	3	-12	-7	-6	-17	-6	-6	-6
acceptor asym. str.	-11	4	-12	-10	-27	-9	10	-11
rmsd		15	8	12	26	6	16	6

4 Conclusions

We have developed a flexible-monomer (12-dimensional) potential for the water dimer. The *ab initio* calculations at nearly a quarter million dimer grid points were performed using the same level of SAPT and the same basis set as in Ref. [14]. The flexible-monomer potential reduces to the rigid-monomer SAPT-5s potential of Ref. [14] when both monomers are undeformed (although we used a slightly modified rigid-monomer fit form, SAPT-5s′). The SAPT-5s′f fit reproduces the computed *ab initio* points to within 0.2 kcal/mole for negative interaction energies. We have computed the second virial coefficients from SAPT-5s′f [32]. The monomer-flexibility correction added to the best available rigid-monomer coefficients computed from SDFT-5s [19] significantly improved the agreement with experiment in the range 300-400 K where the experimental data are most accurate. For large temperatures, the correction worsens the agreement, but this may be due to the fact that the experimental analysis producing these data utilized the information based on the SAPT-5s virial coefficients.

The SAPT-5s′f potential did not perform too well in calculations of dimer spectra, in particular it produced poor IR shifts. This problem was traced to insufficient accuracy of the fit at the bottom of the potential energy well, as determined by comparing the harmonic shifts computed *ab initio* and using SAPT-5s′f. The problem was overcome by fitting the SAPT-5s′fIR potential to a data set which in addition to the points used for SAPT-5s′f included, with very large weights, new points computed in the region very near the minimum and the Hessians. The resulting anharmonic shifts agree with experiment better than any previously published *ab initio* results. In fact, the accuracy is sufficiently high to recommend one set of experimental data over another. The low-energy VRT region of the spectrum is very similar for both SAPT-5s′f and SAPT-5s′fIR potentials and the effects of monomer flexibility are insignificant compared to the differences between the SAPT-5s results and experiment. Although SAPT-5s′fIR predicted the water dimer spectra very well, the virial coeffients computed from this potential do not represent the flexibility effect properly [32]. Thus, it appears that one has to use a more elaborate form of the fitting function to simultaneously reproduce both types of properties.

Acknowledgment

This research was supported by the NSF grants CHE-0239611 and CHE-0555979.

References

[1] D. A. McQuarrie. *Statistical Mechanics*. Harper, New York, 1976.

[2] J. O. Hirschfelder, C. F. Curtiss, and R. B. Bird. *Molecular Theory Of Gases And Liquids*. Wiley, New York, 1954.

[3] U. Buck and F. Huisken. Infrared spectroscopy of size-selected water and methanol clusters. *Chem. Rev.*, 100:3863, 2000.

[4] G. Chalasinski and M. M. Szczesniak. State of the art and challenges of the *ab initio* theory of intermolecular interactions. *Chem. Rev.*, 100:4227, 2000.

[5] B. Jeziorski, R. Moszynski, and K. Szalewicz. Perturbation theory approach to intermolecular potential energy surfaces of van der Waals complexes. *Chem. Rev.*, 94:1887, 1994.

[6] K. Szalewicz, K. Patkowski, and B. Jeziorski. Intermolecular interactions via perturbation theory: from diatoms to biomolecules. In D. J. Wales, editor, *Intermolecular Forces and Clusters*, volume 116 of *Structure and Bonding*, pages 43–117. Springer, 2005.

[7] K. Szalewicz, R. Bukowski, and B. Jeziorski. On the importance of many-body forces in clusters and condensed phase. In C.E. Dykstra, G. Frenking, K.S. Kim, and G.E. Scuseria, editors, *Theory and Applications of Computational Chemistry: The First 40 Years. A Volume of Technical and Historical Perspectives*, chapter 33, pages 919–962. Elsevier, Amsterdam, 2005.

[8] P. Jankowski and K. Szalewicz. A new *ab initio* interaction energy surface and high-resolution spectra of the H_2–CO van der Waals complex. *J. Chem. Phys.*, 23:104301, 2005.

[9] C. Millot, J. C. Soetens, M. T. C. M. Costa, M. P. Hodges, and A. J. Stone. Revised anisotropic site potentials for the water dimer and calculated properties. *J. Phys. Chem. A*, 102:754, 1998.

[10] A. Wallqvist and R. D. Mountain. Molecular models of water: derivation and description. In K. B. Lipkowitz and D. B. Boyd, editors, *Reviews in Computational Chemistry*, volume 13, page 183. Wiley-VCH, New York, 1999.

[11] B. Guillot. A reappraisal of what we have learnt during three decades of computer simulations on water. *J. Mol. Liquids*, 101:219, 2002.

[12] E. M. Mas, K. Szalewicz, and B. Jeziorski. Pair potential for water from symmetry-adapted perturbation theory. *J. Chem. Phys.*, 107:4207, 1997.

[13] G. C. Groenenboom, E. M. Mas, R. Bukowski, K. Szalewicz, P. E. S. Wormer, and A. van der Avoird. Water pair and three-body potential of spectroscopic quality from ab initio calculations. *Phys. Rev. Lett.*, 84:4072, 2000.

[14] E. M. Mas, R. Bukowski, K. Szalewicz, G. C. Groenenboom, P. E. S. Wormer, and A. van der Avoird. Water pair potential of near spectroscopic accuracy. I. Analysis of potential surface and virial coefficients. *J. Chem. Phys.*, 113:6687, 2000.

[15] G. C. Groenenboom, P. E. S. Wormer, A. van der Avoird, E. M. Mas, R. Bukowski, and K. Szalewicz. Water pair potential of near spectroscopic accuracy. II. Vibration-rotation-tunneling levels of the water dimer. *J. Chem. Phys.*, 113:6702, 2000.

[16] M. J. Smit, G. C. Groenenboom, P. E. S. Wormer, A. van der Avoird, R. Bukowski, and K. Szalewicz. Vibrations, tunneling, and transition dipole moments in the water dimer. *J. Phys. Chem. A*, 105:6212, 2001.

[17] E. M. Mas, R. Bukowski, and K. Szalewicz. *Ab initio* three-body interactions for water. I. Potential and structure of water trimer. *J. Chem. Phys.*, 118:4386, 2003.

[18] E. M. Mas, R. Bukowski, and K. Szalewicz. *Ab initio* three-body interactions for water. II. Effects on structure and energetics of liquid. *J. Chem. Phys.*, 118:4404, 2003.

[19] Bukowski, K. Szalewicz, G. C. Groenenboom, and A. van der Avoird. Interaction potential for water dimer from symmetry-adapted perturbation theory based on density functional description of monomers. *J. Chem. Phys.*, 125:044301, 2006.

[20] R. Bukowski, K. Szalewicz, G. C. Groenenboom, and A. van der Avoird, manuscript in preparation.

[21] H. L. Williams, E. M. Mas, K. Szalewicz, and B. Jeziorski. On the effectiveness of monomer-, dimer-, and bond-centered basis functions in calculations of intermolecular interaction energies. *J. Chem. Phys.*, 103:7374, 1995.

[22] R. S. Fellers, C. Leforestier, L. B. Braly, M. G. Brown, and R. J. Saykally. Spectroscopic determination of the water pair potential. *Science*, 284:945, 1999.

[23] F. N. Keutsch, N. Goldman, H. A. Harker, C. Leforestier, and R. J. Saykally. Complete characterization of the water dimer vibrational ground state and testing the VRT(ASP-W)III, SAPT-5st, and VRT(MCY-5f) surfaces. *Mol. Phys.*, 101:3477, 2003.

[24] P. F. Bernath. *Spectra of Atoms and Molecules*. Oxford, New York, 1995.

[25] O. L. Polyansky, P. Jensen, and J. Tennyson. The potential energy surface of $H_2{}^{16}O$. *J. Chem. Phys.*, 105:6490, 1996.

[26] K. T. Tang and J. P. Toennies. An improved simple model for the van der Waals potential based on universal damping functions for the dispersion coefficients. *J. Chem. Phys.*, 80:3726, 1984.

[27] C. Leforestier, F. Gatti, R. S. Fellers, and R. J. Saykally. Determination of a flexible (12d) water dimer potential via direct inversion of spectroscopic data. *J. Chem. Phys.*, 117:8710, 2002.

[28] A. J. Misquitta, R. Podeszwa, B. Jeziorski, and K. Szalewicz. Intermolecular potentials based on symmetry-adapted perturbation theory including dispersion energies from time-dependent density functional calculations. *J. Chem. Phys.*, 123:214103, 2005.

[29] G. S. Tschumper, M. L. Leininger, B. C. Hoffman, E. F. Valeev, H. F. Schaefer III, and M. Quack. Anchoring the water dimer potential energy surface with explicitly correlated computations and focal point analyses. *J. Chem. Phys.*, 116:690, 2002.

[30] S. S. Xantheas and T.H. Dunning. *Ab initio* studies of cyclic water clusters $(H_2O)_n$, n=1–6. I. Optimal structures and vibrational spectra. *J. Chem. Phys.*, 99:8774, 1993.

[31] Z. S. Huang and R. E. Miller. High-resolution near-infrared spectroscopy of water dimer. *J. Chem. Phys.*, 91:6613, 1989.

[32] G. Murdachaew *et al.* Manuscript in preparation.

Brill Academic Publishers
P.O. Box 9000, 2300 PA Leiden,
The Netherlands

Lecture Series on Computer and Computational Sciences
Volume 6, 2006, pp. 492-519

The life and work of Vedene H. Smith, Jr.

Ajit J. Thakkar[1]

Department of Chemistry,
University of New Brunswick,
Fredericton, New Brunswick E3B 6E2,
Canada

Received 31 July, 2006

Abstract: A tribute to the life and work of Vedene H. Smith, Jr. is presented in the form of a brief biographical sketch, some personal reminiscences, and an extensive bibliography.

Keywords: biography, tribute, bibliography

PACS: 01.60.+q

1 Biographical sketch

Vedene Smith was born on April 19, 1935 in Syracuse, N.Y. and raised in New York, Alabama, Georgia and Florida. He graduated from two high schools, one in Georgia and one in Florida, in the same year. As a youth, he bred and exhibited poultry and played competitive golf. In 1955, he earned a B.A. from Emory University. The program's distribution requirements enabled him to study a number of subjects outside the sciences. Among his hobbies at the time was attending poetry readings by Robert Frost, and sometimes transcribing new poems that Frost had not published. He then moved to the Georgia Institute of Technology. He was influenced there by a number of master teachers including a theoretical physicist Harold A. Gersch, a molecular spectroscopist William H. Eberhardt, and two applied mathematicians, John A. Nohel and Marvin B. Sledd. His enthusiasm for mathematics was such that he completed all the course-work for a master's degree in applied mathematics in 1957. He then worked with Gersch and earned a physics Ph. D. in 1960 for a thesis on the theory of many-boson systems. There followed a stint outside academia during which he worked as an applied mathematician. Next, Vedene went to Per-Olov Löwdin's Quantum Chemistry Institute at Uppsala University, Sweden and earned first the Fil. Lic. and then the Fil. Dr. degree with a thesis on independent particle models and electron correlation.

Vedene was very proud of his academic genealogy and enjoyed relating his work to that of his academic forebears. Figure 1 displays his "Ph. D. tree" in which descent is traced through formal Ph. D. advisors. It shows that Vedene's intellectual heritage is remarkably wide-ranging; the tree includes theoretical physicists such as Boltzmann, Ehrenfest and Uhlenbeck, physical chemists such as Ostwald, Fajans and von Liebig, and mathematicians like Petzval and Malmsten.

Vedene was named Docent in Quantum Chemistry at Uppsala in 1967. Later that year, Harold McIntosh and John Coleman persuaded Vedene to join the chemistry faculty of Queen's University in Kingston, Ontario as an Associate Professor. He was promoted to Professor just three years later. He was Head of the Queen's Chemistry Department from 1979 to 1989. He was given the

[1]E-mail: ajit@unb.ca, Web page: http://www.unb.ca/chem/ajit/

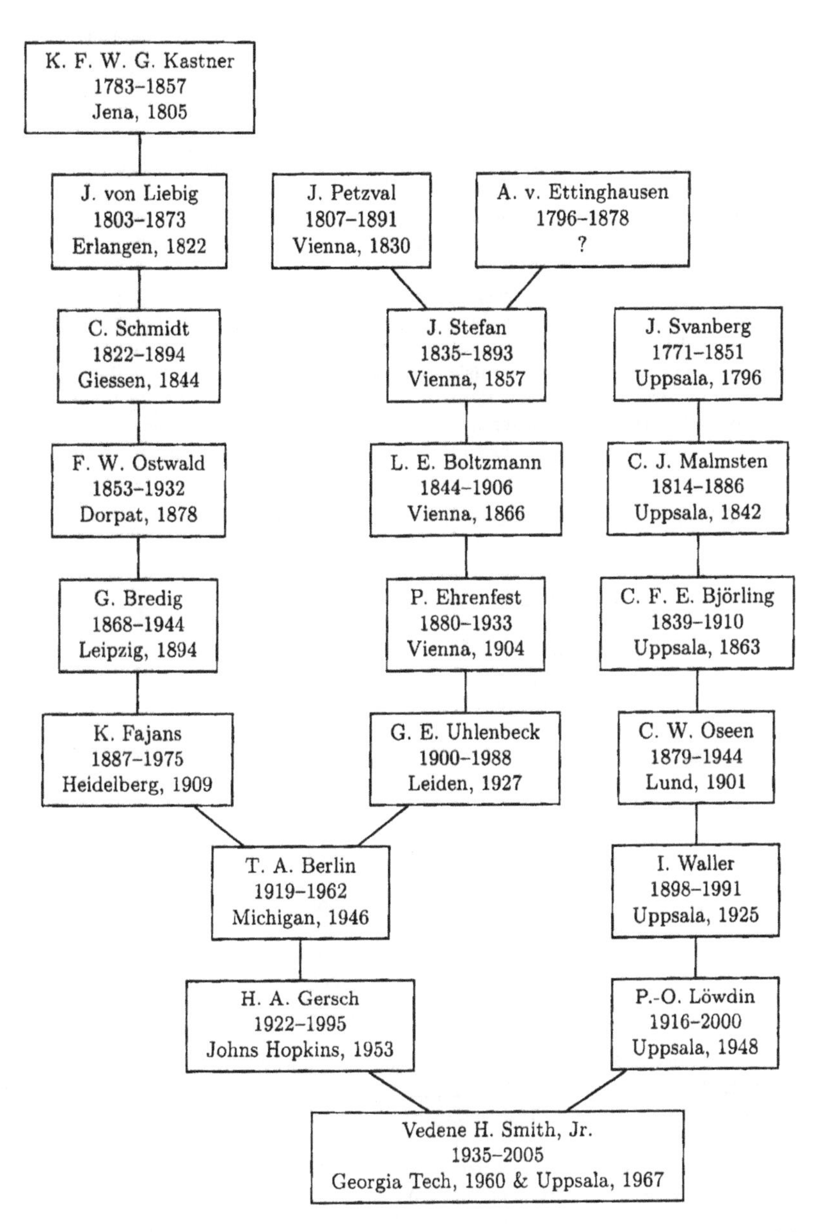

Figure 1: Vedene Smith's academic genealogy traced through Ph. D. advisors. The third line in each box gives the place and date of the Ph. D. degree.

charge to improve its research profile and the number and quality of the graduate students, without sacrificing the high reputation of its undergraduate Honours programs in Chemistry and Engineering Chemistry. Despite the tight fiscal constraints of the time, nine new faculty of high quality were hired during his tenure as Head. Major equipment was obtained for research. An application to the Natural Sciences and Engineering Research Council (NSERC) for a major instrument to be used by Queen's researchers and industrial users from Dupont with partial funding from Dupont and NSERC led to the introduction of the NSERC University-Industry Collaboration program. He was chosen by his peers to chair the Committee of Chairs of Departments of Chemistry of Ontario Universities (CCDCOU) from 1982 to 1986. Upon his retirement in 2000, Queen's University made him a Professor Emeritus.

At Queen's, he served as research supervisor to 18 Ph.D. students; see Table 1. Some of them are now carrying on his academic legacy as professors themselves: Bob Benesch at the Royal Military College, Mike Whangbo at North Carolina State University, Ajit Thakkar at the University of New Brunswick, Alfredo Simas at the Universidade Federal de Pernambuco, Sirkka-Liisa Korppi-Tommola at the University of Jyvaeskylae, Robert Woods at the University of Georgia, and Robin Sagar at the Universidad Autonoma Metropolitana Iztapalapa. Vedene always showed his pleasure when he met their Ph.D. students referring to them as his academic grandchildren.

Table 1: Vedene Smith's academic descendants (Ph. D. students).

Name	Ph. D. year
Robert Benesch	1970
Myung Hwan Whangbo	1974
Ajit J. Thakkar	1976
Aniko Foti	1978
Alfredo M. Simas	1982
Sirkka-Liisa Korppi-Tommola	1982
Alan D. Cameron	1985
Maged Khalil	1989
Robert J. Woods	1990
Jiahu Wang	1994
Robin P. Sagar	1994
Hartmut Schmider	1994
Jeno Nagy	1995
Beiyan Jin	1998
Minhhuy Hô	1998
Alexei Frolov	1999
Guilin Duan	2000
Qizhi Cui	2002

He also supervised ten M. Sc. students including Carolyn Elson, W. John Janis, Albert Byrne, William Westgate, Wendy Pell, Sylviana Constas, B. James Clark, and three others (Foti, Sagar, Hô) who are already listed in Table 1 because they went on to complete a Ph. D. under his supervision.

He was a mentor to many visiting researchers and post-doctoral fellows who worked in his laboratory. Their numbers include Sven Larsson, W. David Moseley, Richard Brown, Kenneth Brubaker, Jang-Wang Liu, Per Kaijser, Piotr Petelenz, Peter Price, Janusz Mrozek, Ewa Broclawik, Abbas Farazdel, Johannes Kaspar, Blair McMaster, Awadh Tripathi, Barbara Szpunar, Kenneth Edgecombe, Changjiang Mei, Rodolfo Esquivel, Anthony Brown, Q. Jin, Juan Carlos Angulo, and Garry Smith as well as several of his former Ph. D. students.

He was a Richard Merton Visiting Professor at the Technical University of Munich in 1971–72 and, on several occasions, a visiting professor at the University of Konstanz and a senior visiting scientist at the Max-Planck-Institute for Chemical Physics of Solids in Dresden. He was the Fraser W. Birss Memorial Lecturer at the University of Alberta in 1993. He gave invited lectures at many international conferences.

Vedene Smith's research was very productive. He published 376 papers and reports on a large variety of topics including electron densities and density matrices, X-ray and Compton scattering, liquid state theory, superconductivity, carbohydrate chemistry, and drug design. His wide scope is best appreciated by perusing his list of publications. He sometimes said that asymptotic analysis and Fourier transforms were the threads that held his work together!

The breadth of his work makes it difficult to pinpoint the area for which he was most appreciated. I would guess that he will probably be remembered most for his work on electron densities and density matrices. His influence had a lot to do with Canada becoming a power house of density functional theory. A bibliometric perspective of the impact of his work is provided by Table 2 which reveals that his ten most-cited publications are on a generator coordinate method for dealing with few-body systems [81, 82], the cusp condition for the interelectronic distribution function [79], obtaining atomic charges from electrostatic potentials [206], a survey of electron densities [86], shell structure in atoms [183], evaluation of $\langle S^2 \rangle$ in Hartree-Fock and density functional theory [280], bonding in chalcogen hexafluorides [60], structural changes upon oxidation and reduction of $S_n, n = 3, 4, 6, 8$ [105], and reciprocal form factors [126].

Table 2: Vedene Smith's 10 most-cited publications as of August 1, 2006

Citations	Reference
205	[81]
93	[79]
91	[206]
87	[82]
84	[86]
80	[183]
63	[280]
56	[60]
54	[105]
54	[126]

Vedene Smith served the Canadian chemistry community in many ways. He held the critical conference portfolio on the first Board of Directors of the Canadian Society for Chemistry (CSC) in 1985-88. He later became the Vice-President (1995-96), and then the President (1996–97) of the CSC. He served on the CSC Executive committee from 1995 to 1998. He pushed for the CSC to be more active in a number of areas including the lobbying of governments on science and technology issues. He became Vice-Chair (1999–2000) and Chair (2000–01) of the Chemical Institute of Canada (CIC).

Vedene Smith was a member of NSERC's Chemistry Grant Selection Committee from 1978 to 1981, NSERC's Committee on Research Computing (1981–82), and Chair of the NSERC Committee on Supercomputing (1983–85). He served from 1998 to 2005 on the Awards Committee of the Canadian National Committee for the International Union of Pure and Applied Chemistry (CNC/IUPAC). He was the Chair of the Board of Trustees of the Canadian Institute of Neutron Scattering (1994–96).

Theoretical Chemistry (1974), a founding co-chair (with Philip Coppens) of the inaugural Gordon Conference on Electron Distributions and Chemical Bonding (1978), the organizing chair of Sagamore VI (1979), the organizing chair of the 68th Canadian Chemical Conference and Exhibition (1985), and the vice-chair of the organizing committee of the 6th International Congress of Quantum Molecular Science (1985).

Vedene edited the proceedings of at least four international meetings, and served on the editorial boards of the *Canadian Journal of Chemistry*, the *International Journal of Quantum Chemistry*, and *Condensed Matter and Materials Communications.*

He was honored by a sixtieth birthday symposium in Kingston, a special volume entitled "Novel Insights on the Electronic Structure of Atoms, Molecules and Solids" (No. 527) of the *Journal of Molecular Structure (Theochem)* on the occasion of his 65th birthday, and posthumously by memorial sessions at PacificChem 2005 and the Sanibel symposium in 2006, a special gathering at Queen's in April 2006, and by a memorial symposium at the ICCMSE 2006 held in Crete.

In 1996, Vedene was diagnosed with a rare form of blood cancer, multiple myeloma. He fought it valiantly for nearly a decade and steadfastly refused to allow it to interfere with his life, his research, and his travels. Many of us cannot forget the sight of Vedene showing up at conferences with two walking sticks and a back support. His fortitude was and continues to be an inspiration to all those who knew him. Vedene Smith died in his sleep on September 30, 2005 at the Helen Henderson Care Centre in Kingston. He is survived by his daughter Stefanie and his wife Regina.

2 Personal reminiscences

I first heard about Vedene from a fellow student who described him as a "high-powered guy" who liked to wear sturdy flip-flops and had a Swedish wife. I was first introduced to Vedene in April 1971 by Jim McCowan. On the latter's recommendation, Vedene offered me a summer job as a research assistant. Even though I had only just completed the second year of a B.Sc. program, Vedene allowed me to work on significant research problems with a great deal of independence. Given my temperament and my emotional state at the time, this was exactly what I needed. The following summer he gave me the same job in Munich where he was on sabbatical. He also encouraged me to take a month out of that summer to attend Per-Olov Löwdin's quantum chemistry summer school in Uppsala and Beitostølen. These experiences hooked me to theoretical and computational chemistry. Vedene started me off on a career that I have pursued since then.

I did my graduate work with Vedene and we continued to collaborate on and off long after that. Our first joint paper was published in 1972 and the last one in 1992. I admired Vedene's broad interests and knowledge of the literature, and we would sometimes phone each other to discuss things we had read about. As the years passed on, we often discussed other issues such as the running of chemistry departments, organizing conferences, and managing difficult people.

Vedene loved travel. Long before the proliferation of airline schedules on the web, it was often profitable and informative to consult Vedene about travel routes to any place that you happened to be going to. He attended countless conferences all over the world, and would often arrive with a suitcase full of paper work. I recall sharing many enjoyable dinners of red snapper accompanied by fine wine at various restaurants close to the locations of the Sanibel symposia.

I developed lymphoma in 2003, and Vedene's brave stand against his cancer encouraged me in my own struggle. The last time I saw Vedene was in August 2005 when my wife Baukje and I visited him at the Helen Henderson Care Centre in Kingston. We shared what memories we could. Since it was evident that the end was near for him, we struggled to find the right words when it was time for me to leave. Ultimately, he apologized for not having organized a dinner for me. I replied that I would take him up on it the next time I visited. He looked straight at me, smiled, and said "You bet." Then I left. I will always remember him that way.

3. Vedene Smith's publications

[1] H. A. Gersch and V. H. Smith, Jr. Cluster integrals and the ground state of bosons with repulsive interactions. *Phys. Rev.*, 119:886–891, 1960.

[2] M. L. Allen and V. H. Smith, Jr. An algorithm for the division of power series. *Comm. A. C. M.*, 5:551, 1961.

[3] D. H. Hay and V. H. Smith, Jr. Integral equations and virial coefficients. *Bull. Am. Phys. Soc.*, 7:569, 1962.

[4] V. H. Smith, Jr. and H. A. Gersch. Theory of many-boson systems: Pair approximations and the Jastrow trial pair function. *Progr. Theor. Phys.*, 29:605–606, 1963.

[5] V. H. Smith, Jr. and H. A. Gersch. Cluster expansions and the theory of many-boson systems. *Progr. Theor. Phys.*, 30:421–434, 1963.

[6] W. Kutzelnigg and V. H. Smith, Jr. The independent particle model for many-electron systems: Comparison of different independent particle approximations, 1964. Technical Note 130, Quantum Chemistry Group, Uppsala University.

[7] V. H. Smith, Jr. Construction of exact spin eigenfunctions. *J. Chem. Phys.*, 41:277–278, 1964.

[8] W. Kutzelnigg and V. H. Smith, Jr. On different criteria for the best independent particle model approximation. *J. Chem. Phys.*, 41:896–897, 1964.

[9] W. Kutzelnigg and V. H. Smith, Jr. The independent particle model for many-electron systems: Symmetry properties and criteria for the validity of the independent particle model, 1964. Technical Note 138, Quantum Chemistry Group, Uppsala University.

[10] P. G. Simpson and V. H. Smith, Jr. Crystallite size distributions from X-ray powder line profiles. *J. Applied Phys.*, 36:3285–3287, 1965.

[11] W. Kutzelnigg and V. H. Smith, Jr. Lower bounds for the eigenvalues of first-order density matrices. *J. Chem. Phys.*, 42:2791–2795, 1965.

[12] V. H. Smith, Jr. On the natural analysis of polyelectronic wavefunctions, 1965. Technical Note 163, Quantum Chemistry Group, Uppsala University.

[13] V. H. Smith, Jr. On the criteria for the best Bogoliubov quasi-particle. *Nuovo Cimento*, 48B:443–446, 1967.

[14] V. H. Smith, Jr. Approximate natural orbitals for carbon ^{1}S. *Theor. Chim. Acta*, 7:245–248, 1967.

[15] V. H. Smith, Jr., P. Bochieri, and C. A. Orzalesi. The ground-state three-body correlation function in a dilute boson gas with hard sphere interaction. *Nuovo Cimento*, 52B:18–27, 1967.

[16] V. H. Smith, Jr. Some aspects of the quantum theory of matter: Independent particle models and correlation. *Acta Univ. Uppsaliensis*, 90:1–13, 1967.

[17] W. Kutzelnigg and V. H. Smith, Jr. Approximation by a single Slater determinant. *Arkiv Fysik*, 38:309–315, 1968.

[18] W. Kutzelnigg and V. H. Smith, Jr. Open and closed shell states in few-particle quantum mechanics I. *Int. J. Quantum Chem.*, 2:531–552, 1968.

[19] V. H. Smith, Jr. and W. Kutzelnigg. Open and closed shell states in few-particle quantum mechanics II. *Int. J. Quantum Chem.*, 2:553–562, 1968.

[20] S. Larsson and V. H. Smith, Jr. One- and two-matrix eigenvalues of the lithium atom. *Queen's Papers on Pure and Applied Mathematics*, 11:296 –300, 1968.

[21] F. E. Harris and V. H. Smith, Jr. Projection of exact spin eigenfunctions. *J. Math. Phys.*, 10:771–782, 1969.

[22] S. Larsson and V. H. Smith, Jr. Analysis of ^{2}S ground state of lithium in terms of natural and best overlap (Brueckner) spin orbitals with implications for the Fermi contact term. *Phys. Rev.*, 178:137–152, 1969.

[23] D. J. McNaughton and V. H. Smith, Jr. The relationship of the quasi-independent model to the Hartree-Fock model. *J. Phys. B*, [1] 2:1138–1141, 1969.

[24] R. Benesch and V. H. Smith, Jr. Correlation effects in X-ray and electron scattering. *Int. J. Quantum Chem.*, 3S:413–421, 1970.

[25] D. J. McNaughton and V. H. Smith, Jr. An investigation of the Kohn-Sham and Slater approximations to the Hartree-Fock exchange potential. *Int. J. Quantum Chem.*, 3S:775–788, 1970.

[26] D. J. Stukel, R. N. Euwema, T. C. Collins, and V. H. Smith, Jr. Exchange study of atomic krypton and tetrahedral semiconductors. *Phys. Rev. B*, 1:779–790, 1970.

[27] R. Benesch and V. H. Smith, Jr. Correlation and X-ray scattering. I. Density matrix formulation. *Acta Cryst.*, A26:579–586, 1970.

[28] R. Benesch and V. H. Smith, Jr. Correlation and X-ray scattering. II. Atomic beryllium. *Acta Cryst.*, A26:586–594, 1970.

[29] R. E. Brown, S. Larsson, and V. H. Smith, Jr. Correlation effects on hyperfine structure expectation values for the boron ^{2}P-ground state. *Phys. Rev. A*, 2:593–599, 1970.

[30] R. Benesch and V. H. Smith, Jr. Scattering factors for atomic lithium from correlated and independent particle model wavefunctions. *J. Chem. Phys.*, 53:1466–1470, 1970.

[31] S. R. Singh and V. H. Smith, Jr. Comparison of two perturbation methods. *Int. J. Quantum Chem.*, 4:519–527, 1970.

[32] R. Benesch and V. H. Smith, Jr. Natural orbitals in momentum space and correlated radial momentum distributions. I. The ^{1}S ground state of Li^+. *Int. J. Quantum Chem.*, 4S:131–138, 1970.

[33] J. E. Harriman and V. H. Smith, Jr. A proposal concerning terminology in reduced density matrix theory, 1970. Report WIS-TCII-379. Theoretical Chemistry Institute, University of Wisconsin.

[34] R. Benesch and V. H. Smith, Jr. Radial momentum distributions for the ^{2}S ground state of the lithium atom. *Chem. Phys. Lett.*, 5:601–604, 1970.

[35] V. H. Smith, Jr. Asymptotic behaviour of incoherent scattering functions. *Chem. Phys. Lett.*, 7:226–228, 1970.

[36] R. E. Brown and V. H. Smith, Jr. One-electron reduced density matrices for Li(^{2}P) and B(^{2}P). *Phys. Rev. A*, 3:1858–187, 1971.

[37] R. Benesch and V. H. Smith, Jr. Radial electron-electron distributions and the Coulomb hole for beryllium. *J. Chem. Phys.*, 55:482–488, 1971.

[38] V. H. Smith, Jr. Cusp conditions for natural functions. *Chem. Phys. Lett.*, 9:365–368, 1971.

[39] S. R. Singh and V. H. Smith, Jr. Perturbation treatment of the Hartree-Fock-Slater(Xα) equations for the three-electron ions. *Int. J. Quantum Chem.*, 5S:421–428, 1971.

[40] R. Benesch, S. R. Singh, and V. H. Smith, Jr. On the relationship of the X-ray form factor to the first-order density matrix in momentum space. *Chem. Phys. Lett.*, 10:151–153, 1971.

[41] R. Benesch and V. H. Smith, Jr. Exact and asymptotic intensity values from 20-parameter Hylleraas-type wave-functions, He and He-like ions. *Int. J. Quantum Chem.*, 5S:35–45, 1971.

[42] S. R. Singh and V. H. Smith, Jr. Perturbation treatment of the Hartree-Fock Slater (Xα) equations. *Phys. Rev.*, A 4:1774–1774, 1971.

[43] R. Benesch and V. H. Smith, Jr. Radial momentum distributions $I_0(p)$ and Compton profiles $J(q)$. The ^{1}S Be atom ground state. *Phys. Rev.*, A 5:114–125, 1972.

[44] R. E. Brown and V. H. Smith, Jr. Discrepancy between theory and experiment for the Compton profiles of molecular hydrogen. *Phys. Rev.*, A 5:140–143, 1972.

[45] S. Larsson and V. H. Smith, Jr. Natural spin orbitals and geminals for the lithium ^{2}S ground state. *Int. J. Quantum Chem.*, 6:1019–1043, 1972.

[46] R. Benesch, S. R. Singh, and V. H. Smith, Jr. The Z^{-1} expansion of the nuclear magnetic shielding constant and the X-ray form factor for the three- and four-electron ions. *Can. J. Phys.*, 50:947–952, 1972.

[47] S. R. Singh and V. H. Smith, Jr. Behaviour of Hartree-Fock and Hartree-Fock-Slater ($X\alpha$) atomic form factors and electron momentum distributions for two-, three- and four-electron ions. *Z. Phys. Chem.*, 255:83–92, 1972.

[48] V. H. Smith, Jr. and A. J. Thakkar. Examination of a new intermolecular potential function. *Chem. Phys. Lett.*, 17:274–276, 1972.

[49] R. E. Brown, S. Larsson, and V. H. Smith, Jr. Correlation effects on hyperfine structure expectation values for atoms with an open 2*p*-subshell. *Phys. Rev. A*, 6:1375–1391, 1972. Erratum *ibid.*, **8**, 2765E (1973).

[50] R. E. Brown, A. J. Thakkar, and V. H. Smith, Jr. Natural analysis of the ^{2}S and ^{2}P states of the lithium-like ions. *Phys. Rev. A*, 7:1192–1195, 1973.

[51] R. E. Brown and V. H. Smith, Jr. Electron correlation and the Compton profile of atomic neon. *Chem. Phys. Lett.*, 20:424–428, 1973.

[52] A. J. Thakkar and V. H. Smith, Jr. Fourier transform of the Morse-V_{dd} potential. *J. Low Temp. Phys.*, 13:331–335, 1973.

[53] L. B. Mendelsohn, B. Block, and V. H. Smith, Jr. Incident X-ray energy dependence of neon Compton profiles. *Phys. Rev. Lett.*, 31:266–269, 1973.

[54] J. W. Liu and V. H. Smith, Jr. The accurate elastic differential cross-section for electron scattering by H_2 in the first Born approximation. *J. Phys. B*, 6:L275–279, 1973.

[55] R. Benesch and V. H. Smith, Jr. Density matrix methods in X-ray scattering and momentum space calculations. In W. C. Price, S. S. Chissick, and T. Ravensdale, editors, *Wave Mechanics—The First Fifty Years*, pages 357–377. Butterworths, London, 1973.

[56] A. J. Thakkar and V. H. Smith, Jr. Atomic interactions in the heavy noble gases. *Mol. Phys.*, 27:191–208, 1974.

[57] A. J. Thakkar and V. H. Smith, Jr. Atomic interactions in neon and helium. *Mol. Phys.*, 27:593–600, 1974.

[58] A. J. Thakkar and V. H. Smith, Jr. MIST - A new interatomic potential function. *Chem. Phys. Lett.*, 24:157–161, 1974.

[59] V. H. Smith, Jr. and M. H. Whangbo. Localized molecular orbital studies in momentum space. I. Compton profiles of $1s$ core electrons and hydrocarbons. *Chem. Phys.*, 5:234–243, 1974.

[60] N. Rösch, M. H. Whangbo, and V. H. Smith, Jr. SCF-$X\alpha$ scattered-wave studies on bonding and ionization potentials. I. Hexafluorides of group VI elements. *J. Am. Chem. Soc.*, 96:5984–5989, 1974.

[61] R. E. Caligaris, O. H. Scalise, J. R. Grigera, V. H. Smith, Jr., and A. J. Thakkar. Comments on the 14-12-8-6 potential function. *Can. J. Chem.*, 52:2444–2448, 1974.

[62] A. J. Thakkar and V. H. Smith, Jr. On a representation of the long-range interatomic potential. *J. Phys. B*, 7:L321–325, 1974.

[63] R. E. Brown and V. H. Smith, Jr. Supergeminals and the applicability of the strongly-orthogonal geminal ansatz: Be(^{1}S) and C(^{1}S). *Queen's Papers on Pure and Applied Mathematics*, 40:75–84, 1974.

[64] Y. Öhrn and V. H. Smith, Jr. On a formula for electron binding energies. *Queen's Papers on Pure and Applied Mathematics*, 40:193 200, 1974.

[65] M. H. Whangbo, E. Clementi, G. H. Diercksen, W. von Niessen, and V. H. Smith, Jr. Hydrogen bonding and the Compton profile of water. *J. Phys. B*, 7:L427–430, 1974.

[66] M. H. Whangbo, W. von Niessen, and V. H. Smith, Jr. Localized molecular orbital studies in momentum space. II. Correspondence between coordinate and momentum space properties of various electron pairs. *Chem. Phys.*, 6:282–290, 1974.

[67] J. W. Liu and V. H. Smith, Jr. Elastic and inelastic scattering intensities for electrons from H_2^+ in the first Born approximation. *Chem. Phys. Lett.*, 30:63–68, 1975.

[68] A. J. Thakkar and V. H. Smith, Jr. Molecular interactions in nitrogen and oxygen. *Mol. Phys.*, 29:731–744, 1975.

[69] F. Bernardi, I. Csizmadia, V. H. Smith, Jr., M. H. Whangbo, and S. Wolfe. A relationship between correlation energies and sizes: the helium-like ions. *Theoret. Chim. Acta*, 37:171–176, 1975.

[70] A. J. Thakkar and V. H. Smith, Jr. Analytic approximations to the vibrational eigenvalues of the ST potential. *Mol. Phys.*, 29:1283–1285, 1975.

[71] J. P. Colpa, A. J. Thakkar, P. Randle, and V. H. Smith, Jr. An analysis of energy differences in atomic multiplets in connection with the inequality formulation of Hund's rules: I. Frozen versus relaxed SCF orbitals and energies. *Mol. Phys.*, 29:1861–1875, 1975.

[72] A. J. Thakkar, D. C. Chapman, and V. H. Smith, Jr. A new analytic approximation to atomic incoherent scattering intensities. *Acta Cryst. A*, 31:391–392, 1975.

[73] A. J. Thakkar and V. H. Smith, Jr. A strategy for the numerical evaluation of Fourier sine and cosine transforms to controlled accuracy. *Comp. Phys. Comm.*, 10:73–79, 1975.

[74] A. J. Thakkar and V. H. Smith, Jr. Ring and other contributions to the higher virial coefficients. *Comp. Phys. Comm.*, 10:80–85, 1975. Erratum *ibid.*, **10**, 438E (1975).

[75] G. H. F. Diercksen, W. P. Kraemer, and V. H. Smith, Jr. The influence of electron correlation on the Compton profile of H_2O. *Phys. Lett.*, 54A:319–320, 1975.

[76] P. Kaijser and V. H. Smith, Jr. Directional Compton profiles for molecular systems: N_2. *Mol. Phys.*, 31:1557–1568, 1976.

[77] N. H. Rösch, M. H. Whangbo, and V. H. Smith, Jr. SCF-Xα scattered-wave studies on bonding and ionization potentials. II. Higher valence state sulfur compounds. *Inorg. Chem.*, 15:1768–1775, 1976.

[78] P. Kaijser and V. H. Smith, Jr. On inversion symmetry in momentum space. In J. L. Calais, O. Goscinski, J. Linderberg, and Y. Öhrn, editors, *Methods and Structure in Quantum Science*, pages 417–426. Plenum Press, New York, 1976.

[79] A. J. Thakkar and V. H. Smith, Jr. The electron-electron cusp condition for the spherical average of the intracule matrix. *Chem. Phys. Lett.*, 42:476–481, 1976.

[80] L. Mendelsohn and V. H. Smith, Jr. Atoms. In B. G. Williams, editor, *Compton Scattering: The investigation of electron momentum distributions*, pages 102–138. McGraw-Hill, London, 1977.

[81] A. J. Thakkar and V. H. Smith, Jr. Compact and accurate integral-transform wavefunctions. I. The 1^1S state of the helium-like ions from H^- through Mg^{10+}. *Phys. Rev. A*, 15:1–15, 1977.

[82] A. J. Thakkar and V. H. Smith, Jr. Compact and accurate integral-transform wavefunctions. II. The 2^1S, 2^3S, 2^1P and 2^3P states of the helium-like ions from He through Mg^{10+}. *Phys. Rev. A*, 15:16–22, 1977.

[83] J. W. Liu and V. H. Smith, Jr. A critical study of high energy electron scattering from H_2. *Chem. Phys. Lett.*, 45:59–63, 1977.

[84] P. Kaijser and V. H. Smith, Jr. Evaluation of momentum distributions and Compton profiles for atomic and molecular systems. *Adv. Quantum Chem.*, 10:37–76, 1977.

[85] A. J. Thakkar and V. H. Smith, Jr. Compact and accurate integral-transform wavefunctions III. Radially correlated wavefunctions for the ground state of the lithium atom. *Phys. Rev. A*, 15:2143–2146, 1977.

[86] V. H. Smith, Jr. Theoretical determination and analysis of electronic charge distributions. *Physica Scripta*, 15:147–162, 1977.

[87] A. J. Thakkar and V. H. Smith, Jr. A suggestion concerning modification of the $n(R) - 6$ potential model. *Mol. Phys.*, 34:597–599, 1977.

[88] A. J. Thakkar and V. H. Smith, Jr. Accurate charge densities and two-electron intracule functions for the helium-like ions. *J. Chem. Phys.*, 67:1191–1196, 1977.

[89] V. H. Smith, Jr. Theoretical studies of the momentum distribution of molecular hydrogen. In D. W. Devins, editor, *Momentum Wave Functions - 1976*, pages 145–150. Am. Inst. Phys., New York, 1977.

[90] R. E. Brown and V. H. Smith, Jr. Compton profiles of first row atoms: First order profiles and correlation effects. *Mol. Phys.*, 34:713–729, 1977.

[91] P. F. Price, I. Absar, and V. H. Smith, Jr. Representations of the electron density and its topographical features. *Isr. J. Chem.*, 16:187–197, 1977.

[92] V. H. Smith, Jr., A. J. Thakkar, W. H. Henneker, J. W. Liu, B. Liu, , and R. E. Brown. Accurate Compton profiles for H_2 and D_2 including the effects of electron correlation and molecular vibration and rotation. *J. Chem. Phys.*, 67:3676–3682, 1977. Erratum *ibid.*, **73**, 4150E (1980).

[93] I. Absar and V. H. Smith, Jr. Basic concepts of quantum chemistry for electron density studies. *Isr. J. Chem.*, 16:87–102, 1977.

[94] D. R. Salahub, A. Foti, and V. H. Smith, Jr. Electron deformation density distribution for cyclic octasulfur by the SCF-$X\alpha$-SW method. *J. Am. Chem. Soc.*, 99:8067–8, 1977.

[95] J. W. Liu and V. H. Smith, Jr. Theoretical Compton profiles and momentum distributions for $H_2^+(^2\Sigma_g^+)$and $H_2(^1\Sigma_g^+)$. *Mol. Phys.*, 35:145–154, 1978.

[96] A. Foti, S. Kishner, M. S. Gopinathan, M. A. Whitehead, and V. H. Smith, Jr. Comparison of SCF-Xα-SW and CNDO-BW calculations on $S_4N_4H_4$ and $F_4S_4N_4$. *Mol. Phys.*, 35:111–127, 1978.

[97] J. R. Sabin and V. H. Smith, Jr. The Compton profile as a criterion for the choice of the local exchange parameter α. *J. Phys. B*, 11:385–389, 1978.

[98] W. J. Janis, P. Kaijser, M. H. Whangbo, and V. H. Smith, Jr. Directional Compton profiles, the $J(\vec{0})$ surface, and scattering factors for benzene. *Mol. Phys.*, 35:1237–1245, 1978.

[99] E. Broclawik, A. E. Foti, and V. H. Smith, Jr. SCF-$X\alpha$-SW studies of octahedral clusters in molybdenum oxides. I. Simple octahedra. *J. Catalysis*, 51:380–385, 1978.

[100] P. Petelenz and V. H. Smith, Jr. Effective electron-hole interaction potentials and the binding energies of exciton-ionized donor complexes. *Phys. Rev. B*, 17:3253–3261, 1978.

[101] A. E. Foti, D. R. Salahub, and V. H. Smith, Jr. Explanation for the structural differences of S_4^{2+},S_4^0 and S_4^{2-}. *Chem. Phys. Lett.*, 57:33–36, 1978.

[102] A. J. Thakkar and V. H. Smith, Jr. N-electron zero-momentum energy expression: A criterion for assessing the accuracy of approximate wavefunctions. *Phys. Rev. A*, 18:841–844, 1978.

[103] A. J. Thakkar and V. H. Smith, Jr. Form factors and total scattering intensities for the helium-like ions from explicitly correlated wavefunctions. *J. Phys. B*, 11:3803–3820, 1978.

[104] W. J. Janis, P. Kaijser, J. R. Sabin, and V. H. Smith, Jr. The influence of polarization functions on the directional Compton profiles of water. *Mol. Phys.*, 37:463–472, 1978.

[105] D. R. Salahub, A. E. Foti, and V. H. Smith, Jr. Molecular orbital study of structural changes on oxidation and reduction of S_3, S_4, S_6 and S_8. *J. Am. Chem. Soc.*, 100:7847–7859, 1978.

[106] V. H. Smith, Jr. Electron densities and reduced density matrices. In P. Becker, editor, ***Electronic and Magnetic Distributions in Molecules and Crystals***, pages 3–25. Plenum Press, New York, 1979.

[107] V. H. Smith, Jr. On the calculation and accuracy of theoretical electron densities. In P. Becker, editor, ***Electronic and Magnetic Distributions in Molecules and Crystals***, pages 27–46. Plenum Press, New York, 1979.

[108] P. Petelenz and V. H. Smith, Jr. Binding energy of the Wannier exciton-ionized donor complex in the CdS crystal. *Can. J. Phys.*, 57:2126–2131, 1979.

[109] P. Petelenz and V. H. Smith, Jr. Binding energies of the lowest excited states of Wannier exciton and Wannier exciton-ionized donor complex in CdS. *J. Phys. C*, 13:47–56, 1980.

[110] E. Broclawik, A. E. Foti, and V. H. Smith, Jr. On the relationship between binding energy and oxidation state in molybdenum oxides. *J. Catalysis*, 62:185–186, 1980.

[111] P. Petelenz and V. H. Smith, Jr. Wannier exciton-ionized donor complexes in thallium halides. *J. Low Temp. Phys.*, 38:413–420, 1980.

[112] P. Petelenz and V. H. Smith, Jr. Binding energies of ionized donor-exciton complexes in polar crystals. *Int. J. Quantum Chem.*, 18:583–586, 1980.

[113] P. Petelenz and V. H. Smith, Jr. Binding energies of D^- ions in CdS. *Phys. Rev. B*, 21:4884–4886, 1980.

[114] A. E. Foti, V. H. Smith, Jr., and S. Fliszar. Charge distributions and chemical effects. XV. SCF-$X\alpha$-SW studies of alkanes. *J. Mol. Struct.*, 68:227–234, 1980.

[115] J. Mrozek, V. H. Smith, Jr., D. R. Salahub, P. Ros, and A. Rozendaal. Electron density in Si_2 and Cl_2. *Mol. Phys.*, 41:509–517, 1980.

[116] E. Broclawik and V. H. Smith, Jr. Electronic structure of selenium chains by the SCF-SW-$X\alpha$ method. *Int. J. Quantum Chem.*, 14S:395–403, 1980.

[117] P. Petelenz and V. H. Smith, Jr. An integral transform method for solving a quantum mechanical three-body problem with exponential-type potentials. *Int. J. Quantum Chem.*, 14S:83–92, 1980.

[118] A. J. Thakkar, A. Simas, and V. H. Smith, Jr. Extraction of momentum expectation values from Compton profiles. *Mol. Phys.*, 41:1153–1162, 1980.

[119] L. Scheire, P. Phariseau, R. Nuyts, A. E. Foti, and V. H. Smith, Jr. On the electronic structure of the xenon fluorides. *Physica*, 101A:22–48, 1980.

[120] P. Kaijser, V. H. Smith, Jr., and A. J. Thakkar. Isotropic and directional Compton profiles for N_2, CO and BF. *Mol. Phys.*, 41:1143–1151, 1980.

[121] A. J. Thakkar and V. H. Smith, Jr. Statistical electron correlation coefficients for the five lowest states of the helium-like ions. *Phys. Rev. A*, 23:473–478, 1981.

[122] E. Broclawik, A. E. Foti, and V. H. Smith, Jr. SCF-SW-Xα studies of octahedral clusters in molybdenum oxides. II. Double octahedra. *J. Catalysis*, 67:103–109, 1981.

[123] P. Petelenz and V. H. Smith, Jr. Mass-ratio dependence and critical binding of exciton-ionized donor complexes in polar crystals. *Phys. Rev. B*, 23:3066–3070, 1981.

[124] J. N. Migdall, M. A. Coplan, D. S. Hench, J. H. Moore, J. A. Tossell, V. H. Smith, Jr., and J. W. Liu. The electron momentum distribution of molecular hydrogen. *Chem. Phys.*, 57:141–146, 1981.

[125] P. Petelenz and V. H. Smith, Jr. Triplet exciton mobilities in weak donor-acceptor complexes. *Chem. Phys. Lett.*, 82:430-433, 1981.

[126] A. J. Thakkar, A. M. Simas, and V. H. Smith, Jr. Internally folded densities. *Chem. Phys.*, 63:175–183, 1981.

[127] A. Simas, A. J. Thakkar, and V. H. Smith, Jr. Momentum space properties of various basis sets used in quantum chemical calculations. *Int. J. Quantum Chem.*, 21:419–429, 1982.

[128] A. E. Foti, V. H. Smith, Jr., and M. A. Whitehead. Bonding in titanium chloride, $TiCl_4$. *Mol. Phys.*, 45:385–396, 1982.

[129] J. Mrozek, P. Petelenz, and V. H. Smith, Jr. Vibronic coupling in excited charge transfer states of solid state charge transfer complexes. *Chem. Phys. Lett.*, 85:245–248, 1982.

[130] E. Broclawik, J. Mrozek, and V. H. Smith, Jr. A quantum chemical investigation of the ammonium radical. *Chem. Phys.*, 66:417–423, 1982. Erratum *ibid.*, **83**, 490E (1984).

[131] B. N. McMaster, V. H. Smith, Jr., and D. R. Salahub. SCF-$X\alpha$-SW electron densities with the overlapping sphere approximation. *Mol. Phys.*, 46:449–463, 1982.

[132] P. Petelenz and V. H. Smith, Jr. Stability of excited states of Wannier exciton-ionized donor complexes. *J. Phys. C*, 15:3721–3724, 1982.

[133] V. H. Smith, Jr. Concepts in charge density analysis: The theoretical approach. In P. Coppens and M. B. Hall, editors, *Electron Distributions and the Chemical Bond*, pages 3–59. Plenum Press, New York, 1982.

[134] B. N. McMaster, J. Mrozek, and V. H. Smith, Jr. An ab initio molecular orbital study of the ammonium radical. *Chem. Phys.*, 73:131–143, 1982.

[135] A. Cameron, V. H. Smith, Jr., and M. C. Baird. Oxidatively-induced migration of hydrogen from metal to carbon monoxide. *Organometallics*, 2:465–467, 1983.

[136] A. Harmalkar, A. M. Simas, V. H. Smith, Jr., and W. M. Westgate. Momentum space properties of atoms. *Int. J. Quantum Chem.*, 23:811–819, 1983.

[137] A. M. Simas, A. J. Thakkar, and V. H. Smith, Jr. Basis set quality. II. Information theoretic appraisal of various s-orbitals. *Int. J. Quantum Chem.*, 24:527–550, 1983.

[138] J. Mrozek, P. Petelenz, and V. H. Smith, Jr. Widths of excited states of Wannier-exciton-ionized donor complex in CdS. *J. Phys. C.*, 16:3131–3135, 1983.

[139] V. H. Smith, Jr., J. R. Sabin, E. Broclawik, and J. Mrozek. The electronic structure of the trisulfur trinitride anion. *Inorg. Chim. Acta*, 77:L101–104, 1983.

[140] V. H. Smith, Jr. Density functional theory and local potential approximations from momentum space considerations. In J. Avery and J.-P. Dahl, editors, *Local Density Approximations in Quantum Chemistry and Solid State Theory*, pages 1–19. Plenum Press, New York, 1984.

[141] A. M. Simas, V. H. Smith, Jr., and P. Kaijser. The nodal structure of the momentum distributions of molecules. *Int. J. Quantum Chem.*, 25:1035–1044, 1984.

[142] A. J. Thakkar, A. N. Tripathi, and V. H. Smith, Jr. Anisotropic electronic intracule densities for diatomics. *Int. J. Quantum Chem.*, 26:159–166, 1984.

[143] W. A. Szarek, S.-L. Korppi-Tommola, H. F. Shurvell, V. H. Smith, Jr., and O. R. Martin. A Raman and infrared study of crystalline D-fructose, L-sorbose and related carbohydrates. Hydrogen bonding and sweetness. *Can. J. Chem.*, 62:1512–1518, 1984.

[144] A. Simas, W. Westgate, and V. H. Smith, Jr. The shell structure of atoms and ions in momentum space. *J. Chem. Phys.*, 80:2636–2642, 1984.

[145] W. A. Szarek, S.-L. Korppi-Tommola, O. R. Martin, and V. H. Smith, Jr. Examination of the molecular properties of D-fructose and L-sorbose by ab initio LCAO-MO calculations. *Can. J. Chem.*, 62:1506–1511, 1984.

[146] A. J. Thakkar, A. N. Tripathi, and V. H. Smith, Jr. Molecular X-ray and electron scattering intensities. *Phys. Rev. A*, 29:1108–1113, 1984.

[147] A. Simas, V. H. Smith, Jr., and A. J. Thakkar. Partial wave analysis of the momentum densities of 14-electron diatomics. *Int. J. Quantum Chem.*, S18:385–392, 1984.

[148] A. Simas, J. Mrozek, and V. H. Smith, Jr. Appraisal of computer hardware and software for quantum chemical calculations. *Int. J. Quantum Chem.*, S18:619–638, 1984.

[149] A. J. Thakkar, A. Simas, and V. H. Smith, Jr. Partial wave analysis of the momentum density. *J. Chem. Phys.*, 81:2953–2961, 1984.

[150] A. M. Simas and V. H. Smith, Jr. Relativistic integrals over Breit-Pauli operators. *J. Chem. Phys.*, 81:5219–5221, 1984.

[151] J. Kaspar and V. H. Smith, Jr. Rydberg transitions in the ammonium radical: Scattered wave-local spin density calculations. *Chem. Phys.*, 90:47–54, 1984.

[152] A. M. Simas, J. Mrozek, and V. H. Smith, Jr. Appraisal of computer hardware and software for scientific calculations. *Das Rechenzentrum*, 7:230–252, 1984.

[153] A. M. Simas, V. H. Smith, Jr., and A. J. Thakkar. Substituent effects in alkynes and cyanimides: A momentum density perspective. *Can. J. Chem.*, 63:1412–1417, 1985.

[154] J. W. Liu and V. H. Smith, Jr. Effect of ground state electron correlation on the (e,2e) reaction spectroscopy of H_2 ($^1\Sigma_g^+$). *Phys. Rev. A*, 31:3003–3011, 1985. Erratum *ibid.*, **39**, 13703–13704E (1989).

[155] J. Kaspar, V. H. Smith, Jr., and B. N. McMaster. The stability and spectra of isotopic forms of the ammonium radical, an ab initio CI study of the dissociation barrier. *Chem. Phys.*, 96:81–96, 1985.

[156] A. Farazdel, W. M. Westgate, A. M. Simas, R. P. Sagar, and V. H. Smith, Jr. Validity of the mass-velocity term in the Breit-Pauli Hamiltonian. *Int. J. Quantum Chem.*, S19:61–68, 1985.

[157] A. M. Simas, V. H. Smith, Jr., and A. J. Thakkar. A momentum density analysis of strong hydrogen bonding. *J. Mol. Struct. (Theochem)*, 123:221–229, 1985.

[158] A. Tripathi and V. H. Smith, Jr. Scattering of X-rays and high-energy electrons from molecules: Comparison of ab initio calculations with experiment. In R. J. Bartlett, editor, *Comparison of Ab Initio Quantum Chemistry with Experiment: State-of-the-Art*, pages 439–462. Reidel, Dordrecht, 1985.

[159] W. M. Westgate, A. M. Simas, and V. H. Smith, Jr. The non-monotonic behaviour of and the presence of slow and fast maxima in the momentum densities of atoms and ions. *J. Chem. Phys.*, 83:4054–4058, 1985.

[160] A. Farazdel and V. H. Smith, Jr. Invalidity of the ubiquitous mass-velocity operator in quasi-relativistic theories. *Int. J. Quantum Chem.*, 29:311–314, 1986.

[161] A. D. Smith, M. A. Coplan, D. J. Chornay, J. H. Moore, J. A. Tossell, J. Mrozek, and V. H. Smith, Jr. Distortion effects in the (e,2e) spectroscopy of helium at high momentum. *J. Phys. B*, 19:969–980, 1986.

[162] M. Kumar, A. N. Tripathi, and V. H. Smith, Jr. Scattering of high energy electrons and X-rays from molecules: The 10-electron series Ne, HF, H_2O, NH_3 and CH_4. *Int. J. Quantum Chem.*, 29:1339–1349, 1986.

[163] J. Kaspar, V. H. Smith, Jr., and B. N. McMaster. Theoretical investigations of the ammonium radicals NH_4, ND_4 and NT_4. In V. H. Smith, Jr., H. F. Schaefer III, and K. Morokuma, editors, *Applied Quantum Chemistry*, pages 403–420. Reidel, Dordrecht, 1986.

[164] D. D. Robertson, V. H. Smith, Jr., A. M. Simas, and A. J. Thakkar. The quality of s-orbitals determined by least squares fitting and constrained variational methods. *Int. J. Quantum Chem.*, 30:717–735, 1986.

[165] A. D. Cameron, V. H. Smith, Jr., and M. C. Baird. On the mechanism of activation of coordinated olefins towards nucleophilic attack. *Int. J. Quantum Chem.*, 20S:657–663, 1986.

[166] W. M. Westgate, A. D. Byrne, V. H. Smith, Jr., and A. M. Simas. Electron-nuclear cusp check, for self-consistent field wavefunctions for the neutral atoms from helium to uranium. *Can. J. Phys.*, 64:1351–2, 1986. Erratum *ibid.*, **67**, 543E (1989).

[167] P. Petelenz and V. H. Smith, Jr. Vibronic coupling in charge transfer complexes II. *Chem. Phys.*, 111:37–46, 1987.

[168] A. M. Tanner, A. J. Thakkar, and V. H. Smith, Jr. Interrelationships between various representations of one-matrices and related densities: A road map and an example. In R. M. Erdahl and V. H. Smith, Jr., editors, *Density Functions and Density Matrices*, pages 327–337. Reidel, Dordrecht, 1987.

[169] D. G. Kanhere, A. Farazdel, and V. H. Smith, Jr. Positron annihilation from F-centres of alkali halide crystals. *Phys. Rev. B*, 35:3131–3137, 1987.

[170] B. Petelenz, P. Petelenz, H. F. Shurvell, and V. H. Smith, Jr. Reconsideration of the electroabsorption spectrum of the anthracene crystal. *Chem. Phys. Lett.*, 133:157–161, 1987.

[171] P. Petelenz, V. H. Smith, Jr., J. Klein, P. Partin, and R. Volte. Recombination of charge carriers in anthracene. *Chem. Phys.*, 112:457–462, 1987.

[172] P. Kaijser, A. N. Tripathi, V. H. Smith, Jr., and G. H. F. Dierksen. Ab initio calculations of one-electron scattering properties of methane. *Phys. Rev. A*, 35:4074–4084, 1987.

[173] R. J. Woods, V. H. Smith, Jr., W. A. Szarek, and A. Farazdel. Ab initio LCAO-MO calculations on α-D-glucopyranose and β-D-fructopyranose, and their thiopyranoid-ring analogues. application to a theory of sweetness. *Chem. Comm.*, pages 937–939, 1987.

[174] P. Petelenz and V. H. Smith, Jr. Binding energies of the muonium and positronium negative ions. *Phys. Rev. A*, 36:5125–5126, 1987.

[175] A. D. Cameron, D. E. Laycock, V. H. Smith, Jr., and M. C. Baird. Hydride addition reactions of the olefin complexes [Fe(η^5-C_5H_5) $(CO)_2$(olefin)-BF_4]: Formyl formation as the kinetically preferred process; regio- and stereo- selectivity during addition to the co-ordinated olefins. *J. Chem. Soc. Dalton Trans.*, pages 2857–2861, 1987.

[176] P. Petelenz and V. H. Smith, Jr. Binding energies of muonic molecules. *Phys. Rev. A*, 36:4078–4080, 1987.

[177] D. D. Robertson, V. H. Smith, Jr., and A. J. Thakkar. More on basis set quality tests. *Int. J. Quantum Chem.*, 32:427–434, 1987.

[178] A. K. Jain, A. N. Tripathi, V. H. Smith, Jr., and A. J. Thakkar. Scattering of fast electrons and X-rays from CO_2 molecules. *Int. J. Quantum Chem. Symp.*, 21:217–227, 1987.

[179] P. Petelenz and V. H. Smith, Jr. Exact matrix elements of the Uehling potential in a basis of explicitly correlated two-particle functions. *Phys. Rev. A*, 35:4055–4059, 1987. Erratum *ibid.*, **36**, 4529E (1988).

[180] B. Petelenz, P. Petelenz, H. F. Shurvell, and V. H. Smith, Jr. Reconsideration of the electroabsorption spectrum of the tetracene and pentacene crystals. *Chem. Phys.*, 119:25–39, 1988.

[181] A. D. Cameron, V. H. Smith, Jr., and M. C. Baird. A reconsideration of the role of slippage in the activation of coordinated olefins towards nucleophilic attack. *J. Chem. Soc. Dalton Trans.*, pages 1037–1073, 1988.

[182] B. Szpunar and V. H. Smith, Jr. Electronic properties and magnetism in high t_c superconductors. *Phys. Rev. B*, 37:2338–2340, 1988.

[183] R. P. Sagar, A. C. T. Ku, V. H. Smith, Jr., and A. M. Simas. The Laplacian of the charge density and its relationship to the shell structure of atoms and ions. *J. Chem. Phys.*, 88:4367–4374, 1988.

[184] B. Szpunar, V. H. Smith, Jr., and R. W. Smith. Electronic structure of antiferromagnetic $YBa_2Cu_3O_6$. *Physica C*, 152:91–98, 1988.

[185] R. P. Sagar, A. C. T. Ku, and V. H. Smith, Jr. An examination of the shell structure of atoms and ions as revealed by the one-electron potential, $\nabla^2\rho(r)/\rho(r)$. *Can. J. Chem.*, 66:1005–1012, 1988.

[186] B. Szpunar and V. H. Smith, Jr. Linear muffin-tin-orbital atomic sphere calculations of the electronic structure of $YBa_2Cu_3O_7$ and $YBa_2Cu_3O_6$. *Phys. Rev. B*, 37:7525–7528, 1988.

[187] A. M. Simas, R. P. Sagar, A. C. T. Ku, and V. H. Smith, Jr. The radial charge distribution and the shell structure of atoms and ions. *Can. J. Chem.*, 66:1923–1930, 1988.

[188] A. Farazdel, A. M. Simas, and V. H. Smith, Jr. Accurate numerical evaluation of infinite integrals of oscillatory functions: Spherical-Bessel transforms. *Acta Phys. Polonica A*, 74:179–191, 1988.

[189] P. Petelenz and V. H. Smith, Jr. Excited charge transfer states of the anthracene crystal. *Acta Phys. Polonica*, A74:307–316, 1988.

[190] B. Szpunar and V. H. Smith, Jr. The use of supercomputers for the study of magnetism in high T_c superconductors, 1988.

[191] B. Szpunar and V. H. Smith, Jr. Itinerant magnetism in high T_c superconductors: YBa_2Cu_3 oxides and related structures. *Int. J. Quantum Chem.*, S22:23–42, 1988.

[192] B. Szpunar and V. H. Smith, Jr. The influence of inequivalent cobalt sites on the spin and orbital magnetic moments in YCo_5. *Portugaliae Physica*, 19:225–228, 1988.

[193] B. Szpunar and V. H. Smith, Jr. Electronic and magnetic structure of superconducting copper oxides. *Portugaliae Physica*, 19:1229–232, 1988.

[194] A. N. Tripathi, V. H. Smith, Jr., and A. J. Thakkar. Scattering of fast electrons and X-rays from moelcules: CH_4 and C_2H_2. *Int. J. Quantum Chem.*, 35:869–885, 1989.

[195] V. H. Smith, Jr. and P. Petelenz. Integral transform wavefunctions in the solution of the quantum mechanical three body problem. In S. E. Jones, J. Rafelski, and H. J. Monkhorst, editors, *AIP Conference Proceedings*, volume 181, pages 295–302, New York, 1989. American Institute of Physics.

[196] P. Petelenz and V. H. Smith, Jr. Vacuum-polarization correction in muonic molecules. *Phys. Rev. A*, 39:1016–1019, 1989.

[197] R. P. Sagar, A. C. T. Ku, V. H. Smith, Jr., and A. M. Simas. Topographical features of the Laplacian of the momentum density of atoms and ions most indicative of shell structure. *J. Chem. Phys.*, 90:6520–6527, 1989.

[198] J. Mrozek and V. H. Smith, Jr. Ab initio computations of the structure and properties of moniliformin, a mycotoxin and related compounds. *Acta Phys. Polonica*, 76:579–589, 1989.

[199] B. Szpunar, V. H. Smith, Jr., and R. W. Smith. Magnetism and electronic structure in the (123) and (2212) copper oxides. *J. Mol. Struct. (Theochem)*, 202:347–361, 1989.

[200] B. Szpunar and V. H. Smith, Jr. Antiferromagnetism in copper oxide planes. *J. Mol. Struct. (Theochem)*, 199:313–326, 1989.

[201] P. Petelenz and V. H. Smith, Jr. Theoretical interpretation of the electroabsorption spectrum of the anthracene crystal. *Chem. Phys.*, 131:409–421, 1989.

[202] V. H. Smith, Jr. On a functional relationship between Compton profiles and stopping powers. *Int. J. Quantum Chem.*, S23:553–553, 1989.

[203] B. Szpunar, V. H. Smith, Jr., and J. Spalek. Electronic structure of antiferromagnetic La_2NiO_4 and $La_{1.5}Sr_{0.5}NiO_4$ systems. *Physica C*, 161:503–511, 1989.

[204] A. K. Jain, A. N. Tripathi, and V. H. Smith, Jr. Momentum densities, Compton profiles, and B functions for the alkaline earth oxides. *Can. J. Chem.*, 67:1886–1891, 1989.

[205] B. Szpunar and V. H. Smith, Jr. Antiferromagnetism and semiconductivity. *Physica B*, 163:29–343, 1990.

[206] R. J. Woods, M. Khalil, W. Pell, S. Moffat, and V. H. Smith, Jr. Derivation of net atomic charges from molecular electrostatic potentials. *J. Comp. Chem.*, 11:297–310, 1990.

[207] A. N. Tripathi, V. H. Smith, Jr., P. Kaijser, and G. H. F. Diercksen. Ab initio calculations of one-electron-scattering properties of ethyne (acetylene) and ethylene molecules. *Phys. Rev. A*, 41:2468–2481, 1990.

[208] R. J. Woods, W. A. Szarek, and V. H. Smith, Jr. An investigation of the relationship between sweetness and intramolecular hydrogen-bonding networks in hexuloses using the semi-empirical molecular orbital method, AM1. *J. Am. Chem. Soc.*, 112:4732–4741, 1990.

[209] V. H. Smith, Jr., D. D. Robertson, and A. N. Tripathi. Monotonicity of the electron momentum density for atomic closed shells in a bare Coulomb field. *Phys. Rev. A*, 42:61–62, 1990.

[210] A. K. Jain, A. N. Tripathi, and V. H. Smith, Jr. Elastic, inelastic and total cross sections for X-ray and electron scattering from molecular silane. *J. Phys. B*, 23:2869–2878, 1990.

[211] B. Szpunar and V. H. Smith, Jr. Calculation of orbital moments in solids. *J. Solid State Chem.*, 88:217–228, 1990.

[212] H. Schmider, V. H. Smith Jr., and W. Weyrich. Determination of electron densities and one-matrices from experimental information. *Trans. Am. Cryst. Assoc.*, 26:125–145, 1990.

[213] D. F. Weaver, M. Khalil, and V. H. Smith, Jr. An AM1 semi-empirical quantum mechanical study of the polyhydration of amino acid side chains. *J. Mol. Struct. (Theochem)*, 226:73–86, 1991.

[214] R. P. Sagar, V. H. Smith, Jr., and A. M. Simas. A Gaussian quadrature for the optimal evaluation of integrals involving Lorentzians over a semi-infinite interval. *Comp. Phys. Comm.*, 62:16–24, 1991.

[215] B. Szpunar and V. H. Smith, Jr. Electronic and magnetic structure of high-T_c cuprates and nickelates. *Studies in High Temperature Superconductors*, 7:135–161, 1991.

[216] M. Khalil, R. J. Woods, D. F. Weaver, and V. H. Smith, Jr. An examination of intermolecular and intramolecular hydrogen bonding in biomolecules by AM1 and MNDO/M semi-empirical molecular orbital studies. *J. Comp. Chem.*, 12:584–593, 1991.

[217] R. J. Woods, W. A. Szarek, and V. H. Smith, Jr. A comparison of semi-empirical and ab initio methods for the study of structural features of relevance in carbohydrate chemistry. *J. Chem. Soc. Chem. Comm.*, 1991:334–337, 1991.

[218] H. Schmider, R. P. Sagar, and V. H. Smith, Jr. Does the atomic structure factor show atomic structure? *J. Chem. Phys.*, 94:4346–4351, 1991.

[219] W. M. Westgate, R. P. Sagar, A. Farazdel, V. H. Smith, Jr., A. M. Simas, and A. J. Thakkar. Momentum-space properties of the neutral atoms from H through U. *Atomic Data and Nuclear Data Tables*, 48:213–229, 1991.

[220] R. J. Woods, W. A. Szarek, and V. H. Smith, Jr. The protonation affinities and deprotonation enthalpies of β-D-fructopyranose and α-L-sorbopyranose. *Can. J. Chem.*, 69:1917–1928, 1991.

[221] H. Schmider, R. P. Sagar, and V. H. Smith, Jr. On the determination of atomic shell boundaries. *J. Chem. Phys.*, 94:8627–8629, 1991.

[222] B. Szpunar and V. H. Smith, Jr. Magnetism and electronic structure in $YBa_2Cu_3O_6$ and $YBa_2Cu_3O_7$. *Phys. Rev. B*, 44:7034–7041, 1991.

[223] B. Szpunar, V. H. Smith, Jr., and N. Hamada. Self-energy corrections for NiO. In G. C. Vezzoli and J. Ashkenazi, editors, *Electronic Structure and Mechanisms for High Temperature Superconductivity*, pages 591–596. Plenum Publishing Company, New York, 1991.

[224] R. P. Sagar, H. Schmider, and V. H. Smith, Jr. Evaluation of Fourier transforms by Gauss-Laguerre quadratures. *J. Phys. A*, 25:189–195, 1992.

[225] H. Schmider, R. P. Sagar, and V. H. Smith, Jr. An investigation of atomic structure in position and momentum space by means of ideal shells. *Can. J. Chem.*, 70:506–512, 1992.

[226] R. P. Sagar and V. H. Smith, Jr. On the calculation of Rys polynomials and quadratures. *Int. J. Quantum Chem.*, 42:827–836, 1992.

[227] A. N. Tripathi, R. P. Sagar, R. O. Esquivel, and V. H. Smith, Jr. Electron correlation in momentum space: The beryllium atom isoelectronic sequence. *Phys. Rev. A,*, 45:4385–4392, 1992.

[228] R. P. Sagar, H. Schmider, and V. H. Smith, Jr. Asymptotic approximations from quadrature rules. *Phys. Rev. A*, 45:6253–6258, 1992.

[229] B. Szpunar and V. H. Smith, Jr. Electronic structure of $Bi_4Sr_4Ca_{2-x}Y_xCu_4O_{16}$ $(x=0,1)$. *Phys. Rev. B*, 45:10616–10621, 1992.

[230] A. J. Thakkar and V. H. Smith, Jr. High-accuracy ab initio form factors for the hydride anion and isoelectronic species. *Acta Cryst.*, A48:70–71, 1992.

[231] H. F. M. da Costa, A. M. Simas, V. H. Smith, Jr., and M. Trsic. The generator coordinate Hartree-Fock method for molecular systems. Near Hartree-Fock limit calculations for N_2, CO and BF. *Chem. Phys. Lett.*, 192:195–198, 1992.

[232] R. O. Esquivel, A. N. Tripathi, R. P. Sagar, and V. H. Smith, Jr. Accurate one-electron momentum-space properties for the lithium isoelectronic sequence. *J. Phys. B*, 25:2925–2941, 1992.

[233] D. F. Weaver, K. E. Edgecombe, V. H. Smith, Jr., and M. N. Anderson. Applications of large-scale computational techniques to the quantum pharmacologic design of drugs for the treatment of epilepsy. In K. R. Billingsley, H. U. Brown III, and E. Dershanes, editors, *Scientific Excellence in Supercomputing: The 1990 IBM Contest Prize Papers*, pages 533–593. The Baldwin Press, Athens, 1992.

[234] H. Schmider, K. E. Edgecombe, V. H. Smith, Jr., and W. Weyrich. One-particle density matrices along the molecular bonds in linear molecules. *J. Chem. Phys.*, 96:8411–8419, 1992.

[235] H. Schmider, V. H. Smith, Jr., and W. Weyrich. Reconstruction of the one-particle density matrix from expectation values in position and momentum space. *J. Chem. Phys.*, 96:8986–8994, 1992.

[236] K. E. Edgecombe, R. O. Esquivel, V. H. Smith, Jr., and F. Müller-Plathe. Pseudoatoms of the electron density. *J. Chem. Phys.*, 97:2593–2599, 1992.

[237] C.-J. Mei and V. H. Smith, Jr. Electronic structure of γ carbon. *Phys. Rev. B.*, 46:7179–7181, 1992.

[238] J. Wang, A. N. Tripathi, and V. H. Smith, Jr. The influence of electron correlation on anisotropic electron intracule and extracule densities. *J. Chem. Phys.*, 97:9188–9194, 1992.

[239] J. Wang, R. P. Sagar, H. Schmider, and V. H. Smith, Jr. X-ray elastic and inelastic scattering factors for neutral atoms Z=2–92. *Atomic Data and Nuclear Data Tables,*, 53:233–269, 1993.

[240] H. Schmider, V. H. Smith, Jr., and W. Weyrich. On the inference of one-particle density matrices from form factors in position and momentum space. *Z. Naturforsch.*, 48a:211–220, 1993.

[241] K. E. Edgecombe, V. H. Smith, Jr., and F. Müller-Plathe. Non-nuclear maxima in the charge density. *Z. Naturforsch.*, 48a:127–133, 1993.

[242] H. Schmider, V. H. Smith, Jr., and W. Weyrich. Atomic orbitals from Compton profiles. *Z. Naturforsch.*, 48a:221–226, 1993.

[243] J. Wang and V. H. Smith, Jr. Ab initio study of the spin density of nitroxide radicals. *Z. Naturforsch.*, 48a:109–116, 1993.

[244] H. Schmider and V. H. Smith, Jr. A spherically averaged representation of the atomic one-particle reduced density matrix. *Theor. Chim. Acta*, 86:115–127, 1993.

[245] J. Wang, A. N. Tripathi, and V. H. Smith, Jr. Electron correlation shifts on pair densities of first-row hydrides. *J. Phys. B*, 26:205–219, 1993.

[246] R. O. Esquivel, J. Chen, M. J. Stott, R. P. Sagar, and V. H. Smith, Jr. Pseudoconvexity of the atomic electron density: lower and upper bounds. *Phys. Rev. A*, 47:936–943, 1993.

[247] R. P. Sagar, R. O. Esquivel, H. Schmider, A. N. Tripathi, and V. H. Smith, Jr. Non-monotonicity of the atomic-electron momentum density. *Phys. Rev. A*, 47:2625–2627, 1993.

[248] R. O. Esquivel, R. P. Sagar, V. H. Smith, Jr., J. Chen, and M. J. Stott. Pseudoconvexity of the atomic electron density: A numerical study. *Phys. Rev. A*, 47:4735–4748, 1993.

[249] C.-J. Mei and V. H. Smith, Jr. Electronic structure of $SrCuO_2$ doped with Ca. *Physica C*, 209:389–392, 1993.

[250] J. Wang and V. H. Smith, Jr. Ab initio studies of the electric field gradients in SCl_2. *Z. Naturforsch.*, 49a:699–702, 1993.

[251] C. Mei and V. H. Smith, Jr. Towards an understanding of the electronic structure of Mott-insulating transition metal oxides. *Int. J. Quantum Chem., Quantum Chem. Symp.*, 27:187–194, 1993.

[252] A. S. Brown and V. H. Smith, Jr. Ab initio electron density distributions in molecules containing sulfur-sulfur bonds. *J. Chem. Phys.*, 99:1837–1843, 1993.

[253] C.-J. Mei and V. H. Smith, Jr. The ab initio electronic structure and stability of the Bucky tube. *Physica C*, 213:157–160, 1993.

[254] H. Schmider, R. O. Esquivel, R. P. Sagar, and V. H. Smith, Jr. Spin magnetic form factors for lithium and its isoelectronic series in position and momentum space. *J. Phys. B.*, 26:2943–2955, 1993.

[255] J. Wang and V. H. Smith, Jr. One-electron and electron pair densities of first-row hydrides in momentum space. *J. Chem. Phys.*, 99:9745–9755, 1993.

[256] A. M. Frolov, V. H. Smith, Jr., and J. Komasa. On the algebraic solution of the non-relativistic three-body problem: Bound states. *J. Phys. A.*, 26:6507–6515, 1993.

[257] C. Mei, K. E. Edgecombe, V. H. Smith, Jr., and A. Heilingbrunner. Topological analysis of the charge density of solids: bcc sodium and lithium. *Int. J. Quantum Chem.*, 48:287-293, 1993.

[258] A. Michalak, J. Mrozek, V. H. Smith, Jr., and D. F. Weaver. A molecular orbital study of two vasoconstricting mycotoxins: Butenolide and ergotamine. *J. Mol. Struct. (Theochem)*, 305:1-7, 1994.

[259] J. Wang and V. H. Smith, Jr. Statistical electron correlation-coefficients and -holes in molecules. *Theor. Chim. Acta,*, 88:35–46, 1994.

[260] J. Nagy, V. H. Smith, Jr., and D. F. Weaver. Calculation of second virial coefficients of alkanes with the MM2 and MM3 force fields. *Mol. Phys.*, 81:1039–1064, 1994.

[261] J. C. Angulo, H. Schmider, R. P. Sagar, and V. H. Smith, Jr. Nonconvexity of the atomic charge density and shell structure. *Phys. Rev. A*, 49:726–728, 1994.

[262] A. M. Frolov, V. H. Smith, Jr., and D. M. Bishop. Bound P-states (L=1) and bound D-states (L=2) of three-body muonic molecular ions. *Phys. Rev. A*, 49:1686–1692, 1994.

[263] H. Schmider, R. P. Sagar, and V. H. Smith, Jr. Atomic moment densities in position and momentum space. *Proc. Indian Acad. Sci.(Chem. Sci.)*, 106:133–142, 1994.

[264] K. E. Edgecombe, D. F. Weaver, and V. H. Smith, Jr. Electronic structure analysis of compounds of interest in drug design. I. Mono- and dicarboxylated pyridines. *Can. J. Chem.*, 72:1388–1403, 1994.

[265] A. M. Frolov and V. H. Smith, Jr. One-photon annihilation in the Ps^- ion and the angular correlation in two-electron ions. *Phys. Rev. A.*, 49:3580–3585, 1994.

[266] H. Schmider, R. P. Sagar, and V. H. Smith, Jr. Density differences for near-Hartree-Fock wavefunctions. *Phys. Rev. A.*, 49:4229–4231, 1994.

[267] M. Hô, R. P. Sagar, J. M. Perez-Jorda, V. H. Smith, Jr., and R. O. Esquivel. A numerical study of molecular information entropies. *Chem. Phys. Lett.*, 219:15–20, 1994.

[268] J. Wang and V. H. Smith, Jr. Electron-pair distributions and chemical bonding. *Int. J. Quantum Chem.*, 49:147–157, 1994.

[269] J. Wang and V. H. Smith, Jr. Evaluation of electron pair densities and their Laplacians in atomic systems. *Chem. Phys. Lett.*, 220:331–336, 1994.

[270] J. Wang, A. N. Tripathi, and V. H. Smith, Jr. Chemical binding and electron correlation effects in X-ray and high energy electron scattering. *J. Chem. Phys.*, 101:4842–4854, 1994.

[271] M. Hô, H. Schmider, K. E. Edgecombe, and V. H. Smith, Jr. Topological analysis of valence electron charge distributions from semiempirical and ab-initio methods. *Int. J. Quantum Chem., Quantum Chem. Symp.*, 28:215–226, 1994.

[272] C.-J. Mei and V. H. Smith, Jr. On the role of doping in high-T_c superconductors. *Int. J. Quantum Chem., Quantum Chem. Symp.*, 28:687–693, 1994.

[273] J. Wang and V. H. Smith, Jr. Evaluation of cross sections for X-ray and high energy electron scattering from molecular systems. *Int. J. Quantum Chem.*, 52:1145–1151, 1994.

[274] J. Wang and V. H. Smith, Jr. Spherically averaged molecular electron densities and radial moments in position and momentum space. *J. Phys. B*, 27:5159–5173, 1994.

[275] M. Hô, R. P. Sagar, V. H. Smith, Jr., and R. O. Esquivel. Atomic information entropies beyond the Hartree-Fock limit. *J. Phys. B*, 27:5149–5157, 1994.

[276] A. M. Frolov and V. H. Smith, Jr. On stimulated nuclear fusion in the cold generated DT hydrides of fissionable elements. *Phys. Lett. A.*, 196:217–222, 1994.

[277] A. M. Frolov and V. H. Smith, Jr. On bound states in two-body systems. *Int. J. Quantum Chem.*, 53:9 -14, 1995.

[278] M. Hô, R. P. Sagar, H. Schmider, D. F. Weaver, and V. H. Smith, Jr. Measures of distance for atomic charge and momentum densities and their relationship to physical properties. *Int. J. Quantum Chem.*, 53:627–633, 1995.

[279] A. M. Frolov and V. H. Smith, Jr. Statistical nuclear correlation coefficients for ^{3}H and ^{3}He nuclei. *Phys. Rev. C.*, 51:423–426, 1995.

[280] J. Wang, A. D. Becke, and V. H. Smith, Jr. Evaluation of $\langle S^2 \rangle$ in restricted, unrestricted Hartree-Fock and density functional based theories. *J. Chem. Phys.*, 102:3477–3480, 1995.

[281] A. M. Frolov, S. I. Kryuchkov, and V. H. Smith, Jr. (e^-, e^+) pair annihilation in the positronium molecule Ps_2. *Phys. Rev. A*, 51:4514–4519, 1995.

[282] J. Wang and V. H. Smith, Jr. Fermi and Coulomb holes of molecules in excited states. *Int. J. Quantum Chem.*, 56:509–519, 1995.

[283] V. H. Smith, Jr. and A. M. Frolov. On properties of the helium-muonic and helium-antiprotonic atoms. *J. Phys. B*, 28:1357–1368, 1995.

[284] A. W. Ross, M. Fink, R. Hilderbrandt, J. Wang, and V. H. Smith, Jr. Complex scattering factors for the diffraction of electrons by gases. In *International Tables of Crystallography vol. C (revised)*, pages 245–338. Kluwer, Dordrecht, 1995.

[285] D. M. Bishop, A. M. Frolov, and V. H. Smith, Jr. Properties of the bound S states ($L = 0$) in Coulomb three-body systems with unit charges. *Phys. Rev. A*, 51:3636–3644, 1995.

[286] J. Wang, R. O. Esquivel, V. H. Smith, Jr., and C. Bunge. Accurate elastic and inelastic scattering factors from He to Ne using correlated wavefunctions. *Phys. Rev. A.*, 51:3812–3818, 1995.

[287] J. Nagy, D. F. Weaver, and V. H. Smith, Jr. A comprehensive study of alkane, non-bonded empirical force fields. Suggestions for improved parameter sets. *J. Phys. Chem.*, 99:8058–8065, 1995.

[288] J. E. Graham, D. C. Ripley, J. T. Smith, V. H. Smith, Jr., and D. F. Weaver. Theoretical studies applied to drug design: Ab initio electronic distributions in bioisosteres. *J. Mol. Struct. (Theochem)*, 343:105–110, 1995.

[289] M. Hô, R. P. Sagar, D. F. Weaver, and V. H. Smith, Jr. An investigation of the dependence of Shannon information entropies and distance measures on molecular geometry. *Int. J. Quantum Chem., Quantum Chem. Symp.*, 29:109–115, 1995.

[290] Q. Jin, C. Mei, and V. H. Smith, Jr. Comparison study of the electronic structure of high T_c superconductors. *Int. J. Quantum Chem., Quantum Chem. Symp.*, 29:189–195, 1995.

[291] J. Nagy, D. F. Weaver, and V. H. Smith, Jr. Ab initio methane dimer intermolecular potentials. *Mol. Phys.*, 85:1179–1192, 1995.

[292] J. Wang and V. H. Smith, Jr. 1/Z expansions for isoelectronic systems from He through Ar. *Phys. Rev. A.*, 52:1060–1066, 1995.

[293] A. M. Frolov and V. H. Smith, Jr. Universal variational expansions for three-body systems. *J. Phys. B*, 28:L449 – L455, 1995.

[294] J. Nagy, V. H. Smith, Jr., and D. F. Weaver. Validation of a reparameterized MM3 non-bonded force field for hydrocarbons: Crystal lattice studies of C_{60} and C_{70} and adsorption of hydrocarbons on graphite. *J. Mol. Struct. (Theochem)*, 358:71–78, 1995.

[295] J. Nagy, D. F. Weaver, and V. H. Smith, Jr. A critical evaluation of benzene analytical non-bonded force fields. reparameterization of the MM3 potential. *J. Phys. Chem.*, 99:13868–13875, 1995.

[296] H. Schmider and V. H. Smith, Jr. Computation of Compton profiles in a weak laser field. *Phys. Rev. A*, 53:3295–3301, 1996.

[297] A. M. Frolov and V. H. Smith, Jr. Bound states with arbitrary angular momenta in non-relativistic three-body systems. *Phys. Rev. A*, 53:3853–3864, 1996.

[298] J. Wang, B. J. Clark, H. Schmider, and V. H. Smith, Jr. Topological analysis of electron momentum densities and the bond directional principle. The first-row hydrides, AH, and homonuclear diatomic molecules A_2. *Can. J. Chem.*, 74:1187–1191, 1996.

[299] J. Wang, V. H. Smith, Jr., C. F. Bunge, and R. Jauregui. Relativistic X-ray scattering factors for He through Ar from Dirac-Hartree-Fock wave functions. *Acta Cryst.*, 52:649–658, 1996.

[300] A. M. Frolov and V. H. Smith, Jr. The CMP invariance for symmetrical Coulomb four-body systems with unit charges. *J. Phys. B*, 29:L433–L440, 1996.

[301] R. O. Esquivel, A. L. Rodriguez, R. P. Sagar, M. Hô, and V. H. Smith, Jr. Physical interpretation of information entropy: Numerical evidence of Collins' conjecture. *Phys. Rev. A*, 54:259–265, 1996.

[302] A. N. Tripathi, V. H. Smith, Jr., R. P. Sagar, and R. O. Esquivel. Electron correlation in momentum space for the neon atom isoelectronic sequence from F^- through Ar^{8+}. *Phys. Rev. A*, 54:1877–1883, 1996.

[303] A. M. Frolov and V. H. Smith, Jr. The properties of the helium-muonic atoms in their electron excited states and four-body helium-muonic ions. *J. Phys. B*, 29:5969–5980, 1996.

[304] A. M. Frolov and V. H. Smith, Jr. Exact finite series for the few-body auxiliary functions. *Int. J. Quantum Chem.*, 63:269–278, 1997.

[305] J. Wang and V. H. Smith, Jr. Electrostatic potential at the nucleus in atoms from 1/Z expansion. *Mol. Phys.*, 90:1027–1029, 1997.

[306] A. M. Frolov and V. H. Smith, Jr. Positronium hydrides and the Ps_2 molecule: Bound state properties, positron annihilation rates and hyperfine structure. *Phys. Rev. A*, 55:2662–2673, 1997.

[307] A. M. Frolov and V. H. Smith, Jr. Nuclear reaction rates in four-body muon molecules. *Phys. Rev. A*, 55:2435–2437, 1997.

[308] A. M. Frolov and V. H. Smith, Jr. The ground state of positronium hydride. *Phys. Rev. A*, 56:2417–2420, 1997.

[309] J. C. Ramirez, C. Soriano, R. O. Esquivel, R. P. Sagar, M. Hô, and V. H. Smith, Jr. Jaynes information entropy of small molecules: Numerical evidence of Collins' conjecture. *Phys. Rev. A*, 56:4477–4482, 1997.

[310] M. Hô, V. H. Smith, Jr., D. F. Weaver, C. Gatti, R. P. Sagar, and R. O. Esquivel. Molecular similarity based on information entropies and distances. *J. Chem. Phys.*, 108:5469–5475, 1998.

[311] M. Hô, D. F. Weaver, V. H. Smith, Jr., R. P. Sagar, and R. O. Esquivel. Calculating the logarithmic mean excitation energy from the Shannon information entropy of the electronic charge density. *Phys. Rev. A*, 57:4512–4517, 1998.

[312] A. M. Frolov and V. H. Smith, Jr. Inner radiation emitted during $\beta^{\pm}$ decay and related problems. *Phys. Rev. A*, 58:1212–1220, 1998.

[313] J. C. Ramirez, J. M. Hernandez Perez, R. P. Sagar, R. O. Esquivel, M. Hô, and V. H. Smith, Jr. On the amount of information present in the one-particle density matrix and the charge density. *Phys. Rev. A*, 58:3507–3515, 1998.

[314] M. Hô, D. F. Weaver, V. H. Smith, Jr., R. P. Sagar, R. O. Esquivel, and S. Yamamoto. An information-entropic study of correlated densities of the water molecule. *J. Chem. Phys.*, 109:10620–10627, 1998.

[315] R. O. Esquivel, A. Vivier-Bunge, and V. H. Smith, Jr. Accurate determination of Fermi-contact interaction in atomic lithium. *J. Mol. Struct. (Theochem)*, 433:43–50, 1998.

[316] F. Sauriol, Y. S. Park, D. F. Weaver, V. H. Smith, Jr., and G. Lord. Combined NMR spectroscopic and quantum mechanical investigation of a biologically relevant hydrogen bonded model system. I. NMR experimental results and estimation of transition state structures and energies. *Can. J. Anal. Sci. Spectrosc.*, 43:137–143, 1998.

[317] P. Ziesche, V. H. Smith, Jr., M. Hô, S. P. Rudin, P. Gersdorf, and M. Taut. The He isoelectronic series and the Hooke's law model. Correlation measures and modifications of Collins' conjecture. *J. Chem. Phys.*, 110:6135–6142, 1999.

[318] B. J. Clark, H. L. Schmider, and V. H. Smith, Jr. On hybrid orbitals in momentum space. In Z. B. Maksic and W. J. Orville-Thomas, editors, *Pauling's Legacy: Modern Modelling of the Chemical Bond*, pages 213–229. Elsevier, Amsterdam, 1999.

[319] G. T. Smith, A. N. Tripathi, and V. H. Smith, Jr. X-ray and electron scattering intensities of molecules calculated using density functional theory. *J. Chem. Phys.*, 110:9390–9400, 1999.

[320] G. Duan, V. H. Smith, Jr., and D. F. Weaver. An ab initio and data mining study on aromatic-amide interactions. *Chem. Phys. Lett.*, 310:323–332, 1999.

[321] Q. Z. Cui and V. H. Smith. Analysis of solvation structure and thermodynamics of methane in water by reference interaction site model theory using an all-atom model. *J. Chem. Phys.*, 113:10240–10245, 2000.

[322] G. Duan, V. H. Smith, and D. F. Weaver. A data mining and ab initio study of the interaction between the aromatic and backbone amide groups in proteins. *Int. J. Quantum Chem.*, 80:44–60, 2000.

[323] J. H. Wang, H. L. Schmider, and V. H. Smith. Comment on "Cusp relations for local strongly decaying properties in electronic systems". *Phys. Rev. A*, 62:016501, 2000.

[324] G. Duan, V. H. Smith, and D. F. Weaver. Characterization of aromatic–amide(side-chain) interactions in proteins through systematic ab initio calculations and data mining analyses. *J. Phys. Chem. A*, 104:4521–4532, 2000.

[325] M. Hô, B. J. Clark, V. H. Smith, D. F. Weaver, C. Gatti, R. P. Sagar, and R. O. Esquivel. Shannon information entropies of molecules and functional groups in the self-consistent reaction field. *J. Chem. Phys.*, 112:7572–7580, 2000.

[326] M. Hô, H. L. Schmider, D. F. Weaver, V. H. Smith, R. P. Sagar, and R. O. Esquivel. Shannon entropy of chemical changes: SN2 displacement reactions. *Int. J. Quantum Chem.*, 77:376–382, 2000.

[327] R.-H. Xie and V. H. Smith. Generation of higher-order atomic dipole squeezing in a high-Q micromaser cavity. (II) Effect of nonlinear one-photon processes. *Physica A*, 301:114–128, 2001.

[328] R.-H. Xie and V. H. Smith. Generation of higher-order atomic dipole squeezing in a high-Q micromaser cavity. (III) Effect of the dynamic Stark shift in two-photon processes. *Physica A*, 301:129–149, 2001.

[329] V. H. Smith. Position and momentum space variables. *Z. Phys. Chem.*, 215:1237–1241, 2001.

[330] G. L. Duan, V. H. Smith, and D. F. Weaver. Characterization of aromatic-thiol π-type hydrogen bonding and phenylalanine-cysteine side chain interactions through ab initio calculations and protein database analyses. *Mol. Phys.*, 99:1689–1699, 2001.

[331] M. Hô, V. H. Smith, R. P. Sagar, and R. O. Esqivel. Asymptotic behaviour of the ratio of density gradient to electron density for atomic systems. *Mol. Phys.*, 99:1727–1728, 2001.

[332] Q. Z. Cui and V. H. Smith. Analysis of solvation structure and thermodynamics of ethane and propane in water by reference interaction site model theory using all-atom models. *J. Chem. Phys.*, 115:2228–2236, 2001.

[333] A. M. Frolov and V. H. Smith. Generalization of the exponential variational ansatz in relative coordinates for bound state calculations in four-body systems. *J. Chem. Phys.*, 115:1187–1196, 2001.

[334] Neerja, A. N. Tripathi, and V. H. Smith. Directional Compton profiles, valence orbital momentum distributions and B(r) functions for the silane molecule. *J. Phys. B*, 34:1233–1243, 2001.

[335] R. P. Sagar, J. C. Ramirez, R. O. Esquivel, M. Hô, and V. H. Smith. Shannon entropies and logarithmic mean excitation energies from cusp- and asymptotic-constrained model densities. *Phys. Rev. A*, 6302:022509, 2001.

[336] M. Hô, W. A. Szarek, and V. H. Smith. Theoretical studies of unusually short bond lengths in oxirane and derivatives. *J. Mol. Struct. (Theochem)*, 537:253–264, 2001.

[337] G. T. Smith, V. H. Smith, and A. M. Frolov. Structures and properties of the hydrides of light elements. *Int. J. Quantum Chem.*, 90:1421–1427, 2002.

[338] Q. Z. Cui and V. H. Smith. K^+/Na^+ selectivity of KcsA potassium channel analyzed by reference interaction site model (RISM) integral equation theory. *Chem. Phys. Lett.*, 365:110–116, 2002.

[339] G. L. Duan, V. H. Smith, and D. F. Weaver. Data mining, ab initio, and molecular mechanics study on conformation of phenylalanine and its interaction with neighboring backbone amide groups in proteins. *Int. J. Quantum Chem.*, 90:669–683, 2002.

[340] Q. Z. Cui and V. H. Smith. Solvation structure, thermodynamics, and molecular conformational equilibria for n-butane in water analyzed by reference interaction site model theory using an all-atom solute model. *J. Phys. Chem. B*, 106:6554–6565, 2002.

[341] R.-H. Xie and V. H. Smith. Generation of higher-order atomic dipole squeezing in a high-Q micromaser cavity. (IV) Discussion of the nondegenerate two-photon Jaynes-Cummings model. *Physica A*, 307:207–220, 2002.

[342] R. P. Sagar, J. C. Ramirez, R. O. Esquivel, M. Hô, and V. H. Smith. Relationships between Jaynes entropy of the one-particle density matrix and Shannon entropy of the electron densities. *J. Chem. Phys.*, 116:9213–9221, 2002.

[343] G. T. Smith, H. L. Schmider, and V. H. Smith. Electron correlation and the eigenvalues of the one-matrix. *Phys. Rev. A*, 65:032508, 2002.

[344] A. M. Frolov, V. H. Smith, and G. T. Smith. Deuterides of light elements: low-temperature thermonuclear burn-up and applications to thermonuclear fusion problems. *Can. J. Phys.*, 80:43–64, 2002.

[345] A. M. Frolov and V. H. Smith. Natural orbital expansions of highly accurate three-body wavefunctions. *J. Phys. B*, 36:4837–4848, 2003.

[346] A. N. Tripathi, H. L. Schmider, and V. H. Smith. Directional Compton profiles and reciprocal form factors for the isoelectronic hydrides PH_3, H_2S and HCl. *J. Phys. B*, 36:4581–4593, 2003.

[347] F. E. Harris, A. M. Frolov, and V. H. Smith. Exponential variational expansion in relative coordinates for highly accurate bound state calculations in four-body systems. *J. Chem. Phys.*, 119:8833–8841, 2003.

[348] A. M. Frolov and V. H. Smith. Bound state properties and astrophysical applications of negatively charged hydrogen ions. *J. Chem. Phys.*, 119:3130–3137, 2003.

[349] A. M. Frolov and V. H. Smith. Exponential variational expansion in relative coordinates and scalar coupling coefficients for three-body systems. *J. Phys. B*, 36:1739–1753, 2003.

[350] R.-H. Xie, L. Jensen, G. W. Bryant, J. J. Zhao, and V. H. Smith. Structural, electronic, and magnetic properties of heterofullerene $C_{48}B_{12}$. *Chem. Phys. Lett.*, 375:445–451, 2003.

[351] R.-H. Xie, G. W. Bryant, J. J. Zhao, V. H. Smith, A. Di Carlo, and A. Pecchia. Tailorable acceptor $C_{60-n}B_n$ and donor $C_{60-m}N_m$ pairs for molecular electronics. *Phys. Rev. Lett.*, 90:206602, 2003.

[352] R.-H. Xie, G. W. Bryant, and V. H. Smith. Raman scattering in C_{60} and $C_{48}N_{12}$ aza-fullerene: First-principles study. *Phys. Rev. B*, 67:155404, 2003.

[353] R.-H. Xie, G. W. Bryant, L. Jensen, J. J. Zhao, and V. H. Smith. First-principles calculations of structural, electronic, vibrational, and magnetic properties of C_{60} and $C_{48}N_{12}$: A comparative study. *J. Chem. Phys.*, 118:8621–8635, 2003.

[354] R.-H. Xie, G. W. Bryant, and V. H. Smith. Electronic, vibrational and magnetic properties of a novel $C_{48}N_{12}$ aza-fullerene. *Chem. Phys. Lett.*, 368:486–494, 2003.

[355] Q. Z. Cui and V. H. Smith. Solvation structure, thermodynamics, and conformational dependence of alanine dipeptide in aqueous solution analyzed with reference interaction site model theory. *J. Chem. Phys.*, 118:279–290, 2003.

[356] G. L. Duan, V. H. Smith, and D. F. Weaver. Validation of the applicability of force fields to reproduce ab initio noncovalent interactions involving aromatic groups. *Adv. Quantum Chem.*, 47:65–92, 2004.

[357] F. E. Harris, A. M. Frolov, and V. H. Smith. New methods for old Coulomb few-body problems. *Int. J. Quantum Chem.*, 100:1086–1091, 2004.

[358] F. E. Harris, A. M. Frolov, and V. H. Smith. Singular and nonsingular three-body integrals for exponential wave functions. *J. Chem. Phys.*, 121:6323–6333, 2004.

[359] A. M. Frolov and V. H. Smith. Exponential representation in the Coulomb three-body problem. *J. Phys. B*, 37:2917–2932, 2004.

[360] R.-H. Xie, G. W. Bryant, C. F. Cheung, V. H. Smith, and J. J. Zhao. Optical excitation and absorption spectra of $C_{50}Cl_{10}$. *J. Chem. Phys.*, 121:2849–2851, 2004.

[361] R.-H. Xie, G. W. Bryant, G. Y. Sun, T. Kar, Z. F. Chen, V. H. Smith, Y. Araki, N. Tagmatarchis, H. Shinohara, and O. Ito. Tuning spectral properties of fullerenes by substitutional doping. *Phys. Rev. B*, 69:201403, 2004.

[362] F. E. Harris, A. M. Frolov, and V. H. Smith. Comment on "Analytic value of the atomic three-electron correlation integral with Slater wave functions". *Phys. Rev. A*, 69:056501, 2004.

[363] F. E. Harris, A. M. Frolov, and V. H. Smith. Highly accurate evaluation of atomic three-electron integrals of lowest orders. *J. Chem. Phys.*, 120:9974–9983, 2004.

[364] R.-H. Xie, G. W. Bryant, G. Y. Sun, M. C. Nicklaus, D. Heringer, T. Frauenheim, M. R. Manaa, V. H. Smith, Y. Araki, and O. Ito. Excitations, optical absorption spectra, and optical excitonic gaps of heterofullerenes. 1. C_{60}, $C_{59}N^+$, and $C_{48}N_{12}$: Theory and experiment. *J. Chem. Phys.*, 120:5133–5147, 2004.

[365] M. R. Manaa, R.-H. Xie, and V. H. Smith. Structure, vibrational, and electronic spectra of heterofullerene $C_{48}(BN)_6$. *Chem. Phys. Lett.*, 387:101–105, 2004.

[366] F. E. Harris, A. M. Frolov, and V. H. Smith. Comment on "Analysis of some integrals arising in the atomic four-electron problem". *J. Chem. Phys.*, 120:3040–3041, 2004.

[367] F. E. Harris and V. H. Smith. Highly compact wave functions for He-like systems. *J. Phys. Chem. A*, 109:11413–11416, 2005.

[368] Q. W. Shi, F. Sauriol, Y. Park, V. H. Smith, G. Lord, and L. O. Zamir. First example of conformational exchange in a natural taxane enolate. *Mag. Reson. Chem.*, 43:798–804, 2005.

[369] Q. Z. Cui and V. H. Smith. Analysis of K^+/Na^+ selectivity of KcsA potassium channel with reference interaction site model theory. *Mol. Phys.*, 103:191–201, 2005.

[370] F. E. Harris and V. H. Smith. Highly compact wavefunctions for two-electron systems. *Adv. Quantum Chem.*, 48:407–419, 2005.

[371] F. E. Harris, A. M. Frolov, and V. H. Smith. Singular integrals and their application to a hypervirial theorem. *Phys. Rev. A*, 72:012511, 2005.

[372] F. E. Harris, V. H. Smith, and A. M. Frolov. Correlated exponential-basis integrals with logarithmic integrands. *Mol. Phys.*, 103:2047–2054, 2005.

[373] R.-H. Xie, G. W. Bryant, J. J. Zhao, T. Kar, and V. H. Smith. Tunable optical properties of icosahedral, dodecahedral, and tetrahedral clusters. *Phys. Rev. B*, 71:125422, 2005.

[374] F. E. Harris, A. M. Frolov, and V. H. Smith. Re: Comment on "New methods for old Coulomb few-body problems". *Int. J. Quantum Chem.*, 106:552–553, 2006.

[375] D. Churchill, J. C. F. Cheung, Y. S. Park, V. H. Smith, G. van Loon, and E. Buncel. Complexation of diazinon, an organophosphorus pesticide, with α-, β-, and γ-cyclodextrin NMR and computational studies. *Can. J. Chem.*, 84:702–708, 2006.

[376] J. R.-H. Xie, V. H. Smith, and R. E. Allen. Spectroscopic properties of dipicolinic acid and its dianion. *Chem. Phys.*, 322:254–268, 2006.

Brill Academic Publishers
P.O. Box 9000, 2300 PA Leiden
The Netherlands

Lecture Series on Computer and Computational Sciences
Volume 6, 2005, pp. 520-525

Dynamics of Protons in Hydrogen Bonds Studied by Theoretical Methods and Vibrational Spectroscopy

Marek J. Wójcik[1]

Faculty of Chemistry,
Jagiellonian University,
30-060 Kraków, Ingardena 3, Poland

Abstract: Dynamics of protons in hydrogen bonds is determined by complex interplay of vibrational interactions. This mechanism is responsible for the complex structure of their infrared and Raman spectra, and for the dynamics of proton tunneling. We present experimental and theoretical evidence for such mechanisms in several hydrogen-bonded systems.

Keywords: Hydrogen bond, vibrational spectra, theoretical models, proton transfer

1. Introduction

Dynamics of protons in hydrogen bonds and proton tunneling in chemical and biological systems plays important role in vital functions and chemical and physical processes. Proton tunneling has fundamental role in many chemical reactions. It has potential application in artificial memory. The phenomenon of potential barrier penetration is also important in many branches of physics: quantum field theory, fission of atomic nuclei, scanning tunneling microscopy, and solid state physics. Dynamics of protons is a complex multidimensional process. In this article we present results of theoretical and spectroscopic studies of vibrational spectra and proton tunneling in several important hydrogen-bonded systems in the gaseous, liquid and solid state.

Infrared spectra of hydrogen-bonded systems have been a subject of numerous experimental and theoretical studies (for reviews see [1-7]). Formation of hydrogen bonds brings striking changes in the X-H (X = O, N, S, F, Cl) stretching bands. These bands are shifted to lower frequencies by the amount which reflects the strength of hydrogen bonds, and their widths and total intensities increase by an order of magnitude.

Modern theories of vibrational spectra of hydrogen bonds [8-11] attribute these spectral effects to multidimensional dynamics of proton motion due to vibronic-type couplings between high-frequency proton and low-frequency hydrogen bond vibrations, combined with Davydov and Fermi resonances. There is an analogy between separation of electronic and vibrational motions, and separation of the fast proton vibrations from the slow hydrogen bond vibrations. Davydov resonance describes interactions between identical X-H groups in crystals or cyclic hydrogen-bonded dimers of carboxylic acids; and Fermi resonance - anharmonic interactions between degenerate (or quasi-degenerate) X-H stretching vibrations and overtone or combination vibrations of other modes in hydrogen-bonded system. These theories allow calculation of infrared and Raman spectra of hydrogen-bonded systems. The effect of deuterium/hydrogen substitution on the spectra and the temperature effect are explained by the theories.

2. Results and Discussion

[1] Corresponding author. E-mail: wojcik@chemia.uj.edu.pl

Infrared O-H stretching band of the cyclic dimer of acetic acid, presented in Fig. 1, constitutes an example of a spectrum for systems of two identical interacting hydrogen bonds. It exhibits irregular progression of sub-bands of varying intensity ratios. After deuteration the band is shifted to lower frequencies and its width and progression are diminished. To model the spectra we assumed the coupling between the O-H(D) and O...O stretching vibrations, resonance coupling between two moieties, and damping mechanisms. The theory presented in Ref. [11] correctly fits the experimental lineshape of the hydrogenated compound and predicts satisfactorily the evolution in the line shape, to the deuterated one by reducing simply the angular frequency of the H-bond bridge and the anharmonic coupling parameter by the factor $\sqrt{2}$.

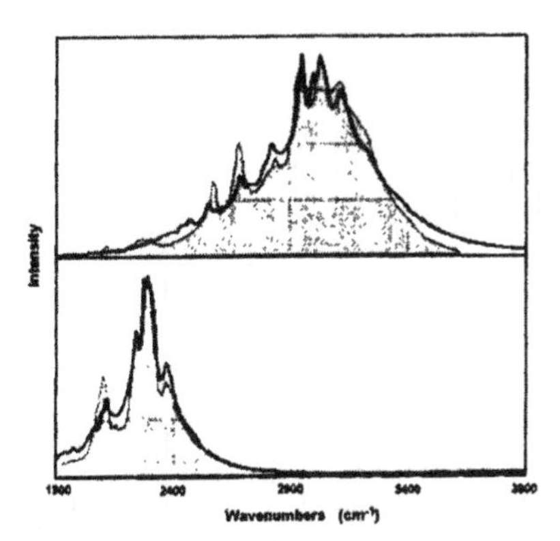

Figure 1: IR line shapes of gaseous cyclic acetic acid dimers and isotope effect. Top $(CD_3CO_2H)_2$: grayed -experiment, line - theory. Bottom $(CD_3CO_2D)_2$: grayed - experiment, line - theory. T = 300 K.

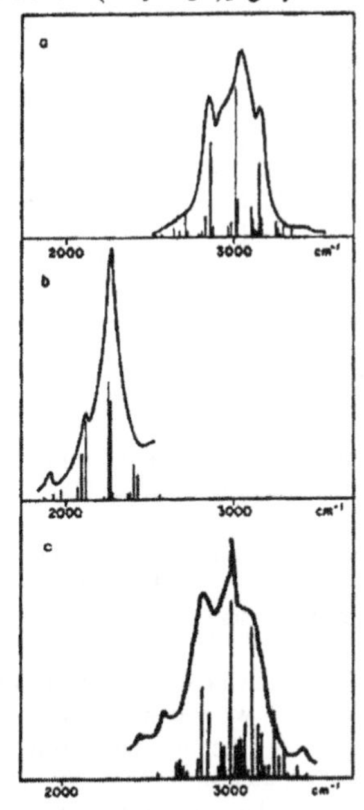

Figure 2: Comparison between the experimental (solid line) and theoretical (δ functions) IR spectra for (a) the 1-methylthymine crystal, (b) the deuterated crystal and (c) IR spectra polarized along the *b* axis of the crystal.

Fig. 2 presents comparison between experimental and calculated spectra for 1-methylthymine crystal, for the deuterated crystal, and for spectra polarized along the *b* axis of the crystal. Calculated spectra are shown as δ functions representing calculated transition energies and relative intensities. Main features of the spectra with regard to the energy and intensity distributions of the fine structure and the width are reproduced in good agreement with the experimental measurements. Substantially different structure and width of the infrared spectrum of the deuterated crystal is the result of replacement of the mass of hydrogen by the mass of deuterium in equations provided by the theory. The polarised spectra have been calculated by taking into account interactions between hydrogen bonds within unit cell of the crystal. They also well reproduce experimental spectra. The details of calculations for 1-methylthymine have been presented in Ref. [10].

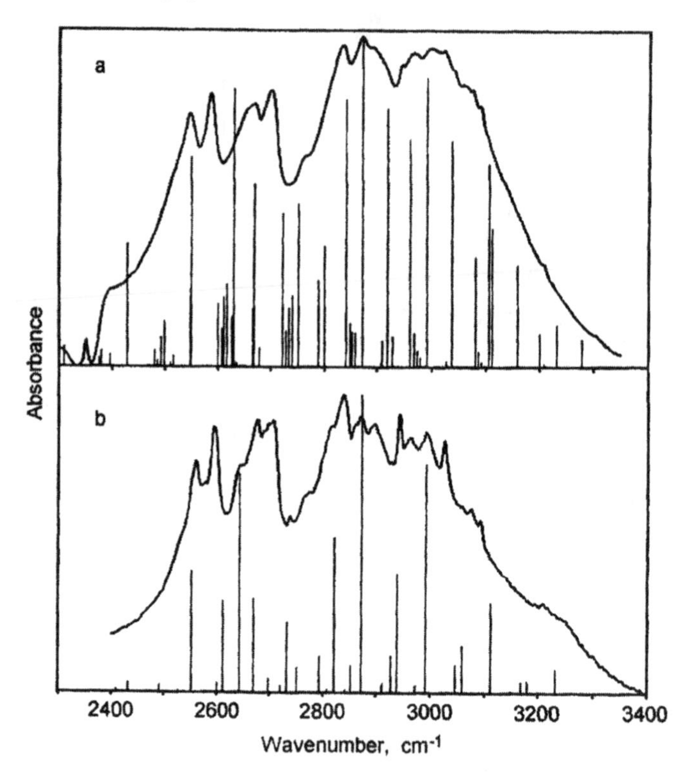

Figure 3: Comparison between the experimental (solid line) and theoretical (δ functions) IR spectra for aspirin crystal at (a) 300K and (b) 77K.

Fig. 3 presents spectra for aspirin at room and liquid nitrogen temperatures [12]. Lowering temperature causes sharpening and narrowing of the O-H stretching bands which is reflected in decreasing number of transitions from excited O...O states (hot transitions).

Quantum model calculations combined with molecular dynamic (MD) simulations allowed us to obtain spectra of water in ionic shells. In Fig. 4 we present calculated infrared bands for the HDO molecules in the first hydration shell of Li^+, for the different types of water-ion geometries found in the second hydration shell of Li^+, and in the first hydration shell of formate anion (13,14). Water molecules in the first shell of cation can have either tetrahedral or trigonal orientation. First type of coordination is prevalent for mono- and divalent hydrated cations from potassium and calcium groups, while the second prevails for divalent first row transition-metal cations and trivalent rare-earth cations. Theoretical simulations in this case give information inaccessible by experimental techniques. Quantum and model studies of ionic solutions complete experimental studies.

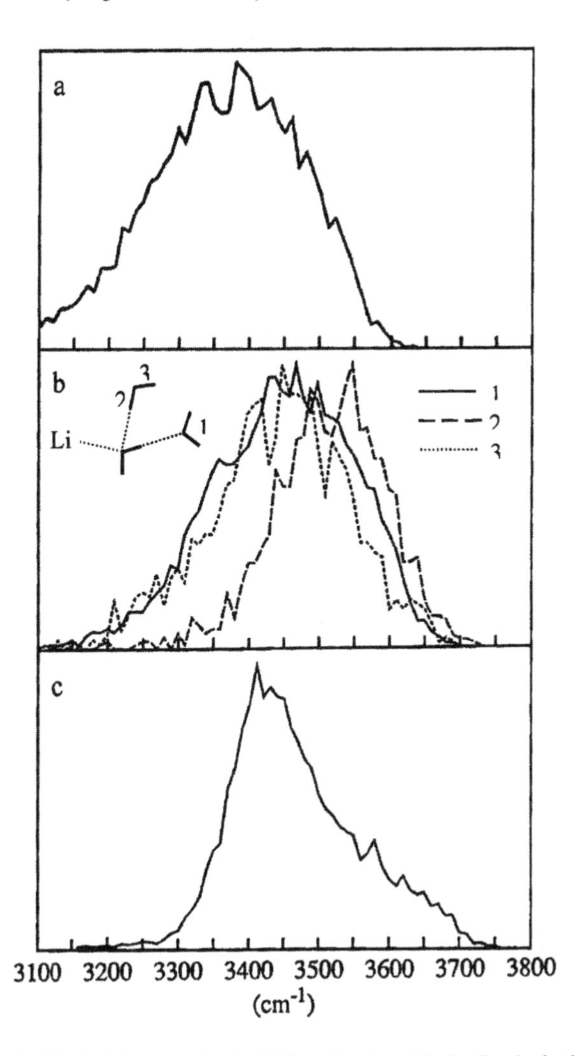

Figure 4: Calculated infrared spectra for the HDO molecules (a) in the first hydration shell of Li^+, (b) for the different types of water-ion geometries found in the second hydration shell of Li^+, and (c) in the first hydration shell surrounding the COO^- end of $HCOO^-$.

In Fig. 5 we present experimental and calculated infrared and Raman spectra of hexagonal ice [15], and in Fig. 6 - experimental and calculated polarized Raman spectra of isotopic mixtures in cubic ice [16]. Theoretical spectra have been calculated for ice clusters containing respectively 727 or 688 water molecules. The spectra in Fig. 6 demonstrate effect of the distribution of clusters of the D_2O molecules embedded in a matrix of H_2O molecules on the vibrational spectra. In the limit of a pure D_2O ice, the excitations are delocalized over the entire crystal, and in the low concentration of D_2O, the spectra are determined by dominant presence of monomers and low-seize clusters.

Vibrational interactions are also present in tunneling splittings accompanying proton tunneling. Proton tunneling in hydrogen-bonded systems is a multidimensional process involving couplings between proton tunneling and low-frequency planar and non-planar vibrations, similar to those responsible for the complex structure of their vibrational spectra. Excitation of vibrational modes promotes or suppresses the tunneling. Strong dependence of tunneling splittings on the excitation of the low-frequency vibrations has been observed in the laser fluorescence excitation experiments [17]. These splittings have been explained by ab initio and model calculations [18].

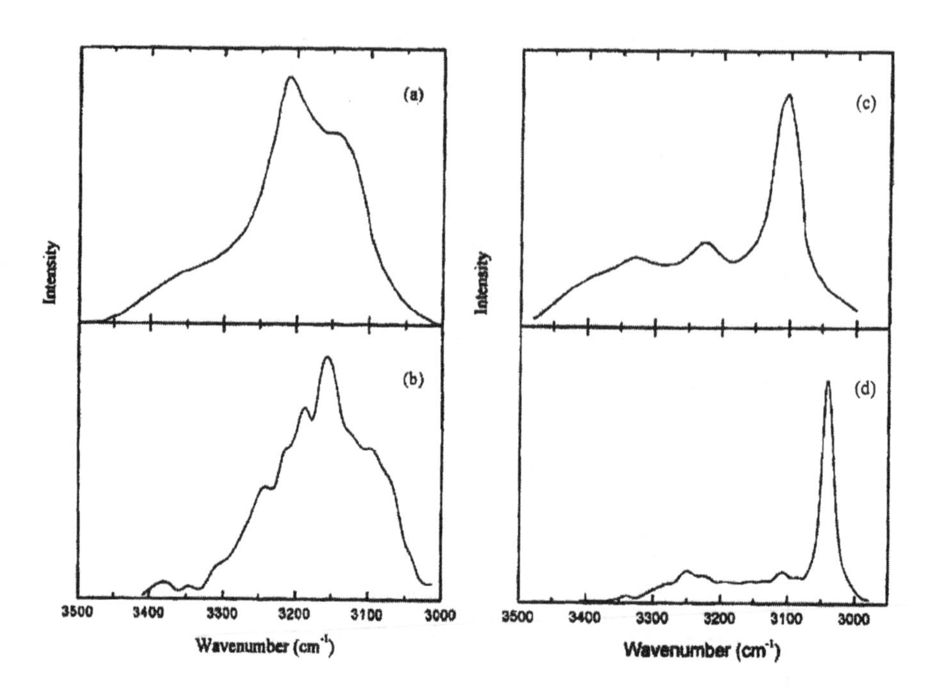

Figure 5: Comparison between experimental and calculated vibrational spectra for the OH stretching region of ice Ih: (a) IR experimental, (b) IR calculated, (c) Raman experimental, (d) Raman calculated.

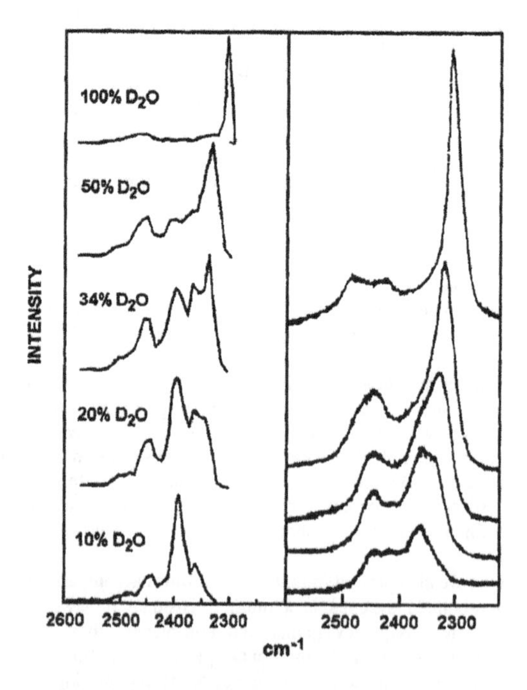

Figure 6: Paralell-polarized Raman spectra for isotopic cubic ice mixtures in the OD stretching region (calculated left side, experimental right side).

Spectroscopic and theoretical studies presented in this article describe complex dynamics of protons in hydrogen bonds. This dynamics determines structures of infrared and Raman spectra of hydrogen-bonded systems, the deuteration and temperature effects, and splittings accompanying multidimensional proton tunneling. We hope these studies allow to better understand the important physical phenomenon of hydrogen bonding.

Acknowledgments

Fig. 1 is reprinted with permission from Ref. (11), Fig. 5 from Ref. (15), and Fig. 6 from Ref. (16), copyrights 1993, 2002 and 2005, American Institute of Physics; Fig. 2 is reprinted from Ref. (10), with the permission of John Wiley & Sons; Fig. 3 from Ref. (12), Fig. 4a from Ref. (13), with the permission of Elsevier; and Figs. 4b and 4c from Ref. (14), with the permission of American Chemical Society. Technical assistance of Mr. Lukasz Boda is kindly acknowledged.

References

[1] D. Hadzi, H.W. Thompson, Eds., *Hydrogen Bonding*, Pergamon Press, London 1959.

[2] G.C. Pimentel, A.L. McClellan, *The Hydrogen Bond*, Freeman Co, San Francisco 1960.

[3] P. Schuster, G. Zundel, C. Sandorfy, Eds., *The Hydrogen Bond. Recent Developments in Theory and Experiments*, Vol. I-III, North Holland 1976.

[4] G.A. Jeffrey, W. Sanger, *Hydrogen Bonding in Biological Structures*, Springer-Verlag, Berlin 1990.

[5] S. Scheiner, Ed., *Hydrogen Bonding. A Theoretical Perspective*, Oxford University Press, Oxford 1997.

[6] G.A. Jeffrey, *Introduction to Hydrogen Bonding*, Oxford University Press, Oxford 1997.

[7] D. Hadzi, Ed., *Theoretical Treatments of Hydrogen Bonding*, J. Wiley, Chichester 1997.

[8] Y. Maréchal, A. Witkowski, *J. Chem. Phys.* **48** 3697 (1968).

[9] A. Witkowski, M. Wójcik, *Chem. Phys.* **1** 9 (1973).

[10] M.J. Wójcik, *Int. J. Quant. Chem.* **10** 747 (1976).

[11] P. Blaise, M.J. Wójcik, O. Henri-Rousseau, *J. Chem. Phys.* **122** 064306 (2005).

[12] M. Boczar, M.J. Wójcik, K. Szczeponek, D. Jamróz, A. Zieba, B. Kawalek, *Chem. Phys.* **286** 63 (2003).

[13] M.J. Wójcik, K. Hermansson, J. Lindgren, L. Ojamäe, *Chem. Phys.* **171** 189 (1993).

[14] J. Lindgren, K. Hermansson, M.J. Wójcik, *J. Phys. Chem.* **97** 5254 (1993).

[15] M.J. Wójcik, K. Szczeponek, S. Ikeda, *J. Chem. Phys.* **117** 9850 (2002).

[16] M.J. Wójcik, V. Buch, J.P. Devlin, *J.Chem. Phys.* **99** 2332 (1993).

[17] H. Sekiya, Y. Nagashima, Y. Nishimura, *J. Chem. Phys.* **92** 5761 (1990).

[18] M.J. Wójcik, H. Nakamura, S. Iwata, W. Tatara, *J. Chem. Phys.* **112**, 6322 (2000

Brill Academic Publishers
P.O. Box 9000, 2300 PA Leiden
The Netherlands

Lecture Series on Computer and Computational Sciences
Volume 6, 2006, pp. 526-539

Multicriteria Choice of Enzyme Immobilization Process for Biosensor Design and Construction

F.A. Batzias[1]

Department of Industrial Management and Technology,
University of Piraeus,
GR-185 34 Piraeus, Greece

Reecived 18 August, 2006; accepted 20 August, 2006

Abstract: A methodological framework, under the form of an algorithmic procedure, is presented for multicriteria choice of enzyme immobilization process (EIP) for biosensor design and construction. An implementation of this procedure, including 21 activity stages and 5 decision nodes, is presented with the following alternatives/criteria, extracted by experts' opinion through a modified Delphi method (in a fuzzy version to count for uncertainty) in such a way that can be easily utilized in special cases of R&D or commercial interest. Alternatives: entrapment/encapsulation, covalent binding, entrapment/crosslinking, crosslinking/electrodeposition, adsorption/electrodeposition, pure adsorption, entrapment/electrodeposition. Criteria: EIP controlability, immobilized enzyme mass loss (due to decreasing diffusional resistance during service time), contribution to the ability of the substrate surface to be regenerated, immobilized enzyme resistance to environmental conditions expected to prevail during its operation, enzyme activity decrease due to immobilization, estimated total cost, EIP simplicity. Detailed monoparametric sensitivity analysis, as regards the impact of each criterion's weight on the suggested/preferred alternative, is performed, and incomparabilities within certain pairs of alternatives are investigated. Process reversibility is also examined on both, theoretical and experimental basis.

Keywords: Enzyme immobilization, biosensor, reversibility, taxonomy, cost, multicriteria choice.

1. Introduction

The objective of enzymes immobilization is their economic application in a way that enhances also controllability of bioprocess and uniformity of convention/production. Consequently, this method is especially recommended to be used with expensive enzymes or/and high added value products to counterbalance the immobilization process cost. Nevertheless, an inexpensive enzyme can be employed as an immobilized agent if another substantial advantage, such as avoidance of contamination or immune response is attained. In the case of production systems, these advantages include also higher efficiency and product purity (in comparison with other production systems), especially in the pharmaceutical/cosmetics industry; in the case of biosensors, these advantages include better values of metrological parameters (in comparison with conventional methods of measurement), especially when monitoring in situ is required.

The classification of enzyme immobilization processes by induction (i.e., by categorizing the techniques used either in industrial practice or in R&D) is very difficult for two reasons: (i) there is not a unique hierarchical system of *criteria divisionis*, established by deduction or at least accepted as a standard/recommended practice, for discrimination/clustering/grouping these processes, and (ii) several unit/partial processes may take place during the whole process. The technique we have adopted to overcome the first difficulty incorporates (i) collecting all cases (from a large sample) with a common

[1] Corresponding author. E-mail: fbatzi@unipi.gr

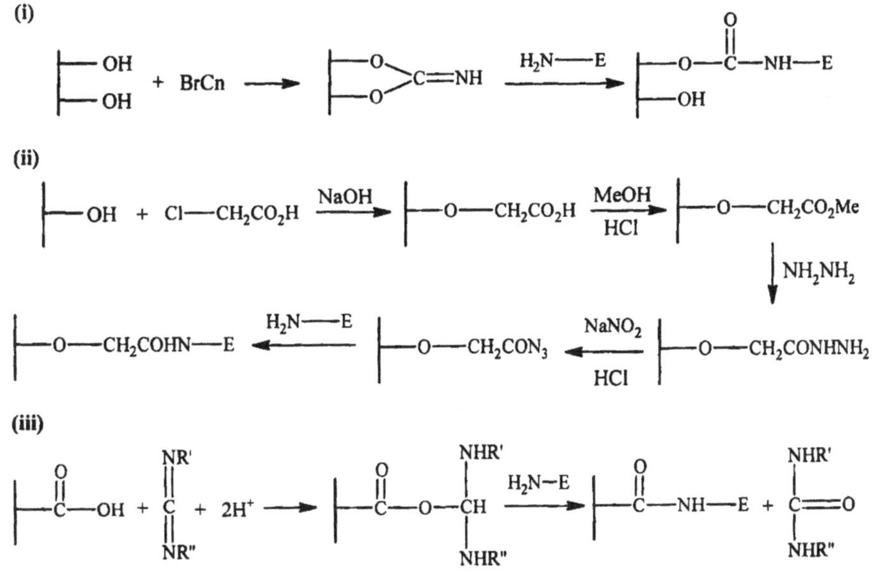

(i)

$$\text{—OH, —OH} + \text{BrCn} \longrightarrow \text{—O, —O}{>}\text{C=NH} \xrightarrow{H_2N\text{—}E} \text{—O—}\overset{O}{\overset{\|}{C}}\text{—NH—E, —OH}$$

(ii)

$$\text{—OH} + \text{Cl—}CH_2CO_2H \xrightarrow{NaOH} \text{—O—}CH_2CO_2H \xrightarrow[HCl]{MeOH} \text{—O—}CH_2CO_2Me$$

$$\downarrow NH_2NH_2$$

$$\text{—O—}CH_2COHN\text{—E} \xleftarrow{H_2N\text{—}E} \text{—O—}CH_2CON_3 \xleftarrow[HCl]{NaNO_2} \text{—O—}CH_2CONHNH_2$$

(iii)

$$\text{—}\overset{O}{\overset{\|}{C}}\text{—OH} + \overset{NR'}{\underset{NR''}{C}} + 2H^+ \longrightarrow \text{—}\overset{O}{\overset{\|}{C}}\text{—O—}\overset{NHR'}{\underset{NHR''}{CH}} \xrightarrow{H_2N\text{—}E} \text{—}\overset{O}{\overset{\|}{C}}\text{—NH—E} + \overset{NHR'}{\underset{NHR''}{C}}\text{=O}$$

Figure 1. The enzyme immobilization onto carbohydrate substrate can be made through (i) the cyanogens bromide technique, (ii) the azide coupling technique, (iii) the carboxyl group (formed on cellulose) technique.

low level characteristic, C_i (e.g., 'oxidation'), i =1, 2,..., n, regardless of overlapping and substrate utilized for immobilization, (ii) selecting the C_i-terms that form a higher level characteristic L_j, j = 1,2, ..., m, where $m \leq n$, regardless of physicochemical homogeneity, e.g., 'chemical reaction' and 'encapsulation' may belong to the same phenomenological level, provided that he connotation of these terms enables the disjunction between alternative whole processes, (iii) comparing these terms with the corresponding classifiers reported in technical literature, (iv) continuing this procedure until all whole cases have been classified at the lowest level with the minimum number of classifiers providing the maximum degree of discrimination (minimax criterion), under the disjunctive clause, i.e., each whole case can be characterized by only one main classifier. For example, the enzyme immobilization onto carbohydrate substrate can be made through (i) the cyanogens bromide technique [1, 2], (ii) the azide coupling technique [3], (iii) the carboxyl group (formed on cellulose, [4]) technique, as shown in Fig.1.

According to the procedure described above, all these whole cases can be classified as 'covalent binding' on the basis of the chemical mechanism with the hydroxyl group playing the central role in this binding; neglecting the role of carbohydrate substrate as common characteristic, this allows for incorporating other techniques of chemical attack either (i) directly to hydroxyl groups of carbohydrates like the cyanuric chloride technique and the transition metal-activation technique or (ii) to protein-and amine-bearing substrates (e.g., utilization of diazobenzidine and its derivatives for selective binding to tyrosyl groups), or (iii) to silane through an inorganic functional group leaving the organic moiety, such as an amine, available for covalent attachment of the enzyme, the most frequently used silane being γ-aminoporpyltriethoxysilane, or (iv) to other inorganic substrates through polymeric bonding like in the case of isocyanate, where carbamate bond is formed between the inorganic surface and the isocyanate group (see [5] and Appendix I).

The technique we have adopted to overcome the second difficulty, i.e., that of unit/partial processes co-existing within the same whole process, includes two actions: (i) to accept as preliminary/transient classifier the partial process which the majority of authors of reporting on the whole process relevant papers accepts directly or indirectly (i.e., by denoting or connoting a classifying term within the text, respectively), (ii) to transfer the case into the second stage of the procedure described above. Applying this technique by means of a data mining intelligent agent, it is probable that when scanning a text (referring to the same immobilization technique by the same author) two terms, belonging to different classifiers are met. For example, Sheppard and Guiseppi-Elli [6] write "The reaction of glutaraldehyde with these primary amines during the immobilization step leads to covalent attachment of the cross-linked enzyme gel to the sensor substrate, ...These reactions lead to the formation of a cross-linked network of silanes containing primary amines, covalently attached to the transducer surface...The most challenging procedure in this protocol is the preparation of the sensing layer by cross-linking the enzyme and albumin on the surface of the conductimetric transducer." When such a conflict (at least at intelligent agent level and consequently at any automatic search/retrieval level) the expert (either system or human) responsible for final decision on classifying should refer to the physicochemical mechanism in order to make the final decision. In this case, the mechanism is the following reaction leading to 'covalent binding', not 'entrapment/cross-linking' (both terms used as classifiers herein – see Implementation), of 3-aminopropyl trimethoxy silane to oxidized silicon surface.

NH_2 H_2C CH_2 H_2C H_3CO—Si—OCH_3 OCH_3 OH Si SiO_2

$\xrightarrow{-3CH_3OH}$

NH_2 H_2C CH_2 H_2C —O—Si—O— O Si SiO_2

The preceding introductory analysis can be used as a basis for the design/development of a framework under the form of an algorithmic procedure for the multicriteria choice of enzyme immobilization process (EIP), when adequate data are provided.

2. Methodology

The optimal choice of EIP depends heavily on adequacy of data provided for running the algorithmic procedure mentioned above. Fuzzy multicriteria analysis (FMCA) has been adopted to count for uncertainty and the output is given in fuzzy and crisp (after defuzzification) numbers, allowing for ranking of alternatives according to preference order. The procedure designed/developed for the needs of the present study includes the following 21 activity stages and 5 decision nodes (for their interconnection, see Fig. 2).

1. Determination of analyte species to be detected/measured and the conditions expected to prevail.
2. Identification of variable/parameter to be measured as mostly representative of the analyte species.
3. Determination of the metrological parameter values according to the specification set a priori.
4. Collection and registration (in lexicographic order) of the enzyme-substrate combinations (ESCs) that can be used for measuring the suggested variable/parameter.
5. Collection of the EIPs corresponding to each ESC.
6. Selection of proper criteria for optimal EIP/ESC choice.
7. Selection enrichment with missing criteria after their expression in controlled vocabulary terms.
8. Grouping/integration of criteria in forms scientifically acceptable, recognizable in technical literature and experts-friendly.
9. Elimination of partial criterion from the whole/integrated criterion with the most weak relevance.
10. Examination of the ESC ordered first among the ESCs that have not been examined yet.
11. FMCA of the EIPs corresponding to the ESC under consideration.
12. Sensitivity analysis to investigate the robustness of solution.
13. Change of the FMCA generalized criterion to obtain sufficient resolution.
14. Evaluation/comparison of the EIPs ranked first in preference order.
15. FMCA of this set of alternative EIPs.
16. Sensitivity analysis of all alternatives.

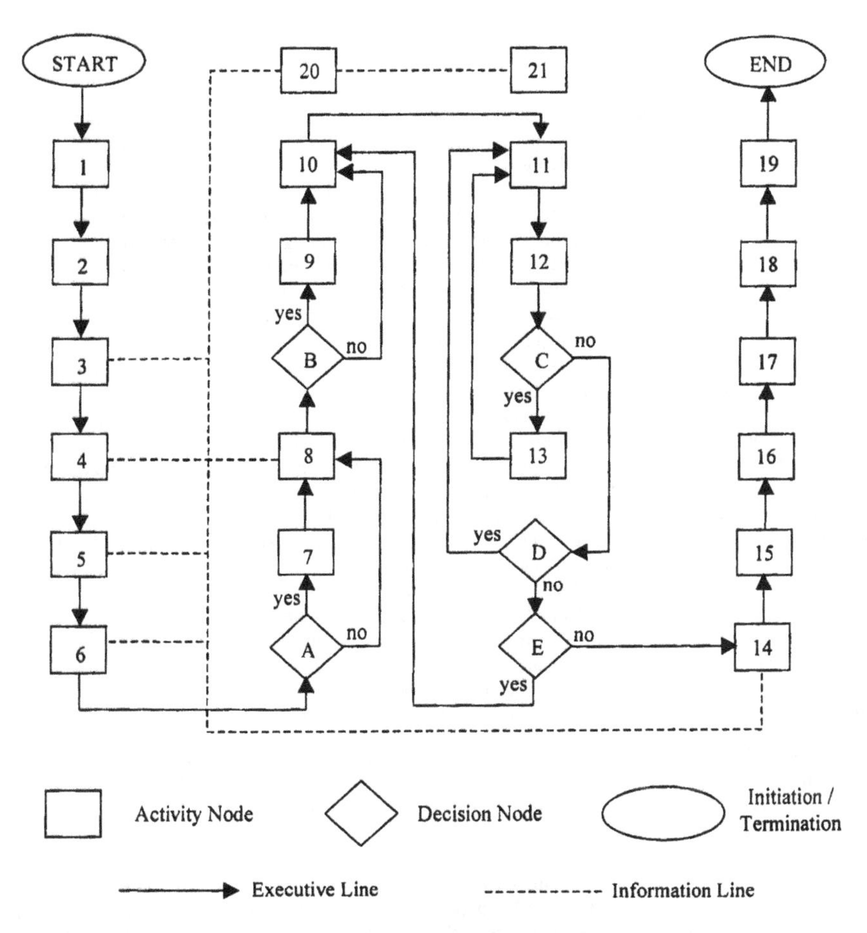

Figure 2. Flowchart of the algorithmic procedure designed/developed for the multicriteria choice of enzyme immobilization process (EIP) for biosensor design and construction.

17. Restructuring of the FMCA matrix by taking into consideration an additional criterion for sensitivity.
18. Re-running of FMCA and final sensitivity analysis performance.
19. Robustness investigation and proposal of solution.
20. KB operations (enrichment, information processing consultation)
21. Intelligent agent performance for knowledge acquisition for external Bases according to an operational interface described in [7].

A. Is there any criterion, taken into consideration (explicitly or implicitly) in empirical choices stored in the KB, not included?
B. Is there any overlapping between criteria?
C. Is the solution sensitive according to specifications, set in stage 3?
D. Is there any need for revising the EIP-set used as input for FMCA, under the light of the EIP-results obtained so far?
E. Is there another ESC, which has not been examined yet?

It is worthwhile noting that the specifications set *a priori* in stage 3 as well the criteria selected in stage 6 are heavily depended on the conditions expected to prevail during service life and the demands of the wider system within which the biosensors subsystem will be integrated. E.g. in the novel system for environmental monitoring through a cooperative/synergistic scheme between bioindicators and biosensors we have investigated recently [8], the operations of the biosensors subsystem are in close interaction with the bioindicators subsystem and all of them depend on the local natural environment.

3. Implementation

The methodology described above was implemented in the case of EIPs entrapment/encapsulation, covalent binding, entrapment/crosslinking, crosslinking/electrodeposition, adsorption/electrodeposition, pure adsorption, entrapment/electrodeposition, named A_j (j=1,2,...,7), respectively. The criteria used were EIP controlability, immobilized enzyme mass loss (due to decreasing diffusional resistance during service time), contribution to the ability of the substrate surface to be regenerated, immobilized enzyme resistance to environmental conditions expected to prevail during its operation, enzyme activity decrease due to immobilization, estimated total cost, EIP simplicity, named f_i (i=1,2,...,7), respectively. The evaluation of the criteria weights vector and the multicriteria preference matrix was performed according to a three-stage Delphi method with consensus in input. The values assigned, under fuzzy form to count for uncertainty, are shown in Appendix II.

The fuzzy multicriteria analysis algorithm followed herein is described in [9], including a piecewise linear reference function $P = H(d) \in [0, 1]$ as a generalized criterion, where d is the difference of the evaluation of two alternatives, a, b; the parameters of H(d) are an indifference threshold q, the greatest value of d, below which there is indifference, and a preference threshold p the lowest value of d above which there is strict preference – the interval between q and p being considered as the region of linear change of preference. The original deterministic model used as basis was Promethee [10], where fuzzy algebra was incorporated according to the rules established by Dubois and Prade [11] while pairwise comparisons between fuzzy sets where performed according to the Cheng and Klein [12]. The results are shown in Fig. 3, where partial ranking of alternatives (PRA), sensitivity analysis for each criterion (SAC) and total ranking of alternatives (TRA), at (a) low preferability resolution obtained for medium q, p values (q = 1.5, p = 3.0) and (b) high preferability resolution obtained for low q, p values (q=0.5, p=1.0). According to TRA-results it seems that 'adsorption' either with electrodeposition or in pure form (A5 and A6, respectively) dominates among the alternatives; entrapment/encapsulation (A1) is the second best while crosslinking/ electrodeposition (A4) appears to be rather sensitive in preference order. The solutions at the (a), (b) levels of preferability resolution are as follows:

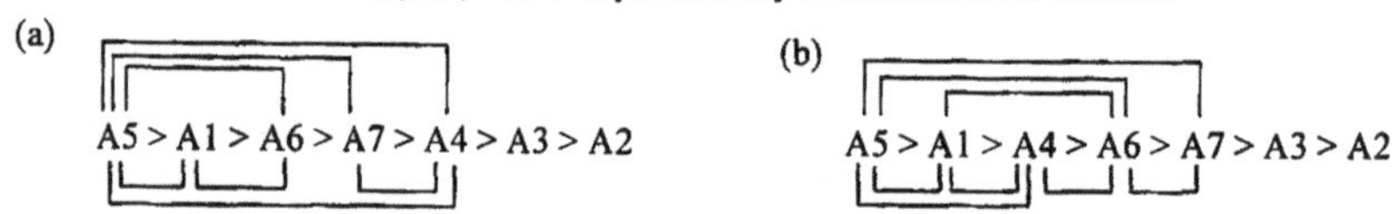

Where the symbol '>' means 'better than' and coupling lines indicate incomparability of the joint alternatives, as it is also shown in the PRA-diagrams of Fig. 3; in the same diagram, the area of each circle equals the relative value (as a crisp number S_j) of the corresponding alternative in the final rating after the Cheng and Klein evaluation while two cycles, not directly connected, characterize the corresponding alternatives as 'incomparable'.

Detailed monoparametric sensitivity analysis as regards the impact of each criterion's weight on the suggested/preferred alternative A_5 (adsorption/electrodeposition) indicates that only for (i) very high deviations from the central weight of criteria f_4, f_5, f_6 (immobilized enzyme resistance to environmental conditions expected to prevail during its operation, enzyme activity decrease due to immobilization, estimated total cost, respectively) and (ii) high preferability resolution obtained for low q, p values, the difference $S_5 - S_1$ becomes negative, implying that the alternative A_1 (entrapment/encapsulation) is suggested; consequently, the solution indicating A_5 as dominant is adequately robust, according to total ranking, although certain pairwise incomparabilities appear, according to partial order.

4. Other biochemical taxonomic approaches

The collection of alternative EIPs used herein for FMCA has been formed by induction, according to the algorithmic procedure shown in Fig. 2. Nevertheless, there are other classification schemes that serve taxonomic purposes. E.g., according to a technical report published by IUPAC [13] EIPs can be classified on the basis of the nature of their attachment onto the support/substrate surface as follows:

1. Covalent bonding of the enzyme to a derivatized, water-insoluble matrix.
2. Intermolecular cross-linking of enzyme molecules using multi-functional reagents.
3. Adsorption of the enzyme onto a water-insoluble matrix.
4. Entrapment of the enzyme inside a water-insoluble polymer lattice or semi-permeable membrane.

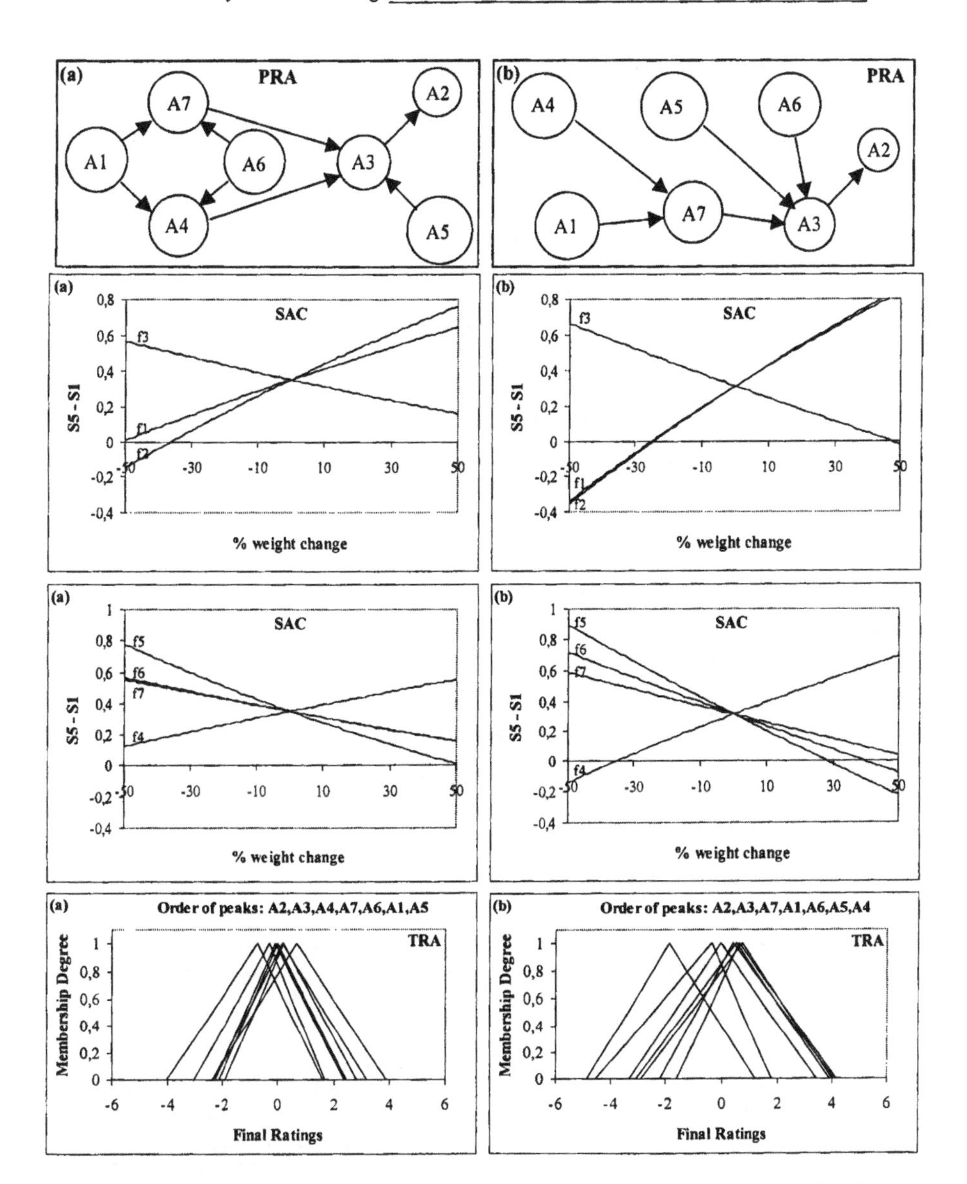

Figure 3. Partial ranking of alternatives (PRA), sensitivity analysis for each criterion (SAC) and total ranking of alternatives (TRA), at (a) low preferability resolution obtained for medium q, p values (q = 1.5, p = 3.0) and (b) high preferability resolution obtained for low q, p values (q=0.5, p=1.0). The arrow → means 'better than'.

On the other hand, the immobilized configuration of an enzyme can be used to form the following taxonomy;

1. Solid-phase immobilized enzyme reactors (packed bed and open tubular) for use in continuous flow techniques such as flow injection analysis and post-column derivatization in liquid chromatography.
2. Immobilized enzyme membranes incorporated into sensors such as potentiometric enzyme electrodes and optical sensors.
3. Solid-phase immobilized enzyme films for use in disposable, dry reagent kits with photometric detection.

The type of support/substrate can also be used to form another kind of taxonomy as follows:

1. Hydrophilic biopolymers based on natural polysaccharides, such as agarose, dextran, cellulose.

2. Lipophilic synthetic organic polymers, such as polyacrylamide, polystyrene, nylon.
3. Inorganic materials, such as controlled pore glass and iron oxide.
Last, hierarchical taxonomies can be structured by deduction, on the basis of logical categorization, as in the following scheme:

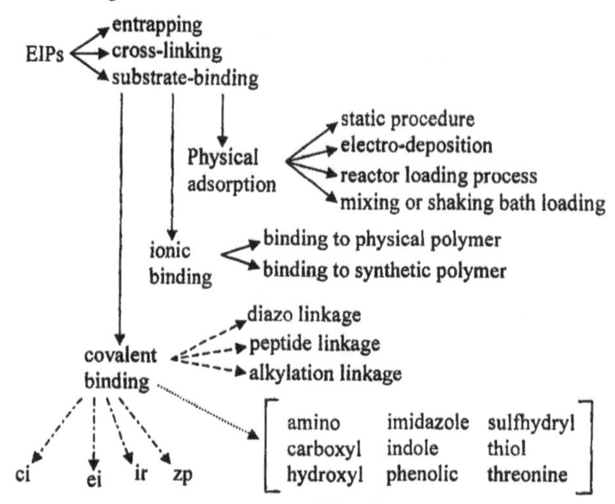

ci: covalent attachment of the enzyme in the presence of a competitive inhibitor or a substrate.
ei: formation of a reversible, covalently linked enzyme-inhibitor complex.
ir: activation of a chemically modified soluble enzyme whose covalent linkage to the matrix is achieved by newly incorporated residues.
zp: activation of a zymogen precursor.

Examples

Diazotization:	Su--N=N--En	Schiff's base formation:	Su--CH=N--En
Amide bond formation:	Su--CO-NH--En	Amidation reaction:	Su--CNH-NH--En
Alkylation and Arylation:	Su--CH_2-NH-En	Thiol-Disulfide interchange:	Su--S-S--En
	Su--CH_2-S—En	Mercury-Enzyme interchange	
UGI reaction (see Appendix III)		Gamma-Irradiation induced coupling	
Su- binding with bifunctional reagents:		Su-O$(CH_2)_2$N=CH$(CH_2)_3$CH=N-En	

where Su = substrate/support, En = Enzyme

For practical reasons, some kind of mixing of certain sub-categories may be desired to obtain a product which satisfies the specifications set *a priori*. On the other hand, two or more EIPs in series may be necessary to obtain such a product. E.g., physical adsorption of the enzyme may be necessary to facilitate a subsequent covalent reaction while, in several cases, effective stabilization of enzymes temporarily adsorbed onto a substrate has been achieved by cross-linking the protein in a chemical reaction subsequent to its physical adsorption. Therefore, although the above mentioned ontological taxonomic approach is based on biochemical reasoning and formal logic, the limitation of the alternative EIPs, used as candidates for optimal selection, by adopting an *ad hoc* phenomenological level, is unavoidable; at any case, this collection is the result of the preliminary round of the Delphi method performed to assign values on the weights vector and the preference matrix, and such as is *a priori* acceptable and consistent to the subsequent stages of the procedure adopted herein.

It is worthwhile noting that, once the EIP has been chosen, further classification is possible through the geometry of the substrate, which usually has the form of particles, films/membranes, tubes, fibers. Last, the substrates, especially the organic ones, constitute another important classifier, as they may play a central role in the chemical mechanism in the case of either industrial production or in citu measurements (via bioreactors or biosensors, respectively); the most usual substrates that can be used in such a classification are polysaccharides, proteins, carbon, polystyrenes, polyacrylates, maleic anhydride based copolymers, polypeptides, vinyl and allyl polymers, and polyamides; obviously, each one of these organic substrates is sub-divided into the next level categories according to the usual taxonomic order of Biology and Organic Chemistry. The final combination of all these classifiers gives the specific EIP, which can be the alternative in multicriteria ranking or/and the object of biochemical R&D. Especially for biosensors the substrate production process, as well as the physico-chemical characterization of the interface and its coupling/combination with any auxiliary item for measuring the analyte, should be incorporated into the final combination mentioned above to formulate a code, which may facilitate storing/retrieval/processing within a KB. For example, a code on immobilization of glucose oxidase on conducting poly (2-fluoroaniline) films by physical adsorption, should also include 'electrochemical deposition' (as substrate production process), 'cyclic voltammetry' (as physico-chemical characterization of the interface) and 'Ag/AgCl reference electrode' (as auxiliary item), if the technique quoted in [14] is adopted.

5. Discussion and experimental evidence

A major advantage of the adsorption, as one of the preferred EIP, is that usually no additional reagents and only a minimum of activation steps are required. Adsorption tends to be less disruptive to the enzymatic protein than chemical means of attachment because the binding is mainly by hydrogen bonds, multiple salt linkages, and Van der Waal's forces. In this respect, the method bears the greatest similarity to the situation found in natural biological membranes and has been used to model such systems. Because of the weak bonds involved, the sorption of the protein resulting from changes in temperature, pH, ionic strength or even the mere presence of substrate, is often observed. Another disadvantage is none-specific, further adsorption of other proteins or other substances as the immobilized enzyme is used. This may alter the properties of the immobilized enzyme or, if the substance adsorbed is a substrate itself for the enzyme, the rate will probably decrease, depending on the surface mobility of enzyme and substrate.

One of the most critical variables in EIP choice including the determination of optimal conditions for effective operation, is reversibility, R, being also an independent variable common to two or more conflicting dependent cost-variables: high R-value means (i) high reactivity with low preparation/construction cost (of biocatalyst or biosensor) and low probability of enzyme to be modified, due to very weak binding to substrate surface, but also (ii) high desorption rate, leading to enzyme mass loss and consequently to unreliable operation (catalysis or measurement, in the case of bioreactor or biosensor, respectively) and decrease of useful lifetime, implying an increased operating cost due to frequent replacement and high maintenance expenses generally. Therefore, the optimal value R_{opt} can be determined by finding out the abscissa of the sum $(RC + CC)_{min}$, where RC and CC are the biocatalyst/biosensor replacement and the construction/preparation cost, respectively. If cheaper technology is invented/applied, the CC-curve moves downwards to CC′ and becomes more flat, implying decrease of R_{opt} to R'_{opt} at $(RC + CC')_{min}$, as shown in Fig. 4a. If skilled labour becomes more expensive, then the RC-curve moves upwards to RC′ (as replacement, especially of spatially distributed biosensors, is more labour intensive in comparison with their construction, which is more capital intensive) and becomes steeper, implying also decrease of R_{opt} to R'_{opt} at $(RC' + CC)_{min}$, as shown in Fig. 4b. If requirements, as regards product/measurement quality, increase, then the MC-curve moves upwards to MC′ and becomes steeper, implying also decrease of R_{opt} at $(MC + PC)_{min}$ to R'_{opt} at $(MC' + PC)_{min}$, as shown in Fig. 4c, where MC and PC are the biocatalysis/measurement cost (increasing with R since reliability decreases due to enzyme mass loss) and the cost due to enzyme properties change (increasing with irreversibility, due to selectivity/sensitivity decrease, as a result of enzyme's modification because of the strong chemical binding onto the substrate surface), respectively. Since all these changes, i.e. cheaper technology, more expensive skilled labour, increase of quality requirements, are in accordance with the general tendency observed in most industrialized countries, we can synthesize the corresponding trends to obtain the overall significant shifting of R_{opt} to R'_{opt} as shown in Fig. 4d. Each cost-diagram is coupled with the corresponding marginal cost diagram, R_{opt} is the abscissa of the point where the two conflicting marginal costs are equal.

Reversibility, as a quasi-value representative of weakness/strength of enzyme binding-on-substrate (or probability of keeping each original structure/configuration while bound and consequently its capability of acting as if it was free/dispersed in solution), can be measured as the index/ratio RI; (active enzyme adsorbed, in mass units) / (total mass of enzyme adsorbed). As the nominator of this ratio is very difficult to be measured, we may replace it with a variable proportional to this mass, since RI is useful for comparison purposes, not as an absolute magnitude; such a variable is the amount desorbed under the conditions expected to prevail during service time, giving the enzyme back to the solution in its original/active form. Since each enzyme-substrate combination is a special case, it is worthwhile examining via RI the same combination under different conditions (i) of substrate

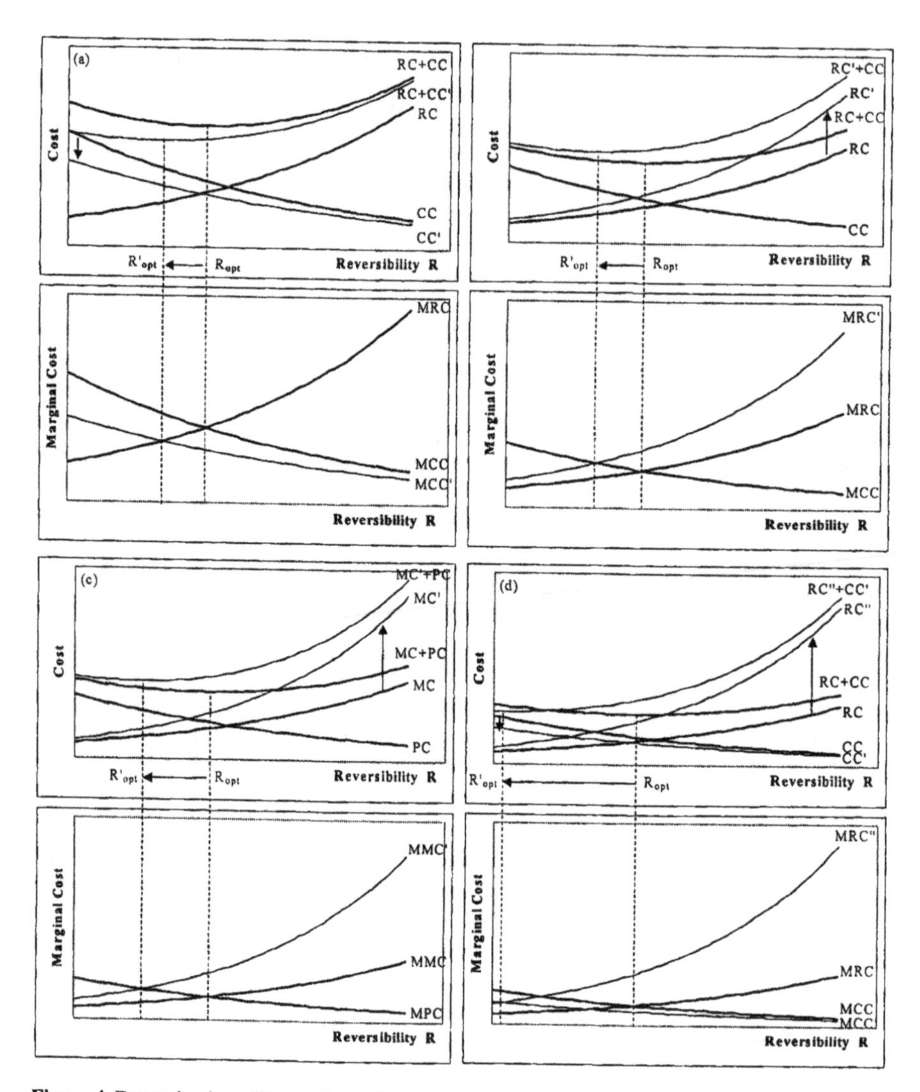

Figure 4. Determination of Ropt, where cheaper technology, more expensive skilled labour, increase of quality requirements, are introduced into the enzyme immobilization process (cases a, b, c, respectively, while case d represents a synthesis giving R_{opt} as result of the cumulative effect).

production and (ii) expected to prevail during service time. An approximation of the first case examination can be achieved by means of adsorption of another molecule, selected among certain substances which have been extensively used to investigate the properties of various substrates. Such a substance is methylene blue (MB) that can be activated by light to an excited state, which in turn activates oxygen to yield oxidizing radicals; these radicals can cause cross linking of amino-acid residues or proteins and hence achieve certain degree of cross linking. It is worthwhile noting that cross linking an enzyme to itself is both expensive and insufficient, as some of the protein material will inevitably be acting mainly as a support; this will result in relatively low enzymatic activity. For this reason cross linking is best used in conjunction with one of the other EIPs, i.e. mostly as a means of stabilizing already adsorbed enzymes and also from preventing leakage for polyacrylamide gels. When the enzyme is also covalently linked to the original substrate, very little desorption is likely by using this method; e.g. carbamy phosphokinase cross-linked to alkylamine glass with glutaraldehyde looses less than 20% of its original activity after continuous performance within a column at room temperature in a two weeks time period.

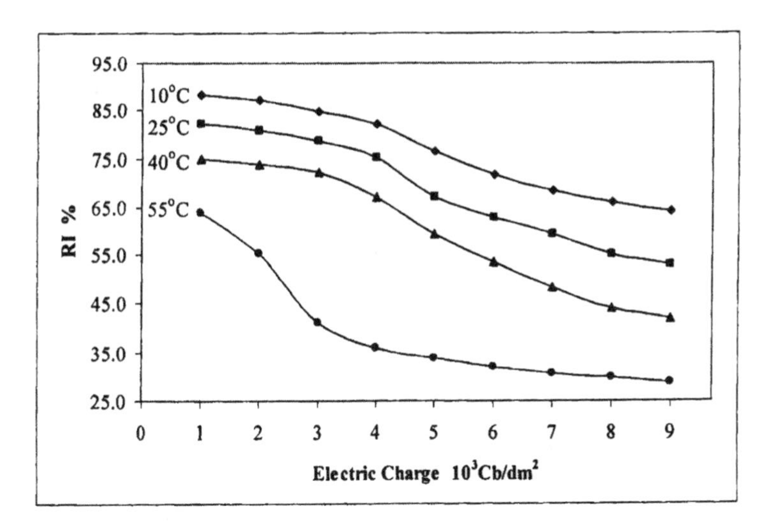

Figure 5. The dependence of reversibility index RI (%) on anodic oxidation electric charge, forming electrochemical porous film of aloumina at various temperatures (electrolyte H_2SO_4 15% w/w).

We have measured RI (%) for MB physically adsorbed on anodically oxidised pure Al 99.7% (0.3% for traces primarily of Si, Fe, Ti, Cu, Zn and secondarily of other elements) as a function of electric charge (10^3 Cb/dm^2) of anodization, which is a measure of the anodic film thickness (alumina is one of the first inorganic substrates used for enzyme immobilization by physical adsorption). The RI-denominator was determined by measuring MB-concentration in the aquatic bath of the dye before and after adsorption at zero point charge pH by means of a HACK DR4000U UV-visible spectrophotometer at λ_{max} = 664 nm (see Appendix IV). The RI-nominator was similarly detected by desorbing the dye in a weak alkaline environment, which was slightly changing because of the $Al(OH)_3$ formation. The original MB-absorption spectrum was used as a reference to check any change in the spectrum curve corresponding to the aquatic solution obtained after desorption. The inverse sigmoid curve obtained for RI (see Fig. 5) indicates that higher electric charge, implying higher anodic oxide thickness, causes higher irreversibility (i.e. lower RI-values); this can be attributed to (i) higher porosity, which can be estimated by means of the classic Keller-Hunter-Robinson model for anodizing with elecrtrolyte H_2SO_4 w/w and (ii) higher Al-dissolution during anodizing resulting to lower crystallinity and higher amorphous/unstructured hydrated aloumina production; this latter product is more likely to cause irreversible coagulation with MB oligomers, a hypothesis which is further supported by (i) the distortion of the reference absorption MB-spectrum, observed when the solution obtained after desorption is examined, and (ii) the downwards shifting of the RI-curve when anodization temperature is increasing, implying higher dissolution rate.

6. Concluding Remarks

The methodological framework, we have developed under the form of an algorithmic procedure, for multicriteria choice of enzyme immobilization process (EIP) for biosensor design and construction, is proved to be functional. The implementation used to prove this functionality uses a 7x7 multicriteria matrix with the following input. Alternatives: entrapment/encapsulation, covalent binding, entrapment/crosslinking, crosslinking/electrodeposition, adsorption/electrodeposition, pure adsorption, entrapment/electrodeposition. Criteria: EIP controlability, immobilized enzyme mass loss (due to decreasing diffusional resistance during service time), contribution to the ability of the substrate surface to be regenerated, immobilized enzyme resistance to environmental conditions expected to prevail during its operation, enzyme activity decrease due to immobilization, estimated total cost, EIP simplicity. The monoparametric sensitivity analysis performed has shown that the suggested solution, as regards the preferred alternative is robust. The specifications set a priori and the criteria selected in certain stages are heavily depended on the conditions expected to prevail during service life and the demands of the wider system within which the biosensors subsystem will be integrated. E.g. in the novel system for environmental monitoring through a cooperative/synergistic scheme between bioindicators and biosensors we have investigated recently [8], the operations of the biosensors subsystem are in close interaction with the bioindicators subsystem and all of them depend on the local natural environment. Consequently, experimental work should be oriented to further investigating the functionality of this equipment under real measurement conditions in situ and the relevant properties of the materials in the laboratory in a real and/or simulated analyte micro-environment; the study on reversibility of adsorption studied here in is a representative sample of such a procedure.

Acknowledgments

The project is co-funded by the European Social Fund & National Resources – EPEAEK II-PYTHAGORAS (Environment: design/ development/ implementation of bioindicators/biosensors).

References

[1] R. Axen, J. Porath and S. Ernback, Chemical coupling of peptides and proteins to polysaccharides by means of cyanogen halides, *Nature (London)* **214** 1302 (1967).

[2] J. Porath, R. Axen and S. Ernback, Chemical coupling of proteins to agarose, *Nature (London)* **215** 1491 (1967).

[3] M.A. Mitz and L.J. Summaria, Synthesis of biologically active cellulose derivatives of enzymes, *Nature (London)* **189** 576 (1961).

[4] N. Weliki and H.H. Weetall, The chemistry and use of cellulose derivatives for the study of biological systems, *Immunochemistry* **2** 293 (1965).

[5] R.A. Messing, Immobilization techniques-enzymes, in *Comprehensive Biotechnology* (C.L. Cooney and A.E. Humphrey Eds.), Pergamon Press, **2** 191-201 (1985).

[6] N.F. Sheppard, Jr. and A. Guiseppi-Elie, Enzyme sensors based on conductimetric measurement, in *Methods in Biotechnology* (A. Mulchandani and K.R. Rogers Eds.), Humana Press, **6** 157-173 (1998).

[7] F.A. Batzias and E.C. Markoulaki, Restructuring the keywords interface to enhance CAPE knowledge via an intelligent agent, *Computer Aided Chemical Engineering* **10**, 829 (2002).

[8] F. Batzias and C.G. Siontorou, A novel system for environmental monitoring through a cooperative/synergistic scheme between bioindicators and biosensors, *Journal of Environmental Management* In press (2006).

[9] A.F. Batzias and F.A. Batzias, Computational and experimental process control by combining neurofuzzy approximation with multicriteria optimization, in *Computational Methods and Experimental Measurements* **11** 107 (2003).

[10] J.P. Brans, Ph. Vincke and B. Mareschal, How to select and how to rank projects: The PROMETHEE method, *European Journal of Operational Research* **24** 228 (1986).

[11] D. Dubois and H. Prade, Operations on fuzzy numbers, *Int. J. Systems Sci.*, **9(6)** 613 (1978).

[12] T.Y. Tseng and C.M. Klein, New algorithm for the ranking procedure in fuzzy decisionmaking, *IEEE Trans. On Systems, Man and Cybernetics*, **19(5)**, 1289 (1989).

[13] P.J. Worsfold, Classification and chemical characteristics of immobilized enzymes, *Pure & Appl. Chem.* **67**(4) 597 (1995).

[14] G.S. Wilson and D.R. Thevenot, Unmediated amperometric enzyme electrodes, in *Biosensors: A Practical Approach* (A.E.G. Cass, Ed.), IRL, New York, 287 (1990).

Appendix I

Example cases of the taxonomic approach adopted herein: attacking γ-aminopropyltriethoxy silane through an inorganic functional group, leaving the organic moiety (amine) available for covalent attachment of the enzyme, where the terminal amine can be fictionalized by functional coupling agents or by protein-/amine-bearing substrates; polymeric isocyanate bonding by forming carbamate bond between the inorganic surface and the isocyanate group with two options; if the enzyme is attached under alkaline conditions, a substituted urea bond is formed between an amine on the protein surface and the isocyanate; if moderately acid conditions are employed, then the isocyanate reacts with a hydroxyl group on the enzyme and a urethane bond is formed.

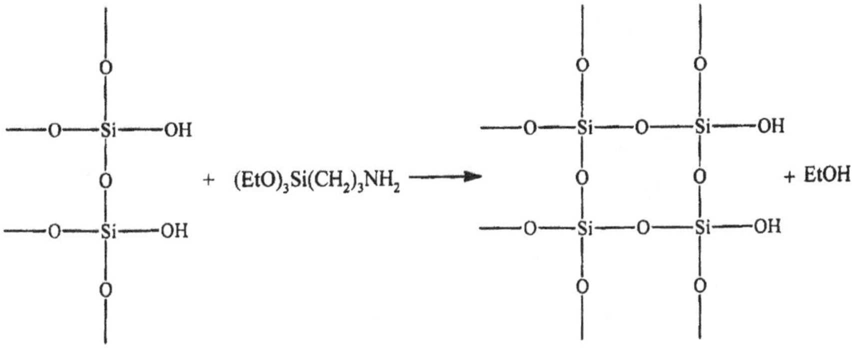

$$\text{—O—Si—OH} \;/\; \text{—O—Si—OH} + (EtO)_3Si(CH_2)_3NH_2 \longrightarrow \text{—O—Si—O—Si—OH} \;/\; \text{—O—Si—O—Si—OH} + EtOH$$

Figure 6. Silane coupling of enzyme to inorganic substrate

(a)

$$O{=}C{=}N\text{—}C_6H_4\text{—}CH_2\text{—}C_6H_3(N{=}C{=}O)\text{—}CH_2\text{—}C_6H_4\text{—}N{=}C{=}O$$

(b)

$$\text{—}C_6H_4N{=}C{=}O + \text{—Ti—OH} \longrightarrow \text{—}C_6H_4N\text{—}C({=}O)\text{—O—Ti—}$$

(c)

$$\text{—}C_6H_4N{=}C{=}O + RNH_2 \longrightarrow \text{—}C_6H_4NH\text{—}C({=}O)\text{—NHR}$$

(d)

$$\text{—}C_6H_4N{=}C{=}O + ROH \longrightarrow \text{—}C_6H_4NH\text{—}C({=}O)\text{—OR}$$

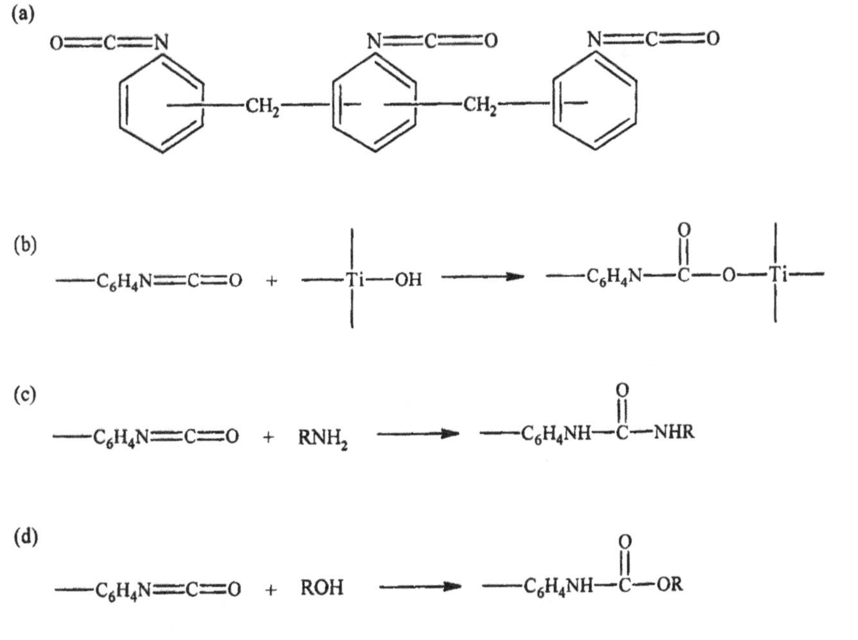

Figure 7. Polymeric isocyaniate bonding of enzyme to inorganic substrate; (a) polymeric isocyanate, (b) carbamate bond formation, (c) substituted urea bond formation, (d) urethane bond formation.

Appendix II

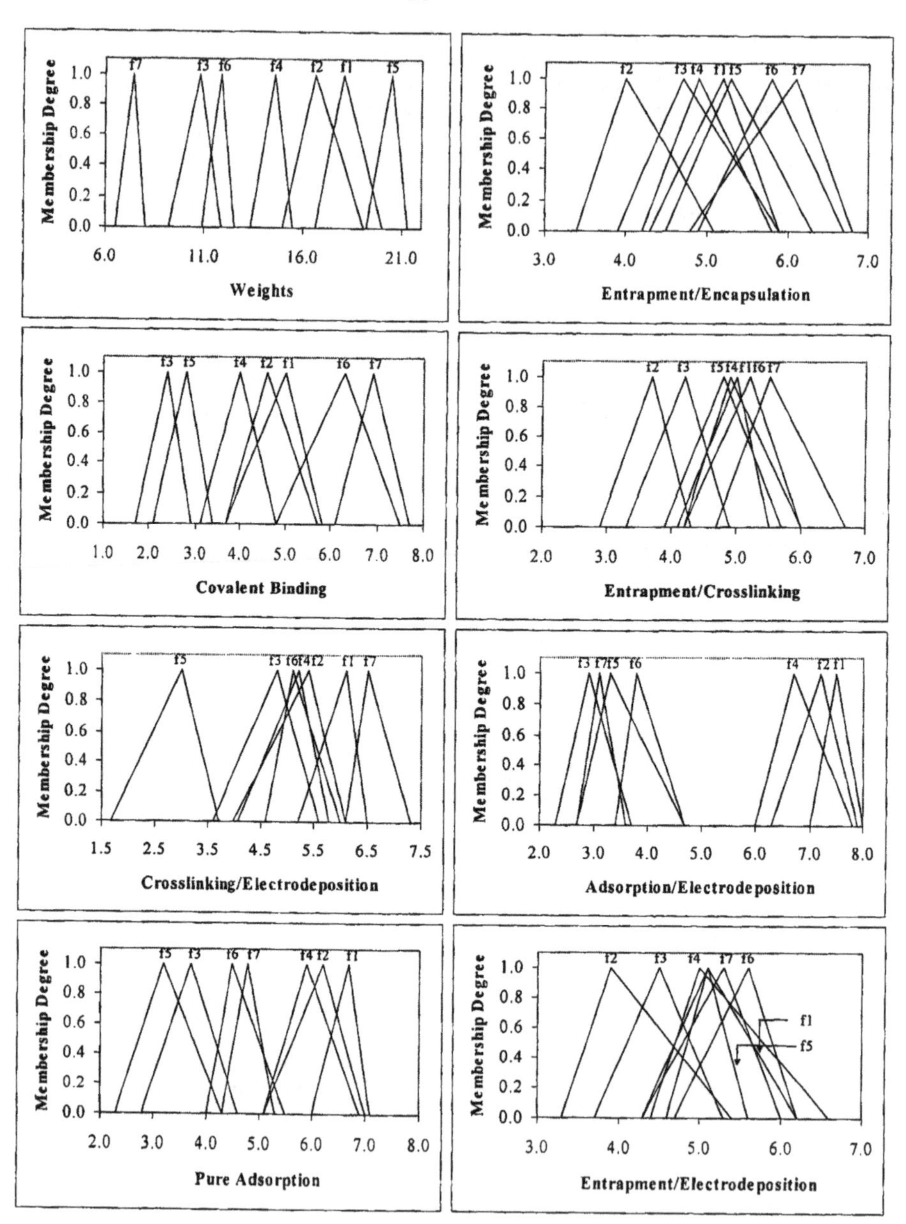

Figure 8. Fuzzy input to (a) weights vector and (b) multicriteria preference matrix as depicted per alternative EIP, according to the stages 10, 11 of the algorithmic procedure shown in Fig. 2.

Appendix III

The UGI reaction, used as a paradigm in the Section on Taxonomy, begins with the reaction of carbonyl with the amine to give imine **1**; condensation of the imine with the isocyanide and carboxylic acid gives intermediate **2**; Acyl transfer of intermediate **2** quickly gives the final bis-amide **3**.

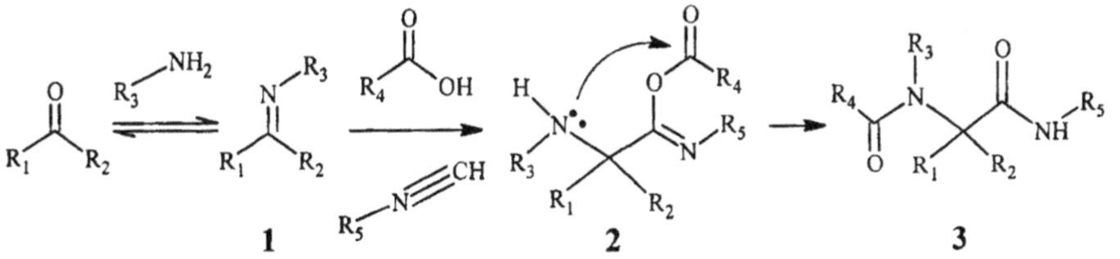

Appendix IV

methylene blue

λ_{max}=664 nm

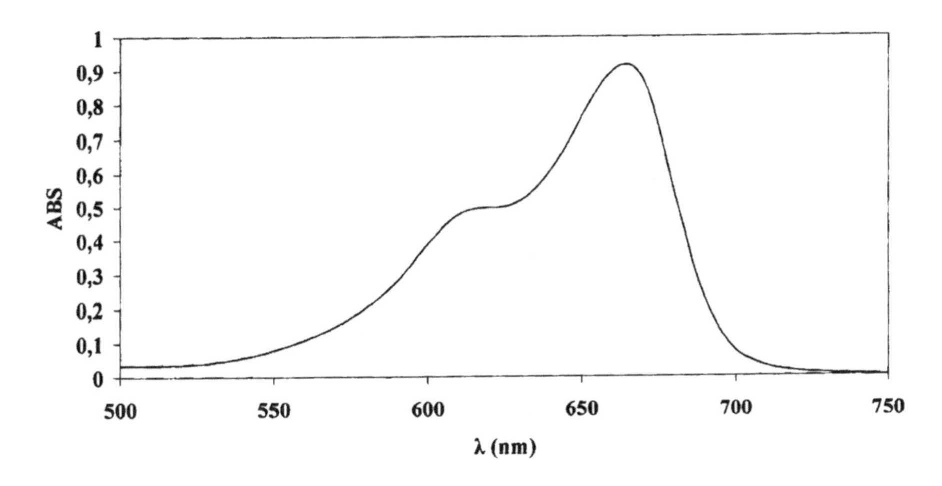

Figure 9. The methylene blue reference absorption spectrum obtained by means of the HACK DR4000U UV-visible spectrophotometer we have used for studying the reversibility index RI (see Fig. 5); irreversible coagulation of MB with anodic Al_2O_3 distorts this spectrum and the deviation identified/quantified by a differential pattern recognition technique gives a measure of the irreversibility corresponding to RI, after desorption has reached equilibrium.

Brill Academic Publishers
P.O. Box 9000, 2300 PA Leiden,
The Netherlands

Lecture Series on Computer and Computational Sciences
Volume 6, 2006, pp. 540-560

Modelling dunes with vegetation and dunes on Mars.

H.J. Herrmann[a,b 1], O. Durán[c], E.J.R. Parteli[c]

a. Institut für Baustoffe, ETH Zürich, Schafmattstr. 6, CH-8093 Zürich, Switzerland.
b. Departamento de Física, Universidade Federal do Ceará, 60455-970, Fortaleza, CE, Brazil.
c. Institut für Computerphysik, ICP, Pfaffenwaldring 27, 70569 Stuttgart, Germany.

Received July 13, 2006; accepted July 20, 2006

Abstract: Sand dunes are found in a wide variety of shapes in deserts and coasts, and also on the Planet Mars. The basic mechanisms of dune formation could be incorporated into a continuum saltation model, which successfully reproduced the shape of the barchan dunes and has been also applied to calculate interaction of barchans in a field. We have recently extended our dune model to investigate other dune shapes observed in nature. Here, we present the first numerical simulation of the transformation of barchan dunes, under influence of vegetation, into parabolic dunes, which appear frequently on coasts. Further, we applied our model to reproduce the shape of barchan dunes observed on Mars, and we found that an interesting property related to the martian saltation is relevant to predict the scale of dunes on Mars. Our model could also reproduce unusual dune shapes of the Martian north polar region, like rounded barchans and elongated linear dunes. Our results support the hypothesis that these dunes are indurated.

Keywords: Sand dunes, aeolian transport, vegetation, Mars.

PACS: 45.70.Qj,05.65.+b,92.10.Wa,92.40.Gc,92.60.Gn

1 Introduction

Sand dunes develop wherever sand is exposed to an agitated medium that lifts grains from the ground and entrains them into a surface flow. On coasts, dunes are constituted of sand which comes from the sea and is therafter deposited onto the beach. Once the grains are exposed to the air, they dry and some can be carried by the wind, initiating sand transport.

The mechanism responsible for dune formation is called *saltation* [1, 2, 3, 4, 5, 6, 7, 8, 9, 10]. Grains are lifted from the ground by the wind, are next accelerated downwind, and after a certain distance ℓ they collide back onto the ground. Each impact grain may eject further grains — mechanism called "splash" [11, 12] — and thus a cascade process occurs, until the saltation flux achieves saturation, since the air transfers momentum back to the grains. Dunes appear because an exposed sand sheet is unstable, evolves into hills, and next into dunes with a slip face. The existence of a saturation transient of the sand flux is reponsible for the existence of a minimum dune size.

1.1 The Dune Model

We have recently proposed a minimal dune model which provides an understanding of important features of real dunes, such as their longitudinal shape and aspect ratio, the formation of a slip

[1] Corresponding author. E-mail: hans@fisica.ufc.br

face, the breaking of scale invariance, and the existence of a minimum dune size [13, 14]. This model combines an analytical description of the turbulent wind velocity field above the dune with a continuum saltation model that allows for saturation transients in the sand flux. The fundamental idea of the dune model is to consider the bed-load as a thin fluid-like granular layer on the top of an immobile sand bed. The model has been latter improved by Schwämmle and Herrmann (2005) [15] to take into account lateral sand transport due to perturbations transverse to the main wind direction.

In previous works we studied the formation and evolution of barchans and transverse dunes. In particular we could successfully reproduce the crescent shape of dunes which develop in areas of unidirectional winds, their rate of motion, stability and interaction with each other. There are, indeed, several different sand patterns which can be observed on coasts according to wind condition, sand availability, humidity level and vegetation. Modellization of sand dunes which accounts for such variables yields an advantageous tool in the understanding of coastal processes and morphology.

Here we present some new results we obtained from an extension of our model to study dunes with vegetation, which are ubiquitous on coasts, and dunes on Mars. In this Section, a summary of the main ingredients of the dune model is presented.

1.1.1 *Wind shear stress*

Sand transport takes place near the surface, in the turbulent boundary layer of the atmosphere [16]. In this turbulent layer, the wind velocity $v(z)$ at a height z may be written as:

$$v(z) = \frac{u_*}{\kappa}\ln\frac{z}{z_0}, \tag{1}$$

where $\kappa = 0.4$ is the von Kármán constant, u_* is the wind shear velocity, which is used to define, together with the fluid density ρ_{fluid}, the shear stress $\tau = \rho_{\text{fluid}}u_*^2$, and z_0 is called aerodynamic roughness.

A dune or a smooth hill can be considered as a perturbation of the surface that causes a perturbation of the air flow onto the hill. In the model, the shear stress perturbation is calculated in the two dimensional Fourier space using the algorithm of Weng *et al.* (1991) [17] for the components τ_x and τ_y, which are, respectively, the components parallel and perpendicular to wind direction. In what follows, we present the sand transport equations introduced in Sauermann *et al.* (2001) [13] and refer to Schwämmle and Herrmann (2005) [15] for a two-dimensional extension of the model.

1.1.2 *Continuum saltation model for sand transport*

The sand transport model uses the air shear stress in the boundary layer to calculate the sand flux. The equation for the sand flux [13] is a differential equation that contains the saturated flux q_s at the steady state, and the characteristic length ℓ_s that defines the transients of the flux:

$$\frac{\partial}{\partial x}q = \frac{1}{\ell_s}q\left(1 - \frac{q}{q_s}\right). \tag{2}$$

In eq. (2), the saturation length ℓ_s is defined as:

$$\ell_s = \frac{1}{\gamma}\left[\frac{\ell}{(u_*/u_{*t})^2 - 1}\right], \tag{3}$$

where ℓ is the average saltation length and u_{*t} is the threshold wind shear velocity for sustained saltation, which defines the threshold shear stress $\tau_t = \rho_{fluid} u_{*t}^2$. The parameter γ models the average number n of grains dislodged out of equilibrium, and is written as:

$$\gamma = \frac{dn}{d(\tau_a/\tau_t)}, \tag{4}$$

where τ_a is the air born shear stress, which is distinguished from the grain born shear stress, i.e. the contribution of the saltating grains near the ground to the total shear stress τ due to their impacts onto the surface. The parameter γ gives thus the amount of grains launched into the saltation sheet when the wind strength deviates from the threshold by an amount τ_a/τ_t. The air born shear stress τ_a is lowered if the number of grains in the saltation layer increases and vice versa, which is called the "feedback" effect: At threshold, the wind just has a strength $\tau_a \approx \tau_t$ sufficient to maintain saltation. The parameter γ depends on microscopic quantities such as the time of a saltation trajectory or the grain-bed interaction, which are not available in the scope of the model.

The steady state is assumed to be reached instantaneously, since it corresponds to a time scale several orders of magnitude smaller than the time scale of the surface evolution. Thus, time dependent terms are neglected.

1.1.3 Surface evolution

The time evolution of the topography $h(x,t)$ is given by the mass conservation equation:

$$\frac{\partial h}{\partial t} = -\frac{1}{\rho_{sand}} \frac{\partial q}{\partial x} \tag{5}$$

where $\rho_{sand} = 0.62\rho_{grain}$ is the mean density of the immobile dune sand [13], while ρ_{grain} is the grain density. If sand deposition leads to slopes that locally exceed the angle of repose $\theta_r \approx 34°$ of the sand, the unstable surface relaxes through avalanches in the direction of the steepest descent. Avalanches are assumed to be instantaneous since their time scale is negligible in comparison with the time scale of the dunes.

For a dune with slip face, flow separation occurs at the brink, which represents a discontinuity of the surface. The flow is divided into two parts by streamlines connecting the brink with the ground. These streamlines define the separation bubble, inside which eddies occur and the flow is often re-circulating [14]. In the model, the dune is divided into several slices parallel to wind direction, and for each slice, one separation streamline is defined at the lee side of the dune. Each streamline is fitted by a third order polynomial connecting the brink with the ground, at the reattachment point. The distance between the brink and the reattachment point, and thus the length of the separation bubble is determined by the constraint of a maximum slope of 14° for the streamlines [14]. Inside the separation bubble, the wind shear stress and sand flux are set to zero.

Simulation steps may be summarized as follows: (i) the shear stress over the surface is calculated using the algorithm of Weng *et al.* (1991) [17]; (ii) from the shear stress, the sand flux is calculated with eq. (2); (iii) the change in the surface height is computed from mass conservation (eq. (5)) using the calculated sand flux; and (iv) if the inclination of the surface gets larger than θ_r, avalanches occur and a slip face is developed. Steps (i) – (iv) are iteratively computed until the steady state is reached.

1.2 Barchan dunes and barchan fields

In spite of its apparent complexity, the procedure presented above is so far the simplest method for calculating the evolution of sand dunes. In particular for barchan dunes the dune model has been extensively tested and its results have been found to be quantitatively in good agreement with field measurements [18]. Moreover, the model has been applied to calculate interaction of barchan dunes in a field, and revealed that dunes may interact in different ways, depending on their sizes. Since smaller dunes have higher velocities, they may easily "collide" with larger, slower wandering ones during their downwind motion in a field. Our simulations show that very small dunes in this case may be "swallawed up" by larger dunes downwind. But if the size difference between the interacting dunes is not too large, then the smaller dune upwind gains sand from the larger one, which then becomes smaller and may wander away (fig. 1). Effectively, it is like if the smaller dune were crossing over the larger one [19], as proposed recently by Besler (1997) [20]. A systematic study of dune interaction has been presented in Durán *et al.* (2005) [21].

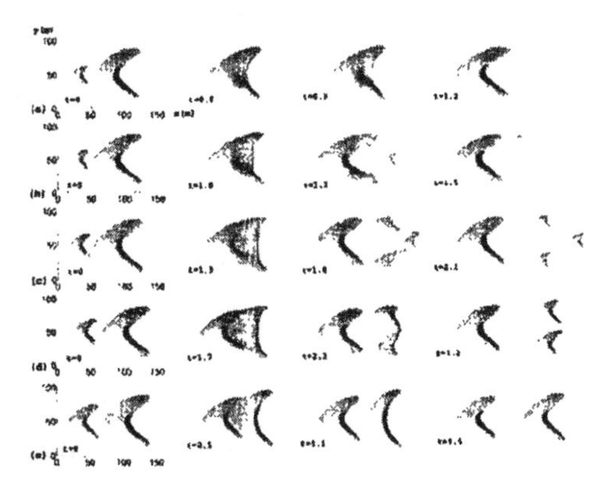

Figure 1: Four simulation snapshots of barchan dune interaction. Cases (a) — (e) correspond to different initial heights for the upwind, smaller barchan (Durán et al. 2005).

Many times, however, dunes meet obstacles and cannot evolve free on bedrocks or sand sheets under the action of wind. A classical example is the case of most coastal dunes. Dunes wandering onto the continent often compete with vegetation, which grows close to the sea where it finds favorable, humid conditions. We all know that plants fix the sand on the ground, but we also expect that the efficiency to stop sand transport should depend for instance on vegetation amount. We investigated with our dune model what happens with a barchan dune when it encounters a vegetated area. Tsoar and Blumberg (2002) [22] proposed that barchans should transform into parabolic dunes under the influence of vegetation. We simulated vegetation following a picture recently introduced by Nishimori and Tanaka (2001) [23], and we could perform the first computational simulation of this transition.

Furthermore, we have applied our dune model to calculate dunes on Mars. We find from our dune model that dune shapes similar to the ones observed on the floor of martian craters could have been achieved by winds of the scarce atmosphere of Mars in the present. And we also found that a property of the martian saltation must be taken into account for correctly predicting the size of dunes on Mars. Another interesting issue we investigated is whether induration of dunes covered by CO_2 frost — in a similar way to an experiment carried out by Kerr and Nigra (1952) [24] where dunes have been covered with oil — could be the explanation for unusual dune shapes observed on the martian north polar region. Supported by the results from our dune model, we concluded that, while vegetation transforms barchans into parabolic dunes on Earth, induration due to CO_2 frost could yield dome-shaped, elliptical and straight, linear dunes on Mars.

2 Transformation of barchans into parabolic dunes: Sand mobility competing with vegetation

One particular factor which can have a significant influence on the dune shape is the presence of vegetation on or around the dunes. A recent investigation of aerial photographs covering a time span of 50 years Tsoar and Blumberg (2002) [22] found that barchans can invert their shape to form parabolic dunes and vice versa when the amount of vegetation changes. Parabolic dunes are U-shaped dunes the arms of which point toward the direction of the prevailing wind. The amount of vegetation varied over that period because of human activities (such as grazing or stabilization) or because of wind power.

The formation of parabolic dunes has been modeled numerically with a lattice model [23, 25]. Though the authors find intermediate formation of parabolic dunes, this does not constitute the transition between full-sized barchan and parabolic dune found in McKee and Douglas (1971) [26] and in Tsoar and Blumberg (2002) [22]. This effect has not been investigated theoretically before. We propose a model for vegetation growth taking into account sand erosion and deposition and use the saltation model [13] for simulating the sand transport which determines the evolution of the dunes.

2.1 Vegetation models

2.1.1 Vegetation Growth:

We propose a continuous model for the rate of vegetation growth, following the idea of Nishimori and Tanaka (2001) [23]. Vegetation is characterized by its local height h_v. We suppose that vegetation can grow until it reaches a maximum height H_v, and that the growth process has a characteristic time t_g, which may be enhanced or inhibited depending on the climatic conditions.

Moreover, the vegetation growth rate should be a function of the time rate of sand surface change $(\partial h/\partial t)$. After any temporal change of the sand surface h, vegetation needs time to adapt to the new conditions. We introduce this phenomenological effect as a delay in the vegetation growth. In this way, the following equation holds:

$$\frac{dh_v}{dt} = \frac{H_v - h_v}{t_g} - \left|\frac{\partial h}{\partial t}\right|. \tag{6}$$

If the second term on the right-hand side of eq. (6) is larger than the first one, i.e. if the erosion rate is sufficiently high, then dh_v/dt becomes negative, which means that vegetation dies since its roots become exposed [22].

2.1.2 Shear Stress Partitioning:

The shear stress partitioning is the main dynamical effect of the vegetation on the flow field and, hence, on the sand transport. The vegetation acts as roughness that absorbs part of the momentum transfered to the soil by the wind. Thus, the total surface shear stress is divided into two components, one acting on the vegetation and the other on the sand grains. The fraction of the total shear stress acting on the sand grains can be described by the expression [27]:

$$\tau_s = \tau\left(1 - \frac{\rho_v}{\rho_c}\right)^2 \tag{7}$$

where τ is the total surface shear stress, τ_s is the shear stress acting on the non-vegetated ground, ρ_v is the vegetation density, defined as $(h_v/H_v)^2$, and ρ_c is a critical vegetation density that depends mainly on the geometric properties of the vegetation [16].

Equation (7) represents a reduction of the shear stress acting on the sand grains, which also yields a reduction of the sand flux. Both equations (6) and (7) contain the interaction between the vegetation and the sand surface. One implication of eq. (6) is that at those places where sand erosion or deposition is small enough vegetation grows. Afterwards, the consequent decrease of the shear stress (eq. (7)), and also of the sand flux, yields sand deposition, which in turn slows down the vegetation growth.

2.2 Results

We performed simulations placing a 4.2 m high barchan dune on a rock bed and then allowing the vegetation to grow. A zero influx and a 0.5 m/s upwind shear velocity are set. We also fixed $\rho_c = 0.5$, a typical value for spreading herbaceous dune plants [16] and $H_v = 1.0$ m. Finally, the dune model parameters are as in Sauermann *et al.* (2001) [13].

We studied the influence of the vegetation characteristic growth time t_g, which contains the information of the growth rate and which, hence, controls the strength of the interaction between the dune and the vegetation.

The upper part of Figure 2 shows snapshots of the evolution of a barchan dune under the influence of vegetation with a characteristic growth time of 7 days. Calculations were performed with open boundaries and a zero influx, which means vegetation is high enough to inhibit sand transport on the flat ground. The evolution of the vegetation density is shown in Figure 3. Initially, the vegetation invades those places where sand erosion or deposition is small, the horns, the crest and the surroundings of the dune except upwind. There, the soil was covered by sand, and, as a consequence of our model, it needs a time t_g to recover. As the vegetation grows, it traps the sand, which then cannot reach the lee side. There, the vegetation cover increases. On the other hand, the vegetation on the windward side is eliminated because its roots are exposed as the dune migrates. However, at the horns the vegetation growth is fast enough to survive the low sand deposition and, thus, sand accumulates there. Hence, whereas the central part of the dune moves forward a sand trail is left behind at the horns. This process leads to the stretching of the windward side and the formation of a parabolic dune. This picture is in accordance to a recent conceptual model to explain such transformation based on field observations [22].

The bottom of figure 2 shows the inverse process. After eliminating all vegetation and setting a constant influx of 0.005 kg/m·s, the parabolic dune is fragmented into small barchanoidal forms which nucleate into a final barchan dune. Although the transformation from parabolic into barchan or transverse dunes have been observed [28], this fragmentation does not occur in nature because the vegetation is eliminated gradually from the parabolic dune — it leaves the central part of the dune first, but still remains at the lateral trails.

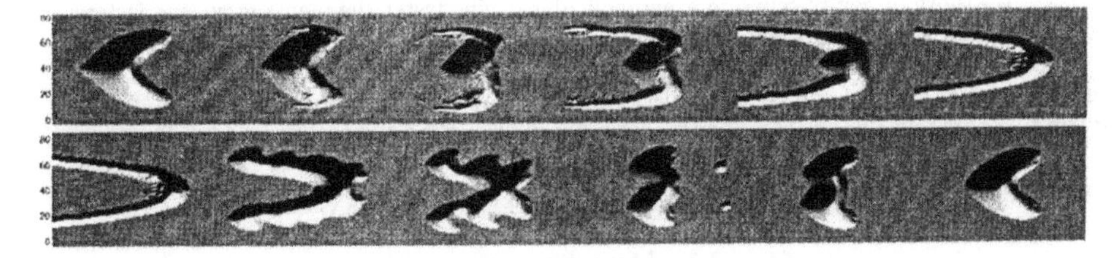

Figure 2: Evolution of an initial barchan dune to a fixed parabolic shape and vice versa. Top: Transformation of barchan into parabolic dune (from the left to the right) with a characteristic time t_g = 7 days (the evolution of vegetation appears in figure 3). Bottom: After vegetation is removed, the opposite process occurs (from the left to the right) and a barchan shape is achieved.

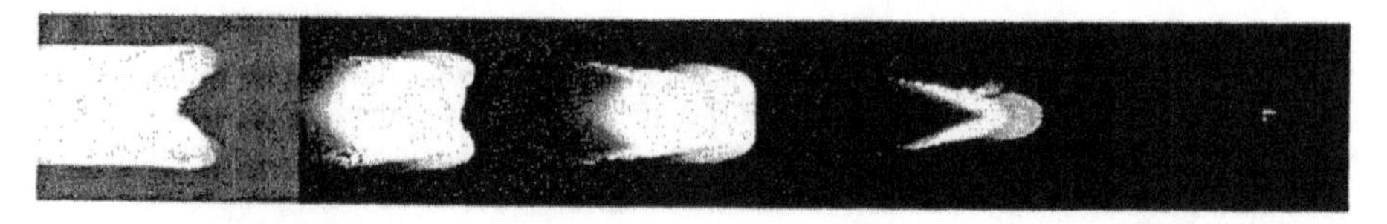

Figure 3: Evolution of the vegetation density in a barchan-parabolic transformation for a vegetation characteristic growth time of 7 days (see upper part of Figure 2). Black means complete covering by vegetation and white means no vegetation.

Both the transition from a barchan into a parabolic dune and the parabolic shape obtained strongly depend on the vegetation growth rate. Figure 4(b), (c) and (d) show the final parabolic dune for three values of t_g (see figure caption). As expected, a longer parabolic dune emerges for a smaller vegetation growth rate, i.e. a higher t_g. On the other hand, if t_g is too high, the vegetation cover is not enough to complete the inversion process (Figure 4(a)). In this case the barchan keeps its shape but leaves lateral sand trails covered by plants. However, due to the constant loss of sand at the arms, the barchan decreases in size until it gets finally stabilized by vegetation.

Notice that the parabolic dunes are slightly asymmetric (Figure 4) despite the initial condition being symmetric. This surprising result is the consequence of the interaction of the vegetation with the sand bed. Once vegetation grows, it protects the soil from erosion. This enhances the growth process, which in turn, increments the soil protection and so on. This mechanism amplifies small asymmetries in the vegetation cover. In our simulation, the initial small asymmetries are due to numerical inaccuracies, in nature, they are a consequence of random factors influencing vegetation growth, like fluctuations in the wind strength, animals, among others.

2.3 Summary

We performed a numerical simulation of the influence of varying amounts of vegetation on dune shapes. We proposed a continuum model describing vegetation growth (eq. (6)). Taking into account the partitioning of the shear stress between the plants and the ground, we used the continuum saltation model to simulate the evolution of the dune shape.

We have reproduced the observed effect of the transition between barchans and parabolic dunes. After a parabolic dune is formed, it is completely covered by vegetation and rendered inactive. So far we could not find any prototype data supporting the time-scales indicated by the model. We have found that the final shape of the parabolic dune evolving from a barchan depends strongly on the growth rate of the vegetation. Slow-growing plants only slow down the arms of the barchan and do not transform it completely into a parabolic dune. The faster the plants grow, the faster the transformation is completed and the shorter are the arms of the resulting parabolic dune.

A quantitative comparison of the results from our vegetation model with data on parabolic dunes measured in the field will be the subject of future work. This will be carried out after our field trip to a parabolic dune field in Northeastern Brazil during the current year.

3 The shape of the barchan dunes in the Arkhangelsky Crater on Mars

3.1 Mars dunes

There are dunes also on the Planet Mars. Sand dunes have been found on Mars for the first time in Mariner 9 images of the Proctor Crater, within the southern highlands of Mars [29, 30]. Since then, other Missions sent to Mars provided us with thousands of images which revealed a rich diversity of dune shapes including barchans, transverse dunes, and also linear and star dunes [31]. With the

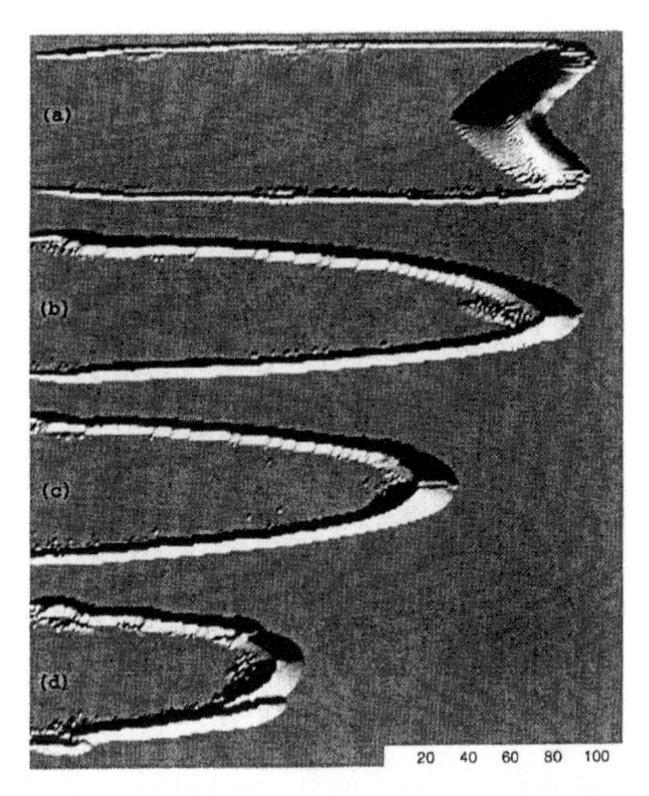

Figure 4: Final or stationary states starting with one barchan dune under the influence of vegetation with different growth rates.The vegetation characteristic growth times are: (a) 12, (b) 8, (c) 7 and (d) 4.6 days. The parabolic dunes are (b) 260 m, (c) 200 m and (d) 130 m long. Note that in (a) the vegetation cover is not dense enough to transform a barchan dune into a parabolic one.

high resolution images available of Mars, e.g. those from the Mars Global Surveyor (MGS) Mars Orbiter Camera (MOC), the study of the Martian surface has become an issue essentially geologic, rather than a subject of the Space and Planetary Science.

While it appears evident that the martian surface has been steadily modified by aeolian processes, it is difficult to know whether martian dunes are in general active (moving) in the present. Sporadic coverage by orbiting spacecraft could generally not detect dune motion in the past few Martian decades [32]. This could be a result for instance of currently insufficient wind strength and/or too low rate of dune motion to be measured. In some locations on Mars, it is even plausible that dunes are immobilized or indurated as commented later in Section 4.

But if Mars dunes are in fact inactive today, what about their age? Atmospheric conditions on Mars may have changed dramatically since the origin of the planet. Mars has been gradually losing atmosphere over the last billions of years mainly due to the impacts of meteorites and the low planetary gravity, $g = 3.71\text{m/s}^2$, which is almost one third of the earth's gravity, $g = 9.81\text{m/s}^2$. While the atmosphere of Mars 4 billions of years ago might have been as dense as the terrestrial atmosphere in the present [33], $1.225 kg/m^3$, the density of the martian air is today only 0.016 kg/m^3 (CO_2 at temperature [34] of 200K and pressure of 6.0mb), and the time at which this density value has been achieved is not known. One motivation to study dunes on Mars with our model is that a study of martian dunes may contribute data on the geological history of the planet, and also help the understanding of aeolian systems and climatic conditions of Mars [35].

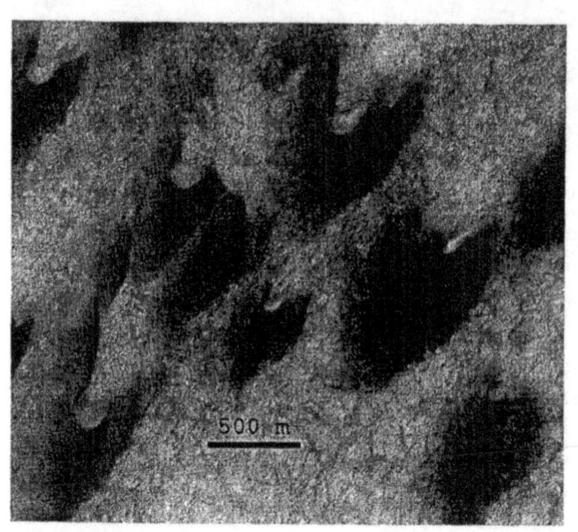

Figure 5: Barchan dunes in the Arkhangelsky Crater, Mars (41.2° S, 25.0° W). Image: MGS Mars Orbiter Camera (MOC), 2004.

Barchan dunes are certainly the most simple and best understood type of dune. Therefore, we started our exploration of Mars dunes with the investigation of a field of barchan dunes located in the Arkhangelsky Crater on Mars (41.2° S, 25.0° W), one image of which is shown in fig. 5. Our goal is to use the dune model to reproduce the dune shape observed in fig. 5 using parameters for Mars. We are interested in answering the following question: Has it been possible that these dunes formed under the atmospheric conditions of Mars in the present? Using the values of martian atmospheric density and gravity, our first aim is to find the conditions of wind and sand flux which define the dune shape observed in fig. 5.

3.2 Model parameters for Mars

Many of the model parameters we need to calculate dunes on Mars are known from the literature. Some of these parameters have been mentioned above — the gravity g and atmospheric density ρ_{fluid}, pressure P and temperature T. Recent Mars Missions obtained important data on sand and wind conditions on Mars, as detailed below, which we use in our calculations.

3.2.1 The sand of Mars dunes

Martian barchans display some essential differences in relation to their terrestrial counterparts. Thermal inertia studies, for example, have shown that the sand of martian dunes is coarser, with a mean grain diameter $d = 500 \pm 100\mu m$ [36], while the size distribution of sand grains of terrestrial dunes presents a maximum around $d = 250\mu m$ [16]. Furthermore, it is known that dunes on Mars are made of grains of basalt [37], while terrestrial dunes are constituted by quartz grains. The angle of repose θ_r of the sand on Mars, or the inclination of the dune slip face has been found to be close to the terrestrial value, $\theta_r \approx 34°$ [32, 38].

3.2.2 Threshold wind velocity for saltation

The dynamics of saltating grains on Mars has been extensively studied by many authors, both using numerical calculations and from wind tunnel experiments on martian atmospheric conditions [3, 38, 39]. The threshold wind shear velocity on Mars has been predicted for a wide range of grain sizes [40, 41, 42]. For grains of $d = 500\mu$m, for example, a value of u_{*t} around 2.0 m/s is predicted for Mars. This is ten times larger than the terrestrial $u_{*t} = 0.2$ m/s [16]. Notwithstanding, wind friction speed values of this order have been reported several times from Mars Missions observations [43, 44, 45].

3.2.3 One unknown saltation parameter

From the saltation model parameters listed in Section 1, there is one, γ (eq. (4)), whose value for Mars we did not calculate. While Sauermann *et al.* (2001) [13] have obtained γ for the case Earth from reported measurements of the saturation time of terrestrial saltation, here we proceeded in a different way to estimate the entrainment rate of grains into saltation, γ, on Mars.

3.3 Simulation of the Arkhangelsky Dunes

We performed simulations with open boundaries and a small influx q_{in} at the inlet, with values typically less than 30% of the saturated flux q_s. Such low values for q_{in}/q_s are characteristic of the interdune in terrestrial dune fields developing on bedrock [46]. A small value of q_{in}/q_s is also reasonable for the Arkhangelsky dunes since sediment access in craters is very commonly restricted [35]. In this way, we need to find for which values of γ and the shear velocity u_* — which is set of course larger than $u_{*t} = 2.0$ m/s — the model reproduces the dune shape in the Arkhangelsky Crater. The initial surface is a gaussian sand hill (fig. 6) with dimensions of the order of the dune sizes observed in fig. 5.

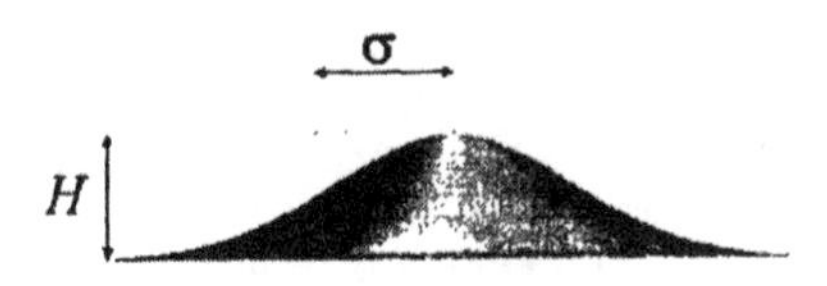

Figure 6: The initial surface is a gaussian hill of height H and characteristic length σ.

We proceed in the following manner. First, we set the value of γ for Mars equal to the earth's [13] and we see if it is possible to find the Arkhangelsky dunes for some value of the wind shear velocity u_*. We obtain the surprising result that no dune appears for any value of u_* in the range of estimated wind friction speeds on the Mars Exploration Rover Meridiani Planum [47] and Viking 1 [44] landing sites on Mars, i.e. for u_* between 2.0 and 4.0 m/s. In other words, the gaussian hill does not develop slip face and disappears because its size is below the critical one for dune formation (fig. 7).

However, if the parameter γ, i.e. the number of grains entrained by the air is of the order of 10 times the earth's then dunes are obtained. Furthermore, as shown in fig. 8, we find that the elongated shape of the Arkhangelsky dunes is in general obtained for a shear velocity close to the threshold. As u_* increases, for a constant value of γ, the dune shape deviates from the Arkhangelsky dunes. One conclusion we get from our calculations, thus, is that the shear velocity in the Arkhangelsky Crater must be close to the threshold friction speed, and probably around 3.0 m/s. Figure 9 shows one dune calculated with this u_* value together with a MOC image of one Arkhangelsky dune.

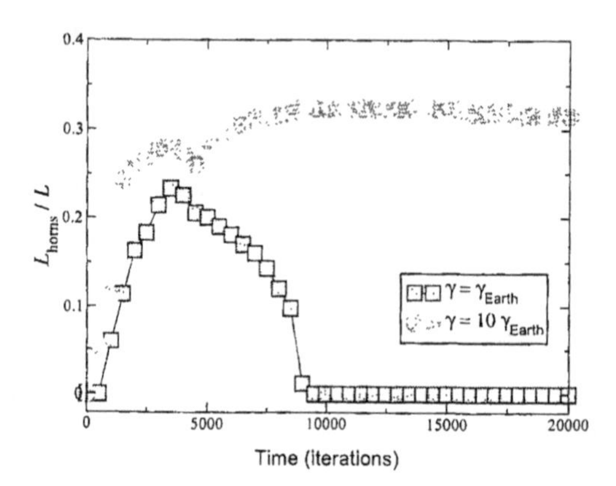

Figure 7: Time evolution of the dune horns length L_{horns} relative to the dune length L using $\gamma = \gamma_{\text{Earth}}$ (squares). In this case, the gaussian in fig. 6 does not achieve an equilibrium barchan shape, and disappears after the sand is blown away, while barchan dunes are obtained with $\gamma = 10\,\gamma_{\text{Earth}}$ (circles). Calculation was performed with $u_*/u_{*t} = 1.75$.

3.3.1 Interpretations and summary

We remark that our value of u_* is an intermediate value between estimated friction speeds at Pathfinder [45] and Viking 1 [44] lander sites on Mars. On the other hand, why should γ be larger on Mars? As mentioned above, the larger the value of γ, the larger the number of grains a percentual increase of the wind strength relative to the threshold succeeds in entraining into the saltation layer. On the other hand, the grains which are entrained by the wind come mainly from the splash after grain-bed collisions. Therefore, we interpret the larger value of γ on Mars as a result of the larger amount of splashed grains, which may be understood as a consequence of the larger velocity reached by saltating grains on Mars.

The mean velocity of the saltating grains calculated with the dune model is around 16.0 m/s, while for Earth the grain velocity is approximately 1.6 m/s. It follows that the momentum transferred by the impacting grains to the sand bed is much larger on Mars than on Earth, what is associated with much larger splashes after impact on Mars. This picture is in agreement with the observation that the number of ejecta after splash is proportional to the velocity of the impacting grain [11]. In fact, it has been already observed that, while on Earth a single grain-bed collision results in ejection of only a few grains, on Mars the number of splashed grains may be of the order of one hundred [48].

What is the consequence of a larger entrainment rate of grains on Mars? A faster population increase of saltating particles amplifies the "feedback effect", which is the decceleration of the wind due to the momentum transfer of the air to the grains [2]. Consequently, a larger splash has an astonishingly important implication for the macroscopic scale of sand patterns: It reduces the distance after which the maximum amount of sand which can be transported by the wind is reached, and the minimum size below which sand hills are continuously eroded and will disappear.

In the model the information of the splash is incorporated in the parameter γ. We notice that an increase in the value of γ means a decrease in the saturation length of the flux (eq. (3)), as expected — the more grains the wind succeeds entraining saltation, the faster the flux should saturate. Since the saturation length is the relevant length scale of barchan dunes, our calculations reveal that this property of the martian splash must be taken into account for correctly predicting the size of dunes on Mars. We have recently found that if the larger splash on Mars is accounted for in the calculations, we can reproduce not only the dune shape as reported here but also the minimal dune size and the dependence of the dune shape with the size [49]. While the saturation length ℓ_s on Earth is around 50 cm, we have obtained $\ell_s \approx 17$ m for Mars using the parameters mentioned above, with which we reproduced the Arkhangelsky dunes. However, if the value of γ on Mars were the same as on Earth, the saturation length and therefore the minimal dune scale on Mars would be accordingly 10 times larger. It would be interesting to use this new insight to make a full microscopic simulation for the saltation mechanism of Mars similar to the one that was recently achieved by [10].

The wide range of barchan forms observed on Mars is a consequence of the different local physical conditions which dictate sand transport [35]. We are presently carrying out a study of the dune shape as a function of the sand flux under different atmospheric and wind conditions valid for Mars. In Section 4 we present another mechanism that appears to be relevant to explain exotic dune forms on the Martian north polar region.

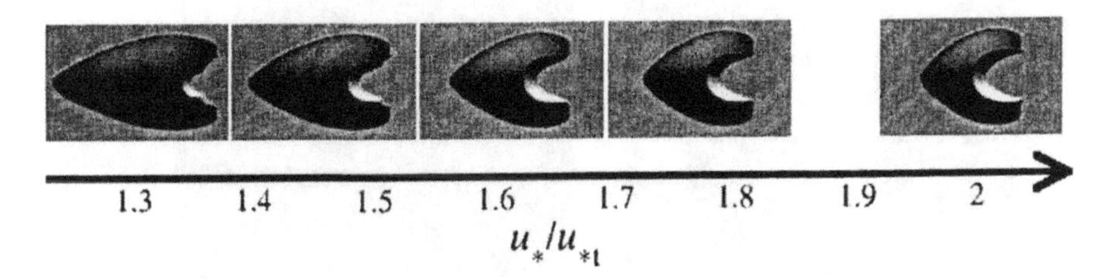

Figure 8: Barchan dunes of width $W = 650$ m calculated using parameters for Mars, with $\gamma = 10\,\gamma_{\mathrm{Earth}}$, and different values of wind shear velocity u_* relative to the threshold u_{*t}. We find $u_*/u_{*t} \approx 1.45$ or u_* around 3.0 m/s in the Arkhangelsky Crater.

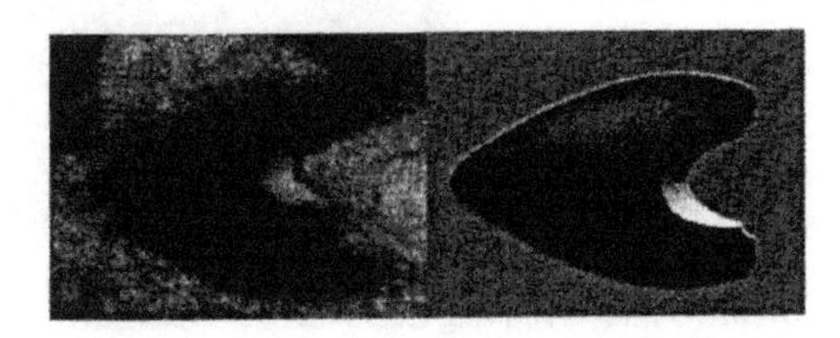

Figure 9: MOC image (left) and calculation (right) of barchan dune of width 650 m in the Arkhangelsky Crater on Mars. Calculation was performed with $u_* = 3.0$ m/s and $\gamma = 10\,\gamma_{\mathrm{Earth}}$.

4 Evidence for indurated sand dunes in the Martian north polar region

The morphology and proximity of differing morphologies of a suite of eolian dunes in the Martian north polar region defy traditional explanation. Unusual features occur in a dune field in Chasma Boreale, in the Martian north polar region. Figure 10 shows examples of the dunes, as seen by the Mars Global Surveyor (MGS) Mars Orbiter Camera (MOC). In this dune field, linear and barchan dunes occur together, side-by-side and in some cases are merged to create a single bedform. The linear dunes lack the sinuosity commonly associated with terrestrial seif dunes [50] — in other words, these linear features are straighter than their counterparts on Earth. Furthermore, the dunes present an additional puzzle -- the barchans are elongated into elliptical forms, and the slip faces are typically small or nonexistent (Figure 11).

4.1 Dune movement and induration

Figure 10: Sub-frames of MGS MOC images showing examples of the elongated, rounded barchans and linear dunes in Chasma Boreale. Both images cover areas about 3 km wide. The picture on the left is from image E01-00104, located near 83.9° N 40.5° W. The right image is from MOC image S02-00901, and is located near 84.2° N 37.9° W. In both cases, sand transport has been from the upper right toward the lower left.

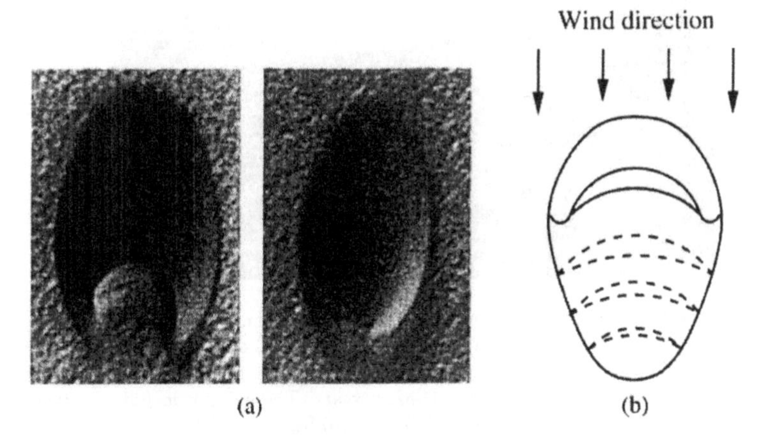

Figure 11: (a) MGS MOC images of a rounded barchan and a dome dune located near 84.9° N 26.6° W, in image M02-00783. The shown parts of the image are each approximately 240 meters wide. (b) Sketch of the deposition in the lee of an oil-soaked barchan (after Kerr and Nigra (1952)). The initial barchan and the final shape are drawn; the dashed lines show the positions of the slip face in intermediate stages, until an elliptical, dome-like shape is produced.

While Martian dunes generally appear to be fresh and have no superimposed impact craters [51] — the presence of which would imply antiquity — repeated imaging by MOC has shown no clear evidence for translation of eolian dunes across the Martian surface. The dunes in the Chasma Boreale region were not imaged by Mariner 9 or Viking at a spatial resolution adequate to see them and measure their movement by comparison with MOC images. Some of the dunes have, however, been repeatedly imaged by MOC during the past four Martian years, and, again, show no evidence for movement (Figure 12). The Chasma Boreale dunes also do not exhibit grooves or steep-walled avalanche chutes. But, the rounded barchans do resemble the oil-soaked dunes of Kerr and Nigra (1952) [24], and the linear dunes simply should not occur with the barchans, unless they are unlike terrestrial linear dunes, and formed—somehow—in a unidirectional wind regime.

We investigate whether it is plausible that the rounded, elliptical dunes in Chasma Boreale could be the product of a process similar to that which was created by successive soaking of dunes in Saudi Arabia with crude oil, and whether the explanation might somehow extend to the occurrence of linear forms in the same dune field. Rather than anthropogenic processes halting dune movement in successive stages, as in the Kerr and Nigra (1952) [24] example, we examine whether induration of dune sand may lead to production of dune forms in Chasma Boreale.

4.1.1 Rounded barchans

Elongated, elliptical, dome-shaped dunes such as those in Figures 10 and 11, might be the product of successive induration as the slip face advances forward and becomes smaller and smaller, until it disappears, as was observed by Kerr and Nigra (1952) [24] for oil-soaked barchans on Earth. In the Saudi Arabia case, the dunes were sprayed with oil, then additional sand arrived at the fixed dune, blew over its top, and became deposited on the lee side, whereupon the dune was again sprayed with oil. This process continued and, over time, the slip face of each dune became successively smaller, and the dune more rounded and elliptical. This is not typical behavior for

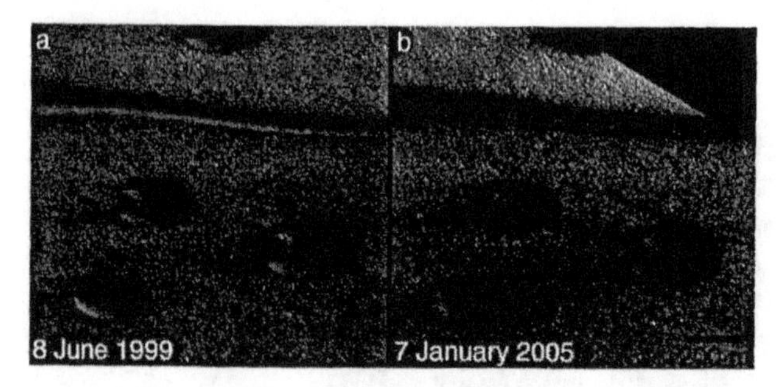

Figure 12: The rounded barchans of Chasma Boreale have been imaged repeatedly by the MGS MOC during the past several Martian years. Relative to features on the substrate across which they are moving, the dunes do not appear to have moved during the mission. The interval between acquisition of the two images shown here is 2.97 Mars years. These dunes include the two shown in Figure 11. (a) Sub-frame of MOC image M02-00783. (b) Sub-frame of MOC image S02-00302. The dunes are illuminated by sunlight from the lower left and are located near 84.9° N, 26.6° W.

lee-side deposition of sand in a dune field, but is a product of the process of successively inhibiting sand movement — in this case, using oil — on the main body of each dune.

The similarity of the oil-soaked dunes to the rounded forms on Mars suggests that the Martian examples could have formed in a similar way, with ice, frozen carbon dioxide, or mineral salts taking the place of oil as the cause for induration. To test the hypothesis under conditions suitable for Mars, we simulated the process of forming a successively-indurated barchan using the appropriate parameters for Mars [52]. Figure 13 shows the successive stages of deposition in the lee of the fixed barchan. The shape of the initial dune and the latter successions were all simulated with the same atmospheric conditions, particle properties, and wind speed; the only change was to simulate induration (non mobility) of each successive stage of the dune development, while at the same time adding new sand from the upwind direction.

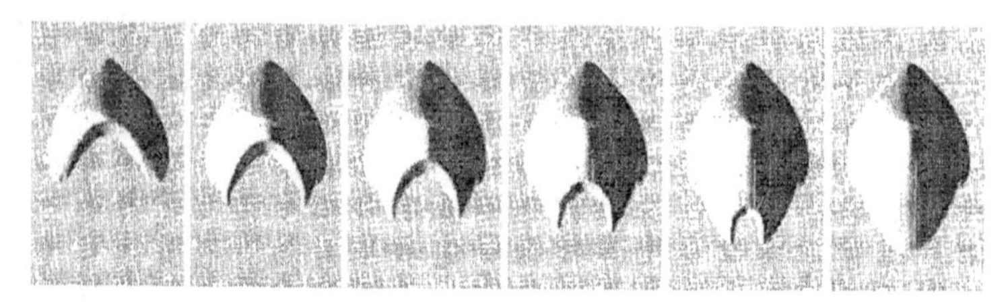

Figure 13: The successive stages of evolution of deposition in the lee of an indurated barchan on Mars. The length of the shown region is 400 meters, the width 240 meters. The first picture shows the initial barchan.

Some small differences between the simulation in Figure 13 and the satellite image (Fig. 11a) have to be discussed. In the simulations, the brink of the initial barchan is still visible at later stages. In reality, this visible ridge would be slowly eroded even if the dune was indurated. The second discrepancy is that the slip face in the photo appears to be very long, but the foot of the slip face is hard to discern, and a slip face extending nearly to the end of the horns would imply a very high slip face brink, which would be in contrast with the flat rounded shape suggested by the shadows and with the height of other Martian dunes described by Bourke *et al.* (2004) [53]. The aspect ratio (height divided by length) of our simulated dunes is consistent with the results of Bourke *et al.* (2004) [53].

4.1.2 Chasma Boreale Linear dunes

In addition to the dome-shaped barchans, the region near the Martian north pole exhibits linear dunes that are very straight, unlike their meandering counterparts on earth [50, 54, 55]. These linear dunes occur with the domes and rounded barchans, suggesting that they form in winds flowing parallel to the linear dunes—yet terrestrial experience requires two oblique winds, not a single, unidirectional wind, to create such bedforms.

Figure 14: A field of linear dunes in Chasma Boreale. This is a part of MOC image E15-00784, located at 84.24° N 39.95° W. It covers an area about 3 km across. The bright outlines of the dunes and of the ridges in the surrounding region indicate that they are covered by seasonal frost, most likely frozen carbon dioxide. The black rectangle indicates the region selected for our linear dune simulation (see Figure 15).

MGS MOC narrow angle images show that the upwind end of the linear dunes, immediately downwind of the domes, shows a knotted structure (e.g., Figure 14). We have performed a computer simulation of such dunes, starting from a straight ridge in wind direction located just downwind of an unerodable dome. As can be seen in Figure 15, the knots at the upwind end are reproduced in early steps of the simulation. Their formation is analogous to the instability of a sand bed under unidirectional winds (e.g., [56, 57]). Later on, however, the simulated linear dune decays into a string of barchans. That this does has not happened in Chasma Boreale can be explained if the linear dunes are indurated or were formed by erosion in the first place. Although our simulations do not show how the linear forms were initiated, they do suggest—as in the case of the elongated, dome-shaped barchans — that the sands are presently or very recently indurated.

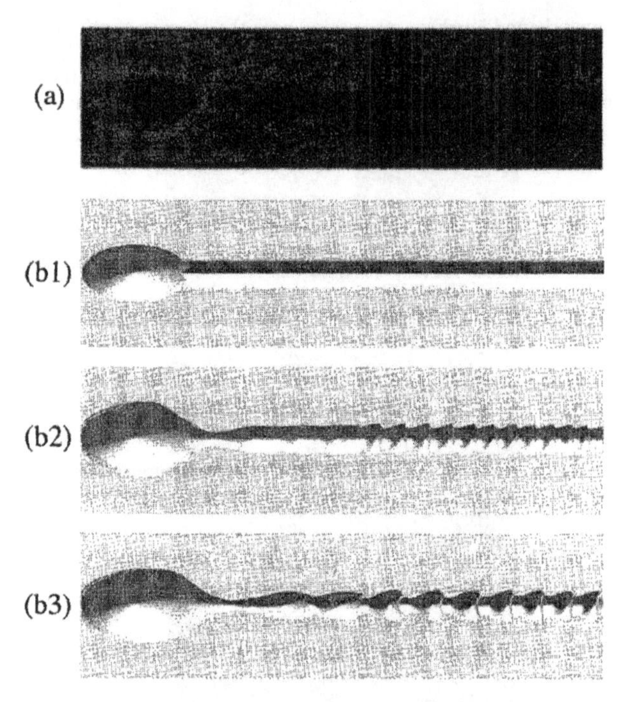

Figure 15: (a) A detail of Figure 14. (b) Simulation of the evolution of a straight linear dune downwind of a dome. Besides small variations in thickness, small barchans can be seen to develop, which are not observed on Mars. From this we conclude that the linear dunes in Figure 14 are indurated.

4.2 Summary

The results of our dune model support, in addition to the observations of artificially indurated dunes by Kerr and Nigra (1952) [24], the hypothesis regarding induration of dunes in Chasma Boreale. While basic, typical dune morphologies are a product of wind regime — with, on Earth, contributions from vegetation and moisture — we conclude that induration may be an additional factor that influences morphology of dunes on the Martian north polar region. In the case of Chasma Boreale, induration has led to creation of elongated, elliptical, dome-shaped dunes; elongated barchans with proportionally tiny slip faces; and straight (as opposed to sinuous, like seif dunes) linear dunes that co-exist with barchan forms in a unidirectional, rather than bidirectional, wind regime.

5 Conclusions

In this work, we have presented extensions of the dune model originally introduced by Sauermann *et al.* (2001) [13] and later improved by Schwämmle and Herrmann (2005) [15] to study dunes with vegetation, and we tested the model to calculate dunes on Mars. We could simulate the transformation of barchans into parabolic dunes, as observed by Tsoar and Blumberg (2002) [22], where we proposed a continuous model for the rate of vegetation growth, following the idea of Nishimori and Tanaka (2001) [23].

Furthermore, we have applied our dune model to calculate the shape of the barchan dunes in the Arkhangelsky Crater on Mars. We found that martian dunes may have been formed by winds close to the threshold, under the atmospheric conditions of Mars in the present, and our calculations reveal that larger splashes on Mars are relevant to explain the size of martian dunes. Finally, application of our model to study north polar dunes on Mars support the hypothesis that the unusual dune shapes observed in the Chasma Boreale region are a result of induration.

ACKNOWLEDGMENTS

This work was supported in part by Volkswagenstiftung and the Max-Planck Price. We acknowledge H. Tsoar and K. Edgett for discussions. E. J. R. Parteli acknowledges support from CAPES - Brasília, Brazil.

References

[1] R.A. Bagnold, *The physics of sand dunes.* Methuen, London, (1941).

[2] P.R. Owen, Saltation of uniformed sand grains in air, *Journal of Fluid Mechanics* **20**: 225 - 242 (1964).

[3] B.R. White, Soil transport by Winds on Mars, *Journal of Geophysical Research* **84 (B9)**, 4643-4651 (1979).

[4] J.E. Ungar and P.K. Haff, Steady state saltation in air, *Sedimentology* **34**: 289–299 (1987).

[5] R.S. Anderson, and P.K. Haff, Wind modification and bed response during saltation in air, *Acta Mechanica (Suppl.)* **1**, 21-51 (1991).

[6] I.K. McEwan and B.B. Willetts, Numerical model of the saltation cloud, *Acta Mechanica (Suppl.)* **1**, 53–66 (1991).

[7] K.R. Rasmussen, J.D. Iversen, and P. Rautahemio, Saltation and wind-flow interaction in a variable slope wind tunnel, *Geomorphology* **17**, 19–28 (1996).

[8] J.M. Foucaut and M. Stanilas, Experimental study of saltating particle trajectories, *Experiments in Fluids* **22**, 321-326 (1997).

[9] Z. Dong, N. Huang, X. Liu, Simulation of the probability of midair interparticle collisions in an aeolian saltating clou, *J. Geophys. Res.* **110**, D24113, doi:10.1029/2005JD006070 (2005).

[10] M.P. Almeida, J.S. Andrade Jr. and H.J. Herrmann, Aeolian transport layer, *Physical Review Letters* **96**, 018001 (2006).

[11] R.S. Anderson, and P.K. Haff, Simulation of aeolian saltation, *Science* **241**, 820-823 (1988).

[12] P. Nalpanis, J.C.R. Hunt, and C.F. Barrett, Saltating particles over flat beds, *Journal of Fluid Mechanics* **251**, 661-685 (1993).

[13] G. Sauermann, K. Kroy, and H.J. Herrmann, A continuum saltation model for sand dunes, *Physical Review E* **64**, 31305 (2001).

[14] K. Kroy, G. Sauermann, and H.J. Herrmann, Minimal model for aeolian sand dunes, *Physical Review E* **66**, 031302 (2002).

[15] V. Schwämmle, and H.J. Herrmann, A model of Barchan dunes including lateral shear stress, *The European Physical Journal E* **16**, 57-65 (2005).

[16] K. Pye, and H. Tsoar, **Aeolian sand and sand dunes**, Unwin Hyman, London (1991).

[17] W.S. Weng, J.C.R. Hunt, D.J. Carruthers, A. Warren, G.F.S. Wiggs, I. Livingstone, I., I. Castro, Air flow and sand transport over sand-dunes, *Acta Mechanica (Suppl.)* **2**, 1-22 (1991).

[18] G. Sauermann, J.S. Andrade Jr., L.P. Maia, U.M.S. Costa, A.D. Araújo, and H.J. Herrmann, Wind velocity and sand transport on a barchan dune, *Geomorphology* **54**, 245-255 (2003).

[19] V. Schwämmle, and H.J. Herrmann, Solitary Wave Behaviour of Dunes, *Nature* **426**, 619-620 (2003).

[20] H. Besler, Eine Wanderdüne als Soliton? *Physikalische Blätter* **53(10)**, 983-985 (1997). (in German).

[21] O. Durán, V. Schwämmle, and H.J. Herrmann, Breeding and solitary wave behaviour of dunes, *Physical Review E* **72**, 021308 (2005).

[22] H. Tsoar and D.G. Blumberg, Formation of parabolic dunes from barchan and transverse dunes along Israel's mediterranean coast, *Earth Surf. Process. Landforms* **27**: 1147–1161 (2002).

[23] H. Nishimori and H. Tanaka, A simple model for the formation of vegetated dunes, *Earth Surf. Process. Landforms* **26**: 1143 - 1150 (2001).

[24] R.C. Kerr, and J.O. Nigra, Eolian sand control, *Bull. Am. Assoc. Petrol. Geol.* **36**: 1541-1573 (1952).

[25] A.C.W. Baas, Chaos, fractals and self-organization in coastal geomorphology: Simulating dune landscapes in vegetated environments, *Geomorphology* **48**: 309-328 (2002).

[26] E.D. McKee and J.R. Douglas, Growth and movement of dunes at White Sand National Monument, New Mexico, *United State Geological Survey* **750**, 108–114 (1971).

[27] R. Buckley, The effect of sparse vegetation on the transport of dune sand by wind, *Nature* **325**, 426-428 (1987).

[28] D. Anton and P. Vincent, Parabolic dunes of the Jafurah desert, eastern province, Saudi Arabia, *Journal of Arid Environments* **11**: 187–198 (1986).

[29] J.F. McCauley, M.H. Carr, J.A. Cutts, W.K. Hatmann, H.Masursky, D.J. Milton, R.P. Sharp, and D.E. Wilhelms, Preliminary Mariner 9 report on the geology of Mars, *Icarus* **17**, 289-327 (1999).

[30] C. Sagan, J. Veverka, P. Fox, L. Quam, R. Tucker, J.B. Pollack, and B.A. Smith, Variable features on Mars: Preliminary Mariner 9 television results, *Icarus* **17**, 346-372 (1972).

[31] K.S. Edgett and Dan. G. Blumberg, Star and linear dunes on Mars, *Icarus* **112**(2), 448-464 (1994).

[32] K.K. Williams, R. Greeley, and J.R. Zimbelman, Using Overlapping MOC Images to Search for Dune Movement and To Measure Dune Heights, *Lunar and Planetary Science* **XXXIV** (2003).

[33] H.J. Melosh, and A.M. Vickery, Impact erosion of the primordial atmosphere of Mars, *Nature* **338**, 487-489 (1989).

[34] Mars Global Surveyor Radio Science Team (2005). http://www-star.stanford.edu/projects/mgs/ ("Late Mars Weather").

[35] M.C. Bourke, M. Balme, and J. Zimbelman, A Comparative Analysis of Barchan Dunes in the Intra-Crater Dune Fields and The North Polar Sand Sea, *LPSC* **XXXV**, 1453 (2004a).

[36] K.S. Edgett, and P.R. Christensen, The Particle Size of Martian Aeolian Dunes, *Journal of Geophysical Research 96 (E5)*, 22765-22776 (1991).

[37] L.K. Fenton, J.L. Bandfield, and A.W. Ward, Aeolian processes in Proctor Crater on Mars: Sedimentary history as analysed from multiple data sets, *Journal of Geophysical Research* **108 (E12)**, 5129 (2003).

[38] R. Greeley, and J. Iversen, **Wind as a Geological Process.** Cambridge Univ. Press, New York (1984).

[39] B.R. White, R. Greeley, J.D. Iversen, and J.B. Pollack, Estimated Grain Saltation in a Martian Atmosphere, *Journal of Geophysical Research* **81 (32)**, 5643-5650 (1976).

[40] R. Greeley, R. Leach, B. White, J. Iversen, and J. Pollack, Threshold windspeeds for sand on Mars: Wind Tunnel Simulations, *Geophysical Research Letters* **7(2)**, 121-124 (1980).

[41] J.D. Iversen, J. B. Pollack, R. Greeley, R., and B.R. White, Saltation Threshold on Mars: The Effect of Interparticle Force, Surface Roughness, and Low Atmospheric Density, *Icarus*, 381-393 (1976).

[42] J.D. Iversen, and B.R. White, Saltation threshold on Earth, Mars and Venus, *Sedimentology* **29**, 111-119 (1982).

[43] B.A. Cantor, P.B. James, M. Caplinger and M.J. Wolff, Martian dust storms: 1999 Mars Orbiter Camera observations, *Journal of Geophysical Research* **106**(E10), 23,653 - 23,687 (2001).

[44] H.J. Moore, The Martian Dust Storm of Sol 1742, *Journal of Geophysical Research* **90**(Supplement): D163-D174 (1985).

[45] R. Sullivan, R. Greeley, M. Kraft, G. Wilson, M. Golombek, K. Herkenhoff, J. Murphy, and P. Smith, Results of the Imager for Mars Pathfinder windsock experiment, *Journal of Geophysical Research* **105 (E10)**, 547-562 (2000).

[46] S.G. Fryberger, A.M. Al-Sari, T.J. Clisham, S.A.R. Rizvi, and K.G. Al-Hinai, Wind sedimentation in the Jafurah sand sea, Saudi Arabia, *Sedimentology* **31**, 413-431 (1984).

[47] R. Sullivan, *et al.* Aeolian processes at the Mars Exploration Rover Meridiani Planum landing site, *Nature* **436**, 58-61 (2005).

[48] J. Marshall, J. Borucki, and C. Bratton, Aeolian Sand Transport in the Planetary Context: Respective Roles of Aerodynamic and Bed-Dilatancy Thresholds, *LPSC* **XXIX**, 1131 (1998).

[49] E.J.R. Parteli, O. Durán, and H.J. Herrmann, Minimal size of a barchan dune, Submitted to Physical Review E (2006).

[50] H. Tsoar, Linear dunes - forms and formation, *Progress in Physical Geography* **13**: 507–528 (1989).

[51] A.G. Marchenko, and A.A. Pronin, Study of relations between small impact craters and dunes on mars, abstracts of the 22nd russian-american microsymposium on planetology, *Tech. Rep. 63-64*, Vernadsky Institute, Moscow (1995).

[52] V. Schatz, H. Tsoar, K.S. Edgett, E.J.R. Parteli, and H.J. Herrmann, Evidence for indurated sand dunes in the Martian north polar region, *Journal of Geophysical Research*, **111(E4)**, E04006, doi:10.1029/2005JE002514 (2006).

[53] M. Bourke, M. Balme, R.A. Beyer, K.K. Williams, J. Zimbelman, How high is that dune? a comparison of methods used to constrain the morphometry of aeolian bedforms on Mars, *LPSC* **XXXV**, 1713 (2004b).

[54] H. Tsoar, Internal structure and surface geometry of longitudinal seif dunes, *Journal of Sedimentary Petrology* **52**, 823–831 (1982).

[55] H. Tsoar, Dynamic processes acting on a longitudinal (seif) sand dune, *Sedimentology* **30**: 567–578 (1983).

[56] B. Andreotti, P. Claudin, and S. Douady, Selection of dune shapes and velocities. Part 1: Dynamics of sand, wind and barchans, *The European Physical Journal B* **28**, 321-329 (2002).

[57] V. Schwämmle, and H.J. Herrmann, Modelling transverse dunes, *Earth Surf. Process. Landforms* **29**(6), 769-784 (2004).

Brill Academic Publishers
P.O. Box 9000, 2300 PA Leiden,
The Netherlands

Lecture Series on Computer and Computational Sciences
Volume 6, 2006, pp. 561-570

High performance computation and numerical validation of e-collision software

N.S. Scott[1]||, L. Gr. Ixaru‡, C. Denis♮ , F. Jézéquel♮, J.-M. Chesneaux♮ and M.P. Scott†

||School of Electronics, Electrical Engineering & Computer Science, †School of Mathematics & Physics,
The Queen's University of Belfast,
Belfast, BT7 1NN, UK

‡Institute of Physics and Nuclear Engineering,
Magurele, Bucharest, R-76900 Romania

♮Laboratoire d'Informatique de Paris 6,
Université Pierre et Marie Curie - Paris 6,
4 place Jussieu, 75252 Paris Cedex 05, France

Received July 13, 2006; accepted July 20, 2006

Abstract: We describe a computational science research programme primarily aimed at engineering numerically robust software that can exploit high performance and distributed computers in the study of electron collisions with atoms and ions. We identify a serious bottleneck in the associated software suite and describe how hand crafted algorithms that exploit a problem's characteristics can be used to resolve the problem. Finally, we draw attention to the problem of round-off error propagation that is inherent in scientific computation and advocate the use of the CADNA library to assist in its understanding and management.

Keywords: atomic collision processes; CADNA; CESTAC method; CP methods; discrete stochastic arithmetic; dynamical control; exponential fitting; floating-point arithmetic; frequency dependent weights; quadrature; round-off error propagation; R-matrix; Slater integrals.

PACS: 02.60.-x; 02.60.Jh; 02.70.-c; 02.50.Ey.

1 Introduction

In this paper we describe a collaborative computational science research programme that is primarily aimed at engineering numerically robust software that can exploit high performance and distributed computers to study electron collisions with atoms and ions.

We begin by briefly introducing the physics of electron collisions (e-collisions). Next we sketch a mathematical model that is based on R-matrix theory[1]. A computational realization of this model that can exploit high performance computers, 2DRMP, is presented in Sec. 4. Here we identify a computational bottleneck associated with the computation of the two-dimensional radial

[1] Corresponding author.E-mail: ns.scott@qub.ac.uk

integrals. We pose and answer the question, "Do the constituent components of the integrands have characteristics that can be exploited to enable the construction of hand crafted quadrature formula that will both improve the accuracy and significantly reduce the computation time of the Slater integrals?"

Scientific computation is inherently flawed with approximations in evidence at every level extending from the physical world to the mathematical model, then to the computational model and finally to the computer implementation. One important source of error that is both esoteric and difficult to manage is that of round-off error propagation which originates from the use of finite precision arithmetic. This theme is explored in the final section where we describe the use of the CADNA library[2], a software tool based on discrete stochastic arithmetic that is used to validate the stability of numerical software.

2 Physics: e-Collisions

Electron collisions with atoms and ions have been the subject of international interest for many years. Data from these processes are of importance in the analysis of physical phenomena in many scientific and technological areas including: aeronomy, astrophysics, biomedicine, gaseous electronics, surface physics, industrial plasmas, environmental, fusion, semiconductor and other technologies[3].

Despite the importance of these applications, relatively little accurate cross section data is known for many of the processes involved. For example, no accurate cross section data, involving high lying excited states, have been calculated for electron collisions with the simplest atomic target, hydrogen. Accordingly, the focus of this paper is on electron impact excitation of H-like atoms and ions at these, so called, intermediate scattering energies.

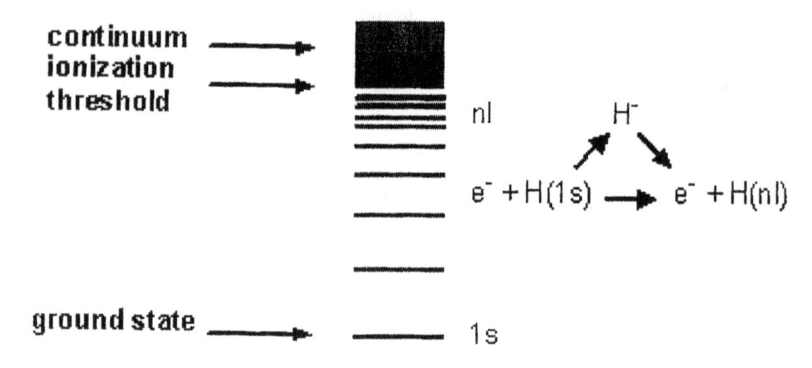

Figure 1: Electron impact excitation. Here a free electron collides with an H-like atom or ion in its ground state resulting in an excited H-like atom and a free electron. This final state can be reached directly or via an intermediate resonance state.

This processes is illustrated in Fig. 1. The process can occur directly or via an intermediate resonance state i.e. a quasi-bound state with a long lifetime. Resonance analysis is important in elucidating the collision process and in determining sensitive parameters against which theoretical computations and experimental results may be compared. Modelling this processes is computationally demanding because account must be taken of the infinite number of continuum states of the ionized target and of the infinite number of target bound states lying below the ionization threshold.

3 Mathematical Model: R-matrix theory

R-matrix theory was first introduced in nuclear physics by Wigner[5, 6]. Around the 1960s it was realized that this approach could also be used in atomic and molecular physics [7]. Since then the R-matrix method has proved to be a remarkably stable, robust and efficient technique for solving the close-coupling equations that arise in electron collision theory[1]. We begin by sketching general R-matrix theory for $(N+1)$-electron targets.

3.1 General R-matrix theory

General R-matrix theory starts by dividing the configuration space describing the collision process into two regions by a sphere of radius $r = a$, where a is chosen so that the charge distribution of the target atom or ion is contained within the sphere. In the internal region ($r \leq a$) exchange and correlation effects between the scattering electron and the target electrons must be included, whereas in the external region such effects can be neglected thereby considerably simplifying the problem.

In the internal region the $(N+1)$-electron wavefunction at energy E is expanded in terms of an energy independent basis set, ψ_k, as

$$\Psi_E = \sum_k A_{Ek}\psi_k. \tag{1}$$

The basis states, ψ_k, are themselves expanded in terms of a complete set of numerical orbitals, u_{ij}, constructed to describe the radial motion of the scattered electron. The expansion coefficients of the u_{ij} set can be determined by diagonalizing

$$(\psi_i|H_{N+1}|\psi_j) = E_k^{N+1}\delta_{ij} \tag{2}$$

where H_{N+1} is the $(N+1)$-electron Hamiltonian operator. This in turn allows the construction of the R-matrix

$$R_{ij} = \frac{1}{2a}\sum_k \frac{\omega_{ik}(a)\omega_{jk}(a)}{E_k^{N+1} - E} \tag{3}$$

at $r = a$, where the amplitudes $\omega_{ik}(a)$ and the poles E_k^{N+1} of the R-matrix are obtained directly from the eigenvectors and eigenvalues of Eq.(2).

In the outer-region the equations reduce to coupled second-order ordinary differential equations. Using a variety of means, such as the 1-D propagation technique implemented in the FARM package [4], these equations can be integrated outwards, subject to the R-matrix boundary conditions at $r = a$, and fitted to an asymptotic expansion. This determines the K-matrix from which the scattering observables including inelastic cross sections and resonance positions and widths can be derived. While the R-matrix is determined by a single diagonalization in the inner-region, the coupled equations in the outer-region must be solved for each scattering energy of interest.

It can be shown that dimension of the Hamiltonian matrix, n, within the internal region is proportional to a^2. Since matrix diagonalization is an n^3 process the size of the inner-region computation rapidly increases when transitions to higher level states are computed. For example, in electron scattering by atomic hydrogen a radius of $a = 60a.u.$ is needed to envelop the $n = 4$ target states while a radius of $360a.u.$ is required to envelop the $n = 10$ target states. This results in an increase in diagonalization time by a factor of approximately 46,000. Solution of the corresponding dense Hamiltonian matrices, which are typically of the order of 50,000 to 100,000, places considerable demands on both computer time and numerical robustness.

In the following subsection we outline an alternative approach that is designed for scattering from H-like systems. It has the advantage of extending the boundary radius of the inner-region far beyond that which is possible using the traditional 'one-sector' technique.

3.2 2D R-matrix theory

As in general R-matrix theory the two-electron configuration space (r_1, r_2) is divided into two regions by a sphere of radius a centered on the target nucleus[8]. However, in the 2-D variant the inner-region is further divided into sub-regions as illustrated in Figure 2.

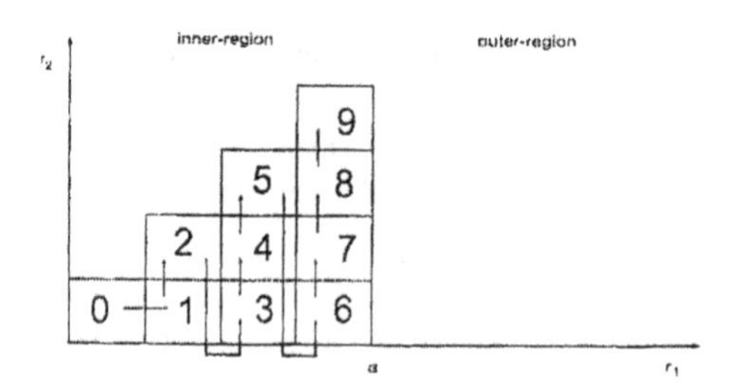

Figure 2: Subdivision of the inner-region configuration space (r_1, r_2) into a set of connected sub-regions labelled 0..9

Within each sub-region energy independent R-matrix basis states, $\theta_k^{LS\pi}(\mathbf{r}_1, \mathbf{r}_2)$, are expanded in terms of one-electron basis functions, ν_{ij}, whose radial forms are solutions of the Schrödinger equation. The expansion coefficients of the ν_{ij} set are obtained by diagonalizing the corresponding 2-electron Hamiltonian matrix. The expansion coefficients and the radial basis functions are then used to construct surface amplitudes, $\omega_{inl_1l_2k}$, associated with each sub-region edge $i \in \{1, 2, 3, 4\}$.

For each incident electron energy a set of local R-matrices (R_{ji}) can be constructed from the surface amplitudes as follows:

$$(R_{ji})_{n'l_1'l_2'nl_1l_2} = \frac{1}{2a_i} \sum_k \frac{\omega_{jn'l_1'l_2'k}\omega_{inl_1l_2k}}{E_k - E}, j, i \in \{1, 2, 3, 4\} \tag{4}$$

Here a_i is the radius of the ith edge, E is the total energy of the two-electron system and E_k are the eigenenergies obtained by diagonalizing the two-electron Hamiltonian in the sub-region. By using the local R-matrices, the R-matrix on the boundary of the innermost sub-region can be propagated across all sub-regions, working systematically from the r_1-axis at the bottom of each strip to its diagonal as illustrated in Figure 2, to yield the global R-matrix, $\Re$, on the boundary of the inner- and outer-region $(r_1 = a)$.

Finally, the global R-matrix $\Re$ is transformed onto an appropriate basis for use in the outer-region. The resulting outer-region equations are identical in form to those described in Section 3.1.

4 Computational Model: 2DRMP

The 2D propagator model described in the previous section has been implemented as the suite of seven programs depicted in Fig. 3. These programs have been designed to operate on serial machines and to exploit distributed memory and shared memory parallelism found on tightly coupled high performance clusters.

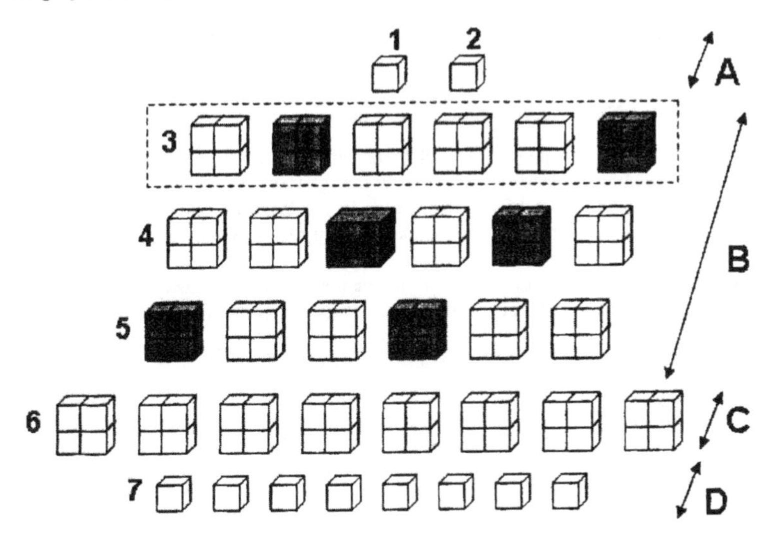

Figure 3: The 2DRMP package. Blocks A and B are independent of the collision energy and need only be performed once while blocks C and D are dependent on the collision energy and must be repeated hundreds of times.

4.1 2DRMP

We begin by sketching the 2DRMP suit. Each program belongs to one of four functional blocks: A, B, C or D. The blocks must be performed sequentially and communication between programs is through files.

Block A contains two independent programs that are not computationally intensive. Program 1 constructs the atomic basis functions used in the transformation of the Global R-matrix, $\Re$, while program 2 computes radial integrals to be used in the construction of the Hamiltonian matrix in off-diagonal sub-regions.

In block B, program 3 constructs a sub-region Hamiltonian matrix, program 4 diagonalizes the matrix, and program 5 constructs the corresponding surface amplitudes from the matrix's eigenvalues and eigenvectors. Each column in this block corresponds to an independent sub-region.

Block C uses program 6 to propagate the global R-matrix, $\Re$, across all the sub-regions of the inner-region. Each element in this block corresponds to a series of propagations, one for each scattering energy.

Block D corresponds to an outer-region program such as FARM. Again each element in this block corresponds to a range of scattering energies.

Programs 3,4,5 and 6 can execute in serial or parallel. For example, program 3 can construct a matrix in serial or in parallel using either MPI or OpenMP. Programs 4,5 and 6 have a similar capability. On a tighly coupled supercomputer, such as HPCx[9], parallelism tends to be horizontal with, for example, many matrices being computed simultaneously (controlled by MPI) and each being spread over several processors (controlled by either MPI or OpenMP). This is illustrated by the dotted rectangle in Fig. 3 where each matrix construction is itself spread across a 2×2 grid of processors.

4.2 High performance computation of the Slater integrals

In large scale virtual experiments using 2DRMP efficient exploitation is severely impeded by a significant load imbalance between the construction of the Hamiltonian matrix on diagonal and off-diagonal sectors. The root of the bottleneck is the large number of two dimensional radial integrals, the so called Slater integrals, that are required on each diagonal sector. The Slater integrals are defined as follows.

Let $X_i = iH$, $i = 0, 1, \ldots, i_{max}$ be equidistant points on the halfaxis $x \geq 0$. On each interval $I_i = (X_i, X_{i+1})$ we consider the eigenvalue problem for the radial Schrödinger equation

$$y''_{nl} = (V(x) - E_{nl})y_{nl}(x), \quad x \in I_i, \tag{5}$$

in a Coulomb field,

$$V(x) = \frac{l(l+1)}{x^2} - \frac{2Z}{x}, \tag{6}$$

where l and Z represent the orbital angular momentum (a nonnegative integer) and the electric charge (a positive real number), respectively. The boundary conditions are as follows:

$$i = 0: \qquad y_{nl}(0) = 0, \qquad a_1 y_{nl}(X_1) + b_1 y'_{nl}(X_1) = 0,$$

$$i > 0: \quad a_0 y_{nl}(X_i) + b_0 y'_{nl}(X_i) = 0, \quad a_1 y_{nl}(X_{i+1}) + b_1 y'_{nl}(X_{i+1}) = 0,$$

where a_0, b_0, a_1 and b_1 are constants dictated by the physics of the problem but subject to the conditions $|a_0| + |b_0| \neq 0$ and $|a_1| + |b_1| \neq 0$. Throughout this paper we use $a_0 = a_1 = 0$ and $b_0 = b_1 = 1$.

The eigenfunctions, y_{nl}, normalized as

$$\int_{X_i}^{X_{i+1}} y_{nl}(x) y_{n'l}(x) dx = \delta_{nn'}, \tag{7}$$

are then used to compute two-dimensional radial integrals of the form

$$R = \int_{X_j}^{X_{j+1}} dx_2 \int_{X_i}^{X_{i+1}} dx_1 \, y_{n_3 l_3}(x_2) y_{n_4 l_4}(x_2) \Delta(x_1, x_2; \lambda) y_{n_1 l_1}(x_1) y_{n_2 l_2}(x_1), \tag{8}$$

where

$$\Delta(x_1, x_2; \lambda) := \begin{cases} \frac{x_1^\lambda}{x_2^{\lambda+1}} & \text{if } x_1 < x_2 \\ \frac{x_2^\lambda}{x_1^{\lambda+1}} & \text{if } x_1 \geq x_2 \end{cases}.$$

Integrals of this form are sometimes called Slater integrals and have to be evaluated in each H by H square sector identified by the pair i, j , where $i = 0, 1, \ldots, i_{max}$ and $j = 0, \ldots, i$. An illustration for $i_{max} = 4$ is shown in Fig.2. Using standard "off the shelf" quadrature formula, Hamiltonian matrix construction on each diagonal sector, to approximately 7 figures of accuracy, can take of the order of hours while each off-diagonal sector takes only tens of seconds.

Efficient computation of the Slater integrals depends primarily on the number of points at which the values of the integrand are needed in order to achieve some predefined accuracy. For example, with a sector size of 15 atomic units, Newton-Cotes formulae typically require more than one thousand mesh points to achieve about eight figures of accuracy. In our research we targeted ten figures of accuracy.

A significant reduction in the number of mesh points is possible but this requires an investigation of the two stages of the process i.e. both the numerical solution of the Schrödinger equation and the actual computation of the Slater integrals. Our aim, therefore, was to design a sequence of numerical methods which advantageously exploited as many characteristic features of the problem as possible. In particular, we wished to devise a two stage computational strategy where the second is influenced and informed by the first.

Our detailed analysis, which is fully described in [10], focused on the l-independence and n-dependence of the frequencies of the oscillatory eigenfunctions that are solutions of the Schrödinger equation. This resulted in the development of the following two stage computational strategy.

Stage 1: Solution of the Schrödinger equation based on piecewise perturbation theory. In practice we opted for CPM{12, 10} of [11] which is a highly accurate version in the class of the CP methods.

Stage 2: Computation of Slater integrals using sets of extended frequency-dependent quadrature rules(EFDQR).

In the first stage the CP{12,10} method enabled us to produce, to an accuracy of about 12 figures, eigenvalues, frequencies, normalized eigenfunctions and the first derivative of the eigenfunctions in a few tens of steps using a fixed step size which is independent of n. By using this information and by exploiting the independence of the frequencies on l, we were able to construct extended frequency dependent quadrature rules for the Slater integrals *using the same mesh points.*

Table 1: Case 1. $L = 0$, $S = 0$, $\Pi = even$, $n_{max} = 20$, $l_{max} = 7$. Case 2. $L = 2$, $S = 0$, $\Pi = even$, $n_{max} = 20$, $l_{max} = 9$. Case 3. $L = 4$, $S = 0$, $\Pi = even$, $n_{max} = 20$, $l_{max} = 11$.

Case	Sector where $i > 0$	No. of Slater integrals	Matrix size	HPCx(secs) SR	HPCx(secs) EFDQR	Speedup
1	0,0	8187600	1680	943	38	25
	i,i	8187600	1680	943	30	31
2	0,0	85741800	5090	9880	163	61
	i,i	85741800	5090	9880	128	77
3	0,0	302869500	8900	34899	472	74
	i,i	302869500	8900	34899	373	94

The new algorithm was incorporated into the construction of the diagonal sector Hamiltonian matrices where time taken for construction is dominated by the computation of the Slater integrals. In table 1 we present three typical cases with fixed n but increasing l that result in 8187600, 85741800 and 302869500 Slater integrals on the respective diagonal sectors[2]. We compare the new EFDQR strategy against the Simpson's Rule (SR) approach implemented in the original 2DRMP code. Sector (0,0) takes longer because of the special treatment required at the origin.

Overall, we have found that the new computational strategy is more accurate, giving an accuracy of 10 rather than 7 figures, and is between one and two orders of magnitude faster than the original implementation. This represents a real and very significant reduction in cpu cycles, particularly for larger cases.

The original 2DRMP code stored the diagonal sector matrices on disk after construction (program 4 in Fig. 3). In a subsequent job using one LPAR[3] per sector, all diagonal sector matrices

[2]The computation on off diagonal sectors is much faster. For example, Case 3 on HPCx takes under 25 secs.
[3]16 processor node on HPCx

were read from disk, block cyclically distributed across their respective LPARs and diagonalized using ScaLAPACK (program 5 in Fig. 3). We have now parallelized the new construction method also using one LPAR per sector. We now combine programs 4 and 5 and build each diagonal sector matrix immediately in its required block cyclic distribution, thereby allowing diagonalization to follow within the same job and avoiding disk IO and storage. From preliminary tests performed we estimate that for case 3 the wall clock time for matrix construction will decrease from 34899 seconds (9.6 hours) to about 45 seconds.

We have therefore transformed matrix construction from a long capacity job to a short capability job very significantly reducing both cpu cycles and wall clock time and increasing machine utilization. However, how can we be sure that the new algorithm gives acceptably accurate results? This is the subject of the next section.

5 Scientific computation: numerical music or numerical noise?

A novel aspect of this collaborative project is the rigorous investigation of the numerical validation of large scale high performance and distributed scientific computation. This is a topic of considerable importance, but one that has received relatively little practical attention in the computational science community. It is well known that the floating point arithmetic commonly used in scientific computing only approximates exact arithmetic. In consequence, each arithmetic statement generates a round-off error. It is not uncommon to find that the same code, using the same data, produces different results when executed on different platforms[4]. Indeed, the same code, using the same data, on the same computer can produce different results when the rounding mode is changed. Potentially these errors can occur at each variable assignment and each arithmetic operation.

The propagation of these errors must be investigated to avoid the production of computed results with no significant digits. In a supercomputing environment where trillions of floating-point operations may be performed per second or in a Grid environment, where the overall computation may involve contributions from many heterogeneous platforms, controlled and rigorous numerical verification is essential to give confidence that the computed results are acceptable.

5.1 The CADNA library

Using Discrete Stochastic Arithmetic[14, 15] the CADNA[5] library[2] is a tool designed to estimate precisely the computing error in computer generated results i.e. to estimate the number of common significant figures between the computed result and the exact result.

The basic idea is to perform the same computation several times propagating the round-off error differently each time. The computer's deterministic arithmetic is replaced by a stochastic arithmetic where each elementary operation is performed N times before the next instruction is executed. For each operation N samples are obtained. The mean value and standard deviation characterize the corresponding stochastic number. The value of this number is the mean value of the different samples. The number of exact significant digits in the number is estimated using the mean value and the standard deviation. If all the samples are zero or if the number of exact significant digits is less than one the number is defined as a computational zero. This means that a computational zero is either a mathematical zero or a number without any significance *i.e. numerical noise.*

The CADNA Fortran implementation is a set of data types, functions and subroutines that may be easily incorporated into any Fortran program. In essence, Fortran types are simply replaced by the corresponding stochastic types. The stochastic numbers are N-tuples containing the perturbed

[4]even when each platform conforms to the IEEE 754 floating-point standard. Variations can occur because of compiler optimization, use of extended precision and behaviour of intrinsic functions.

[5]Control of Accuracy and Debugging for Numerical Applications

floating point values. Arithmetic operators, logical operators and intrinsic functions have been overloaded so that when an operator is used the operands are N-tuples and the returned result is an N-tuple. The library, therefore, enables a scientific code to be executed using random arithmetic without having to make major changes to the original code. During execution when a numerical anomaly is detected, dedicated counters specific to CADNA are incremented. At the end of the run, all these counters with their signification are printed on standard output. A nice feature is that intermediate and final results are output only to the exact number of significant digits.

In this project we are using CADNA systematically throughout 2DRMP to estimate the computing error of intermediate and final results. In addition, we have used CADNA to establish *benchmark results* for the Slater integrals against which the EFDQR method can be compared. Here, when using Newton-Cotes formulae, we employ CADNA to control dynamically the optimum integration step used in the computation of the Slater integrals *i.e.* the step for which the global error, consisting of both the truncation error and the round-off error, is minimal[16]. This is briefly described in the following section. This work will be described further at the conference and is described in detail in [18].

5.2 Benchmark results using CADNA

When an approximate numerical method is used, a truncation error, $e(h)_m$, which depends on the chosen step size h, is generated in addition to the computing error, $e(h)_c$, which is due to round-off error propagation. From the perspective of a computational scientist, the practical error of interest is the global error $e(h)_g$ which combines $e(h)_m$ and $e(h)_c$. As h increases $e(h)_m$ increases and $e(h)_c$ decreases, but when h decreases $e(h)_m$ decreases and $e(h)_c$ increases. These two sources of error counterbalance each other. Ideally we wish to minimize the global error which occurs when $e_c(h) \approx e_m(h)$. With normal floating-point arithmetic this is not possible as we have no estimate of $e_c(h)$, however with CADNA it is possible.

It can be shown that if $I(h)$ is an approximation of order p of an exact value I, *i.e.* $I(h) - I = Kh^p + \mathcal{O}(h^q)$ with $1 \leq p < q$, $K \in \mathbb{R}$ then if computations are performed, in the convergence zone, until the difference between two successive approximations is a computational zero then the significant bits of the last approximation not affected by round-off errors are in common with the *exact value* of the integral, I, up to one bit [17]. In effect, we generate a sequence of approximations, I_n, halving the step size each time. When the difference between two successive iterations is a computational zero no further meaningful computation can occur and we have achieved the optimal computed result for the method using the finite precision available. Since Newton-Cotes type formulae are of the appropriate approximation form we have established using 2^{17} integration steps benchmark results for a range of Slater integrals accurate to 10 exact significant digits. The new EFDQR agree fully with the benchmark results.

6 Concluding remarks

In this paper we have described a computational science project involving researchers from France, Romania and the UK. The primary aim being to engineer numerically reliable software that can exploit high performance and distributed computers in the study of electron collisions with atoms and ions. However, we have also presented some ideas of more general interest.

First, we have advocated the construction of hand crafted algorithms that exploit a problem's characteristics rather than using general "off the shelf" algorithms. In doing so we have dramatically improved performance and made more efficient use of valuable machine cycles.

Second, we have drawn attention to the fact that scientific computation is inherently flawed. We have focused on the propagation of round-off error due to finite precision floating-point arithmetic.

In particular, we have advocated and demonstrated the use of the CADNA library as a diagnostic tool to help appreciate and control these errors.

Acknowledgments

The authors are grateful to the UK EPSRC for their support through the grants GR/M01784/01 and GR/R89073/01. LGrI is indebted to the UK EPSRC for a Visiting Fellowship to Queen's University Belfast and NSS is grateful to Laboratoire d'Informatique de Paris 6, l'Université Pierre et Marie Curie, Paris for a Visiting Professorship.

References

[1] P. G. Burke, K. A. Berrington(eds), Atomic and Molecular Processes and R-matrix Approach. IOP Publishing, Bristol (1993).

[2] CADNA website - http://www.lip6.fr/cadna.

[3] K. H. Becker, C. W. McCurdy, T. M. Orlando and T. N. Rescigno, Workshop on "Electron-Driven Processes:Scientific Challenges and Technological Opportunities", http://attila.stevens-tech.edu/physics/People/Faculty/Becker/EDP/ EDP-FinalReport.pdf.

[4] V. M. Burke and C. J. Noble, Comput. Phys. Commun. (1995) **85**, 471–500.

[5] E. P. Wigner, Phys. Rev. (1946) **70**, 15–33.

[6] E. P. Wigner, Phys. Rev. (1946) **70**, 606–618.

[7] P. G. Burke, Adv. Atom. Mol. Phys. (1968) **4**, 173–219.

[8] P. G. Burke, C. J. Noble and M. P. Scott, Proc. Roy. Soc. A (1987) **410**, 287–310.

[9] HPCx, UK National Supercomputing Centre - http://www.hpcx.ac.uk/.

[10] L. Gr. Ixaru, N. S. Scott and M. P. Scott, SIAM Journal on Scientific Computing, (2006), accepted for publication.

[11] L. Gr. Ixaru, H. De Meyer, G. Vanden Berghe, Journal Comp. and Appl. Math., (1998) **88**, 289–314.

[12] R. Alan et al, UK e-Science All Hands Meeting, (2003) 556–561 (http://www.nesc.ac.uk/events/ahm2003/AHMCD/pdf/125.pdf).

[13] Virginia Faro-Mazo, P. Preston, T. Harmer and N.S. Scott, UK e-Science All Hands Meeting (2004) 1007–1010 (http://www.allhands.org.uk/2004/proceedings/papers/134.pdf).

[14] J. Vignes, Math. Comput. Simulation, (1993) **35**, 233–261.

[15] J. Vignes, Num. Algo. (2004) **37**, 377–390.

[16] J.-M. Chesneaux, C. Denis, F. Jézéquel and N. S. Scott, Scan2006 Duisburg, Germany.

[17] F. Jézéquel, C. R. Acad. Sci. Paris - Mécanique, (2006) **334**, 362.

[18] N. S. Scott, F. Jézéquel, C. Denis and J.-M. Chesneaux, Comput. Phys. Commun. to be published (2007).

Brill Academic Publishers
P.O. Box 9000, 2300 PA Leiden
The Netherlands

Lecture Series on Computer and Computational Sciences
Volume 6, 2006, pp. 571-580

An Alternative Proof of the Onsager Reciprocal Relations for Multi-component Diffusion

George D. Verros[1] and Athena K. Testempasi[2]
Department of Electrical Engineering
Technological & Educational Institute (T.E.I.) of Lamia
GR-35100 Lamia, Greece

Received July 13, 2006; accepted July 20, 2006

Abstract: In this work an alternative proof of the Onsager Reciprocal Relations (ORR) for simultaneous multi-component diffusion and heat transfer in the presence of external forces is given. This proof is based on sound principles such as the Galilean Invariance thus eliminating any doubt about the ORR. It is believed that this proof may be generalized to other processes beyond multi-component diffusion thus leading to a generalized framework for transport phenomena and irreversible thermodynamics.

Keywords: Multi-component Diffusion, Heat Transfer, Friction Coefficients, Onsager Reciprocal Relations

PACS : 0570Ln ; 6610Cb ; 8715Vv ; 44

1. Introduction

The field of irreversible thermodynamics provides us with a general framework for the macroscopic description of processes. [1]-[3] More specifically, irreversible thermodynamics is an indispensable part of many computational models describing transport phenomena, material manufacture, etc. [4]-[9]. Irreversible thermodynamics is based on four postulates above and beyond those of equilibrium thermodynamics [1]-[3]:

1. The equilibrium thermodynamic relations apply to systems that are not in equilibrium, provided that the gradients are not too large (quasi-equilibrium postulate)
2. All the fluxes ($\mathbf{J}_i$) in the system may be written as linear relations involving all the thermodynamic forces, $\mathbf{X}_i$. (linearity postulate, $J_i = \sum_{j=1}^{n} L_{ij} X_i$; i =1,2...n)
3. No coupling of fluxes and forces occurs if the difference in the tensorial order of the flux and force is an odd number (Curie's postulate)
4. In the absence of magnetic fields and assuming linearly independent fluxes or thermodynamic forces the matrix of coefficients in the flux-force relations is symmetrical. This postulate is known as the Onsanger Reciprocal Relations (ORR):$L_{ij} = L_{ji}$.

Onsager derived these relations for the heat transfer case in 1931[10]-[11]. He used the principle of microscopic reversibility by applying the invariance of the equations of motion for atoms and molecules with respect to time reversal (the transformation t→-t). This means that the mechanical equations of the motion (classical as well as quantum mechanical) of the particles are symmetrical with respect to the time. In other words, the particles retrace their former paths if all velocities are reversed[12]. Onsager also made a principal decision: the transition from molecular reversibility to microscopic reversibility can be made. It is important to remark that Onsager did not use a particular molecular model. As a consequence the results and limitations of the theory are valid for all materials, so that the theory can be related to continuum theory[12]. Casimir further developed this theory[13].

In the literature there appear to be two groups of derivations of Onsager reciprocal relations (ORR). In the first of these, it is assumed that the macroscopic laws of motion hold for the averages of the macroscopic coordinates (such as temperature gradient, concentration gradient, etc) even if their values

---

[1]Corresponding author. E-mail: verros@eng.auth.gr, gdverros@otenet.gr

[2]Also Environmental Education Center of Edessa, 58200 Edessa, GREECE

are microscopic. The second group assumes a definite statistical law for the path representing the system in phase space[12]. Although there is experimental evidence for the validity ORR [14]-[16], as was noticed by Prigogine and Kondepudi[17] in a recent review the theoretical basis of ORR requires careful considerations. Moreover, doubts about the ORR have been raised in the literature[18]-[20]; For example, In *Rational Thermodynamics*[19] Truesdell remarks, "Onsager's and Casimir's claims that their assertions follow from *the principle of microscopic reversibility* have been accepted with little question... the reversibility theorem and Poincare's recurrence theorem make irreversible behavior impossible for dynamical systems in a classical sense. Something must be added to the dynamics of conservative systems, something not consistent with it, in order to get irreversibility at all." [20] Moreover, it is believed that if equilibrium relations such as the Gibbs-Duhem equations are added in the mathematical model for multi-component diffusion by simply applying the quasi equilibrium postulate, then the Onsager Reciprocal Conditions are not necessarily fulfilled (Ref [2] p.67-69); The aim of this work is to investigate the theoretical basis of this principle for the multi-component diffusion case.

Multi-component diffusion is the main issue in many industrial processes including metallurgy, ceramics manufacture and chemical engineering process such as drying, extraction, membrane separation etc.[3], [21]-[24]. The industrial importance of multi-component diffusion has led to numerous studies. Most of these studies are based on the Onsager model for multi-component diffusion.[4]-[9] According to this model the molar fluxes $\mathbf{J}_i$ in a N-component system can be expressed as a linear combination of the vector of chemical potential gradients $\mathbf{X}_i$ and a **NxN** matrix of Onsager $\mathbf{L}_{ij}$ phenomenological mobility coefficients:

$$J_i = \left\|L_{ij}\right\| X_i \; ; \; X_i = -grad\mu_i \; ; \; J_i = c_i(v_i - v) \tag{1}$$

where $\mathbf{v}$ refers to the velocity of the centre of mass, $\boldsymbol{\mu}$ represents the chemical potential and $\mathbf{v}_i$ stands for the local velocity defined as $\mathbf{J}_i/\mathbf{c}_i$, where $\mathbf{c}_i$ is the molar concentration of the i-th substance. The Onsager reciprocal relations (ORR) for multi-component diffusion state that $L_{ij} = L_{ji}$; i,j =1,2...n-1. These relations were applied by several workers in the field to reduce the number of unknown mobility coefficients or in other words to reduce the degrees of freedom [4]-[9] The aim of this work is to eliminate any doubt about ORR by giving a simple proof for the isothermal and non-isothermal multi-component diffusion case in the presence of external forces. This is achieved by studying the transformation laws between fluxes defined relative to an arbitrary velocity and by applying the Gibbs-Duhem theorem for multi-component diffusion. In what follows, the derivations of Onsager reciprocal relations for the multi-component isothermal diffusion and simultaneous heat & mass transfer are given, the main issues of this work are analyzed and finally conclusions are drawn.

2. Derivation of Onsager Reciprocal Relations for Multi-component Isothermal Diffusion

The uncompensated heat produced by an irreversible process is given by the dissipation function. The dissipation function is derived from an entropy balance [1]-[3]. The starting point of this work is the definition of the dissipation function Ψ in the absence of viscous flows for a non-elastic, non-reacting, isothermal & isotropic fluid containing n diffusing species [1]-[3], [12]:

$$\Psi = T\sigma = \sum_{i=1}^{n} J_i X_i \; ; i = 1,2...n \tag{2}$$

where σ is the rate of production of entropy per unit volume, T stands for the thermodynamic temperature and the molar flux $\mathbf{J}_i$ is measured relative to the velocity $\mathbf{v}$ of the centre of mass :

$$J_i = c_i(v_i - v) \; ; \; v = \sum_{i=1}^{n} M_i c_i v_i / \rho \; ; \; \sum_{i=1}^{n} J_i M_i = 0 \tag{3}$$

c_i is the molar concentration, M_i stands for the molar mass of the i-th species, the density ρ is given as: $\rho = \sum_{i=1}^{n} M_i c_i$ and the thermodynamic forces X_i are given as $X_i = -(grad\mu_i)_T + F_i$, where $(grad\mu_i)_T = (grad\mu_i)_{T,P} + V_i grad(p)$ is the gradient of i-th substance molar chemical potential, V_i stands for the partial molar volume of the i-th substance, p is the hydrostatic pressure and F_i represents the external force per mole of each substance. In this work, it is assumed that external forces act on the system or in other words there is no mechanical equilibrium.

At this point it is assumed that the quasi-equilibrium postulate holds true (see Introduction); Consequently, equilibrium thermodynamic relations such as the Gibbs-Duhem equation, can be applied to the system. Schmitt and Graig[25] have shown that if the transformed thermodynamic forces $x_i^{'} = X_i + M_i \left(grad(p) - \sum_{j=1}^{n} c_j F_j \right) \Big/ \rho$ are introduced in the dissipation function, then by using the Gibbs - Duhem equation $\sum_{i=1}^{n} c_i (grad\mu_i)_{T,P} = 0$ the following equation is derived:

$$\sum_{i=1}^{n} c_i x_i^{'} = 0 \tag{4}$$

It is also necessary to introduce new fluxes, defined relative to an arbitrary reference velocity $v^{\neq}$:

$$j_i^{\neq} = c_i (v_i - v^{\neq}) \; ; \; v^{\neq} = \sum_{i=1}^{n} w_i v_i \tag{5}$$

where w_i are weighting factors whose sum is unity: : $\sum_{i=1}^{n} w_i = 1$

If these new fluxes are introduced in Eq. (2), then the dissipation function is invariant under the transformation to the new set of fluxes as well as to the thermodynamic forces [25, 26]:

$$\Psi = \sum_{i=1}^{n} j_i^{\neq} x_i^{'} \tag{6}$$

The above equation derived by Schmitt and Craig [25] is a generalization of the Prigogine theorem [27] for multi-component diffusion in the absence of the mechanical equilibrium.

Please note, that the fluxes $j_i^{\neq}$ are also linearly dependent, since from eq (5)

$$\sum_{i=1}^{n} w_i j_i^{\neq} / c_i \tag{7}$$

By using Eq (7) the dissipation function (Eq. 6) is written as:

$$\Psi = \sum_{i=1}^{n} j_i^{\neq} x_i^{'} = \sum_{i,j=1}^{n-1} (\delta_{ij} + w_i c_j / c_i w_n) j_i^{\neq} x_j^{'} \text{ or } \Psi = \sum_{i,j}^{n-1} A_{ij} j_i^{\neq} x_j^{'} \tag{8}$$

where $A_{ij} = \delta_{ij} + w_i c_j / c_i w_n$; i,j = 1,2..,n-1 (9)

Eq (8) can be regarded as the sum of fluxes and transformed dynamic forces $\Psi = \sum_{i=1}^{n-1} j_i^{\neq} \left(\sum_{j=1}^{n-1} A_{ij} x_j^{'} \right)$

or, also as the sum of products of transformed fluxes and thermodynamic forces:

$$\Psi = \sum_{j=1}^{n-1}\left(\sum_{i=1}^{n-1} A_{ij} j_i^{\neq}\right) x_j' \tag{10}$$

The main idea of irreversible thermodynamics is to derive fundamental macroscopic laws from the dissipation function[1]-[3]; For this purpose the linearity postulate (see Introduction) is applied to the fluxes and driving forces as these appear in the dissipation function. In our case there are n-1 independent fluxes and driving forces (see Eq. 10). Consequently, by using the linearity postulate (see Introduction) the following equations between fluxes and thermodynamic driving forces are derived:

$$j_i^{\neq} = \sum_{j,k=1}^{n-1} l_{ij}^{\neq}\left(A_{jk} x_k'\right) = \sum_{j,k=1}^{n-1}\left(l_{ik}^{\neq} A_{kj}\right) x_j' \quad \text{or}$$

$$x_i' = \sum_{j,k=1}^{n-1} r_{ij}^{\neq}\left(A_{kj} j_k^{\neq}\right) = \sum_{j,k=1}^{n-1}\left(r_{ik}^{\neq} A_{jk}\right) j_j^{\neq} \tag{11}$$

The quantities $l_{ij}^{\neq}$ are the mobility coefficients and $r_{ij}^{\neq}$ are the friction coefficients for diffusion respectively. Since the fluxes and the thermodynamic forces in the above equations are linearly independent, the Onsager Reciprocal Relations (ORR) state that

$$l_{ij}^{\neq} = l_{ji}^{\neq} \quad \text{or} \quad r_{ij}^{\neq} = r_{ji}^{\neq} \quad ; i,j = 1,2 \ldots n-1 \tag{12}$$

At this point the mobility coefficients ($L_{ij}^{\neq}$) and the friction coefficients ($R_{ij}^{\neq}$) are introduced by using the linearity postulate[1]-[3], [26],[28]:

$$j_i^{\neq} = \sum_{j=1}^{n} L_{ij}^{\neq} x_j' \quad ; \quad x_i' = \sum_{j=1}^{n} R_{ij}^{\neq} j_j^{\neq} \quad ; i = 1,2 \ldots n \tag{13}$$

In the above equations it is assumed that the mobility coefficients and the friction coefficients are different to those defined in Eq (11). By using Eq. (7) and eliminating $j_n^{\neq}$ from Eq. (13), the following equation is derived:

$$x_i' = \sum_{j=1}^{n-1}\left(R_{ij}^{\neq} - w_j c_n R_{in}^{\neq} / c_j w_n\right) j_j^{\neq} \quad ; i=1,2..n \tag{14}$$

From Eq. (4), (9) and (11b) it follows that

$$x_n' = -\sum_{i,j=1}^{n} c_i \left\{ r_{ij}^{\neq} + \left(w_j / c_j w_n\right) \sum_{k=1}^{n-1} c_k r_{ik}^{\neq} \right\} j_j^{\neq} / c_n \tag{15}$$

A comparison of Eq. (11b) and (15) with (14) gives

$$x_n' = -\sum_{i,j=1}^{n} c_i \left\{ r_{ij}^{\neq} + \left(w_j / c_j w_n\right) \sum_{k=1}^{n-1} c_k r_{ik}^{\neq} \right\} j_j^{\neq} / c_n \quad ; i,j = 1,2..n-1 \tag{16}$$

$$R_{ij}^{\neq} - w_j c_n R_{in}^{\neq} / c_j w_n = r_{ij}^{\neq} + \left(w_j / c_j w_n\right) \sum_{k=1}^{n-1} c_k r_{ik}^{\neq} \quad ; j=1,2 \ldots n-1 \tag{17}$$

Following Tyrrell and Harris [26] the friction coefficients (see Eq. 13) are introduced into Eq (4):

$$\sum_{k=1}^{n} c_k \dot{x}_k = \sum_{k=1}^{n} c_k \sum_{i=1}^{n} R_{ki}^{\neq} j_i^{\neq} = \sum_{i=1}^{n} j_i^{\neq} \sum_{k=1}^{n} R_{ki}^{\neq} c_k^{\neq} = 0 \text{ or}$$

$$\sum_{j=1}^{n} c_j R_{ij}^{\neq} = \sum_{j=1}^{n} c_j R_{ji}^{\neq} = 0 \text{ ; i =1,2...n} \tag{18}$$

The above equation holds true due to the fact that the fluxes are defined relative to an arbitrary velocity and in the most general case, $\sum_{i=1}^{n} j_i^{\neq} \neq 0$ By introducing Eq. (18) into Eq. (17) it follows that

$$\left(\sum_{i=1}^{n-1} R_{ij}^{\neq} c_i - \sum_{i=1}^{n-1} c_i r_{ij}^{\neq}\right) - \left(w_j / c_j w_n\right)\left(R_{nn}^{\neq} - \sum_{j=1}^{n-1} c_i c_j r_{ij}^{\neq} / c_n^2\right) = 0 \text{ ; j = 1,2..n-1} \tag{19}$$

As Eq. (19) holds for arbitrary concentrations, one could derive the following equations:

$$\sum_{i=1}^{n-1} c_i R_{ij}^{\neq} = \sum_{i=1}^{n-1} c_i r_{ij}^{\neq} \text{ ;j = 1,2 .. n-1} \tag{20}$$

$$R_{nn}^{\neq} = -\sum_{j=1}^{n-1} c_i c_j r_{ij}^{\neq} / c_n^2 \tag{21}$$

From equation (20) it follows that $R_{ij}^{\neq} = r_{ij}^{\neq}$; i,j = 1,2...n-1 (22)

The above equation could be viewed as a consequence of the Galilean invariance. Galilean transformations describe the change from one reference system to another by means of a uniform translation. In the classical theory, the physical laws and equations have to be invariant with respect to reference systems that are in relative translation at a constant velocity. They are said to be invariant Galilean transformations[12]. In our case, the fluxes defined in Eq. (11) and (13) could be viewed as fluxes defined with respect to different reference systems. More specifically, the fluxes in Eq. (11) could be viewed as defined relative to the velocity of the n-th substance v_n, while the fluxes in Eq (14) are defined relative to an arbitrary velocity $v^{\neq}$. According to Galilean transformation the physical laws and equations have to be invariant regarding the different system of reference. The mobility and the friction coefficients are physical quantities characterizing matter; Consequently, the mobility as well as the friction coefficients are independent of the fluxes reference system. Therefore, Eq. (22) can be directly derived from the Galilean invariance. In a similar way, the invariance of the dissipation function with respect to the flux reference velocity can be viewed as a consequence of the Galilean invariance.By using Eq. (22) one can eliminate the first term on the left and the right hand-side of Eq (16) and the following equation is directly derived:

$$R_{in}^{\neq} = -\sum_{j=1}^{n-1} c_j r_{ij}^{\neq} / c_n \quad \text{or} \quad \sum_{j=1}^{n} c_j R_{ij}^{\neq} = 0 \quad \text{; i = 1,2...n-1} \tag{23}$$

By using Eq. (18), (21) and Eq. (23) it follows that[26]:

$$\sum_{j=1}^{n} c_j R_{ij}^{\neq} = \sum_{j=1}^{n} c_j R_{ji}^{\neq} = 0 \text{ or } R_{ij}^{\neq} = R_{ji}^{\neq} \text{ ; i, j = 1,2 ...n} \tag{24}$$

Given the above equality, by taking into account Eq. (22) the Onsager reciprocal relations for the linearly independent friction coefficients are derived:

$$r_{ij}^{\neq} = r_{ji}^{\neq} \text{ ; i,j =1,2...n-1} \tag{25}$$

As the fluxes and the driving forces defined in Eq. (11) are linearly independent then there is the inverse matrix of the friction coefficients $r_{ij}^{\neq}$; This inverse matrix is the matrix of mobility coefficients $l_{ij}^{\neq}$. The matrix of friction coefficients is symmetrical (see Eq. 25) and also the inverse matrix, defined as the matrix of mobility coefficients has to be symmetrical [16]:

$$l_{ij}^{\neq} = l_{ji}^{\neq} \; ; \; i,j = 1,2...n-1 \tag{26}$$

The above equations are the Onsager Reciprocal Relations for isothermal diffusion. The analysis presented in this work is not a new analysis; The idea of using Eq. (24) for proving ORR can be found in the work of Miller[16]. Moreover, the methodology for deriving Eq. (24) can be found in Lorimer's [26] work for the frame of reference for diffusion in membranes and liquids. However, Lorimer [26] used the equality between friction coefficients (Eq. 22) as an arbitrary assumption. In this work Eq. (22) was derived either as a consequence of the Galilean invariance or as a consequence of the relation between driving thermodynamic forces (see Eq. 4 and Eq. 18)

3. Derivation of Onsager Reciprocal Relations for Simultaneous Heat Transfer & Multi-Component Diffusion

The uncompensated heat produced by an irreversible process is given by the dissipation function. The dissipation function is derived from an entropy balance [1]-[3]. The starting point of this work is the definition of the dissipation function Ψ in the absence of viscous flows for a non-elastic, non-reacting, isotropic fluid containing n diffusing species [28]:

$$\Psi = T\sigma = \sum_{i=1}^{n} j_i^{\neq} x_i' + j_q' x_u \; ; x_u = -gradT/T \; ; \; j_q' = j_q - \sum_{i=1}^{n} H_i j_i \; ; \; -\nabla . j_q = \rho \frac{dq}{dt} \; ; i=1,2..n \tag{27}$$

where σ is the rate of production of entropy per unit volume, T stands for the thermodynamic temperature, H_i is the partial molar of species i, q represents heat, ρ is the density.

The molar flux $j_i^{\neq}$ is given by Eq. (5) and the thermodynamic forces x'_i are defined as

$x_i' = X_i + M_i \left(grad(p) - \sum_{j=1}^{n} c_j F_j \right) / \rho$ By using Eq (7) the dissipation function (Eq. 27) is written

$$\text{as: } \Psi = \sum_{i,j}^{n-1} A_{ij} j_i^{\neq} x_j' + j_q' x_u \; ; \quad A_{ij} = \delta_{ij} + w_i c_j / c_i w_n \; ; i,j = 1,2..,n-1 \tag{28}$$

Eq (28) can be regarded as the sum of fluxes and transformed dynamic forces:

$$\Psi = \sum_{i=1}^{n-1} j_i^{\neq} \left(\sum_{j=1}^{n-1} A_{ij} x_j' \right) + j_q' x_u = \sum_{j=1}^{n-1} \left(\sum_{i=1}^{n-1} A_{ij} j_i^{\neq} \right) x_j' + j_q' x_u \tag{29}$$

By using the linearity postulate (see Introduction) the following equations between fluxes and thermodynamic driving forces are directly derived:

$$j_i^{\neq} = \sum_{j,k=1}^{n-1} l_{ij}^{\neq} \left(A_{jk} x_k' \right) + l_{iu}^{\neq} x_u = \sum_{j,k=1}^{n-1} \left(l_{ik}^{\neq} A_{kj} \right) x_j' + l_{iu}^{\neq} x_u$$

$$j_q^{\neq} = \sum_{j,k=1}^{n-1} l_{uj}^{\neq} \left(A_{jk} x_k' \right) + l_{uu}^{\neq} x_u = \sum_{j,k=1}^{n-1} \left(l_{ik}^{\neq} A_{kj} \right) x_j' + l_{uu}^{\neq} x_u$$

$$x_i' = \sum_{j,k=1}^{n-1} r_{ij}^{\neq}\left(A_{kj} j_k^{\neq}\right) + r_{iu}^{\neq} x_u = \sum_{j,k=1}^{n-1}\left(r_{ik}^{\neq} A_{jk}\right) j_j^{\neq} + r_{iu}^{\neq} x_u$$

$$x_u = \sum_{j,k=1}^{n-1} r_{uj}^{\neq}\left(A_{kj} j_k^{\neq}\right) + r_{uu}^{\neq} x_u = \sum_{j,k=1}^{n-1}\left(r_{ik}^{\neq} A_{jk}\right) j_j^{\neq} + r_{uu}^{\neq} x_u \tag{30}$$

The quantities $l^{\neq}$ are the conductivity coefficients and $r^{\neq}$ are the friction coefficients for diffusion or heat transfer, respectively. Since the fluxes and the thermodynamic forces are linearly independent (Eq. 29-30), the Onsager reciprocal relations (ORR) state that

$$l_{ij}^{\neq} = l_{ji}^{\neq}\,,\ l_{uj}^{\neq} = l_{ju}^{\neq} \text{ or } r_{ij}^{\neq} = r_{ji}^{\neq}\,,\ r_{uj}^{\neq} = r_{ju}^{\neq} \qquad \text{i,j} = 1,2\ldots\text{n-1} \tag{31}$$

By introducing of the friction coefficients $R^{\neq}$ and by using the linearity postulate [2], [26] it follows that:

$$x_i' = \sum_{j=1}^{n} R_{ij}^{\neq} j_j^{\neq} + R_{iu}^{\neq} j_q \ ; \ x_u = \sum_{j=1}^{n} R_{uj}^{\neq} j_j^{\neq} + + R_{uu}^{\neq} j_q \ ; \quad \text{i} = 1,2\ldots.\text{n} \tag{32}$$

In the above equations it is assumed that the the friction coefficients are different than these defined in Eq (30). By using Eq. (7) and eliminating $j^{\neq}_n$ from Eq. (32), the following equation is derived:

$$x_i' = \sum_{j=1}^{n-1}\left(R_{ij}^{\neq} - w_j c_n R_{in}^{\neq} / c_j w_n\right) j_j^{\neq} + R_{iu}^{\neq} j_q \quad ; \quad \text{i}=1,2..\text{n}$$

$$x_u = \sum_{j=1}^{n-1}\left(R_{uj}^{\neq} - w_j c_n R_{un}^{\neq} / c_j w_n\right) j_j^{\neq} + R_{uu}^{\neq} j_q \tag{33}$$

From Eq. (4), (28b) and (30) it follows that

$$x_n' = -\sum_{i,j=1}^{n} c_i \left\{ r_{ij}^{\neq} + \left(w_j / c_j w_n\right) \sum_{k=1}^{n-1} c_k r_{ik}^{\neq} \right\} j_j^{\neq} / c_n + \sum_{i=1}^{n-1} c_i r_{iu}^{\neq} / c_n \tag{34}$$

Comparison of Eq. (30) and (34) with Eq. (33) gives:

$$R_{ij}^{\neq} - w_j c_n R_{in}^{\neq} / c_j w_n = r_{ij}^{\neq} + \left(w_j / c_j w_n\right) \sum_{k=1}^{n-1} c_k r_{ik}^{\neq} \ ; \ R_{iu}^{\neq} = r_{iu}^{\neq} \ \text{i,j} = 1,2..\text{n-1} \tag{35}$$

$$R_{nj}^{\neq} - w_j c_n R_{nn}^{\neq} / c_j w_n = -\sum_{i=1}^{n-1} c_i \left\{ r_{ij}^{\neq} + \left(w_j / c_j w_n\right) \sum_{k=1}^{n-1} c_k r_{ik}^{\neq} \right\} j_j^{\neq} / c_n \tag{36}$$

$$\sum_{i=1}^{n-1} c_i r_{iu}^{\neq} + c_n R_{nu}^{\neq} = 0 \ \text{ or by using (35b) } \sum_{i=1}^{n} c_i R_{iu}^{\neq} = 0 \tag{37}$$

$$R_{uj}^{\neq} - w_j c_n R_{un}^{\neq} / c_j w_n = r_{uj}^{\neq} + \left(w_j / c_j w_n\right) \sum_{k=1}^{n-1} c_k r_{uk}^{\neq} \qquad ; \quad R_{uu}^{\neq} = r_{uu}^{\neq} \tag{38}$$

By substituting Eq. (32) into Eq. (4) the following equation is derived [21]: $\sum_{i=1}^{n} c_i \sum_{j=1}^{n} R_{ij}^{\neq} j_j^{\neq} + \sum_{i=1}^{n} c_i R_{iu}^{\neq} j_q = 0$ From this equation and from Eq. (37b) by taking into account that the total sum of molar fluxes relative to an arbitrary velocity , in the most general case, is not equal to zero it follows that $\sum_{i=1}^{n} c_i R_{ij}^{\neq} = 0$ (39)

By introducing Eq. (39) into Eq. (36) it follows that

$$\left(\sum_{i=1}^{n-1} R_{ij}^{\neq} c_i - \sum_{i=1}^{n-1} c_i r_{ij}^{\neq}\right) - \left(w_j / c_j w_n\right)\left(R_{nn}^{\neq} - \sum_{j=1}^{n-1} c_i c_j r_{ij}^{\neq} / c_n^2\right) = 0 \;\; ; j = 1,2..n\text{-}1 \quad (40)$$

As Eq. (40) holds for arbitrary concentrations, one could derive the following equations:

$$\sum_{i=1}^{n-1} c_i R_{ij}^{\neq} = \sum_{i=1}^{n-1} c_i r_{ij}^{\neq} \;\; ; j = 1,2 .. n\text{-}1 \qquad R_{nn}^{\neq} = -\sum_{j=1}^{n-1} c_i c_j r_{ij}^{\neq} / c_n^2 \quad (41)$$

From the above equation it follows that : $R_{ij}^{\neq} = r_{ij}^{\neq}$; i,j = 1,2…n-1 (42)

By using Eq. (41) one can eliminate the first term of the left and the right hand-side of Eq (35) and the following equation is directly derived:

$$R_{in}^{\neq} = -\sum_{j=1}^{n-1} c_j r_{ij}^{\neq} / c_n \quad ; \quad i = 1,2\ldots n\text{-}1 \quad \text{or} \quad \sum_{j=1}^{n} c_j R_{ij}^{\neq} = 0 \quad (43)$$

By comparing Eq. (43) and Eq. (39) it follows that [26],[28]:

$$\sum_{j=1}^{n} c_j R_{ij}^{\neq} = \sum_{j=1}^{n} c_j R_{ji}^{\neq} = 0 ;\ i, j = 1,2 \ldots n \quad \text{or} \quad R_{ij}^{\neq} = R_{ji}^{\neq} \;\; ; i, j = 1,2 \ldots n \quad (44)$$

These are the Osanger reciprocal relations for the diffusion friction coefficients.

From equation (38a) it follows directly that $R_{uj}^{\neq} = r_{uj}^{\neq}$. By substituting this equation into Eq. (38) and by taking into account Eq.(35b) and Eq. (37)it follows that $\sum_{i=1}^{n} c_i R_{iu}^{\neq} = \sum_{i=1}^{n} c_i R_{ui}^{\neq} = 0$ or $R_{uj}^{\neq} = R_{ju}^{\neq}$.

Given this equality, by taking into account Eq. (35b) the Onsager reciprocal relations for the $r_{uj}^{\neq}$ are derived: $r_{uj}^{\neq} = r_{ju}^{\neq}$; $l_{uj}^{\neq} = l_{ju}^{\neq}$; j =1,2…n-1. It should be noted here that this work follows closely the work of Lorimer [28]. However, in this work the equality $R_{uj}^{\neq} = r_{uj}^{\neq}$ and the equality $R_{ij}^{\neq} = r_{ij}^{\neq}$ (Eq. 42) are derived directly from the Gibbs-Duhem theorem.

4. Results & Conclusions

In the present work an alternative proof of the Onsager reciprocal relations (ORR) for multi-component diffusion in the presence of external forces (absence of mechanical equilibrium) is given. This proof is based on the transformation laws between fluxes defined relative to an arbitrary velocity (Galilean Invariance, GI). The main characteristic of this proof is its simplicity eliminating thus any doubt about the ORR for the multi-component diffusion case. Consequently, every model for multi-component diffusion should satisfy the Onsager reciprocal relations. This analysis shows that that if the Gibbs-Duhem equations are added in the mathematical model for multi-component diffusion by simply applying the quasi equilibrium postulate, the ORR are also fulfilled. Therefore, it is not necessary to seek for specific conditions for the friction coefficients to satisfy the ORR. The friction coefficients characterize energy and matter; Therefore according to the Galilean Invariance as shown in this work, these coefficients are independent of the fluxes reference system. Moreover, this work shows that the ORR for this particular case could be viewed as a consequence of the Galilean Invariance. It is believed that this proof may be generalized to other processes beyond multi-component diffusion thus leading to a generalized framework for irreversible thermodynamics and transport phenomena.

Acknowledgment. The authors thank Ms Kate Somerscales for her help in preparing the manuscript

References

[1] J.O. Hirschfelder, C.F. Curtiss and R.B. Bird, *Molecular Theory of Gases and Liquids*, Wiley, New York, 1964.

[2] S.R. de Groot and P. Mazur, *Non-Equilibrium Thermodynamics*, Dover Publications, New York, 1984.

[3] R.B. Bird, W.E. Stewart and E. N. Lightfoot, *Transport Phenomena*, John Wiley & Sons, New York, 2002.

[4] J.S. Vrentas, J.L. Duda, H.C Ling, *J. Appl. Pol. Sci.*, Vol. *30*, 4499-4516 (1985).

[5] C.F. Curtiss, R.B. Bird, *Nat. Acad. Sci. USA*, Vol. *93*, 7440-7445 (1996).

[6] P.E. Price Jr, I.H. Romdhane, *AIChE J.*, Vol. *49(2)*, 309-322 (2003).

[7] G.D. Verros, N.A Malamataris, *Polymer*, Vol. *46*, 12626-12636 (2005).

[8] G.D. Verros, N.A Malamataris, *In Lecture Series on Computer and Computational Sciences, Vol. 3: In the Frontiers of Computational Science*, G. Maroulis, T.E. Simos (Eds), Brill Academic Publishers: Leiden, The Netherlands, pp. 360 – 383 (2005).

[9] P. J. A. M. Kerkhof, M. A. M. Geboers, *AIChE J.* , Vol. *51*(1), 79-121 (2005).

[10] L. Onsager, *Physical Review*, Vol. 37, 405-426 (1931).

[11] L. Onsager, *Physical Review*, Vol. 38, 2265-2279 (1931).

[12] G.D.C. Kuiken, *Thermodynamics of Irreversible Processes-Applications to Diffusion and Rheology*, John Wiley & Sons, New York, 1994.

[13] H.B.G. Casimir, *Rev. Mod. Phys*, Vol. 17(2-3), 343-350 (1945).

[14] D.G. Miller, *Chem. Rev.*, Vol. 60(1), 16-37 (1960).

[15] D.G. Miller, *J. Phys. Chem.*, Vol. *63(8)*, 570-578 (1959).

[16] D.G. Miller, *J. Phys. Chem.*, Vol. *70(8)*, 2639-2659 (1966).

[17] B. D. Coleman and C. Truesdell, *J. Chem. Phys*, Vol. 33(1), 28-31 (1960).

[18] C. Truesdell,. *Rational Thermodynamics*, Springer-Verlag, New York, 1984.

[19] J. M. Zielinski, S. Alsoy, *J. Polym. Sci.. Part B: Polym. Phys.*, Vol *39*, 1496 (2001).

[20] D.P. Kondepudi and I. Prigogine, *Thermodynamics, Nonequilibrium*, in Encyclopedia of Applied Physics (Ed. G. L. Trigg), Wiley-VCH, 2003.

[21] H.J.V. Tyrell and K.R. Harris, *Diffusion in Liquids A Theoretical and Experimental Study*, Butterworths, London, 1984.

[22] R. Taylor and R. Krishna, *Multicomponent Mass Transfer*. Wiley, New York, 1993.

[23] E.L. Cussler, *Diffusion Mass Transfer in Fluid Systems*, Cambridge Univ. Press, Cambridge, 1997.

[24] J.C. Slattery, *Advanced Transport Phenomena*, Cambridge University Press, Cambridge, 1999.

[25] A. Schmitt, J.B. Craig, *J. Phys. Chem.*, Vol. 81(13), 1338-1342 (1977).

[26] J.W. Lorimer *J. Chem. Soc. Faraday Trans. II*, Vol. *74*, 75-83 (1978).

[27] I. Prigogine, *Etude Thermodynamique des Phι nomθnes Irrι versibles*, Desoer, Liège, 1947.

[28] J.W. Lorimer, *J. Chem. Soc. Faraday Trans. II*, Vol. 74, 84-92 (1978).

Brill Academic Publishers
P.O. Box 9000, 2300 PA Leiden
The Netherlands

Lecture Series on Computer and Computational Sciences
Volume 6, 2006, pp. 581-591

Cellular Automata (CA) As a Basic Method for Studying Network Dynamics

[1,a,b] D. Bonchev, [a]L. B. Kier and [a,c]C.-K. Cheng

[a]Center for the Study of Biological Complexity,
[b]Department of Mathematics and pplied Mathematics,
[c]Department of Computer Sciences
Virginia Commonwealth University
Richmond, VA, 23284-2030, USA

Received July 25, 2006; accepted July 30, 2006

Abstract: Macromolecular networks in living cell are the main focus of postgenomic biology and medicine. While network structure is well characterized by means of graph theory and information theory, problems are still encountered in describing and modeling dynamics of cellular networks. The present work summarizes the studies initiated at the Center for the Study of Biological Complexity at VCU, Richmond, aiming at establishing cellular automata as a standard for performing modeling of processes within metabolic, protein, and gene regulatory networks. The CA methodology is presented along with applications to specific pathways of biomedical importance. It is shown that the method enables the identification of characteristic dynamic patterns, and devising of strategies for pathway control. Preliminary results are also shown from a more general approach relating network topology and dynamics.

Keywords: Networks, pathways, cellular automata, modeling, dynamics

Mathematics SubjectClassification: 92B99

PACS: 33.15.-e

1. Introduction

The living cell is an extraordinarily highly organized system. Its major ingredients DNA, proteins, enzymes, and metabolites, take part in a variety of interactions (enzymatic reactions, protein-protein binding, gene/protein regulation, signaling), which connect them in a gigantic network containing tens of thousands of nodes. It is within this network that biological functions are organized, controlled and performed. Departing from reductionism, postgenomic cell biology is gradually transforming into systems biology refocusing its main interests from individual genes, proteins, and metabolites to the functioning of the system as whole. On the present stage of its development, studies of networks involving all types of interactions between the cellular ingredients are still an exception. The high complexity of the global cellular network has necessarily stimulated the analysis of the subnetworks based on interactions of a single specific type. Cell biology is intensively studying the properties of protein-protein interaction networks, metabolic networks, and gene regulatory networks, and the pathways within these networks, which are responsible for the individual biological functions.

It was a huge step toward the unity of science to realize that networks are the universal language of the complex world, and every branch of science can benefit from using networks to represent the systems it studies. Examples extend from the Internet and World Wide Web to financial and market networks to transport and electrical grids to social networks to ecological and biological networks to molecules as

[1] Corresponding author. Email: dgbonchev@vcu.edu

atomic networks and even to the structure of space-time. It was even more surprising to find out that dynamic evolutionary networks have common properties independent on the specifics of the interactions that connect the objects represented by network nodes. The closeness of nodes in these networks (imaginatively termed "small-worldness"), the fact that any two nodes in such a network are only few steps away [1] was the first well established common feature of the complex networks. It was also realized that the links in such networks are distributed among nodes according to power law, producing many nodes having only few neighbors, and only few hubs that are highly connected and are of prime importance for the existence and functioning of the system. [2] The evolution of networks was also found to follow the common pattern of "preferential attachment" of the new nodes to nodes that are highly connected (the "rich-get-richer" rule). [3].

The quantitative description of network *structure* is provided by graph theory [4] and information theory [5], with contributions coming from a variety of sciences, such as applied mathematics, computer sciences, physics, social sciences [6], and mathematical chemistry. [7] Complexity theory was also attracted in the characterization of network topology [8,9], and new network descriptors continue to be developed.[10] The situation with network *dynamics* is more complicated. The classical method for modeling dynamics by systems of differential equations faces a big challenge from the huge size of the cellular networks and the lack of methods for simultaneous measurements of dynamic parameters in the living cell. A viable alternative is to use the method of cellular automata [11], which while not offering explicit mathematical solutions, provides a convenient way to study trends and patterns of network dynamics, which makes it useful in predicting systems behavior. The present article summarizes some results from the efforts of the Center for the Study of Biological Complexity at the Virginia Commonwealth University to establish cellular automata method as a basic tool for studying the dynamics of cellular networks.

2. The Cellular Automata Method

Cellular automata were first proposed by the mathematician Stanislaw Ulam and the mathematical physicist John von Neumann a half century ago, and recently regain popularity after Steven Wolfram advocated the method as a radically new approach to science [11]. The first such system proposed by von Neumann was made up of square cells in a matrix, each with a state, operating with a set of rules in a two-dimensional grid. With the development of modern digital computers it became increasingly clear that these fairly abstract ideas could in fact be usefully applied to the examination of real physical systems. Cellular automata has five fundamental defining characteristics: 1) they consist of a discrete lattice of cells, 2) they evolve in discrete time steps, 3) each site takes on a finite number of possible values, 4) the value of each site evolves according to the same deterministic rules, 5) The rules for the evolution of a site depend only on a local neighborhood of sites around it.

The simulation of a dynamic system using cellular automata requires several parts that make up the process. The cell is the basic model of each ingredient, molecule or whatever constitutes the system. The grid may have boundaries or be part of a topological object that eliminates boundaries. Cellular automata have been designed for one, two or three-dimensional arrays. The most commonly used is the two-dimensional grid. The square cell has been the one most widely used. Each cell in the grid is endowed with a primary state, i.e., whether it is empty or occupied with a particle, object, molecule or whatever the system requires studying the dynamic event. Secondary information is contained in the state description that encodes the differences among cell occupants in a study.

The dynamic character of cellular automata is developed by the simulation of movement of the cells. This may be a simultaneous (synchronous) process or each cell, in turn, may asynchronously execute a movement. Each cell computes its movement based on rules derived from the states of other nearby cells. These nearby cells constitute a neighborhood. The rules may be deterministic or they may be stochastic, the latter process driven by probabilities of certain events occurring. When all cells in the grid have computed their state and have executed their movement (or not) it is one iteration, a unit of time in the cellular automata simulation. Each cell is identified in the program and is drawn randomly for the choice of movement or not. The question of which type of movement to use depends upon the system being modeled and the information sought from the model but it should reflect reality.

Cell movement is governed by rules called transition functions. The rules involve the immediate environment of each cell called the neighborhood. The most common neighborhood used in two-dimensional cellular automata is called the von Neumann neighborhood after the pioneer of the method. A cell, i, is in the center of four cells, j, adjoining the four faces of i. The rules governing cell movement may be deterministic or probabilistic. Deterministic cellular automata use a fixed set of rules, the values of which are constant and uniformly applied to the cells of each type. In probabilistic cellular automata, the movement of i is based on a probability-chosen rule where a certain probability to move or not to move is established for each type of i cell at its turn. Its state, (empty or occupied) is determined, and then its attribute as an occupant is determined. The probability of movement is next determined by a random number selection.

The movement of cells is based upon rules governing the events inherent in cellular automata dynamics. These are rules that describe the probabilities of two adjacent cells separating, two cells joining at a face, two cells displacing each other in a gravity simulation or a cell changing its state after joining with another cell. The first rule is the movement probability, P_m. This rule involves the probability that an occupant in an unbound cell, i, will move to one of four adjacent cells, j, if that space is unoccupied. If it moves to a cell whose neighbor is an occupied cell, k, then a bond will form between cells i and k.

A joining trajectory parameter, J(AB), describes the movement of a molecule at A, to join with a molecule at B, when an intermediate cell is vacant. This rule follows the rule to move or not to move described above. It thus involves the extended von Neumann neighborhood of ingredient A, and has the effect of adding a short-range attraction or repulsion component to the interaction between ingredients A and B. J is a non-negative real number. When $J = 1$, the molecule A has the same probability of movement toward or away from B as for the case when the B cell is empty. When $J > 1$, molecule A has a greater probability of movement toward an occupied cell B than when cell B is empty. When $0 < J < 1$, molecule A has a lower probability of such movement, and this can be considered as a degree of mutual repulsion. When $J = 0$, molecule A can not make any movement (all neighboring cells are occupied).

Just as two cells can join together, so their tessellated state can be broken. The first of two trajectory or interaction rules is the breaking probability, P_B. The $P_B(AB)$ rule is the probability for a molecule, A, bonded to molecule, B, to break away from, B. If molecule A is bonded to two molecules, B and C, the simultaneous probability of a breaking away event from both B and C is $P_B(AB)*P_B(AC)$. If molecule A is bound to three other molecules (B, C, and D), the simultaneous breaking probability of molecule A is $P_B(AB)*P_B(AC)*(P_B(AD)$. Of course, if molecule, A, is surrounded by four molecules it cannot move.

3. Specifics of the Cellular Automata Application to Cellular Networks

Enzymatic reactions are the basic biochemical reactions in the living cells. The intimate mechanism of these reactions includes the formation of an enzyme-substrate complex SE, which transforms into enzyme-product complex, PE, and dissociates to free the reaction product P:

$$\mathbf{S + E \rightarrow SE \rightarrow PE \rightarrow P + E}$$

Enzymes in the CA modeling are usually considered stationary, and the movement probability for such an enzyme cell is set at $P_M = 0$. The substrates and products of reactions are modeled as highly mobile with $P_M = 1$. The CA model selected is asynchronous, i.e., the cells change their states and compute their states one at a time. The joining and breaking probabilities may vary within the 0 to 1 range depending on the specifics of interactions. In protein-protein interaction networks the formation of a protein dimer is reflected by assuming $P_B = 0$. In the case where there are no special requirements, the joining trajectory parameter is taken as $J = 1$, and the breaking probability $P_B = 0.5$ to allow sufficient time for interaction. Only the cells involved in a specific state change, i.e., enzyme-substrate → enzyme-product, are endowed with a state change probability rule, defined by transition probability, P_C, which describes the probability of an ES pair of cells changing to an EP pair of cells. This is a basic variable in the CA modeling of enzymatic reactions, which depends on the enzyme activity, the presence of inhibitors and other factors that affect the enzymes. The transition probability is varied in a wide range, except for the case of irreversible reactions for which $P_c = 1$. The collection of rules associated with a network species is thus a profile of the structure of that species and its relationship with other species, within the definition of a molecular system. By systematically varying the rules one can develop a profile of configurations reflecting the influences of different structures. A general method for the CA modeling of a single enzymatic reaction was developed by Kier [12,13].

Concentrations are another group of variables in the CA modeling, represented by a number of cells for each reaction species. Concentration changes during the consecutive iterations of changing cells states constitute the essence of the ***temporal CA model***. Iterations continue until a steady state or (more rarely) an equilibrium state is reached and concentrations do not change further. The steady state concentrations are part of the output data of the CA modeling, and may be plotted against the varied basic parameters to produce the *spatial CA model* of the network under study. Enzyme concentrations are usually not varied. It is assumed that they are in sufficient amount to support the reactions. On the other hand, the potential deficit of a certain enzyme can be reflected in the modeling by a lower transitional probability, and the same can be done in cases of enzyme inhibition.

In what follows, we present some examples of the application of the cellular automata method to model the dynamics of some cellular processes.

4. Cellular Automata Modeling of the Mitogene Activated Protein Kinase (MAPK) Signaling Pathway [14]

The mitogene activated protein kinase is a signaling pathway, relaying signals from the plasma membrane to targets in the cytoplasm and nucleus. The molecular mechanism of this pathway has been studied intensively during the last ten years [15,16]. Our cellular automata modeling was limited to the major cascade part of the pathway. The cascade shown in Figure 1, is presented as given by Huang and Ferrell [17].

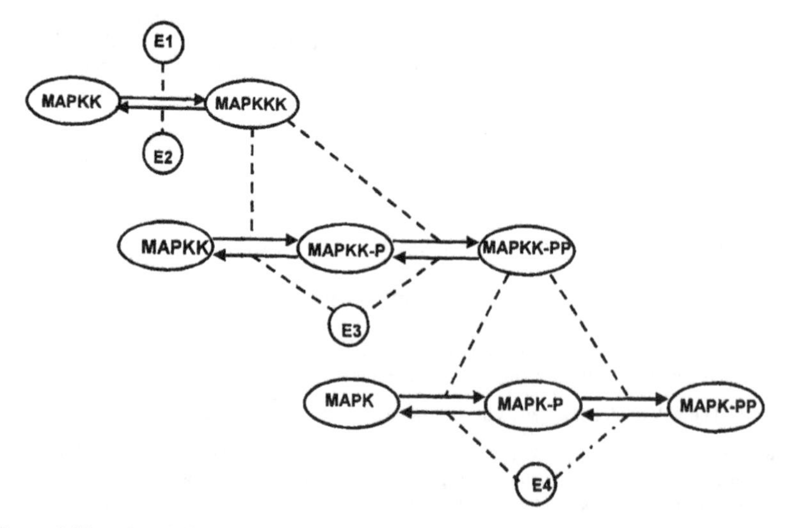

Figure 1. The MAPK signaling cascade. The catalytic action of enzymes is presented by dashed lines. E3 and E4 are MAPKK-protease and MAPK-protease, respectively, and the hypothetical enzymes E1 and E2 affect the reactions of MAPKKK activation and deactivation.. P stands for phosphate, PP for diphosphate. The products of the cascade rows 1 and 2 act as enzymes for the phosphorylation reactions at rows 2 and 3, respectively.

The biochemical reactions involved in the MAPK cascade are described as follows:

$$
\begin{array}{llll}
A + E1 & \rightarrow AE1 & \rightarrow BE1 & \rightarrow B + E1 \\
B + E2 & \rightarrow BE2 & \rightarrow AE2 & \rightarrow A + E2 \\
C + B & \rightarrow CB & \rightarrow DB & \rightarrow D + B \\
D + B & \rightarrow DB & \rightarrow EB & \rightarrow E + B \\
D + E3 & \rightarrow DE3 & \rightarrow CE3 & \rightarrow C + E3 \\
E + E3 & \rightarrow EE3 & \rightarrow DE3 & \rightarrow D + E3 \\
F + E & \rightarrow EF & \rightarrow EG & \rightarrow G + E \\
G + E & \rightarrow EG & \rightarrow EH & \rightarrow H + E \\
G + E4 & \rightarrow GE4 & \rightarrow FE4 & \rightarrow F + E4 \\
H + E4 & \rightarrow HE4 & \rightarrow GE4 & \rightarrow G + E4
\end{array}
$$

The three substrates (MAPKKK, MAPKK, and MAPK) and the four enzymes have some prescribed initial concentration, expressed as a number of CA cells. We varied the initial concentrations of the three substrates (MAPKKK, MAPKK, and MAPK) and the competencies of enzymes E1 to E4. The basic variable was the concentration of MAPKKK (the cascade input signal), which was examined within a 25-fold range from 20 to 500 cells. The concentrations of MAPKK and MAPK were kept constant (500 or 250 cells) in most of the models. The four enzymes E1to E4 were represented by 50 cells each. In one series of models, we kept the MAPKKK initial concentration equal to 50 cells, the transition probabilities of three of the enzymes, $P_c = 0.1$, and varied the probability of the fourth enzyme within the 0 to 1 range. In another series, *all* enzyme transition probabilities were kept constant ($P_c = 0.1$), whereas the concentrations of substrates were varied. A third series varied both substrate concentrations and enzyme propensities. Recorded were the variations in the concentrations of the three substrates MAPKKK, MAPKK, and MAPK, and those of the five products MAPKKK*, MAPKK-P, MAPKK-PP, MAPK-P, and MAPK-PP.

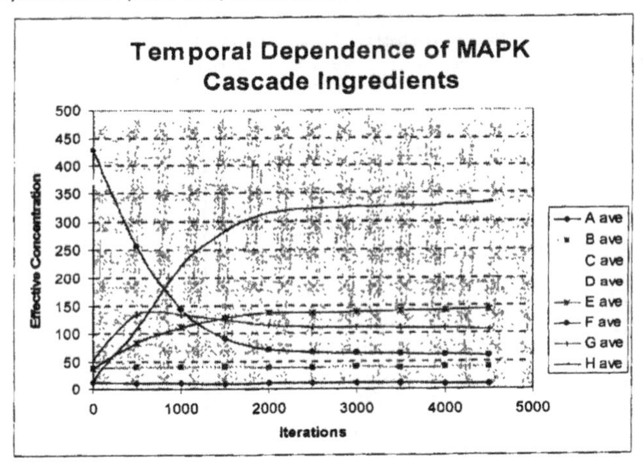

Figure 2. The temporal dependence of the MAPK cascade species (A = MAPKKK, B = MAPKKK*, C = MAPKK, D = MAPKK-P, E = MAPKK-PP, F = MAPK, F = MAPK-P, H = MAPK-PP) for enzyme propensities P_c (E1) = P_c (E2) = P_c (E3) = P_c (E4) = 0.1, and substrates initial concentration $[A_o] = 50$, $[C_o] = [F_o] = 500$.

Fig. 2 shows an example of one of the temporal models obtained for the MAPK cascade. The number of iterations needed to reach a steady state (or much less frequently an equilibrium state) is used as an inverse measure of time (The faster the process the smaller the number of iterations needed to reach the steady-state). It may be seen that at the conditions selected 5000 iterations suffice to reach a steady state at which the cascade output signal (H = MAPK-PP) reaches about 340/500 = 68% of its maximal value.

The results obtained for the steady-state concentrations of a series of models are plotted to represent 2-D or 3-D *spatial* CA model. Fig. 3 illustrates such a model for a series with variable propensity of the E3 enzyme. It is seen that the maximum amplification of the cascade signal (the largest production of the doubly phosphorylated MAPK, denoted in Fig. 3 as H) occurs at an intermediate E3 propensity, due to the reversing of the second row phosphorylation reactions. This result shows the potential of the CA modeling to predict dynamic patterns and to help in finding optimum conditions for of the input signal amplification.

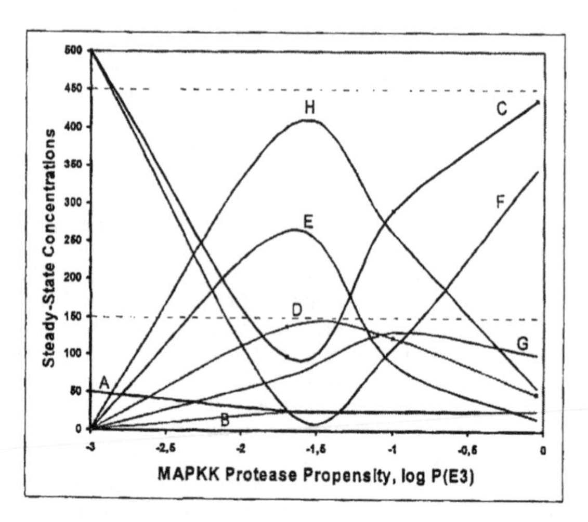

Figure 3. Influence of the MAPKK-protease propensity P(E3) on the steady-state concentrations of the MAPK cascade species. Model parameters used: enzyme propensities P_c (E1) = P_c (E2) = P_c (E4) = 0.1, substrates initial concentration $[A_o]$ = 50, $[C_o]$ = $[F_o]$ = 500. The best amplification of cascade signal (the highest concentration of H) at these conditions is obtained for the enzyme E3 propensity within the range of 0.01 – 0.02. Similar range is obtained for the optimal propensity of enzymes E2 and E4 (not shown).

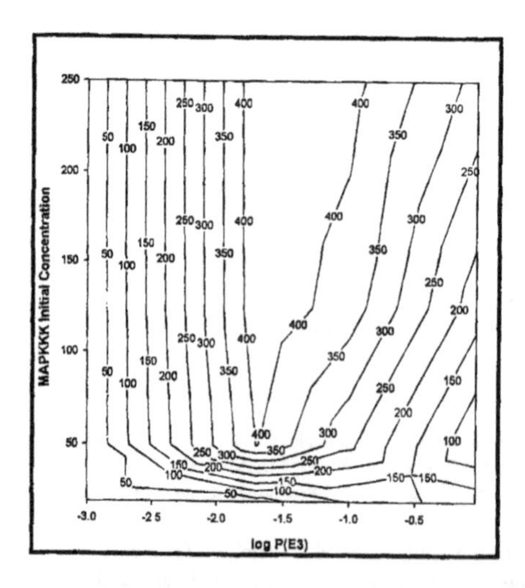

Figure 4. Contour plot of the MAPK-PP steady-state concentration at variable MAPKKK initial concentration and variable MAPKK-protease propensity. Other model parameters used include E1, E2, and E4 enzyme propensities equal to 0.1, and the second and third cascade row initial concentrations $[MAPKK_o]$ = $[MAPK_o]$ = 500. The optimal conditions for obtaining the highest MAPK-PP yield are those within the contour level of 400 cells, and the enzyme E3 propensity near to or higher than 0.02.

The search for optimal results in a CA modeling can be facilitated by the use of 3-D or contour plots. The optimization of the MAPK cascade dynamics can be best attained when plotting the steady-state concentrations against the initial concentration of the input signal (A = MAPKKK) and the propensity of either E3 enzyme or the E4 one. Such a plot is shown in Fig. 4, where the contour line [H4] = 400 indicates that such optimal conditions can be realized for [A] > 50 for a relatively narrow range of moderately inhibited MAPKK-protease.

An important outcome of the CA modeling is the possibility to summarize the patterns of dynamic changes in the network studied in the form of recommendations how to manipulate the network variables in order to reach a certain result. Such a summary for the modeling of the MAPK cascade is shown in Table 1.

Table 1. Inhibiting enzymes E1 to E4 as a tool for controlling the MAPK pathway

Objectives To Accomplish		Do This	Validity Range
Decrease [MAPK]	→	Inhibit E2, E3, E4	P = 0.9 → P = 0.02
Increase [MAPK]	→	Inhibit E1	P = 0.9 → P = 0
Decrease [MAPK-PP]	→	Inhibit E1	P = 0.9 → P = 0
Increase [MAPK-PP]	→	Inhibit E3, E4	P = 0.9 → P =0.02
Decrease [MAPKK]	→	Inhibit E3	P = 0.9 → P =0.02
Increase [MAPKK]	→	Inhibit E1	P = 0.9 → P = 0

5. Cellular Automata Modeling of the FAS-Activated Apoptosis Pathway

The apoptosis signaling pathway is controlling the programmed cell death that can be initiated either to destroy cells that are no more needed in the cell development or to protect the organism in case of pathogenic attack. Due to the variety of input signals, the apoptosis pathway is very complex and

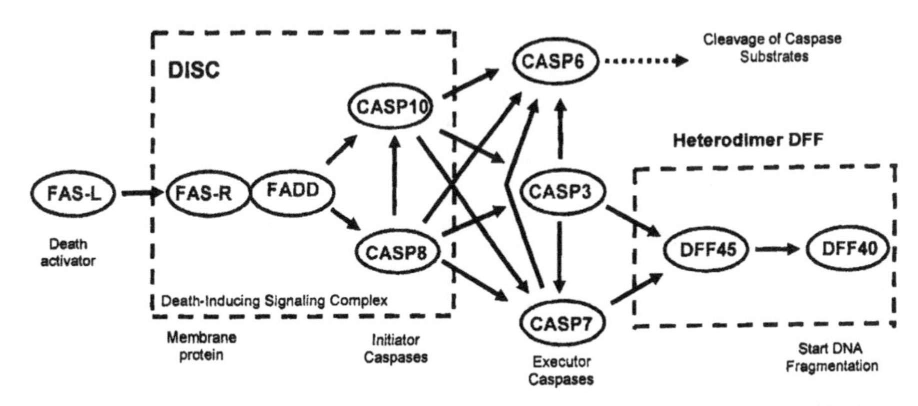

Figure 5. The FAS-activated apoptosis signaling pathway includes a chain of interactions that release and activate caspase enzymes. It ends with the output signal, the release of the the DFF40 protein, which starts the DNA fragmentation.

includes different mechanisms of cell suicide. In Fig. 5, we show one such mechanism activated by the FAS protein. The interaction of the FAS ligand with the corresponding membrane receptor releases the two caspase enzymes CASP-8 and CASP-10, which are bonded in the death-inducing signaling complex DISC. After the release, the two caspases activate on their turn the executor caspases CASP-3, CASP-6, and CASP-7. The chemical signal finally arrives at the DFF heterodimer, which releases the

DFF40 protein, the immediate effector of the DNA fragmentation. The high connectivity of the pathway enables multiple options for an effective transmission of the input signal.

The initial steps of the complex biochemical interactions associated with the signal transmission are shown below. The complete description of these reactions will be presented in a forthcoming publication, along with a detailed analysis of the CA modeling results.

FAS-L + DISC → DISC' + CASP-8*

FAS-L + DISC → DISC'' + CASP-10

CASP-8* + CASP-10 → CASP-8* + CASP-10*

The CA modeling included variations of the species concentrations and activity, the latter expressed as probability within the 0 to 1 range. 3-fold to 5-fold acceleration of the signal transmission was predicted to occur at high activity of participating caspases. Fig. 6 illustrates the types of results obtained by simultaneously varying the activity probability of the executor caspases 3 and 7 (Series 1 through 6 correspond to increasing activity of caspase-3). The number of iterations needed to attain 50% of the maximum output signal was used as an indirect inversely proportional measure of the processes rate. The example illustrates the potential of cellular automata modeling for finding the optimal conditions for either cell death or cell survival, depending on the specific targets of the biomedical research.

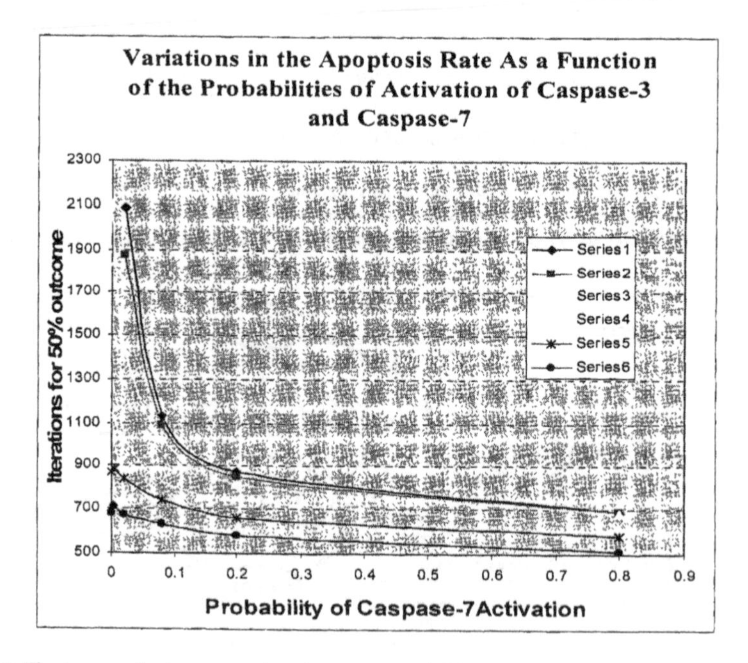

Figure 6. The increase in the caspase-3 and caspase-7 activities' probability reduces considerably the number of CA iterations needed for achieving 50% of the maximum output signal in the FAS-initiated apoptosis. Series 1 through 6 are those with probability 0.001, 0.005, 0.02, 0.08, 0.2, and 0.8, respectively

6. Cellular automata as a tool for identifying and investigating topological patterns that influence network dynamics

The large size of networks in biology is a major challenge to any method for modeling network dynamics. Topological network analysis can save a substantial amount of time and resources by identifying structural trends that influence strongly network dynamics. Studying such topology/dynamics relationships in small network fragments (subgraphs, motifs [18], one can then search for conservation or modification of these patterns after reconstructing the network from its pieces. Such a project is under way in the Center for the Study of Biological Complexity at the Virginia Commonwealth University.

An illustration of our approach is given in the generation series, shown in Fig. 7. It incorporates ten small graphs having three or four vertices, connected by directed single-step graph transformations. One may distinguish in Fig. 7 ring closures (**1** → **3**, **2** → **3**, **4** → **5**, **6**, **6** → **8**, **7** → **9**), adding a node (**2** → **4**) or adding a branch (**3** → **7**), and forming additional cycles (**5** → **8**, **5** → **9**, **8** → **10**, **9** → **10**). The corresponding information on the changes in the dynamics of transforming the input concentration of the species **A** into the products **B**, **C** or **D** is also shown.

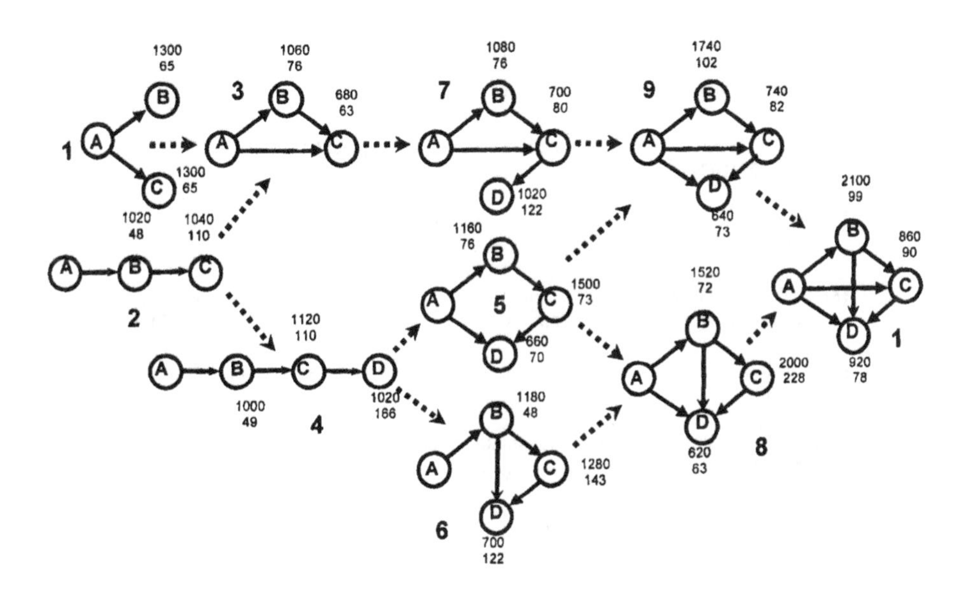

Figure 7. A flow chart of directed topological transformations between small network subgraphs (motifs) and the related information on the change in the dynamics of network flows. The pairs of numbers attached to most nodes show the number of cellular automata iterations needed to obtained a 100% (the upper number) or a 50% (the lower one) conversion of the input concentration of species **A** into the products **B**, **C** or **D**, respectively.

Analyzing Fig. 7, one can readily identify the patterns that speed up the production of the output product **D** or **C**. Thus, the number of CA iterations needed to obtain 100% conversion of A into D (or C for three-vertex graphs) is reduced to almost a half for ring closures: **1** → **3** (1300 to 680), **2** → **3** (1040 to 680), **4** → **5** (1020 to 660), **4** → **6** (1020 to 700), **7** → **9** (1020 to 640). The only exception of this strong quantitative trend is the transformation of **6** into **8** (700 to 620), the small acceleration for which is explained by the high speed of the **A** → **D** conversion already achieved in structure **6** at the cost of the existing feed-forward motif from **B** to **D**. The further examination of Fig. 7 reveals that it is difficult to separate clearly the ring closure pattern from the feed-forward one, which emerges in several different modifications. One may consider the **6** → **8** transition rather as a representative of a double (or secondary) feed-forward pattern, along with the **5** → **8** (660 → 620) and **5** → **9** transitions (660 → 640). The **8** → **10** and **9** → **10** transformations producing a complete graph already slow down the rate of converting **A** into **D**, due to the emerging conflict between two double feed-forward motifs.

A more detailed analysis of the topology/dynamics relationships in network motifs and patterns will be presented in forthcoming publications.

References

[1] D.J. Watts, and S. H. Strogatz, Collective Dynamics of "Small-World" Networks, *Nature* 393 440-442(1998)

[2] A.-L Barabasi and R. Albert, Emergence of Scaling in Random Networks, *Science* 286 509-512 (1999)

[3] A.-L. Barabasi, *Linked: The New Science of Networks*, Perseus, New York, 2002.

[4] F. Harary, *Graph Theory*, 2nd ed., Addison-Wesley, Reading, MA, 1969.

[5] D. Bonchev, *Information-Theoretic Indices for the Characterization of Chemical Structures*, Research Studies Press, Reading, U.K., 1982.

[6] L.C. Freeman, Centrality in Social Networks: Conceptual Clarification, *Social Networks* 1 215-239(1979).

[7] D. Bonchev and D.H., Rouvray, Eds., *Mathematical Chemistry, Vol. VII, Complexity in Chemistry*. Francis and Taylor, London, U. K., 2003.

[8] D. Bonchev. Complexity of Protein-Protein Interaction Networks, Complexes and Pathways, in *Handbook of Proteomics Methods*, M. Conn, ed. Humana, New York, p. 451-462 (2003).

[9] Bonchev, D. (2004) Complexity Analysis of Yeast Proteome Network, *Chem. Biodiversity*, 1, 312-326.

[10] D. Bonchev and G.A. Buck, Quantitative Measures of Network Complexity, in: D. Bonchev and D.H. Rouvray, Eds., *Complexity in Chemistry, Biology, and Ecology*, Springer, New York, p. 191-235, 2005.

[11] S. Wolfram, *A New Kind of Science*, Wolfram Media, Champaign, IL, 2002.

[12] L. B. Kier, C.-K. Cheng, B. Testa and P.-A. Carrupt, .A Cellular Automata Model of Enzyme Kinetics, *J. Molec. Graphics* 14 227-238(1996).

[13] L.B. Kier, P.G. Seybold and C.-K. Cheng, *Cellular Automata Modeling of Chemical Systems*, Springer, Amsterdam, 2005.

[14] L.B. Kier, D. Bonchev, and G.A. Buck, Modeling Biochemical Networks: A Cellular Automata Approach, *Chem. Biodiversity* 2, 233-243(2005).

[15] B.N. Kholodenko, Negative Feedback in the Mitogen-Activated Protein Kinase, *Eur. J. Biochem*. 267 1583-1588(2000).

[16] F.A. Brightman and D.A. Fell, Differential Feedback Regulation on the MAPK Cascades Underlies the Quantitative Differences in EGF and NGF Signaling in PC12 Cells, *FEBS Lett*. 482 169-174(2000).

[17] C.–Y.F. Huang, J.E. Ferell, Ultrasensitivity in the Mitogen-Activated Protein Kinase Pathways *Proc. Natl. Acad. Sci. USA* 93 10078-10083(1996).

[18] R. Milo, S. Itzkovitz, N. Kashtan, R. Levitt, and U. Alon, Network Motifs: Simple Building Blocks of Complex Networks, *Science* 298 824-827(2002).